HIGH EXCITATION AND SHORT PULSE PHENOMENA

Third Trieste ICTP–IUPAP Semiconductor Symposium "High Excitation and Short Pulse Phenomena"
International Centre for Theoretical Physics, Trieste, Italy, 2–6 July 1984

HIGH EXCITATION AND SHORT PULSE PHENOMENA

PROCEEDINGS OF THE THIRD TRIESTE
ICTP–IUPAP SEMICONDUCTOR SYMPOSIUM

INTERNATIONAL CENTRE FOR THEORETICAL PHYSICS
TRIESTE, ITALY
2–6 JULY 1984

Guest Editor:

M.H. PILKUHN

Physikalisches Institut, Stuttgart University, Stuttgart, F.R.G.

1985

NORTH-HOLLAND – AMSTERDAM

Reprinted from Journal of Luminescence Volume 30 (1985)

ISBN: 0 444 86931 X

Published by:

North-Holland Physics Publishing

a division of
Elsevier Science Publishers B.V.
P.O.Box 103
1000 AC Amsterdam
The Netherlands

Sole distributors for the U.S.A. and Canada:

Elsevier Science Publishing Company, Inc.

52 Vanderbilt Avenue
New York, N.Y. 10017
U.S.A.

Library of Congress Cataloging in Publication Data

Trieste ICTP-IUPAP Semiconductor Symposium (3rd : 1984)
High excitation and short pulse phenomena.

"Reprinted from Journal of luminescence, volume 30 (1985)"--T.p. verso.
Includes bibliographies and indexes.
1. Semiconductors--Congresses. 2. Semiconductors--Plasma effects--Congresses. 3. Exciton theory--Congresses. I. Pilkuhn, M. H., 1934- . II. Title. III. Title: Short pulse phenomena.
QC610.9.T75 1984 537.6'22 85-3018
ISBN 0-444-86931-X (U.S.)

Printed in The Netherlands

PREFACE

The Trieste Semiconductor Symposia, sponsored by the Committee for Joint IUPAP–UNESCO Semiconductor Activities in Trieste, take place biennially at the International Centre for Theoretical Physics (ICTP). They are organized in cooperation with the Gruppo Nazionale di Struttura della Materia of the National Research Council, under the auspices of the Italian Ministry for Scientific Research. The symposia are meant to cover focal aspects of semiconductor physics and potential applications at a state-of-the art level. The topic chosen for the first symposium in 1980 was "Amorphous Silicon: Physics and Applications"; the second symposium, in 1982, was devoted to "Surfaces and Interfaces: Physics and Electronics", while the present one deals with "High Excitation and Short Pulse Phenomena".

Each topic is presented and discussed by a body of leading scientists in the field, chosen to provide a thorough and well-balanced coverage of both basic physical aspects and device implications. The symposia extend over a five-day period, with a good fraction of the time devoted to discussion and presentation of latest results by the participants in the audience. The latter includes mainly research scientists already active in the field, but also students and guests of the Centre.

A basic component of the success of the symposia, so far, has been the outstanding performance of the chairmen who have designed the technical program and selected the speakers. In the present case, we wish to express our deep appreciation for the work done by M.H. Pilkuhn from the University of Stuttgart, but also by the meeting secretary, A. Forchel, who have succeeded in producing a very lively and most constructive meeting.

The helpful assistance and the suggestions from our colleagues in the IUPAP–UNESCO Committee – P.N. Butcher, R. Enderlein and F. Stern – should be given full recognition. The continuous encouragement of the IUPAP–Semiconductor Committee, in particular of its chairman C. Hilsum, has also been determinant to the start and the development of the enterprise.

The Trieste Symposia would not have been possible without the generous financial aid provided by our two main supporters, CNR (National Research Council) and IBM–Italy. C. Rizzuto, chairman of the Gruppo Nazionale di Struttura della Materia of CNR, and M. Morelli, director of cultural and academic programs of IBM–Italy, are to be personally credited for their interest in this matter. Support from the Trieste International Centre was also of great importance. Finally, a special acknowledgment should go to the staff of the Centre and also of the International School for Advanced Studies (SISSA), for their capable handling of all local organizational problems.

We close by expressing our hope that future symposia are characterized by as much interest and success as the past ones, so as to be a recognized point of reference for the semiconductor community in Europe and in the rest of the world.

Trieste, 6 July 1984

A. Frova and E. Tosatti
(Members, Committee for Joint
IUPAP–UNESCO Semiconductor
Activities in Trieste)

INTRODUCTION

The Third IUPAP Semiconductor Symposium in Trieste was devoted to "High Excitation and Short Pulse Phenomena", a topic which certainly belongs to one of the most interesting and active fields in semiconductor research. In fact, during more than one decade, the study of many-body phenomena in highly excited semiconductors has developed rapidly into an important aspect of fundamental solid state physics discussed repeatedly at many conferences.

It was soon recognized that the field implied more than the mere study of many-body interactions of carriers at high density. Equally attractive was the exploration of the transition from a "single-particle situation" (e.g. excitons) at low excitation to a "many-body situation" (e.g. electron–hole liquids) at high excitation, and some of the key topics have been:

- excitonic molecules, multiple exciton complexes, exciton screening;
- formation of electron–hole drops, band-gap renormalization, phase diagrams.

The Trieste symposium came at a time when the physical understanding of many of these topics had reached a fairly high level already. For this reason, rather than reviewing "traditional" themes like electron–hole drops, ground-state energies, or phase diagrams, new and often still controversial topics were discussed at the Trieste symposium. The following selection will give a flavour of the issues treated in Trieste:

- non-equilibrium electron–hole plasmas in direct and indirect gap semiconductors, transport of plasmas and of hot carriers, expansion velocities;
- screening effects involved in the exciton–plasma transition;
- optical bistability and nonlinearity, new experimental observations and theoretical treatment;
- new experimental techniques with very short laser pulses, femtosecond spectroscopy, ultrashort pulse investigations of plasmas, quantum wells, or melting processes;
- physical aspects of laser annealing and of melting at the highest excitation levels;
- new results on high excitation and short pulse physics of quantum wells, excitons in quantum wells, quantum well lasers;
- fast relaxation processes, carrier–carrier and carrier–phonon interactions.

The specialists who had convened in Trieste, from many countries all over the world, discussed these novel and other aspects of high excitation and short pulse physics. New results as well as reviews of recent developments and of the present status of our understanding were presented by 41 speakers during one week. The lively and intense discussions, for which much time had been reserved, took place in the very pleasant and stimulating atmosphere of the International Centre for Theoretical Physics (ICTP).

These Proceedings contain the keynote papers of the symposium and also the contributions from the participants. They convey ample information about our present knowledge and understanding of the field of high excitation and short pulse phenomena, and they also give an impression of new trends for future development.

As during the previous semiconductor symposia, the Trieste Centre for Theoretical Physics provided the congenial and most pleasant environment essential for fruitful discussions. The support of the Centre to the symposium is gratefully acknowledged, and also the very important support of the Italian National Research Council (CNR), of IBM–Italy, and of the IUPAP Semiconductor Commission.

It has to be mentioned with special acknowledgement that the continuous initiative and guidance of the Joint IUPAP–UNESCO Committee for the Semiconductor Activities in Trieste have made the symposia in the past and also this present symposium possible. Andrea Frova is the Chairman of this Committee, and the other members are: P.N. Butcher, R. Enderlein, F. Stern and E. Tosatti. Apart from this, the enormous amount of work done by Andrae Frova and by Erio Tosatti in organizing the symposium deserves special gratitude and recognition. Not all the many people who helped or supported the symposium can be mentioned here by name, but they deserve our sincere thanks.

It is hoped that these Proceedings convey the spirit of the lively Third Semiconductor Symposium in Trieste, that they may serve as a source of information and reference, and that they may stimulate new ways for future research.

Stuttgart, FRG
September 1984

Manfred H. Pilkuhn
Chairman and Proceedings Editor

CONTENTS

Journal of Luminescence 30 (1985) 1–17
North-Holland, Amsterdam

ELECTRON–HOLE PLASMA GENERATION AND EVOLUTION IN SEMICONDUCTORS

Monique COMBESCOT and Julien BOK

Groupe de Physique des Solides de l'Ecole Normale Supérieure, 24 rue Lhomond, 75231 Paris Cedex 05, France

We discuss various physical problems related to dense electron–hole (e–h) plasmas in silicon-like semiconductors such as: the e–h plasma energy in the quantum and classical limits including the actual band structure; the dielectric constant using the appropriate plasma frequency mass and the variation of the plasma relaxation time due to e–h collisions; the plasma generation including its nonlinear aspect when the e–h density passes the plasma frequency density of the pump beam; the e–h plasma evolution due to Auger recombination and impact ionization as well as the e–h plasma hydrodynamics; finally, we will discuss also the possibility of a very fascinating melting at $T=0$.

Fifteen years ago, Keldysh [1] suggested that some line in the Germanium luminescence spectrum, below the well-known exciton line, could be due to an electron–hole plasma. At that period, the pump laser could produce only a very dilute exciton gas, but as it turns out that in Ge the lowest energy of the e–h plasma state is below the exciton energy, the exciton gas condenses at low temperature into an e–h plasma of definite density – this density which corresponds to the plasma energy minimum, is 2×10^{17} cm^{-3} in Ge and 2×10^{18} cm^{-3} in Si. Then started the very popular subject of electron–hole drops [2]. Some years later, the use of short pulses allowed one to create directly e–h plasmas with densities slightly above the density of drops, and so to get plasmas at higher densities.

The subject of e–h plasmas has been revived recently by experiments on laser annealing of ion-implanted semiconductors. Very short pulses as well as powerful lasers are used. In silicon typical e–h densities of 10^{19} cm^{-3} are reached with nanosecond pulses. These densities can go up to few 10^{20} cm^{-3} with 10 ps pulses. Very recently the Palaiseau [3] laser, delivering 1 J cm^{-2} in 100 fs, allowed us to reach densities above 10^{22} cm^{-3}, which is extremely high if one remembers that the total number of valence band electrons is 2×10^{23} cm^{-3}.

In this lecture, we will give a review of the most important problems related to e–h plasmas, especially in the very high density limit.

0022-2313/85/\$03.30

We will first discuss the calculation of the energy of an e–h plasma in the quantum and classical limits. A very precise calculation has been done in relation with e–h drops [4, 5] within the parabolic approximation for the band structure. We will discuss the deviations from these results if one takes into account a more realistic band structure at high density, computed by Chelikowski and Cohen [6].

Most of the experiments on e–h plasmas are optical experiments which rely on the response of the crystal in presence of e–h pairs. They are directly related to the dielectric constant. We will discuss the appropriate mass to use in the plasma frequency as well as the plasma relaxation time which controls the interband absorption. We will show that at high density, this relaxation time strongly decreases, as it is dominated by e–h collisions, and no more by e–ph interactions.

In section 3, we treat the generation of the e–h plasma from a laser beam absorption. The problem is usually rather simple as the created e–h density stays low enough for the sample dielectric constant at the pump frequency not to be modified. For densities of the order of the plasma frequency density of the pump, one has to solve a problem of nonlinear optics with a time- and space-dependent dielectric constant.

The time evolution of the e–h plasma is studied in section 4. The e–h pairs mainly recombine via Auger effect, and this mechanism is so efficient that it is the main reason for preventing the existence of very dense plasmas. These e–h pairs also move due to their density gradient, always present as they are created from the absorption of the laser beam, with a certain penetration depth. We will write the hydrodynamic equations, for a very dense plasma, especially in the quantum limit.

Finally, we will discuss the fascinating subject of a melting at $T=0$, which should be obtained when a very large number of covalent bounds are broken by excitation of e–h pairs. We will also point out some experimental problems which arise if one tries to see a decrease of the melting temperature due to additional e–h pairs created by a laser pulse.

1. Electron–hole plasma energy

An analytical expression for the energy of an e–h plasma can be obtained in two limits: the quantum limit and the classical limit.

1.1. Quantum limit

When the temperature T is small compared to the chemical potential, the plasma is degenerate. For $T=0$, the energy is composed of a kinetic, exchange and correlation term,

$$E \sim \int_0^{K_F} \frac{\hbar^2 k^2}{2m} \, d^3k - \int_0^{K_F} \frac{4\pi e^2}{(k-k')^2} \, d^3k \, d^3k' - E_{corr}(n)$$
$$= \alpha K_F^5 - \beta K_F^4 - E_{corr} = n(an^{2/3} - bn^{1/3} - \varepsilon_{corr}(n)). \qquad (1)$$

At large densities (i.e., when the interparticle distance r_s measured in Bohr radius is much less than 1) the correlation energy can be neglected compared to the exchange term.

In relation with e–h drops, the energy of an e–h plasma in silicon has been calculated [4,5] taking into account the full complexity of the silicon band structure. The plasma specific heat [7] has also been determined for silicon, which allows one to obtain the lowest order temperature expansion of the free energy as

$$\frac{F(N, T, V)}{N} = 5.3(10^{-18}n)^{2/3} - 11(10^{-18}n)^{1/3} - 0.12\tilde{T}^2(10^{-18}n)^{-2/3}. \qquad (2)$$

This value [4] is expressed in meV, while N is the total number of e–h pairs in a volume V, with a density $n = N/V$ expressed in cm^{-3} and a temperature $\tilde{T}$ expressed in meV.

Table 1 gives the corresponding values of the kinetic and exchange contributions to the $T=0$ free energy for densities varying from 10^{19} cm^{-3} to 10^{21} cm^{-3}. (In order to have a density scale, it might be useful to recall that for the drop density 2×10^{18} cm^{-3}, r_s is of order 1.)

The temperature T^* is such that, for a given density, the specific heat contribution equals the kinetic energy. It is a gross estimate of the temperature at which the plasma changes from a quantum gas to a classical gas. The exact determination of the crossover temperature is in fact much more complex as, due to the conduction band degeneracy, the electrons become classical before the holes.

Table 1
Kinetic energy ε_{kin}, exchange energy ε_{exch} and chemical potential μ^0 at $T=0$ for various e–h densities n. μ^0_{CN} is calculated within the usual parabolic band structure, while μ^0_{cc} is calculated using Chelikowski and Cohen calculation. T^* is such that ε_{kin} equals the specific heat contribution, while T^{**} is such that $kT^{**} = \mu^0_{cc}$

n	10^{19} cm^{-3}	10^{20} cm^{-3}	5×10^{20} cm^{-3}	10^{21} cm^{-3}	5×10^{21} cm^{-3}	10^{22} cm^{-3}
ε_{kin}	25 meV	0.11 eV	0.33 eV	0.53 eV		
ε_{exch}	−24 meV	−0.051 eV	−0.087 eV	−0.11 eV		
T^*	360 K	1700 K	4800 K	7700 K		
μ^0_{CN}	10 meV	0.12 eV	0.43 eV	0.73 eV		
μ^0_{cc}			0.49 eV	0.68 eV	1.5 eV	2.2 eV
T^{**}				4000 K	8900 K	13000 K

But, the estimate allows one to say that a plasma with few $10^{20}\ \mathrm{cm}^{-3}$ e–h pairs is a quantum gas at room temperature and could be considered as such up to the silicon melting temperature.

Table 1 gives also the e–h pair chemical potential (without the gap) $\mu^0_{\mathrm{CN}} = \partial F(N, 0, V)/\partial N$ using eq. (1), up to $n = 10^{21}\ \mathrm{cm}^{-3}$.

One expects that, with increasing e–h density, the low density parabolic band structure becomes incorrect and consequently, also the expression (2) for the free energy. In order to obtain the chemical potential of a denser plasma, we have considered the silicon band structure of Chelikowski and Cohen [6]. Using their results, it is a good approximation to take for the e and h density-of-states versus energy a linear behaviour

$$\mathrm{d}n_{\mathrm{e}}/\mathrm{d}E = \alpha E, \qquad \mathrm{d}n_{\mathrm{h}}/\mathrm{d}E = \beta E. \tag{3}$$

For silicon, $\alpha \simeq \beta = 1.7 \times 10^{22}\ \mathrm{cm}^{-3}\ \mathrm{eV}^{-2}$. This allows one to calculate the electron chemical potential for $T = 0$, if one neglects the Coulomb contribution (an approximation which is valid in the limit of large n we are considering)

$$n_{\mathrm{e}} = \int_0^{\mu^0_{\mathrm{e}}} \alpha E\ \mathrm{d}E = \frac{\alpha \mu_{\mathrm{e}}^2}{2}. \tag{4}$$

As the e–h pair chemical potential is $\mu^0_{\mathrm{e}} + \mu^0_{\mathrm{h}}$, this finally gives for the chemical potential of a dense e–h plasma of Si in the quantum limit,

$$\mu^0_{\mathrm{cc}} = 2.2(10^{-22} n)^{1/2}, \tag{5}$$

where n is in cm^{-3} and μ in eV. This expression should be valid at high density up to 2 eV as it relies on a linear expression of the density-of-states valid up to 1 eV for e and h.

Table 1 gives the corresponding values of μ^0_{cc} for n between $5 \times 10^{20}\ \mathrm{cm}^{-3}$ and $10^{22}\ \mathrm{cm}^{-3}$. One sees that μ^0_{cc} crosses μ^0_{CN} for $n \sim 7 \times 10^{20}\ \mathrm{cm}^{-3}$. This means that the "low density" expression (2) for the free energy could be used up to densities as high as $10^{21}\ \mathrm{cm}^{-3}$.

Equation (5) for a dense e–h plasma chemical potential allows us also to have an estimate of the temperature T^{**} below which the plasma is degenerate using $kT^{**} \simeq \mu^0_{\mathrm{cc}}$. Table 1 shows the corresponding results. One can note that T^{**} and T^* are in good agreement in the crossover region.

1.2. Classical limit

Up to now, we have discussed the case of an e–h plasma in the quantum limit, and we have seen that for densities above a few $10^{20}\ \mathrm{cm}^{-3}$ this is the case for a plasma in equilibrium with the lattice (as its temperature will be below 1700 K). Let us now turn to the other interesting case of a very dense, but very hot e–h plasma. This will imply one has to study it before it has cooled down, i.e., within 1 ps or less. Using the appropriate value of the density-of-states deduced from

Chelikowsky and Cohen [6], the chemical potential of a classical free electron gas of density n_e verifies

$$n_e = \int_0^{\infty} \alpha E \exp[(\mu_e - E)/kT] \, dE, \tag{6}$$

so that the chemical potential of n e–h pairs in Si reads

$$\mu_{cc}^{T} = 2kT \ln[n/\alpha(kT)^2]. \tag{7}$$

2. Dielectric constant in the presence of an electron–hole plasma

Let us first outline the classical derivation of the intraband contribution to the dielectric constant. An electron with charge q and mass m, in an electric field E, moves according to

$$m_e(\ddot{x} + \dot{x}/\tau) = qE,$$

where τ is a relaxation time which describes some damping process for the e motion (we will say more about that τ later). For $E = E_0 e^{-i\omega t}$, the solution is $x = x_0 e^{-i\omega t}$; and the resulting polarization of N electrons is simply

$$P_e = Nqx = \frac{Nq^2}{m_e(-\omega^2 - i\omega/\tau)} E. \tag{8}$$

Adding the polarization of N holes with mass m_h and supposing that the relaxation time τ is the same, one gets for the polarization of a plasma having N e–h pairs,

$$P = Nq^2 \left(\frac{1}{m_e} + \frac{1}{m_h} \right) \frac{1}{-\omega^2 - i\omega/\tau} E. \tag{9}$$

This gives the so-called intraband contribution to the dielectric constant which comes from the existence of a plasma. One can recall that the presence of a damping process in the plasma motion gives an imaginary part to the dielectric constant, and consequently an absorption of the electromagnetic wave by the e–h plasma.

A more elaborate calculation [8] of the dielectric constant of an e–h plasma has already been done within the RPA approximation. This calculation takes into account the fact that the conduction band is degenerate and that the energy of an electron $\boldsymbol{k}$ has an ellipsoidal symmetry associated with the masses $m_{\parallel}$ and $m_{\perp}$. Moreover, the Coulomb interaction is not diagonal between heavy holes and light holes, which are the eigenstates of the Kohn–Luttinger representation of the valence band; consequently, a heavy hole can become a light hole after scattering, and the two populations of holes are not independent. It turns out that the light holes, which usually do not play an important role as their number is much smaller than the number of the heavy holes, give in fact the main contribution to the intraband dielectric constant [8]. Finally, the proper mass to use in the dielectric constant of a Ge-like semiconductor is

$$\frac{1}{\mu}=\frac{1}{3}\left(\frac{1}{m_{\parallel}}+\frac{2}{m_{\perp}}\right)+\frac{1}{2}\left(\frac{1}{m_{\mathrm{H}}}+\frac{1}{m_{\mathrm{L}}}\right). \tag{10}$$

This gives for silicon $\mu=0.12$ using the usual low density band structure. We have seen that at high density (say above 10^{21} cm^{-3}) one should take a different density-of-states according to Cohen and Chelikowski [6], and μ should increase. For electron and hole effective masses equal to the free electron mass, the plasma frequency effective mass goes to $\mu=\frac{1}{2}$.

Adding the polarizability of the lattice itself to the plasma contribution, the dielectric constant of a silicon sample excited with an e–h plasma of density N finally reads

$$\varepsilon=\varepsilon_0+\mathrm{i}\varepsilon_0'-\frac{Nq^2}{\mu(\omega^2+\mathrm{i}\omega/\tau)}\equiv(n+\mathrm{i}k)^2, \tag{11}$$

where $\varepsilon_0+\mathrm{i}\varepsilon_0'$ is the crystal dielectric constant at frequency ω and $n+ik$ is the sample index in the presence of the e–h plasma.

Let us give some orders of magnitude for the plasma density N_p at resonance, i.e., the density at which the real part of ε becomes zero and consequently the electromagnetic field changes from a propagating wave to a vanishing wave.

$$1/N_\mathrm{p}=4\pi\mathrm{e}^2/[\mu\omega^2(1+1/\omega^2\tau^2)\varepsilon_0]. \tag{12}$$

For a wavelength $\lambda=0.62\,\mu$m, $\varepsilon_0=14.4$ and assuming for simplicity $\omega\tau=0$, $N_\mathrm{p}=5\times10^{21}$ cm^{-3} if one takes $\mu=0.12$. N_p can go up to 2×10^{22} cm^{-3} if $\mu=0.5$. If one uses a 0.31 μm wavelength, the plasma resonance is reached for a density in the range 2–8 $\times10^{22}$ cm^{-3} (depending on the value taken for μ), which is extremely high compared with the total number of valence band electrons, 2×10^{23} cm^{-3}. On the other hand, a plasma density $N=5\times10^{20}$ cm^{-3} would correspond to the resonance for $\lambda\sim4\,\mu$m, and a 10 μm wavelength to $N_\mathrm{p}\sim2\times10^{19}$ cm^{-3}.

Figure (1) gives the reflectivity versus N/N_p for various relaxation times τ, in the case of a sample having a spatially homogeneous dielectric constant given by eq. (11). On the same figure is also shown the penetration depth $X=\lambda/4\pi k$. X rapidly drops when N/N_p approaches 1.

2.1. Electron–hole plasma relaxation time

The relaxation time τ which enters into the dielectric constant describes some damping process for the motion of electrons and holes in the presence of an electrical field. For a pure crystal, this damping comes from collisions between electrons or holes and the lattice and τ is of the order of the electron–phonon

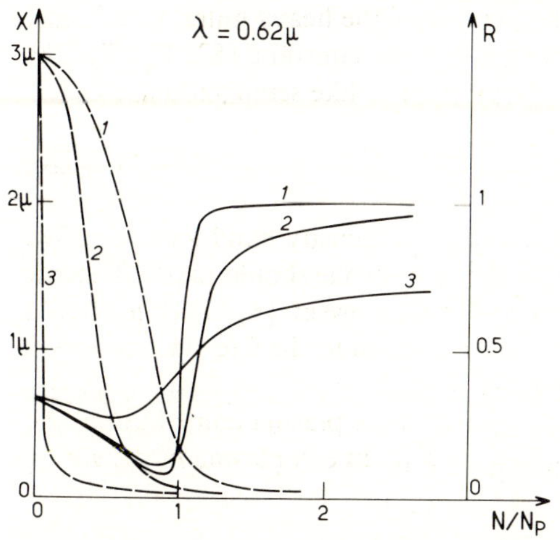

Fig. 1. Reflectivity R and penetration depth X versus e–h density N measured in plasma frequency density N_p for a spatially homogeneous silicon sample, and various values of τ. Curves 1: $\tau = 10^{-13}$ s; curves 2: $\tau = 5 \times 10^{-15}$ s; curves 3: $\tau = 5 \times 10^{-16}$ s.

relaxation time $\tau_{ph} \sim 10^{-13}$ s. But besides collisions with the atoms, there are also collisions between two electrons, two holes and one electron and one hole. In fact, as the collisions between two electrons (or two holes), conserve the total momentum of the two electrons, the total current cannot be changed and no damping results from such collisions. On the other hand, during a collision between one electron and one hole, the total momentum can still be conserved, whereas, the two particles having opposite charges, the total current (due to the electrons and the holes) changes and this gives rise to a new relaxation time τ_{eh} which is expected to decrease with the probability of such collision, i.e., when the e–h density N increases.

The e–h collision time can be easily estimated from the electron–impurity [9] collision time calculation using the screened Coulomb interaction and the Rutherford scattering cross section. One finds that τ_{eh}^{-1} increases linearly with plasma density at low n and saturates at high density (for a degenerate plasma) to a value of the order of $10^{16}\ s^{-1}$.

Adding the damping due to electron–phonons interaction, the total relaxation τ entering the dielectric constant is finally

$$\frac{1}{\tau} = \frac{1}{\tau_{ph}} + \frac{1}{\tau_{eh}}, \tag{13}$$

τ is of order 10^{-13} s for a low density plasma and decreases with increasing N. Indication of such a variation of τ has already been found in heavely doped semiconductors [10]. Very recently, τ as low as 10^{-16} s has been found in silicon excited by femtosecond pulses [3].

3. Generation of an electron–hole plasma by laser beam absorption

The general expression for the variation of the electromagnetic energy is

$$\frac{\partial u}{\partial t}=\frac{E+E^*}{2}\frac{\partial}{\partial t}\left(\frac{D+D^*}{2}\right) \qquad \text{with } \boldsymbol{D}=\varepsilon\boldsymbol{E}. \tag{14}$$

Noting that ε will vary more slowly with time than E, the time average of the energy beam variation becomes

$$\partial\bar{u}/\partial t=\tfrac{1}{4}(-\mathrm{i}\omega\varepsilon+\mathrm{i}\omega\varepsilon^*)EE^*. \tag{15}$$

The energy absorption comes from the imaginary part of the dielectric constant. There are two contributions to this imaginary part: one comes from the e–h plasma absorption (or intraband absorption), via the relaxation time τ and Joule effect, and gives a plasma heating. The second comes from the interband term ε_0' and is due to creation of e–h pairs. Picking only that part in $\partial\bar{u}/\partial t$, one gets the e–h generation rate as each created pair corresponds to the absorption of one photon energy $\hbar\omega$,

$$\hbar\omega\frac{\partial N}{\partial t}=\omega\frac{\varepsilon_0'}{2}EE^*. \tag{16}$$

(1) if the e–h density N stays far from the plasma frequency resonance, the sample dielectric constant is essentially independent of N and so stays constant with space and time $\varepsilon\simeq\varepsilon_0+\mathrm{i}\varepsilon_0'$: the electric field in the sample has then an exponential behaviour

$$E(z,t)=\frac{2E_0}{1+n_0+\mathrm{i}k_0}\exp\left[-\mathrm{i}\omega\left(t-\frac{n_0+\mathrm{i}k_0}{c}z\right)\right], \tag{17}$$

where $\varepsilon_0+\mathrm{i}\varepsilon_0'=(n_0+\mathrm{i}k_0)^2$. (The electric field amplitude results from the continuity of the E and B fields at the irradiated surface $z=0$.)

If one now introduces the Poynting vector or energy flux outside the sample $P_0=|(E_0+E_0^*)/2\wedge(H_0+H_0^*)/2|=\frac{1}{2}cE_0E_0^*$ and inside the sample $P=\frac{1}{2}cn_0EE^*$, the general expression (16) of the e–h density variation can be rewritten in a more intuitive form

$$\begin{aligned}\frac{\partial N}{\partial t}&=\frac{1}{\hbar\omega}\omega\frac{2n_0k_0}{2}E(0)E^*(0)\exp\left(-\frac{2\omega k_0 z}{c}\right)\\&=\frac{(1-R)P_0}{\hbar\omega D_0}\exp(-z/D_0),\end{aligned} \tag{18}$$

where $D_0=c/2\omega k_0$ is the usual laser beam penetration depth and $1-R=P(z=0)/P_0$ is the transmition factor at the irradiated surface. The quantity $(1-R)P_0/\hbar\omega$ is just the number of photons which penetrates into a unit surface and the exponential term takes into account the beam absorption.

(2) Equation (18) is no longer valid if the e–h plasma density becomes of the order of, or larger than, the plasma frequency resonance density. Then the dielectric constant ε depends on space and time via N. In that case, one has to go back to the Maxwell equations and consider a problem of nonlinear optics for a sample having a space- and time-dependent absorption. From curl $E = -\partial B/\partial t$ and c^2 curl $B=\partial D/\partial t$ with $\boldsymbol{D}=\varepsilon\boldsymbol{E}$, one gets for an electromagnetic wave propagating in the z direction,

$$c^2(\partial^2 E_y/\partial z^2) = -\omega^2 \varepsilon E_y, \qquad B_x = (\mathrm{i}/\omega)(\partial E_y/\partial z). \tag{19}$$

The proper solution relies on two boundaries conditions. For $z\to\infty$, the e–h density N goes to zero, the dielectric constant goes to $\varepsilon_0 + \mathrm{i}\varepsilon_0'$ and the electric field recovers the exponential behaviour of an unexcited sample

$$E(\infty, t) = A(t) \exp\left[\mathrm{i}\frac{\omega}{c}(n_0 + \mathrm{i}k_0)z\right]. \tag{20}$$

The coefficient $A(t)$ has to be determined from a second boundary condition. If E_0 and E_1 are the electric fields of the incident and reflected waves, the continuity of $\boldsymbol{E}$ and $\boldsymbol{B}$ at $z=0$ reads

$$E_0 + E_1 = E(0, t), \qquad -\frac{E_0}{c} + \frac{E_1}{c} = \frac{\mathrm{i}}{\omega}\frac{\partial E}{\partial z}\bigg|_{z=0}, \tag{21}$$

so that the electric fields outside and inside the crystal are related by

$$2E_0 = E(0, t) - \frac{\mathrm{i}c}{\omega}\frac{\partial E(z, t)}{\partial z}\bigg|_{z=0}. \tag{22}$$

As the knowledge of the electric field at any time and space allows us to obtain the plasma generation rate via eq. (16), eqs. (11), (16), (19), (20) and (22) form a complete set of equations and its numerical resolution will give the e–h density profile at any time as well as the instantaneous reflectivity,

$$R(t) = |E(0, t) - E_0|^2/|E_0|^2. \tag{23}$$

The evolution of the electronic system during the pulse excitation can be understood qualitatively noting that the leading edge of the pulse finds a virgin silicon sample: it is absorbed over a penetration depth D_0 ($\sim 3\,\mu$m for $\lambda = 0.62\,\mu$m) and creates e–h pairs with an exponential profile over D_0. The sample keeps reacting that way until the surface e–h density reaches the plasma frequency density N_p. Then the real part of the dielectric constant turns negative, the beam transmitted into the sample becomes a vanishing wave and is now absorbed over a few hundred angstroms, creating e–h pairs over that distance only. At the same time, the reflectivity increases as for a metal, and the values reached are closely related to the plasma conductivity, i.e., its relaxation time τ. The nonlinear set of equations has been solved numerically. Figure (2) shows the e–h density profile $N(z)$ at the end of the pulse for various pulse intensities. One

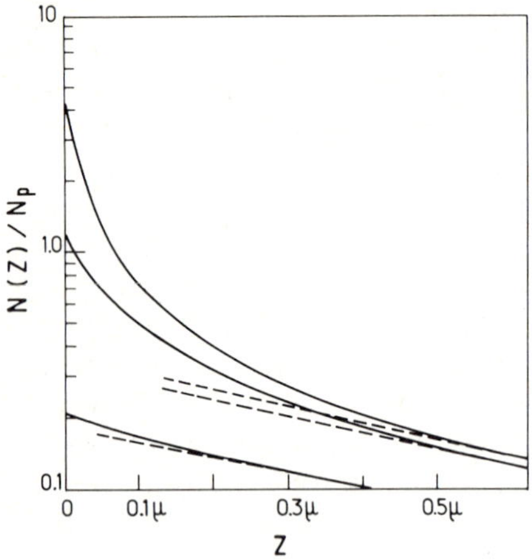

Fig. 2. Electron–hole density profile at the end of the pulse for various pulse intensities: $I/I_0 = 0.2$, 1, 5.

checks that, far from the surface, $N(z)$ has an exponential behaviour and that $N(z)$ increases rapidly above N_p only very close to the surface. This calculation allows one also to get the relaxation time τ by comparison of the average of the instantaneous reflectivity over the pulse duration obtained from eq. (23) with the experimental measurement of the total energy reflected from the entire pulse [3–11]. The best fit gives $\tau \sim 10^{-16}$ s and rules out values of τ of order 10^{-15} s and below. This value is much higher than the usual electron–phonon relaxation time (which anyhow is expected to increase – and not to decrease – with the plasma density due to screening) and indicates that a new damping mechanism appears in a very dense plasma (for $n \sim 10^{21}$–10^{22} cm^{-3}) due to electrons–holes collisions (see Section 3). We want to emphasize that the preceding calculation is based on two main assumptions.

– The interband dielectric constant does not change during the pulse; this implies that the lattice temperature does not change – this should be correct for pulses shorter than 1 ps – but also that $\varepsilon_0 + i\varepsilon'_0$ does not vary very much with plasma density.

– The e–h density results only from pairs created by interband absorption. This implies that the total Auger time (e–h recombination plus impact ionization) is longer than the pulse duration, probably correct for pulses as short as the 100 fs ones. This also implies that the e–h pairs do not move during the pulse. As we will see later about the plasma motion, this is also usually verified for femtosecond pulses.

4. Electron–hole plasma evolution

Once generated by the absorption of the laser beam, the e–h pairs will recombine and move. Let us study these two physical phenomena.

4.1. Auger effect

In silicon, the variation of the e–h pair density comes essentially from the Auger effect. An impressive amount of work on Auger effect has been done by A. Haug and coworkers [12]. We will just outline here the derivation of the Auger lifetime. The number of e–h pairs can change either by a process where one e–h pair ($\boldsymbol{k}_1$, $\boldsymbol{k}'_1$) disappears giving its energy to a second electron $\boldsymbol{k}_2$ which goes up in the conduction band in $\boldsymbol{k}'_2$ – Auger recombination – or by the inverse process where, after a Coulomb interaction one e–h pair is created from an electron $\boldsymbol{k}'_2$ high in the band – impact ionization. The statistical occupation factor for such processes is

$$P = f_e(1)\,f_e(2)\,f_h(1')(1-f_e(2')) - (1-f_e(1))(1-f_e(2))(1-f_h(1'))\,f_e(2'), \tag{24}$$

where f_e and f_h are the Fermi distribution for e and h. As the total energy is conserved in both processes and as $1-f=f\,\exp[(\varepsilon-\mu)/kT]$, P can be rewritten as

$$P = \{1-\exp[-(E_g+\mu_e+\mu_h)/kT]\}\,f_e(1)\,f_e(2)\,f_h(1')(1-f_e(2')).$$

Using now the Fermi golden rule, the transition probability for the change of the number of e–h pairs is

$$W \sim \int \mathrm{d}^3k_1\,\mathrm{d}^3k_2\,\mathrm{d}^3k'_1\,\mathrm{d}^3k'_2\,|M|^2\,\delta(\boldsymbol{k})\,\delta(\varepsilon)P,$$

where M is the transition matrix element and the δ functions insure conservations of impulse and energy. For an indirect gap semiconductor such as silicon, the first-order transition is not possible and the $\boldsymbol{k}$ conservation is insured by an additional phonon. But in any case, for a classical plasma, the Boltzmann distributions in P allow one to rewrite the probability transition as

$$W = \left(1-\exp\left[-\frac{E_g+\mu_e+\mu_h}{kT}\right]\right)\exp\left(\frac{2\mu_e+\mu_h}{kT}\right)A, \tag{25}$$

where A is a factor which does not depend on μ, i.e., on the electron or hole density and contains all the details such as band structure, phonons etc . . . . Finally, as $\mu = kT\ln n$,

$$W = Cn(n^2 - n_0^2(T)) = n/\tau(n), \tag{26}$$

$n_0(T)$ being the density for which the e–h pair chemical potential $E_g+\mu_e+\mu_h=0$, i.e., the equilibrium carrier density for a carrier temperature T. The first term in

W corresponds to the Auger recombination while the second gives the impact ionization. For a quantum e–h gas, there is obviously no impact ionization, and the density dependence of the Auger recombinaison is weaker than in the classical case [12]. Moreover, at very high densities, screening effects should be included, which also will smooth the density dependence.

For silicon $C=4\times10^{-31}$ cm^6 s^{-1}, this gives an Auger recombination time $(Cn^2)^{-1}\sim2\times10^{-12}$ s for $n\sim10^{21}$ cm^3. If the plasma is in equilibrium with the lattice, the carrier temperature will be the lattice temperature, $n_0(T)$ will be much smaller than n and the impact ionization negligible compared with Auger recombination. This Auger recombination is extremely efficient to prevent very high density. Let us consider the Palaiseau femto-second laser which, at highest fluence, delivers ~1 J cm^{-2} in $\tau_p=10^{-13}$ s; if these 2 eV photons spread over a 3 μm penetration depth, this corresponds to a generation rate of $G=10^{22}$ cm^{-3}/ 10^{-13} s. Using the usual Auger coefficient, this gives a density such that $G=Cn^3$, i.e., $n\sim10^{22}$ cm^{-3} only. In fact at those densities, the usual value for C, obtained in the classical low density limit, does not apply, and the effective Auger time will be longer. Moreover, at these e–h densities, the plasma frequency of the pump beam is overpassed, i.e., the laser penetration depth is much smaller, so that highest densities can indeed be reached.

4.2. Plasma hydrodynamics

The study of the e–h plasma evolution corresponds to the determination of the local e–h density, the temperature, and the velocity $\boldsymbol{v}$ at any time. For that one needs three equations. For simplicity, we will assume an uniform excitation such that the only space variable is the distance z to the irradiated surface.

$$\frac{\partial n}{\partial t}+\nabla nv=G(z,t)-\frac{n}{\tau(n)}, \tag{27}$$

G being the laser generation rate and $\tau(n)$ the full Auger lifetime.

The second hydrodynamic equation is the equation of motion of a unit volume of e–h pairs or Euler's equation

$$Mn\left(\frac{\partial v}{\partial t}+\frac{v\partial v}{\partial z}+\frac{v}{\tau^*}\right)=-\nabla p, \tag{28}$$

M is the translational mass of an e–h pair, and v/τ^* is a viscous term which describes phenomenologically the damping of the e–h motion [13]. For simplicity, τ^* is assumed to be constant. Neglecting first the quadratic term in v, this gives, for steady state, an e–h current [14]

$$j=nv=-(\tau^*/M)\nabla p. \tag{29}$$

It is possible to rewrite eq. (29) for the current in the usual Onsager form derived from the Boltzmann equation

$$j = -(\tau^*/M)(n\nabla\mu + s\nabla T), \tag{30}$$

where the Gibbs–Duhem relation has been used and s is the entropy per unit volume.

Equation (29) or (30) gives for the diffusion coefficient ($j = -D\nabla n$)

$$D = \frac{\tau^*}{M}\left.\frac{\partial P}{\partial n}\right|_T \equiv \frac{n\tau^*}{M}\left.\frac{\partial\mu}{\partial n}\right|_T. \tag{31}$$

In the classical limit, the pressure $P = 2nkT$, so that one recovers the usual expression for the ambipolar diffusion coefficient $D = (2\tau^*/M)kT$. For $\tau^* \sim 10^{-13}$ s and $M \simeq 0.5$, the value of D at room temperature is of order 20 $cm^2\ s^{-1}$.

For an e–h gas in the quantum limit, the pressure p is deduced from the free energy (2) as $p = -\partial F/\partial v|_T \simeq \frac{2}{3}an^{5/3}$ and the diffusion coefficient reads

$$D \simeq \tfrac{10}{9}a(\tau^*/M)n^{2/3}. \tag{32}$$

For $n \sim 10^{21}\ cm^{-3}$, D is $10\times$ larger than the classical limit value at room temperature. As discussed in section 1, expression (32) should be valid below $10^{21}\ cm^{-3}$. At higher density, Chelikowski and Cohen's band structure leads to $\mu = a'\sqrt{n}$ (see eq. (5)), so that the diffusion coefficient, $D' \simeq \frac{1}{2}a'(\tau^*/M)\sqrt{n}$, has a weaker dependence in density.

The last and third hydrodynamical equation, in addition to the continuity equation (27) and the Euler's equation (28), is the equation for the entropy or heat equation. This equation should include the heat flux in the e–h system as well as in the phonon system. Moreover, one should write a detailed account of the heat produced by the very fast thermalization of the created e–h pairs, as well as the thermalization of the Auger electrons – or holes. Fortunately, this rather complicated heat equation can be avoided for a very dense e–h plasma in the quantum limit. In that case, the pressure does not depend on T, and the temperature is decoupled from the two first equations, so that the density and velocity distributions can be determined [13] without solving the heat equation.

Let us now give some order of magnitude for the e–h pairs velocity in the quantum limit. For a plasma created by a laser with a penetration depth x_0, the density gradient ∇n is of order n/x_0 so that

$$v = j/n \simeq D\nabla n/n \sim D/x_0. \tag{33}$$

If $n \sim 10^{21}\ cm^{-3}$, x_0 is of order of 2 μm for a 0.62 μm pump beam. Using (32), this gives an e–h pair velocity $v \sim 10^4$ m/s. One may then compare the decrease of the e–h density due to this e–h current with the decrease due to Auger recombination. Going back to the conservation equation (27), the effective lifetime τ_j associated with the current corresponds to

$$n/\tau_j \sim \nabla j \simeq j/x_0 \sim D\nabla n/x_0 \sim Dn/x_0^2. \tag{34}$$

This gives $\tau_j \sim x_0^2/D \sim 2\times 10^{-10}$ s: as τ_j is longer than the Auger lifetime at this density, the evolution of the e–h pairs created over 2 μm is essentially controlled by Auger effect.

For $n \sim 10^{22}\ \mathrm{cm}^{-3}$, for the same 0.62 μm pump beam, the penetration depth falls to $x_0 \sim 1000$ Å. Using again (32) for D, one finds $v \sim 10^6$ m/s. With such a velocity, the plasma expands during a 100 fs pulse over $L \sim v\tau_p \sim 10^{-7}$ m, i.e., over a distance of the order of the penetration depth x_0. Consequently, this velocity will affect the e–h density. The same conclusion is obtained if one considers the effective lifetime τ_j which is then found to be of the order of 10^{-13} s, i.e., the pulse duration time. But velocities as high as 10^6 m/s are larger than the sound velocity: the coupling between the e–h pairs and the phonon system cannot be described as simply as with a viscous friction term v/τ^* in eq. (28). Such very high velocities are found when the e–h density profile is very steep and one may expect in that case a highly nonlinear regime as for shock waves [15].

5. Fascinating melting at $T=0$

A few years ago, Van Vechten [16] proposed a new melting mechanism to explain the laser annealing of ion-implanted semiconductors. Even if it is now well accepted that, with 10^{-9} and even 10^{-11} s laser pulses, the annealing is just due to a simple thermal process, Van Vechten's idea is very nice and one can wonder how to produce and see such a melting. The basic idea is the following: each time one electron is excited from the valence band to the conduction band, one covalent bond is broken and if all the electrons are removed from the valence band, there would be no more bonds between the atoms and surely no more solid phase: this melting, due to the existence of a very dense e–h plasma, does not need any thermal excitation and could happen even at $T=0$.

More precisely, it has been shown that the frequency of the TA phonon, responsible for the crystal stability, decreases with increasing e–h pair density, and cancels when a fraction of the total number of valence band electrons is excited. This softening of the TA mode has already been used 10 years ago by Heine and Van Vechten [17] to explain the observed band-gap variation with T. More recently, [18] we have studied the effect of this softening on the melting temperature of Si or Ge in order to find the precise effect of additional e–h pairs due to a laser excitation. We have found that this softening of the TA mode produces already an instability in the electronic system at a critical temperature T^*. The crystal will undergo a phase transition at T^* if this electronic instability takes place before the usual instability in the atomic lattice itself. If this is the case, the thermal melting of silicon might be already due to e–h pairs. This hypothesis, which is supported by the fact that in silicon T^* is of order 2000 K and that the melting temperature is rather low compared with the vacancy energy, could be easily checked experimentally as the e–h plasma density at

T^* is found to be of order of 10^{21} cm^{-3}, i.e., two orders of magnitude above the value deduced from low density interpolation.

Let us now outline the main steps of the calculation. The free energy F of a crystal having N_a atoms in a volume V, at a temperature T in the presence of N e–h pairs contains an atomic part, a phonon part and an e–h plasma part. The equilibrium plasma density at a temperature T, for an isolated system, i.e., in absence of laser excitation, is obtained by writing that the e–h pair chemical potential $\mu = \partial F/\partial N|_{T,V,N_a} = 0$. μ is composed of an electron term μ_e ($\mu_e \sim kT$) ln n_e for a classical gas and $\mu_e \sim n^{2/3}$ for a high density quantum gas), a hole term μ_h and a contribution coming from the phonon free energy via the softening of the TA mode. Using, as a first approximation, a linear dependence for the TA frequency

$$\omega(q, n) = \omega_0(q)(1 - \alpha n/n_a), \tag{35}$$

where n_a is the density of atoms, the e–h pairs depending part of the phonon free energy

$$\sum kT \ln\left[1 - \exp\left(-\frac{\hbar\omega}{kT}\right)\right] \simeq \sum kT \ln\frac{\hbar\omega}{kT} \sim N_a kT \ln\left(1 - \frac{\alpha n}{n_a}\right) \tag{36}$$

gives a contribution to μ which is negative, increases with T and diverges at $n = n_a/\alpha$. The e–h chemical potential, including this TA mode contribution, reads finally

$$\mu = E_{GO} + \mu_e + \mu_h - \frac{\alpha kT}{1 - \alpha n/n_a}. \tag{37}$$

In the classical limit, one recovers the band-gap variation with temperature, as eq. (37) gives for the equilibrium density (at $\mu = 0$)

$$n \sim \exp[-(E_{GO} - \alpha kT)/2kT], \tag{38}$$

if n stays far below n_a/α. If one now allows n to increases, one finds for $\mu(n, T)$ the set of curves shown in fig. 2. As the phonon contribution to μ increases with T, there is a temperature T^* above which no solution can be found to the equilibrium condition $\mu = 0$ and consequently no e–h density at equilibrium. The highest stable e–h density n^*, reached at the critical temperature T^*, has to be a sizeable fraction of the density scale n_a/α for the softening of the TA mode, as the instability in the electronic system results from this softening. In silicon, $n_a/\alpha \sim 8 \times 10^{21}$ cm^{-3} as deduced from the T dependence of the band gap, and $n^* \sim 10^{21}$ cm^{-3}.

If now the crystal is irradiated by a laser, the e–h density is not any more given by $\partial F/\partial N = 0$ but is imposed by a balance between the laser generation G rate and the full Auger effect (e–h recombinaison and impact ionization). Such a plasma, with density $n(G)$, can be stable at a temperature T only if its compressibility is positive; this imposes $\partial\mu/\partial n$ positive for $n(G)$ and consequently T

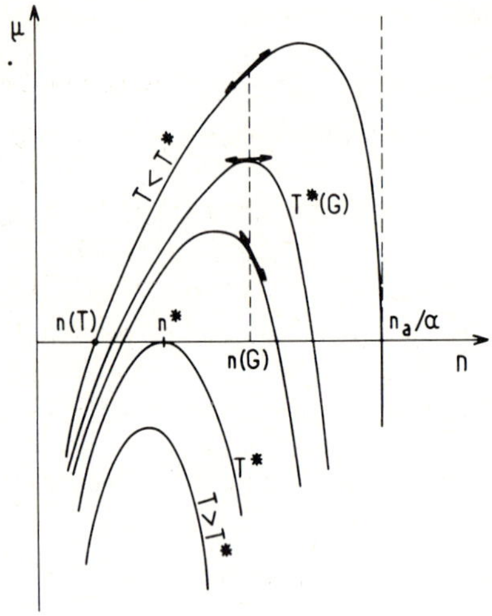

Fig. 3. Electron–hole pair chemical potential μ versus density n for various temperatures. T^* is the instability temperature without laser, while $T^*(G)$ is the instability (or melting) temperature with a laser having a generation rate G, i.e., imposing an e–h density $n(G)$.

below a critical value $T^*(G)$ (see fig. 3). This temperature $T^*(G)$ which is the highest for which a plasma with density $n(G)$ can be stable, is in fact nothing else than the melting temperature in presence of a laser with a generation rate G. Here again, as the density scale for the phonons softening is n_a/α, one expects a sizeable change in the melting temperature, only if the laser can produce a sizeable fraction of n_a/α. At the limit of $n(G)=n_a/\alpha$, the melting temperature drops to zero.

Let us now end by discussing the possibility of seeing this melting at $T=0$. One would need a very high density plasma ($\sim 10^{22}\,\text{cm}^{-3}$) and on the other hand a very cold sample in order to be sure that the fusion is not a simple thermal one. This implies a very powerful laser in order to reach $10^{22}\,\text{cm}^{-3}$, as the Auger processes are extremely fast at these densities, but also, to look at the transition very soon, as the crystal will be heated up very fast, precisely due to the cooling of the Auger electrons by phonons emission. This could be achieved with femtosecond lasers if one looks at the transitions over few hundred femtoseconds. Now, comes the problem of the nucleation time for a cold melting as well as the intrinsic time necessary to see that the crystal is indeed a liquid, i.e., the atoms have moved over a distance of the order of their interatomic distance. In order for this time to be smaller than the heating time $\sim 10^{-13}$ s, the atom velocity has to be larger than 10^3 m/s. In consequence, the domain where it is possible to produce and see melting at zero degree is very tiny and unfortunately will make the experiment very difficult, if not impossible.

References

[1] L.V. Keldysh, 1968, Proc. 9th Int. Conf. Physics Semicon., Moscou (Leningrad, Nauka) pp. 1303–12.
[2] For a review on e–h drops see T.M. Rice, J. Hensel, T. Phillips and G.A. Thomas, Solid State Physics, Vol. 32.
[3] D. Hulin, M. Combescot, J. Bok, A. Migus, J.Y. Vinet and A. Antonetti, Phys. Rev. Lett., March (1984).
[4] M. Combescot and P. Nozières, J. Phys. C5 (1972) 2369.
[5] W.F. Brinkman and T.M. Rice, Phys. Rev. B7 (1973) 1508.
[6] J.R. Chelikowsky and M.L. Cohen, Phys. Rev. B14 (1976) 556.
[7] M. Combescot, Phys. Rev. Lett. 37 (1976) 390.
[8] M. Combescot and P. Nozières, Solid State Commun. 10 (1972) 301.
[9] Landau and Lifchitz, Quantum mechanics, p. 595.
[10] M. Miyao, T. Motooka, N. Natsuaki and T. Tokuyama, Laser and electron-beam solid interactions, eds., J.F. Gibbons et al. (Academic, New York, 1978) p. 163,
[11] See also D. Hulin's contribution to this conference.
[12] A. Haug, D. Kerkhoff and W. Lochmann, Phys. Status Solidi B89 (1978) 357; A. Haug, J. Phys. C16 (1983) 4159 and reference therein.
[13] M. Combescot, Phys. Lett. 85A (1981) 308; Phys. Rev. Lett. 49 (1982) 1413.
[14] In case of a spatially varying band gap E_g, as in presence of an uniaxial pressure or a temperature gradient, one should add $n\nabla E_g$ to the RHS of eq. (29). This term is usually negligible compared with the density gradient due to the pump beam penetration depth.
[15] M. Combescot, Solid State Commun. 81 (1979) 30.
[16] J.A. Van Vechten, J. Phys. (Paris) C41 (1980) C4-15.
[17] V. Heine and J.A. Van Vechten, Phys. Rev. B13 (1976) 1622.
[18] M. Combescot and J. Bok, Phys. Rev. Lett. 48 (1982) 1413; J. Bok, Phys. Lett. 84A (1981) 448.

Journal of Luminescence 30 (1985) 18–36
North-Holland, Amsterdam

NON-EQUILIBRIUM PLASMA: THEORY AND EXPERIMENT

G. MAHLER* and A. FOURIKIS

Institut für theoretische Physik, Universität Stuttgart D-7000 Stuttgart 80, FRG

Transport properties of highly excited semiconductors are investigated. The model under consideration is an interacting electron–hole–acoustic phonon system far from equilibrium, described by parametrized single-particle distribution functions. Attention is focused on ambipolar transport of the electronic subsystem including diffusion, thermodiffusion, and drift. Modifications due to many-body effects are discussed. The theoretical scheme is used to interpret experimental data from various optical measurements.

1. Introduction

Properties of highly excited semiconductors have intensively been studied during the last twelve years or so. From a careful optical analysis a detailed understanding of the phase diagram of the generated electron–hole system has been achieved [1]. However, in spite of the fact that the state induced by the laser is indeed far from equilibrium there seemed to be no need to go beyond more or less phenomenological kinetic considerations [2]. Only recently [3] has it become clear that the generally used concept of quasi-equilibrium is in fact insufficient to describe the highly excited state. This shortcoming is due to the fact that inhomogeneous excitation conditions (coupling to a laser field across a crystal surface) will, in general, even under stationary conditions, lead to an inhomogeneous state which should therefore exhibit internal macroscopic currents. Study of those currents is the subject of transport theory. Consideration of the above situation under high excitation conditions is the subject of the present paper [4].

* Present address: Laboratoire de spectroscopie et optique du corps solide, F67084 Strasbourg Cedex.

2. The system

The system under consideration is defined by its effective Hamiltonian $\hat{H}$ which describes excitations with respect to the so-called "effective vacuum" $|0\rangle$. In case of a simple two-band semiconductor this vacuum corresponds to the rigid periodic lattice and from the electronic-state point of view to a completely filled valence band. This vacuum, thus, defines the zero-point energy $E_0=0$, the zero-point momentum $\boldsymbol{p}_0=0$ (modulo reciprocal lattice vector), the zero point of electron number in conduction band $N_e=0$ and of holes in valence band $N_h=0$. For the sake of simplicity excitations are restricted to electrons e, holes h, and acoustic phonons ph, with given dispersion relations $E_0^\lambda(p^\lambda)$, $\lambda=$e, h, ph. These three subsystems are coupled: electrons and holes by Coulomb interactions, electrons (holes) and phonons by deformation-potential coupling and phonons among themselves by anharmonicity.

$$H=\sum_{\lambda} H_1^\lambda+\sum_{\lambda\neq\lambda'} H_2^{\lambda\lambda'}, \tag{1}$$

$$H_1^\lambda=\sum_{\nu} E_0^\lambda(p_\nu^\lambda), \tag{2}$$

$$H_2^\lambda=\tfrac{1}{2}\sum_{\nu,\nu'} \phi^{\lambda\lambda'}(\boldsymbol{r}_\nu^\lambda, \boldsymbol{r}_{\nu'}^{\lambda'}). \tag{3}$$

Electron–hole bound states are not considered here.

3. The homogeneous state

Classically, the state of a many-particle system may be described by means of the many-particle distribution function $f(\boldsymbol{x}_1^1, \boldsymbol{x}_2^1, \ldots, \boldsymbol{x}_\nu^\lambda, \ldots)$, where

$$\boldsymbol{x}_\nu^\lambda=\{\boldsymbol{p}_\nu^\lambda, \boldsymbol{r}_\nu^\lambda\} \tag{4}$$

denotes momentum and position of particle ν in subsystem λ.

Quantum mechanically $\boldsymbol{x}_\nu^\lambda$ has only a restricted meaning in terms of wave packets [5] (a more rigorous picture is provided by the Wigner matrix). A much reduced but often sufficient information is contained in the one-particle distribution functions $f^\lambda(\boldsymbol{x}^\lambda, t)$.

3.1. Thermal equilibrium

In case of thermal equilibrium the state is spatially homogeneous and uniquely fixed by temperature T,

$$f_{\mathrm{w}}^\lambda(\boldsymbol{p})=\{\exp[E^\lambda(\boldsymbol{p}^\lambda)-\mu^\lambda-\boldsymbol{w}\cdot\boldsymbol{p}^\lambda]/k_{\mathrm{B}}T]\pm 1\}^{-1}, \tag{5}$$

with the constraints

$$n = n^{\mathrm{e}} = n^{\mathrm{h}}, \tag{6}$$

$$\mu^{\mathrm{e}}(n, T) - \mu^{\mathrm{h}}(n, T) = 0, \tag{7}$$

$$\mu^{\mathrm{ph}}(n^{\mathrm{ph}}, T) = 0, \tag{8}$$

$$\boldsymbol{w} = 0. \tag{9}$$

Here, n^{λ} are the particle densities, μ^{λ} the chemical potentials, and $\boldsymbol{w}$ a drift velocity. $\boldsymbol{w}$ must be zero in equilibrium since the momentum is not collisionally invariant [5] (see below).

3.2. Non-equilibrium

More intensive variables are required to specify the homogeneous non-equilibrium state. For the following it is essential that the processes responsible for the evolution of the state occur on two different time scales [1]: While fast processes (e–e, h–h, and e–h scattering) are assumed to still establish a temperature in each subsystem, the slow ones (e–ph, h–ph scattering, e–h recombination) control the deviations from equilibrium. These deviations are incorporated into eq. (5) by relaxing some of the above constraints:

(i) Electronic quasi-equilibrium,

$$\mu^{\mathrm{e}}(n, T) - \mu^{\mathrm{h}}(n, T) \neq 0 \tag{10}$$

(n is no longer a function of T).

(ii) Drift of local system (drag), $\quad \boldsymbol{w} \neq 0. \qquad (11)$

(iii) Non-equilibrium between subsystems

$$T = T^{\mathrm{e}} = T^{\mathrm{h}}, \qquad \Delta T = T - T^{\mathrm{ph}} \neq 0, \tag{12}$$

$$\boldsymbol{w} = \boldsymbol{w}^{\mathrm{e}} = \boldsymbol{w}^{\mathrm{h}}, \qquad \Delta \boldsymbol{w} = \boldsymbol{w} - \boldsymbol{w}^{\mathrm{ph}} \neq 0. \tag{13}$$

We see that the state is now characterized by the parameter set

$$\{T, n, \boldsymbol{w}, T^{\mathrm{ph}}, \boldsymbol{w}^{\mathrm{ph}}\}. \tag{14}$$

As discussed by Keizer [22], such non-equilibrium states will, in general, be connected with unusual fluctuation spectra, which, however, we do not consider here.

4. Influence of interactions

The influence of interactions is twofold [6]: in the first place they contribute to the non-dissipative processes of the system, i.e., to the thermodynamic functions like internal energy, etc.; secondly they are indispensable for dissipative phenomena, i.e., relaxation and transport processes.

4.1. Self-energy corrections

Let us consider

$$U=\langle\langle\hat{H}\rangle\rangle=\sum_{\lambda}\int \mathrm{d}^6x\, E_0^{\lambda}(\boldsymbol{p}^{\lambda})f^{\lambda}(\boldsymbol{x}^{\lambda})+\tfrac{1}{2}\sum_{\lambda,\lambda'}\iint \mathrm{d}x_1^6\,\mathrm{d}x_2^6\phi^{\lambda\lambda'}f_2^{\lambda\lambda'}(\boldsymbol{x}_1^{\lambda},\boldsymbol{x}_2^{\lambda'}). \tag{15}$$

For the thermal equilibrium U was just the usual total internal energy. Quasi-particle energies are then introduced as functional derivatives [7]

$$E^{\lambda}(\boldsymbol{x})=\delta U/\delta f^{\lambda}(\boldsymbol{x}), \tag{16}$$

which give the change of U when a particle λ with momentum $\boldsymbol{p}$ is added to the system at position $\boldsymbol{r}$. Neglecting correlations,

$$f_2^{\lambda\lambda'}=f^{\lambda}f^{\lambda'}, \tag{17}$$

eq. (16) is equivalent to the quasi-particle dispersion relation

$$E^{\lambda}(\boldsymbol{x})=E_0^{\lambda}(\boldsymbol{p})+S^{\lambda}(\boldsymbol{r}), \tag{18a}$$

with

$$S^{\lambda}(\boldsymbol{r})=\sum_{\lambda'}\int \mathrm{d}^6x^{\lambda'}\phi^{\lambda\lambda'}f^{\lambda'}(x^{\lambda'},t). \tag{18b}$$

The self-energy correction S^{λ} (here in the mean-field approximation) depends on state $f^{\lambda'}$.

4.2. Evolution of distribution functions

The Boltzman equation for quasi-particles reads [6, 7]:

$$\hat{\mathscr{D}}^{\lambda}f^{\lambda}(\boldsymbol{x}^{\lambda},t)=\sum_{\lambda'}J^{\lambda\lambda'}(\boldsymbol{x}^{\lambda},t), \tag{19}$$

where

$$\hat{\mathscr{D}}^{\lambda}=\frac{\partial}{\partial t}+\frac{\partial E^{\lambda}}{\partial \boldsymbol{p}^{\lambda}}\cdot\frac{\partial}{\partial \boldsymbol{r}^{\lambda}}-\frac{\partial E^{\lambda}}{\partial \boldsymbol{r}^{\lambda}}\cdot\frac{\partial}{\partial \boldsymbol{p}^{\lambda}}+\boldsymbol{F}_0^{\lambda}\cdot\frac{\partial}{\partial \boldsymbol{p}^{\lambda}} \tag{20}$$

is the streaming operator, $\boldsymbol{F}_0^{\lambda}$ an external force on particles of subsystem λ and $J^{\lambda\lambda'}$ the collision integrals.

5. The inhomogeneous state

To zero order, the inhomogeneous state is characterized by the local non-equilibrium state, i.e., by eq. (5) where all intensive parameters are now allowed to depend on $\boldsymbol{r}^{\lambda}$. In order to obtain the first-order corrections we linearize [8]

$$J^{\lambda\lambda'} = -\frac{f^{\lambda} - f_0^{\lambda}}{\tau_{\mathrm{R}}^{\lambda\lambda'}} - \frac{f^{\lambda} - f_{\mathrm{w}}^{\lambda}}{\tau_{\mathrm{N}}^{\lambda\lambda'}}, \tag{21}$$

where $\tau_{\mathrm{R}}^{\lambda\lambda'}$ is the relaxation time due to resistive scattering (umklapp, impurity, boundary scattering) and $\tau_{\mathrm{N}}^{\lambda\lambda'}$ relaxation time due to momentum conserving N processes. Keeping in (19) only $J^{\mathrm{e,h-ph}}$ and $J^{\mathrm{ph-ph}}$, we get for $\tau_{\mathrm{R}}^{\lambda\lambda'} \gg \tau_{\mathrm{N}}^{\lambda\lambda'} \approx \tau^{\lambda\lambda'}$

$$f^{\lambda}(\boldsymbol{x}^{\lambda}, t) = f_0^{\lambda}(\boldsymbol{x}^{\lambda}, t) - \frac{\partial f_0^{\lambda}}{\partial \boldsymbol{p}^{\lambda}} \tau^{\lambda-\mathrm{ph}} T^{\lambda}\{E^{\lambda}(\boldsymbol{x}^{\lambda})\boldsymbol{X}_u^{\lambda} + \boldsymbol{X}_n^{\lambda}\} - \frac{\partial f_0}{\partial E^{\lambda}} \boldsymbol{w}^{\lambda}\boldsymbol{p}^{\lambda}, \tag{22a}$$

$$\boldsymbol{X}_n^{\lambda} = \nabla(-\mu^{\lambda}/T^{\lambda}) + \boldsymbol{F}_0^{\lambda}/T^{\lambda}, \qquad \boldsymbol{X}_u^{\lambda} = \nabla(1/T^{\lambda}). \tag{22b}$$

It will turn out, however, that $\boldsymbol{w}$ for the electron–hole system is not necessarily small, so that the linear correction becomes insufficient. Fortunately, the effect of a fast drift can easily be included to any order provided

$$\partial E^{\lambda}/\partial \boldsymbol{p}^{\lambda} = \boldsymbol{p}^{\lambda}/m^{\lambda}, \tag{23}$$

which holds for parabolic bands. In this case,

$$f_{\mathrm{w}}(\boldsymbol{p}^{\lambda}, \boldsymbol{r}) = f_0^{\lambda}(\boldsymbol{p}^{\lambda} - m^{\lambda}\boldsymbol{w}, \boldsymbol{r}^{\lambda}). \tag{24}$$

6. Macroscopic variables

Macroscopic variables are obtained by averaging microscopic one-particle variables $\xi^{\lambda}(\boldsymbol{x}^{\lambda})$ over momentum space [6].

6.1. Local fields

For the particle number, $\xi^{\lambda} = 1$, we find

$$\langle 1 \rangle \equiv \frac{1}{n^{\lambda}(\boldsymbol{r}^{\lambda}, t)} \int \frac{\mathrm{d}^3\boldsymbol{p}^{\lambda} g^{\lambda}}{(2\pi\hbar)^3} f^{\lambda}(\boldsymbol{x}^{\lambda}, t) = 1, \tag{25}$$

which defines the local particle density $n^{\lambda}(\boldsymbol{r}^{\lambda}, t)$; g^{λ} is the degeneracy (e.g., $g^{\lambda} = 2$ for spin 1/2 particles). Similarly, we introduce the kinetic energy density

$$u_{\mathrm{kin}}^{\lambda}(\boldsymbol{r}, t) = n^{\lambda}\langle E_0^{\lambda}(\boldsymbol{p}^{\lambda})\rangle, \tag{26}$$

the quasi-particle energy density [7]

$$\delta u^{\lambda}(\boldsymbol{r}, t) = \delta u_{\mathrm{kin}}^{\lambda} + \delta n^{\lambda} S^{\lambda}(\boldsymbol{r}), \tag{27}$$

and the quasi-momentum density

$$\boldsymbol{q}^{\lambda}=n^{\lambda}\langle \boldsymbol{p}^{\lambda}\rangle. \tag{28}$$

6.2. Local currents

The corresponding current densities are

$$\boldsymbol{j}_{\xi}^{\lambda}(\boldsymbol{r}^{\lambda}, t)=n^{\lambda}\langle \xi^{\lambda}(\partial E^{\lambda}/\partial \boldsymbol{p}^{\lambda})\rangle.$$

In particular,

$$\boldsymbol{j}_{\mathrm{kin}}^{\lambda}=n^{\lambda}\langle E_{0}^{\lambda}(\partial E^{\lambda}/\partial \boldsymbol{p}^{\lambda})\rangle \tag{29a}$$

is the local kinetic energy current,

$$\boldsymbol{j}_{u}^{\lambda}=n^{\lambda}\langle E^{\lambda}(\partial E^{\lambda}/\partial \boldsymbol{p}^{\lambda})\rangle=\boldsymbol{j}_{\mathrm{kin}}^{\lambda}+S^{\lambda}\boldsymbol{j}_{n}^{\lambda} \tag{29b}$$

the total energy current, and

$$\boldsymbol{j}_{n}^{\lambda}=n^{\lambda}\langle \partial E^{\lambda}/\partial p^{\lambda}\rangle\equiv n^{\lambda}\boldsymbol{v}_{\mathrm{D}}^{\lambda} \tag{30}$$

the local particle current with v_{D}^{λ} being the macroscopic velocity. For a parabolic dispersion law

$$\boldsymbol{j}_{n}^{\lambda}=\boldsymbol{q}^{\lambda}/m^{\lambda}. \tag{31}$$

Finally,

$$R^{\lambda}=n^{\lambda}\left\langle \boldsymbol{p}^{\lambda}\frac{\partial E^{\lambda}}{\partial \boldsymbol{p}^{\lambda}}\right\rangle \tag{32}$$

is the momentum flux tensor.

Using distribution functions (22), we get in the lattice reference frame: (compare ref. [9])

$$\boldsymbol{j}_{n}^{\lambda}=L_{nn}^{\lambda}\boldsymbol{X}_{n}^{\lambda}+L_{nu}^{\lambda}\boldsymbol{X}_{u}^{\lambda}+n^{\lambda}\boldsymbol{w}, \tag{33}$$

$$\boldsymbol{j}_{u}^{\lambda}=L_{un}^{\lambda}\boldsymbol{X}_{n}^{\lambda}+L_{uu}^{\lambda}\boldsymbol{X}_{u}^{\lambda}+\delta u^{\lambda}\boldsymbol{w}, \qquad \lambda=\mathrm{e, h}, \tag{34}$$

$$\boldsymbol{j}_{u}^{\mathrm{ph}}=L_{uu}^{\mathrm{ph}}\boldsymbol{X}_{u}^{\mathrm{ph}}+\boldsymbol{w}^{\mathrm{ph}}\; n^{\mathrm{ph}}\left\langle E^{\mathrm{ph}}+\boldsymbol{p}_{-}\frac{\partial E^{\mathrm{ph}}}{\partial \boldsymbol{p}_{-}}\right\rangle. \tag{35}$$

These $L_{jk}=L_{kj}$ are the transport tensors, local properties, which can also be expressed as momentum averages with respect to f_{0}^{λ}:

$$L_{nn}^{\lambda}=T^{\lambda}n^{\lambda}\left\langle\frac{\partial}{\partial \boldsymbol{p}^{\lambda}}\left\{\tau^{\lambda-\mathrm{ph}}(\boldsymbol{p}^{\lambda})\frac{\partial E^{\lambda}}{\partial \boldsymbol{p}^{\lambda}}\right\}\right\rangle_{0},$$

$$L_{nu}^{\lambda}=T^{\lambda}n^{\lambda}\left\langle\frac{\partial}{\partial \boldsymbol{p}^{\lambda}}\left\{\tau^{\lambda-\mathrm{ph}}(\boldsymbol{p}^{\lambda})\frac{\partial E^{\lambda}}{\partial \boldsymbol{p}^{\lambda}}E^{\lambda}\right\}\right\rangle_{0},$$

$$L^{\lambda}_{uu} = T^{\lambda} n^{\lambda} \left\langle \frac{\partial}{\partial \boldsymbol{p}^{\lambda}} \left\{ \tau^{\lambda-\mathrm{ph}}(\boldsymbol{p}^{\lambda}) \frac{\partial E^{\lambda}}{\partial \boldsymbol{p}^{\lambda}} (E^{\lambda})^2 \right\} \right\rangle_0 . \tag{36}$$

Using eq. (24) $u^{\lambda}_{\mathrm{kin}}$ and δu^{λ} contain the energy contribution due to the drift $\boldsymbol{w}^{\lambda}$; the proper internal energy is usually [9] defined with respect to the reference frame moving at speed $\boldsymbol{w}$.

Neglecting self-energy correction (second part of eq. (32)), and viscosity, the linear corrections due to inhomogeneities do not contribute to the flux tensor, so that

$$R^{\lambda}_0 = \Pi^{\lambda} \mathbf{1} + n^{\lambda} m^{\lambda} w^{\lambda} w^{\lambda}, \tag{37}$$

where Π^{λ} is the partial hydrodynamic pressure. For the phonon system this holds only to lowest order [10] in $\boldsymbol{w}^{\mathrm{ph}}$ ($n^{\mathrm{ph}} m^{\mathrm{ph}}$ is the phonon mass density).

7. Transport coefficients

7.1. Scattering model

Within the present approximations the transport tensors are uniquely given once the dispersion relations $E^{\lambda}(x)$ and the scattering model are known. In case of phonons, approximating:

$$\tau^{\mathrm{ph-ph}}(\boldsymbol{p}^{\mathrm{ph}}) (\partial E^{\mathrm{ph}} / \partial p^{\mathrm{ph}})^2 \simeq \tau^{\mathrm{ph-ph}}_{\mathrm{eff}} (v^{\mathrm{ph}}_{\mathrm{eff}})^2 \simeq \mathrm{const}, \tag{38}$$

we find the usual result [8] for the diagonal elements,

$$L^{\mathrm{ph}}_{uu} = (T^{\mathrm{ph}})^2 \tau^{\mathrm{ph-ph}}_{\mathrm{eff}} (v^{\mathrm{ph}}_{\mathrm{eff}})^2 c^{\mathrm{ph}} / 3, \tag{39}$$

where c^{ph} is the phonon specific heat per unit volume. For the case of electron (hole)–phonon scattering we start from [11]

$$\tau^{\lambda-\mathrm{ph}}(p^{\lambda}) = A^{\lambda-\mathrm{ph}}_s |p^{\lambda}|^s \tag{40}$$

with $s = -1$ for deformation potential coupling. In this case A_{-1} involves the respective deformation-potential constant [11]. We find

$$L^{\lambda}_{nn} = g^{\lambda} k_{\mathrm{B}} T^2 A^{\lambda-\mathrm{ph}}_{-1} F^{\lambda}_0 / 3\pi^2 \hbar^3, \tag{41}$$

$$L^{\lambda}_{nu} = g^{\lambda} k^2_{\mathrm{B}} T^3 A^{\lambda-\mathrm{ph}}_{-1} F^{\lambda}_1 / 3\pi^2 \hbar^3, \tag{42}$$

$$L^{\lambda}_{uu} = g^{\lambda} k^3_{\mathrm{B}} T^4 A^{\lambda-\mathrm{ph}}_{-1} F^{\lambda}_2 / \pi^2 \hbar^3, \tag{43}$$

where F^{λ}_v denotes the Fermi integrals (see fig. 1)

$$F^{\lambda}_v(\eta^{\lambda}) = \int_0^{\infty} \frac{x^v \,\mathrm{d}x}{\exp(x - \eta^{\lambda}) + 1}, \tag{44a}$$

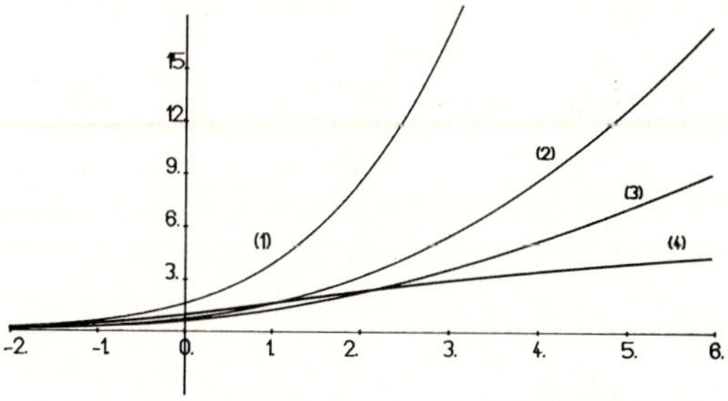

Fig. 1. Fermi integrals $F_v(h)$ as defined in eq. (44): (1) $v=2$, (2) $v=1$, (3) $v=1/2$, (4) $v=-1/2$.

with

$$\eta^\lambda = \mu_0^\lambda / k_B T. \tag{44b}$$

Here the self-energy corrections cancel out as they modify energy levels and μ^2 in the same way.

7.2. *Ambipolar transport*

The two kinetically independent electronic subsystems e, h are in fact strongly correlated macroscopically. This is due to electrical forces [12]

$$\boldsymbol{F}_0^{e,h} = q^{e,h} \cdot \boldsymbol{\varepsilon} \tag{45}$$

arising from the electric field $\boldsymbol{\varepsilon}$ created by local charge imbalance in the system itself. Within the framework of ambipolar transport $\boldsymbol{\varepsilon}$ is determined from the condition

$$\boldsymbol{j}_n = \boldsymbol{j}_n^e = \boldsymbol{j}_n^h, \tag{46}$$

$$\boldsymbol{j}_u = \boldsymbol{j}_u^e + \boldsymbol{j}_u^h, \tag{47}$$

while $n_e = n_h = n$ is still kept. As a result we are left with only one particle and one energy current controlled by the thermodynamic forces

$$\boldsymbol{X}_n = \boldsymbol{X}_n^e + \boldsymbol{X}_n^h, \tag{48}$$

$$\boldsymbol{X}_u = \boldsymbol{X}_u^e = \boldsymbol{X}_u^h, \tag{49}$$

and the effective transport coefficients

$$L_{nn} = L_{nn}^e \cdot L_{nn}^h / (L_{nn}^e + L_{nn}^h), \tag{50}$$

$$L_{nu} = (L_{nu}^e L_{nn}^h + L_{nu}^h L_{nn}^e) / (L_{nn}^e + L_{nn}^h), \tag{51}$$

$$L_{uu} = L_{uu}^e + L_{uu}^h - (L_{nu}^e - L_{nu}^h)^2 / (L_{nn}^e + L_{nn}^h). \tag{52}$$

7.3. Transformation of forces

Changing to the forces ∇T and ∇n we get from eqs. (33) and (34) the ambipolar currents in the more familiar form [13]

$$j_n = -D_T \frac{1}{T} \nabla T - D\nabla n + n\boldsymbol{w}, \tag{53}$$

$$j_u = -K\nabla T - B_T\nabla n + \delta u \cdot \boldsymbol{w} + \boldsymbol{R}\cdot\boldsymbol{w}. \tag{54}$$

Here, D is the ambipolar diffusion coefficient (see fig. 2),

$$D = \frac{L_{nn}}{T}\left(\frac{\partial\mu}{\partial n}\right), \tag{55}$$

D_T is the thermodiffusion coefficient,

$$D_T = \frac{1}{T}\left[L_{nu} - L_{nn}\left(\frac{\partial u}{\partial n}\right)\right], \tag{56}$$

K is the heat conduction coefficient,

$$K = \frac{1}{T^2}\left[L_{uu} - L_{un}\left(\frac{\partial u}{\partial n}\right)\right], \tag{57}$$

and finally, B_T is the Dufour coefficient,

$$B_T = \frac{L_{un}}{T}\left(\frac{\partial\mu}{\partial n}\right). \tag{58}$$

Besides the convective contributions $\sim \boldsymbol{w}$, transport equations (53) and (54) involve three dissipative processes; diffusion, heat conduction and thermodiffusion. The sign of D_T sensitively depends on the scattering model: $D_T > 0$ for $s > -2$. These ambipolar transport coefficients are shown in fig. 2 for silicon. Note that given the material parameters $A_{-1}^{\lambda-\mathrm{ph}}$ these results are completely determined.

7.4. Self-energy corrections for currents

As shown by Khalatnikov [7], the momentum-flux tensor has to be corrected by

$$\boldsymbol{R}^\lambda = \boldsymbol{R}_0^\lambda + \mathbf{1}(n^\lambda\langle E^\lambda\rangle - u^\lambda). \tag{59}$$

The modifications for the other current densities are easily derived from eqs. (22b) and (36). One finds that in the transport equations for $\boldsymbol{j}_n^\lambda$ and $\boldsymbol{j}_{\mathrm{kin}}^\lambda$, the thermodynamic force $\boldsymbol{X}_n$ for $S^\lambda = 0$ is replaced by

$$\boldsymbol{X}_n^\lambda \to \boldsymbol{X}_n^\lambda - \frac{1}{T^\lambda}\nabla S^\lambda. \tag{60}$$

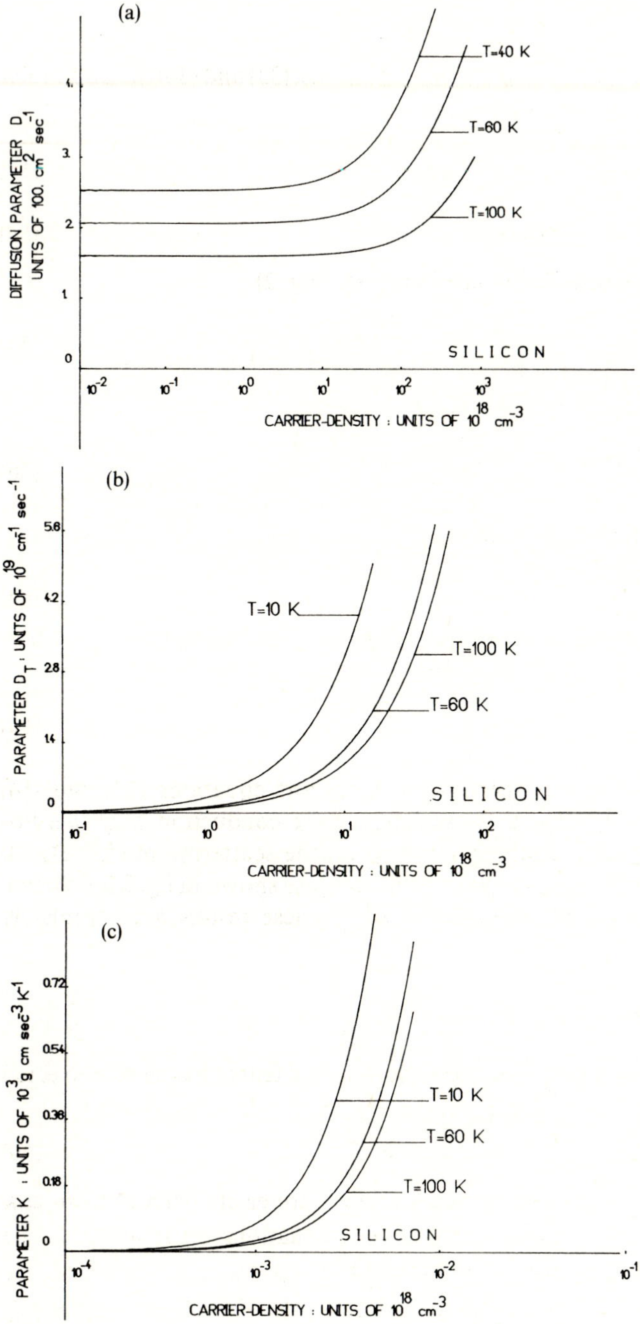
(a)
DIFFUSION PARAMETER D
UNITS OF 100. cm^2 sec^{-1}
T=40 K
T=60 K
T=100 K
SILICON
CARRIER-DENSITY : UNITS OF 10^{18} cm^{-3}
(b)
PARAMETER D_T : UNITS OF 10^{19} cm^{-1} sec^{-1}
T=10 K
T=100 K
T=60 K
SILICON
CARRIER-DENSITY : UNITS OF 10^{18} cm^{-3}
(c)
PARAMETER K , UNITS OF 10^3 g cm sec^{-3} K^{-1}
T=10 K
T=60 K
T=100 K
SILICON
CARRIER-DENSITY : UNITS OF 10^{18} cm^{-3}

Fig. 2(a), (b), (c).

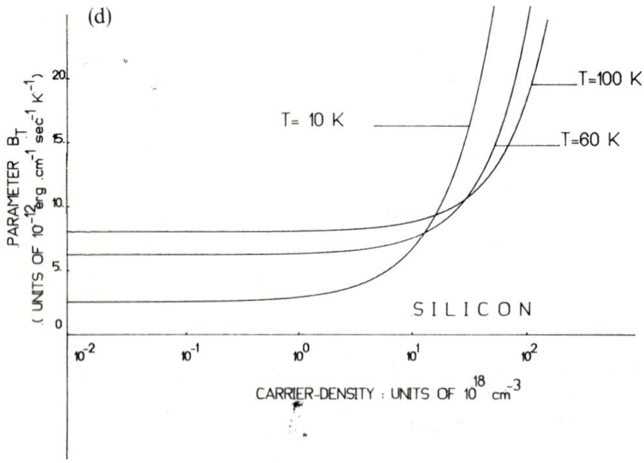

Fig. 2. Ambipolar transport coefficients for Si. Deformation potential parameters have been taken as $E_1^e - E_1^h = 3.5$ eV and $E_1^h/E_1^e = 2$. Effective mass $m_e = 0.32$ and $m_h = 0.56$, sound velocity 9×10^5 cm/s. (a) Isothermal diffusion constant D; (b) Thermodiffusion coefficient D_T; (c) Heat conduction coefficient K; (d) Dufour coefficient B_T.

If we consider, e.g., self-energy corrections due to e–ph coupling:

$$E^\lambda(\boldsymbol{x}) = E_0^\lambda(\boldsymbol{p}^\lambda) + S^{\lambda-\mathrm{ph}}(T^{\mathrm{ph}}(\boldsymbol{r})), \tag{61}$$

we find (for $\boldsymbol{w} = 0$)

$$\boldsymbol{j}_n^\lambda = -D_T^\lambda \frac{1}{T} \nabla T - \frac{L_{nn}^\lambda}{T} \frac{\partial S^{\lambda-\mathrm{ph}}}{\partial T^{\mathrm{ph}}} \nabla T^{\mathrm{ph}} - D^\lambda \nabla n. \tag{62}$$

We see that the second term also relates a temperature gradient to a particle current [14, 15]: The self-energy parts S^λ contribute to the temperature-dependent band structure [15]. For the intraband terms (Fan terms) one finds

$$\partial S^{\lambda-\mathrm{ph}}/\partial T^{\mathrm{ph}} < 0, \qquad \lambda = \mathrm{e, h},$$

so that a negative temperature gradient will oppose diffusion ("carrier confinement"). This self-energy effect, however, should be distinguished from the proper thermodiffusion effect (first-term). One should further note that the invariance for L_{jk} carries over to the ambipolar coefficients but not to those applying to the forces ∇T, ∇n: here we have, e.g. from eq. (55),

$$D = D_0 \left(\frac{\partial \mu}{\partial n}\right)\left(\frac{\partial \mu_0}{\partial n}\right)^{-1}, \tag{63}$$

as studied in ref. [12]. The same relation holds for the Dufour coefficient. We observe that these coefficients vanish at critical points; one should remember, however, that fluctuations have not been taken into account at present.

8. Balance equations

Balance equations are obtained by multiplying the Boltzmann equation with the microscopic variable ξ^λ, and averaging over momentum space [6]:

$$\frac{\partial}{\partial t}\sum_\lambda n^\lambda\langle\xi^\lambda\rangle = -\operatorname{div}\left\{\sum_\lambda n^\lambda\left\langle \xi^\lambda \frac{\partial E^\lambda}{\partial \boldsymbol{p}^\lambda}\right\rangle\right\} + \sum_\lambda n^\lambda \left\langle \frac{\partial E^\lambda}{\partial \boldsymbol{p}} \operatorname{grad} \xi^\lambda \right\rangle + J_{\text{coll}}(\xi^\lambda), \tag{64}$$

where

$$J_{\text{coll}}(\xi^\lambda) = \sum_{\lambda\lambda'} n^\lambda \langle J^{\lambda\lambda'}\xi^\lambda\rangle. \tag{65}$$

ξ^λ is called hydrodynamic variable, if $J_{\text{coll}}(\xi^\lambda)=0$. In the present case, collisional invariance only holds for the particle (electronic subsystem) and the energy balance.

8.1. Particle and energy balance

From eq. (64) we get

$$\frac{\partial n}{\partial t} + \operatorname{div} \boldsymbol{j}_n = \sigma_n, \tag{66}$$

$$\frac{\partial u}{\partial t} + \operatorname{div} \boldsymbol{j}_u - \boldsymbol{j}_n \operatorname{grad}(S^{\text{e}} + S^{\text{h}}) + \sigma^{\text{tr}} = \sigma_u, \tag{67}$$

$$\frac{\partial u^{\text{ph}}}{\partial t} + \operatorname{div} \boldsymbol{j}_u^{\text{ph}} - \boldsymbol{j}_n^{\text{ph}} \operatorname{grad} S^{\text{ph}} - \sigma^{\text{tr}} = \sigma_u^{\text{ph}}. \tag{68}$$

Here the two separate balance equations [16] account for the fact that due to local non-equilibrium, $\Delta T \neq 0$, there is an energy transfer between the two subsystems [17],

$$\sigma_u^{\text{tr}} = \sigma_u^{\text{e-ph}} + \sigma_u^{\text{h-ph}}, \tag{69}$$

$$\sigma_u^{\lambda-\text{ph}} \sim (m^\lambda)^4 T^3 \frac{\Delta T}{T} F_1^\lambda. \tag{70}$$

The source terms on the right-hand sides follow from (hydrodynamic) source models describing the influence of the environment, which is neither contained in the effective Hamiltonian (1) nor in the Boltzmann equation. So we may use, e.g.,

$$\sigma_n = \sigma_n^{\text{laser}} - n/\tau - \gamma n^3, \tag{71}$$

where the first term describes the laser impact, the second term spontaneous recombination and the last term Auger recombination (there is an additional sink term due the tendency of e–h pairs to form excitons).

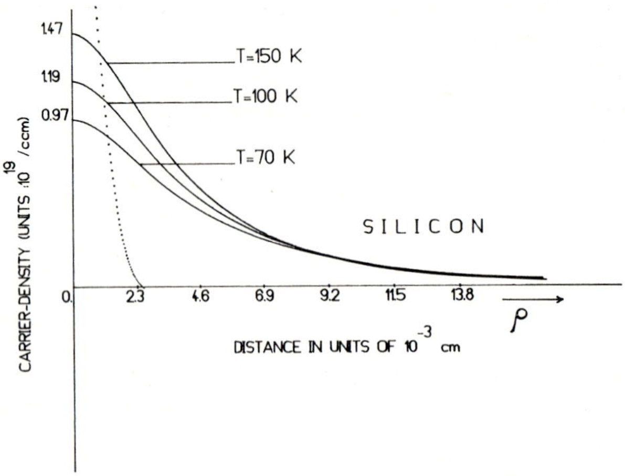

Fig. 3. Stationary density profile for Si; $T = \text{const}$. The laser (dotted line) is directed perpendicular onto the crystal which posesses slab geometry; ρ is the distance from the center of the laser spot, the spot size is 10^{-3} cm, the number of injected e–h pairs $10^{22}\ \text{s}^{-1}$.

In fig. 3 we give an example for a slab geometry (slab thickness is considered small compared to the absorption length) excited by a laser field (frequency ω, intensity I, Gaussian profile of spot-size radius R_0) centered at $\rho = 0$. The problem has cylindrical symmetry so that fields and currents solely depend on the distance ρ from the symmetry axis. Calculations are carried out for $\boldsymbol{w} = \boldsymbol{w}^{\text{ph}} = 0$ and $T = \text{const}$. Results under non-isothermal conditions will be published elsewhere.

8.2. Momentum balance

For the momentum we derive approximately from eq. (21)

$$J_{\text{coll}}(\boldsymbol{p}^{\lambda}) = \sum_{\lambda\lambda'} n^{\lambda} \langle J^{\lambda\lambda'} \boldsymbol{p}^{\lambda} \rangle \simeq -\left(\frac{\boldsymbol{q}^{\text{e}}}{\tau_{\text{R}}^{\text{e-ph}}} + \frac{\boldsymbol{q}^{\text{h}}}{\tau_{\text{R}}^{\text{h-ph}}} + \frac{\boldsymbol{q}^{\text{ph}}}{\tau_{\text{R}}^{\text{ph-ph}}} \right). \tag{72}$$

Again, as $\Delta \boldsymbol{w} \neq 0$, we consider the two momentum balance equations

$$\frac{\partial \boldsymbol{q}}{\partial t} + \text{div}\, R + \frac{-\boldsymbol{q}}{\tau_{\text{R}}} + \sigma_q^{\text{tr}} = \sigma_q, \tag{73}$$

$$\partial \boldsymbol{q}^{\text{ph}}/\partial t + \text{div}\, R^{\text{ph}} + {}_{-}\boldsymbol{q}^{\text{ph}}/\tau_{\text{R}}^{\text{ph-ph}} - \sigma_q^{\text{tr}} = 0, \tag{74}$$

$$\boldsymbol{q}^{\text{e}}/\tau_{\text{R}}^{\text{e-ph}} + \boldsymbol{q}^{\text{h}}/\tau_{\text{R}}^{\text{h-ph}} \simeq -\boldsymbol{q}/\tau_{\text{R}}. \tag{75}$$

σ_q is an external momentum source. The momentum transfer may be written (in linear approximation)

$$\sigma_q^{\text{tr}} = M\, \Delta \boldsymbol{w}/\tau_{\text{eff}}, \tag{76}$$

with $M = m_e + m_h$. These three balance equations control the three unknowns X_n^λ, X_u^λ and w^λ entering f_w^λ. The momentum balance is needed to describe

(i) drag: $w^{ph} \simeq w$ (damping mainly due to umklapp),

(ii) phonon wind: $w^{ph} > w$,

(iii) electronic drift: $w > w^{ph}$.

If

$$\Delta w \simeq w \simeq v_D, \tag{78}$$

so that

$$q = Mnw, \tag{79}$$

eq. (73) with $\sigma_q = 0$ reduces to the damped Euler equation [18]

$$\frac{\partial w}{\partial t} + \frac{1}{nM} \nabla \Pi + (w\nabla)w = -w\left(\frac{1}{\tau_{eff}} + \frac{1}{\tau_R} - \frac{1}{\tau}\right). \tag{80}$$

As we see, damping is generally due to momentum transfer and umklapp, partially compensated by finite-lifetime effects. This equation, in linearized form, should not be confused with the diffusion equation (53) as is occasionally done. For real fluid mixtures under local equilibrium conditions drift is a non-dissipative process, i.e., it does not contribute to entropy production [13].

9. Velocities

9.1. Stationary diffusion velocity

From eqs. (53), (66) and (71) we find for $w = 0$, $T = \text{const.}$, $D = \text{const.}$, $\gamma = 0$ and a half-space geometry, $x > 0$,

$$\partial^2 n/\partial x^2 = -\sigma_n/D = n/D\tau. \tag{81}$$

This equation holds away from the surface where $\sigma_n^{laser} \simeq 0$. The solution is

$$\partial n/\partial x = -(1/\sqrt{D\tau})n, \tag{82}$$

so that from

$$j_n = n v_{diff} = -D(\partial n/\partial x) \tag{83}$$

we obtain

$$v_{diff} = (D/\tau)^{1/2}. \tag{84}$$

On the other hand, in order to have a meaningful definition for local density, temperature, etc., the gradients should be restricted by [9]

$$\left|\frac{\partial n}{\partial x} \cdot \frac{1}{n}\right| \lambda \ll 1, \tag{85}$$

where λ is the mean free path. When combined with eq. (82) the above gives immediately

$$v_{\rm diff} \ll D/\lambda \simeq v^{\rm therm}\,, \tag{86}$$

where $v^{\rm therm}$ is the Fermi velocity in the degenerate case.

9.2. Drift velocity

Conceptually there is no restriction of the kind (86) on the drift velocity $\boldsymbol{w}$ as the drift need not primarily be caused by inhomogeneities. Together with damping the drift rather causes inhomogeneities. The Euler equation in its non-linearized form (80) can account for high velocities; the corresponding equation for phonons, eq. (74), on the other hand, only holds to lower order in $\boldsymbol{w}^{\rm ph}$. Contrary, however, to electrical transport, there is no external force in the present case which may induce high drift. Nevertheless, experiments [3,19,20] seem to indicate even supersonic drift velocities. In this case also the momentum transfer from non-equilibrium phonons ("phonon wind") can be excluded. What remains is the momentum source possibly related to the carrier injection within a thin (1 μm) sheet close to the surface. This injection required a microscopic description which has not yet been carried out. Combined with a momentum damping time [21,25] of 10^{-10} s, we obtain a typical drift length of 10 μm. Supersonic phenomena should roughly be restricted on this length and time scale (region I).

It seems plausible to assume that a phonon wind will be generated in this region, which will subsequently prevent the e–h system from being slowed down below $w^{\rm ph} \lesssim v_{\rm s}$ (sound velocity). As a result in region II we have a common drag of both subsystems with the much smaller damping rate due to umklapp processes only.

9.3. Experimental tests

As the local optical response is controlled by single-particle distribution functions, their deviation from equilibrium can, in principle, be inferred from the respective optical distortion. However, the linear-order corrections due to inhomogeneities do not enter and also the drift velocity $\boldsymbol{w}$ can only be measured in this way if w approaches $v_{\rm F}$. The corresponding experiments are usually carried out under quasi-stationary conditions:

(a) Luminescence lineshape [3];
(b) Gain (ratio of gain to absorption) [23];
(c) Gain as a function of sample thickness [23];
(d) Raman scattering (single-particle/collective modes) [24].

A second group of experiments usually runs under non-stationary conditions; these methods can account also for velocities well below $v_{\rm F}$ and even below the velocity of sound:

(e) Light-scattering [26];
(f) Expansion to Mott transition [27];
(g) Time-of-flight spectroscopy [19];
(h) Optical imaging [25].

Figure 4 shows the influence of drift on the spontaneous emission (a) in direct gap semiconductors, (b) in indirect semiconductors. Due to different selection rules, lines narrow in case (a), while they broaden in case (b). We see that if these lines with drift were erroneously interpreted by a fitting procedure without drift, one would infer in case (a) a large red shift and an underestimate of n, in case (b) a blue shift and an overestimate of n. This is exactly what has been

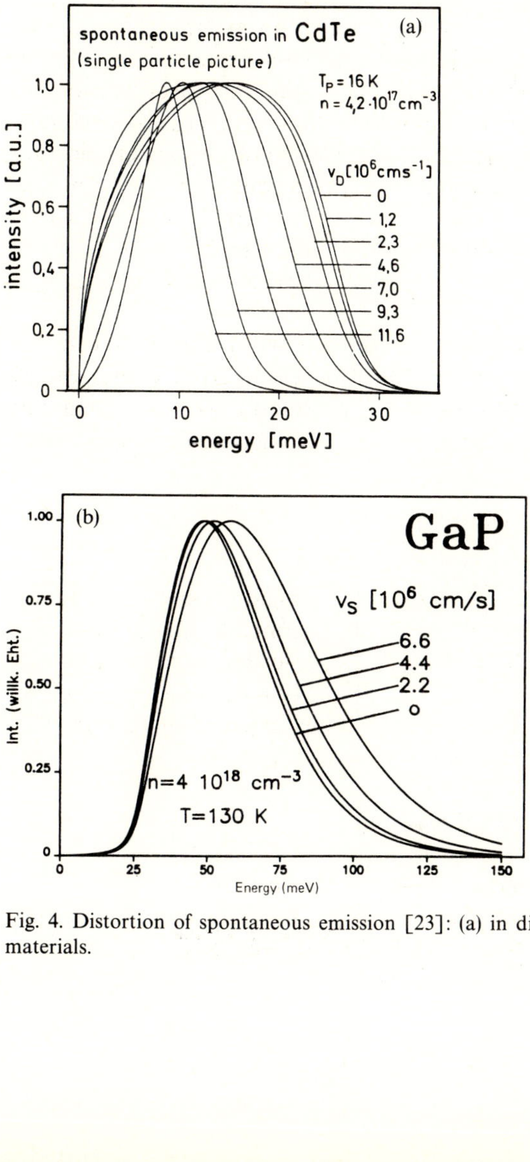

Fig. 4. Distortion of spontaneous emission [23]: (a) in direct-gap materials; (b) in indirect-gap materials.

found previously [3]. It is obvious that, e.g., the narrow line of case (a) shifted by roughly one Rydberg to lower energies (as compared to theory and to those cases, where drift is prevented) cannot be explained by any spatial inhomogeneity.

Figure 5 shows the absorption coefficient (optical response without many-body effects) as it is influenced by drift. In addition to a shift of the zero point we observe a rapid decrease of the gain/absorption ratio.

Figure 6, finally, depicts the measured Raman shift with and without spatial confinement in back-scattering geometry. In this case one does not only find the influence of drift, but even its direction, namely, away from the excited surface.

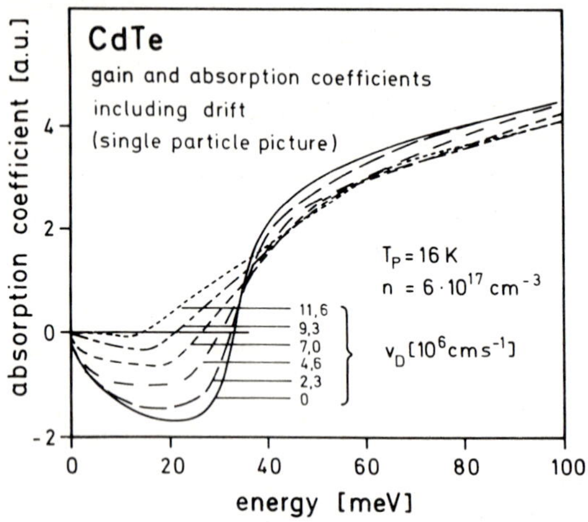

Fig. 5. Distortion of gain [23].

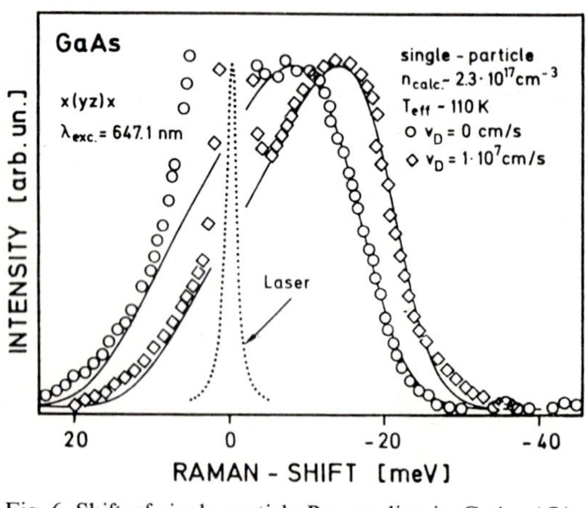

Fig. 6. Shift of single-particle Raman line in GaAs: (○) confined carriers (no drift); (◇) carriers drifting away from the surface (back scattering).

10. Conclusions

We have shown in this paper how drift and diffusive processes can be treated in a system far from equilibrium. Our approach has been based on a parametrization of the non-equilibrium distribution functions followed by an explicit derivation of the macroscopic equations (balance equations). We discussed ambipolar transport, the influence of many-body effects on thermodynamic functions and transport coefficients and non-equilibrium drift effects: phonon wind, electronic (supersonic) drift and common drag. The experimental verifications for non-equilibrium effects are abundant. However, the origin of the supersonic electronic speeds is still not clear. Unfortunately, even if this high speed is incorporated by macroscopic boundary conditions, the theoretical description will, in general, have to include five balance equations (three for the electronic subsystem, two for phonons) in order to get a fairly complete picture. It is also not clear how the previously found $n(T)$ relations, which had been interpreted without drift [28], can be understood under these much more general conditions. On the other hand, one may note from fig. 3 that already for isothermal diffusion the maximum density increases with temperature.

References

[1] T.M. Rice, Solid State Phys. 32 (1977) 1.
[2] Y. Prokrowskii, Phys. Status Solidi (a) 11 (1982) 385.
[3] A. Forchel, H. Schweizer and G. Mahler, Phys. Rev. Lett. 51 (1983) 501.
[4] Preliminary results have been given previously: G. Mahler and A. Forchel, Helv. Phys. Acta 56 (1983) 875.
[5] M. Dresden, Rev. Mod. Phys. 33 (1961) 265.
[6] Y.L. Klimontovich, Kinetic Theory of Non-Ideal Gases and Non-Ideal Plasmas (Pergamon, London, New York, 1982).
[7] I.M. Khalantnikov, in: The Boltzmann Equation, Proc. Int. Symp. Vienna, 1972 (Springer, Wien, 1973).
[8] H. Beck, in: Dynamical Properties of Solids, Vol. 2, eds., G.K. Horton and A. Maradudin (North-Holland, Amsterdam, 1975).
[9] I. Gyarmati, Non-Equilibrium Thermodynamics (Springer, Berlin, 1970).
[10] C.P. Enz, Rev. Mod. Phys. 46 (1968) 705.
[11] B.R. Nag, Electron Transport in Compound Semiconductors (Springer, Berlin, 1980).
[12] J.F. Young and H.M. van Driel, Phys. Rev. B 26 (1982) 2147.
[13] S.R. de Groot and P. Mazur, Non-Equilibrium Theomodynamics (North-Holland, Amsterdam, 1969).
[14] J.A. van Vechten and M. Wautelet, Phys. Rev. B 23 (1981) 5543.
[15] M. Schlueter, G. Martinez and M. Cohen, Phys. Rev. B 12 (1975) 654.
[16] J.H. Herzinger and H.G. Kaper, Mathematical Theory of Transport Processes in Gases (North-Holland, Amsterdam, 1972).
[17] S.M. Kogan, Sov. Phys. Solid State 4 (1963) 1813.
[18] M. Combescot, Phys. Lett. 85A (1981) 308.

[19] A. Forchel, B. Laurich, G. Tränkle, H. Hillmer and G. Mahler, in: Proc. 17th ICPS, San Francisco, 1984 (to be published).
[20] H. Schweizer, E. Zielinski, A. Forchel and G. Mahler, in: Proc. Int. Conf. Lum., Madison 1984 (to be published).
[21] D.S. Pan, D.S. Smith and T.C. McGill, Phys. Rev. B 17 (1978) 3284.
[22] J. Keizer, J. Chem. Phys. 69 (1978) 2609.
[23] H. Schweizer, Thesis Stuttgart 1984; H. Schweizer and E. Zielinski, J. Lumin. 30 (1985) 37.
[24] H. Nather and L. Quagliano, J. Lumin. 30 (1985) 50.
[25] M.A. Tamor, M. Greenstein and J.P. Wolfe, Phys. Rev. B 27 (1983) 7353.
[26] T.C. Damen and J.M. Worlock, in: Proc. 3rd Int. Conf. Light Scattering in Solids, eds., M. Balkanski, R.C. Leite and S.P.S. Porto (Flamarion, Paris, 1976) p. 183.
[27] A. Forchel, H. Schweizer, B. Laurich, G. Tränkle and G. Mahler, MRS Symp. Proc., Boston 1983 (North-Holland, Amsterdam, 1984).
[28] G. Mahler, G. Maier, A. Forchel, B. Laurich, H. Sanwald and W. Schmid, Phys. Rev. Lett. 47 (1981) 1855; in this paper the transport coefficients have not been calculated consistently with respect to a specific scattering model.

Journal of Luminescence 30 (1985) 37–49
North-Holland, Amsterdam

PLASMA EXPANSION AND BAND-GAP RENORMALIZATION IN CdTe AND GaP

H. SCHWEIZER and E. ZIELINSKI
Physikalisches Institut, Universität Stuttgart, D 7000 Stuttgart 80, FRG

Detailed experimental and theoretical investigations, based on direct excite and probe experiments and on line-shape analysis, on the influence of many-body interaction, polar interaction, and transport effects on the energetic position and shape of the electron–hole plasma (EHP) emission are reported. The comparison of optical EHP properties in an indirect (GaP) and a direct (CdTe) semiconductor of similar polarity offers the possibility of a separation of drift effects and polar effects with respect to band-gap renormalization. Direct evidence of a fast plasma drift in CdTe is obtained by thickness sensitive optical excite and probe experiments.

1. Introduction

High-excitation experiments in semiconductors offer the possibility to study dense electron–hole plasma (EHP) systems in their equilibrium [1,2] and nonequilibrium [3] state. The understanding of the equilibrium EHP is nearly perfect [1] whereas the nonequilibrium state is still under investigation. This is mainly due to the higher degree of freedom, which is mirrored by a larger set of macroscopic observables describing the nonequilibrium EHP correctly. Thus the appropriate set of macroscopic observables, e.g. density, temperature, particle, and momentum flux, giving a correct description of the nonequilibrium EHP, is not evident [4,5]. Assuming a particle and momentum flux striking discrepancies [6] could be removed [7] for many direct and indirect gap semiconductors. The striking properties of the expanding plasma are:

– The band-gap renormalization of the expanding plasma seems to be opposite in direct and indirect gap materials [7] related to the well-stated renormalization of the equilibrium plasma of electron–hole droplets (EHD) [1].

– In indirect gap semiconductors the above described apparently distinct band-gap renormalization of nonequilibrium and equilibrium plasma can be observed directly [8] by comparison of EHP emission curves slightly below and above the critical temperature T_c [7] for EHD [9].

0022-2313/85/$03.30

There is no possibility to explain this jump of the renormalized gap by many-body theory, since the temperature dependence of exchange and correlation energy is very weak. A possible explanation of the above stated properties of direct and indirect semiconductors due to their different band structure is ruled out by the result of the screening experiment of the Γ-exciton [10] in Germanium. The exciton is screened by free (L) electrons (indirect gap) and free (Γ) holes (direct gap), which is well described by the above stated many-body theory.

The analysis done in ref. [7] is mainly based on more indirect arguments (line-shape fit, energetic position, complicated comparison with theory). Additionally the direct gap semiconductors investigated up to now usually show a stronger polar interaction than the indirect gap semiconductors (with the exception of GaP, AlAs, and the thallous halides) resulting in a stronger band-gap renormalization. Due to the always present nonequilibrium situation the influence of drift effects and polar effects on band-gap renormalization cannot be separated in direct gap semiconductors. Therefore an uncertainty occurs, whether the experimentally observed band-gap renormalization has been influenced mainly by nonequilibrium effects or rather by polar effects. In the case of CdS the polar correction seems to be sufficient [11], however, in the case of GaAs with slightly weaker polar interaction the drift correction seems to be more appropriate [7].

We therefore tried to separate polar effects and drift effects comparing materials with similar polar interaction but direct and indirect band structure. Using GaP as indirect material, where an equilibrium *and* a nonequilibrium EHP is present [2], and CdTe as direct material, always showing a nonequilibrium EHP we are able to estimate the polar interaction in GaP and CdTe experimentally. Moreover we have performed an optical excite and probe experiment with spatial resolution to demonstrate the plasma expansion effect in CdTe directly.

2. Experimental

The high excitation experiments (excite and probe) were performed with N_2-pumped dye lasers ($\tau_{rep}=100$ Hz, $\tau_p=10$ ns) on pure CdTe samples. The excitation laser was operated as narrow band laser with 1 meV line width tuneable in frequency for resonant excitation, and the probe laser as broad band laser. The maximum of the probe laser emission was slightly above the exciton resonance to avoid additional excitations by the probe laser and to improve the signal-to-noise ratio in the energetic region of the gain. The signal was dispersed by a double monochromator detected by a GaAs photo multiplier and processed by a coupled setup of boxcar integrator and lock-in amplifier. Data processing and storage was obtained by a computer. The experimental conditions and the setup of the GaP experiments are described elsewhere [2].

3. Renormalized band-gap in GaP

In this section we demonstrate that in the polar material GaP a surface excited EHP, at temperatures above T_c, shows a nonequilibrium behaviour very similar to the behaviour of the EHP in the elementary semiconductors Si and Ge [7]. In analogy to Si and Ge a characteristic jump of the renormalized band-gap of the EHP occurs (see fig. 1) in going from temperatures below T_c (open circles in fig. 1) to temperatures above T_c (solid circles in fig. 1), which is ascribed to a strong drift effect of the EHP above T_c [7]. Since the EHP below T_c shows a well-defined equilibrium behaviour [2], drift can be neglected in the line-shape evaluation of the EHD. This means a separation of the drift effect because the renormalized band-gap of the EHD depends only on carrier density and polar interaction. Therefore the experimental band-gap renormalization of the EHD can be used to evaluate the influence of the polar interaction on band-gap shrinkage.

One finds (see fig. 2) that the experimental renormalization values are closer to the theory using the formula of Vashista and Kalia [12], which means weak phonon interaction, than to the theory taking into account the full phonon interaction [13] indicating that the SPPA using the present form of polar interaction [11,13] overestimates the band-gap renormalization. The curvature of the renormalization curve, however, is more appropriate described taking into account phonon interaction, which is indicated by the critical temperature of the EHD in GaP [2]. Because of the relatively small density range in high-excitation experiments the curvature of the experimental renormalization

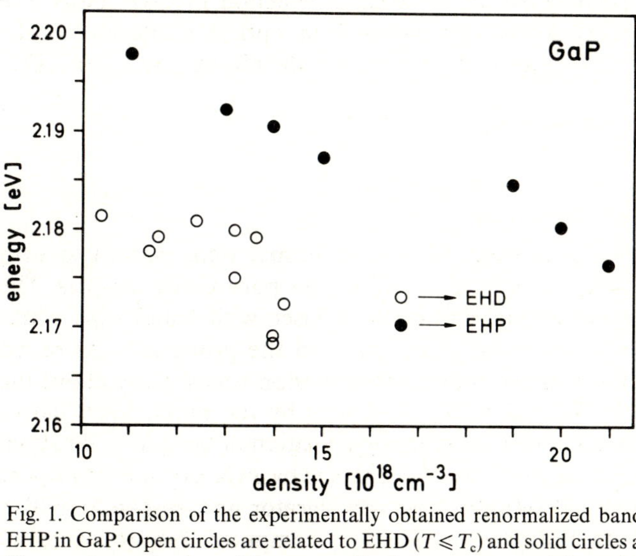

Fig. 1. Comparison of the experimentally obtained renormalized band-gap of the EHD and the EHP in GaP. Open circles are related to EHD ($T \leqslant T_c$) and solid circles are related to EHP ($T \geqslant T_c$). The experimental values of the renormalized gap are corrected due to lattice temperature.

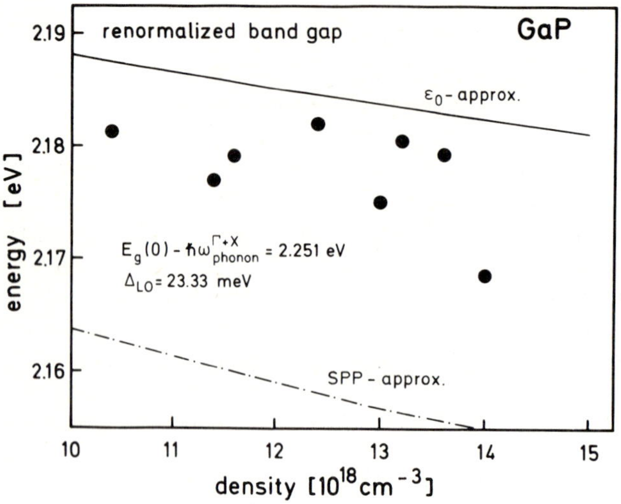

Fig. 2. Comparison of the renormalized band-gap of the EHD (solid circles) with the theoretical band-gap in ε_0-approximation using the formula of Vashista and Kalia [12] and in single-plasmon pole approximation (SPPA) using the formula of Schmitt-Rink et al. [13] including phonon interaction. The zero density value for both theoretical renormalization curves is the experimentally observable band-gap including the phonon replica [2] as indicated in the figure.

curve is not observable. For a computation of the energetic position of the renormalized band-gap only, it is therefore possible to use the formula of Vashista and Kalia [12], which should be also appropriate for the case of CdTe because of the similar polarity compared to GaP as will be seen in section 7.

4. Excite and probe experiment (optical gain)

Overcoming the more indirect observations of the plasma expansion we have performed optical excite and probe experiments on thin CdTe plane parallel platelets of different thicknesses. To avoid a different optical response of the material due to background impurities the samples were cut from the same ingot and then polished and etched to the appropriate thickness. Using optical excite and probe experiments there is a possibility to test the spatial distribution and density of the surface excited electron–hole plasma as it is sketched in fig. 3.

On the left side of fig. 3 a nonexpanding plasma is depicted. For different thicknesses $d \geqslant \delta$ [14] we expect no effect on optical gain for nonexpanding plasmas because the inverted region remains constant at all values of d. On the other hand gain would be affected very strongly if the plasma expands as shown

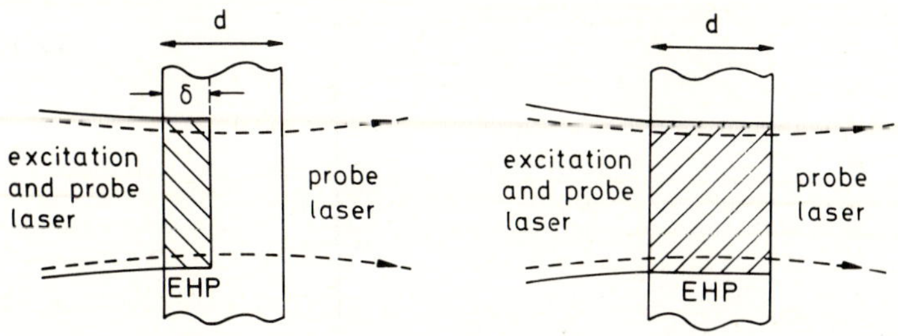

Fig. 3. Excite and probe experiment (optical gain) in a thin semiconductor platelet with plasma expansion (right) and without plasma expansion (left).

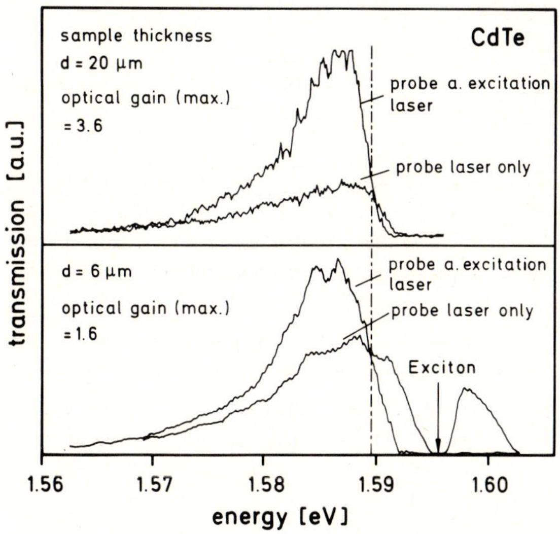

Fig. 4. Linear optical transition of the probe laser with and without excitation. The 6 μm sample shows the exciton state which is not resolved in the 20 μm sample. Further explanation see text.

in the right side of fig. 3. The result of this experiment is shown in fig. 4 for sample thicknesses of 20 and 6 μm.

For both sample thicknesses the gain region starts at nearly the same energetic position (dash-dotted line) and ends roughly for both curves at similar positions (point where the transmission curve with excitation hits the transmission curve without excitation) indicating very similar densities. Above the crossover point (dash-dotted line), which has the meaning of the chemical potential of the EHP in an equilibrium interpretation, the absorption is enhanced due to the band-gap renormalization at high carrier densities. Most important is the fact, that the optical transmission in the gain region increases if the sample thickness increases (fig. 4).

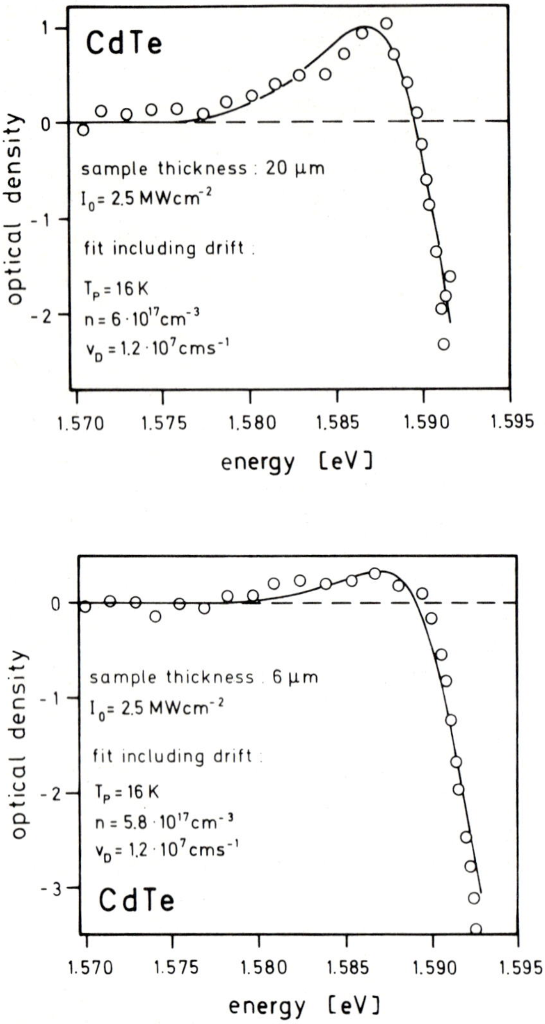

Fig. 5. Comparison of theoretical (solid line) and experimental (open circles) gain spectra for sample thicknesses of 20 μm (top) and 6 μm (bottom).

The gain coefficient for both samples is roughly constant (20 μm: 640 cm^{-1}; 6 μm: 780 cm^{-1}). This indicates that both samples are nearly filled with carriers. The average carrier velocity estimated from the sample thickness (20 μm) and the carrier life time [15] reaches about 6.7×10^6 cm/s.

To get insight in the physical parameters (temperature, density, velocity) of the EHP, we have calculated the optical gain taking into account a rigidly shifted Fermi sphere [7]. As expected from the spectral width and position of the transmission spectra, we find in the line-shape fit similar densities for the 20 μm and 6 μm sample and a carrier velocity which agrees well with the above estimate (see fig. 5). Notice that we have used only average values for density, temperature and velocity. Caused by our poor knowledge about the precise carrier profile which must be, due to the small value of δ compared to d, clearly nondiffusive close to the surface we would overstress the physical interpretation by using a certain profile. The carrier velocity obtained by a line-shape fit may be reduced and become closer to the estimated value, if we would take into account a lifetime broadening process [16].

5. Optical gain and absorption

Due to the progress in calculation of absorption spectra at arbitrary carrier densities [10,17] with the help of the many-body theory, there is more insight in the relation of the precise many-body theory, which needs a high computation expenditure, to the less precise single particle result where computation is very simple. One finds at very high densities that the correct absorption constant factorizes [13] into the simple single-particle function times a many-body correction. Above the crossover point the absorption reaches a value which is near the exciton continuum value at low excitation [17]. Important for the further discussion is the fact, that the gain-to-absorption amplitude ratio, symmetrical to the crossover point, depends only on the degeneracy of the very dense Fermi system of electrons and holes. This last statement is used to support the expansion results obtained in section 4.

In fig. 6 we have calculated the single-particle absorption coefficient at various drift velocities of the carriers. The main result is that drift changes the gain-to-absorption ratio very strongly. Another important result is that the crossover point (change from gain to absorption) has no longer the meaning of the chemical potential of the EHP and depends strongly on carrier drift. Looking up now the ratio of gain-to-absorption in the equilibrium case one finds a value of about 1:3 theoretically.

A comparison with the experiment (fig. 7) leads to a value of 1:20! From the temperature and energetic position (equivalent to the carrier density) of the EHP we know its degeneracy. Due to that degeneracy we can calculate the single-particle absorption coefficient (fig. 7, curve in short dashes). To satisfy the correct absorption value (above 1.59 eV about 15000 cm^{-1}) we have to multiply the single-particle curve with the symmetric many-body enhancement function leading to about 5000 cm^{-1} gain which is never seen, but in consistency with the low energetic position (high density) of the gain curve. Instead of this value we only see 780 cm^{-1} at maximum. Assuming drift, the experimentally observed gain-to-absorption ratio can be reproduced (curve in long dashes), therefore we believe that the experimentally observed strong reduction of the gain shows strong evidence for a fast plasma expansion in CdTe.

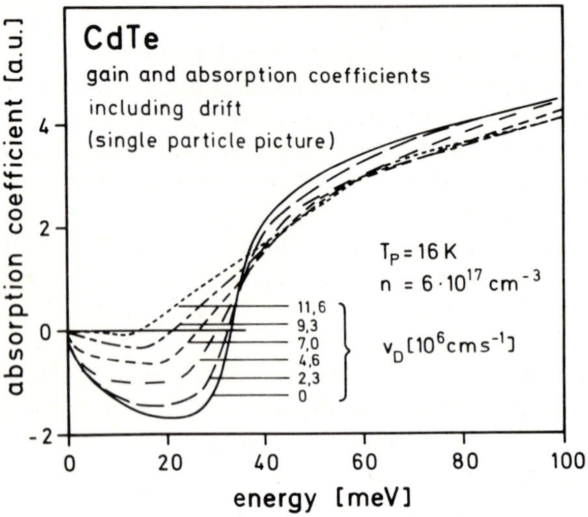

Fig. 6. Single-particle dielectric function at fixed temperature and density for various drift velocities.

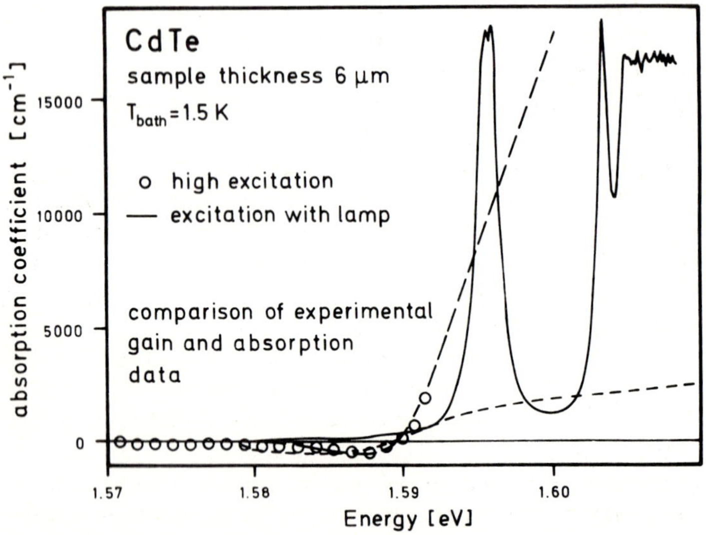

Fig. 7. Comparison of experimental gain values (open circles) to the experimental absorption values of the exciton (solid line) in CdTe. The curve in short dashes is single-particle theory without drift, the curve in long dashes is single-particle theory with drift. The computation of both curves includes k-selection. Further explanation see text.

6. Luminescence experiments in CdTe

On the basis of sections 4 and 5 we have performed spatially resolved luminescence experiments in CdTe. If we have a strong plasma expansion from the surface into the bulk in the high-excitation experiment, we must have different contributions of stimulated and spontaneous emission in the optical backward emission, depending on the average expansion length of the carriers into the bulk.

In fig. 8 we present the results of the spatially resolved luminescence of a CdTe sample with an accurate flat surface. Scanning from the edge to the center of the laser spot we find a line-width narrowing of the emission which we interpret in terms of amplified spontaneous emission (ASE).

Using the formula of the amplified spontaneous emission in one dimension [18] in addition to the measured maximum gain coefficient 90, we find an average

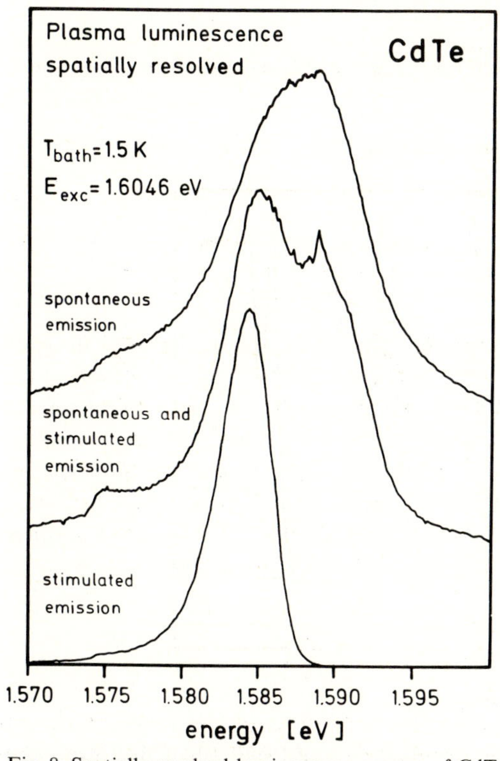

Fig. 8. Spatially resolved luminescence spectra of CdTe in backward geometry. The three curves (top to bottom) are related to the edge, intermediate, and center region of the laser excitation spot. Explanation see text.

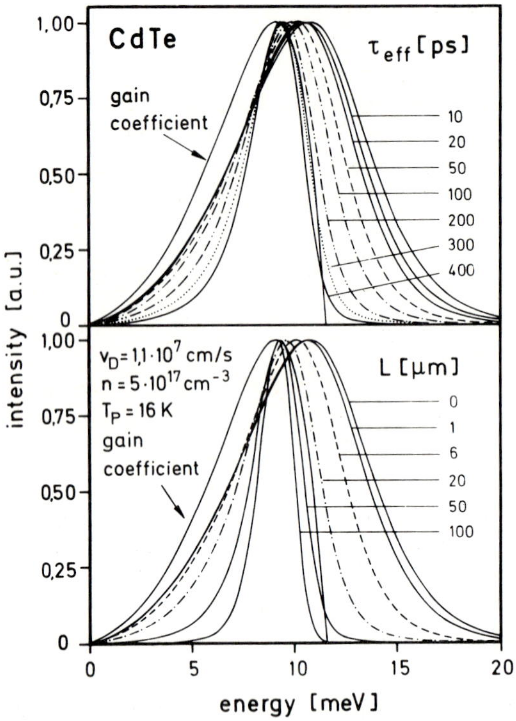

Fig. 9. Theoretical spectra of amplified spontaneous emission (ASE) as a function of the average expansion length (lower part) or the effective life-time [19] (upper part). Further explanation see text.

expansion length of the carriers between 20 and 50 μm as depicted in fig. 9 (lower part). This corresponds to an average effective plasma lifetime [19] (upper part of fig. 9) of 200 to 400 ps, which agrees well with the estimate. Additionally, another small detail supports the drift model. If no plasma expansion would be present the ASE would be centered on the spontaneous emission. With drift the maximum of the ASE considerably shifts to lower energies. A wavelength dependent gain saturation effect [20] can be excluded, since the value of the expansion length is too small.

7. Band-gap renormalization in CdTe

In fig. 10 we have summarized the experimental band-gap renormalization values without drift interpretation (solid circle) and with drift interpretation (open circles) in comparison to the theory [12] in ε_0-approximation and in SPPA including phonon interaction [11,13]. To obtain agreement between the experiment and any theoretical curve one had to take into account the drift

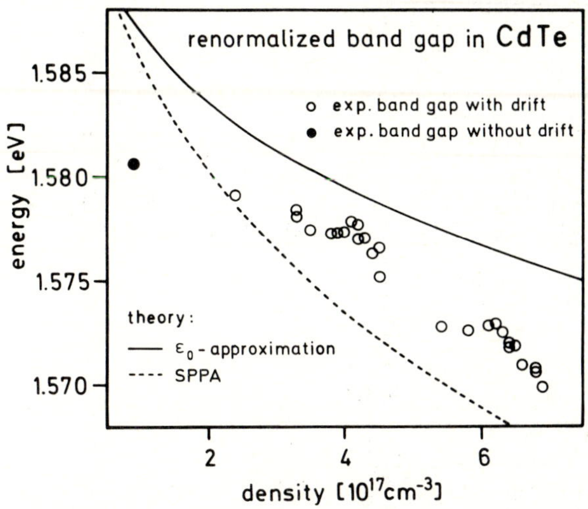

Fig. 10. Comparison of experimental band-gap renormalization with theoretical band-gap renormalization in ε_0-approximation and in SPPA. The experimental value obtained in the equilibrium interpretation of the spectra is represented by the solid circle. The zero density value for both theoretical curves is the experimental band-gap obtained from the exciton spectra (fig. 7) $E_g(0) = 1.6059 \pm 0.0002$ eV [21].

assumption, since the experimental "non-drift value" (solid circle) is situated below both theoretical curves.

On the other hand relying on the ε_0-approximation leads to unreasonable high drift velocities of the carriers overcoming the LO-phonon velocity barrier. We therefore fitted the experimental spectra in a nonlinear regression procedure resulting in an average drift velocity of 1.1×10^7 cm/s and in a band-gap renormalization situated between both theories. Using the SPPA one would have of course lower drift velocities in the case of CdTe ($\sim 5 \times 10^6$ cm/s), which are too low with respect to the excite and probe experiments reported in sections 4 and 5. Moreover due to the inverse behaviour of the renormalization in indirect gap semiconductors compared to the direct one [7] the drift velocity of the carriers in GaP would overcome the LO-velocity barrier, which is clearly ruled out by the position of the renormalized band-gap of the EHD (section 3).

8. Conclusions

In optical excite and probe experiments we have shown in a direct manner the existence of a fast plasma expansion. Further investigations (line shape and energetic position of the gain spectra, ratio of gain-to-absorption, and spatially resolved luminescence) give physical parameters which agree well to those determined by the direct experiment. The average velocity of the plasma expansion near the surface overcomes the speed of sound. The observed high initial velocities at the surface which are of the order of the Fermi velocity are physically not fully understood and are probably due to the preparation of the plasma very close beneath the surface. Due to the nonequilibrium situation in high-excitation experiments the EHP cannot be interpreted in terms of quasi equilibrium. It must be interpreted rather in terms of a nonequilibrium formalism taking into account the particle and energy currents.

References and footnotes

[1] T.L. Reinecke, A. Forchel, H. Schweizer, B. Laurich, A. Hangleiter, J. Wagner and W. Schmid, Physica 117B & 118B (1983) 315–317.

[2] H. Schweizer, A. Forchel and W. Schmid, Physica 117B & 118B (1983) 318–320.

[3] S. Modesti, L.G. Quagliano, A. Frova, J.L. Staehli and M. Guzzi, J. Lumin. 24/25 (1981) 581; K.M. Romanek, H. Nather and E.O. Göbel, ibid., 585;
A. Forchel, H. Schweizer, H. Nather, K.M. Romanek, J. Fischer and G. Mahler Physica 117B & 118B (1983) 336–338.

[4] G. Mahler and A. Fourikis, J. Lumin. 30 (1985) 18.

[5] G. Röpke and F.E. Höhne, Phys. Status Solidi (b) 107 (1981) 603;
G. Röpke, W. Ebeling and W.D. Kraft, Physica 101A (1980) 243.

[6] G. Beni and T.M. Rice, Phys. Rev. B18 (1978) 768;
O. Hildebrand and E.O. Göbel Proc. 13th Int. Conf. Phys. of Semicond. Rome, 1976 (North-Holland, Amsterdam, New York, Oxford, 1976) 942.

[7] A. Forchel, H. Schweizer and G. Mahler, Phys. Rev. Lett. 51 (1983) 501.

[8] In direct gap semiconductor materials a direct experimental observation of the different energetic positions of equilibrium and nonequilibrium plasma, like in indirect semiconductors [7], is impossible because of the absence of an equilibrium plasma (EHD). Due to the short lifetimes of the excess carriers a nonequilibrium plasma is always present.

[9] Indicating the different behaviour of nonequilibrium and equilibrium EHP the following nomenclature is used:
– for the equilibrium EHP (EHD) as introduced earlier,
– for the nonequilibrium EHP (EHP).

[10] H. Schweizer, A. Forchel, A. Hangleiter, S. Schmitt-Rink, J.P. Löwenau and H. Haug, Phys. Rev. Lett. 51 (1983) 698.

[11] M. Rösler and R. Zimmermann, Phys. Status Solidi (b) 83 (1977) 85.

[12] P. Vashista and R.K. Kalia, Phys. Rev. B25 (1982) 6492.

[13] S. Schmitt-Rink, D.B. Tran Thoai and H. Haug, Z. Phys. B39 (1980) 25.

[14] δ depends mainly on the absorption length and the carrier diffusion length (CdTe: $\delta \sim l = \sqrt{D\tau} \leqslant 2\ \mu m$). The value of the diffusion length l is an upper bound due to the lifetime value [15].

[15] The restricted time resolution (300 ps) of our experimental setup (mode locked Ar-laser) prevents us from measuring the EHP lifetime directly, which is equal or even shorter than 300 ps.

[16] P.T. Landsberg, Proc. Phys. Soc. A 62 (1949) 806;
R.W. Martin and H.L. Störmer, Solid State Commun. 22 (1977) 523;
V.D. Kulakovskii, V.B. Timofeev and V.M. Edels'ki, Sov. Phys.-JETP 47 (1978) 193 (Zh. Eksp. Teor. Fiz. 74 (1978) 372);
W. Schmid, Phys. Status Solidi (b) 94 (1979) 413;
C. Klingshirn and H. Haug, Phys. Rep. 70 (1980) 315–398.

[17] H. Haug, Advances in Solid State Physics, *XXII*, 149 (1983).

[18] In the backward scattering geometry we have a solid angle of 7° at maximum, due to the high refractive index of CdTe, which legitimates the one-dimensional consideration for the ASE: $I_{ASE} = [I_{spon}(E, n, v, T)/g_0 g(E, n, v, T)]\{\exp[g_0 g(E, n, v, T)L] - 1\}$.

[19] The effective lifetime corresponding to L is calculated as $\tau = L/v$.

[20] E.O. Göbel, O. Hildebrand and K. Löhnert, IEEE, J. Quant. Electron. QE-13, NO.10 (1977) 73.

[21] E. Zielinski, Thesis, Universität Stuttgart (1983).

Journal of Luminescence 30 (1985) 50–64
North-Holland, Amsterdam

INELASTIC LIGHT SCATTERING IN HIGHLY EXCITED GaAs

Heinz NATHER and Lucia G. QUAGLIANO*
Physikalisches Institut, Universität Stuttgart, Pfaffenwaldring 57, D-7000 Stuttgart 80, FRG

The coupled plasmon–LO-phonon and single-particle excitations of a photocreated electron–hole plasma in high purity GaAs are studied by means of Raman scattering measurements. The experimental results are described by the random-phase approximation with respect to the multicomponent character of optical plasma. From single-particle spectra the plasma distribution function and drift velocities up to 10^7 cm/s have been determined. Using the coupled modes as probe for the plasma density the temperature dependence of the recombination in the EHP is studied.

Introduction

Inelastic light scattering from the electronic plasma in highly doped semiconductors has been studied extensively throughout the last years [1]. Light is scattered by both the single-particle and collective excitations (coupled plasmon–LO-phonon modes). The properties and mechanisms of the different scattering processes are generally well understood [2].

At high optical excitation intensities the formation of a dense, high mobility, electron–hole plasma (EHP) has been observed in many semiconductors [3]. Under these conditions a strong expansion is observed because the carrier density n_{EHP} is higher than the density n^*, i.e. the density that minimizes the plasma free energy [4]. This rapid particle flux and the ensuing density and temperature gradients affect the optical properties of the EHP. The plasma expansion has been shown to be responsible for several discrepancies between the predictions of equilibrium theories and experiments [5–7]: as a consequence of the high velocities the luminescence and gain lineshapes are changed due to a modified distribution function.

On the other hand the experimental data can be described in the framework of an equilibrium model if the carrier expansion is suppressed by restricting the

* Present and permanent address: Istituto di Metodologie Avanzate Inorganiche del C.N.R., Rome, Italy

0022-2313/85/$03.30 © Elsevier Science Publishers B.V.
(North-Holland Physics Publishing Division)

EHP in a fixed volume (3-D carrier confinement) [8, 9]. The present work deals primarily with the contributions to the light scattering spectra from the single-particle excitations and collective plasmon–LO-phonon modes of an uniformly excited EHP. The single-particle spectra which reflect the distribution of the electrons are best suited to serve as a probe for the equilibrium density and temperature as well as for the non-equilibrium distribution in the presence of carrier flux. The collective mode can be used as a measure of the EHP density.

Beside the interest in the possible Raman excitations in a multicomponent plasma, these experiments are particularly important for the study of EHP properties, since from the Raman spectra independent values for the plasma density and temperature can be obtained. Moreover this scattering process permits the determination of these plasma parameters in non-degenerate plasmas where optical gain measurements cannot be used.

This paper reports detailed Raman scattering studies on GaAs taken at different electronic temperatures and an extended range of excitation intensities. The first part describes the experimental details. The second part presents the theoretical model for the Raman spectra. Especially the scattering process in the presence of more than one mobile species will be discussed. A comprehensive survey of the experimental results will be given in section (3). The discussion of these results will illustrate that for uniformly excited plasma the linewidth, the peak positions and the relative intensities of the experimental spectra are well described by the random-phase approximation (RPA). The comparison of single-particle spectra taken at homogeneous excitation conditions with those obtained for inhomogeneous excitation will point out that in high mobility plasma drift velocities up to 10^7 cm/s are achieved. Finally the density–temperature relation for homogeneous EHP is presented. The data are discussed in the frame of a recombination model which accounts very well for the experimental data.

1. Experimental details

The Raman scattering experiments were performed with a Kr^+-laser which could be operated in the cw or cavity dumped mode. The strong emission line at 647.1 nm served to excite the EHP and to scatter from it. Since GaAs is strongly absorbing at this wavelength the penetration depth of the laser light is $\simeq 0.3$ μm. To avoid density gradients due to inhomogeneous excitation, the plasma was created in microstructures as illustrated in fig. 1. These high purity GaAs/GaAlAs heterostructures were grown by molecular beam epitaxy (MBE) on (100) GaAs substrates. The thickness of the "active" GaAs layer ($d \simeq 0.5$ μm) is comparable to the penetration depth of the laser. Circular islands obtained by photolithography and subsequent etching are used to suppress possible lateral carrier diffusion. For the optical experiment, the sample was mounted in a

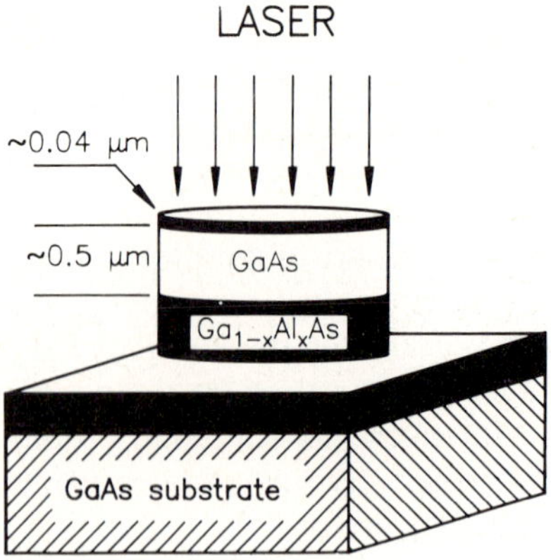

Fig. 1. Schematic design of the sample.

temperature controlled dewar. All experiments were performed in backscattering geometry with the incoming light polarized along $y \parallel (010)$ and the scattered light either along $y \parallel (010)$ or $z \parallel (001)$. The incident and scattered light is directed along $x \parallel (100)$ and $\bar{x} \parallel (\bar{1}00)$. The scattered light was focused on the entrance slit of a 0.85 m double monochromator equipped with two 1200 groove/mm gratings blazed at 1 µm. A RCA 31034A photomultiplier with GaAs cathode was used in a photocounting mode to process the signal.

2. Theoretical background

Light scattering mechanisms and selection rules for electronic excitations have been extensively reviewed by Mooradian [1] and Klein [2]. The scattering spectra of the EHP and lattice excitations in high purity GaAs obtained under the given experimental conditions have resonant character, i.e., the energy of the laser photons is very close to the $E_0 + \Delta_0$ electronic interband energy gap of GaAs. In this case the dominating scattering contributions fall into two categories:

(1) *Single-particle* scattering which couples to the laser radiation via spin-density fluctuations and can be observed in the perpendicular polarization $\bar{x}(yz)x$.

(2) *Collective* excitations due to charge–carrier-density fluctuations with parallel polarization $\bar{x}(yy)x$.

The two different scattering processes are separated by the different selection rules and show an appreciable enhancement of the scattered light intensity if the laser line is close to the $E_0 + \Delta_0$ energy gap.

Collective mode

For polar materials (like GaAs) strong coupling between the longitudinal phonons and the free carrier system exists if the plasma frequency

$$\omega_p = 4\pi e^2 n/m^* \varepsilon_\infty \tag{1}$$

is comparable to the LO-phonon frequency ω_{LO}. Because of this coupling the plasmons and phonons cease to be independent excitations and two mixed plasmon–LO-phonon modes labeled L_- and L_+ are observed. The scattering cross section of a multicomponent plasma due to charge-density fluctuations can be written as [10]

$$\frac{\partial\delta}{\partial\omega\partial\Omega} \propto [1-\exp(\hbar\omega/k_B T)]^{-1} \,\mathrm{Im}[\chi_e(\omega, \boldsymbol{q}) - \chi_e^2(\omega, \boldsymbol{q})/\varepsilon(\omega, \boldsymbol{q})], \tag{2}$$

where $\chi_e(\omega, \boldsymbol{q})$ is the susceptibility of the free electrons and $\varepsilon(\omega, \boldsymbol{q})$ the total dielectric function.

$$\varepsilon(\omega, \boldsymbol{q}) = \varepsilon_\infty + \chi_e(\omega, \boldsymbol{q}) + \chi_h(\omega, \boldsymbol{q}) + \chi_L(\omega), \tag{3}$$

$\chi_h(\omega, \boldsymbol{q})$ is the contribution of the photoexcited holes including inter- and intra-band transitions of light and heavy holes. $\chi_L(\omega)$ represents the lattice susceptibility. The intra-band terms of electrons and holes can be evaluated for finite temperatures with Lindhard–Mermin functions [11,12], whereas the inter-valence-band susceptibility can be approximated as given in ref. [13].

Figure 2 compares numerical calculations of the light scattering spectra for a one-component plasma $\chi_h(\omega, \boldsymbol{q}) = 0$ (fig. 2a) with those of a multicomponent electron–hole plasma $\chi_h(\omega, \boldsymbol{q}) \neq 0$ (fig. 2b) at different plasma densities. Although both calculations show the typical two coupled modes (L_- and L_+), there is a remarkable change in the lineshape if the inter- and intra-band transitions of the photocreated holes are considered in the dielectric function $\varepsilon(\omega, \boldsymbol{q})$ (see fig. 2b). At low Raman energies a new band arises in the spectra which is not present if only the contribution of the electrons is taken into account. This band has to be attributed to the acoustic plasmon mode [10] and is a consequence of the two types of mobile species in the optical plasma. The analysis of the several contributions to the valence-band susceptibility indicates that especially the inter-valence-band transitions of photoexcited holes modify the spectral lineshape. The strong damping of the L_+ mode at high EHP densities is a direct consequence of such transitions and prevents the detection of this mode. The coupled mode energies are given by the zeros of the total dielectric function (eq. (3)). Using the simple Drude expression for the free carrier system and the classical expression for the lattice vibrations, one can derive the EHP density n_{EHP} from the functional dependence of the plasma frequency $\omega_p(n)$ on the frequencies ω_- and ω_+ of the two coupled plasmon–LO-phonon modes [2],

$$\omega_\pm^2 = \tfrac{1}{2}(\omega_p^2 + \omega_{LO}^2) \pm \tfrac{1}{2}[(\omega_p^2 + \omega_{LO}^2)^2 - 4\omega_p^2\omega_{TO}^2]^{1/2}. \tag{4}$$

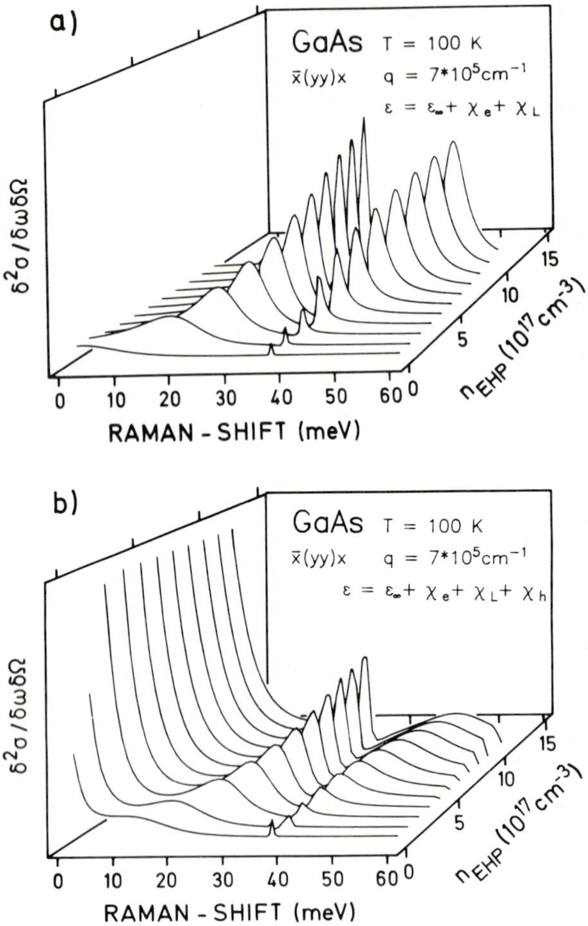

Fig. 2. Calculated coupled plasmon–LO-phonon mode spectra at different plasma densities. (a) Without including the intra- and inter-band susceptibility of the photoexcited holes. (b) With the intra- and inter-band contribution.

The effective mass in the expression of the plasma frequency has to be replaced by the optical mass m_{op} of the e–h pairs in this approximation. Therefore, the carrier density n_{EHP} can be obtained directly from the experimental values of the L_- and L_+ modes.

Single-particle mode

Scattering by individual carriers of the electron–hole plasma is related under resonant excitation conditions to fluctuations in the spin density. Figure 3 shows schematically this scattering process. The incident photon ($\hbar\omega_L$) excites

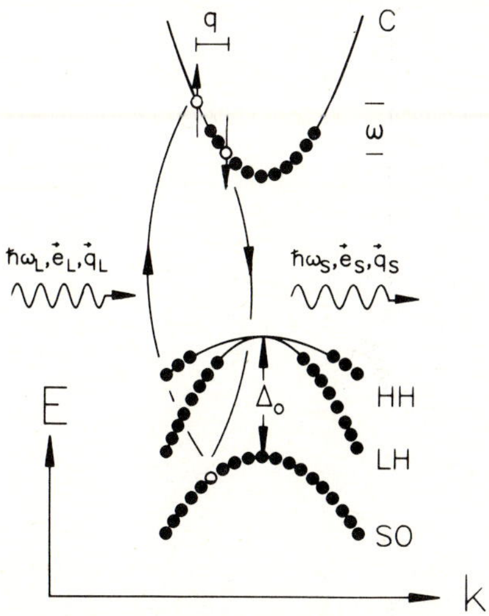

Fig. 3. The dominant resonant single-particle light scattering process if the laser energy is close to the $E_0+\Delta_0$ energy gap. A photon ($\hbar\omega_L$) is absorbed by exciting a electron with spin "up" and simultaneous recombination of an electron with spin "down" and emission of another photon ($\hbar\omega_S$).

an electron with spin "up" from the split-off valence band to the conduction band above the Fermi energy. An electron with opposite spin below the Fermi level recombines with the hole in the valence band and emits a photon ($\hbar\omega_S$). This Raman process is a consequence of the mixed spin character of the split-off band and is dominant for electrons if the exciting energy is close to the $E_o+\Delta_o$ energy gap, because of the non-degenerate character of the split-off band. Ignoring any resonant effects on the spectral lineshape, the Stokes light scattering cross section of spin-density single-particle excitations can be written as [14]

$$\frac{\partial^2\partial}{\partial\omega\partial\Omega}\propto[1-\exp(\hbar\omega/k_BT)]^{-1}\,\mathrm{Im}\,\varepsilon_e(\omega,\boldsymbol{q}), \qquad (5)$$

where ω is the Stokes shift and $\varepsilon_e(\omega, \boldsymbol{q})$ the dielectric function of the electron gas. $\mathrm{Im}\,\varepsilon_e(\omega, \boldsymbol{q})$ reflects the density-of-states associated with single-particle transitions and can be calculated by a finite temperature Lindhard expression [11].

3. Experimental results and discussion

In fig. 4(a) is shown a typical polarized spectrum $\bar{x}(yy)x$ for a high optical excited GaAs sample. The spectrum is assigned to charge-density fluctuations and shows two well-defined bands at the energies of the coupled optic plasmon–LO-phonon modes. No LO-phonon line has been observed from the GaAs.

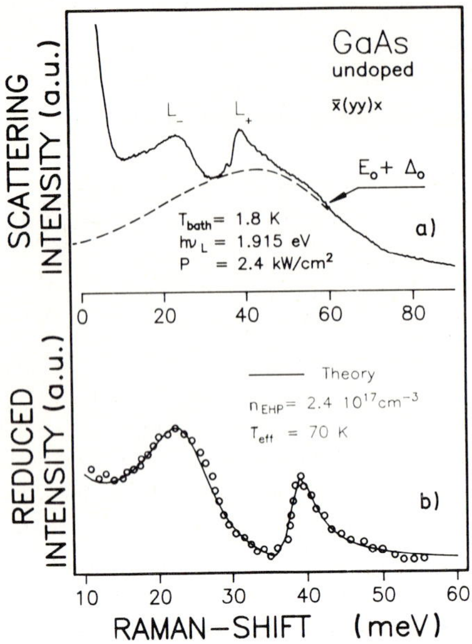

Fig. 4. Polarized light scattering spectra for a photon energy near the $E_0+\Delta_0$ optical gap. (a) Full spectra with $E_0+\Delta_0$ background. (b) Obtained from spectrum (a) after subtraction of the $E_0+\Delta_0$ luminescence background. The solid line is the calculated lineshape according to eq. (2).

This is in contrast to previous works on non-confined plasma [7, 15], where the LO-phonon line has been observed together with the coupled modes because of the plasma inhomogeneity.

The spectral lineshape is particularly pronounced in fig. 4(b) (dots), after subtraction of the broad $E_0+\Delta_0$ luminescence background. Calculation of the coupled mode lineshape requires knowledge of the EHP density and temperature. The density can be estimated in a first approximation by an evaluation of the coupled mode frequencies ω_-, ω_+ (see eq. (4)). Using the well-known optical parameters of GaAs a consistent density of $n_{\mathrm{EHP}} \simeq 2.4\times10^{17}\ \mathrm{cm}^{-3}$ can be obtained independently from both plasmon–LO-phonon modes. The effective carrier temperature can be estimated from the high energy tail of the EHP luminescence to $T_{\mathrm{eff}} \simeq 70$ K. The solid line in fig. 4(b) corresponds to the above plasma data and an electron damping $\Gamma = 2$ meV in the Mermin correction of the Lindhard integral [12]. The agreement with the measured spectra is excellent. The linewidth as well as the relative intensities of the two coupled modes are simultaneously fitted by the theory. It has to be noted that:

(1) No satisfactory agreement with the experimental data can be obtained if the inter-valence-band contribution to the dielectric function is neglected. This shows that the $V_{LH}\rightarrow V_{HH}$ transitions of photoexcited holes have a substantial influence on the damping of the coupled modes.

(2) The electron damping as found from the collision-relaxation time $\tau[\Gamma=1/(2\pi\tau c)]$ is by a factor 3–10 smaller than in the case of equivalently doped n-type GaAs. This is a direct consequence of the high mobility in the optically excited plasma in high purity GaAs and makes the present lineshape fitting more significant than for doped material.

Figure 5 summarizes the experimentally obtained coupled mode energies (dots) versus the EHP density as calculated according to eq. (4), whereas the solid line represents the peak position of these modes as obtained from the full theoretical lineshape calculation after eqs. (2) and (3). Although the Drude formula (eq. (4)) does not include any damping of the EHP there is no substantial difference to the calculation with damping. The slight disagreement at low EHP densities has to be attributed to the fact that both models are high density approximations. Although the lineshape is strongly influenced by the inter- and intra-valence-band transitions of photocreated holes, there is no significant change in the peak position of the two coupled modes. The plasma density can therefore easily be obtained from the functional dependence of the collective mode energies on the plasma frequency ω_p without the complex procedure of a numerical lineshape fitting.

Next we consider the depolarized spectra $\bar{x}(yz)x$. Typical examples are shown in fig. 6 after subtraction of the luminescence background. Similar to the case of n-type GaAs, such spectra are caused by spin-density fluctuations of the free electrons and reflect directly the single-particle excitations of the EHP. As shown above for the collective mode a lineshape calculation requires the knowledge of the plasma density and temperature. Because of the much simpler expression (5), i.e., only the imaginary part of the electronic dielectric function has to be taken into account, one can obtain the two plasma parameter from a fitting procedure. The best fits are shown by the solid line in fig. 6. It should be pointed out that the plasma densities and temperatures determined from the single-particle spectra agree very well with the values obtained from the coupled mode analysis and photoluminescence fittings [9]. The main results can be summarized as follows:

(1) The Raman scattering process allows the detection of the EHP parameters for degenerate as well as for *non*-degenerate plasma with the same high accuracy. This is important to the application of optical gain and transmission measurements which are limited to degenerate EHP.

(2) The electron-damping parameter that give the best fit to the experimental data is found to $\Gamma\simeq 0$ meV, whereas the collective modes reflect $\Gamma\neq 0$ meV. A detailed analysis of this difference is beyond the scope of this work.

The present studies on a photoexcited electron–hole system will be used in the following part in order to investigate the non-equilibrium state and the density–temperature relation of the EHP.

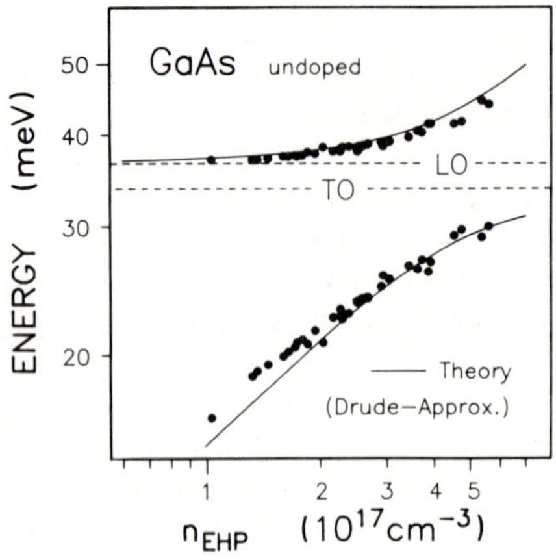

Fig. 5. Plasmon–LO-phonon mode energies as a function of the EHP density calculated from eq. (4). The full line represents the peak position of the coupled mode lineshape as given by eq. (2).

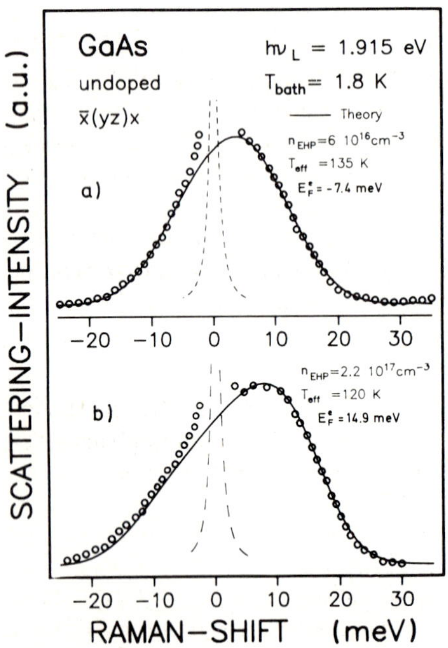

Fig. 6. Single-particle spectra observed in the $\bar{x}(yz)x$ configuration near the resonance gap for a non-degenerate (a) as well as for a degenerate EHP (b). The full lines are theoretical calculations according to eq. (5).

4. Light scattering from non-equilibrium EHP

High photoexcitation of bulk GaAs has been the subject of intensive studies [16]. Although direct-gap materials have a much simpler band structure than indirect-gap semiconductors the agreement between theoretical predictions and experimental data is not satisfactory for the direct-gap materials. The high carrier mobility in conjunction with a high Fermi pressure of the EHP leads to the expansion of the photoexcited plasma far beyond the pumped region [5,6]. If sufficiently high, this particle flux modifies the carrier distribution function [16].

Luminescence and gain experiments under intense photoexcitation show that the lineshape can be explained by taking into account a rigid shift of the Fermi-sphere in the $\boldsymbol{k}$-space relative to the band extrema. Light scattering from single-particle excitations can directly provide detailed information about the plasma distribution function [17].

Mooradian [18] applied this technique at first to study the electron-distribution function induced by an applied high electric field. In order to investigate the non-equilibrium distribution of the EHP a sample like the one of fig. 1 but without the GaAlAs layer between the "active" GaAs layer and the GaAs substrate was used. This structure allows a nearly one-dimensional plasma expansion parallel to the incident laser light without lateral density gradients. Because of the lack of carrier confinement, one needs laser power about 40–100 times higher than the laser power for the 3-D confinement to create the same EHP density. The density was checked by analyzing the coupled mode spectra. The high laser power required, is a first indication of fast plasma expansion effects.

Raman spectra as recorded for a 3-D confined plasma (circles) and a non-confined EHP (squares) are depicted in fig. 7. The laser power was adjusted in both experiments so that the collective modes reflect the same EHP density. Although the lineshape of the non-equilibrium spectra is not significantly changed compared to the equilibrium case, its position is shifted toward higher Stokes energies. In order to fit the experimental single-particle spectra with densities consistent with the coupled mode results we assume in a first approximation a shifted Fermi–Dirac distribution function [19]

$$\tilde{f}(\boldsymbol{k}, \boldsymbol{k}_{\mathrm{D}}, T)=\left[\exp\left\{\left[\frac{\hbar^2}{2m}(\boldsymbol{k}-\boldsymbol{k}_{\mathrm{D}})^2-\mu\right]\Big/k_{\mathrm{B}}T\right\}+1\right]^{-1}, \tag{6}$$

where μ is the chemical potential, $v_{\mathrm{D}}=\hbar k_{\mathrm{D}}/m$ is the drift velocity of the carriers relative to the lattice. In order to express the results in terms of the equilibrium Fermi–Dirac distribution one can show in the frame of the RPA, that the imaginary part of the dielectric function can be expressed as

$$\mathrm{Im}\,\tilde{\varepsilon}(\omega, \boldsymbol{q}, \boldsymbol{k}_{\mathrm{D}})\simeq\int \mathrm{d}^3\boldsymbol{k}\, f(\boldsymbol{k}, T)\,\delta\left[\frac{\hbar^2}{2m}(-k^2-2\boldsymbol{q}\boldsymbol{k}+\boldsymbol{q}\boldsymbol{k}_{\mathrm{D}})+\hbar\omega\right]. \tag{7}$$

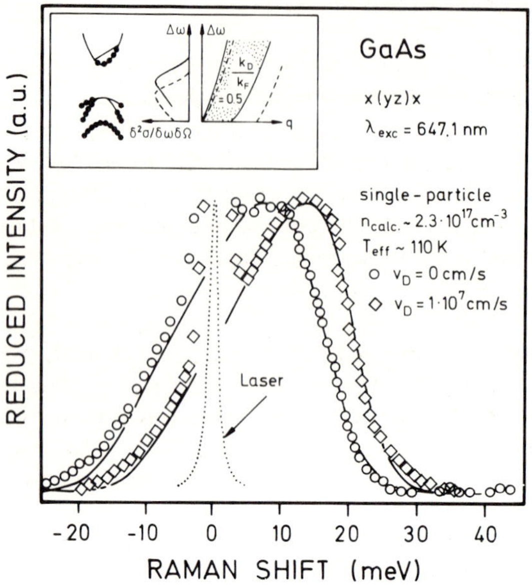

Fig. 7. Comparison between measured single-particle spectra for a 3-D confined plasma (circles) and a non-confined EHP (squares). The dashed line is the calculated lineshape according to eq. (5), whereas the full line is a fit including a drift velocity $v_D = 10^7$ cm/s in eqs. (6) and (7). The plasma density is estimated from the coupled mode spectra. The inset shows schematically the influence of the carrier diffusion on the dispersion and lineshape of the single-particle scattering process.

In the inset of fig. 7 the influence of the carrier drift on the single-particle spectra is schematically shown. The displaced distribution function modifies the single-particle dispersion and thus the lineshape position. This *Doppler* shift due to the expansion of the EHP reflects directly the drift velocity v_D and because of the vector character in eq. (7) also its direction. This is an important difference to the evaluation of gain and luminescence experiments where only the amount of the drift velocity can be measured. A comparison of the experimental single-particle spectra with theoretical ones, based on a dielectric function of the form according to eq. (7), is shown by the solid line. The best fit can be achieved with $v_D = 10^7$ cm/s. This value is consistent with previous experiments on the nano- and picosecond timescale [6,20]. The broken line fits the spectrum for the confined EHP with $v_D = 0$.

An open question at present is the assumption of a rigidly shifted Fermi-sphere. Experiments at very high power levels ($P > 300$ kW/cm^2) show a single-particle lineshape which cannot be explained by a simple Doppler shifted lineshape, indicating that there are further processes in this high excitation field which are worthy to study.

5. Density–temperature relation of quasi-equilibrium EHP

Investigations of the $n-T$ relation of highly excited GaAs have been extensively persued within the last decade [3, 21]. The observed independence of the EHP density on the pump intensity suggests that an initially overdense EHP expands fast to a "final" density. The nearly linear dependence of this "final" density on the carrier temperature was first related to the Mott density [22]. On the basis of a hydrodynamical model, and more basically on a thermodiffusion concept [23], the observed $n-T$ relationship could be theoretically reproduced and is found to be due to the non-equilibrium nature of high excited EHP.

The present study is concerned with an application of Raman spectroscopy to confined low- and high-temperature plasma. The EHP was created by a tuneable laser (710 nm $<\lambda_{exc}<$ 810 nm) operating cw. A second, weak laser ($\lambda_L = 647$ nm) served to probe the optically excited EHP. This novel technique allows one to study the plasma also at low electronic temperatures. Figure 8 shows a series

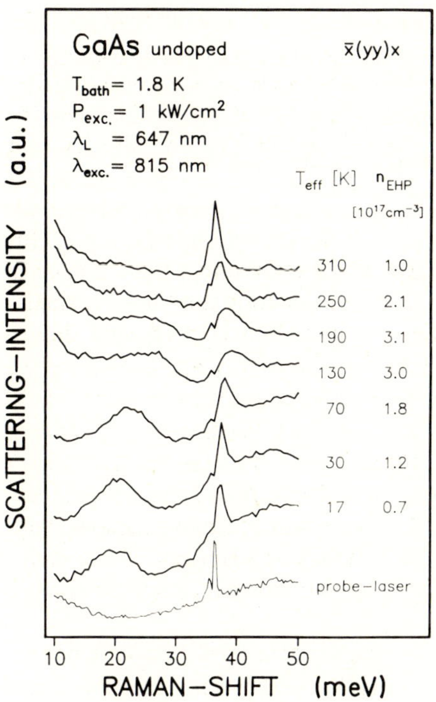

Fig. 8. Coupled plasmon–LO-phonon modes of a photocreated EHP at different electronic temperatures. The lower trace is recorded without the pump laser (λ_L) and reflects only the LO-phonon of the "active" GaAs layer.

of coupled plasmon–LO-phonon spectra recorded at different temperatures and at a constant power density. The lower trace (weak line) was recorded with the probe laser and shows the LO-phonon line of the GaAs layer, whereas the upper traces (probe + pump laser) reflect the spectral features of the coupled plasmon–LO-phonon modes. From the Raman energies one obtains a plasma density which increases from $n_{EHP} \simeq 7 \times 10^{16}$ cm^{-3} at $\simeq 17$ K up to $\simeq 3.2 \times 10^{17}$ cm^{-3} at $\simeq 170$ K. A further increase in temperature reduces the density to 1×10^{17} cm^{-3} at $\simeq 300$ K. To exclude that this change in the carrier density with temperature is caused by a temperature effect of the Raman process itself, we have measured coupled mode spectra of a comparable doped n-type GaAs sample and also calculated the temperature dependence of the Raman spectra. Neither experimentally nor theoretically does one find a significant shift of the coupled modes with the temperature. A further evidence of a strong temperature correlated EHP density at high temperature is found from the integrated EHP luminescence. Such measurements reflect an unexpected strong decrease of the luminescence intensity [24] for $T_{eff} \gtrsim 170$ K.

Figure 9 shows the experimental observed n–T values for three different excitation power densities (symbols). The most striking result is the decrease in plasma density above 170 K at all excitation densities. This is different to high-temperature experiments on non-confined EHP in Si [25] where a $n \sim T^{3/2}$ dependence was found in accordance with predictions of the thermodiffusion model [23]. On the other hand, in the case of an equilibrium EHP the recombination channels such as bimolecular, Auger and interface recombination should determine the free carrier density. Under this assumption one can describe the plasma behaviour in the framework of a balance equation between the generation and recombination of carriers. With a bimolecular recombination coeffi-

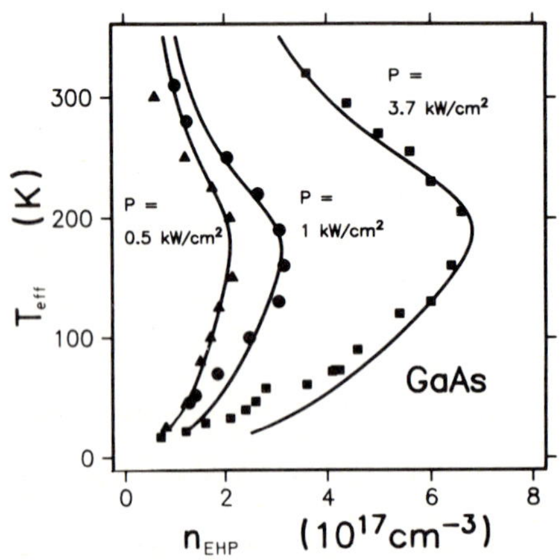

Fig. 9. EHP density and temperature at different power densities. Full lines are calculated as described in the text.

cient $B(T) \sim 1/T$ as expected for the Boltzmann limit and a temperature dependent Auger coefficient as calculated by Haug et al. [26] for the phononless valence-band Auger process (CHSH) one can calculate the n–T relation.

The results of these calculations are shown in fig. (9) for different generation rates. The experimental data are satisfactorily fitted by the model which includes only generation and recombination of carriers. Most striking is the evaluation at high temperatures which is governed by the onset of the strongly temperature dependent Auger recombination process, whereas at low temperatures this process plays no important role on the EHP density.

Summary

In conclusion we can state that the inelastic light scattering by collective and single-particle excitations has proven to be an accurate and versatile tool in measuring the equilibrium as well as the non-equilibrium properties of photo-excited electron–hole plasma.

From the spectral features of high temperatures EHP we deduce a correlation of the EHP density and temperature with recombination channels like bimolecular and Auger recombination.

The following parameters have been used in all calculations:

$$m_e = 0.0665 m_0, \quad m_{lh} = 0.082 m_0, \quad m_{hh} = 0.5 m_0, \quad m_{s0} = 0.15 m_0,$$

$$E_0 = 1.51 \text{ eV}, \quad \Delta_0 = 0.33 \text{ eV}, \quad \omega_{LO} = 36.5 \text{ meV}, \quad \omega_{TO} = 33.8 \text{ meV},$$

$$\varepsilon_\infty = 10.91, \quad \varepsilon_0 = 12.9, \quad q = 6.9 \times 10^5 \text{ cm}^{-1}.$$

Acknowledgements

The authors are grateful to K. Ploog for providing the high purity GaAs/GaAlAs heterostructures. We also thank A. Forchel for valuable discussions.

References

[1] A. Mooradian, in: Advances in Solid State Physics, Vol. 9, H.J. Queisser, ed. (Pergamon-Vieweg, Braunschweig, 1969) p. 74.

[2] For a review, see e.g., M.V. Klein, in: Light Scattering in Solids, M. Cardona, ed. (Springer Verlag, Berlin, 1975) p. 147.

[3] E.O. Göbel and G. Mahler, in: Advances in Solid State Physics, Vol. 14 (Pergamon-Vieweg, Braunschweig, 1979) p. 105;
C. Klingshirn and H. Haug, Phys. Rep. 70 (1981) 315.

[4] S. Modesti, A. Frova, M. Capizzi, L.G. Quagliano, J.-L. Staehli and M. Guzzi, J. Phys. Soc. Japan 49, Suppl. A (1980) 515.
[5] S. Modesti, L.G. Quagliano, A. Frova, J.-L. Staehli and M. Guzzi, J. Lumin. 24/25 (1981) 581.
[6] K.M. Romanek, H. Nather, J. Fischer and E.O. Göbel, J. Lumin. 24/25 (1981) 585.
[7] K.M. Romanek, H. Nather and E.O. Göbel, Solid State Commun. 39, (1984) 23.
[8] M. Capizzi, S. Modesti, A. Frova, J.-L. Staehli, M. Guzzi and R.A. Logan, Phys. Rev. B 28 (1984) 2028.
[9] H. Nather and L.G. Quagliano, Solid State Commun. 50 (1984) 75.
[10] P.M. Platzman and P.A. Wolff, in: Waves and Interactions in Solid State Plasmas (Academic Press, New York, 1973).
[11] J. Lindhard, Dan. Mat. Fys. Medd. 28 (1954) No. 8.
[12] N.D. Mermin, Phys. Rev. B 1 (1970) 2362.
[13] R. März, thesis, Frankfurt (1983).
[14] A. Pinczuk, G. Abstreiter, R. Trommer and M. Cardona, Solid State Commun. 30 (1979) 429.
[15] J.E. Kardontchik and E. Cohen, Phys. Rev. Lett. 42 (1979) 669.
[16] See A. Forchel, H. Schweizer and G. Mahler, Phys. Rev. Lett. 51 (1983) 501 and references therein;
A. Forchel, B. Laurich, H. Hillmer, G. Tränkle and M. Pilkuhn, J. Lumin. 30 (1985) 67.
[17] A. Mooradian, in: Light Scattering Spectra of Solids, G.B. Wright, ed. (Springer Verlag, New York, 1968) p. 285.
[18] A. Mooradian and A.L. McWhorter, in: Proc. 11th Int. Conf. Phys. Semicond., S.P. Keller, J.C. Hensel and F. Stern, eds. (U.S. A.E.C. Div. of Tech. Information, Oak Ridge 1970) p. 380.
[19] Y.C. Lee and N. Tzoar, Phys. Rev. 140 (1965) A396.
[20] C.L. Collins and P.Y. Yu, Solid State Commun. 51 (1984) 123.
[21] M. Combescot, Phys. Status Solidi (b) 86 (1978) 349.
[22] O. Hildebrand, E.O. Göbel, K.M. Romanek, H. Weber and G. Mahler, Phys. Rev. B 17 (1978) 4775.
[23] G. Mahler, G. Maier, A. Forchel, B. Laurich, H. Sanwald and W. Schmid, Phys. Rev. Lett. 47 (1981) 1855.
[24] H. Nather and L.G. Quagliano, in: Proc. 17th Int. Conf. Phys. Semicon. (San Francisco, 1984) in press.
[25] B. Laurich and A. Forchel, J. Phys. C5, 10 (1983) 43.
[26] A. Haug, D. Kerkhoff and W. Lochmann, Phys. Status Solidi (b) 89 (1978) 357.

Journal of Luminescence 30 (1985) 65–66
North-Holland, Amsterdam

ELECTRON–HOLE PLASMA IN $Ga_{1-x}Al_xAs$: EXPANSION AND CONFINEMENT (extended abstract)

A. SELLONI, M. CAPIZZI and S. MODESTI

Dipartimento di Fisica, Universita' La Sapienza, Roma, Italy

It has often been pointed out that the value of the band-gap renormalization ΔE_G obtained from gain and luminescence spectroscopy in direct-gap semiconductors (DGS) were up to 1.5 effective Rydbergs higher than the theoretical estimates for the corresponding plasma density. Recently, these discrepancies have been attributed to: use of questionable phenomenological models to fit the experimental lineshapes [1]; and/or neglect of non-equilibrium effects in the plasma (i.e., the Fermi pressure [1] and the drift velocity [2]). Because of the high excitation intensities which are usually employed to overcome the short carrier lifetime, the plasma density in the optically pumped region exceeds the so-called liquid density n^*. The ensuing Fermi pressure causes a rapid plasma expansion. Contrasting values for the diffusion length L_D have been reported, pointing either to a moderate expansion ($L_0 \sim 10$–20 μm) or to a strong expansion picture. We measured the luminescence – with 20 μm spatial resolution – in $Ga_{1-x}Al_xAs$ samples tailored to insure a "one-dimensional" expansion. Filamentary samples were obtained by photolithographically etching 15 μm wide and 400 μm long islands, consisting of a central active layer of $Ga_{1-x}Al_xAs$ (0.5 μm thick), sandwiched between two alloy layers with wider gap. The local density was determined from fitting of the luminescence lineshape, using the model discussed in ref. [3]. All the plasma profiles are consistently described by a value of $L_D \simeq 25$ μm, which is considered an upper limit because of the finite dimensions of the laser spot and of the presence – at low density – of an underlying bound exciton band. This result has been confirmed by independent measurements of the spatially integrated luminescence from small circular $Ga_{0.61}Al_{0.39}As$ islands of varying diameter d ($25\ \mu m \leqslant d < \infty$).

Plasma expansion strongly influences the luminescence and gain spectra. In order to study the EHP properties, we choose therefore to restrict our attention to the spectra obtained from circular 3-d confined plasma, i.e., from islands with diameter smaller than or equal to the exciting laser spot. Our theoretical analysis is based on the Random Phase Approximation (RPA) and includes k-conservation. Different models have been tested for the electron and hole spectral functions, to evaluate the influence of plasmon replicas on

0022-2313/85/$03.30

lineshapes and band positions. The self-consistent RPA overestimates plasmon sidebands and underestimates the band-gap renormalization. This shows that "beyond RPA" interactions are important in luminescence, even at densities as high as $20n^*$. An excellent fit of the experimental lineshapes, over a wide range of densities and temperatures, is achieved instead using simple spectral functions, accounting only for the single particle energy renormalization and lifetime broadening. Correspondingly, good agreement between theory and experiment is found for the band-gap reduction. An extended account of the work will appear in a special issue of Helvetica Physica Acta during the course of this year.

We are very indebted to our colleagues A.Frova and J.L. Staehli for starting this work and participating in various stages of its development. We thank M. Guzzi for collaboration in the experimental part, R. Logan for custom preparation of the layered epitaxial structures, and G. Chiaretti for careful photolithographic processing of the samples.

References

[1] M. Capizzi, S. Modesti, A. Frova, J.L. Staehli, M. Guzzi and R.A. Logan, Phys. Rev. B29 (1984) 2028.

[2] A. Forchel, H. Schweizer and G. Mahler, Phys. Rev. Lett. 51 (1983) 501.

[3] A. Selloni, S. Modesti and M. Capizzi, Phys. Rev. B30 (1984) 821.

Journal of Luminescence 30 (1985) 67–81
North-Holland, Amsterdam

OPTICAL STUDIES OF FAST PLASMA TRANSPORT IN Si

A. FORCHEL, B. LAURICH, H. HILLMER, G. TRÄNKLE and M. PILKUHN

Physikalisches Institut, Universität Stuttgart, Pfaffenwaldring 57, D 7000 Stuttgart 80, FRG

We present two new optical techniques for the investigation of the transport properties of ambipolar plasmas in semiconductors. The first method is a time-of-flight technique: Using surface doping with shallow impurities which provide a characteristic bound exciton emission we introduce optically active spatial markers into the thin Si wafers investigated. Carrier pairs are excited at the undoped surface of the wafer by short laser pulses. From time-resolved studies of the bound exciton emission we obtain average velocity values for the ambipolar transport through the sample. The most attractive feature of the time-of-flight method compared to other spatially resolved measurements is the combination of very high spatial resolution (submicron range) with a high sensitivity. The second method used to study the plasma transport is based on a time-resolved investigation of the Mott transition between the electron–hole plasma and free excitons in Si. Using well-established values for the Mott transition and the temporal evolution of the plasma and the free exciton emissions we obtain values for the velocity with which the plasma expands to the Mott density as functions of the excitation power and the temperature. The advantage of this method is the lack of any need for spatial resolution. Possible extensions of both methods are discussed.

Introduction

Under strong optical excitation collective phenomena are known to dominate the optical spectra of intrinsic semiconductors. These phenomena range from the condensation of carrier pairs into an electron–hole liquid at very low temperatures to the ultra-high density, high temperature electron–hole plasma obtained in femto-second laser pulse experiments [1,2].

From a thermodynamic point of view the collective properties of the charge carrier system can be divided into equilibrium and non-equilibrium processes. One of the most interesting phenomena observed under quasi-equilibrium conditions is the electron–hole liquid condensation, which has been studied for a wide variety of materials with indirect band structure [3,4]. Experimentally, condensation phase diagrams have been determined from the evaluation of luminescence spectra in unstrained and uniaxially strained Si and Ge as well as for GaP. The ground state ($T=0$) and critical parameters ($T=T_c$) have been

0022-2313/85/$03.30 © Elsevier Science Publishers B.V.
(North-Holland Physics Publishing Division)

found to be in quantitative agreement with the results of microscopic calculations. Another property of the electronic system which can be understood using equilibrium models is the screening of the exciton binding by free carriers ("Mott transition"): Experimental and theoretical results for the Mott density as a function of the carrier temperature are in good agreement [5,6].

In contrast to effects for which the quasi-equilibrium description provides a quantitative explanation, there exists a wealth of data – mostly from direct gap materials and from indirect gap semiconductors above the critical point – where the equilibrium picture leads to inconsistent results. For example, the evaluation of the gain spectra ("lineshape fits") in highly excited direct gap semiconductors provides always plasma densities which are lower by about a factor of two than those obtained from the energetic position of the gain. Due to density dependent many-body effects the energy gap is reduced in the plasma (band renormalization) [7], and therefore the energetic position of the optical response of the plasma is directly related to the plasma density. The inconsistency does not depend on the particular theory used to describe the spectra (e.g., single particle picture without momentum conservation [8], or many-body theory including momentum conservation [9]). The different density values are particularly surprising since in indirect gap materials at temperatures below the critical point the density values determined from the energetic position and from the lineshape agree within a few percent [10]. Above the critical point, on the other hand, one observes a strong disagreement between the densities obtained from both methods. In contrast to the behaviour of the plasma in the direct gap materials, the density obtained from the energetic position is lower by approximately a factor of two than the density derived from a lineshape analysis.

Another persistent puzzle has been observed in the case of Raman scattering studies on GaAs [11] and GaP [12] (above T_c). In principle, the single-particle scattering spectrum as well as the energy of the coupled plasmon–phonon modes ω_+ and ω_- can be used to derive the plasma density. The straightforward evaluation of the data, however, implied different densities from each process.

Forchel and coworkers [13] have proposed a simple model including previously neglected non-equilibrium (transport) effects which is capable to resolve all above mentioned inconsistencies. The model assumes that the ambipolar plasma expands due to, e.g., density gradients into the bulk of the sample. If the average expansion velocity is comparable to the Fermi velocity of either holes or electrons, the distribution function of these carriers will deviate significantly from the equilibrium Fermi distribution. The simple single-particle model explains consistently the measured gain, emission and Raman spectra if one assumes an ambipolar expansion of the non-equilibrium plasma with velocities between 2×10^6 and 1×10^7 cm/s, depending on the particular material under investigation. In addition to the magnitude of the plasma velocity, the evaluation of the single-particle spectra implies that the direction of motion is into the sample, and hence confirms the main assumption of the model [11,13].

The expansion velocities inferred from the evaluation of the spectra are surprisingly high, especially since it is not clear which kind of force accelerates the carrier pairs. In the frame of studies of the "phonon wind" induced motion of electron–hole liquid, for example, velocities limited by the speed of sound are observed. Velocities in the lower and mid 10^6 cm/s range have previously been obtained for unipolar plasmas in high electric fields [14]. In addition to the somewhat indirect spectroscopic velocity determination it is clearly of high interest to determine the ambipolar expansion velocity directly.

Previously a number of experiments have been designed with temporal and spatial resolutions which allow in principle the determination of the expansion velocity. They include pulsed infrared plasma absorption [15] and spatially and time-resolved luminescence spectroscopy [16] and phonon detection [17]. Most of these techniques have been developed initially to study phonon wind transport effects in Ge, and later Si. The relatively low velocities allowed studies with typical temporal resolutions worse than 10 ns and spatial resolutions in the order of 30 μm [16]. The spatial resolution, in particular, is limited by the weak intensities emitted by the indirect gap materials. Furthermore, due to surface/edge scattering effects, measurements of a plasma expansion into the sample are severely restricted.

Optical time-of-flight method

Since we are particularly interested in the fast expansion process into the bulk of the sample, we have developed a new time-of-flight method with sub-micron spatial resolution and with a temporal resolution of about 1 ns. The basic principle of the method is illustrated in fig. 1. A thin high purity Si sample is doped through one surface, most suitably by ion implantation, with impurity atoms which are known to give rise to an effective and characteristic bound exciton luminescence [18]. These impurities serve as spatial markers for the presence of electron–hole pairs. Since we excite carrier pairs within ~1 μm

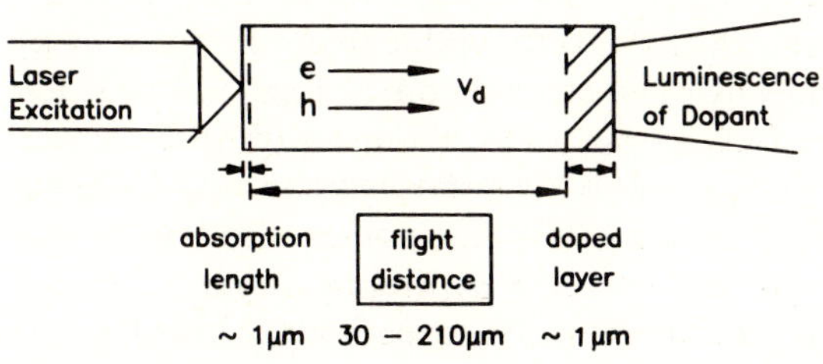

Fig. 1. Schematic design of sample and setup for the time-of-flight, technique.

of the opposite surface only electron–hole pairs which have travelled through the sample can be bound at the impurities. The bound exciton intensity therefore is a measure for the number of carrier pairs which have reached the doped surface layer. From a time-resolved study of the BE intensity one can easily obtain the average velocity with which the carriers have moved through the sample.

The ideal impurity for these time-of-flight experiments has to satisfy the following conditions:

(i) The BE emission should occur at an energy where no other luminescence band as, e.g., free exciton or electron–hole liquid/electron–hole plasma-emission arises.

(ii) The BE recombination should be fast ($\tau \leqslant 1$ ns). Then the BE intensity is directly proportional to the number of carrier pairs in the doped layer.

(iii) The BE should be stable at higher temperatures. For studies of non-equilibrium plasma transport in Si, for example, the emission should be clearly observable up to at least 25 K [19].

Especially at higher temperatures the first and the last requirement are very difficult to reconcile. The thermal stability of a bound exciton increases with its binding energy. Hence the thermally most stable simple BE's in indirect gap semiconductors overlap with the phonon assisted free exciton emission.

We have selected indium as the dopant for the present experiments. It fulfils the above requirements only partially: The In recombination is rather fast ($\tau = 2.7$ ns) [20]. The emission of the In bound excitons occurs at 10868 Å, which corresponds to the high energy tail of the TA-phonon-assisted recombination of the free exciton. At low electronic temperature (below about 4 K) the excitonic contribution at the In wavelength is negligible. At temperatures above about 15 K, however, due to the strong increase of the free exciton intensity and the simultaneous decrease of the In BE luminescence, expansion related signals are very difficult to detect. Another set of samples was prepared by thallium diffusion into Si (layer depth $\cong 8\ \mu$m, [Tl] $\cong 10^{16}$ cm^{-3}). Compared to the case of In the Tl bound exciton has a very short lifetime (270 ps [21]). Due to its higher binding energy it is thermally relatively stable. At low temperature, however, its luminescence overlaps with the TA replica of the electron–hole liquid emission and at high temperature with the strong TO/LO replica of the free exciton.

The samples were mounted in a variable temperature dewar. Either a mode-locked, Q-switched and frequency-doubled Nd–YAG Laser ($\lambda = 532$ nm, repetition rate typically 4 kHz, peak power 1 kW, pulse length $\simeq 0.5$ ns) or a cavity-dumped, mode-locked Ar Laser ($\lambda = 514$ nm, repetition rate 0.4 MHz, peak power 5 W, pulse length $\simeq 0.3$ ns) were used to excite the samples. For the thicker samples the luminescence was detected by a high sensitivity photomultiplier (Sl, time resolution $\simeq 5$ ns), whereas a less sensitive high speed photomultiplier (Sl, time resolution $\leqslant 1$ ns) was used for samples with thicknesses below 100 μm. The signals were processed using fast gated electronics (time-delayed spectra) or time to pulse-height conversion in conjunction with a multichannel analyzer (for studies of temporal changes at a fixed energy).

Figure 2 displays time-delayed spectra from a thallium doped sample at 13 K. The dashed trace is a spectrum recorded immediately after the laser pulse and displays mainly the emission of the electron–hole liquid. In particular, no structure is observed at the wavelength corresponding to the Tl bound exciton (BE) emission. The solid line represents a spectrum recorded with a time delay of 60 ns. Here we observe a strong emission due to recombination of the Tl BE, indicating that a large fraction of the carrier pairs has travelled through the sample within the delay time.

More detailed information about the evolution of the plasma expansion with time can be obtained from the time development of the BE emission. Figure 3 displays the onset and decay of the In bound exciton luminescence in a sample with 95 μm thickness (full line). For a calibration of the time scale, the time dependence of the emission after a direct optical excitation of the In layer is included (dotted line) [22]. During approximately the first 5 ns the direct excitation of the In BE gives rise to a rapidly increasing and steeply decaying emission. The long tail for times in excess of about 12 ns is due to carrier pairs which after diffusion through the sample incidentally recombine at an In site. Compared to the emission after direct optical excitation, the BE signal due to ambipolar transport through the sample (excitation of the unimplanted surface) is significantly different: The maximum of the transport induced emission is delayed by about 20 ns and the onset of the emission occurs about 10 ns after the laser pulse ($t=0$).

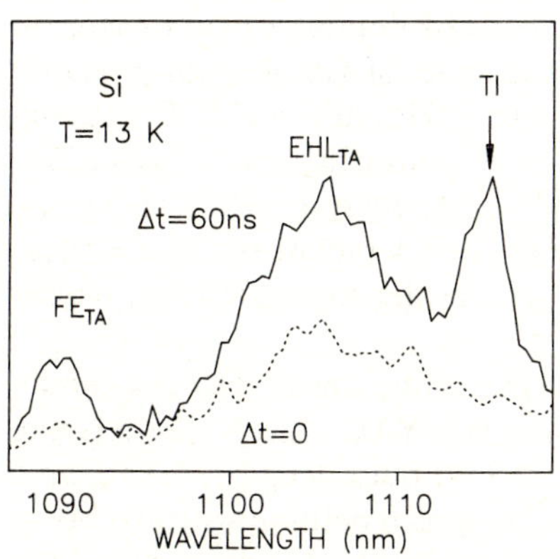

Fig. 2. Time-delayed emission spectra of a Si wafer, which was doped on one surface with thallium, after excitation of the undoped surface. Broken line: spectrum obtained immediately after the laser pulse. Solid line: spectrum recorded with a delay of 60 ns with respect to the excitation. The detection gate width was 10 ns in both cases. See text.

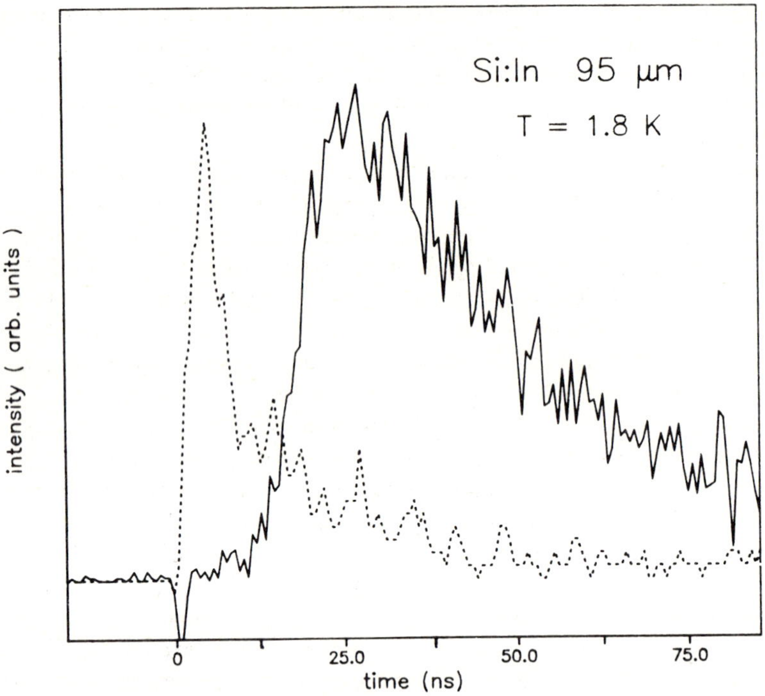

Fig. 3. Temporal changes of the In bound exciton emission intensity, (i) after direct optical excitation of the In implanted surface (dotted line), (ii) after optical excitation of the undoped surface.

Using the measured crystal thickness the average velocity corresponding to the intensity maximum is close to 5×10^5 cm/s. The onset of the signal yields average velocities close to 1×10^6 cm/s. This implies that a significant fraction of the carriers travel through the sample with velocities above the speed of sound [23]. In particular, the initial velocities are inconsistent with a "sound barrier" for the ambipolar transport, which has been inferred from recent spatial imaging studies by Wolfe and coauthors [23]. Note, that our evaluation uses only the minimum distance (wafer thickness) between the excited and unexcited surface and thereby gives a lower limit for the carrier velocities: Due to a small laser spot (diameter of about 30 μm) compared to the sample thickness the plasma will expand into a hemisphere, and hence the average path of a carrier pair necessary to reach the doped surface layer will be longer than the sample thickness.

Great care has been taken to ascertain that the onset of the emission is due to ambipolar transport only. By using appropriate sample holders we exclude any direct optical excitation of the In doped layer, which is certified by the absence of the BE intensity during the first 5 ns. In principle, the near infrared emissions from the free excitons or electron–hole plasma could furthermore excite the In doped layer. Due to the indirect bandgap, however, the intrinsic emissions occur at lower energies and hence cannot be absorbed by the indium. More important, the optical density αd for such a transition is 10^{-6} [24], which is by orders of magnitude too low to make any detectable contribution to the In bound exciton signal.

Using the optical time-of-flight method we have studied the ambipolar transport in highly excited Si at different temperatures below 15 K and for different thicknesses. Figure 4 displays the thickness dependence of the maximum average expansion velocities derived from the bound exciton emission onsets at a bath temperature of 1.8 K. We observe a rapid increase of the maximum average velocity with decreasing thickness: For a sample thickness of 300 μm the peak average velocity v_{max} is 4×10^5 cm/s, whereas 2×10^6 cm/s are obtained in a 78 μm thick wafer.

Figure 5 displays the measured temperature dependence of v_{max} for the 95 μm sample. Between 1.8 K and 16 K the maximum velocity decreases from 1×10^6 to 3×10^5 cm/s, which indicates a strong increase of the damping of the ambipolar transport at high temperatures.

In order to understand qualitatively the observed features of the ambipolar transport we propose the following schematic model (fig. 6): The laser pulse leads to strong local heating of the excited spot compared to the bulk of the sample, which is maintained close to the selected bath temperature. From time-resolved measurements of exciton and plasma spectra, e.g. by Shah and Dayem, initial temperatures as high as 45 K have been observed after exciting Si at 2 K with an Ar-ion laser [25]. For our experiments we therefore assume that the electronic and lattice temperature within the excited spot are above

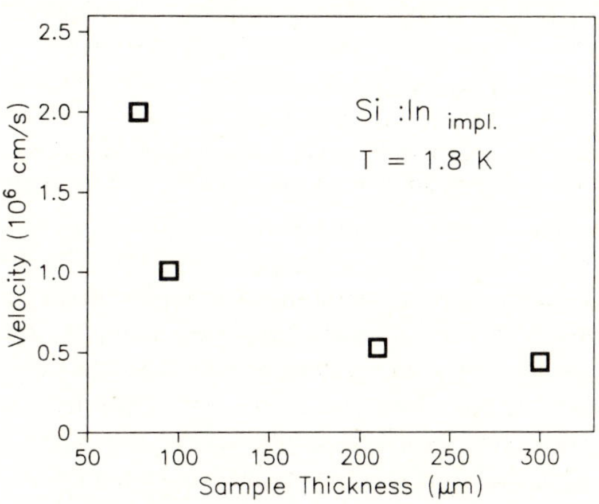

Fig. 4. Dependence of the maximum average velocity of the ambipolar expansion on the sample thickness. The values of v_{max} were obtained from the onsets of the In BE emission in time-resolved measurements similar to the one depicted in fig. 3.

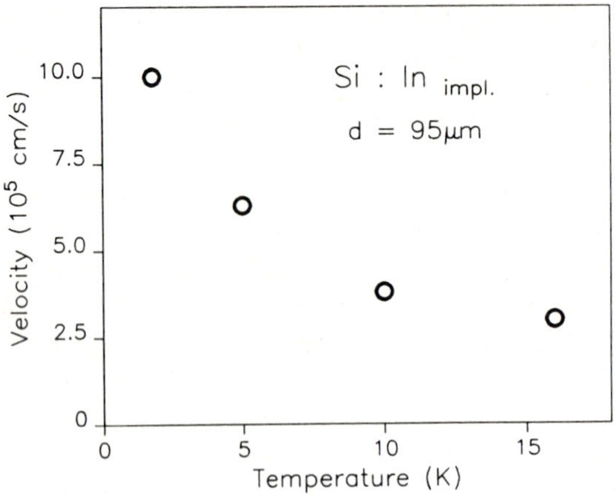

Fig. 5. Temperature dependence of the maximum average velocity for the 95 μm thick sample.

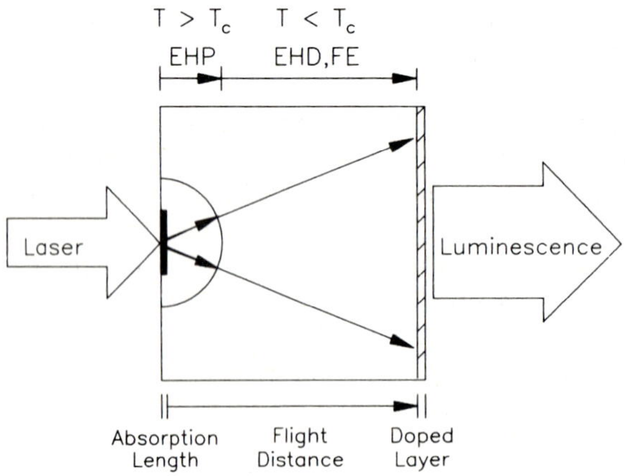

Fig. 6. Schematic illustration of the two-component model proposed to explain the observed transport phenomena. The bath temperature is assumed to be much lower than the critical temperature for electron–hole droplet (EHD) condensation. See text.

T_c. In this spatial region the carriers form a non-equilibrium plasma. At a certain distance from the excitation spot the temperature falls below T_c and for high enough carrier densities electron–hole liquid condensation occurs. To a first approximation we describe the transport using two different but constant values of the velocity for electron–hole pairs within the plasma ($T > T_c$) and the liquid ($T < T_c$). This two-component picture is justified for the following reasons: From the results of the emission spectra of the non-equilibrium plasma and from the Mott transition based transport study we expect plasma velocities in the 10^6 cm/s range (see below). For the motion of the electron–hole liquid, on the other hand, a number of studies implies velocities below the sound velocity (5.9×10^5 cm/s for TA-phonons in Si), which decrease rapidly with increasing temperature. Since the measured velocities are situated above as well as below the sound velocity, both contributions are necessary to explain the data.

In our picture the observed strong increase of the maximum plasma velocity with decreasing sample thickness arises since the dimensions of the heated spot are independent of the sample thickness, whereas the region with electron liquid based carrier transport is reduced. The temperature dependence at a fixed thickness, on the other hand, is determined by the temperature variation of the slower component and hence decreases according to the thermal dependence of the phonon-wind induced motion of the electron–hole liquid. In order to further check the validity of our model, measurements at a bath temperature above T_c are necessary and currently prepared.

Mott transition based investigation of the electron–hole plasma expansion

The transition of carrier pairs from a "metallic" plasma state into the "insulating" phase of free excitons ("Mott transition") has been studied intensively [26], both from an experimental and a theoretical point of view. Theoretically, dynamical screening theories including many-body effects have been developed [27]. The results of static and dynamic theories for the screening of the exciton binding energy have been compared to experiments mostly in indirect gap materials (Si, Ge) [5,6,28]. For the temperature dependent Mott density the different theories give values very close to the results of the classical Debye–Hückel approach. Although based on very different methods for the density determination the experimental data is in good agreement with the theory [5,6,28].

We have used the well-understood properties of the Mott transition in Si to study the plasma expansion at temperatures above the critical point for electron–hole liquid condensation. For the experiments the identical equipment as described in the previous paragraph has been used with the exception of the Nd–YAG laser.

Figure 7 displays the temporal variation of the electron–hole plasma and free exciton emission observed in high purity Si at 30 K together with the laser pulse. For the measurements the spectrometer wavelength was set to 11350 Å (EHP) and to 11250 Å (FE). The free exciton data has been corrected for overlapping contributions from the plasma using the analysis of time-delayed spectra of both emissions.

Figure 7 contains several unexpected features: (i) The plasma emission rises very fast according to the excitation pulse. In contrast, the free exciton displays a very slow increase and reaches its intensity maximum about 30 ns after the end

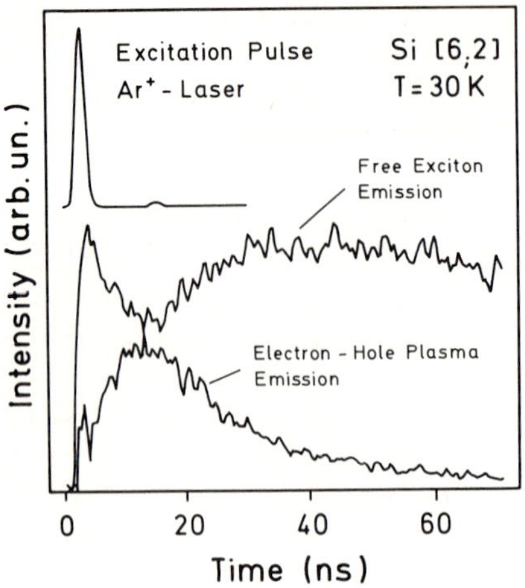

Fig. 7. Time dependence of the electron–hole plasma emission and the free exciton emission in intrinsic Si above the critical point for electron–hole liquid condensation. The topmost trace depicts the time dependence of the laser pulse and displays the time resolution of the setup.

of the excitation pulse. (ii) The decay of the plasma intensity is much too fast to be due to recombination. In our experiment we excite carrier densities in the low 10^{18} cm^{-3} range. From studies of the electron–hole liquid at lower temperature the Auger mechanism has been identified as the dominating recombination channel in high density plasmas. At the present densities the lifetime due to Auger recombination amounts to about 100 ns. (iii) The time constant of the decrease of the plasma emission is strikingly similar to the time constant of the increase of the exciton intensity.

The temporal development of the plasma and exciton emission are explained consistently if the Mott transition is taken into account. The laser pulse creates electron–hole pairs at densities about two orders above the Mott density. At this high density no carriers are in the excitonic state since the Mott density gives the upper limit for the density of a free carrier gas in which excitons are still stable. Due to, e.g., density gradients, the plasma expands into the sample and the plasma density drops locally below the Mott density. In those regions free excitons are formed. Since the exciton and the plasma lifetimes are long compared to the time scale considered, the exciton intensity increases until virtually all carriers are in spatial regions with a density below or equal to the Mott density.

As tests for our interpretation we have studied the temporal behaviour of plasma and exciton at different temperatures and different excitation intensities. Figure 8 displays the onset of the FE intensity (upper part) in comparison with the decay of the plasma emission (lower traces) at temperatures of 50 K and 30 K for a constant excitation intensity. As discussed above for the case of 30 K,

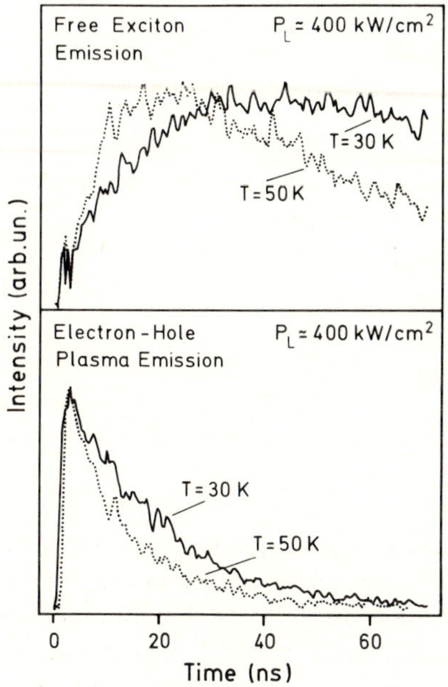

Fig. 8. Temperature dependence of onset and decay of exciton (upper part) and plasma (lower part) emission intensity. Dotted and full traces have been recorded at 50 K and 30 K, respectively.

similar time constants for the plasma decay and the exciton intensity rise are clearly observed at 50 K. More important, we see a significant reduction of this time constant as the temperature is raised. Since the Auger recombination is known to be only weakly temperature dependent [29], this behaviour clearly contradicts an interpretation based on recombination. The Mott transition densities, on the other hand, increase linearly with the temperature. For a fixed excitation intensity our expansion model implies that higher Mott densities will be reached earlier and hence explains the shorter time constants observed at 50 K.

Figure 9 depicts the time Δt within which the free exciton intensity reaches a maximum at a temperature of 30 K as a function of the excitation intensity. We observe a significant increase of Δt with increasing excitation intensity. From our expansion approach this behaviour is expected, since the expansion time down to the Mott density will decrease if the initial density approaches the Mott density. Note, that this behaviour is incomprehensible in terms of recombination of the carriers: The lifetime due to Auger recombination should decrease as the carrier density is increased. Since the observed delay times represent also the plasma decay times the experimental results clearly show an opposite trend.

Due to the fact that recombination can be neglected during the interesting

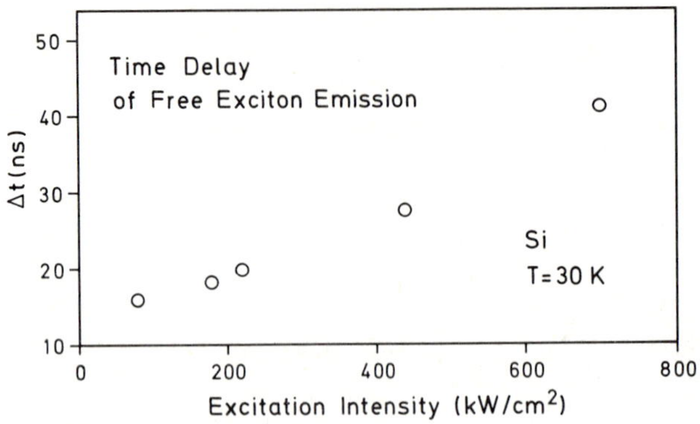

Fig. 9. Excitation intensity dependence of the time by which the intensity maximum of the free exciton is delayed with respect to the laser pulse.

time scale of strong plasma expansion effects, we can obtain good estimates for the plasma velocity from the data. In order to evaluate the plasma expansion velocity we note, that the free exciton emission is due to recombination of non-interacting electron–hole pairs. The emission intensity is therefore proportional to the number of excitons in the sample. Since the probability of recombination is low, we assume that the number of carrier pairs remains constant for the time during which the exciton intensity increases. We further identify the maximum of the FE intensity with the time during which all carriers have reached sample regions where the density is below the Mott density. As can be seen from the temporal development of the plasma intensity (fig. 7) this assumption is well justified. From the laser pulse energy absorbed in the sample we then calculate the number of electron–hole pairs created in the sample and calibrate the FE intensity maximum by this number. Within the above given assumptions, the gradual rise of the exciton intensity can now be used to determine how many carriers have reached the Mott density for any delay time before the maximum occurs. From the number of carrier pairs which form excitons and the Mott density we obtain a minimum volume which these carrier pairs occupy. If we then assume a reasonable geometry for the expansion (e.g. expansion into a hemisphere) we can calculate a typical length for the expansion. Since we measure the delay times with respect to the laser pulse, we can easily derive average expansion velocities of the electron–hole plasma down to the Mott density as a function of the fraction of carrier pairs in the excitonic subsystem $f = N_{\mathrm{Ex}}/N_{\mathrm{Tot}}$, where N_i are numbers of electron–hole pairs in the excitonic state and in the total system, respectively.

Figure 10 displays the velocities obtained from this method as a function of f. The values are in the mid 10^6 cm/s range for the initial expansion and decrease into the low 10^5 cm/s range for the completion of the expansion. The observed trend agrees with the simple expectation, that the initial velocity

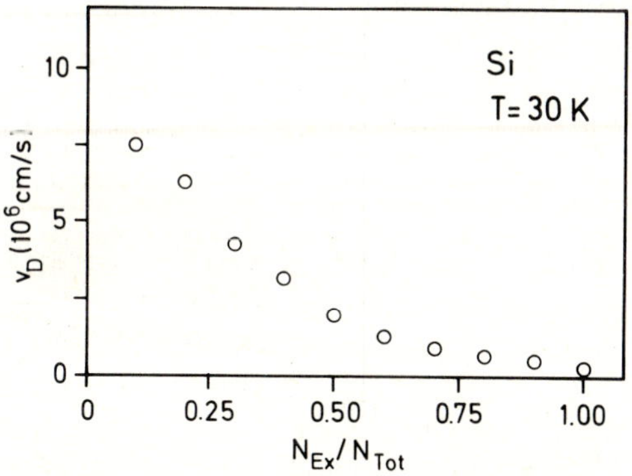

Fig. 10. Expansion velocities for the plasma expansion to the Mott density as derived from the onset of the free exciton emission intensity. See text.

of the dense plasma will be reduced as the average plasma density drops during the expansion.

The velocities from the present method are consistent with those observed in the low temperature time-of-flight experiments and they are furthermore consistent with the velocities derived from the analysis of the spectra of the non-equilibrium plasma. In extension to the other methods, the Mott density approach allows us to estimate typical volumes and hence distances through which the fast expansion of the plasma occurs. For example, the expansion of 50% of the carrier system down to the Mott density yields expansion lengths of about 30 μm.

Conclusions

We have developed two new optical methods to study the transport in ambipolar plasmas and have used them to investigate the plasma expansion in Si with previously unreached precision. The optical time-of-flight technique, in particular, provides direct access to the ambipolar transport with submicron spatial resolution. The high spatial resolution can be used only since the method is extremely sensitive. A further attractive feature of our technique is the full compatibility with fast laser systems, which in conjunction with the possibility to prepare very sharp doping profiles (0.1 μm range and below) is very promising for future direct studies of ballistic transport over short distances. The key problems with the time-of-flight method are associated with the search for a suitable dopant, which should have fast BE recombination, should be thermally stable, and spectroscopically separated from the intrinsic emissions.

The Mott transition of free excitons has been used here to study the plasma expansion above the critical point for the electron–hole liquid condensation.

The most attractive feature of this method is its lack for any spatial resolution. Possible extensions include for example the study of the Mott transition in samples doped with low concentrations of shallow impurities. Due to the different binding energies of the excitons at the impurities different Mott densities are expected, which will allow in principle to monitor the expansion velocity as a function of the carrier density.

Both methods have unambiguously determined that the expansion of ambipolar plasmas includes carrier pairs with very high velocities in the 10^6 cm/s range. Our results contradict the idea of a "sound barrier" introduced by Wolfe and coworkers [23] to describe the lower maximum velocities extracted from spatial imaging studies of EHL and EHP transport. We are convinced, that the results of ref. [23] are limited by the experimental condition, especially when the motion into the sample is addressed. The present direct measurements of the expansion velocity are consistent with the velocities estimated from the evaluation of the plasma spectra using approximate non-equilibrium distribution functions [13]. In the case of the time-of-flight experiments, the results are presently limited, however, by the temperature stability of the impurities. Measurements with thermally more stable dopants are currently performed and will be published elsewhere.

References

[1] J.C. Hensel, T.G. Phillips, G.A. Thomas and T.M. Rice, Solid State Physics 32, eds., F. Seitz, D. Turnbull and H. Ehrenreich (Academic Press, New York, 1977).

[2] C.V. Shank, J. Lumin. 30 (1985) 243.

[3] G.A. Thomas, T.M. Rice and J.C. Hensel, Phys. Rev. Lett. 33 (1974) 219.

[4] T.L. Reinecke, A. Forchel, H. Schweizer, B. Laurich, A. Hangleiter, J. Wagner and W. Schmid, Phyisca 117B&118B (1983) 315.

[5] H. Scheizer, A. Forchel, A. Hangleiter, S. Schmitt-Rink, J. Löwenau and H. Haug, Phys. Rev. Lett. 51 (1983) 698.

[6] G.B. Norris and K.K. Bajaj, Phys. Rev. B 26 (1982) 6706 and references therein.

[7] P. Vashishta, S.G. Das and K.S. Singwi, Phys. Rev. B 10 (1974) 5108.

[8] O. Hildebrand, E.O. Göbel, K.M. Romanek, H. Weber and G. Mahler, Phys. Rev. B 17 (1978) 4775.

[9] S. Schmitt-Rink, D.B. Tran Thoai and H. Haug, Solid State Commun. 36 (1980) 381.

[10] P. Vashishta and R.K. Kalia, Phys. Rev. B 25 (1982) 6492.

[11] K.M. Romanek, H. Nather and E.O. Göbel, Solid State Commun. 39 (1981) 23; H. Nather and G. Quagliano, J. Lumin. 30 (1985) 50.

[12] J.E. Kardontchik and E. Cohen, Phys. Rev. Lett. 42 (1979) 669.

[13] A. Forchel, H. Schweizer and G. Mahler, Phys. Rev. Lett. 51 (1983) 501.

[14] See, e.g., E.M. Conwell, Solid State Physics, Suppl. 9, eds., H. Ehrenreich, F. Seitz and D. Turnbull (Academic Press, New York, 1967).

[15] T.C. Damen and J.M. Worlock, Proc. 3rd Int. Conf. Light Scattering in Solids, eds., M. Balkanski, R.C. Leite and S.P. Porto (Flammarion, Paris, 1976) p. 183.

[16] J.P. Wolfe, Physica 117B&118B (1983) 321; J. Lumin. 30 (1985) 82.

[17] J.C. Hensel and R.C. Dynes, Phys. Rev. Lett. 39 (1977) 969.

[18] P.J. Dean and D.C. Herbert, Topics in Current Physics (Excitons), ed., K. Cho (Springer Verlag, Berlin, Heidelberg, 1979).
[19] A. Forchel, B. Laurich, G. Moersch, W. Schmid and T.L. Reinecke, Phys. Rev. Lett. 46 (1981) 678.
[20] W. Schmid, Phys. Stat. Sol. (b) 84 (1977) 529.
[21] T. Steiner and M.L.W. Thewalt, Solid State Commun. 49 (1984) 1121.
[22] Both measurements were performed under otherwise identical conditions by turning the sample by 180 degrees in order to directly excite the implanted layer.
[23] See, e.g., in ref. [16] and F. Steranka and J.P. Wolfe, to be published.
[24] P.J. Dean, W.F. Flood and G. Kaminski, Phys. Rev. 163 (1967) 721.
[25] J. Shah and A.H. Dayem, Phys. Rev. Lett. 37 (1976) 861.
[26] N.F. Mott, Metal–Insulator Transitions (Barnes and Noble, New York, 1974).
[27] H. Haug, Advances in Solid State Physics 22, ed., P. Grosse (Vieweg, Braunschweig, 1982) p. 149.
[28] J. Shah, M. Combescot and A.H. Dayem, Phys. Rev. Lett. 38 (1978) 1497.
[29] J. Dziewior and W. Schmid, Appl. Phys. Lett. 31 (1977) 346.

Journal of Luminescence 30 (1985) 82–113
North-Holland, Amsterdam

TRANSPORT OF DEGENERATE ELECTRON–HOLE PLASMAS IN Si AND Ge

J.P. WOLFE

Physics Department and Materials Research Laboratory, University of Illinois at Urbana-Champaign, 1110 W. Green Street, Urbana, IL 61801, USA

At low crystal temperatures, pulsed-laser excitation of Si and Ge can produce a mobile electron–hole plasma with a Fermi energy much larger than $k_B T$. The motion of this degenerate plasma away from the excitation surface depends intimately on its interactions with high-frequency phonons. Momentum damping and phonon-wind forces are principal factors which determine the plasma motion on nanosecond and longer time scales. A variety of luminescence and heat-pulse experiments are reviewed here which characterize the transport behavior of photo-generated electron–hole plasma in these indirect-gap semiconductors.

1. Introduction

Intense photoexcitation of pure semiconductors produces electron–hole plasmas which have been the subject of study in a number of crystals. The nature of the plasma depends on a number of factors including the generation rate of non-equilibrium carriers, their initial energies, thermalization times, scattering times and recombination lifetimes. The maximum plasma density and the average kinetic energy of the carriers generally depend on the energy and the length of the excitation pulse. On one extreme, recent experiments have shown that the electron–hole plasma produced by intense, ultrashort (<picosecond) laser pulses can have a density approaching a few percent of the atomic density of the crystal [1,2]. Such high plasma densities can modify the atomic binding in the crystal. The energy relaxation of photoexcited carriers proceeds on a picosecond time scale by generation of optical phonons. In time, the carriers relax to the bottom of their respective bands and eventually recombine.

A large number of studies have been conducted with excitation pulses long compared to the initial relaxation time of the electrons, i.e., pulse lengths in the nanosecond to microsecond range. In such cases, both electronic and thermal

0022-2313/85/$03.30

energy are generated during the excitation pulse. The evolution of this electronic and thermal energy on the nanosecond and greater time scale is the subject of the present paper. I will concentrate principally on the transport of electron–hole plasma at low temperatures in pure crystals of Si and Ge. In these indirect-gap semiconductors recombination lifetimes are relatively long, permitting the plasma to reach a quasi-equilibrium with the lattice. At low temperatures, such plasmas – including an electron–hole liquid phase – can be degenerate, i.e., have a Fermi energy considerably larger than kT.

The transport of photo-generated plasmas has been the subject of recent theoretical conjecture and controversy. On one hand, it has been hypothesized that the electron–hole plasma produced by nanosecond laser sources could be self-confined to the excitation surface, aiding in the annealing of crystalline defects [3]. On the other extreme, it has been postulated that, at least at low temperatures, the plasma drifts away from the surface at velocities approaching the Fermi velocity of the plasma, i.e., in the 10^6 to 10^7 cm/s range [4,5]. This transport problem involves the basic carrier–carrier and carrier–phonon interaction processes in the semiconductor and represents an important challenge to theorists and experimentalists alike.

The basic problem of how a photo-generated plasma moves away from the excitation region after its initial thermalization is hardly a trivial one. Since a considerable amount of electronic and thermal energy is deposited near the crystal surface as a result of the optical excitation, one must deal with inhomogeneities in space, time and energy. Hydrodynamic [6] and thermodiffusion [7] models have been developed to deal with these complexities, and the validity of such models remains to be fully tested. Fortunately, at liquid helium temperatures, experiments have shown that some simplifications occur: (1) The plasma takes the form of droplets of constant-density (but variable volume) electron–hole liquid. (2) The electronic temperature and liquid density are relatively constant following the exciting pulse for the lifetime of the carriers. (3) The thermal energy deposited in the excitation region by carrier thermalization is dissipated to a large degree by emission of non-equilibrium ballistic phonons into the otherwise cold lattice. In other words, the crystal remains cold even very close to the excitation spot, due to the rapid non-diffusive flux of heat from this region. The local temperature of the lattice can be ascertained in some experiments from the spatially-resolved luminescence spectrum of the EHL. The possibility of spatially-extended temperature gradients, of course, is enhanced at higher temperatures, where the ballistic mean-free-path of the non-equilibrium phonons is reduced.

As one can readily guess, a principal process in the evolution of electronic and thermal energy following optical excitation is the interaction of electrons and acoustic phonons. In the first place, the electron–phonon interaction determines the momentum damping time of the electron–hole plasma. This parameter effectively gives the viscosity of the crystal to plasma motion. In the second place, the non-equilibrium phonons can act as an expansive force on the plasma. This novel idea of a "phonon wind" was originally put forth by Keldysh [8] and has been experimentally observed in the transport of electron–hole droplets in Si and Ge.

As an example of these ideas, consider a recent experiment showing the spatial transport of electron–hole liquid in Ge at 2 K following a 120 ns Nd:YAG laser pulse [9]. The absorbed energy is 4.6 μJ, corresponding to 3×10^{13} photons with a wavelength of 1.06 μm. A spatial profile of the liquid is obtained by passing a sharply-focused image of the crystal across the entrance slit of a spectrometer, which selects the EHL recombination luminescence at 1.75 μm wavelength. The light is detected by a cooled Ge photodiode and time selected by a boxcar averager with a 10 ns gate. The radius of the liquid distribution about the excitation point (a 40 μm spot) is plotted as a function of time in fig. 1(a). In fig. 1(b) is plotted luminescence spectra for several times after the start of a laser pulse. The overall time resolution obtained with the Ge photodiode and subsequent electronics is 30 ns.

The density of the electron–hole liquid can be directly ascertained from the energy position of the luminescence emission. The arrow in the figure indicates the renormalized band gap for the liquid (minus an acoustic phonon energy), due to its Coulombic binding energy. A detailed theoretical calculation relates this renormalized gap to the plasma density [10]. Alternatively, the density may be obtained from a fit of the luminescence spectrum to a Fermi distribution of carrier energies, giving the Fermi energy shown in the figure. Both methods give an equilibrium liquid density of about 2.3×10^{17} pairs/cm^{-3}. For the case of this Q-switched laser pulse, the liquid density remains quite constant throughout its lifetime; indicating that the liquid is in the form of constant-density droplets. The open circles show the measured CW spectrum at 2 K.

Now examine the spatial buildup shown in fig. 1(a). As the pulse is turning on, the radius of the luminescence profile (plotted as the dots) increases. The solid line indicates the predicted radius of the distribution assuming that each absorbed photon contributes an electron–hole pair towards a single large hemisphere of liquid at a density of 2.3×10^{17}/cm^3. The agreement at early times suggests that indeed a single large droplet of about 300 μm radius is formed. Thereafter, the distribution continues to expand, indicating that the liquid is being dispersed into smaller droplets of the same density. Indeed, as time continues the distribution expands to over 1 mm in radius, as is shown in fig. 1(c) [11]. One might be naively tempted to conclude that the droplets are projected ballistically into the sample and come to rest with a momentum damping time of several microseconds. However, as will be discussed in the next section of this paper, the momentum damping time of droplets is of order one nanosecond. This implies that the liquid droplets are being pushed by a continuous force of several microseconds duration. Experiments have shown that this force arises from the non-equilibrium flux of ballistic phonons originating at the excitation

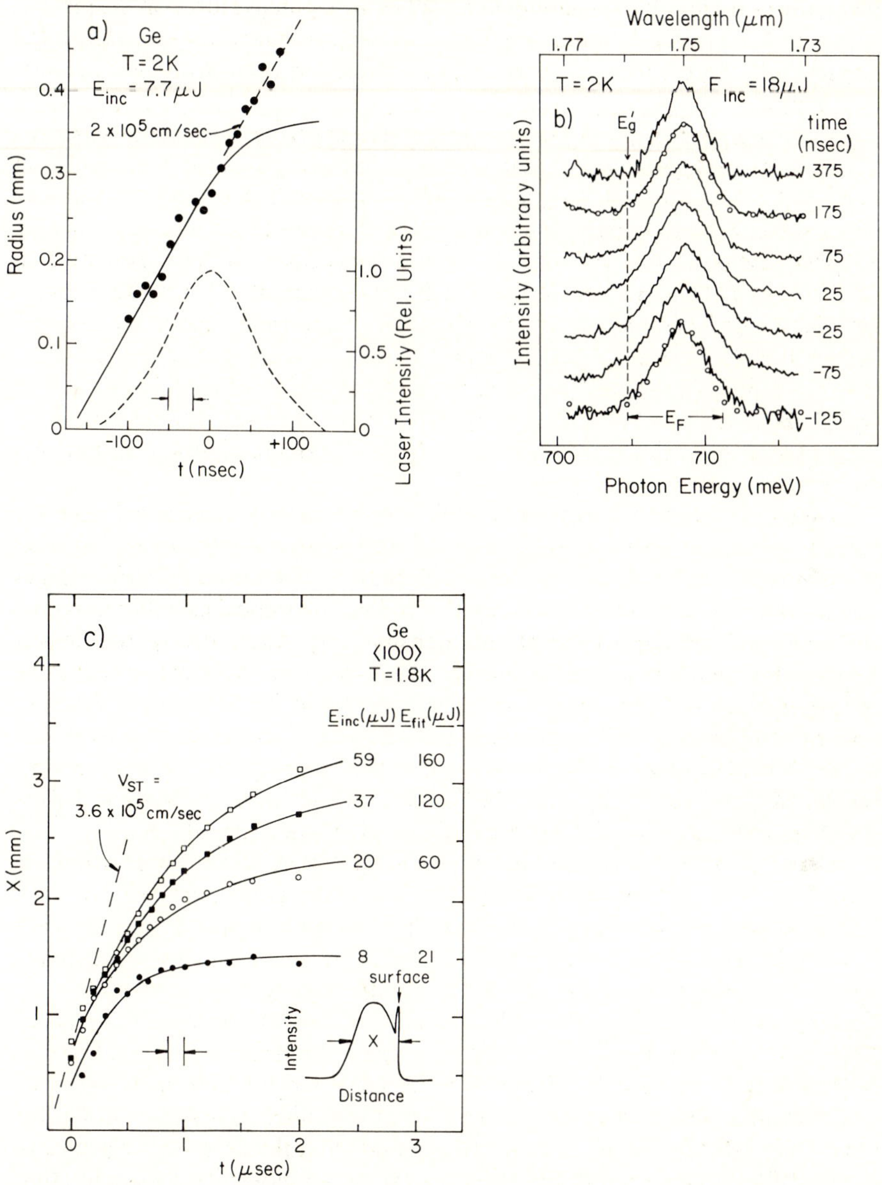

Fig. 1. (a) Radius of the spatial profile of EHL luminescence in Ge following a Q-switched Nd:YAG laser pulse, as described in the text. From ref. [9]. (b) Time-resolved luminescence spectra for EHL at 2 K. E_g' is the renormalized energy gap for the electron–hole liquid [9]. The open circles show the EHL spectrum for CW excitation at 2 K. (c) Time evolution of the penetration depth of EHL luminescence at 1.8 K from ref. [11]. Solid lines are theoretical fits to the data to be described in a later section. For (c) the laser spot radius was 0.9 mm.

point, as a by-product of the carrier thermalization process.

Before pursuing these interesting issues, consider what the plasma expansion is like above the critical temperature of the liquid, which occurs at about 7 K in Ge. Figure 2 shows the time-resolved expansion and spectra at 10 K for a similar Q-switched laser pulse [9]. Unlike the liquid case, the plasma density

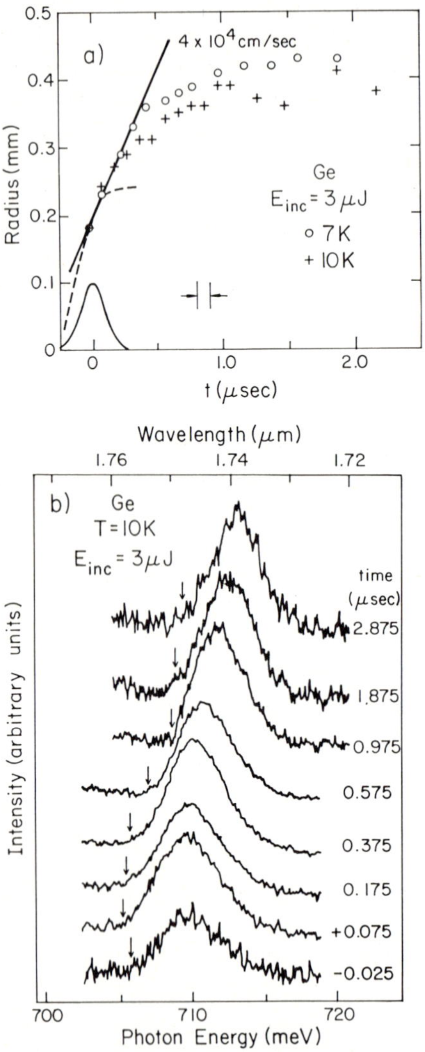

Fig. 2. (a) Radius of the luminescence profile of the EHP in Ge at 7 K and 10 K. Spatial resolution $= 60\,\mu m$ [9]. (b) Time-resolved luminescence spectra for the electron–hole plasma (EHP) in Ge at $T = 10$ K following a 200 ns Q-switched laser pulse with peak at $t \approx 0$ ns. The shift in renormalized gap E'_g with time indicates a reduction in plasma density. Time resolution $= 100$ ns [9].

decreases continuously in time, as indicated by the shift of the renormalized band gap (arrows). Again, the plasma is pushed far into the crystal (presumably by a phonon wind), but the velocities are somewhat slower, due largely to the decreased momentum damping time of carriers at this higher temperature.

Various techniques have been developed in our laboratory over the last several years to characterize the transport of photoexcited phases in semiconductors. By combining spectroscopic and imaging techniques and exploiting the stress properties of these materials, detailed information has been gained about the interaction of electrons and phonons. In addition, techniques for generation and detection of non-equilibrium phonons have proven valuable in the study of electron–hole liquid. Phonon spectroscopy and phonon imaging are two of these methods. I will first describe the measurement of momentum damping time for the electron–hole liquid. It is found that the Fermi statistics of this electron–hole plasma give rise to some interesting effects. Secondly, I will describe the striking effects produced by the phonon wind. Finally, I will show some very recent measurements of phonon-wind driven plasmas in Si.

2. Momentum damping of electron–hole liquid

The scattering of carriers in an electron–hole plasma by phonons shows some interesting features due to the Fermi nature of the particle statistics. To appreciate these properties, consider first the electron–phonon scattering expected for a single bound electron–hole pair, i.e., an exciton. The basic method for calculating the carrier-phonon scattering rate in a semiconductor is the deformation-potential method formulated by Bardeen and Shockley in 1950 [12]. The coupling between electron and phonon is through the deformation potential, or energy shift in the conduction band per unit strain. The signature of deformation-potential scattering is the $T^{3/2}$ temperature dependence of the scattering rate as contained in the basic result:

$$\tau^{-1} = \frac{3}{2\pi\sqrt{2}} \frac{m^{3/2} D^2}{\rho s^2 h^4} (k_B T)^{3/2}, \tag{1}$$

where ρ is the crystal density, s is the sound velocity, m is the carrier mass, and D is the deformation potential. This form of the scattering time has been verified over a rather broad temperature range for the exciton in ultrapure silicon at low temperatures [13], as shown in fig. 3. In the above formula, m is taken to be the sum of the electron and hole masses, and D is taken to be the sum of the electron and hole deformation potentials, giving a predicted scattering rate within about a factor of 2 of the observed rate. The measurement of the exciton scattering rate was accomplished by a strain-gradient method to be described below for the electron–hole liquid. Note that the drift mobility of the excitons, plotted on the right scale, is extremely high at low temperatures.

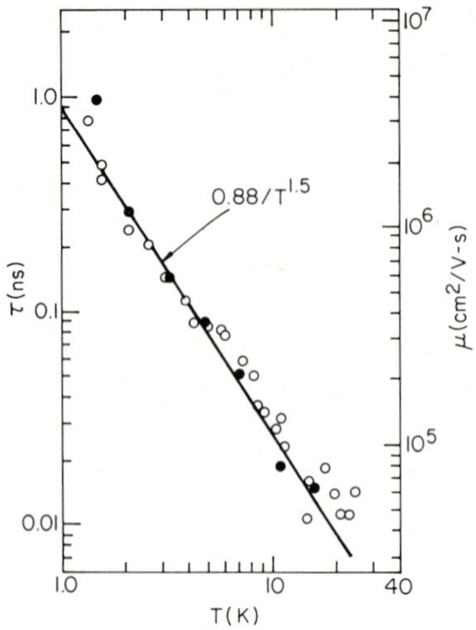

Fig. 3. Temperature dependence of the momentum damping time of excitons in Si, as determined by measurement of their drift velocity in a strain gradient. From ref. [13].

In contrast, the scattering of a carrier in a plasma must take into account the degenerate statistics which limits the number of states into which the carrier can scatter. The basic method for calculating τ for phonon scattering in an electron–hole plasma is analogous to the derivation of the Bloch conductivity for a simple metal. In this case the electron and hole systems are interpenetrating degenerate Fermi-liquid systems. The motion of an electron–hole droplet corresponds to the spatial superposition of the electron and hole currents with no net charge transported. Again, the carrier-phonon interaction is treated in the deformation potential approximation [14]. Applying this method to electron–hole liquid, Keldysh et al. [15] obtained the following result for the scattering rate:

$$\tau^{-1} = \frac{2}{3(2\pi)^3} \frac{m_e^2 D_e^2 + m_h^2 D_h^2}{\hbar^2 (m_e + m_h) \rho s n} \cdot I\left(\frac{T}{T_0}\right) \cdot \left(\frac{k_B T}{\hbar s}\right)^5, \tag{2}$$

where $m_{e,h}$ are the electron and hole masses, $D_{e,h}$ are the deformation potentials, n is the electron–hole pair density, ρ is the crystal density, and s is the sound velocity. The integral I depends on the crystal temperature and the parameter $T_0 = 2\hbar k_F s/k_B$, which is a characteristic temperature of the particular plasma system. By setting either the electron or hole mass to zero in eq. (2), one recovers the classical Bloch conductivity result. This formula neglects the effects of screening and anisotropy, which can be accounted for by more detailed calculations. Also, it assumes that the carrier drift velocity is much less than the sound velocity. Corrections due to near-sonic velocities will be discussed shortly.

Figure 4 shows the temperature dependence of the carrier-phonon scattering rate – i.e., the momentum-damping rate – predicted by eq. (2). Three different Fermi systems, with different parameters T_0, are compared in this figure. The temperature scale appropriate to each is shown. One can see that for temperatures above about 1 K, electron–hole droplets in Ge should display an approximately linear variation of scattering rate with temperature. Droplets in Si, however, due to the larger Fermi wavevector and sound velocity, are expected to display the classic T^{-5} Bloch prediction for temperatures below about 2 K. For reference, one can see that the corresponding transition temperature for pure copper is in the 20 K to 30 K range. The dashed line at the top indicates the mobility saturation in ordinary copper metal. In fact, this T^{-5} dependence has never been conclusively demonstrated in a metal, presumably because of the stringent requirements on impurity and defect concentration. From this diagram, the electron–hole liquid in high-purity Si seems to be a promising candidate for testing the Bloch scattering prediction at low temperatures.

One might rightfully ask what the effects of screening and anisotropies in masses and deformation potentials are. The scattering of a carrier in the electron–hole liquid by phonons will tend to be screened out as the result of the Coulomb interaction with other carriers in the plasma. These effects have been calculated by several workers [16,17] and it is found that the damping rate is modified by typically less than a factor of 2.

Let us now consider the measurement of momentum damping time for the

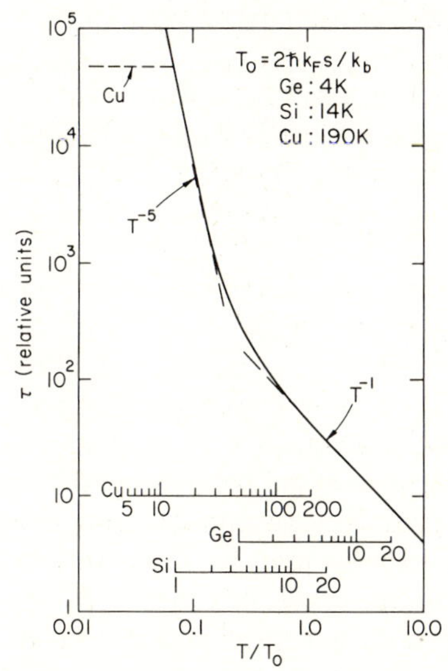

Fig. 4. Temperature dependence of the Bloch conductivity relation, eq. (2). Relative scattering time is plotted versus reduced temperature T/T_0 for the various degenerate Fermi systems. From ref. [11].

electron–hole liquid. In the drift regime, the momentum damping time is simply related to the mobility of the carriers, i.e., $\mu = \tau/m$, where m is the carrier mass. Experimentally, the mobility is the ratio of the drift velocity to the applied force, $\mu = v_d/F$. Generally, the force is applied by an electric field, but in the case of neutral entities such as excitons and electron–hole liquid, this is not possible. Instead, it has been shown that a strain-gradient produces an effective force on these particles [18,19]. This is because the energy gap of a semiconductor can be significantly altered by an applied stress. A gradient in stress produces a gradient in electron–hole pair energy, or a force on the pair. The use of calibrated strain gradients was first demonstrated for excitons in silicon [13], and the result is shown in fig. 3. In this case, the momentum damping time is simply given by $\tau = mv_d/F$.

Figure 5(a) shows the experimental arrangement used to determine the mobility of electron–hole droplets in silicon. An inhomogeneous stress is applied to the silicon sample by the rounded stress rod and steel pad (which helps to distribute the gradient). This Hertzian contact geometry actually produces a strain maximum, or energy minimum, inside the crystal. Droplets produced at the surface of the crystal are pulled into this strain well. The strain well actually provides a convenient means for collecting and studying a con-controlled-volume of electron–hole liquid or excitons [19], but we are presently interested only in their drift towards this energy minimum. The drift of electron–hole droplets towards this energy minimum is shown dramatically in fig. 5(b), which is a vidicon TV image of the electron–hole droplet luminescence ($\lambda \approx 1.2$ μm). In this experiment, the stress maximum below the stresser was approximately 50 kgf/mm^2. The temperature of the crystal was 2.0 K and the excitation was produced by a continuous argon ion laser with absorbed power of about 300 mW. The strain-induced force F, or energy gradient, is determined from the spectral shift of the droplet luminescence as a function of position x along the droplet streamer. Controlled gradients with values up to about 0.6 eV/cm were produced in this manner.

The drift velocity v_d of the excitons or droplets is determined by measuring the onset time of the luminescence at various points in the crystal after the excitation pulse. Of course, care must be taken to ensure that the strain-gradient force is the only force on the carriers, i.e., phonon wind forces are negligible. In fig. 6 are plotted the measured drift velocities for electron–hole droplets as a function of the strain-gradient force for two different crystal temperatures [20]. These results were obtained by space and time resolved photon counting. The horizontal scale is divided by the charge of an electron in order to convert the units of force (eV/cm) to the conventional motive force (electric field in V/cm). The data in fig. 6 show that the velocities increase with increasing motive force. At the smallest forces, the velocity increases approximately linearly with increasing force, as was the case for excitons. However, as the force is increased, the drift velocities begin to show a non-linear behavior. Indeed, at the lowest

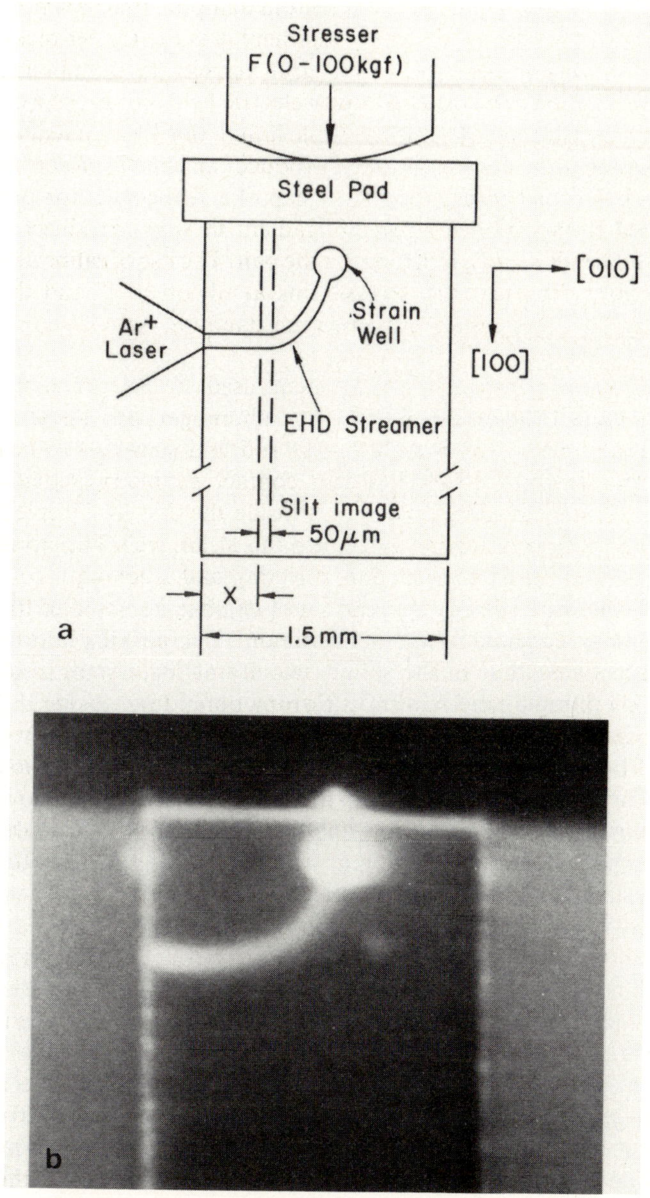

Fig. 5. (a) Experimental configuration for the drift experiments. Spatially-resolved luminescence profiles can be obtained by projecting an image of the crystal on the slit plane of the spectrometer. Drift velocity is determined by pulsing the laser and measuring the onset time of luminescence signal at various slit positions. (b) Vidicon image of an EHD streamer in Si at 2 K with 300 mW CW excitation. The stress maximum in the strain well is about 50 kgf/mm^2. From ref. [20].

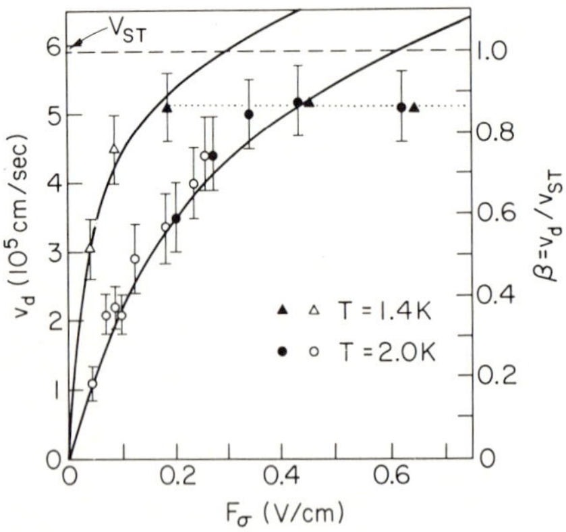

Fig. 6. Force dependence of the EHD drift velocity at 1.4 K and 2.0 K. The open symbols indicate use of the steel pad (fig. 5(a)) to reduce the strain gradient, and solid symbols denote data taken without the pad. The solid lines are the scaled results of the theory described in the text. Notice that the 1.4 K data exhibits a strong saturation in v_d above 0.2 V/cm, as shown by the dotted line. From ref. [20].

temperature of 1.4 K, the drift velocity remains *constant* above a force of 0.2 V/cm as the motive force is increased by another factor of 3. This striking saturation of the drift velocity near that of the sound velocity of the crystal is an indication that a strong damping mechanism is turning on and increasing the scattering rate of carriers in proportion to the motive force. It was simply not possible to push the droplets faster than the sound velocity in these experiments. The vertical scale on the right of fig. 6 is the ratio of the drift velocity to the slow transverse acoustic phonon velocity in the crystal. It appears that the droplets saturate at a velocity slightly *less* than the sound velocity, an effect which may be related to the focusing of phonons in this crystal.

How can we understand this non-linear relationship between drift velocity and the motive force for electron–hole droplets? The electron–hole liquid is a degenerate Fermi fluid of electrons and holes. At low droplet velocities the scattering of carriers in the liquid by phonons primarily involves carriers near the Fermi surface, since those deep in the Fermi sea are unable to scatter. One consequence of this is that, just as in a conducting metal, the high density of carriers and exclusion principle greatly enhances the average carrier mobility. In fact at low temperatures and low velocities the EHD are found to be more mobile than free excitons. The effect of the finite drift speed on the Fermi system is to shift the center of the Fermi surface away from $v=0$. This is shown schematically in fig. 7. This diagram shows that at zero temperature the liquid cannot emit phonons until it reaches the sound velocity, but at finite temperatures the emission of phonons can occur below the sound velocity, due to the thermally excited carriers. In addition to this emission process, one must

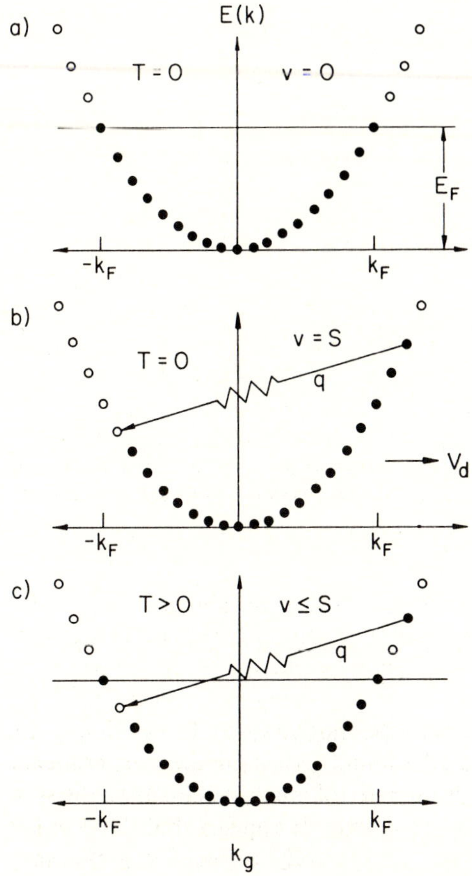

Fig. 7. Schematic momentum–energy diagram for EHL. Solid symbols represent occupied electronic states. At $T=0$, phonons can be emitted when $v_d > s$, the sound velocity. At finite temperatures, phonon emission can occur at $v_d < s$. From ref. [20].

consider the redirection of thermal phonons in the crystal by the electron–hole liquid. A detailed theory of the phonon damping process has been presented by D'yakonov and Subashiev [21], who extend the original calculation of Keldysh to include the effects of finite velocity. The solid lines in fig. 6 show an application of their theory to the droplet transport in Si [20]. The fit of this theory to the experimental data is seen to be quite good up to the saturation velocity of the droplets, however, it does not explain the abrupt saturation effect.

The temperature dependence of the droplet drift velocity is also predicted by this theory, and a comparison with experimental results is shown in fig. 8. At the highest motive forces ($F = 0.63$ V/cm) the drift velocity is almost independent of temperature. This is clearly because the droplets are in the saturated

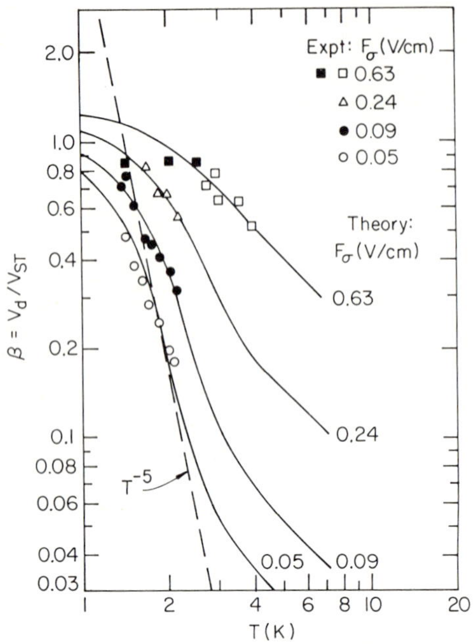

Fig. 8. Temperature dependence of the drift velocity for several strain gradient forces. Solid lines are the theory of ref. [21], using the same scale factor as in fig. 6. From ref. [20].

velocity regime. For smaller motive forces, the temperature dependence is more rapid, and the theory and data begin to approach the simple T^{-5} Bloch result. The finite-velocity damping theory was used to generate the family of curves of v_d versus T at selected values of F, using a single scale factor which determines the strength of the damping. These curves are shown as the solid lines in fig. 8. At 2 K, the theory predicts an absolute value for the momentum damping time of $\tau = 1.0$ ns, whereas $\tau = 0.6$ ns was observed experimentally. Considering the complexity of the electron–hole liquid system, it may be said that theory and experiment are in very good accord for drift velocities less than the observed saturation velocity. Finally, the mobility and damping time for electron–hole droplets in silicon may be plotted as a function of temperature and compared to the simple Bloch result (dashed curve) in fig. 9. The mobility approaches 10^7 cm^2/Vs at the lowest temperatures. The damping time is obtained from the measured mobility $\mu = v_d/F$ using $\tau = \mu m$ and the pair mass $0.45\, m_0$ [20]. In fig. 6 the experimental data points which were taken at low drift velocities agree quite well with this simple relation.

The saturation of electron–hole droplet drift velocities at high motive forces and low temperatures remains as an interesting effect to be studied in more quantitative detail. Tamor and I suggested that this striking effect can be accounted for by a "Cherenkov" emission of phonons as the liquid attempts to

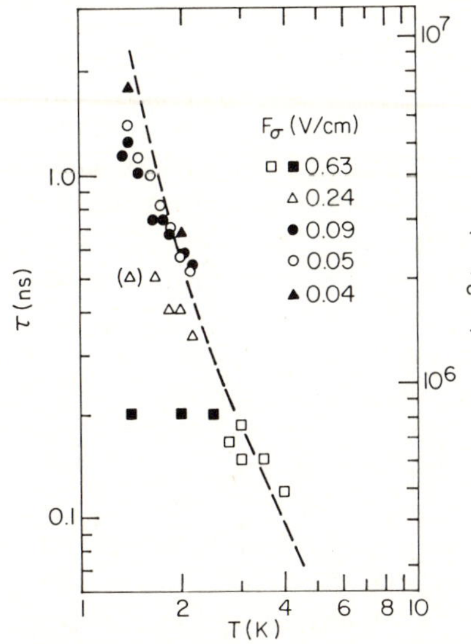

Fig. 9. Scattering time and mobility of the EHL versus temperature. The points that lie far off of the Bloch theory (dashed curve) correspond to saturated (near-sonic) velocities. From ref. [20].

exceed the sound velocity in the crystal. Attempts have been made to calculate the increased momentum damping due to this effect [21]. A simple explanation for the subsonic velocity limit ($\beta=0.9$ rather than $\beta=1.0$) may lie in the anisotropy of phonon velocities in the silicon crystal. One must take into account the effects of phonon focusing, which implies that the phonon wavevector and group velocity vector are not collinear.

For completeness, let me mention that measurements of droplet mobilities have also been made for electron–hole liquid in germanium. The experiments are technically more difficult in this crystal, because a solid state detector is required to measure the recombination luminescence near 1.75 μm, and the sensitivity and time resolution of such detectors are not as good as photo tubes for which photon counting techniques can be used. For this reason, it was not possible to observe velocity saturation effects, but the droplet mobility at smaller motive forces was determined [11], and as expected from the discussion of fig. 4, the electron–phonon scattering time was found to vary inversely with temperature over the observed range of 1.4–2.0 K. As for silicon, the measured droplet mobilities in Ge are extremely high: $\mu=4.2\times10^6/T$ (cm^2/Vs).

Many of the ideas applied to electron–hole liquid in these studies should be applicable to electron–hole plasmas at somewhat higher temperatures. In section 4 we will examine the transport of plasma above the critical temperature for EHL. At this point, we turn to another interesting aspect of electron–hole plasma transport in semiconductors, namely the effect of phonon wind forces produced by the thermalization of the photoexcited carriers.

3. The phonon wind

The momentum damping times of electron–hole liquid are clearly much shorter than the time observed for droplets to come to rest after their creation by a laser pulse. Figure 1 shows that this latter time is of order of a few microseconds. Therefore, a force with a duration of a few microseconds is required to explain these results. The requisite force is found to be a non-equilibrium flux of phonons which are absorbed by the carriers, thereby transferring their momentum to the droplets. It should be noted that diffusion is a very slow process for the droplets, owing to their large mass: the diffusion length of an electron–hole droplet containing 10^7 pairs in Ge at a temperature of 2 K is predicted to be of order 1 μm. Clearly the droplets must be drifting in a force field in order to move millimeter distances during their lifetime.

One of the most interesting aspects of this phonon wind process is that the phonons doing the pushing are ballistic in nature. This fact is dramatically demonstrated by a photograph of the electron–hole droplet cloud in germanium. Figure 10(a) shows the 1.75 μm luminescence as obtained by a special infrared imaging system [22]. The cloud is highly anisotropic, showing that the droplets are pushed along certain crystalline directions more strongly than along other directions. The sharp features (flares) emanate from the excitation point, which

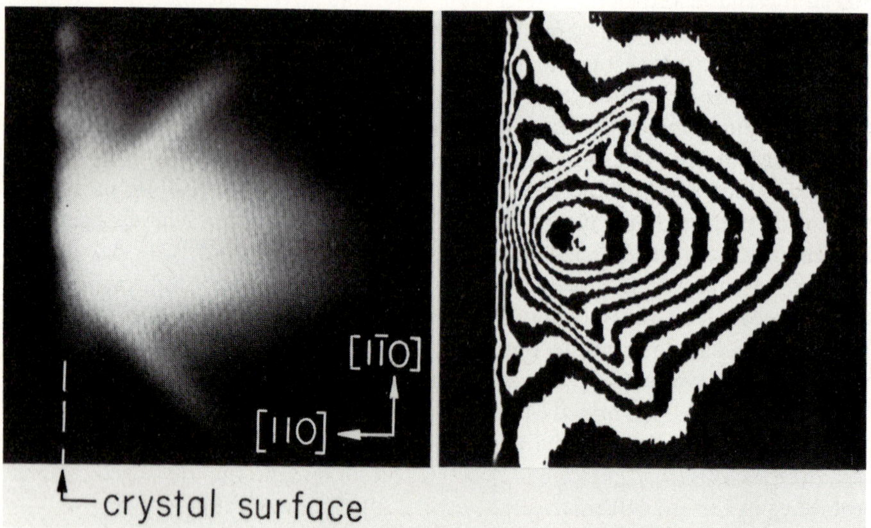

Fig. 10. (a) Cloud of electron–hole droplets in Ge at 1.8 K as viewed along the (110) excitation surface, which is seen as the vertical line of scattered luminescence light. The plane of the photo is (001). Photo width = 1.6 mm on crystal. The CW Nd:YAG excitation, focused to a 100 μm spot, is absorbed within 1 μm of the surface. From ref. [24]. (b) Contours of constant luminescence intensity showing the shape of the EHL cloud: sharp flares in the (100) planes and broad lobes along ⟨111⟩ directions [24].

indicates that the source of the phonon-wind force is rather localized near this point. The photo also shows that there must be a concentrated flux of phonons propagating *ballistically* for more than a millimeter into the crystal. If the phonons were scattered or the phonon source were more distributed throughout the crystal the cloud features would certainly be much more diffuse. A more quantitative picture of the cloud shape can be gained by plotting contours of constant luminescence intensity as is done in fig. 10(b). This is possible because the image was recorded digitally.

The fact that *phonons* emanating from a point source are strongly directed in certain crystalline directions is a subject of study in its own right. This effect – known as "phonon focusing" – results from the basic elastic anisotropy of the crystal itself and is common to all crystals. One imagines that the carrier thermalization near the excitation point produces a more or less random distribution of phonon wavevectors. An isotropic distribution of wavevectors does not lead to an isotropic distribution of energy flux because, in an anisotropic medium, the phonon wavevector $\boldsymbol{k}$ and phonon group velocity vector V are not in general collinear. Along certain crystalline directions a large solid angle of k-space gets mapped into a very small solid angle of group velocity space, causing an increase in phonon energy flux along that group velocity direction. Furthermore, along certain crystalline directions there are mathematical singularities in flux.

The observation of large electron–hole droplet cloud anisotropies in Ge motivated G.A. Northrop and I to devise a means for imaging the ballistic flux of phonons directly. Our "phonon imaging" method employs a laser beam as a heat source and a tiny superconducting bolometer as a thermal detector [23]. By adjusting the angle between the source and detector (raster-scanning the laser spot), it is possible to get a spatial map of the phonon flux anisotropy in a crystal. Figure 11(a) shows the ballistic heat flux for a Ge crystal oriented as shown in fig. 11(b). There is an intense concentration of phonons within a $\pm 2°$ angle of the $\{100\}$ planes of the crystal, and it is clearly this intense concentration of phonons which produces the flares in the cloud of droplets.

An understanding of the phonon focusing pattern can be gained by mapping the singular flux directions onto the surface of the crystal. This calculation is described in ref. [23]. The dashed lines in fig. 11(b) represent the fast transverse phonon mode and the solid lines represent the slow transverse mode. The longitudinal phonons display no singular structure, however, they are broadly focused about the $\langle 111 \rangle$ directions in the crystal. These longitudinal phonons give rise to the broad lobe in the cloud which protrudes into the crystal in fig. 10.

In addition to the large anisotropies induced by the phonon focusing effects, there are additional anisotropies due to the coupling of electrons and phonons. Recall that the electron energy surface in Ge consists of four ellipsoids along the $\langle 111 \rangle$ directions. The electron–phonon coupling depends on the phonon wavevector and polarization directions with respect to the ellipsoid axis. In addition,

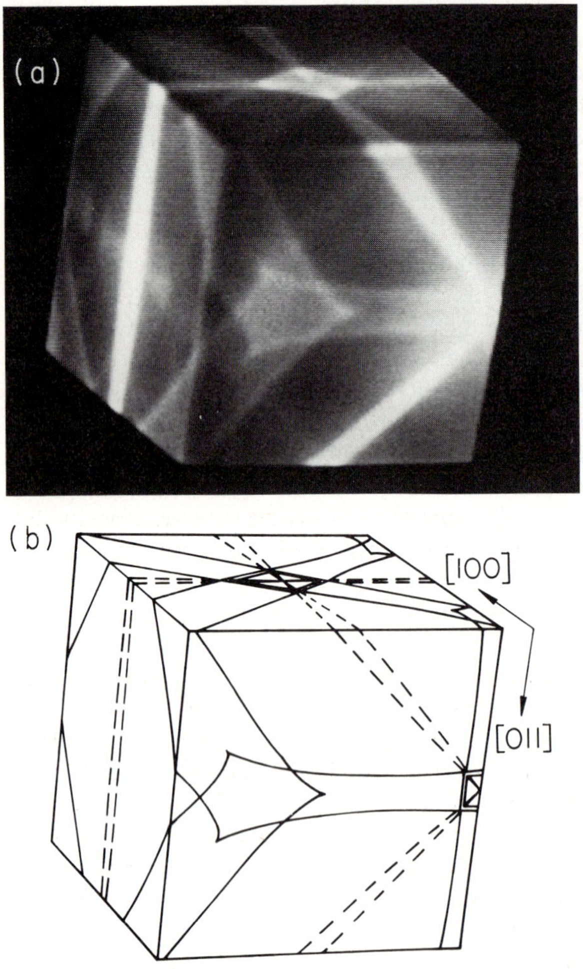

Fig. 11. (a) Ballistic phonon image of a Ge crystal showing the cause of the sharp anisotropies in the EHL cloud. This may be viewed as a picture of the ballistic phonon flux emanating from the center of the right rear ($01\bar{1}$) surface. From ref. [24]. (b) Theoretical calculation of the singular flux directions for the experiment in (a). Solid lines are for the slow transverse mode and dashed lines are for the fast transverse mode. The fast transverse flux gives rise to the flares in the droplet cloud, fig. 10. From ref. [23].

the degree of anisotropy in the droplet cloud depends upon the frequency distribution of the ballistic phonons due to the ellipsoidal shape of the electronic Fermi surfaces: the degree of anisotropy is predicted to increase as the temperature of the phonon source is increased. A further, interesting complication is that the phonon wind force acting on a given droplet in the crystal is not radially directed from the excitation point – an effect due to the fact that the phonon wavevector and group velocity vector are not collinear. These effects have been considered in calculations presented in refs. [25] and [22]. It is found that an overall satisfactory agreement is reached between theory and experiment with regard to the anisotropic cloud shape in Ge.

The frequency distribution of the non-equilibrium phonons turns out to be an important issue. By comparison with theory, it is found that the anisotropy of the cloud in Ge is explained reasonably well assuming a Planck distribution of phonons with temperature $T=6$ K [22]. It is theoretically expected that phonons with wavevectors greater than twice the Fermi wavevector will not be absorbed by the liquid [8]. A novel "phonon spectroscopy" experiment has demonstrated the existence of this $2k_F$ cutoff. It happens that a frequency-tunable source of phonons with wavevectors about equal to $2k_F$ can be produced by superconducting tunnel junctions. The experiment consists of observing the absorption of a beam of these quasi-monochromatic phonons by a cloud of electron–hole droplets [26]. The results are shown in fig. 12. Part (a) shows a linearly increasing absorption of phonons up to the point where the $2k_F$ cutoff was expected. As luck would have it, the sensitivity of the experiment ran out for phonon frequencies above the predicted cutoff (dashed line). Fortunately it was realized that the cutoff frequency could be moved downward by stressing the crystal along the $[11\bar{1}]$ direction and projecting the phonons along the [111] direction. This particular stress raises the [111] electron ellipsoid in energy, thereby decreasing the Fermi wavevector along its long axis. (The electrons preferentially fill the lower $[11\bar{1}]$ valley.) Parts (b) through (e) of fig. 12 show the phonon absorption spectrum for increasing stress. The result is a striking demonstration that the liquid transmits phonons above the cutoff frequency, as indicated by the knee in the dashed theoretical curves. The practical implication of this result is that, for phonon sources with temperatures above a few degrees Kelvin, most of the high-frequency phonons generated will actually not provide an effective wind for the droplets. A 10 K black body distribution has a peak at 600 GHz, from $h\nu_{peak}=2.8\ kT$.

Consider now the droplet transport for the case of *pulsed* laser excitation. It is reasonable to expect some quite different effects in this case. After all, the thermalization of hot carriers takes place on a subnanosecond scale, implying that a strong phonon-wind source is produced almost simultaneously with the optical excitation pulse. The high-frequency longitudinal optical phonons produced by the initial carrier thermalization down-convert on a subnanosecond scale into acoustic phonons. Further down-conversion into low-frequency acoustic phonons is possible by anharmonic processes. Down-converted phonons with wavevectors less than twice the Fermi wavevector of this electron–hole plasma exert a phonon wind force on it. However, since the droplets travel at speeds less than the sound velocity (see fig. 1), a one-shot source of ballistic phonons would quickly leave the droplets behind. Yet, fig. 1 also indicates that there is a continuous drift of droplets into the sample for a few microseconds,

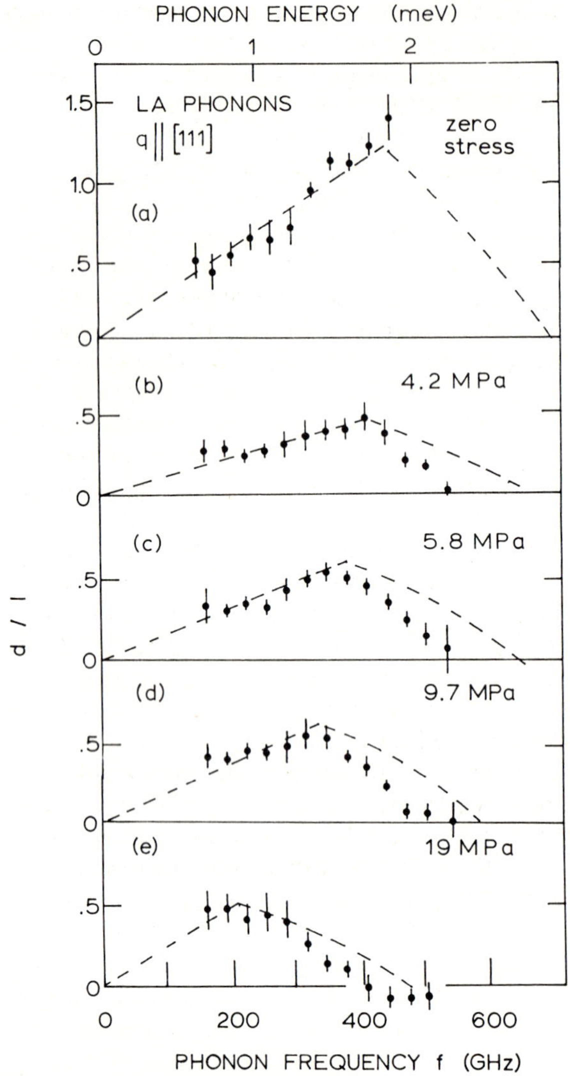

Fig. 12. Phonon spectroscopy of the EHL in Ge. The monochromatic phonon intensity at the detector is given by $I = I_0 \exp(-d/l)$ where l is the mean-free-path of phonons through pure EHL and d is the total path length through liquid. Plotted is $-\ln(I/I_0)$. Phonons with wavevectors above $2k_F$, which is varied by applying stress to the crystal, are transmitted more readily through the liquid. For phonons below the cutoff (i.e., at the knee in the dashed theoretical curves) the mean-free-path of phonons through pure EHL is in the 5–10 μm range. From ref. [26].

implying that there is a source of phonon wind for times well after the phonon pulse [11]. The solid curves in fig. 1(c) are the predicted penetration depth of the cloud assuming the existence of a long-lived hot spot at the excitation point. The lifetimes required for this hot spot are between 0.3 and 1.5 μs for the range of laser energies shown in fig. 1(c). These times are much longer than the laser pulse of 100 ns and indicate that the thermal energy is being delayed near the excitation point due to energy down-conversion and scattering processes.

It is worth considering for a moment the evolution of thermal energy near the excitation point. Phonon imaging experiments have been performed to determine the size of the hot spot created by pulse laser excitation [27, 28]. It is found that the heat pulses arriving at the bolometer detector on the opposite side of the crystal display a sharp pulse at the ballistic time-of-flight followed by a long (about 10 μs) tail, caused by phonons which are delayed by scattering. By observing the sharp features in the phonon focusing pattern as a function of arrival time. It is possible to separate out phonons which originate from the localized hot spot and those which are delayed by diffusive scattering in the bulk of the crystal. This type of "spatial filtering" experiment corroborated the result obtained in the cloud buildup experiment – namely that a hot spot is produced which has a lifetime order of 1 μs. The idea of a hot spot was originally proposed by Hensel and Dynes [29] but they concluded that its lifetime was of order 10 μs and its size was as large as 600 μm. The later spatial filtering and cloud-buildup experiments indicate a considerably shorter lifetime and size – less than about 100 μm. The 10 μs long tail following the ballistic heat pulses is actually due to phonon scattering in the bulk.

An idea of the processes involved in the transport of thermal energy away from the excitation point can be gained from fig. 13. Plotted are Planck distributions for several "heater" temperatures. (Actually, the frequency distribution of phonons produced by photoexcitation of a semiconductor is not necessarily a Planck distribution.) The fate of phonons emitted from the hot spot depends radically on their frequency, because the mean-free-path of phonons at low temperatures is limited by isotope scattering. An accepted theory for the scattering rate of phonons of frequency ν is simply given by $\tau^{-1} = A\nu^4$, where $A = 4.49 \times 10^{-41}$ s^3 for Ge. In fig. 13 the dashed curves are the absorption factor,

$$\alpha(\nu) = 1 - \exp(-A\nu^4 l/s), \tag{3}$$

for two path lengths, $l = 1$ cm and 100 μm, with s the phonon velocity. Phonons with frequency greater than about 10^{12} Hz scatter within only 100 μm of the source and hence are quite localized. Down-conversion of these high-frequency phonons near the excitation point produces low-frequency phonons which escape the heated region and are detected as the delayed ballistic signal. Phonons with frequencies in the range 0.2 to 1 THz have mean-free-paths between 100 μm and 1 cm. They leave the source region and scatter and/or down-convert before reaching the detector, producing the long 10 μs tail on the heat pulses. If the

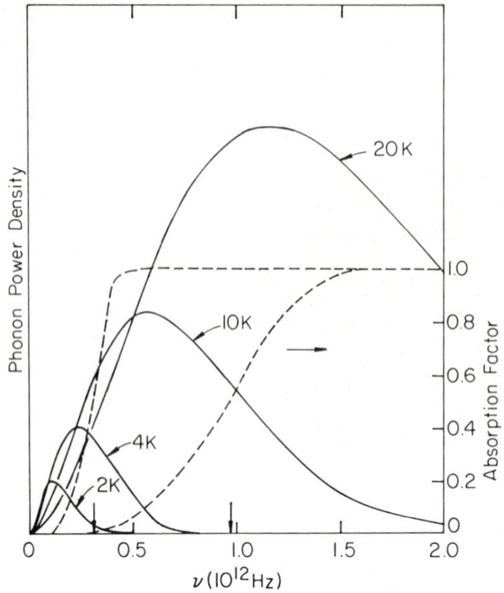

Fig. 13. Planck distributions for heater sources of several temperatures. The dashed lines represent the absorption due to isotope scattering, as described in the text. From ref. [27].

temperature of the heated region is above about 4 K, the number of these diffusive phonons exceeds the number of low-frequency ballistic phonons. This picture provides a qualitative understanding of both the cloud expansion in Ge and the ballistic heat pulse signals. It should be noted that the highest energies used in the cloud experiments are approaching the damage threshold for the crystal itself. Thus we have the interesting situation where the local crystal temperature at the surface is approaching the melting temperature (1200 K), and yet the temperature of the crystal in the bulk remains close to the lattice temperature, as indicated by the luminescence spectrum for the electron–hole droplets (see fig. 1(b)).

Not surprisingly, the cloud of electron–hole droplets produced by pulsed laser excitation is quite different from that produced by continuous excitation. Figure 14(a) shows a luminescence image of the cloud of electron–hole droplets produced by a 300 ns Nd:YAG laser pulse. For comparison fig. 14(b) shows the cloud of droplets produced by continuous excitation. The difference in these two images is striking: one notable feature is that the sharp flares in the CW image are replaced by dark regions in the pulsed image. The cloud produced by pulsed excitation has the appearance of a shell of droplets which has been punctured along the symmetry planes. In fact, it is expected that pulsed excitation should produce a shell of electron–hole liquid: the carriers are all produced during the laser pulse, and they will be pushed into the crystal together by the resulting phonon wind from the hot spot. The strong fast-transverse (FT) wind along the (100) planes has the effect of dispersing the droplets further into the sample than for directions of less intense phonon flux, which seems to explain the holes in the shell of droplets. This intense FT phonon wind is present in the

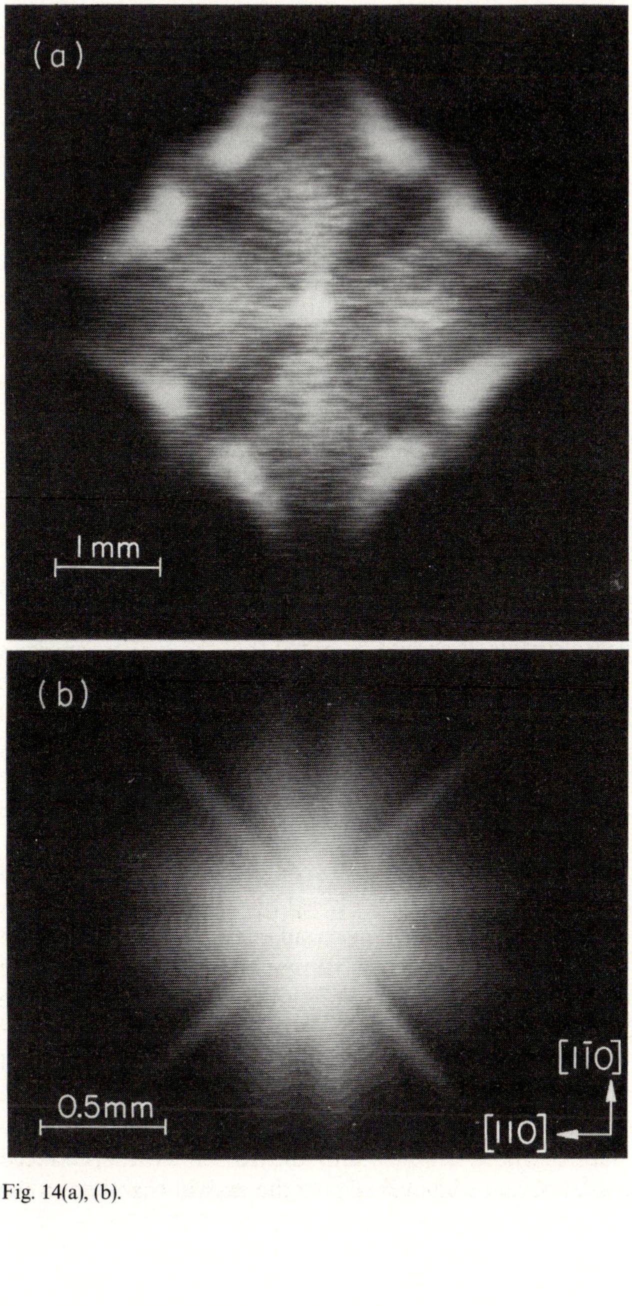

Fig. 14(a), (b).

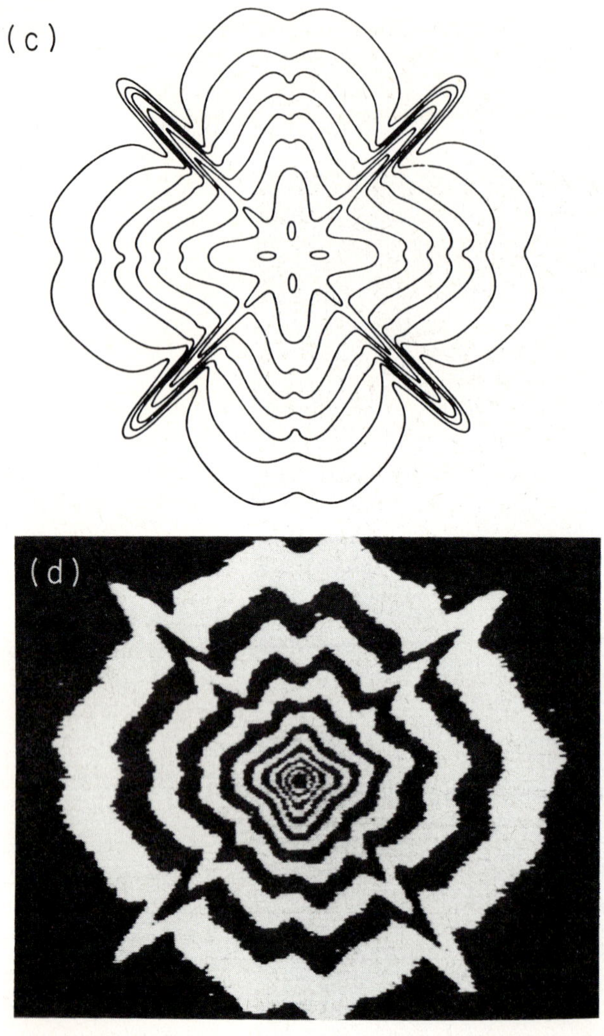

Fig. 14. (a) Time-resolved luminescence image of the EHL cloud in Ge taken 5 μs after a 20 μJ Q-switched Nd:YAG laser pulse of 300 ns duration. $T = 1.8$ K. This is a "face view" looking through the crystal into the laser beam, which is absorbed within ~ 1 μm of the surface. From ref. [11]. (b) CW cloud of EHL in Ge for the same sample geometry as above. (c) Calculated contours of constant luminescence intensity expected for CW excitation. From ref. [25]. (d) Contours of constant luminescence intensity of the EHL cloud in (b). From ref. [25].

CW case, and yet the cloud is more intense along these directions, indicating that there are more droplets channeled into these flares.

Thus, the CW image suggests that near the excitation point there must be a "fill-in" effect which channels more droplets into this region as it is depleted by the intensely focused phonons. This effect is contained in the theoretical assumption that near the excitation point there is a uniform distribution of liquid which, under the influence of the phonon wind, is distributed into larger regions in the crystal. Figure 14(c) shows the calculation of the CW cloud shape as a function of distance from the excitation point using this assumption and the ideas previously outlined above. Both the anisotropies in the phonon wind flux and the electronic deformation potential have been accounted for. In addition, this calculation allows for the attenuation of the phonon wind strength due to absorption by electron–hole droplets. In particular, one can see that the broad ⟨111⟩ lobes due to LA phonons are not so prominent at large radii as at small radii. These changes in the cloud shape with radius from the excitation point can be observed experimentally by producing a contour plot (fig. 14(d)) of the luminescence image in fig. 14(b). The experimental contours agree remarkably well with those calculated theoretically, indicating that the phonon wind theory, with all its complications, is in reasonable accord with experiment.

So far, we have considered only those phonons localized near the excitation region which are a direct result of the thermalization of hot carriers. (These have been termed "T-phonons".) In addition to these phonons, there are phonons emitted during the recombination of carriers in the droplets. Such "R-phonons" have a distributed source, and hence would tend to wash out any anisotropy generated in the cloud by the highly-directed T-phonons. There have been several studies of the relative importance of T- versus R-phonons in Ge [30,31].

The effect of R-phonons seems to be much more pronounced in the cloud of electron–hole droplets in Si. In fact, very little anisotropy is observed in the electron–hole droplet cloud, as shown in fig. 15. Due to the different effective masses and dielectric constant, the electron–hole liquid in Si is considerably denser than in Ge. The liquid density in Si is 3.3×10^{18} cm^{-3}, compared to 2.3×10^{17} cm^{-3} in Ge. Since the recombination rate increases with electron–hole pair density, the lifetime of the liquid in Si is only 140 ns, compared to 40 μs in Ge. From measurements of quantum efficiency of luminescence [32], it is known that the recombination is primarily by an Auger process, whereby the energy of a recombining pair is converted into kinetic energy of a remaining carrier. The thermalization of this hot carrier produces phonons. Thus, due to the short droplet lifetime and rapid Auger recombination, the cloud of droplets in Si is relatively small, and its shape and size is determined predominantly by the intense concentration of R-phonons. The face view image shown in fig. 15 shows none of the anisotropy observed in the cloud of droplets in Ge, fig. 14(b). Still, the phonon wind is the predominant transport force for these droplets [33]. The cloud expansion process is analogous to the expansion of a cloud of charged particles. In this case, the droplets are emitting phonons which are absorbed by other droplets, causing an effective repulsive interaction between the droplets [34]. As with the T-phonon source, the size of the cloud of droplets

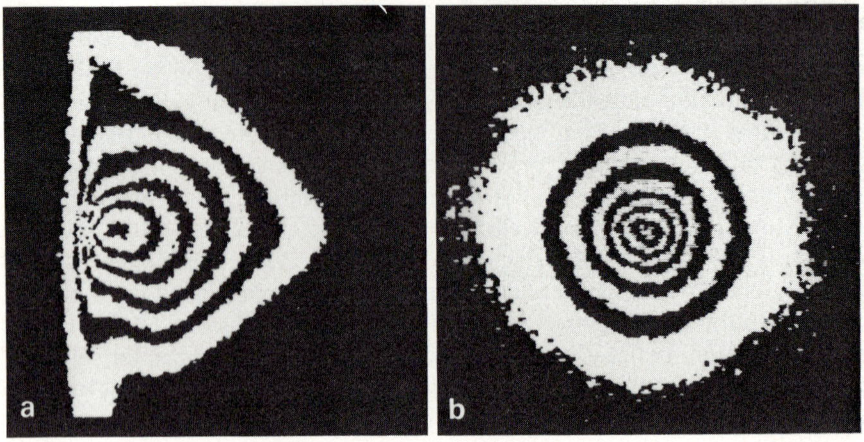

Fig. 15. Contour map of the EHL cloud luminescence in Si viewing parallel (a) and perpendicular (b) to the excitation surface. The cloud was produced by 220 mW CW excitation with an Ar^+ laser. The photographs span about 600 μm across on the crystal. From ref. [33].

depends upon the excitation power which increases the total number of droplets in the cloud.

Recent time-resolved imaging experiments have shown the expansion of electron–hole droplet cloud in Si following intense pulse excitation [35]. Figure 16(a) shows a shell of electron–hole droplets at a time 300 ns after the excitation pulse. In this case, the intense laser radiation, due to a 50 μJ, 5 ns N_2 laser pulse at $\lambda = 337$ nm wavelength, produces a strong T-phonon wind which blows the droplets at near-sonic velocities away from the excitation point. Profiles of this expansion process are obtained by slitscans of the droplet luminescence at various times after the pulse, using an experimental detection system similar to that used for the momentum-damping measurements. The laser was focused to a $100 \times 600\,\mu m^2$ spot on the sample. Even at this high excitation level the droplet cloud is observed to move with subsonic velocity. The width of the shell expands as it moves into the crystal, indicating a repulsive phonon wind force between the droplets themselves. Again, as in Ge, an understanding of this transport process requires a thermal hot-spot which lives for approximately 1 μs after the laser pulse.

4. Phonon-wind driven plasma in silicon

As the temperature of a crystal of Si or Ge is raised, a point is reached at which electron–hole droplets will not form. This critical temperature is at about 7 K in Ge and about 22 K in Si. It is still possible to form electron–hole plasma above this temperature, however, the increased entropy in the system

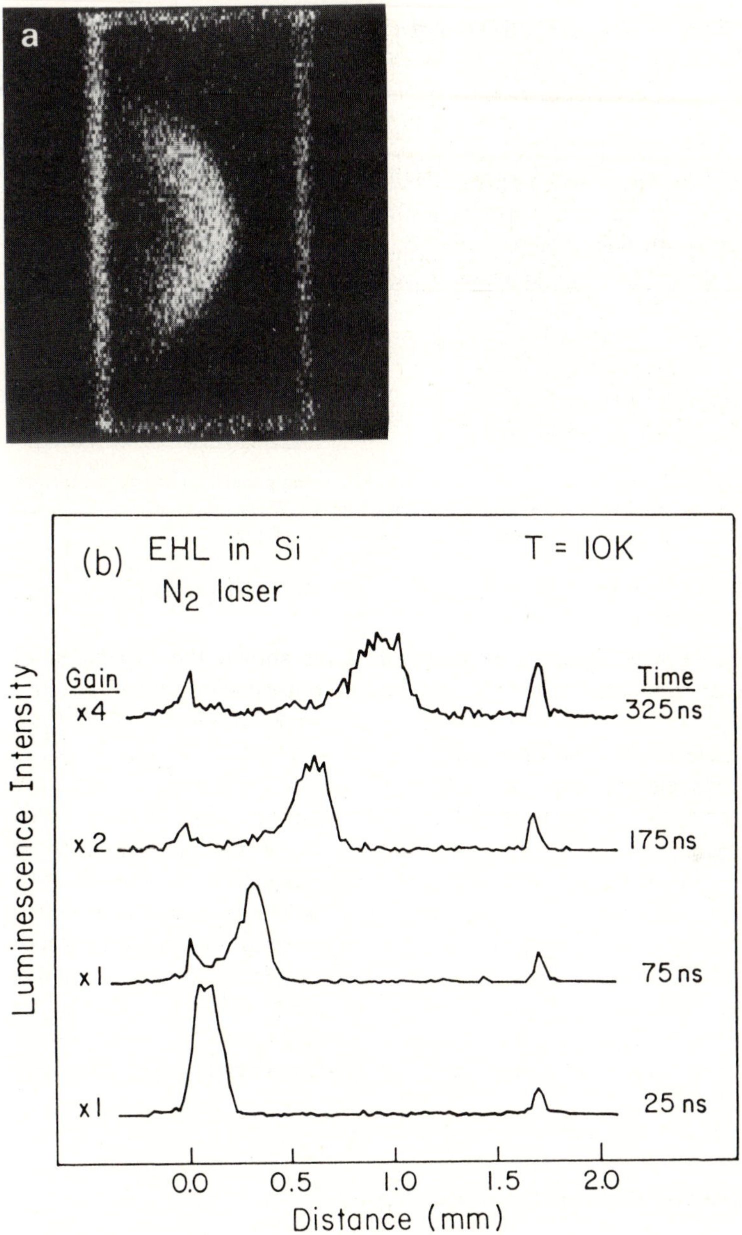

Fig. 16. (a) Time-resolved image of an expanding EHL cloud in Si taken 300 ns after a 5 ns, 50 μJ N_2 laser pulse, incident at the left. The crystal dimensions are $1.7 \times 2.9 \times 3.0$ mm^3. From ref. [35]. (b) Profiles of the expanding cloud in Si for various times after the laser pulse, showing near-sonic drift of the shell of droplets [35].

does not favor formation of a constant-density fluid. Indeed, there is nothing intrinsically different between a plasma at the liquid density and electron–hole liquid: the plasma formed above the critical temperature is simply not thermodynamically stable. This instability is displayed rather dramatically in fig. 2(b), which shows the density of the electron–hole plasma decreasing as the electron–hole pairs decay, in direct contrast to the electron–hole liquid.

The time-resolved spectra of electron–hole plasma in Si are shown in fig. 17(a). These are luminescence spectra from a pure Si crystal which has been excited with a cavity-dumped Argon-ion laser pulse of 8 ns duration and 0.4 μJ

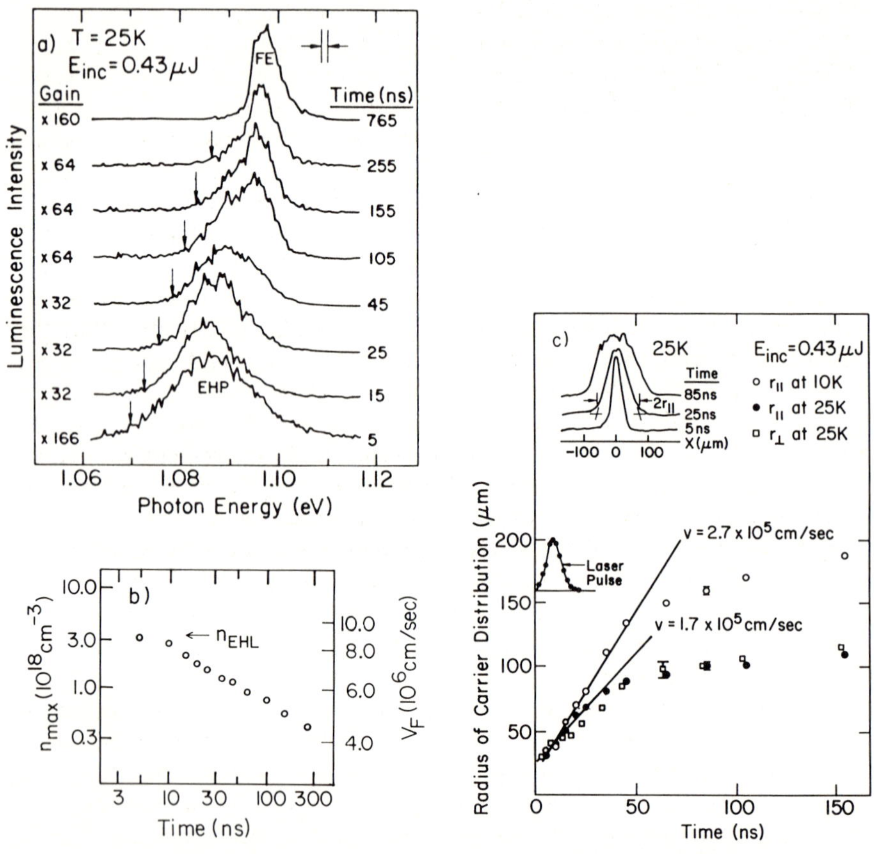

Fig. 17. (a) Time-resolved spectra of the EHP in Si following a cavity-dumped Ar^+ laser pulse. A 5 ns gate was used for the first 30 ns and, to optimize signal, was gradually opened to 50 ns at later times. The arrows indicate the position of the low-energy edge of the plasma luminescence. From ref. [35]. (b) Maximum plasma density as a function of time as estimated from the position of the low-energy edge of the plasma luminescence. (c) Radii of the carrier distribution parallel ($r_{\parallel}$) and perpendicular ($r_{\perp}$) to the crystal surface versus time. Both EHL at 10 K and EHP at 25 K are shown. The inset shows time-resolved slitscans of EHP looking into the laser beam direction.

energy. Initially, a plasma with density close to that of EHL is formed, and as the pairs recombine this density decreases, leaving only excitons at the longest times. The arrows at the low-energy edge of the spectrum approximate the renormalized band gap. Using the theoretical relationship between renormalized gap and plasma density [10], the maximum density of the plasma versus time is plotted in fig. 17(b).

Luminescence spectra of electron–hole plasma in semiconductors have caused a recent controversy in terms of the spatial transport of this plasma. Forchel et al. have noticed an interesting temperature dependence of the plasma density [36]. They found that the density of the plasma actually increases with increasing temperature, contrary to what they expected. To explain this behavior, Mahler et al. [7] have devised a thermodiffusion theory which incorporates a spatially varying plasma density and crystal temperature away from the excitation point. With this model they found that the density of the plasma is a maximum away from the excitation surface and this maximum density can be regarded as the predominant density of pairs in the plasma. This model was able to explain the increasing plasma density with temperature observed by Forchel et al. In these data, Forchel et al. [36] determined the density of the plasma from the width of the plasma emission line. More recently they recognized that the density obtained in this way is not consistent with the density obtained from the renormalized band gap. Indeed, the densities that they obtained in Si from the shape of the spectrum were about a factor of 3 higher than derived from the energy position of the luminescence; i.e., the experimental spectra appeared to be too broad. To explain this apparent anomaly, Forchel et al. [5] have postulated that the plasma drifts at a velocity comparable to its Fermi velocity. They concluded that the plasma densities were actually lower than those originally reported and that the spectral widths can be explained by a drifted Fermi distribution. The drift velocities required to explain their data were in the range 10^6 to 10^7 cm/s. These velocities were clearly much higher than the velocities obtained for electron–hole liquid at lower temperatures, which motivated us to examine the transport of electron–hole plasma at these higher temperatures.

Time-resolved spatial profiles of the electron–hole plasma in Si following the 0.4 μJ laser pulse are shown in the inset of fig. 17(c). These are spatial profiles looking into the laser beam direction. Additional profiles were obtained viewing parallel to the excitation surface. Both $r_\parallel$ and $r_\perp$ are plotted as a function of time during and after the laser pulse. The time resolution in this experiment was 5 ns and the spatial resolution was 30 μm. From these data, one can readily get an estimate of the average velocity during the first 20 ns following the start of the laser pulse. This initial average velocity for the electron–hole plasma at 25 K is seen to be about 1.7×10^5 cm/s.

Clearly, the drift velocities obtained on these time and space scales are much less than the Fermi velocities of the electron–hole plasma, as indicated on the right scale of fig. 17(b). If, for example, the plasma were drifting at a velocity of 4×10^6 cm/s – i.e., at half of the electronic Fermi velocity deduced from the spectral shift at 20 ns – the plasma would have traveled 800 μm during the span of 20 ns. It is concluded therefore that the broadening of the electron–hole plasma luminescence line cannot be due to extremely high drift velocities as postulated by Forchel et al. [5]. It is much more likely that these broader luminescence lines are a result of the basic inhomogeneities in the plasma. The density of the plasma is a function of both space and time, and an inhomogeneous distribution in density could easily result in a broadened luminescence line. Also, at the early times (e.g., during and shortly after the laser pulse) the plasma may have a temperature above the lattice temperature.

Additional experiments have been performed on the plasma expansion in Si using much higher excitation energies [35]. A 5 ns N_2 laser having up to 50 μJ energy was used to excite the crystal, as in the electron–hole droplet experiment described above. Time-resolved photon-counting images of the electron–hole plasma luminescence are shown in fig. 18. The plasma is seen to expand as a shell into the crystal, although this shell is somewhat smaller than that of the electron–hole liquid due to the increased momentum damping rate expected at these higher temperatures. The fact that the plasma distribution takes the shape of a shell indicates that it is *driven* into the crystal by a phonon wind source, as observed previously in the electron–hole liquid experiments.

The similarity of the transport processes for electron–hole liquid and electron–hole plasma is further underscored by the temperature dependence of the initial expansion velocity, as shown in fig. 19. The expansion velocity of the plasma increases with decreasing temperature, as expected from a temperature dependence of the momentum damping time. There is virtually no detectable

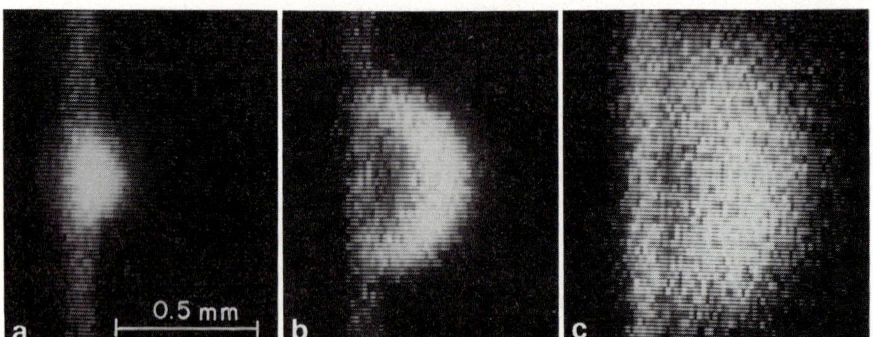

Fig. 18. Time-resolved images of EHP in Si at 32 K following excitation with a 50 μJ N_2 laser pulse. The laser is incident from the left and the white vertical line is scattered luminescence light from the crystal surface. Images (a) (EHP at 50 ns) and (b) (EHP at 550 ns), taken at $h\nu = 1.090$ eV, show an expanding shell of plasma. Image (c) (Fe at 3 μs) shows the free exciton distribution ($h\nu = 1.098$ eV) well after the EHP has decayed.

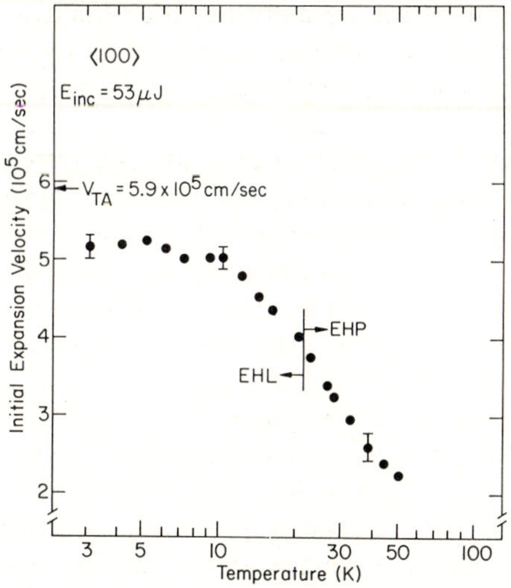

Fig. 19. Initial expansion velocities, obtained from slitscans as in fig. 16(b), versus crystal temperature. Excitation is with a pulsed N_2 laser. Data below 22 K correspond to EHL and those measured above 22 K correspond to EHP.

difference between the expansion velocity of the plasma above 22 K and the electron–hole liquid below 22 K. As the temperature is lowered further the expansion rate of the electron–hole liquid begins to saturate, at a velocity slightly below that of the sound velocity in the crystal. This saturated velocity is identical to that observed in the drift experiments of Tamor, as described earlier. It is suggested that a strong phonon wind of LA phonons pushes the droplets up to speeds of the slow transverse phonons in the crystal, at which point a strong damping process due to the emission of these (TA) phonons is present. This highly non-linear damping as an interesting aspect of the plasma transport process.

Clearly, more experimental and theoretical work is needed to characterize the transport of the electron–hole plasmas in these semiconductors. The previous thermodiffusion and hydrodynamic models proposed to explain these phenomena do not take into account the non-equilibrium phonon wind which, from the spatially-resolved measurements, is seen to play such an important role. A description of the plasma expansion in some ways is more complicated than that of electron–hole droplets at lower temperatures. Most importantly, the plasma density is non-uniform, so the momentum damping time and the absorption of phonons will be spatially dependent. Nevertheless, a reasonably quantitative description of the expanding shell of plasma, as demonstrated in the time-resolved images of fig. 18, is possible in terms of the phonon wind and momentum damping processes reviewed in the body of this paper. Interesting experimental challenges remain in the characterization of plasma expansion at higher temperatures and in direct-gap semiconductors, in which the recombination lifetimes are considerably shorter. The long-lived electron–hole plasmas in Si and Ge have provided some basic insights into the interaction of carriers and phonons in semiconductors.

This work has been supported by the National Science Foundation under Grant DMR–80–24000 and Materials Research Laboratory Grant DMR–80–20250.

References

[1] C.V. Shank, R. Yen and C. Hirbmann, Phys. Rev. Lett. 50 (1983) 454.
[2] D. Hulin, M. Combescot, J. Bok, A. Migus, J.Y. Vinet and A. Antonetti, Phys. Rev. Lett. 52 (1984) 1998.
[3] M. Wantelet and J.A. Van Vechten, Phys. Rev. B 23 (1981) 5551.
[4] A. Forchel, H.S. Schweizer, H. Nather, K.M. Romanek, J. Fischer and G. Mahler, Physica 117B (1983) 336.
[5] A. Forchel, H. Schweizer and G. Mahler, Phys. Rev. Lett. 51 (1983) 501.
[6] M. Combescot, Solid State Commun. 30 (1979) 81.
[7] G. Mahler, G. Maier, A. Forchel, B. Laurich, H. Sanwald and W. Schmid, Phys. Rev. Lett. 47 (1981) 1855.
[8] L.V. Keldysh, JETP 23 (1976) 86.
[9] S.J. Kirch and J.P. Wolfe, to be published.
[10] P. Vashista and R. Kalia, Phys. Rev. B 25 (1982) 6492;
J.C. Hensel, T.G. Phillips, G.A. Thomas and T.M. Rice, Solid State Physics, eds., H. Enrenreich, F. Seitz and D. Turnbull (Academic, New York, 1977) Vol. 32.
[11] M. A. Tamor, M. Greenstein and J.P. Wolfe, Phys. Rev. B 27 (1983) 7353.
[12] J. Bardeen and W. Shockley, Phys. Rev. 80 (1950) 72.
[13] M.A. Tamor and J.P. Wolfe, Phys. Rev. Lett. 44 (1980) 1703.
[14] C. Herring and E. Vogt, Phys. Rev. 101 (1956) 944.
[15] L.V. Keldysh and S.G. Tikhodeev, JETP Lett. 21 (1975) 273.
[16] D.S. Pan, D.L. Smith and T.C. McGill, Phys. Rev. B 17 (1978) 3284.
[17] R.S. Markiewicz, Phys. Status Solidi B 90 (1978) 585.
[18] See the recent review by V.S. Bagaev, T.I. Galkina and N.N. Sibeldin, in: Electron-Hole Droplets in Semiconductors, eds., C.D. Jeffries and L.V. Keldysh (North-Holland, New York, 1983) p. 276.
[19] See the recent review by J.P. Wolfe and C.D. Jeffries, ibid., p. 431.
[20] M.A. Tamor and J.P. Wolfe, Phys. Rev. B 26 (1982) 5743.
[21] M.I. D'yakonov and A.V. Subashiev, JETP 48 (1978) 980;
A.V. Subashiev, Sov. Phys. Solid State 22 (1980) 431.
[22] M. Greenstein and J.P. Wolfe, Phys. Rev. B 24 (1981) 3318; Phys. Rev. Lett. 41 (1978) 715.
[23] G.A. Northrop and J.P. Wolfe, Phys. Rev. B 22 (1980) 6196; Phys. Rev. Lett. 43 (1979) 1424.
[24] J.P. Wolfe, M. Greenstein and G.A. Northrop, J. Phys. Soc. Jpn. 49 (1980) Suppl. A, p. 491.
[25] R.S. Markiewicz, M. Greenstein and J.P. Wolfe, Solid State Commun. 35 (1980) 339.
[26] W. Dietsche, S.J. Kirch and J.P. Wolfe, Phys. Rev. B 26 (1982) 780.
[27] M. Greenstein, M.A. Tamor and J.P. Wolfe, Phys. Rev. B 26 (1982) 5604.
[28] S.J. Kirch and J.P. Wolfe, Phys. Rev. B 29 (1984) 3382.
[29] J.C. Hensel and R.C. Dynes, Phys. Rev. Lett. 39 (1977) 969.
[30] R.S. Markiewicz, Solid State Commun. 33 (1980) 701.

[31] M. Greenstein and J.P. Wolfe, Solid State Commun. 33 (1980) 309.
[32] Ya. Pokrovskii, Phys. Status Solidi B 11 (1972) 385;
See also, P.L. Gourley and J.P. Wolfe, Phys. Rev. B 24 (1981) 5970.
[33] M.A. Tamor and J.P. Wolfe, Phys. Rev. B 21 (1980) 739.
[34] See also for Ge: J. Doehler and J.M. Worlock, Phys. Rev. Lett. 41 (1978) 980.
[35] F.M. Steranka and J.P. Wolfe, to be published.
[36] A. Forchel, B. Laurich, W. Schmid and M.H. Pilkuhn, J. Phys. Soc. Jpn. 49 (1980) 487.

Journal of Luminescence 30 (1985) 114–119
North-Holland, Amsterdam

ON THE INVESTIGATION OF THE DIFFUSION PROCESSES OF PHOTOEXCITED CARRIERS IN SILICON BY ps-REFLECTIVITY MEASUREMENTS

H. BERGNER, V. BRÜCKNER and M. SCHUBERT

Physics Department of the Friedrich-Schiller University Jena, DDR–6900 JENA, GDR

The optical excitation of non-equilibrium charge carriers in semiconductors results in a carrier distribution which gives rise to a diffusion process into the bulk of the semiconductor. The diffusion time depends on the absorption coefficient and on the ambipolar diffusion constant. These diffusion times are on a subnanosecond time scale for our experimental conditions. By time-resolved reflectivity measurements we have studied the dependence of the ambipolar diffusion constant on the density of excited carriers. Experimental results were compared with theoretical ones.

Time-resolved reflectivity measurements can be used to study the temporal behaviour of optically generated non-equilibrium charge carriers in semiconductors, e.g., in Ge [1] and Si [2]. The mainly used excite-and-probe technique is based on the excitation of carriers into the absorption band. After excitation a test by a time delayed pulse is performed. We used a ps Nd:YAG spectrometer that yields pulses down to about 20 ps.

The temporal behavior of the carrier density is in principle determined by the superposition of various processes. Such processes are the linear and nonlinear recombination, the diffusion, the Auger recombination and the surface recombination. In the present paper the influence of linear recombination and diffusion on the temporal behaviour is investigated by means of time-resolved reflectivity measurements. Particularly the influence of the generated carrier concentration on the ambipolar diffusion is studied.

In fig. 1 the experimental arrangement is shown. The diameter of the probe beam considerably falls below the diameter of the exciting beam. Therefore, the influence of the surface coordinates y and z can be neglected and only the dependence on the coordinate x (perpendicular to the surface) occurs. It can be assumed that the sample surface is carefully prepared, so that the surface recombination rate takes a relatively low value and therefore, this process can be neglected. The energies of the probe beam, the reflected beam, and the exciting beam were measured by a laser energy monitoring system. To omit

0022-2313/85/$03.30

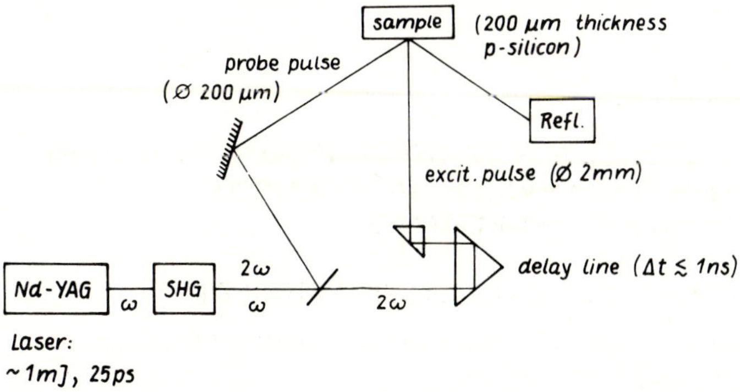

Fig. 1. Experimental arrangement of time-resolved reflectivity measurements.

the excitation energy fluctuations of the pulses an averaging is performed by a microcomputer K 1510, which also evaluates the characteristic parameters of the processes involved.

The carrier density $N(t)$ at time t in a surface layer can be obtained from the refractive index $n(t)$. Following the Drude model the difference between $n(t)$ and the refractive index without excitation n_0 is given by

$$n(t)-n_0=-[e^2/2\varepsilon_0 n_0 m^* \omega_p^2]N(t). \tag{1}$$

Here e is the electron charge, ε_0 is the permittivity of vacuum, m^* is the optical mass assigned to the electron–hole pairs, and ω_p is the frequency of the probe pulse.

Under physically reasonable assumptions [3] the temporal and spatial behaviour of the non-equilibrium charge carrier density $N(x, t)$ can be described by

$$\frac{\partial}{\partial t} N(x, t)-D\frac{\partial^2}{\partial x^2} N(x, t)+\frac{N(x, t)}{\tau}=G(x, t), \tag{2}$$

where D is the ambipolar diffusion constant, τ is the carrier lifetime, and $G(x, t)$ describes the generation by the laser pulse. To gain a physical insight under simple conditions we first will discuss a delta-function-like pulse excitation. After excitation we have to find the solution of a homogeneous differential equation where, due to the excitation, the initial distribution $N(x, 0)=N(0, 0)$ $\exp(-\alpha x)$ holds. α is the linear absorption coefficient of the medium. The carrier density at the surface $N(0, t)$ is measurable via reflectivity that is determined by the behaviour of N in a thin layer at the surface. The solution is

$$N(0, t)=N(0, 0)\exp(-t/\tau)\{\exp(D\alpha^2 t)[1-\operatorname{erf}\sqrt{D\alpha^2 t}]\}. \tag{3}$$

The expression in the braces takes on the value $\frac{1}{2}$ at a characteristic diffusion time $t_D \approx 0.59/D\alpha^2$. In fig. 2 the function $N(0, t)$ is plotted versus t for several

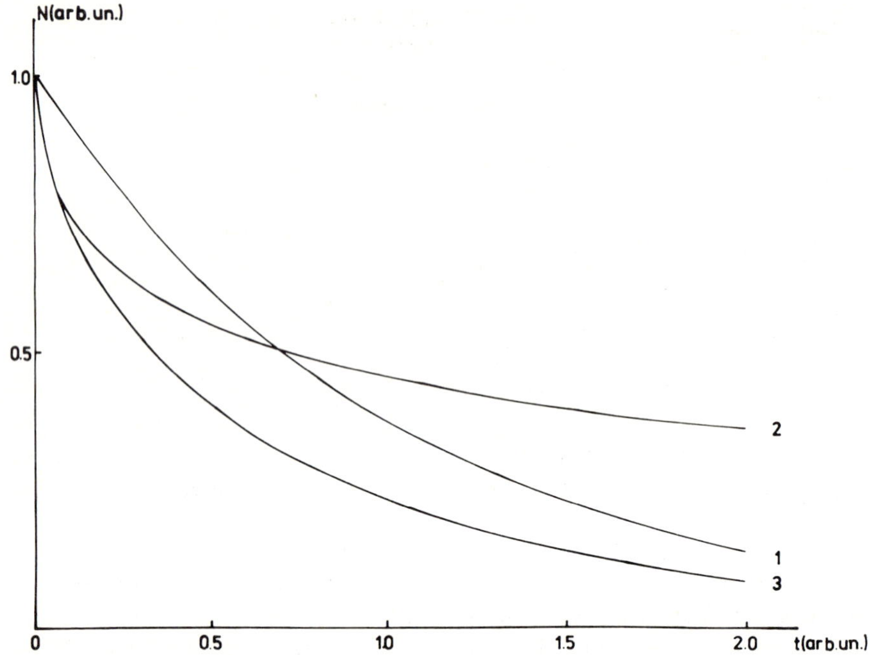

Fig. 2. Time dependence of carrier density at the surface. Curve (1): recombination only, diffusion neglected; curve (2): diffusion only, recombination neglected; curve (3): diffusion and linear recombination present.

cases. In the 'pure' recombination case ($D=0$) an exponential decay occurs. If only diffusion acts and recombination can be neglected ($\tau\rightarrow\infty$) we have a relatively rapid change for small t. The third curve shows the behaviour caused by diffusion as well as linear recombination. As a general result one may say that a diffusion-like behaviour occurs for small t, whereas a recombination-like behaviour appears for large t.

In our experiments the excitation is effected by a Gaussian-shaped pulse with respect to the dependence on time. Our numerical solutions of the inhomogeneous continuity equation (2) show the strong influence of the pulse width on the diffusion process, even if the pulse duration is small in comparison with the characteristic diffusion time t_D [3]. These numerical solutions of the continuity equation (2) were used to fit our experimental results.

Time-resolved reflectivity measurements have been made in monocrystalline p-doped silicon. We have investigated two types of silicon: the sample S1 with an acceptor density of $n_a < 10^{13}$ cm^{-3} and the sample S2 with $n_a \approx 1.5\times10^{15}$ cm^{-3}. In fig. 3 we have plotted the normalized reflectivity of both samples versus the delay time and also the 'best' numerical curves. For sample S1 (+) an ambipolar diffusion constant $D=44$ cm^{2}s^{-1} was obtained. The recombination

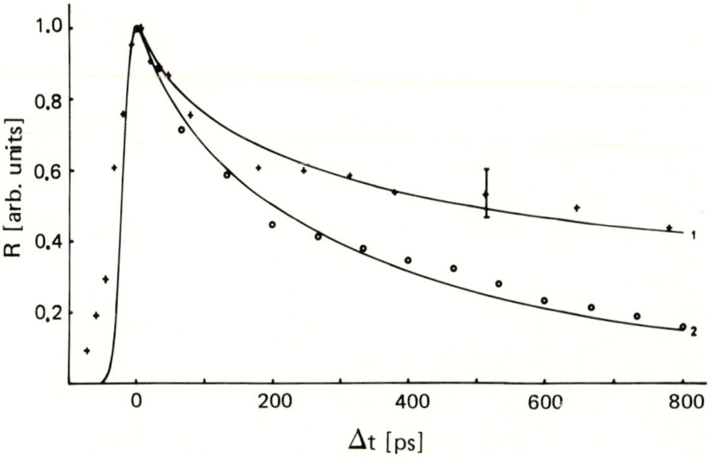

Fig. 3. Normalized reflectivity as a function of delay time. Curve (1): acceptor concentration $n_a < 10^{13}$ cm^{-3}; curve (2): $n_a \approx 1.5 \times 10^{15}$ cm^{-3}. Solid lines are best numerical fits.

constant lies in the region $\tau \approx 10$ ns. In this region the alternation of τ does not lead to a significant change of the theoretical curve. By fitting the numerical curve to the experimental results of sample S2 (○) we found the values $D = 44$ cm^{2}s^{-1} and $\tau = 750$ ps. So one can say that a mainly diffusion process takes place in sample S1, whereas the relaxation process of the carriers in the sample with higher acceptor concentration is controlled by diffusion as well as linear recombination. The ambipolar diffusion constant was found to be larger than the usual value of silicon at room temperature and at low carrier concentration.

To study the dependence of the ambipolar diffusion constant on the generated carrier concentration the time-resolved reflectivity of the sample S1, where only diffusion takes place, was studied at different excitation energy levels (fig. 4). Comparing these experimental results with the numerical ones the best fit was found at diffusion constants D of 18 cm^{2}s^{-1} (curve 1 in fig. 4), 30 cm^{2}s^{-1} (curve 2), and 60 cm^{2}s^{-1} (curve 3) at excitation energy densities of 1.4 mJ/cm^2 (+), 2.9 mJ/cm^2(△) and 6.6 mJ/cm^2 (○), respectively. At an excitation energy density of 0.3 mJ/cm^2 no differences in the temporal behaviour in comparison with 1.4 mJ/cm^2 excitation was found.

In order to compare our experimental results with the theoretical ones we will use the dependence of the ambipolar diffusion coefficient on the carrier density as presented in [1,4]. At high excitation ($N \approx 10^{20}$ cm^{-3}) the semiconductor is degenerated. If the carriers can be considered to move independently, the Boltzmann transport theory predicts for strong degeneracy [1] $D_{deg}/D_{nond} \propto N^{1/3}$. In fig. 5 we compare the calculations of Young and Van Driel [4] with our experimental results. The dashed curve in fig. 5 corresponds to the theory of independent carriers, whereas the full line represents the model

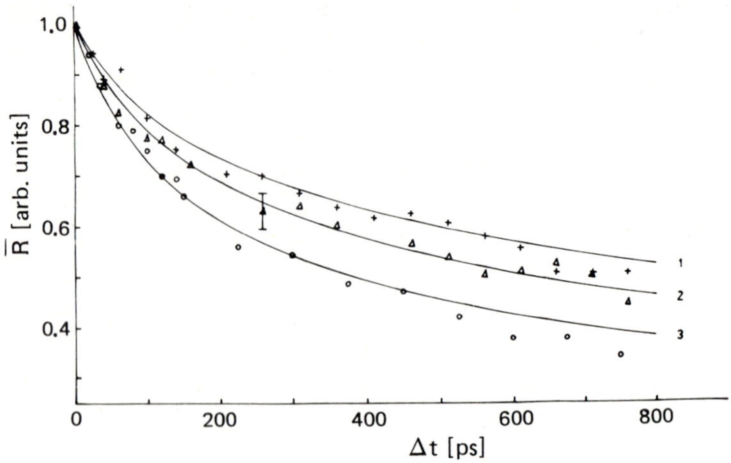

Fig. 4. Normalized reflectivity as a function of delay time for a sample where only diffusion takes place. Curves (1)–(3) refer to different excitation levels (see text).

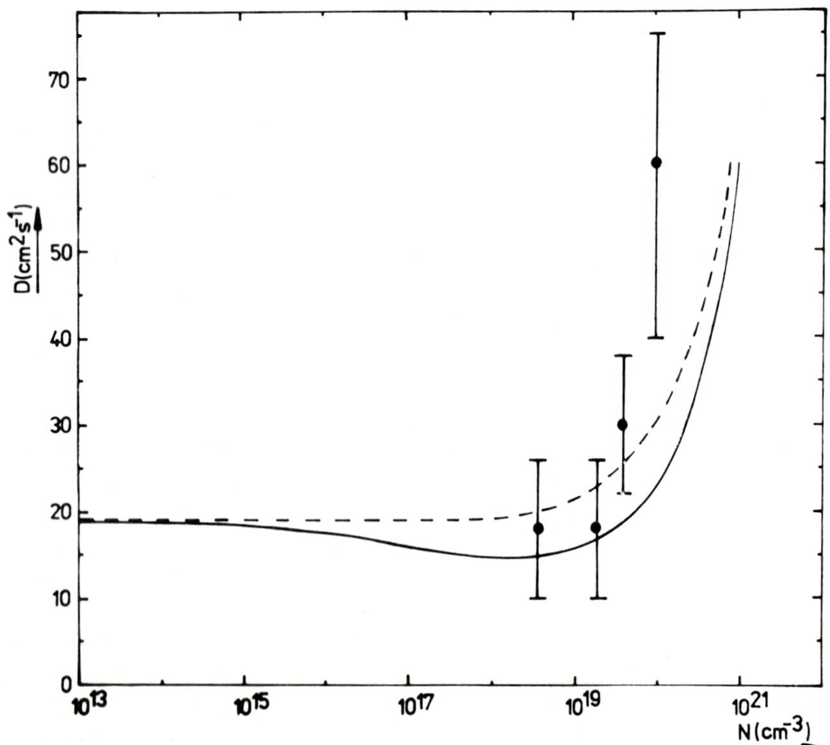

Fig. 5. Ambipolar diffusion coefficient as a function of carrier density: comparison of experimental data with theoretical curves [4].

regarding many-body effects. The generated carrier concentrations in our experiments were calculated from the excitation energy density taking into account Fresnel reflection. The experimentally obtained data are in a rather good agreement with the theoretical curves [5].

Finally it should be mentioned that diffusion occurs in the presence of density gradients, and a given experiment will not be able to yield the diffusion coefficient for a particular density. Therefore, our calculated ambipolar diffusion coefficient D is an effective value.

References

[1] D.H. Auston and C.V. Shank, Phys. Rev. Lett. 32 (1974) 1120.
[2] H. Bergner, V. Brückner and B. Schröder, Kvant. Elektr. 10 (1983) 1150.
[3] H. Bergner and V. Brückner, Opt. Quant. Electr. 15 (1983) 477.
[4] J.F. Young and H.M. van Driel, Phys. Rev. B26 (1982) 2147.
[5] H. Bergner and V. Brückner, Phys. Status Solidi (a) 79 (1983) K85.

Journal of Luminescence 30 (1985) 120–143
North-Holland, Amsterdam

HOT-CARRIER TRANSIENT TRANSPORT

Carlo JACOBONI

Dipartimento di Fisica, Universita' di Modena, Via Campi, 213/A, 41100 Modena, Italy

In this article a review is presented of the present knowledge on the problem of transient transport of charge carriers in semiconductors with high applied electric fields. After a summary of the fundamentals of charge transport the Monte Carlo method is presented as applied to such problems. Transient transport is then presented with special emphasis on the most modern results related to autocorrelation function of velocity fluctuations.

1. Introduction

In this lecture we shall try to give a review of the present knowledge on the problem of transient transport of charge carriers in semiconductors with high applied electric fields. This problem has recently received great interest in connection with the fabrication of ultrasmall devices (submicron), made possible by molecular beam epitaxy. Furthermore, modern laser technology allows for the experimental analysis of ultrafast transport processes, so that it is now useful to have a theoretical knowledge of charge transport at very short times and distances.

Charge transport in semiconductors is in general a difficult problem, both from the mathematical and the physical point of view. In fact, the integro-differential equation (the Boltzmann equation) which describes the problem does not offer simple, or even complicated, analytical solutions except for very few cases, usually not applicable to real systems. Furthermore, since transport quantities are derived from averages over many physical processes whose relative importance is not known a priori, the formulation of reliable microscopic models for the physical system under investigation is difficult. When one moves from linear to nonlinear response conditions, the difficulties become even greater.

The situation has greatly improved after the introduction of the Monte Carlo (MC) technique [1]. MC is a statistical numerical means used, in general,

0022-2313/85/$03.30 © Elsevier Science Publishers B.V.
(North-Holland Physics Publishing Division)

for solving mathematical problems; as such, it was born well before its application to transport in semiconductors [2] and has been applied to a large number of scientific fields. In the case of charge transport, however, the statistical numerical approach to the solution of the Boltzmann equation proves to be a direct simulation of the dynamics of charge carriers inside the crystal, so that, while the solution of the equation is being built up, any physical information required can be easily extracted.

In what follows we shall first review the fundamentals of charge transport (section 2) and of the hot-electron problem (section 3). The MC method will be briefly summarized in section 4. Transient transport will be treated in the two final sections (sections 5 and 6) with special emphasis to the most modern results related to the autocorrelation function of velocity fluctuations.

2. Fundamentals of charge transport. The Boltzmann equation

The approach commonly used in almost all treatments of transport properties in semiconductors is semiclassical in nature. The scattering cross sections for electronic transitions due to imperfections of the host crystal are derived from the quantum theory of scattering, while charge carriers are considered to move between two successive scattering events according to classical mechanics. The electron momentum $h\boldsymbol{k}$ changes with time according to

$$\hbar \frac{\mathrm{d}\boldsymbol{k}}{\mathrm{d}t} = e\left(\boldsymbol{E} + \frac{1}{c}\,\boldsymbol{v}\times\boldsymbol{B}\right), \tag{1}$$

$$\boldsymbol{v} = \frac{1}{\hbar}\nabla_{\boldsymbol{k}}\varepsilon, \tag{2}$$

where e, $\boldsymbol{k}$, $\boldsymbol{v}$, and ε are the electron charge (with its sign), wavevector, velocity, and energy, respectively; $\boldsymbol{E}$ and $\boldsymbol{B}$ are applied electric and magnetic fields, and h and c are the Planck constant and the velocity of light in vacuum respectively. In the following only electric fields will be considered.

The fundamental quantity in classical transport theory is the carrier distribution function $f(\boldsymbol{r}, \boldsymbol{k}, t)$, defined as proportional to the density of electrons in the sixth-dimensional space $(\boldsymbol{r}, \boldsymbol{k})$, where $\boldsymbol{r}$ indicates the position in the crystal. The normalization constant can be chosen in such a way that

$$\frac{2}{(2\pi)^3}\int \mathrm{d}^3k \int \mathrm{d}^3r\, f(\boldsymbol{r}, \boldsymbol{k}, t) = N,$$

where N is the number of electrons in the crystal.

Since spin interactions are usually not involved in transport phenomena in presence of electric fields alone, the spin variables are not explicitely considered.

The equation which describes transport phenomena in the semiclassical approach is the Boltzmann equation [3]:

$$\frac{\partial f}{\partial t} + \boldsymbol{v} \cdot \nabla_r f + \dot{\boldsymbol{k}} \nabla_k f = \left. \frac{\partial f}{\partial t} \right|_{\text{coll}}, \tag{3}$$

where the right-hand side is the rate of change of f due to collisions.

The assumption of eq. (3) at the basis of the transport theory implies the hypothesis of pointlike particles in phase space against the uncertainty principle, and instantaneous scattering process. When these two hypotheses are not justified, quantum transport equations, based on the density matrix concept, should be used [4, 5].

Once the distribution function is known, all quantities of interest, such as carrier drift velocity, mean energy and diffusion coefficient, can be obtained as functions of time, applied fields, temperature, and/or carrier concentration gradient.

As can be seen from inspection of the Boltzmann equation and of eqs. (1) and (2), the basic knowledge required for the analysis of any transport phenomenon is concerned with the band structure $\varepsilon(\boldsymbol{k})$ of the material and the scattering mechanisms which the electrons undergo within the crystal. For reasons of brevity we shall assume in what follows that both the band structure and the scattering mechanisms undergone by charge carriers in the most important semiconductors are known, at least to some extent [6,7].

3. Nonlinear transport – hot carriers

Nonlinear transport traditionally deals with problems which arise when a sufficiently strong electric field $\boldsymbol{E}$ is applied to a material, so that the current density deviates from the linear response.

Carrier heating and its relation to nonlinearity can be easily understood in a simple model semiconductor (spherical and parabolic band). The presence of a field inside the crystal modifies the free paths of the electrons between two successive scattering events by adding to the velocity a component in the direction of the field:

$$\boldsymbol{v}^{(i)} = \boldsymbol{v}_0^{(i)} + \frac{e\boldsymbol{E}}{m} t^{(i)}, \tag{4}$$

where $\boldsymbol{v}^{(i)}$ is the instantaneous velocity of the ith electron, $t^{(i)}$ is the time elapsed after the last scattering event of the ith electron and $\boldsymbol{v}_0^{(i)}$ is the initial value of $\boldsymbol{v}^{(i)}$ at $t^{(i)} = 0$.

The drift velocity of the electron gas is obtained by averaging $\boldsymbol{v}^{(i)}(t)$ over all electrons at a given time t, corresponding to a different values of $t^{(i)}$ for each electron. If there is no accumulation of the effect of the electric field on the initial velocities $\boldsymbol{v}_0^{(i)}$ of the successive free electron paths, all these initial velocities reproduce the electron distribution function in the absence of field. The mean value of the electron energy differs from its zero-field value only by the amount gained during each single flight. As long as this energy is negligible with respect to the electron energy, since the electron scattering probabilities are functions of the electron energies, the distribution of the $t^{(i)}$'s at any given time t is independent of $\boldsymbol{E}$. The drift velocity is therefore linear in the electric field according to Ohm's law.

On the other hand, if the field is sufficiently high, the energy gained by the electrons during each single flight is not negligible. The electrons cannot fully dissipate this energy at each scattering event and the effect of the field accumulates on the initial velocities $\boldsymbol{v}_0^{(i)}$. The distribution of the $t^{(i)}$'s also depends upon the field, and the linearity of Ohm's law is lost in averaging eq. (4). The mean energy of the carriers then increases, and it would continue to increase up to the breakdown of the sample were it not for scattering mechanisms whose dissipation capacity increases at increasing electron energies. These scattering

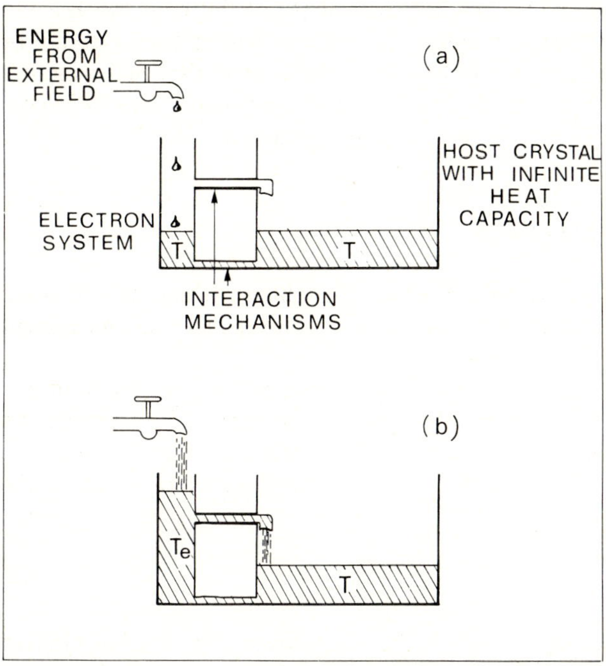

Fig. 1. Hydraulic analogue of the concept of hot electrons. Energy is transferred from the external electric field to the electron system and to the host crystal. In the low-field case (a) some interaction mechanisms are capable to maintaining the temperature of the electron system equal to that of the thermal bath. At high fields (b) the energy supplied to the electrons is higher, and a new stationary state is attained by an increased efficiency of the scattering mechanisms already active at low fields and, sometimes, by the onset of new mechanisms.

mechanisms establish a new stationary state in which the average electron energy is higher than at equilibrium. This is the origin of the expression "hot-electron effects" applied to nonlinear transport phenomena in semiconductors. Figure 1 shows a hydraulic analogue of the concept of hot electrons.

A general picture of hot-electron effects in pure semiconductors can be given as follows (see fig. 2). Let us first divide all semiconductors into two categories depending on whether the most effective scattering mechanisms are due to electrostatic fields (polar crystals) or to deformation potential (covalent semiconductors). Let us then note that as the external field increases above the Ohmic region the mean energy of the electrons in steady-state conditions increases.

For covalent materials the scattering mechanisms become more effective as the electron energy increases, so that (i) the mobility decreases as the field increases, and (ii) steady state is attained with a small energy rise above thermal equilibrium: the electron mean energy increases slowly with field at the beginning of the deviation from linear-response regime, as shown in fig. 2.

For polar materials the efficiency of the scattering mechanisms does not

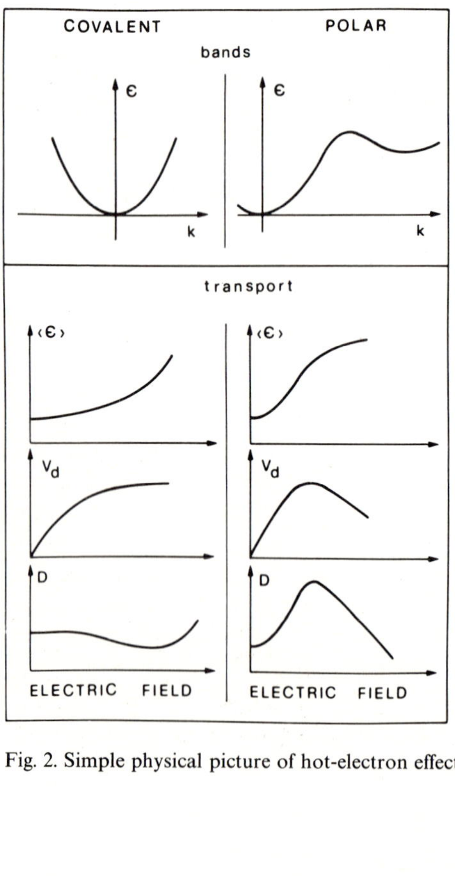

Fig. 2. Simple physical picture of hot-electron effects in semiconductors (see text).

effectively increase (and may even decrease) as the electron mean energy increases. Thus, the mean electron energy increases very fast to reach steady-state conditions as the applied field increases (eventually polar runaway is attained), while the mobility does not decrease. In a parabolic band the drift velocity controlled by polar scattering may even be super-Ohmic in the warm-electron region [8].

At higher fields in covalent semiconductors the mobility keeps decreasing toward saturation of the drift velocity and the electric mean energy keeps increasing. Eventually upper valleys are reached; this happens in Ge but not in Si, since in Si the upper valleys are so high in energy that breakdown occurs first. At low temperatures in Ge, the population of the upper valleys gives rise to a negative differential mobility, but the effect is quite small since the mean energy of the electrons increases slowly in increasing fields and the upper valleys are populated slowly.

In polar semiconductors the mean energy of the electrons increases much faster, and above threshold field for polar runaway it would increase indefinitely, giving rise to dielectric breakdown. Electrons then populate the upper valleys in the band and the population of these upper valleys increases very rapidly as the external field increases. At this point the new mechanism of nonequivalent intervalley scattering comes into play. Since this is very effective in dissipating both energy and momentum, above threshold field for polar runaway the energy increase with field is slowed down and the drift velocity decreases with increasing field, giving rise to the well-known phenomenon of negative differential mobility at the basis of the Gunn effect [9,10] (see fig. 2).

If we now consider the diffusion of electrons, we must remember that at high fields the Einstein relation does not hold. However, we may consider qualitatively a very rough generalization of that equation:

$$D=\tfrac{2}{3}\mu(E)\langle\varepsilon\rangle/e \tag{5}$$

and, by taking into account the considerations made above on $\mu(E)$ and $\langle\varepsilon\rangle$, we obtain the general behavior shown in the lowest part of fig. 2. It is determined, in general, by the competing effects of reducing $\mu(E)$ and increasing $\langle\varepsilon\rangle$ as field increases. In covalent semiconductors the two effects are almost balanced, the former being slightly prevalent, so that D decreases slowly with increasing fields. In polar semiconductors we have seen to the contrary that there is first a sharp increase of $\langle\varepsilon\rangle$ not accompanied by a decrease in mobility, so that D increases very rapidly with $\boldsymbol{E}$. Then we saw that the increase in energy is slowed down very much while the mobility decreases very sharply: a sharp reduction of D follows.

For holes, similar considerations hold except for the absence of upper valleys.

Almost all hot-electron effects can be described as details of the above general picture.

4. Solution of transport problems – the Monte Carlo method

From the analytical point of view, transport phenomena in nonlinear regimes would be completely described by solving the Boltzmann equation without linearizing it with respect to the external force. This is a formidable mathematical problem and is still unsolved today.

A big step forward in the solution of the Boltzmann equation has been however accomplished with the introduction of numerical techniques. The two most important of them, which date from 1966 (Semiconductor Kyoto Conference), are the iterative technique and the Monte Carlo (MC) method.

The iterative technique [11–13] yields a solution of the Boltzmann equation by means of an iterative procedure, and processes the whole distribution function at each step of the procedure. For this reason it can be useful when we deal with physical phenomena which depend on details of the distribution function.

The MC method [1,7,14–16] is by far the most popular of the two techniques. This method, as applied to the study of stationary transport properties of homogeneous semiconductors, consists of a simulation of the motion of one electron subject to the accelerative force of external fields and to given scattering mechanisms. By selecting random numbers, we make these mechanisms act stochastically in accordance with the probabilities determined by the differential and total cross sections, known from the theory of the corresponding processes. The electron path is recorded for a time sufficiently long to be representative, on the basis of ergodicity, of the behavior of all the carriers which contribute to the transport process. Figure 3 illustrates the principles of the MC method (see the figure caption).

When the transport process under investigation is not homogeneous or not stationary, it is necessary to simulate a large number of electrons and follow them during their dynamic histories, in order to obtain the desired information on the process of interest.

The MC technique, as a direct simulation of the dynamics of charge carriers inside the crystal, enables us to extract any required physical information while the solution of the transport equation is being built up. Furthermore, it enables us to simulate particular physical situations unattainable in experiments, or even to investigate on nonexisting materials, in order to emphasize special features of the phenomenon under study. This use of the MC makes it similar to an experimental technique: the simulated "experiment" can in fact be compared with analytically formulated theories.

5. Transient transport

The transient dynamic response of a physical system to given external forces is the response to these forces during the evolution of the system from given

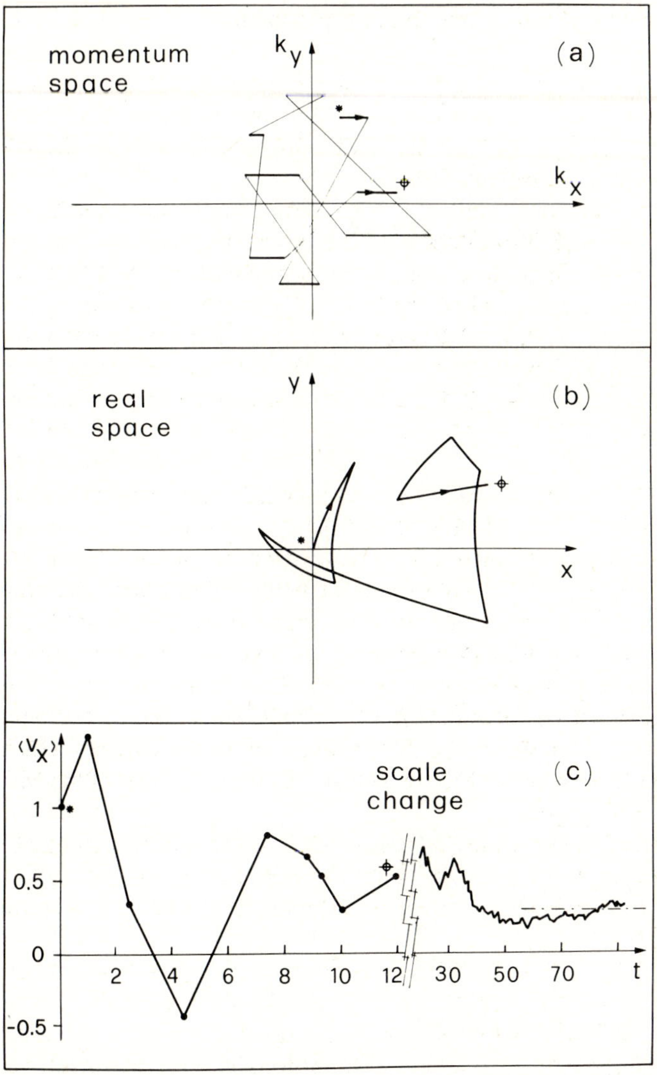

Fig. 3. The principles of the MC method; for simplicity a two-dimensional model is considered here. Part (a) of the figure shows the simulation of the sampling electron in momentum space, subject to an accelerating force due to an electric field oriented along the positive x direction. The heavy segments are due to the effect of the field during free flights, while light lines represent discontinuous variations of k due to scattering processes. Part (b) shows the path of the particle in real space. It is composed of eight fragments of parabolas corresponding to the eight free flights in part (a) of the figure. Part (c) shows the average velocity of the particle obtained as a function of the simulation time. The left section of the curve ($t < 12$) is obtained by the simulation illustrated in the parts (a) and (b) of the figure. The horizontal dot–dashed line represents the "exact" drift velocity obtained with a very long simulation time. Special symbols indicate corresponding points in the three parts of the figure (* is the starting point). All units are arbitrary.

initial conditions to the steady state. The duration of the transient response is not known a priori and, in general, it will be at least of the order of the longest of the characteristic times of the carrier system. This time may be called the "transient-transport time" and it depends upon the values of the applied field and temperature. In the case of high-field transport in semiconductors it may roughly correspond to the energy relaxation time or to the time necessary for the repopulation of the different valleys. Furthermore the transient response depends strongly upon the initial conditions of the carriers.

Transient transport may occur in different physical systems. The first rather "academic" situation we may think of is that of an infinite homogeneous system with a sudden change of the applied field, for example, from zero to a high value: the electron distribution requires some time to reach its steady-state nonequilibrium value. The second system of interest is a much more realistic one and can be present in ultrasmall devices: in a steady-state situation particles may enter a small region with a very high applied field and reach the end of this region so quickly that the space dependent distribution function does not reach the form equivalent to a steady-state homogeneous system.

The two situations mentioned above can be combined together when the system of interest is subject to alternating fields with frequency higher than the inverse of some electronic relaxation time ($\omega \cdot \tau > 1$) and wavevector higher than the inverse of the electronic mean free path ($q \cdot l > 1$).

For space or time-dependent problems the analytical solution of the Boltzmann equation is even more difficult than for homogeneous and stationary problems, while for MC programs little work needs to be added to attack such problems. This confirms again the usefulness of the MC method, in particular for the analysis of small devices, where it is often necessary to consider both the transient dynamic response to voltage changes and the electronic behavior at different points of the device with different field or material properties [17]. For such problems we cannot rely on the ergodic property of the system, and an ensemble of particles must be explicitly simulated [18,19]. Provided the number of simulated particles is sufficiently large, the average value of a quantity of interest, obtained on this sample ensemble as a function of time, will be representative for the average on the entire gas.

To determine the precision of the results obtained, the entire ensemble can be separated into a certain number of sub-ensembles; for each of them an estimate of the quantity of interest will be performed. Then their average value and standard deviation can be taken as the most probable value and the statistical uncertainty.

The transient dynamic response which will be obtained by means of the simulation, will of course depend upon the initial conditions of the carriers, and these must be assumed according to the situation which is to be explored.

As examples of application of the MC method to transient transport, we show in figs. 4 and 5 the mean velocity $\langle v \rangle$ and the mean energy $\langle \varepsilon \rangle$ of an en-

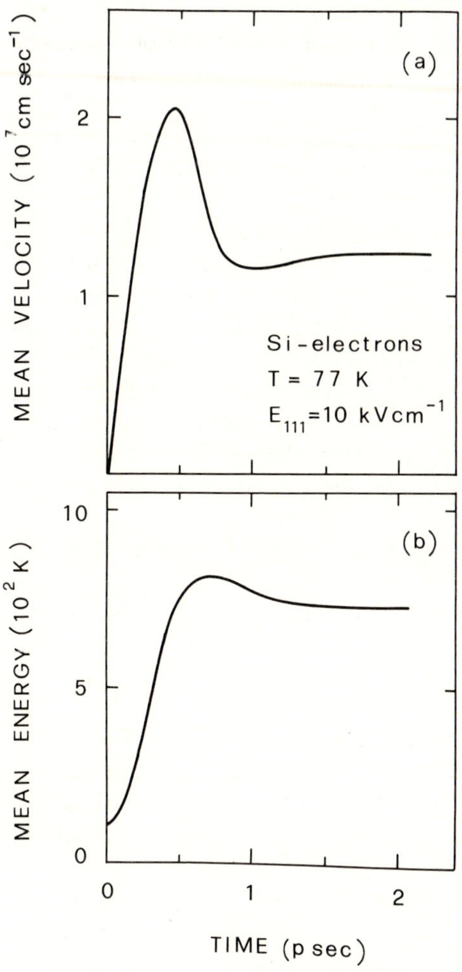

Fig. 4. MC results for the mean velocity (a), and the mean energy (b) for an ensemble of electrons in Si at 77 K, as functions of the time elapsed after the application of a field $E = 10$ kV/cm along a $\langle 111 \rangle$ direction.

semble of carriers in Si and GaAs as functions of the time elapsed after the instantaneous application of a static field. The silicon model used in the calculations is the same as given in ref. [20]. The GaAs model is that given in ref. [14]. In both cases electrons have initially the equilibrium Maxwellian distribution in $\boldsymbol{k}$-space. For the case of Si they are randomly distributed among the six valleys, while in GaAs they are all in the central valley.

As can be seen from figs. 4 and 5, a general feature of high-field transient transport, starting from equilibrium conditions, is an overshoot of $\langle v \rangle$ and $\langle \varepsilon \rangle$

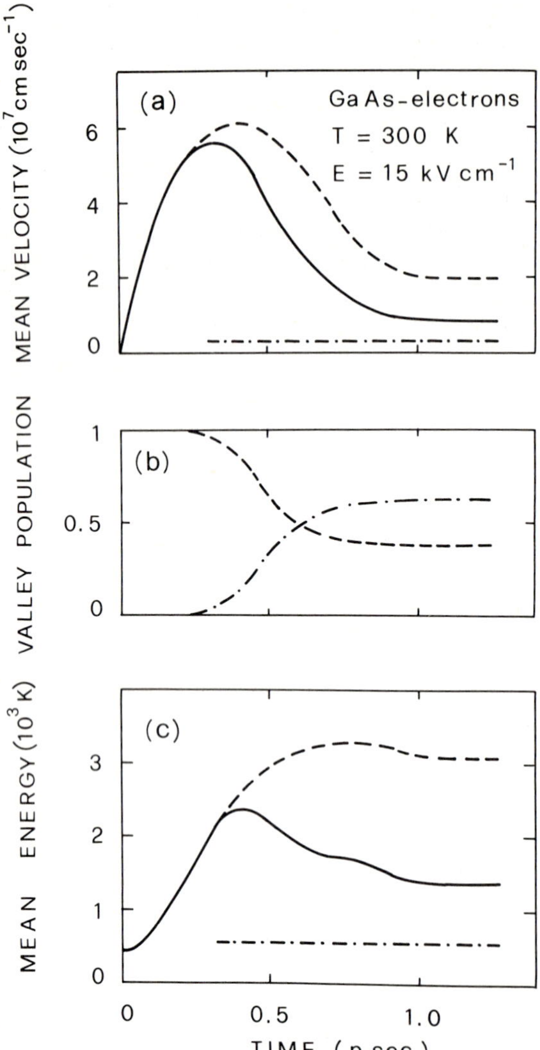

Fig. 5. MC results for the mean velocity (a), the valley populations (b), and the mean energy (c), for an ensemble of electrons in GaAs at 300 K as functions of the time elapsed after the application of a field $E = 15$ kV/cm. Dashed and dot–dashed curves refer to quantities averaged over the central and upper valleys, respectively. Continuous curves refer to quantities averaged over all the particles.

before reaching the steady-state value. The overshoot of $\langle v \rangle$ is associated with larger momentum relaxation times of "cold" electrons [13,21–23]. The overshoot of mean energy in Si, which is found at times slightly larger than those of the overshoot of mean velocity, is associated with an initial excess of absorbed power of the electron gas, reduced later by the setup of intervalley scattering.

In the case of GaAs the transient transport is also influenced by the electron transfer to upper valleys, which further reduces the average velocity and kinetic energy. In fact, as shown in fig. 5, in correspondence with the population of the upper valleys, the mean energy and velocity in the central valley become higher than the corresponding total quantities because in the upper valleys $\langle v \rangle$ and $\langle \varepsilon \rangle$ have values approximately constant and much lower than those of the central valley.

The very first part of transient transport is determined by the acceleration due to the electric force, before the effect of scattering; this regime is called "ballistic transport". In such conditions, if the average initial velocity of the particles is zero, then the average velocity $\langle v_z(t) \rangle$ along the field direction and

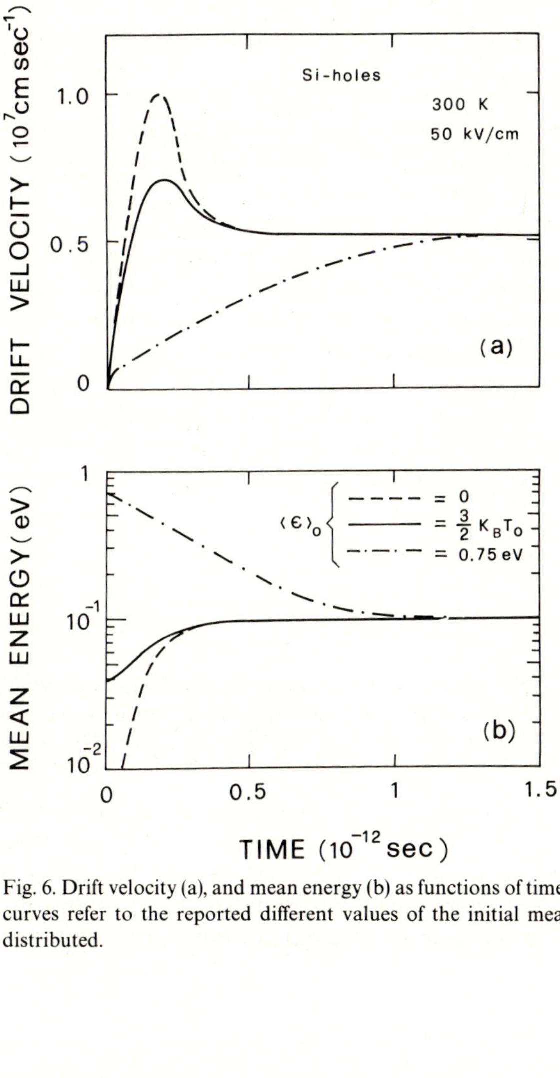

Fig. 6. Drift velocity (a), and mean energy (b) as functions of time for the case of holes in Si. Different curves refer to the reported different values of the initial mean energy with velocities randomly distributed.

the average energy $\langle \varepsilon(t) \rangle$ change with time, in a simple model, as:

$$\langle v_z(t) \rangle = at, \tag{6}$$

$$\langle \varepsilon(t) \rangle = \langle \varepsilon(0) \rangle + \tfrac{1}{2} m a^2 t^2, \tag{7}$$

where a is the acceleration eE/m. The linear and squared dependencies predicted by the above equations are actually present as initial behavior of the results reported in figs. 4 and 5.

In order to show the importance of initial conditions, fig. 6 reports ensemble MC calculations for the case of holes in Si at $T = 300$ K and $E = 50$ kV/cm [24]. Here the transient behavior of the drift velocity and the mean energy is investigated at different initial mean energies. An overshoot in the drift velocity can be seen that is larger for the colder initial condition and disappears for the hottest case. When the electron energy is higher in the initial conditions than in the steady state, in fact, the drift velocity is at first lowered by a higher scattering and reaches its stationary value more slowly.

Space- and time-dependent phenomena may present similar characteristics and they have sometimes been confused in simulation problems. For instance the drift velocity overshoot seen above in the time-dependent homogeneous case, is often used in discussing the behavior of an electron stream coming from a low-field region and entering a region with a high applied field.

If this reasoning is correct from a qualitative point of view, in rigorous calculations the kind of average that must be done is different whether time- or space-dependent phenomena are considered. In a steady-state phenomenon with a space dependence of the applied field it is in general necessary to consider the electron properties at given positions, so that the simulation must perform recording and averages for given points at any times, while when the object of the investigation is the evolution in time of a homogeneous system, average quantities must be evaluated at given times independent of the particle position.

6. Diffusion, velocity autocorrelation, and noise

6.1. Steady state

Diffusion is described, at a phenomenological level, by Fick's first law. If an electric field $\boldsymbol{E}$ is present, which would produce, in absence of diffusion, a drift velocity $\boldsymbol{v}_d$, and if diffusion and drift do not influence each other, the current density is given by

$$J_i = e\left\{ m(\boldsymbol{r}) v_{d_i} - D_{ij} \frac{\partial n(\boldsymbol{r})}{\partial x_j} \right\}, \tag{8}$$

where $\boldsymbol{r}$ is the space position with components x_j, $n(\boldsymbol{r})$ is the particle density, D_{ij} is the diffusion coefficient tensor; sum over repeated indices is implied.

By combining this equation with the continuity equation, we obtain the diffusion equation

$$\frac{\partial n}{\partial t} = -v_{d_i}\frac{\partial n}{\partial x_i} + D_{ij}\frac{\partial^2 n}{\partial x_i \partial x_j}, \tag{9}$$

where $\boldsymbol{v}_d$ and D_{ij} have been assumed to be space independent.

For vanishing small electric field strengths, diffusivity D and mobility μ are field independent and verify the Einstein relation

$$D = \mu k_0 T/E. \tag{10}$$

At high fields the Einstein relation fails, and the study of diffusion is generally performed through eq. (9) by introducing a field-dependent diffusion coefficient $D_{ij}(E)$.

From the diffusion equation it is easy to obtain

$$\frac{d}{dt}\langle (z - \langle z \rangle)^2 \rangle = 2D_\ell \tag{11}$$

where D_ℓ is the diagonal component of D along $\boldsymbol{v}_d$.

From eq. (11) another expression for D can be obtained by differentiation:

$$D = \langle \delta z(t) \delta v(t) \rangle, \tag{12}$$

where $\delta z(t) = z(t) - \langle z \rangle$ and $\delta v(t) = v(t) - \langle v \rangle$ are the fluctuations of the particle position and velocity with respect to their ensemble averages [16,25]. Equation (12) gives a clear physical insight into the diffusion phenomenon: the spreading of a carrier ensemble comes from the particle space–velocity correlations which arise during the time evolution of the system. Particles with velocity higher than average, having accumulated this positive fluctuation in the near past, are, in general, in the front of the carrier space distribution and, similarly, the slower particles are in the rear of the distribution. Thus particles in the front will tend to run away, and particles in the rear will tend to be left behind, which causes a further spreading of the distribution.

Equation (11) can be used to determine the diffusion coefficient by means of a MC simulation: a number of particles is independently simulated, and their positions are recorded at fixed times. At large enough times the second central moment shows the linearity predicted by eq. (11), and from its slope D is obtained. At short times eq. (11) does not yield the correct steady-state diffusivity because correlations are not fully developed (see below).

Other components of D can be obtained in a similar way with the MC method, since, in general, we get, with analogous calculations:

$$D_{ij} = \frac{1}{2}\frac{\mathrm{d}}{\mathrm{d}t}\langle (x_i - \langle x_i \rangle)(x_j - \langle x_j \rangle) \rangle. \tag{13}$$

The diffusion phenomenon is strictly related to noise. This is so because both of them are essentially due to the stochastic velocity fluctuations $\delta v(t)$ over the drift value $\boldsymbol{v}_d$. The mathematical quantity which describes the common origin of diffusion and noise is the autocorrelation function of velocity fluctuations, which carries the information on how large these fluctuations are and how they decay in time.

If we consider, for simplicity, a one-dimensional situation, the autocorrelation function of velocity fluctuations is defined as

$$C(t) = \langle \delta v(t') \delta v(t'+t) \rangle, \tag{14}$$

where brackets indicate ensemble average, and the mean value in steady-state conditions is independent of t'. $C(t)$ is related to the diffusion D and to the noise spectrum of velocity fluctuations $S_v(\omega)$ through the equations [3,26]

$$D = \int_0^{+\infty} C(t)\, \mathrm{d}t, \tag{15}$$

$$D(\omega) \equiv \int_0^{+\infty} C(t) e^{i\omega t}\, \mathrm{d}t = \tfrac{1}{2} S_v(\omega). \tag{16}$$

Thus, D can be determined, after the evaluation of $C(t)$, by eq. (15).

Thus a determination of the noise spectrum of the velocity fluctuations is equivalent to the determination of a frequency-dependent diffusion coefficient $D(\omega)$, entering into problems in which the time during which the gradient $\partial n/\partial r$ varies appreciably is small compared with the electron relaxation time ($\omega \cdot \tau > 1$). In these situations the diffusion equation in the form (8) does not hold, and it must be generalized by using Fourier analysis [29–32].

The evaluation of the autocorrelation function of δv in steady-state conditions can be easily performed from a MC simulation [7,27].

Figure 7 shows the autocorrelation function of velocity fluctuations for electrons in Si at 77 K with an applied field $E = 10$ kV/cm. The two curves refer to $\boldsymbol{E} \| \langle 100 \rangle$ and $\boldsymbol{E} \| \langle 111 \rangle$. In the first case an intervalley fluctuation [28] is present due to the different partial drift velocities in different valleys of the silicon conduction band.

The negative part of $C(\tau)$ is related to the threshold nature of intervalley scattering which introduces a partial streaming motion, such that positive velocity fluctuations tend to be followed after some time by negative fluctuations and vice versa.

6.2. Transient

As we have seen, the diffusion process of a carrier ensemble comes from the particle space–velocity correlations which arise during the evolution in time of the system.

Starting from an initial condition in which the particle positions and velocities

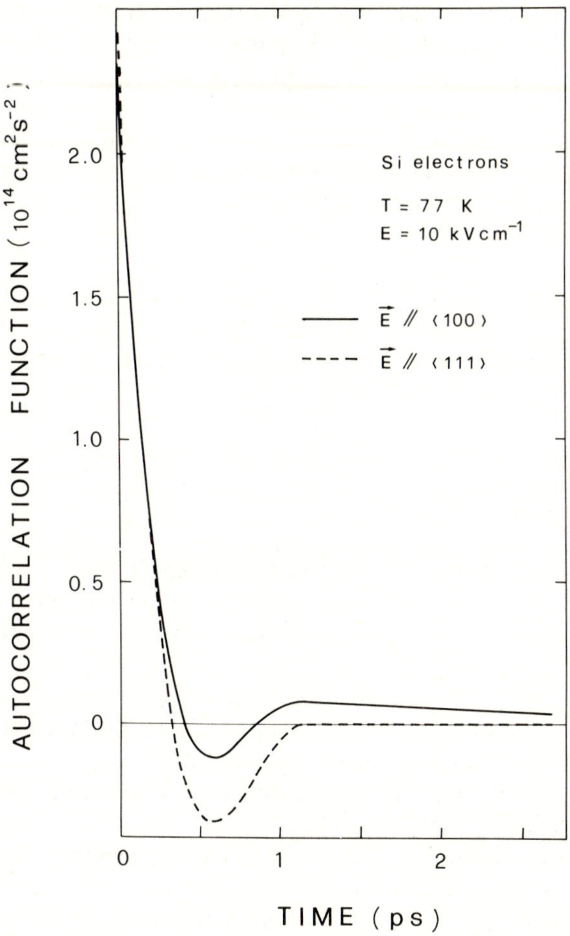

Fig. 7. Velocity autocorrelation function for electrons in Si, as obtained from a MC simulation, at 77 K for $E=10$ kV/cm applied along ⟨100⟩ (continuous line) and ⟨111⟩ (dashed line) directions, respectively.

are totally uncorrelated, the process which occurs during the time necessary for setting up the correlations will be defined as the "correlation transient".

Furthermore, when a high electric field is applied at a certain time to the electron ensemble, the transport process itself has to pass through the transient regime, discussed above, which is necessary for attaining the stationary distribution $f(\boldsymbol{k})$ in $\boldsymbol{k}$-space.

In what follows we shall discuss how these two different transients can be analyzed separately, but simultaneously, and how their effect influences the transient of the diffusivity of the electron ensemble.

The definition of the transient diffusion coefficient has been given by a generalization of eq. (11) to arbitrary small times [33–35]:

$$D(t)=\frac{1}{2}\frac{\mathrm{d}}{\mathrm{d}t}\langle(z(t)-\langle z\rangle_t)^2\rangle, \tag{17}$$

where $z(t)$ is the space position of a carrier at time t along the z direction parallel to $\boldsymbol{v}_{\mathrm{d}}$. If there is no correlation between the initial positions and velocities of the particles eq. (17) leads to [27]:

$$D(t)=\int_0^t \mathrm{d}\tau\, C_t(\tau), \tag{18}$$

with

$$C_t(\tau)\equiv\langle\delta v(t)\delta v(t+\tau)\rangle, \qquad 0\leqslant\tau\leqslant t. \tag{19}$$

Equation (18) reduces to eq. (15) in steady-state conditions ($t\rightarrow\infty$).

By comparison of these two expressions, we see that in the present case:

–the integration interval does not extend to $+\infty$, but toward the past, back only to the initial conditions; this finite integration brings about the effect of the correlation transient;

–the autocorrelation function to be integrated in the transient analysis (eq. (19)) is not time independent; it is given by the specific ensemble average at a particular time and its shape provides information about the transport transient.

The transient autocorrelation function $C_t(\tau)$ can be calculated with a simple many-particle MC procedure [27].

If the field is switched on at a time t_{s} larger than the time $t=0$ of the initial conditions [36] the effect of the transport transient is separated in time from the initial correlation transient of the zero-field diffusion.

Figures 8 and 9 present MC results for the second central moment and for the transient longitudinal diffusion coefficient obtained with eq. (17) as a function of time for electrons in Si and in GaAs.

For the case of Si the following initial conditions have been taken: electrons are randomly situated in one of the six valleys with equal probability, the velocities are chosen according to an equilibrium Maxwellian distribution, and the electrons are all positioned at $r=0$. Figure 8 shows that the second central moment first increases quadratically with time, as predicted by ballistic behavior. At intermediate times (0.2–1.0 ps) an irregular behavior is exhibited by the diffusivity: a tendency to level off, followed by an overshoot ($t\simeq 0.8$ ps). Then, at sufficiently long times the second central moment shows the linear dependence on time, with a slope corresponding to steady-state diffusion. This behavior is due essentially to the combined action of the acceleration impressed by the external field and to the dissipation of energy and momentum associated with intervalley scattering [27].

For the case of GaAs, shown in fig. 9, in the initial conditions all the particles

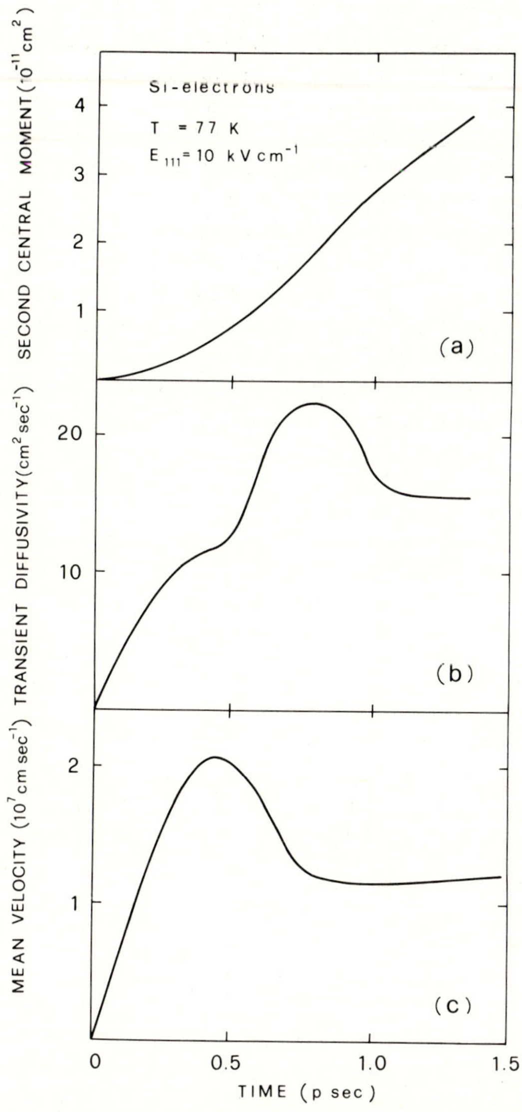

Fig. 8. Second central moment (a) and longitudinal diffusion coefficient (b) as functions of time for electrons in Si at 77 K and $E = 10$ kV/cm along a $\langle 111 \rangle$ direction. The final slope of the curve in (a) corresponds to the steady-state value $D = 15$ cm^2/s shown in (b) at sufficiently long times.

have been situated in the central valley. Figure 9 shows that the second central moment at short times exhibits, as in the previous case, the quadratic ballistic behavior; it achieves a maximum value ($t \simeq 0.5$ ps), then it decreases, attains a minimum ($t \simeq 0.8$ ps), and finally increases monotonically. The asymptotic behavior corresponds to the steady-state value of $D=20\ \mathrm{cm}^2/\mathrm{s}$. In the range of times between 0.6 and 0.8 ps the negative slope exhibited by the second central moment versus time yields a negative transient diffusivity region, with a minimum value of D of about $-20\ \mathrm{cm}^2/\mathrm{s}$ at $t=0.65$ ps [23]. The behavior of the diffusivity shown in fig. 9 can again be described in terms of the energy dependence of the scattering mechanisms; now an essential role is played by intervalley scattering between central and satellite valleys [37].

The results for transient diffusion can be analyzed and understood in terms of the transient autocorrelation function $C_t(\tau)$ [37]. Figure 10 shows the transient autocorrelation function for electrons in Si at $T=77$ K and $\boldsymbol{E}_{100}=10$ kV/cm; the same initial conditions used in fig. 8 have been taken. The curves are interrupted at $\tau=t$ when the correlation with the initial conditions is reached. The area under each curve gives the corresponding value of the transient diffusion coefficient at time t. At very short times this area is very small both because

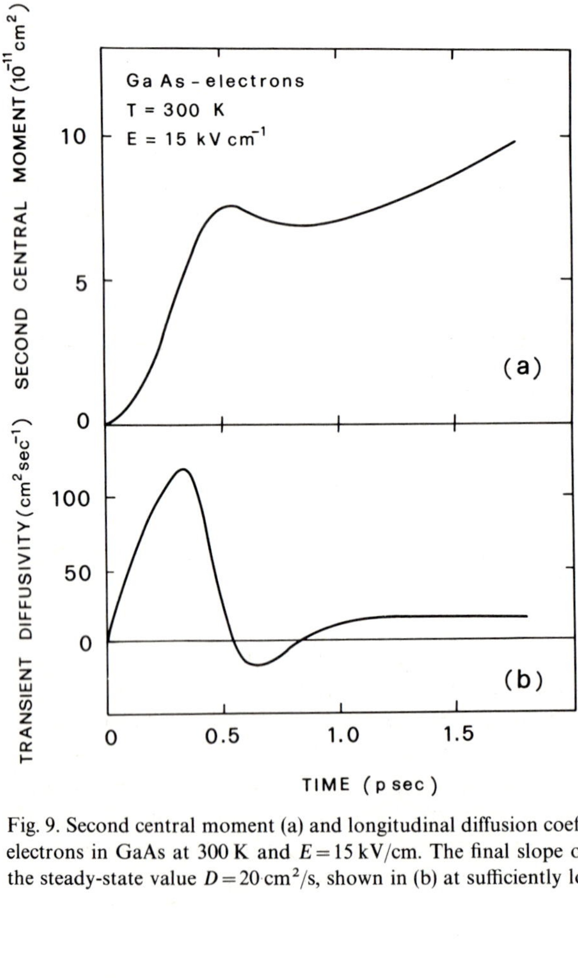

Fig. 9. Second central moment (a) and longitudinal diffusion coefficient (b) as functions of time for electrons in GaAs at 300 K and $E=15$ kV/cm. The final slope of the curve in (a) corresponds to the steady-state value $D=20\ \mathrm{cm}^2/\mathrm{s}$, shown in (b) at sufficiently long times.

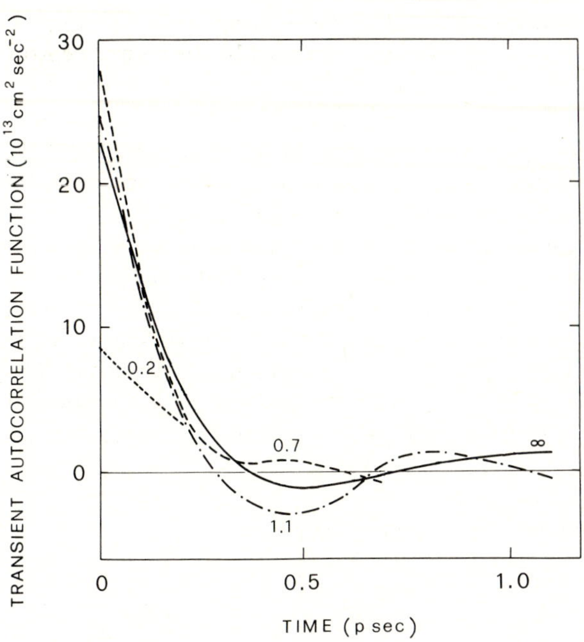

Fig. 10. Transient velocity autocorrelation function as a function of time at fixed times t (reported in ps over each curve in the figure) for electrons in Si at 77 K and $E=10$ kV/cm along a ⟨100⟩ direction. Each curve is interrupted at $\tau=t$ when the correlation with the initial conditions is reached. The stationary autocorrelation function (continuous line) is shown for comparison.

$C_t(\tau)$ starts from low values, and because it is interrupted at short τ. The first effect is a consequence of the transient transport, the electrons being still cold, while the second effect is due to the fact that at small times the correlations are not yet established (correlation transient). As t increases, the area reaches larger values, and it is maximum when the positive part of $C_t(\tau)$ is totally present, and negative correlations are not yet present. This leads to the overshoot of the transient diffusivity. When negative correlations are established, $D(t)$ decreases toward the steady-state value, which is attained when the shape of $C_t(\tau)$ reproduces the steady-state function and the integral of the autocorrelation function is extended to all significant values of τ.

As a further example, which may be of interest for the case of optically excited electrons, we show in fig. 11 the transient transport quantities for the case of electrons in Si at $T=77$ K and $E_{100}=10$ kV/cm, with the initial conditions corresponding to electron energies much larger than both thermal equilibrium and steady-state values. It can be seen that the mean energy decreases monotonically to its steady value; since the electrons are not initially cold, the drift velocity reaches the steady-state value without overshoot and the initial overshoot of the diffusivity is enhanced, with respect to the case corresponding to the

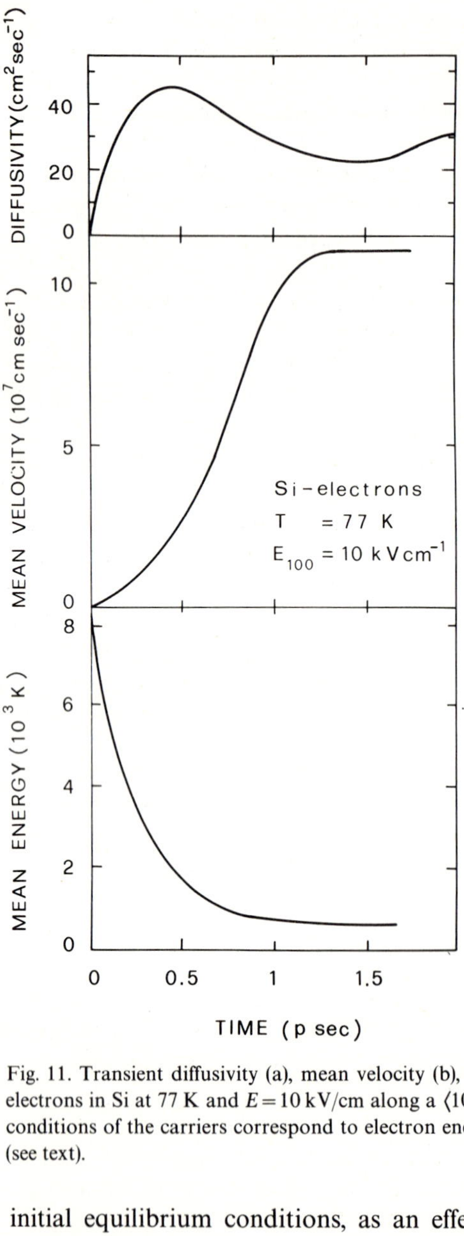

Fig. 11. Transient diffusivity (a), mean velocity (b), and mean energy (c) as functions of time for electrons in Si at 77 K and $E = 10$ kV/cm along a $\langle 100 \rangle$ direction in a situation in which the initial conditions of the carriers correspond to electron energies much larger than the steady-state value (see text).

initial equilibrium conditions, as an effect of the large initial energy of the carriers.

As a final result, we report in fig. 12 an analysis of the transient diffusivity in a case in which the electric field is applied after the onset of the zero-field correlations. Before application of the field, the diffusivity slowly reaches the steady-state value. When the electric field is applied a transport transient occurs with heating of the carriers on a shorter time scale; during this time new correlations are established and the diffusivity reaches the new lower steady-state value passing through a minimum with a region of negative values. This negative region is in great contrast with the overshoot of D seen for the case in fig. 8. This can be explained by considering that the first electrons which undergo intervalley scattering after the field application are the fastest electrons in the direction of the field; during the first part of the process (zero-field correlation transient) the distributions of carriers have fully developed the space–velocity correlations at the basis of the diffusion process, so that a slowing down of the fastest particles due to intervalley scattering produces a shrinking of the space distribution corresponding to the negative D. The comparison between the second part of fig. 8 and fig. 12 may be useful to appreciate the influence of initial conditions in the transient diffusion.

In conclusion, we may summarize the main ideas contained in this review as follows.

(i) Hot-electron transient transport is described by the time-dependent Boltzmann equation without linearization with respect to the external field.

(ii) Analytical solutions to this problem are not available, while many-particle MC techniques yield a numerical solution for any specific physical situation.

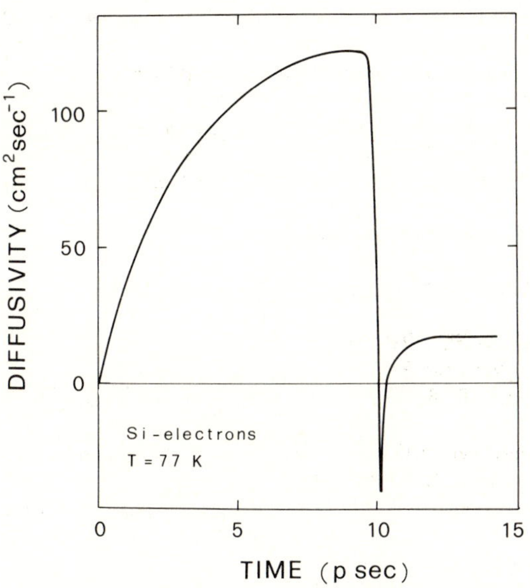

Fig. 12. Transient diffusivity as a function of time for electrons in Si at 77 K, in a physical situation in which a field $E_{111} = 10$ kV/cm is switched on at time $t_s = 9.57$ ps after the initial conditions ($t = 0$) of uncorrelated particles.

(iii) Transient transport depends strongly upon the initial conditions and lasts a time of the order of few picoseconds.

(iv) When the initial electron distribution is cold during the transient drift velocity, diffusivity and mean energy present in general an overshoot before reaching the steady-state value.

(v) Transient diffusivity is influenced by both the transport transient and the correlation transient, and can be analyzed by the time-dependent autocorrelation function of velocity fluctuations.

References

[1] T. Kurosawa, Proc. 8th Int. Conf. Phys. Semic., Kyoto, J. Phys. Soc. Japan, Suppl. 21 (1966) 424.

[2] N.P. Buslenko, D.I. Golenko, Y.A. Shreider and V.G. Sragovich, in: The Monte Carlo Method, ed., Y.A. Shreider (Pergamon, London, 1966).

[3] F. Reif, Fundamentals of Statistical and Thermal Physics (McGraw Hill, New York, 1965).

[4] W. Kohn and J.M. Luttinger, Phys. Rev. 108 (1957) 590.

[5] J.R. Barker, J. Phys. C 6 (1973) 2663.

[6] K. Seeger, Semiconductor Physics (Springer, Vienna, 1973).

[7] C. Jacoboni and L. Reggiani, Rev. Mod. Phys. 55 (1983) 645.

[8] C. Jacoboni and L. Reggiani, Phys. Lett. A 33 (1970) 333.

[9] H. Frolich, Proc. Roy. Soc. A 188 (1947) 521, 532.

[10] R. Stratton, Proc. R. Soc. A 246 (1958) 406.

[11] H. Budd, J. Phys. Soc. Japan, Suppl. 21 (1966) 420.

[12] P.J. Price, Proc. 5th Int. Conf. Phys. Semic., ed., S.M. Ryvkin (Nauka, Leningrad, 1968) p. 273.

[13] H.D. Rees, J. Phys. Chem. Solids 30 (1969) 643.

[14] W. Fawcett, D. A. Boardman and S. Swain, J. Phys. Chem. Solids 31 (1970) 1963.

[15] W. Fawcett and E.G.S. Paige, J. Phys. C 4 (1971) 1801.

[16] P.J. Price, in: Semiconductors and Semimetals, Vol. 14 (Academic Press, New York, 1979) p. 279.

[17] R.W. Hockney and J.W. Eastwood, Computer Simulation using Particles (Pergamon, London, New York, 1982).

[18] P.A. Lebwhol and P.J. Price, Solid State Commun. 9 (1971) 1221.

[19] P.A. Lebwhol and P.J. Price, Appl. Phys. Lett. 19 (1971) 530.

[20] R. Brunetti, C. Jacoboni, L. Reggiani, G. Bosman and R.J.J. Zijlstra, J. Appl. Phys. 52 (1981) 6713.

[21] G. Ruch, IEEE Trans. Electron. Devices ED–19 (1972) 652.

[22] G. Baccarani, C. Jacoboni and A.M. Mazzone, Solid State Elect. 20 (1977) 5.

[23] B. Boittiaux, E. Constant, L. Reggiani, R. Brunetti and C. Jacoboni, Appl. Phys. Lett. 40 (1982) 407.

[24] L. Reggiani, J.C. Vaissiere, J.P. Nougier and D. Gasquet, Proc. 3rd Int. Conf. on Hot Carriers in Semiconductors, J. Physique Coll., Tome 42 C7 (1981) 357.

[25] P.J. Price, in: Fluctuation Phenomena in Solids, ed., R.E. Burgess (Academic Press, New York, 1965).

[26] P.J. Price, J. Appl. Phys. 31 (1960) 124.

[27] C. Jacoboni and R. Brunetti, Phys. Rev. B 15 29 (1983) 5739.

[28] R. Brunetti, C. Jacoboni and L. Reggiani, Proc. 3rd Int. Conf. on Hot Carriers in Semiconductors, J. Physique Coll., Tome 42 C7 (1981) 117.

[29] W.A. Shlup, Phys. Kondens. Materie, 13 (1971) 307.

[30] S.V. Gantsevich, V.L. Gurevich and R. Katilius, Phys. Cond. Matter 18 (1974) 165.
[31] C. Jacoboni, Phys. Status Solidi (b) 65 (1974) 61.
[32] C. Jacoboni, L. Reggiani and R. Brunetti, Proc. 3rd Int. Conf. on Hot Carriers in Semiconductors, J. Physique Coll., Tome 42 C7 (1981) 123.
[33] A. Alberigi-Quaranta, V. Borsari, C. Jacoboni and G. Zanarini, Appl. Phys. Lett. 22 (1973) 103.
[34] D. K. Ferry and J. R. Barker, J. Appl. Phys. 52 (1981) 818.
[35] L. Reggiani, R. Brunetti and C. Jacoboni, Proc. 3rd Int. Conf. on Hot Carriers in Semiconductors, J. Physique Coll., Tome 42 C7 (1981) 111.
[36] E. Constant and B. Boittiaux, Proc. 3rd Int. Conf. on Hot Carriers in Semiconductors, J. Physique Coll., Tome 42, C7 (1981) 73.
[37] R. Brunetti and C. Jacoboni, in: Semiconductors Processes Probed by Ultrafast Laser Spectroscopy, ed., R.R. Alfano (Academic Press, New York, to be published).

Journal of Luminescence 30 (1985) 144–153
North-Holland, Amsterdam

NEW TRANSPORT PHENOMENA IN VARIABLE GAP SEMICONDUCTORS AND THEIR DEVICE APPLICATIONS

F. CAPASSO and B.F. LEVINE

AT&T Bell Laboratories, Murray Hill, New Jersey 07974, USA

This article reviews recent work on electron transport in graded gap semiconductors. Many new applications are also discussed such as staircase devices, heterojunction bipolar transistors and sawtooth superlattices.

1. Electron velocity measurements in variable gap semiconductors

Kroemer first proposed the use of a graded gap p-type layer for the base of a bipolar transistor to reduce the minority carrier (electron) transit time in the base [1]. Figure 1 shows the energy band diagram of such a transistor under equilibrium conditions (no injection). When the base-emitter and base-collector junctions are respectively forward – and reverse-biased, electrons are injected from the emitter into the base and move over to the collector layer.

With no grading in the heavily doped p-base minority carriers (electrons) are transported by diffusion, a relatively slow process. In addition, a fraction

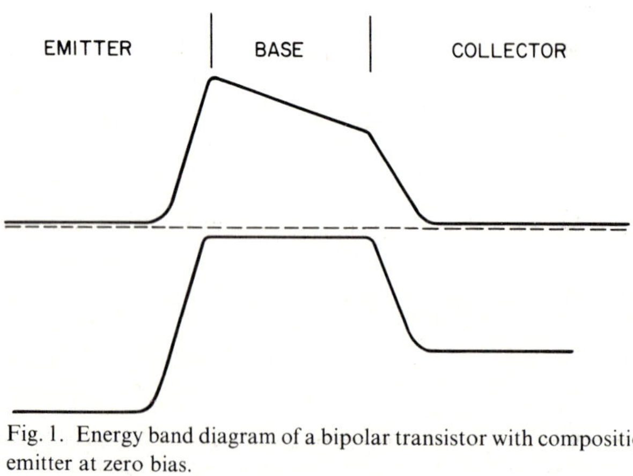

Fig. 1. Energy band diagram of a bipolar transistor with compositionally graded base and wide gap emitter at zero bias.

0022-2313/85/$03.30

of the injected electrons recombine with holes in the base, thus decreasing the base transport factor.

The presence of bandgap grading in the base creates a quasi-electric field acting on the electrons. These therefore move predominantly by drift (if the field is sufficiently strong) leading to a higher velocity in the base. This increases the cutoff frequency or, alternatively can be traded-off against the base resistance [2].

Recently Levine et al. [3,4], using an all optical method measured for the first time the electron velocity in a heavily p^+ doped compositionally graded $Al_xGa_{1-x}As$ layer, similar to the base of the bipolar transistor illustrated in fig. 1.

The energy band diagram of the sample is sketched in fig. 2 along with the principle of the experimental method. The measurement technique is a "pump and probe" scheme. The pump laser beam, transmitted through one of the AlGaAs window layers is absorbed in the first few thousand Å of the graded layer. Optically generated electrons, under the action of the quasi-electric field drift towards the right in fig. 2 and accumulate at the end of the graded layer. This produces a refractive index change at the interface with the second window layer. This refractive index variation produces a reflectivity change that can be probed with the counter propagating probe laser beam. This reflectivity change is measured as a function of the delay between pump and

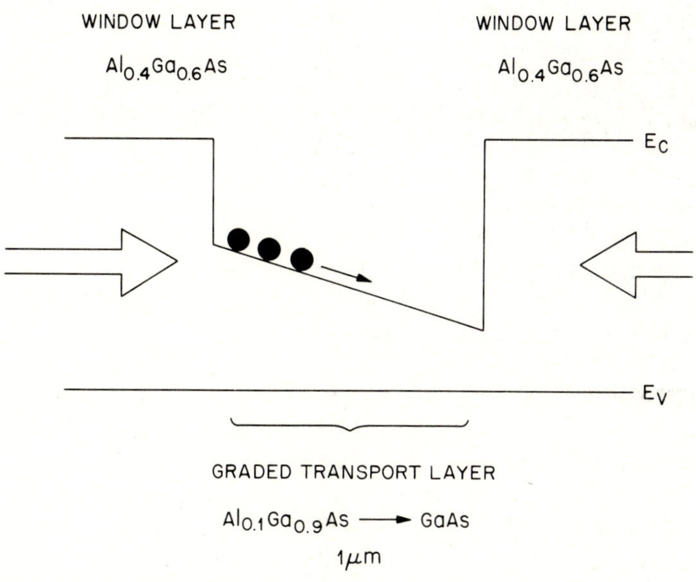

Fig. 2. Band diagram of graded gap layer covered with high gap window layers. The technique used to measure the electron drift velocity is also schematically illustrated.

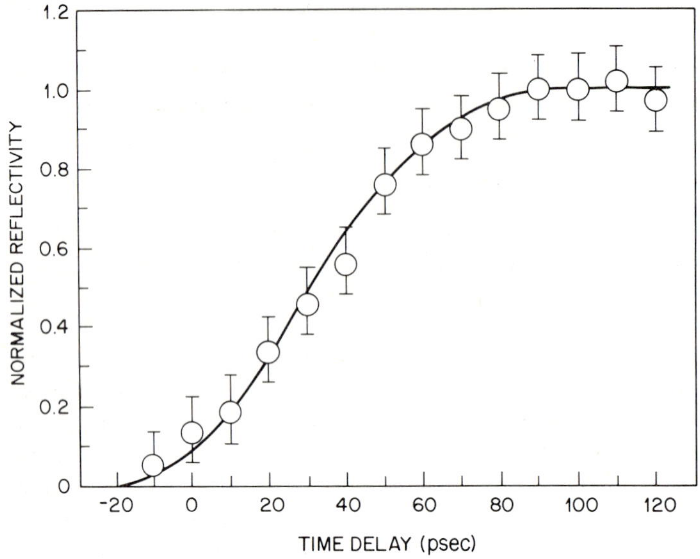

Fig. 3. Normalized experimental results for pump induced reflectivity change vs. time delay obtained in 1 μm thick graded gap p^+ AlGaAs, at a quasi-electric field $F=1.2$ kV/cm.

probe beam using phase sensitive detection techniques. The reflectivity data are shown in fig. 3 for a sample with a 1 μm thick transport layer graded from $Al_{0.8}Ga_{0.9}As$ to GaAs and doped to $p \cong 2\times10^{18}/cm^3$. This corresponds to a quasi-field of 1.2 kV/cm. The laser pulse width was 15 ps and the time 0 in fig. 3, represents the center of the pump pulse as determined by two-photon absorption in a GaP crystal cemented near the sample. The transit time is approximately given by the shift of the half height of the reflectivity curve from zero, which is $\tau=33$ ps. Taking as the drift length the graded layer thickness minus the absorption length of the pump beam ($1/\alpha \cong 2500$ Å), one estimates a minority carrier velocity 2.3×10^6 cm/s.

In these relatively thick samples diffusion effects are important and cause a spread in the arrival time of electrons at the end of the sample, given roughly by the 10–90% rise time of the reflectivity curve, i.e. 63 ps.

It is interesting to note that the drift mobility obtained from the measurement is $\mu_d = v_e/F = 1900$ cm^2 V/s, which is comparable with the usual mobility of 2200 cm^2 V/s at the doping level of the graded layer in GaAs.

Electron velocity measurements were also made in a 0.42 μm thick strongly graded ($F_e=8.8$ kV/cm) highly doped ($p=4\times10^{18}$ cm^{-3}) $Al_xGa_{1-x}As$ layer graded from $Al_{0.3}Ga_{0.7}As$ to GaAs [4]. A transit time of only 1.7 ps was measured, more than an order of magnitude shorter than that for $F=1.2$ kV/cm, as shown in fig. 4, corresponding roughly to a velocity $v \cong 2.5\times10^7$ cm/s. The velocity can be obtained rigorously and accurately (±10% error)

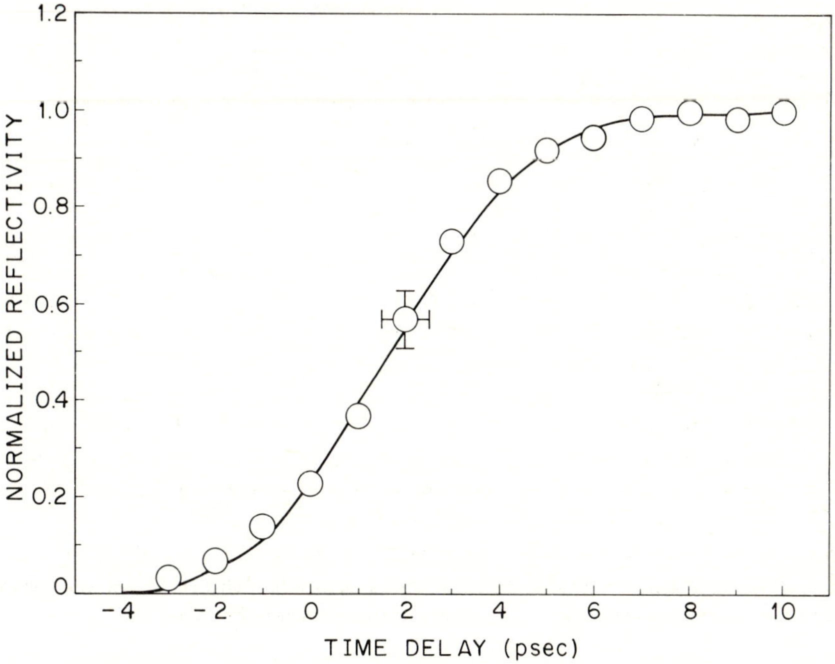

Fig. 4. Normalized experimental reflectivity change vs. time delay measured in a 0.4 μm thick graded gap p^+ AlGaAs layer at a quasi-electric field $F = 8.8$ kV/cm.

from the reflectivity data, by solving the drift diffusion equation and taking into account the effects of the pump absorption length (especially important in the thin sample), the partial penetration of the probe beam in the graded material. Including all these effects one finds that reflectivity data can be fitted using only one adjustable parameter, the electron drift velocity [4]. This velocity is $v_e = 2.8 \times 10^6$ cm/s for $F = 1.2$ kV/cm and $p = 2 \times 10^{18}/\mathrm{cm}^3$, and $v_3 = 1.8 \times 10^7$ cm/s for $F = 8.8$ kV/cm and $p = 4 \times 10^{18}/\mathrm{cm}^3$.

We see that when we increased the quasi-field from 1.2 to 8.8 kV/cm (a factor of 7.3) the velocity increased from 1.8×10^6 cm/s to 1.8×10^7 cm/s (a factor of 6.5). That is, we observed the approximate validity of the relation $v = \mu F$. In fact, using $\mu = 1700$ cm^2/s (for $p = 4 \times 10^{18}$ cm^{-3}) we calculate $v = 1.5 \times 10^7$ cm/s for $F = 8.8$ kV/cm in reasonable agreement with the experiment. It is worth noting that this measured velocity of 1.8×10^7 cm/s (in the quasi-field) is significantly larger than that for pure undoped GaAs where $V = 1.2 \times 10^7$ cm/s for an ordinary electric field of 8.8 kV/cm. In fact the measured high velocity is comparable to the peak velocity reached in GaAs for $F = 3.5$ kV/cm before the intervalley transfer occurs from the central to the L valley. It is noteworthy that our measured velocity is also comparable to the maximum possible phonon limited velocity in the central valley of GaAs. This is given by $V_{max} = [(E_p/m^*)\tanh(E_p/2kT)]^{1/2} = 2.3 \times 10^7$ cm/s, where $E_p = 35$ meV is the optical phonon energy and the effective mass $m^* = 0.67m_0$).

This high velocity can be understood without reference to transient effects since the transit time is much larger than the momentum relaxation time of 0.3 ps. The large velocity results from the fact that the electrons spend most of their time in the high velocity central valley rather than in the low velocity L valley. This may result from the injected electron density being so much less than the hole doping density that the strong hole scattering can rapidly cool the electrons without excessively heating the holes [5]. Furthermore, the electrons remain in the central valley throughout their transit across the graded layer since the total conduction band edge drop ($\Delta E_g = 0.37$ eV) is comparable to the GaAs Γ–L separation ($\Delta E_{\Gamma L} = 0.33$ eV) and therefore they do not have sufficient excess energy for significant transfer to the L valley.

Response time and bandwidth measurements in an AlGaAl/GaAs phototransistor with a graded gap base also independently confirmed that the base transit time is greatly reduced by the quasi-field in the base [6].

More recently three terminal graded-gap transistors have been reported [2]. Cutoff frequencies $f_T = 16$ GHz and current gains $\beta \gtrsim 150$ have been achieved [7,8].

2. Impact ionization in graded gap semiconductors and the staircase solid-state photomultipler

A large difference between the ionization rates for electron (α) and holes (β) (ionization rate ≡ number of secondary pair created per unit length) is an essential requirement for low noise avalanche photodiodes [9]. The necessary asymmetry between α and β can be enhanced using a graded gap semiconductor as illustrated in fig. 5. This represents the energy diagram of such a semiconductor under a very high electric field.

It is seen that electrons experience a higher effective electric field and lower ionization energy than holes, since they are moving towards lower gap regions. The effect is a significant increase in the α/β ratio, when the grading exceeds 1 eV/μm. This enhancement has been demonstrated by Capasso et al. in a p–i–n avalanche photodiode with a graded gap AlGaAs 0.5 μm wide region [10].

By applying a suitable bias to a sawtooth superlattice structure one can achieve the solid-state equivalent of a photomultiplier [11–13] (fig. 6).

Figure 6(a) shows the band diagram of the graded gap multilayer material (assumed intrinsic) at zero applied field. Each stage is linearly graded in composition from a low (E_{g1}) to a high (E_{g2}) bandgap, with an abrupt step back to low bandgap material. The conduction-band discontinuity shown accounts

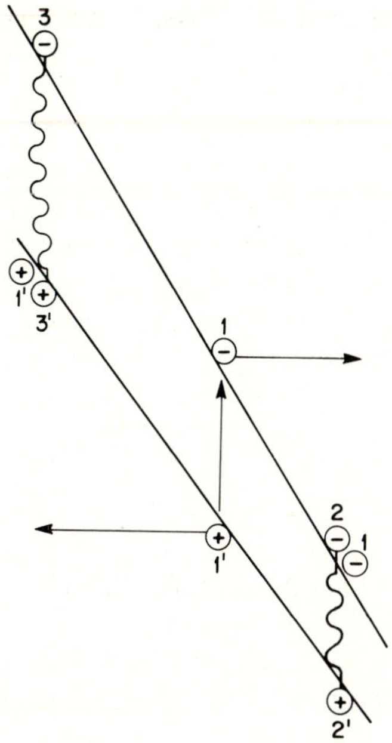

Fig. 5. Schematic of impact ionization in a graded gap semiconductor. The 1–1′ e–h is the initial electron–hole pair. Note the lower ionization energy and higher effective electric field seen by the electron.

for most of the bandgap difference, as is typical of many III–V heterojunctions. The materials are chosen for a conduction-band discontinuity comparable to or greater than the electron ionization energy E_{ie} in the low-gap material following the step. The biased detector is shown in fig. 6(b). Holes do not ionize since $\Delta E_v \simeq 0$ and the field is too low to cause ionization.

Consider a photoelectron generated near the p^+ contact; the electron does not impact ionize in the graded region before the conduction-band step because the effective field is too low. At the step, however, the electron ionizes since $\Delta E_c > E_{ie}$ and the process is repeated at every stage. The steps correspond to the dynodes of a phototube.

Note the low bias voltage V of this structure; for 5 stages with $\Delta E_c = 1$ eV, the gain is $\simeq 2^5 = 32$ and V slightly greater than 5 V. The reason that in the staircase APD the avalanche noise is lower than in the best conventional APD (one in which only one type of carrier can ionize) can be understood as follows. In a conventional APD the avalanche is more random because carriers can ionize everywhere in the avalanche region, while in the staircase APD electrons ionize at well-defined positions in space (i.e., the multiplication process is more deterministic). Note that similarly, in a photomultiplier tube, the avalanche is essentially noise free.

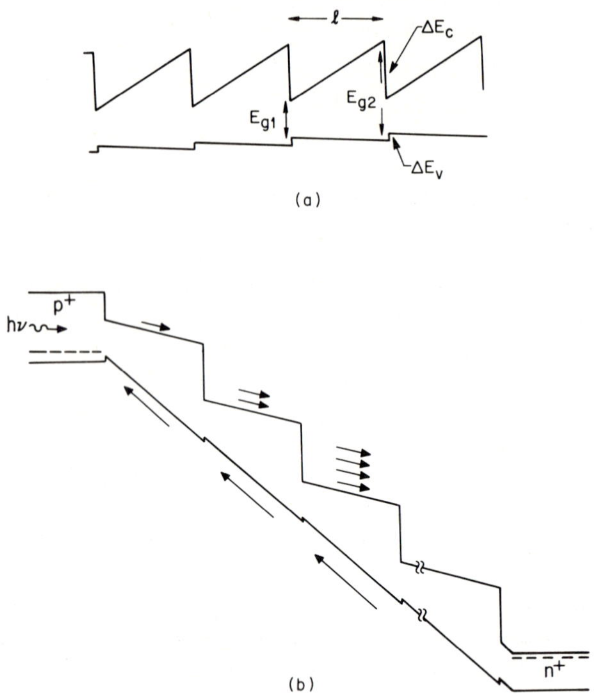

Fig. 6. Energy band diagram of (a) unbiased-graded multilayer region and (b) complete staircase detector under bias. The arrows in the valence bands simply indicate that holes do not impact ionize.

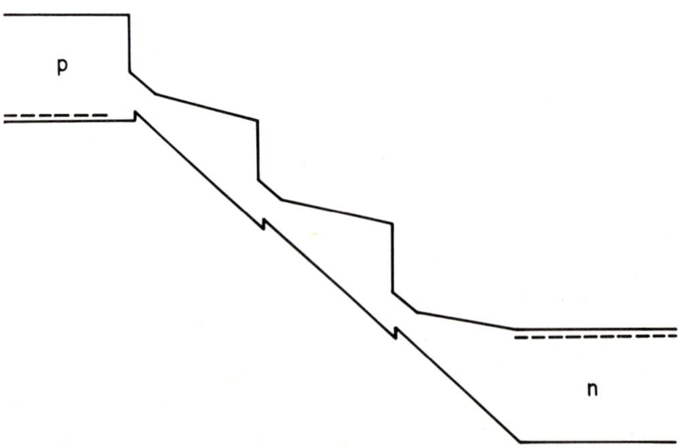

Fig. 7. Band diagram of a modified staircase APD. The finite ionization mean-free-path above the ionization threshold is taken into account by introducing ungraded regions after the steps.

A possible material system for the implementation of this device in the 1.3–1.6 μm region is AlGaAsSb/GaSb as discussed in ref. [12]. In practical cases a thin ungraded layer (100–200 Å) should be left after the conduction-band step to account for the finite ionization mean-free-path ($\lambda_{ii} \approx 50$ Å) (fig. 7).

3. Repeated velocity overshoot devices and electrical polarization effect in sawtooth superlattices

Other interesting applications of staircase potentials have been proposed. We shall discuss here the repeated velocity overshoot device. This structure offers the potential for achieving average drift velocities well in excess of the maximum steady-state velocity over distances greater than 1 μm [14].

Figure 8(a) shows a general type of staircase potential structure. The corresponding electric field, shown in fig. 8(b), consists of a series of high field regions of value E_1 and width d superimposed upon a background field E_0. The background field E_0 is chosen so that the steady-state electron energy distribution is not excessively broadened beyond its thermal equilibrium value, but at the same time the average drift velocity is still relatively high. Electrons, upon entering the high field regions, gain rapidly energy and momentum, so that the drift velocity overshoots the steady-state value. The energy and momentum are then allowed to relax in the subsequent low field region. With the staircase structure repeated velocity overshoot can be achieved (fig. 8(c)). A graded gap AlGaAs structure with steps $\Delta w = 0.2$ eV could be used for this purpose (fig. 9).

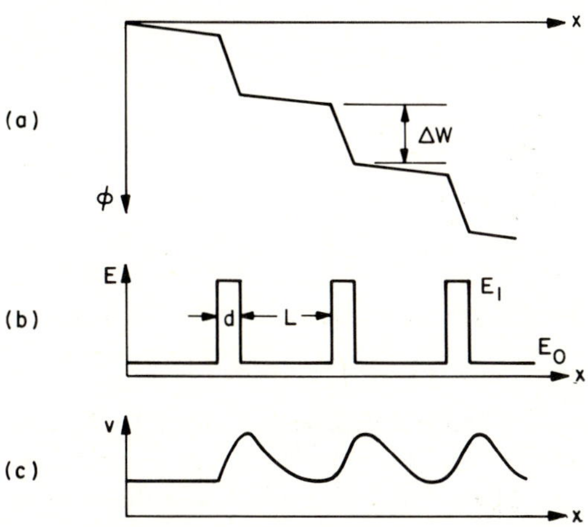

Fig. 8. Principle of repeated velocity overshoot. Stairstep potential configuration (a) and the corresponding electric field (b). The ensemble velocity as a function of position is illustrated schematically in (c).

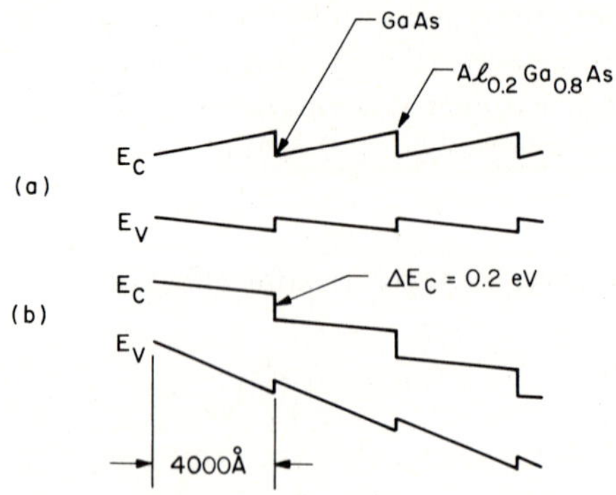

Fig. 9. Band diagram of graded gap repeated velocity overshoot device.

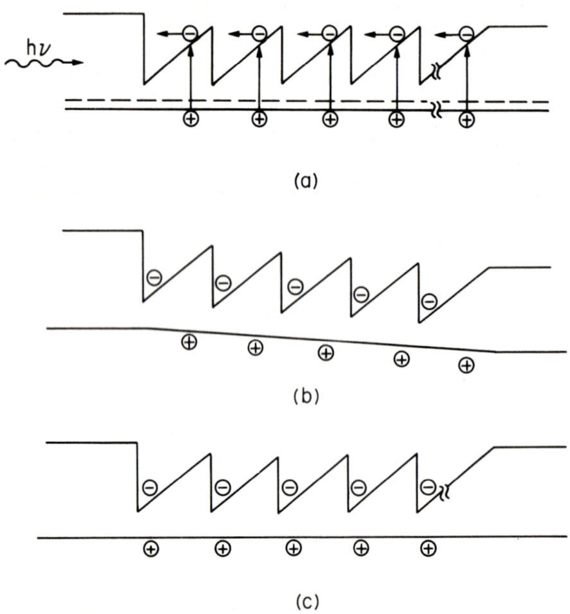

Fig. 10. Illustration of the formation and decay of the macroscopic electrical polarization in the sawtooth superlattice structure.

Sawtooth superlattices have also other intriguing physical properties such as the possibility of generating a transient macroscopic electrical polarization extending over many periods of the superlattice. This effect is a direct consequence of the lack of reflection symmetry [15]. The energy band diagram of a sawtooth p-type superlattice is sketched in fig. 10(a). The layer thicknesses are typically a few hundred angstroms and a suitable material is graded gap

$Al_xGa_{1-x}As$. The superlattice is sandwiched between two highly doped p^+ contact regions. Let us assume that electron–hole pairs are excited by a very short pulse as shown in fig. 10(a). Electrons experience a high quasi-electric field (typically $\simeq 10^5$ V/cm) due to the grading whereas the total force acting on holes is smaller because of the valence band edge lineup in p-type materials. Therefore electrons separate from holes and reach the low-gap side in a sub-picosecond time ($<10^{-13}$ s). This sets up an electrical polarization in the sawtooth structure which results in the appearance of a voltage across the device terminals (fig. 10(b)). This macroscopic dipole moment and its associated voltage subsequently decay in time by a combination of (a) dielectric relaxation and (b) hole drift under the action of the internal electric field produced by the separation of electrons and holes (Fig. 10(c)). This polarization phenomenon has recently been observed in AlGaAs sawtooth superlattices. The transient photovoltage showed a decay time of ≈ 150 ps [15].

In conclusion the new devices and phenomena discussed in this article indicate that with appropriate combinations of heterojunctions, superlattice structures and graded gap materials one can modify continuously and in a nearly arbitrary fashion the energy band diagram of a semiconductor (bandgap engineering) [8,16]. This permits one to modify to an unprecedented degree the transport properties and to tailor them to a given device application. Many new exciting applications are expected in the near future from this new approach to materials and device design.

References

[1] H. Kroemer, RCA Rev. 18 (1957) 332.
[2] J.R. Hayes, F. Capasso, A.C. Gossard, R.J. Malik and W. Wiegmann. Electron. Lett. 13 (1983) 410.
[3] B.F. Levine, W.T. Tsang, C.G. Bethea and F. Capasso, Appl. Phys. Lett. 41 (1982) 470.
[4] B.F. Levine, C.G. Bethea, W.T. Tsang, F. Capasso, K.K. Thornber, R.C. Fulton and D.A. Kleinman, Appl. Phys. Lett. 42 (1983) 769.
[5] J. Shah, R.F. Leheny, R.E. Nahory, J. Degani and A.E. DiGiovanni Appl. Phys. Lett. 39 (1981) 569.
[6] F. Capasso, W.T. Tsang, C.G. Bethea, A.L. Hutchinson and B.F. Levine, Appl. Phys. Lett. 42 (1983) 93.
[7] D.L. Miller, P.M. Asbeck, R.J. Anderson and F.H. Eisen, Electron. Lett. 19 (1983) 367.
[8] F. Capasso, Surf. Sci. 142 (1984) 513.
[9] R.J. McIntyre, IEEE Trans. Electron Dev. ED-13 (1966) 164.
[10] F. Capasso, W.T. Tsang, A.L. Hutchinson and P.W. Foy. Inst. Phys. Conf. Ser. No. 67 (1982) 473.
[11] F. Capasso and W.T. Tsang, Proc. Int. Electron Devices Meeting, San Francisco 1982 (IEEE, New York, 1982) p. 334.
[12] F. Capasso, W.T. Tsang and G.F. Williams, IEEE Trans. Electron Devices, ED-30 (1983) 381.
[13] F. Capasso, W.T. Tsang and G.F. Williams, Proc. SPIE, 340 (1982) 50.
[14] J.A. Cooper, F. Capasso and K.K. Thornber IEEE Electron. Dev. Lett. EDLI-3 (1982) 497.
[15] F. Capasso, S. Luryi, W.T. Tsang, C.G. Bethea and B.F. Levine, Phys. Rev. Lett. 51 (1983) 2318.
[16] F. Capasso, J. Vac. Sci. Technol. B1 (1983) 457.

Journal of Luminescence 30 (1985) 154–161
North-Holland, Amsterdam

EXCITONIC VERSUS PLASMA SCREENING IN HIGHLY EXCITED GALLIUM ARSENIDE

G.W. FEHRENBACH, W. SCHÄFER and R.G. ULBRICH

Institut für Physik der Universität Dortmund, 46 Dortmund 50, FRG

We report band edge absorption spectra of GaAs under conditions of high electron–hole pair excitation in the range of 10^{13}–10^{17} cm^{-3} achieved with picosecond light pulses tuned (i) to the 1s exciton energy and (ii) to the continuum ($\hbar\omega > E_g$), where critical densities for bleaching of the exciton absorption lines are much lower. A substantial blue shift of the 2s absorption line relative to the 1s exciton was found, the latter remained constant in energy up to complete bleaching.

1. Introduction

The transient absorption spectrum near the band edge of highly excited semiconductors depends on the electron–hole pair density [1,2], the distribution function of the two-particle excitations, and the degree of correlation between the single particles, i.e. in lowest order the exciton effect. As indicated in fig. 1 and as discussed earlier we have to distinguish between two basically different situations: the photon energy $\hbar\omega$ of the excite pulse equals the 1s exciton energy, or $\hbar\omega$ is larger than E_g [3,4]. In the former case the existence and polarizability of the excitons themselves causes screening effects of essentially dielectric nature. In the latter case a plasma-type screening occurs due to the abundance of populated continuum states. This paper reports the extension of our earlier work to the $n=2$ discrete exciton state.

2. Experiment

In order to prepare a nonequilibrium e–h pair excitation in the direct gap semiconductor GaAs we have used ultrathin samples and picosecond light pulses to avoid the two basic problems:

(i) spatial inhomogeneity of the excitation;

(ii) indirect determination of the e–h pair density.

The GaAs samples of controlled thickness $t=0.5$–2 μm were prepared by

0022-2313/85/$03.30

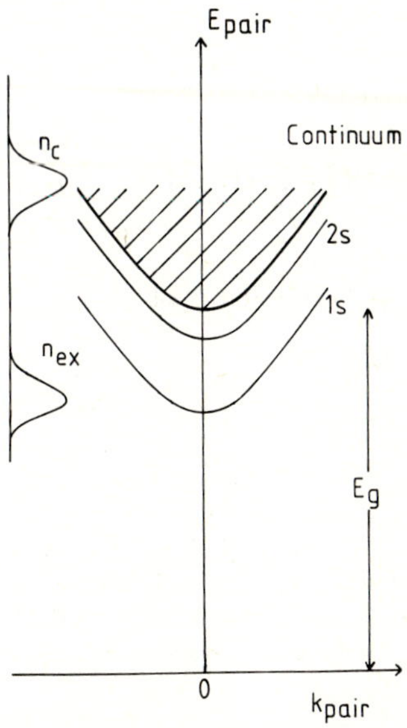

Fig. 1. Two-particle excitations in GaAs. Energetic positions of the excite pulses are shown for resonant 1s exciton (n_{ex}) and continuum excitation (n_c).

standard lapping, etching, and ion milling techniques from high purity liquid-phase-epitaxy material. Sample thicknesses in the range of 5 μm were measured directly with a microscope. The thickness of thinner samples was then determined from their continuum absorption ($0.85–1.0 \times 10^4\ cm^{-1}$). We have been able to prepare 0.5 μm thick GaAs crystals showing a well resolved 1s exciton resonance (see fig. 3). The discrete 2s exciton resonance could be observed only in samples thicker than 2 μm (see fig. 4). The samples were held freely immersed in liquid He II at 1.2 K.

The experiments were performed in the excite-and-probe configuration shown in fig. 2. Using two independently tunable, synchronously mode-locked dye lasers we are able to excite and probe at different energies and to compare

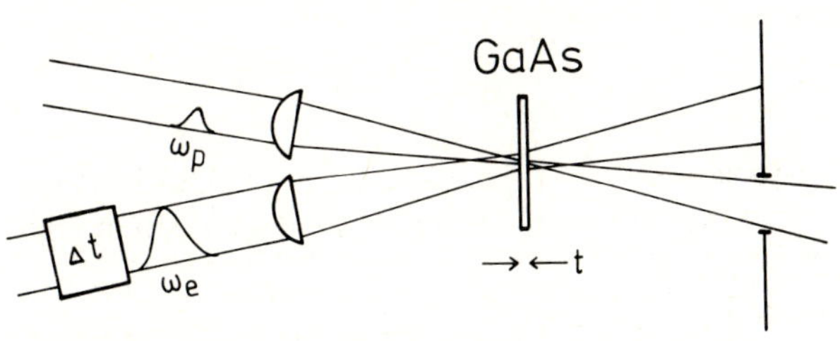

Fig. 2. Experimental setup: excite and probe configuration with two independently tunable synchronously mode-locked dye lasers with 10 ps pulse length and 1 meV spectral width.

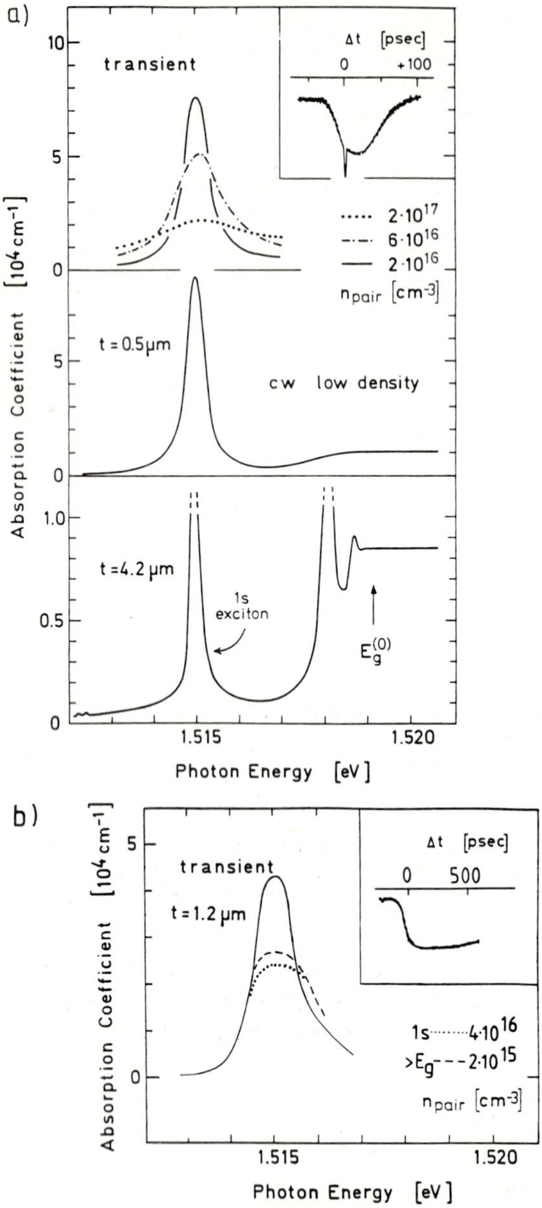

Fig. 3. The cw low density and transient high density optical absorption spectra for: (a) resonant excitation with light pulses centered on the 1s exciton line; (b) resonant 1s exciton vs. continuum excitation. Inset: temporal behavior of the probe beam absorption for fixed photon energy 1.515 eV after deposition of $n_{ex} = 6 \times 10^{16}$ cm^{-3} by the excite pulse. The lower and middle trace of fig. 3(a) show the cw low density absorption spectrum of an optically thick, respectively, ultrathin sample for comparison.

directly the transient change of the absorption under 1s exciton or continuum excitation conditions. The spectral width of the 10 ps light pulses varied from 1 to 1.5 meV. Cross-correlation measurements gave 5–10 ps relative jitter between the two pulse trains.

The excite and the probe beams were focussed to a 30, respectively, 20 μm spot size at the sample surface. A thighter focus of the probe beam was chosen in order to keep the effects of the transverse inhomogeneity due to the Gaussian beam profiles small. By changing the probe beam position on the sample we also found that lateral diffusion is negligible on a 20 μm length scale, even at our highest densities.

In general the probe pulses were delayed by 15 to 20 ps to ensure that the probe pulses reached the sample after complete deposition of the excite pulse energy. Since all measured recovery times were longer than 30 ps we could safely determine the actual pair densities in the sample from the deposited excite-pulse energies. The average power of the excite-pulse train was measured with a calibrated power meter. The pulse shape was monitored with background-free second-harmonic autocorrelation. The intensity level between the light pulses was checked with a fast d.c. photodiode (risetime 100 ps) and a single photon timing setup (prompt response 600 ps). These measurements gave a precise determination of the energy of a single excite pulse. The fraction of the excite-pulse energy that was absorbed in the sample was then determined by careful measurements of the absolute spectra of the incident, the transmitted, and the reflected excite pulse with a 0.85 m double-grating spectrometer (resolution 0.1 meV). Probe pulse generated densities were always below 10^{11} cm^{-3}.

Figure 3(a) (lower trace) shows for reference the cw low-density absorption spectrum of an optically thick sample with the exciton Rydberg series up to $n=3$. The corresponding spectrum of the ultrathin sample (fig. 3(a), middle trace) shows a well resolved 1s exciton resonance. The top traces of fig. 3(a) are

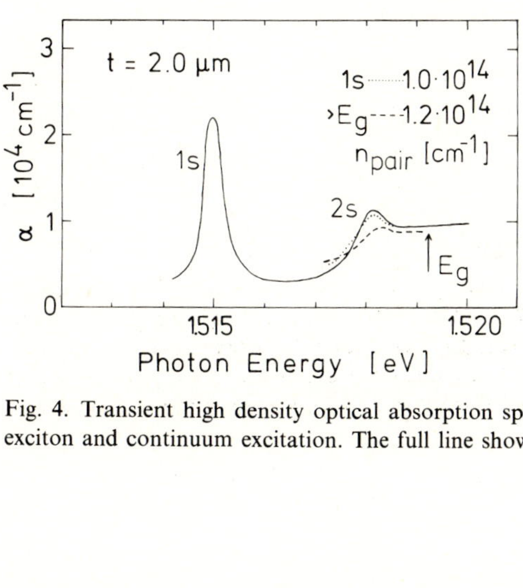

Fig. 4. Transient high density optical absorption spectra of the 2s exciton line for resonant 1s exciton and continuum excitation. The full line shows the cw low density absorption spectrum.

typical transient absorption spectra of the ultrathin sample for three different exciton densities, taken at $\Delta t = 8$ ps. In fig. 3(b) we have plotted for direct comparison the transient absorption of a slightly thicker sample after resonant 1s exciton, respectively, continuum pumping (10 meV above the band gap). In both cases the 1s exciton line remains stable in energy up to complete bleaching. For continuum pumping we observe the same qualitiative features, but at more than 10 times lower densities: the onset of additional broadening occurs at exciton densities of 2×10^{16} cm^{-3} and at continuum pair densities of 10^{15} cm^{-3}. In figs. 5 and 6 we have plotted the measured energetic positions (circles) and excess halfwidths (vertical bars) of the 1s exciton line.

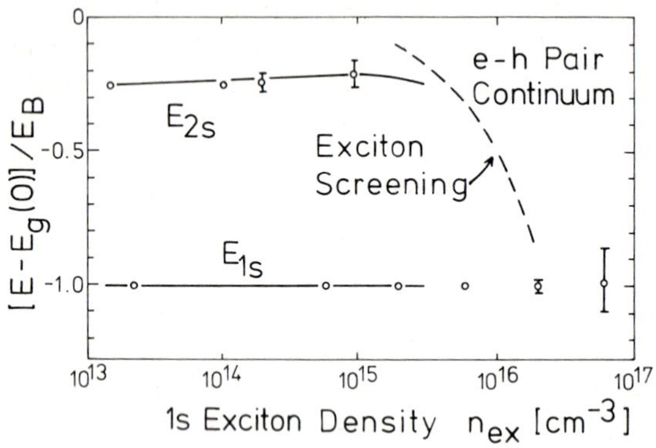

Fig. 5. Experimentally determined peak positions (circles) and additional broadening (vertical bars) of the 1s and 2s exciton absorption structure for different exciton densities. The full lines give the calculated two lowest exciton energies, the broken line the theoretical continuum edge $E_g(n_{ex})$ as a function of exciton densities.

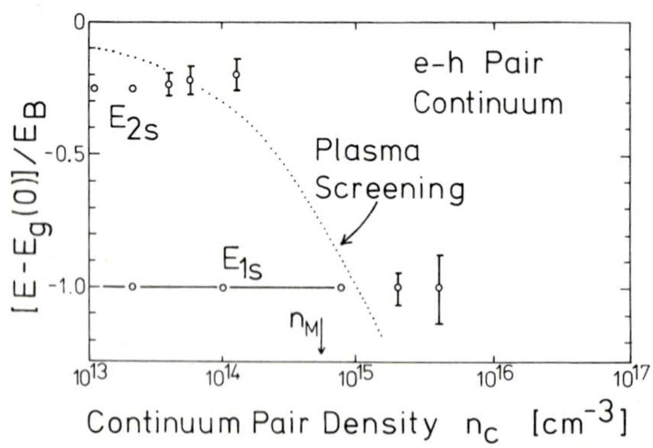

Fig. 6. Same as fig. 5 for different continuum pair densities. The full line indicates the calculated 1s exciton energy and the dotted line shows the theoretical dependence of the continuum band edge $E_g(n_c)$ for the plasma screening mechanism taken from ref. [5]. The arrow marks the "classical" Mott density n_M.

The bleaching effect has an extremly fast recovery time in thin ($t<1$ μm) samples: with 8 ps excitation light pulses we observed the restoration of the unperturbed absorption spectrum within 50 ps, as shown in the inset of fig. 3(a). In a thicker sample ($t=1.2$ μm) of the same crystal quality we found an appreciably longer decay time of 500 ps for the bleaching effect (see inset of fig. 3(b)). In even thicker samples ($t=4$ μm) the recovery time of ~ns is the same as the radiative lifetime of the corresponding bulk crystal.

These findings and additional quantum efficiency measurements lead us to the conclusion that the bleaching decay time constant in thin samples is mainly determined by surface recombination: the free surface acts as an efficient nonradiative recombination channel. The interplay between sample thickness t, and the bulk exciton diffusion length, which is in the order of 1 μm, determines whether capture into shallow impurities with subsequent radiative (or nonradiative) bulk recombination annihilates the excitons or whether the surface is the dominant "sink" for the extraction of energy. As soon as diffusion and surface recombination are the dominant recombination channel the recovery time becomes proportional to t^2. In this simple model a diffusion constant $D=10\ cm^2/s$ results in a recovery time of 50 ps for the ultrathin sample ($t=0.5$ μm) and 300 ps for a 1.2 μm thick sample.

In fig. 4 we present the transient probe absorption spectrum with the probe pulse tuned to the 2s exciton resonance, taken at $\Delta t=10$ ps. Since we have not succeeded in preparing 1 μm thick samples showing a well resolved 2s exciton line we have to take into account that in a 2 μm thick sample the excitation density is inhomogeneous, especially in the case of resonant 1s exciton excitation where the peak absorption value in this sample was $2\times10^4\ cm^{-1}$. The three key features observed are as follows:

(i) the onset of additional broadening for resonant 1s exciton excitation occurs at $n_{ex}=2\times10^{14}\ cm^{-3}$;

(ii) the onset of additional broadening for continuum excitation occurs at $n_c=3\times10^{13}\ cm^{-3}$, a density one order of magnitude lower compared with (i);

(iii) the 2s exciton line surprisingly shows a substantial blue shift of 0.2 of its binding energy for both pumping conditions.

In figs. 5 and 6 we have also plotted the measured energetic positions (circles) and excess halfwidths (vertical bars) of the 2s exciton line for resonant pumping of the 1s exciton and for continuum excitation. The onset of broadening of both, the 1s and the 2s exciton absorption line is in good agreement with the theoretical dependence of the continuum band edge for plasma screening (dotted curve), taken from ref. [5] and for exciton screening (broken curve), taken from ref. [3]. There is also an excellent agreement of the experimentally observed absorption peak energies with the theoretical calculations (full lines).

The energy scheme shown in fig. 7 compares the 1s and 2s exciton state for different excitation levels. The effective potential, changed due to exciton–

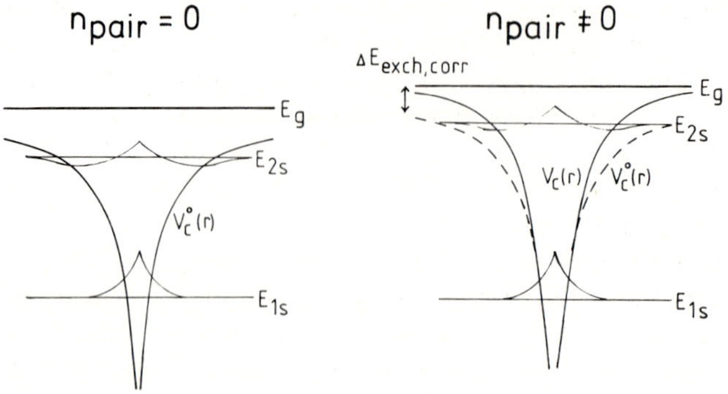

Fig. 7. Energy scheme showing the 1s, 2s exciton, and the band gap energy E_g. The e–h Coulomb potential $V_c^0(r)$ of the unexcited crystal is changed to $V_c(r)$ in the presence of excited e–h pairs due to screening and exchange. The experimentally observed blue shift of the 2s exciton indicates a blue shift of the energy gap E_g.

exciton interaction, is also shown qualitatively as a function of the relative electron–hole distance. The blue shift of the 2s exciton relative to the stable 1s exciton is to our knowledge the first experimental indication of a blue shift of the energy gap, defined by

$$E_g(n_p) = \lim_{n\to\infty} E_{ns}(n_p),$$

$$n_p := n_{ex}, \quad \text{resp.} \quad n_p := n_c,$$

under high excitation conditions in the 10^{15} cm^{-3} pair density range.

3. Theory

In order to explain the experimental findings, we have previously [3,4] calculated to lowest order in n_{ex} the critical density (corresponding to the Mott density) in either case. In the present, more detailed calculation for the case of 1s excitation, three essential contributions enter the effective excitonic Hamiltonian in addition to the usual hydrogenic part:

(i) a dynamic self-energy contribution due to the additional excitonic screening;

(ii) an induced electron–hole interaction potential, which tends to decrease the excitonic binding energy;

(iii) exchange type contributions to the self-energy (including the screened interaction) which are due to the fermionic character of the two-particle excitations.

In contrast to the treatment in [3,4] a mean gap approximation is used for the excitonic screening function and its dynamical character is treated in the Shindo approximation. Incorporation of the three contributions up to infinite order in the density of excited pairs leads to a nonlinear integral equation [6], which is transformed to a matrix equation using a Gaussian quadrature. This matrix equation is solved by iteration.

The results of these calculations up to densities $n_{ex} = 4 \times 10^{15}\,\mathrm{cm}^{-3}$ (calculations for higher densities are still in progress) are shown as full lines in fig. 5. There is a complete cancellation for the energy of the 1s exciton, whereas the 2s exciton is shifted to higher energies. This is in accordance with experiment. The dependence of the continuum edge (broken line) is taken from the lowest order calculation of ref. [3]. It should be modified through exchange type contributions to the self-energy, which presumably shifts the critical densities to still higher values. Up to now there is no quantitative explanation for the blue shift of the 2s exciton in the presence of continuum pairs. An adequate treatment of this case has to take e–h correlation into account. This will, in contrast to plasma screening, stabilize the binding of the pairs.

4. Conclusion

Band edge absorption spectra of optically excited GaAs show empirically (i) a basic difference in critical density (respectively, screening behavior) for resonant 1s excitation compared with continuum excitation, (ii) a constant 1s exciton energy and a novel blue shift of the 2s level in both cases. For the excitonic screening case, the theoretical treatment including e–h correlation and exchange describes correctly the observed 1s and 2s exciton energies.

References

[1] E.O. Göbel and G. Mahler, Festkörperprobleme: Advances in Solid State Physics (Vieweg, Braunschweig, 1979) Vol. XIX, p. 109.

[2] C. Klingshirn and H. Haug, Phys. Rep. 70 (1981) 315.

[3] G.W. Fehrenbach, W. Schäfer, J. Treusch and R.G. Ulbrich, Phys. Rev. Lett. 49 (1982) 1281.

[4] G.W. Fehrenbach, W. Schäfer, J. Treusch and R.G. Ulbrich, Physica 117B and 118B (1983) 303.

[5] R. Zimmermann, K. Kilimann, W.D. Kraeft, D. Kremp and G. Röpke, Phys. Status Solidi (b) 90 (1978) 175.

[6] W. Schäfer, Z. Phys., to be published.

Journal of Luminescence 30 (1985) 162–170
North-Holland, Amsterdam

PHOTOLUMINESCENCE OF GERMANIUM NEAR THE SCREENING IONIZATION LIMIT OF EXCITONS

I. BALSLEV

*Fysisk Institut, Odense Universitet, D-5230 Odense M, Denmark**

and

Max-Planck-Institut für Festkörperforschung, Heisenbergstrasse 1, 7000 Stuttgart 80, FRG

Photoluminescence spectra from highly excited germanium at 25–30 K are presented. The density of free electrons and holes is determined from induced free carrier absorption and achieved values up to 10^{17} cm^{-3}. At such densities the electron–hole system is pumped to well above the screening ionization limit of excitons. The band-to-band luminescence, however, is strongly Coulomb enhanced. The spectral shape of the measured luminescence fits well with calculations based on a Yukawa interaction potential.

1. Introduction

The electron dynamics in highly photoexcited semiconductors have been studied in great detail in the seventies and eighties [1–4]. Among the most spectacular effects in this field is the formation of the electron–hole liquid [4,5]. In this respect the best studied electron–hole system is that of photoexcited germanium having a very long life time of excitons and free carriers. However, even in this material there is some uncertainty about electron dynamics at temperatures above 5 K. Below 5 K there is a well-established phase diagram involving coexistence of excitons and metallic droplets, but in the range 5–20 K there are experimental [6] and theoretical [7] indications of a very complex phase diagram. The complexity is due to competition between the gas–liquid and the Mott-like transition.

The experimental results to be discussed here are optical spectra of indirect transitions in photoexcited germanium above 25 K. In this case there is no phase separation of any kind, and the Mott-like transition of the electron–hole system is described as a screening ionization of excitons in a homogeneous

* Present address.

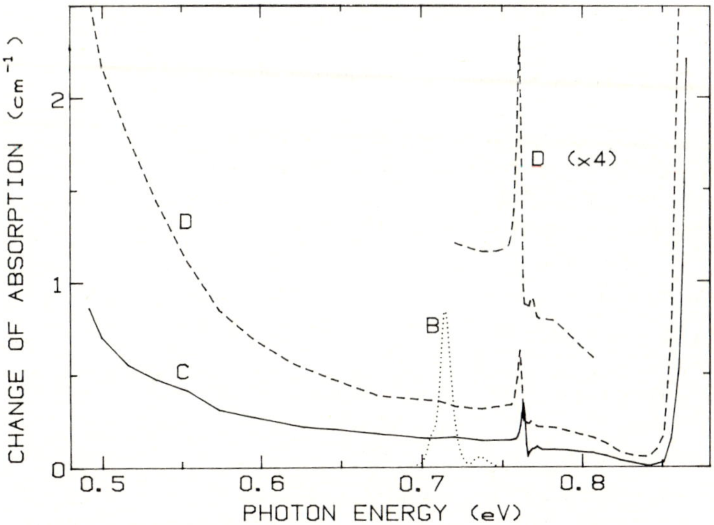

Fig. 1. Experimental spectra recorded for highly excited germanium. Curves C and D are spectra of induced absorption at two different excitation levels at 45 K and curve B is the luminescence spectrum at 30 K.

phase. In the experiments samples of pure Ge is illuminated by a Nd:YAG laser. Sample heating is reduced by a 10:1 chopper. The induced absorption as well as the luminescence is detected simultaneously as discussed in detail in ref. [8]. Typical spectra obtained with a 0.6 mm thick sample are shown in fig. 1.

For photon energies $\hbar\omega$ above 0.85 eV the shift of the direct absorption edge is seen, while the free carrier absorption is detected for $\hbar\omega < 0.75$ eV. The latter contribution provides accurate information on the density n_f of free pairs. The structure of the induced absorption between 0.75 and 0.85 eV is discussed in detail in ref. [8].

In the present article the attention will be focussed on the luminescence, which is observed in the range 0.7–0.75 eV (see fig. 1). This luminescence is due to phonon assisted recombination of excitons and free pairs. The excitation condition (temperature and excitation level) reported here is similar to that reported by Schweizer et al. [9], but the interpretation in the present work is different.

In section 2 the expected line shape of the indirect luminescence will be derived from a simple model involving free carrier screening of the electron–hole interaction. In section 3 it is demonstrated that the spectra calculated for systems well above the screening ionization limit fit well with the observed luminescence spectra. The implications of the interpretation and the remaining problems on the electron–hole system in Ge are discussed in the concluding section 4.

2. Theory of phonon assisted luminescence

Let us describe the electron–hole interaction in a photo-excited semiconductor by a Yukawa potential

$$V(r) = -(2E_{x0}a_B/r)\exp(-k_s r), \tag{1}$$

where r is the electron–hole separation and k_s is the screening constant. Without screening the excitons are characterized by the Rydberg energy E_{x0} and the Bohr radius a_B. In a static Debye–Hückel limit and with equal electron and hole masses, k_s is given by

$$k_s^2 = 16\pi n_f a_B E_{x0}/kT, \tag{2}$$

where kT is the thermal energy and n_f is the common density of free electrons and holes. The Mott criterion $k_s a_B = 1.2$ [10] defines the Mott density n_M

$$n_M = \frac{(1.2)^2 kT}{16\pi E_{x0}} a_B^{-3}. \tag{3}$$

When discussing direct and indirect interband transitions it is convenient to introduce a function $D(E_r, k_s)$ which is proportional to the dipole transition probability of a specific pair state. The pair considered is characterized by the energy E_r of the relative electron–hole motion. Thus $E_r > 0$ for the pair continuum, and $E_r = -E_{x0}$ for $n=1$ excitons. In case of direct transitions across a band gap E_g^0, $D(\hbar\omega - E_g^0, k_s)$ is proportional to the imaginary part of the interband susceptibility. With $k_s = 0$, Elliott's result [11] can be expressed as

$$D(E_r, 0) = \sum \frac{1}{n^3}\,\delta\left(\frac{E_r}{E_{x0}} + \frac{1}{n^2}\right) + \frac{\theta(E_r)/2}{1 - \exp(-2\pi\sqrt{E_{x0}/E_r})}, \tag{4}$$

where θ is the unit step function. Note that $D(E_r, k_s)$ is normalized so that $D(0,0) = 1/2$. In case of infinite screening

$$D(E_r, \infty) = \frac{1}{\pi}\,\theta(E_r)\sqrt{E_r/E_{x0}}\,. \tag{5}$$

The function $D(E_r, k_s)$ has been studied by Dow [12] who replaces the Yukawa potential with the slightly different Hültén potential. A simple alternative calculation of $D(E_r, k_s)$ is based on Stahl's polariton equations [13] as described in appendix I. Typical spectra with a life time broadening of $\Gamma = 0.1\ E_{x0}$ are shown in fig. 2. These results are similar to those obtained from many-body calculations [14]. Note that $D(E_r, k_s)$ is close to a suitably broadened step function

$$D(E_r, k_s) \approx \theta(E_r)/2 \tag{6}$$

in a wide range above screening ionization limit ($1.2 < k_s a_B \lesssim 2$).

Turning to allowed phonon assisted transitions the spontaneous lumine-

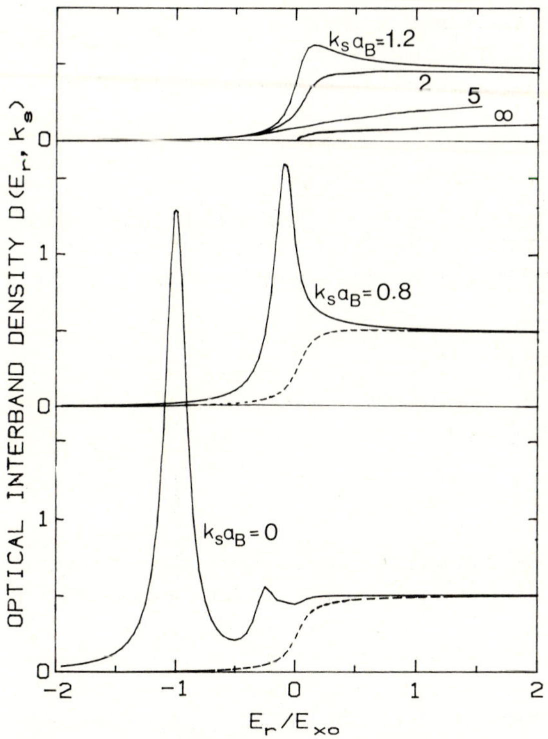

Fig. 2. Calculated optical interband density $D(E_r, k_s)$ for various values of $k_s a_B$. All curves are calculated as explained in appendix I ($\Gamma = 0.1\ E_{x0}$) except the curve for $k_s \to \infty$. This has $\Gamma = 0$ (cf. eq. (5)).

scence from an electron–hole system is proportional to [15]

$$I(\omega) \propto \int_0^\infty D(E_r, k_s)\sqrt{E_t} f(E_t, E_r)\delta(\hbar\omega - E_g - E_r - E_f + E_p)\,dE_t, \tag{7}$$

where E_g is the renormalized indirect gap, E_p is the emitted phonon energy, E_t and $f(E_t, E_r)$ is the occupation probability of a pair state with E_r and E_t as relative and translational energy, respectively. For fully screened excitons ($k_s a_B > 1.2$) and Maxwell–Boltzmann statistics

$$f(E_t, E_r) \propto n_f \exp(-(E_r + E_t)/kT). \tag{8}$$

Assuming further the restricted range $1.2 < k_s < 2$ in which eq. (6) is valid, one obtains

$$I(\omega) \infty \Delta E^{3/2} \theta(\Delta E) \exp(-\Delta E/kT), \tag{9}$$

with

$$\Delta E = \hbar\omega - E_g + E_p. \tag{10}$$

Note that the power 3/2 in eq. (9) should be replaced by 2 in the limit $k_s \to \infty$ and by 1/2 in case of a pure exciton gas ($k_s = 0$, $kT \ll E_{x0}$).

3. Line shape fitting

Typical spectra of luminescence of Ge at ~ 30 K are shown in fig. 3. Here spectrum A is recorded as described in section 1. For this spectrum the density

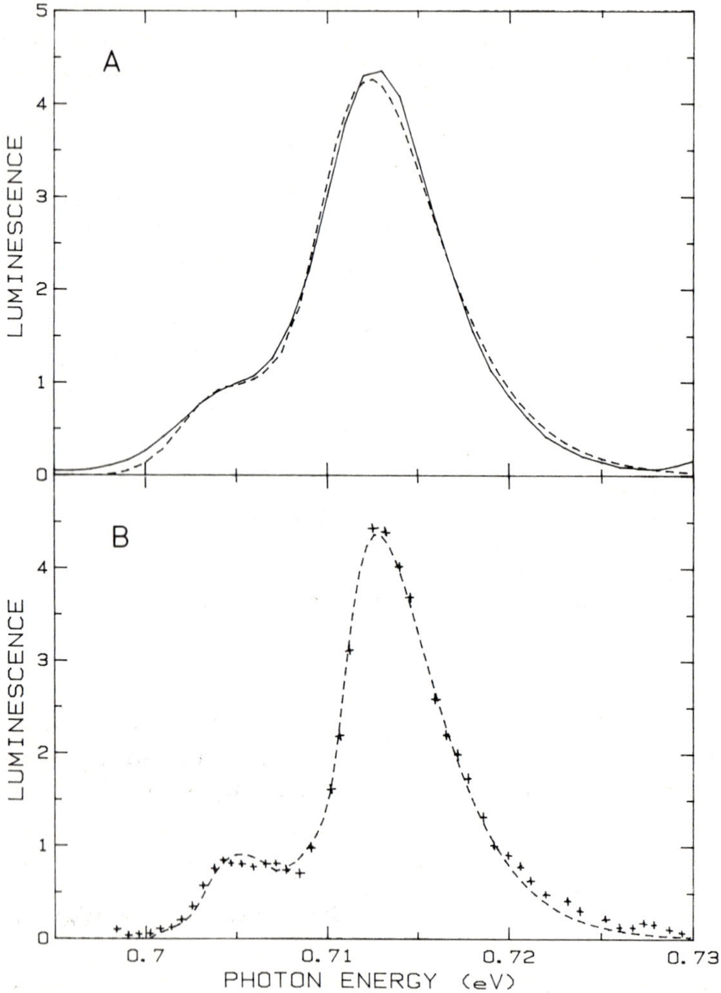

Fig. 3. Luminescence spectra of highly excited germanium. Full curve in the upper figure is the presently reported experiment, while the points in the lower figure are experimental results of ref. [9]. Dotted curves are calculated results discussed in the text. Data for the spectra are given in table 1.

Table 1
Results of fitting components based on eq. (9) to luminescence spectra of this work and of ref. [9].

	Sample temperature (K)	Free pair density (cm^{-3})	Fit values T (K)	ΔE_g (meV)
Fig. 3A (present work)	27	1.1×10^{17}	29	-7.9
Fig. 3B (ref. [9])	~ 30	3×10^{16}	23	-6.1

of free pairs is determined from free carrier absorption, and the sample temperature is determined from the zero excitation energy gap (derived from transmission modulation spectra 90° out of phase with the laser excitation [8]). The spectrum B, fig. 3, is taken from ref. [9]. The values of n_f and T for the spectra in fig. 3 are given in table 1.

The line shape observed is well described as a superposition of two phonon components ($E_p = 27.3$ meV and 35 meV for LA and TO phonons respectively) each representing band-to-band luminescence from a system without excitons. The calculated curves in fig. 3 are based on eqs. (9) and (10) with a Lorentzian broadening of 0.5 meV included. In the fits the sample temperature and the band renormalization $\Delta E_g(n_f) = E_g(n_f) - E_g(0)$ as primary fitting parameters. Accurate values for $E_g(0) - E_{x0}$ at the relevant temperatures are taken from ref. [15] and E_{x0} is taken to be 4.0 meV [16]. The results of the fits are given in table 1.

It is seen that the fits are satisfactory with reasonable fit values for the temperature. The results for ΔE_g can be used for estimating the Mott density. For this we need a functional dependence of ΔE_g on n/n_M. In the range $n < n_M$ the theoretical results reported by Haug and Schmitt-Rink [17] and by Zimmermann et al. [18] is well represented by

$$\Delta E_g \approx -E_{x0}(n/n_M)^{1/2}. \tag{11}$$

Extending this dependence up to $\sim 3n_M$, the experimental Mott density derived from table 1 is about 2×10^{16} cm^{-3} at 25 K. This agrees well with the Debye–Hückel value (1.2×10^{16} cm^{-3} for $T = 25$ K, $a_B = 11$ nm and $E_{x0} = 4.0$ meV [16]).

The interpretation given here is significantly different from that of ref. [9], in which the spectra are decomposed into two electronic contribution one from free pairs without Coulomb enhancement (power 2 in eq. (9)) and one from excitons (power $\frac{1}{2}$ in eq. (9)).

4. Discussion and conclusion

This work provides a consistent description of photoexcited germanium at 25–30 K in the pair density range 2×10^{16}–10^{17} cm^{-3}. The electron–hole system is above the screening ionization limit of excitons, but the band-to-band transitions has considerable Coulomb enhancement. The system just below the Mott density has spectra very similar to that above, but is difficult to treat theoretically. Here, the excitons have a density $n_x = 0$. Rough estimates of the ionization degree $n_f/(n_x + n_f)$ yield values of the order 0.7 just below the Mott density at 25 K. Very little is known about the properties of excitons which are almost ionized by screening. Such excitons have very large radii and could contribute to band renormalization, screening, and intraband absorption almost as free pairs. If this is the case all observable consequences of the screening ionization transition are all continuous. The excitonic oscillator strength goes gradually to zero (see fig. 2) and the many-body effects caused by large orbit excitons are more or less identical to those of pair states in the continuum.

Appendix I

As mentioned in section 2 the expression $D(\hbar\omega - E_g, k_s)$ is proportional to the imaginary part of the susceptibility χ associated with direct transitions. In the present work χ is calculated by numerical integration of Stahl's polariton equations for a simple pair of band extrema [13]

$$\left[E_g - \hbar\omega - \frac{\hbar^2}{2\mu}\nabla_r^2 - \frac{\hbar^2}{2m}\nabla_R^2 + V(\bar{r})\right] Y(\bar{r}, \bar{R}) = M(\bar{r})E(R), \tag{A1}$$

$$(\varepsilon\omega^2 + c^2\nabla_R^2)E(\bar{R}) = -\frac{2\omega^2}{\varepsilon_0}\int M(\bar{r})Y(\bar{r}, \bar{R})\mathrm{d}^3r = \chi\omega^2 E(R), \tag{A2}$$

$Y(\bar{r}, \bar{R})$ is the coherent electron–hole amplitude, $\bar{R}$, $\bar{r}$ the center-of-mass and relative coordinate, respectively, $E(\bar{R})$ the electric field, ε the background dielectric constant, m, μ the total and reduced effective mass, respectively, E_g the energy gap, ω the polariton frequency, $V(r)$ the electron–hole interaction, and $M(r)$ the transition dipole density. Neglecting spatial dispersion ($m \to \infty$) it is possible to separate variables. Then

$$\left[E_g - \hbar\omega - \frac{\hbar^2}{2\mu}\nabla_r^2 + V(\bar{r})\right] Y(\bar{r}, \bar{R}) = \frac{2M(\bar{r})\int M(\bar{r})Y(\bar{r}, \bar{R})\mathrm{d}^3r}{\varepsilon_0\chi}. \tag{A3}$$

We insert for $V(\bar{r})$ the Yokawa potential (eq. (2)) and use for $M(\bar{r})$ the shell structure as in ref. [19]:

$$M(\bar{r}) = M_0\delta(\rho - \rho_0)/4\pi\rho_0^2, \tag{A4}$$

$$\rho = |\bar{r}|/a_B, \tag{A5}$$

where $\rho_0 \ll 1$. With the substitution

$$Y(\bar{r}, \bar{R}) = e^{ikR}\frac{1}{\rho}w(\rho) \tag{A6}$$

and a damping Γ one obtains

$$\frac{E_g-\hbar(\omega-i\Gamma)}{E_{x0}}w(\rho)-w''(\rho)-\frac{2}{\rho}\exp(-k_s a_B\rho)w(\rho)=\frac{2M_0^2\delta(\rho-\rho_0)w(\rho_0)}{4\pi\rho_0^2\chi E_{x0}\varepsilon_0}. \quad \text{(A7)}$$

Here it is used that $E_{x0}=\hbar^2/2\mu a_B^2$. As ρ_0 is small $w(\rho)$ is essentially proportional to ρ for $0<\rho<\rho_0$. Hence

$$\frac{\partial w(\rho_0+)}{\partial\rho}-\frac{\partial w(\rho_0-)}{\partial\rho}\simeq\frac{\partial w(\rho_0+)}{\partial\rho}-\frac{w(\rho_0)}{\rho_0}=-\frac{2M_0^2w(\rho_0)}{4\pi\rho_0^2\chi E_{x0}\varepsilon_0}, \quad \text{(A8)}$$

or approximately

$$\chi=\frac{-w(\rho_0)}{\partial w(\rho_0+)/\partial\rho}\,\frac{2M_0^2}{4\pi\rho_0^2E_{x0}\varepsilon_0}. \quad \text{(A9)}$$

There is no source for $w(\rho)$ at infinity, so with $\Gamma>0$ it is possible to generate meaningful solutions for $w(\rho)$ by applying the boundary condition

$$\frac{\partial w(\rho_{max})}{\partial\rho\cdot w(\rho_{max})}=-\kappa,$$

$$\kappa=(\hbar(\omega_g-\omega+i\Gamma)/E_{x0})^{1/2},$$

$$\rho_{max}=(\mathrm{Re}(\kappa))^{-1}>0.$$

With this condition it is very simple to integrate numerically the equation

$$\left[\frac{E_g-\hbar(\omega-i\Gamma)}{E_{x0}}-\frac{2}{\rho}\exp(-k_s a_B\rho)\right]w(\rho)-w''(\rho)=0 \quad \text{(A11)}$$

from ρ_{max} to ρ_0. The logarithmic derivative of $w(\rho)$ at $\rho=\rho_0$ is then inversely proportional to χ (see eq. (A9)).

Identifying $\hbar\omega-E_g$ as the energy E_r of the relative motion it is seen that

$$D(E_r,k_s)=C\,\mathrm{Im}\left(\frac{w(\rho_0)}{\partial w(\rho_0+)/\partial\rho}\right)\Bigg|_{\hbar\omega=E_g+E_r}, \quad \text{(A12)}$$

where the proportionality constant C can be determined from the fact that $D(0,0)=1/2$ (see eq. (4)).

For $\Gamma=0.1\ E_{x0}$ the spectra of $D(E_r, k_s)$ calculated in this way are reasonably independent of the spatial parameters ρ_0 and ρ_{max} in a wide range about the values $\rho_0=0.1$ and $\rho_{max}=50$. A suitable step size in the numerical integration is $\Delta\rho=0.05$.

References

[1] C.D. Jeffries, Science 189 (1975) 955.

[2] Physics of Highly Excited States in Solids, Lecture Notes in Physics 57 (Springer, Berlin, 1976).

[3] C. Klingshirn and H. Haug, Phys. Rep. 70 (1981) 315.
[4] L.V. Keldysh, in: Proc. 9th Int. Conf. Physics of Semiconductors, Moscow, 1968, eds., S.M. Ryvkin and Yu. V. Shmartsev (Nauka, Leningrad, 1968) p. 1303.
[5] J.C. Hensel, T.G. Phillips and G.A. Thomas, in: Solid State Physics, eds., H. Ehrenreich, F. Seitz and D. Turnbull (Academic Press, New York, 1977) Vol. 32, p. 88.
[6] L.J. Schowalter, F.M. Sterenka, M.B. Salamon and J.P. Wolfe, Phys. Rev. B 29 (1984) 2970.
[7] I. Balslev, Phys. Status Solidi (b) 101 (1980) 749.
[8] I. Balslev to be published in Physical Review.
[9] H. Schweitzer, A. Forchel and W. Klingenstein, Phys. Status Solidi. (b) 102 (1980) 343.
[10] N.F. Mott, Metal–Insulator Transitions (Taylor and Francis, Ltd., London, 1974) p. 124.
[11] R.J. Elliott, Phys. Rev. 108 (1957) 1384.
[12] J.D. Dow, in: Proc. 12th Int. Conf. Physics of Semiconductors, Stuttgart, ed., M.H. Pilkuhn (Teubner, Stuttgart, 1974) p. 957.
[13] A. Stahl, Phys. Status Solidi (b) 106 (1981) 575.
[14] H. Schweizer, A. Forchel, A. Hangleiter, S. Schmitt-Rink, J.P. Löwenau and H. Haug, Phys. Rev. Lett. 51 (1983) 698.
[15] G.G. Macfarlane, T.P. McLean, J.E. Quarrington and V. Roberts, Phys. Rev. 108 (1957) 1377;
T.P. McLean, Progress in Semiconductors, ed., A.F. Gibson (Heywood and Co., Ltd., London, 1960) Vol. 5, p. 53.
[16] M. Altarelli and N.O. Lipari, Phys. Rev. Lett. 36 (1976) 619.
[17] H. Haug and S. Schmitt-Rink, to be published in Progress in Quantum Electronics.
[18] R. Zimmermann, K. Kilimann, W.D. Kraeft, D. Kremp and G. Röpke, Phys. Status Solidi (b) 90 (1978) 175.
[19] A. Stahl and I. Balslev, Phys. Status Solidi (b) 113 (1983) 583.

Journal of Luminescence 30 (1985) 171–187
North-Holland, Amsterdam

NONEQUILIBRIUM MANY-BODY THEORY OF OPTICAL NONLINEARITIES OF SEMICONDUCTORS

H. HAUG*

Department of Physics, Indiana University, Bloomington, Ind. 47405, USA

Laser excited semiconductors are in a state far from equilibrium. Close to intrinsic resonances they show large optical nonlinearities due to many-body effects. Nonequilibrium Green's functions are best suited to describe highly excited semiconductors, because they allow a consistent determination of the eigenmodes and the occupation numbers of the excitations involved.

1. Introduction

The traditional spectroscopy of semiconductors studies the linear optical response in the frequency region close to the band-gap, where the absorption sets in. These spectra are determined by the interaction of a single electron–hole pair with phonons and impurities. With the development of lasers, the excitation became so strong that the interaction between the large number of excited electron–hole pairs are much more important than the interaction of a single pair with the vibrating lattice and the lattice imperfections. This became strikingly manifest when at low temperatures in several indirect-gap semiconductors a condensation of the excitation into a plasma-liquid was observed.

The theoretical explanation of the properties of an electron–hole plasma-liquid could make use of the highly developed electron gas theory with only minor modifications. The attractive electron–hole Coulomb interaction turned out to be not of great importance in this phase. Furthermore, due to the long lifetimes of the electron–hole pairs in indirect-gap semiconductors, non-equilibrium effects played a minor role.

The situation changes if one considers direct-gap semiconductor materials [1]. Due to the short lifetimes of the excitations the laser intensities have to be higher in order to come into the region which is dominated by many-body effects. As a consequence, one has to use pulse excitations to avoid overheating

* Permanent address: Institut für Theoretische Physik, Universität Frankfurt, FRG.

and damage. Thus nonequilibrium effects become more pronounced. Temporal and spatial variations of the excitation density have in general to be taken into account in order to understand the optical properties of such systems [2]. At intermediate excitation levels the attractive Coulomb potential between electrons and holes is still strong enough to form bound states, i.e., excitons and biexcitons. In this region many-exciton effects are important.

A very challenging task for the theory is the description of the region in which a transition from an exciton gas into a plasma occurs. This problem is solved only partly. Several groups showed how this transition can be described in a $T=0$ pair formalism. The Frankfurt group [3] has put forward a solution which is valid at high temperatures, where the screening of the Coulomb forces is due to free carriers. The theory describes the experimental findings within the limitations of the model [4]. The study of this intermediate regime is not at all only a high-brow academic problem. It is obvious, that large absorptive and dispersive changes are linked with the vanishing of the sharp exciton resonance. For illustration, we include the calculated spectra [3] of GaAs for various electron–hole pair densities. Figure 1 shows how the screening

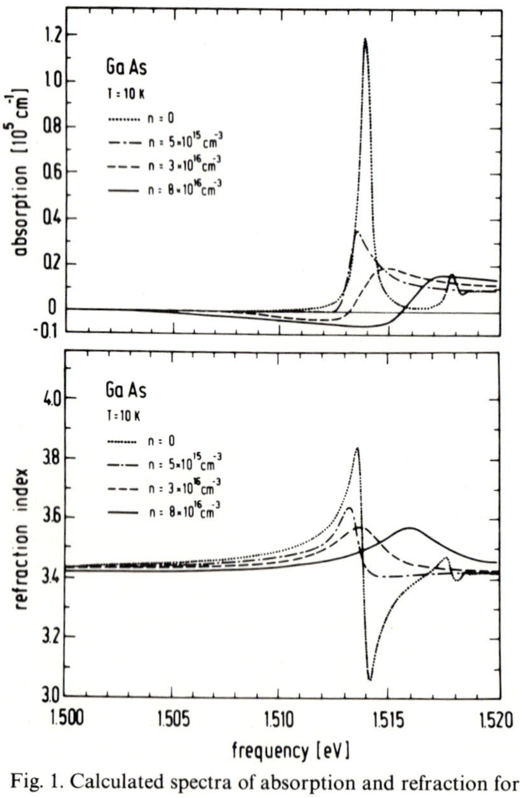

Fig. 1. Calculated spectra of absorption and refraction for GaAs [3].

reduces the oscillator strength of the exciton and at the same time reduces the effective band-gap. The changes in the optical spectra under excitation can be used to build bistable optical devices, which are needed for direct optical data processing. Impressive progress in the development of technologically interesting semiconductor devices has been made in the last few years. I refer to the contributions of S.D. Smith [5] and D.S. Chemla [6]. Another example for resonance-enhanced nonlinearities is the two-photon resonance of the biexciton. In this case, the resonance increases with increasing laser intensity, causing again enough dispersive changes to generate optical bistability.

Highly excited semiconductors are and have to be treated as nonequilibrium systems. In the present contribution, I will advertise the use of nonequilibrium Green's functions which have been introduced by Keldysh [7]. The formalism of Keldysh allows a consistent calculation of the single-particle energies and of the distribution function for carriers, even under pulse excitation, where nonthermal distributions are generated. A further advantage is that the Keldysh formalism provides us with strict diagrammatic rules for the calculation of higher-order many-body contributions under nonequilibrium conditions. The formalism has been used recently by Schäfer and Treusch [8] as well as by Iwanow and Keldysh [9] to study the properties of highly excited exciton systems.

2. Nonequilibrium Green's functions

The response of a system to external fields is described by retarded Green's functions (for simplicity I assume local homogeneity, but the formalism allows also the study of transport problems in locally inhomogeneous situations [10, 11])

$$G^{\mathrm{r}}(k, t, t') = -\mathrm{i}\theta(t-t')\langle[a_k(t); a_k^\dagger(t')]_\pm\rangle. \tag{1}$$

The signs $\pm$ of the commutator refer to Fermions and Bosons, respectively. If the system is in thermal equilibrium, temperature Green's functions are very handy to calculate retarded functions. If the system is not in equilibrium, one Green's function is not sufficient to describe the system. Instead one has to introduce four Green's functions which form a two by two matrix $G^{ij}(k, t, t')$. The matrix elements are:

$$G^{++} = -\mathrm{i}\langle Ta_k(t)a_k^\dagger(t')\rangle = G, \qquad G^{+-} = \pm\mathrm{i}\langle a_k^\dagger(t')a_k(t)\rangle = G^<,$$
$$G^{-+} = \mathrm{i}\langle a_k(t)a_k^\dagger(t')\rangle = G^>, \qquad G^{--} = -\mathrm{i}\langle \tilde{T}a_k(t)a_k^\dagger(t')\rangle = \tilde{G}. \tag{2}$$

Here, G is the usual time-ordered Green's function, $G^<$ is the particle propagator, because $G^<(k, t, t) = \pm\mathrm{i}n_k(t)$. Similarly, $G^>$ is the hole propagator with $G^>(k, t, t) = -i(1 \mp n_k(t))$. Finally, $\tilde{G}$ is the antitime-ordered Green's function, the antitimeordering operator $\tilde{T}$ orders an operator with an earlier time argument to the right. The matrix element G^{ij} obeys a Dyson equation

$$G^{ij}=G^{0ij}+G^{0ik}\Sigma^{kl}G^{kj}, \tag{3}$$

where the summation convention is assumed. The self-energy matrix Σ^{kl} can be evaluated by using the familiar Feynman techniques, summing at internal vertices also over the Keldysh index.

From the definitions of the G^{ij} it follows that

$$G^{\mathrm{r}}=G^{++}-G^{+-}=G^{-+}-G^{--}, \qquad G^{\mathrm{a}}=G^{++}-G^{-+}=G^{+-}-G^{--}. \tag{4}$$

Using (4) one gets from the matrix Dyson equation the Dyson equation for the retarded Green's function:

$$G^{\mathrm{r}}=G^{0\mathrm{r}}+G^{0\mathrm{r}}\Sigma^{\mathrm{r}}G^{\mathrm{r}}, \tag{5}$$

where

$$\Sigma^{\mathrm{r}}=\Sigma^{++}+\Sigma^{+-}=-(\Sigma^{--}+\Sigma^{+-}), \qquad \Sigma^{\mathrm{a}}=\Sigma^{++}+\Sigma^{-+}=-(\Sigma^{--}+\Sigma^{-+}). \tag{6}$$

The nonequilibrium Green's functions depend on two time arguments t, t'. It is advantageous to introduce $\sigma=t-t'$ and $\tau=(t+t')/2$ and Fourier transforming with respect to the relative time coordinate σ. Thus, one gets, e.g., $G^{\mathrm{r}}(k, \omega, \tau)$. One sees, that these functions allow one quite naturally to calculate spectra which can change in time. Similar arguments also hold for the space coordinates, if local homogeneity is not assumed, i.e., r, $r'\to\rho$, $R\to k$, R. In the expression for the retarded self-energy numbers of quasiparticles will generally appear. These numbers can be calculated from the equation for $G^{<}$ following a procedure by Kadanoff and Baym [10]. We will illustrate the use of nonequilibrium Green's functions for two nearly exactly solvable models. A two-band free-carrier system and an exciton–biexciton system, both coupled to a strong coherent classically treated laser field as well as to a continuum of bath variables.

3. Band-to-band transitions [12]

In order to obtain an exactly solvable model, we disregarded for the time being the Coulomb interactions and treat simply laser-induced and spontaneous transitions between two bands. The Hamiltonian is ($\hbar=1$).

$$H=\sum_{i,k} e_{ik}a_{ik}^{\dagger}a_{ik}+\sum_{\lambda}\Omega_{\lambda}b_{\lambda}^{\dagger}b_{\lambda}-[E^{(-)}\exp(+\mathrm{i}\omega_0 t)P^{(+)}+E^{(+)}\exp(-\mathrm{i}\omega_0 t)P^{(-)}]$$

$$+\sum_{\lambda,k} g_{\lambda k}(b_{\lambda}^{\dagger}a_{2k}^{\dagger}a_{1k}+b_{\lambda}a_{1k}^{\dagger}a_{2k}), \tag{7}$$

where

$$e_{1k}=E_g+k^2/2m_e, \qquad e_{2k}=-k^2/2m_h, \qquad P^{(+)}=\sum p_k a_{2k}^\dagger a_{1k}=(P^{(-)})^+. \tag{8}$$

The interaction of the classical laser fields $E^{(\pm)}$ is written in the rotating wave approximation. Ω_1 is the energy of the bath quanta. The bath consists of the vacuum of the continuum of all photon modes. The laser is resonant at the energy $k^2/2m+E_g=\omega_0$, where m is the reduced electron–hole mass. The resonant energies in both bands are $\varepsilon_1^r=(\omega_0-E_g)m/m_e+E_g$ and $\varepsilon_2^r=-(\omega_0-E_g)m/m_h$. Adding and subtracting

$$H_0=\sum \varepsilon_i^r a_{ik}^\dagger a_{ik}+\sum \omega_0 b_\lambda^\dagger b_\lambda,$$

one gets $H=(H-H_0)+H_0=H'-H_0$. In the interaction representation $H=\exp(\mathrm{i}H_0t)H'\exp(-\mathrm{i}H_0t)$ we find

$$H=\sum_{i,k}\varepsilon_{ik}a_{ik}^\dagger a_{ik}-\sum_\lambda \omega_\lambda b_\lambda^\dagger b_\lambda-(E^{(-)}P^{(+)}+E^{(+)}P^{(-)})+\sum_{\lambda,k}g_{\lambda k}(b_\lambda^\dagger a_{2k}^\dagger a_{1k}+\text{h.c.})$$

$$\varepsilon_{1k}=\Delta_k\frac{m}{m_e}, \qquad \varepsilon_{2k}=-\Delta_k\frac{m}{m_h}, \quad \text{with} \quad \Delta_k=\frac{k^2}{2m}+E_g-\omega_0, \qquad \omega_\lambda=\Omega_1-\omega_0. \tag{9}$$

For simplicity, we assume $E^{(-)}=E^{(+)}=E$(real). Because of the presence of a coherent laser beam, we have to use nondiagonal single-particle Green's functions $G_{\rho\mu}^{ij}(t,t')$, where the lower indices refer to the bands: $(\rho,\mu)=(1,2)$.

Taking the time derivations of the retarded Green's function, we get from (3) and its Hermitian conjugate with the use of (4) and (6)

$$G_\rho^0(t)^{-1}G_{\rho\rho}^r(t,t')=\delta(t-t')+pEG_{\nu\rho}^r(t,t')+\int \mathrm{d}t''\,\Sigma_{\rho\rho}^r(t,t'')G_{\rho\rho}^r(t'',t'),$$

$$(G_\rho^0(t')^{-1})^*G_{\rho\rho}^r(t,t')=\delta(t-t')+pEG_{\rho\nu}^r(t,t')+\int \mathrm{d}t''\,\Sigma_{\rho\rho}^r(t,t'')\Sigma_{\rho\rho'}^r(t'',t'), \tag{10}$$

and for $\rho\neq\nu$,

$$G_\rho^0(t)^{-1}G_{\rho\nu}^r(t,t')=pEG_{\nu\nu}^r+\int \mathrm{d}t''\,\Sigma_{\rho\rho}^r(t,t'')G_{\rho\nu}^r(t'',t'),$$

$$(G_\rho^0(t')^{-1})^*G_{\rho\nu}^r(t,t')=pEG_{\rho\rho}^r+\int \mathrm{d}t''\,G_{\rho\nu}^r(t,t'')\Sigma_{\nu\nu}^r(t'',t'), \tag{11}$$

where $G_\rho^0(t)^{-1}=\mathrm{i}\partial/\partial t-\varepsilon_\rho$.

We have already evaluated the self-energy due to the interaction with the field E. Now, we introduce relative and central coordinates

$$\sigma=t-t', \qquad \tau=(t+t')/2,$$

so that

$$\partial/\partial t=\partial/\partial\sigma+\tfrac{1}{2}\partial/\partial\tau, \qquad \partial/\partial t'=-\partial/\partial\sigma+\tfrac{1}{2}\partial/\partial\tau.$$

Adding the pairs of equations and taking the Fourier transform, we obtain

$$G^{0\mathrm{r}}_{\rho}(\omega)^{-1}G^{\mathrm{r}}_{\rho\rho}(\omega,\tau)=1+\tfrac{1}{2}pE[EG^{\mathrm{r}}_{\nu\rho}(\omega,\tau)+EG^{\mathrm{r}}_{\rho\nu}(\omega,\tau), \tag{12}$$

$$\left[2\omega-\sum_{\mu}(\varepsilon_{\mu}+\Sigma_{\mu\mu}(\omega,\tau))\right]G^{\mathrm{r}}_{\rho\nu}(\omega,\tau)=pE\sum_{\mu}G^{\mathrm{r}}_{\mu\mu}(\omega,\tau), \tag{13}$$

with the solutions

$$G^{\mathrm{r}}_{\rho\rho}(\omega,\tau)=\frac{\omega-\varepsilon_{\rho}-\Sigma^{\mathrm{r}}_{\rho\rho}(\omega,\tau)}{(\omega-\varepsilon_{\rho}-\Sigma^{\mathrm{r}}_{\rho\rho}(\omega,\tau))(\omega-\varepsilon_{\nu}-\Sigma^{\mathrm{r}}_{\nu\nu}(\omega,\tau))-p^{2}E^{2}}, \tag{14}$$

$$G^{\mathrm{r}}_{\rho\nu}(\omega,\tau)=\frac{pE}{(\omega-\varepsilon_{\rho}-\Sigma^{\mathrm{r}}_{\rho\rho}(\omega,\tau))(\omega-\varepsilon_{\nu}-\Sigma^{\mathrm{r}}_{\nu\nu}(\omega,\tau))-p^{2}E^{2}}. \tag{15}$$

If one neglects the radiative self-energies, the eigenfrequencies (poles of (14) and (15)) [13,14] are

$$\omega_{1,2}=\tfrac{1}{2}((\varepsilon_{1}+\varepsilon_{2})\pm[(\varepsilon_{1}-\varepsilon_{2})^{2}+4p^{2}E^{2}]^{1/2}. \tag{16}$$

Thus, the spectrum exhibits in this strong field limit a gap (see fig. 2). The radiative self-energies are given by (neglecting vertex corrections, which are non-resonant)

$\Sigma^{ij}_{\rho\rho}(\omega)=$

$$=-\mathrm{i}g^{i}\sum_{\lambda}\int\frac{\mathrm{d}\omega'}{2\pi}\,D^{0ij}_{\lambda}(\omega')G^{ij}_{\nu\nu}(\omega-\omega')g^{j}. \tag{17}$$

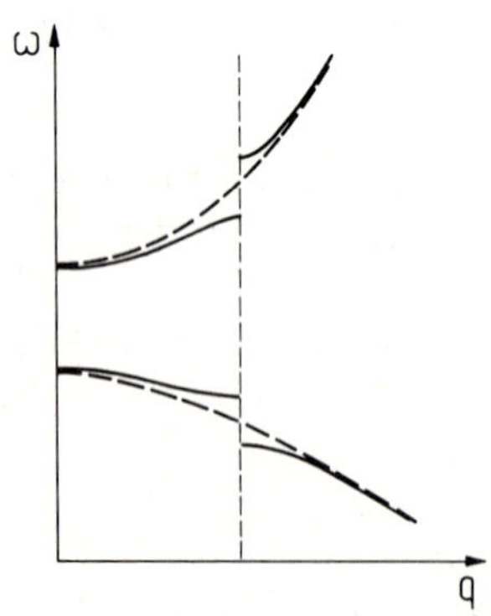

Fig. 2. Schematical plot of the dispersion (eq. (16)).

With the Keldysh rule $g^{\pm}=\pm g$, we get with (5) and (6)

$$\Sigma^{\mathrm{r}}_{\rho\rho}(\omega)=+\mathrm{i}g^2\sum_{\lambda}\int\frac{\mathrm{d}\omega'}{2\pi}\left(D^{0++}_{\lambda}(\omega')G^{++}_{\nu\nu}(\omega-\omega')-D^{0+-}_{\lambda}(\omega')G^{+-}_{\nu\nu}(\omega-\omega')\right)$$

$$=\mathrm{i}g^2\sum_{\lambda}\int\frac{\mathrm{d}\omega'}{2\pi}\left(D^{0\mathrm{r}}_{\lambda}(\omega')G^{\mathrm{r}}_{\nu\nu}(\omega-\omega')+D^{0\mathrm{r}}_{\lambda}(\omega')G^{+-}_{\nu\nu}(\omega-\omega')+D^{0+-}_{\lambda}(\omega')\times G^{\mathrm{r}}_{\nu\nu}(\omega-\omega')\right), \tag{18}$$

where

$$D^{0\mathrm{r}}_{\lambda}(\omega')=\frac{1}{\omega'-\omega_{\lambda}+\mathrm{i}\delta}-\frac{1}{\omega'+\omega_{\lambda}+\mathrm{i}\delta},$$

and

$$D^{0+-}_{\lambda}(\omega')=-2\pi\mathrm{i}[N_{\lambda}\delta(\omega'-\omega_{\lambda})+(1+N_{\lambda})\delta(\omega'+\omega_{\lambda})]$$
$$=2\pi\mathrm{i}\delta(\omega'+\omega_{\lambda}),\quad \text{because } N_{\lambda}=0. \tag{19}$$

In order to calculate the self-energies, we have to know the particle propagator $G^{+-}_{\nu\nu}$, i.e., the number of electrons and holes. From the matrix Dyson equation (15) one gets

$$G^{+-}=G^{0+-}+G^{0++}(\Sigma^{++}G^{+-}+\Sigma^{+-}G^{--})+G^{0+-}(\Sigma^{-+}G^{+-}+\Sigma^{--}G^{--}). \tag{20}$$

Operating from the left with $G^{0}(t)^{-1}=\mathrm{i}\partial/\partial t-\varepsilon_{\nu}$, on eq. (20), we get

$$G^{0^{-1}}(t)G^{+-}(t,t')=\Sigma^{++}G^{+-}+\Sigma^{+-}G^{--}$$
$$=\Sigma^{\mathrm{r}}G^{<}-\Sigma^{<}G^{\mathrm{a}}, \tag{21}$$

where we used the identities

$$G^{0^{-1}}(t)G^{0++}(t,t')=\delta(t-t'),\qquad G^{0^{-1}}(t)G^{0+-}(t,t')=0.$$

Operating with $(G^{0^{-1}}(t'))^{*}$ on the second argument of $G=G_0+G\Sigma G_0$ one gets

$$(G^{0^{-1}}(t'))^{*}G^{+-}(t,t')=-G^{\mathrm{r}}\Sigma^{<}+G^{<}\Sigma^{\mathrm{a}}, \tag{22}$$

where we used $(G^{0^{-1}}(t'))^{*}G^{0--}(t,t')=-\delta(t-t')$. Introducing again relative and central coordinates, one gets by subtracting (22) from (21)

$$\mathrm{i}\frac{\partial}{\partial\tau}G^{<}(\omega,\tau)=G^{<}(\omega,\tau)(\Sigma^{\mathrm{r}}(\omega,\tau)-\Sigma^{\mathrm{a}}(\omega,\tau))-\Sigma^{<}(\omega,\tau)(G^{\mathrm{r}}(\omega,\tau)-G^{\mathrm{a}}(\omega,\tau)), \tag{23}$$

or with the use of (5) and (6),

$$\mathrm{i}\frac{\partial}{\partial\tau}G^{<}(\omega,\tau)=G^{<}(\Sigma^{<}-\Sigma^{>})+\Sigma^{<}(G^{>}-G^{<})$$
$$=G^{>}\Sigma^{<}-G^{<}\Sigma^{>}. \tag{24}$$

Applying this procedure to our two-band model, we find ($\nu \neq \rho$)

$$i\frac{\partial}{\partial\tau} G^<_{\rho\rho}(\omega,\tau) = pE(G^<_{\nu\rho} - G^<_{\rho\nu}) + G^>_{\rho\rho}\Sigma^<_{\rho\rho} - G^<_{\rho\rho}\Sigma^>_{\rho\rho}, \tag{25}$$

$$\left(i\frac{\partial}{\partial\tau} + \varepsilon_\rho - \varepsilon_\nu\right) G^<_{\rho\nu}(\omega,\tau) = pE(G^<_{\nu\nu} - G^<_{\rho\rho}) + G^<_{\rho\nu}(\Sigma^r_{\rho\rho} - \Sigma^a_{\nu\nu}) + G^r_{\rho\nu}\Sigma^<_{\nu\nu} - \Sigma^<_{\rho\rho}G^a_{\rho\nu}. \tag{26}$$

In the following, we will assume that the electrons and holes (26) are in a quasi-equilibrium, so that we have to calculate only the quasichemical potentials μ_ρ. μ_ρ can be calculated from the total number of particles ρ, which is given by

$$i\sum_k \int \frac{d\omega}{2\pi} G^<_{\rho\rho}(\omega,\tau) = N_\rho, \tag{27}$$

while the polarization is given by

$$i\sum \int \frac{d\omega}{2\pi} pG^<_{12}(\omega,\tau) = P^{(+)}. \tag{28}$$

Thus, one finds from (25)

$$\frac{\partial N_1}{\partial\tau} = -i(P^{(-)} - P^{(+)})E - \sum_k \int \frac{d\omega}{2\pi} (G^>_{11}\Sigma^<_{11} - G^<_{11}\Sigma^>_{11}). \tag{29}$$

In the first term one obtains

$$-i(P^{(-)} - P^{(+)}) = \mathrm{Im}\, P = \varepsilon''(\omega_0)E/4\pi = n(\omega_0)c\alpha(\omega_0)E/4\pi.$$

From the averaged Poynting vector one finds the intensity of the laser field $I = n(\omega_0)cE^2/4\pi$, so that the first term of (29) is equal to $\alpha(\omega_0)I/\omega_0$, which is just the absorption rate. The self-energies are

$$\begin{aligned}\Sigma^<_{11} = \Sigma^{+-}_{11} &= -ig^+ \sum_\lambda \int \frac{d\omega'}{2\pi} D^{0+-}(\omega')G^{+-}_{22}(\omega-\omega')g^- \\ &= 2\pi g^2 \int \frac{d\omega'}{2\pi} (N_\lambda\delta(\omega'-\omega_\lambda) + (1+N_\lambda)\delta(\omega'+\omega_\lambda))G^{+-}_{22}(\omega-\omega') \\ &= g^2 G^{+-}_{22}(\omega+\omega_\lambda),\end{aligned}$$

because $N_\lambda = 0$, and $\Sigma^>_{11} = \Sigma^{-+}_{11} = g^2 G^{-+}_{22}(\omega-\omega_\lambda)$. Notice, that $\omega_\lambda = \Omega_\lambda - \omega_0$ can be negative. Thus, the rate equation for the total number of electrons in one band is

$$\partial N_1/\partial\tau = \alpha(\omega_0)I/\omega_0 - R = -\partial N_2/\partial\tau, \tag{31}$$

where the rate of spontaneous emission is given by

$$R = g^2 \sum_{k,\lambda} \int \frac{d\omega}{2\pi} (G^>_{22}(\omega{}^+\omega_\lambda)G^<_{11}(\omega) - G^<_{22}(\omega-\omega_\lambda)G^>_{11}(\omega)). \tag{32}$$

Let us now turn to the dipole equation (26). It reads for $\rho=1$, $\nu=2$

$$\left(\mathrm{i}\frac{\partial}{\partial\tau}-\varepsilon_1+\varepsilon_2\right)G^{<}_{12}(\omega,\tau)=pE(G^{<}_{22}-G^{<}_{11})+G^{<}_{12}(\Sigma^{\mathrm{r}}_{11}-\Sigma^{\mathrm{a}}_{22})+G^{\mathrm{r}}_{12}\Sigma^{<}_{22}-\Sigma^{<}_{11}G^{\mathrm{a}}_{12}. \tag{33}$$

The stationary solution of (33) with (28) is

$$P^{(+)}=\mathrm{i}\int\frac{\mathrm{d}\omega}{2\pi}\,\frac{p^2E(G^{<}_{22}-G^{<}_{11})+G^{\mathrm{r}}_{12}\Sigma^{<}_{22}-\Sigma^{<}_{11}G^{\mathrm{a}}_{12}}{(G^{r\,-1}_{11}-G^{a\,-1}_{22})}\,. \tag{34}$$

In order to discuss the result (34), we show first, that it includes the known low and high field limits. Under quasi-equilibrium conditions we have

$$G^{<}_{\rho\rho}=\mathrm{i}f_\rho(\omega)A_{\rho\rho}(\omega),\qquad G^{>}_{\rho\rho}=-\mathrm{i}(1-f_\rho(\omega))A_{\rho\rho}(\omega),$$

where $f_\rho(\omega)=1/[\exp(\beta(\omega-\mu_\rho))+1]$, so that

$$G^{\mathrm{r}}_{\rho\rho}-G^{\mathrm{a}}_{\rho\rho}=G^{>}_{\rho\rho}-G^{<}_{\rho\rho}=-\mathrm{i}A_{\rho\rho}. \tag{35}$$

Similarly, one has for the self-energies

$$\Sigma^{<}_{\rho\rho}=-\mathrm{i}f_\rho(\omega)\Gamma_{\rho\rho}(\omega),\qquad \Sigma^{>}_{\rho\rho}=\mathrm{i}(1-f_\rho(\omega))\Gamma_{\rho\rho}(\omega), \tag{36}$$

so that $\Sigma^{\mathrm{r}}_{\rho\rho}-\Sigma^{\mathrm{a}}_{\rho\rho}=\Sigma^{<}_{\rho\rho}-\Sigma^{>}_{\rho\rho}=-\mathrm{i}\Gamma_{\rho\rho}$. (Note the difference in signs as compared to the Green's functions.)

(A) *Low-field limit*

In this limit the result (34) has to reduce to the well-known linear response result. It follows from (15) that for $E\to0$,

$$G^{\mathrm{r}}_{12}(\omega)=pEG^{\mathrm{r}}_{11}(\omega)G^{\mathrm{r}}_{22}(\omega). \tag{37}$$

Inserting (35)–(37) into (34), we find

$$G^{<}_{12}(\omega)=\mathrm{i}pE\,\frac{f_2(A_{22}+G^{\mathrm{r}}_{11}(\omega)G^{\mathrm{r}}_{22}(\omega)\Gamma_{22})-f_1(A_{11}+G^{\mathrm{a}}_{11}G^{\mathrm{a}}_{22}\Gamma_{11})}{G^{r\,-1}_{11}-G^{a\,-1}_{22}}\,. \tag{38}$$

From (36) one gets $\Gamma_{\rho\rho}=-\mathrm{i}(G^{\mathrm{a}\,-1}_{\rho\rho}-G^{\mathrm{r}\,-1}_{\rho\rho})=-A_{\rho\rho}/G^{\mathrm{r}}_{\rho\rho}G^{\mathrm{a}}_{\rho\rho}$, therefore

$$G^{*}_{12}(\omega)=\mathrm{i}pE(f_2A_{22}G^{\mathrm{r}}_{11}+f_1A_{11}G^{\mathrm{a}}_{22}). \tag{39}$$

From (28) we find

$$\begin{aligned}P^{(+)}&=-\sum_k\int\frac{\mathrm{d}\omega}{2\pi}\frac{\mathrm{d}\omega'}{2\pi}\,p^2E\left(\frac{f_2(\omega)A_{22}(\omega)A_{11}(\omega')}{\omega-\omega'+\mathrm{i}\delta}+\frac{f_1(\omega)A_{11}(\omega)A_{22}(\omega')}{\omega-\omega'-\mathrm{i}\delta}\right)\\&=-\sum_k\int\frac{\mathrm{d}\omega}{2\pi}\int\frac{\mathrm{d}\omega'}{2\pi}\,p^2E\,\frac{f_2(\omega)-f_1(\omega')}{\omega-\omega'+\mathrm{i}\delta}\,A_{11}(\omega)A_{22}(\omega'),\end{aligned} \tag{40}$$

so that

$$\chi(\omega_0)=P^+/E^+ \\ =-\sum_k \int \frac{d\omega}{2\pi} \int \frac{d\omega'}{2\pi} p^2 \frac{f_2(\omega)-f_1(\omega')}{\omega-\omega'+i\delta} A_{11}(\omega')A_{22}(\omega). \tag{41}$$

Approximating the spectral functions by δ-functions

$$A_{\rho\rho}=2\pi\delta(\omega-\omega_\rho),$$

one gets the usual linear response formula

$$\chi=-\sum_k p^2[f_2(\varepsilon_2)-f_1(\varepsilon_1)]/(\varepsilon_2-\varepsilon_1+i\delta), \tag{42}$$

or

$$\chi''=\sum_k \pi p^2(f_2(\varepsilon_2)-f_1(\varepsilon_1))\delta(\varepsilon_2-\varepsilon_1), \tag{43}$$

where $\varepsilon_2-\varepsilon_1=E_{kv}+\hbar\omega_0-E_{kv}$.

(B) High-field limit

In this case, we can neglect the self-energies due to spontaneous emission and find

$$P^{(+)}=-\sum_k \int \frac{d\omega}{2\pi} p^2 Ef(\omega) \left(\frac{A_{22}(\omega)-A_{11}(\omega)}{\varepsilon_2-\varepsilon_1}\right). \tag{44}$$

The stationarity in the rate equation requires $\mu_1=\mu_2=0$, i.e., the bands are filled with electrons up to the energies resonant with the laser field (total bleaching). Using the eigenfrequencies $\omega_{1,2}$ (16), we find

$$P^+=-\sum_k p^2 E[f(\omega_2)-f(\omega_1)]/(\omega_2-\omega_1+i\delta), \tag{45}$$

which is of the same form as the linear response result (42) except that ε_1, ε_2 are replaced by the renormalized energies ω_1, ω_2.

In general, $\mu_1=(m/m_e)\mu$ and $\mu_2=-(m/m_h)\mu$ have to be determined numerically by solving self-consistently the polarization equation (34) and the rate equation (31).

4. The biexciton two-photon resonance [15]

The two-photon resonance of the biexciton is another source of large optical nonlinearities. We developed a simple theory of the nonlinear optical response by taking into account only the most resonant contribution [16]. Raman scattering experiments [17], transmission measurements [18], and direct measurements of the changes of the index of refraction [19] were in good qualitative agreement with the theory [20]. However, recently the theory has been extended by taking into account further less resonant terms [21]. The results, so seemingly exact, are quite conflicting as will be discussed in more detail at the end of this chapter. The nonlinear dielectric function has a pole in the upper half of the complex plane, and the resulting polariton dispersion is – to say the least – very strange.

We will derive the extended results again by the nonequilibrium Green's function technique and point out the conflicting features of the result.

In order to proceed in close analogy with section 3, we will again assume a coherent classical field. The Hamiltonian is

$$\tilde{H}=\sum \varepsilon_{xk}a_k^\dagger a_k+\sum \varepsilon_{mk}B_k^\dagger B_k+\sum \Omega_\lambda b_\lambda^\dagger b_\lambda-\mu_x(E^{(+)}a_q^\dagger+E^{(-)}a_q)$$

$$-\mu_{xm}\sum (E^{(-)}B_{q+q'}^\dagger a_{q'}+E^{(+)}a_{q'}^\dagger B_{q+q'})+\sum_{k,\lambda} M_{k'\lambda}(B_{k'}^\dagger b_\lambda+B_{k'}b_\lambda^\dagger), \tag{46}$$

where $\varepsilon_{xk}=\omega_{xk}-\omega_0$, $\varepsilon_{mk}=\omega_{mk}-2\omega_0$, $\Omega_\lambda=\omega_\lambda-2\omega_0$ and a_k, B_k, b_λ are the exciton, biexciton and bath quantum operators. The polarization P^+ is given by

$$P^{(+)}=\mu_x\langle a_q\rangle+\mu_{xm}\sum_{q'}\langle a_q^\dagger B_{q+q'}\rangle. \tag{47}$$

The main difference with the Fermi system of section 3 is that in the Boson system of excitons and biexcitons the one-point functions $\langle a_q\rangle$, $\langle B_{2q}\rangle$ are finite under a coherent excitation. We split the last term for convenience into a factorized term and the rest:

$$P^{(+)}=\mu_x\langle a_q\rangle+\mu_{xm}\langle a_q^\dagger\rangle\langle B_{2q}\rangle+\mu_{xm}\sum(\langle a_q^\dagger B_{q+q'}\rangle-\langle a_q\rangle\langle B_{2q}\rangle)$$

$$=\sum_{i=1}^{3} P_i^{(+)}, \tag{48}$$

and calculated the three terms separately.

The equations of motion for the one-point functions are

$$G_{xq}^0(t)^{-1}\langle a_q\rangle=-\mu_x E^{(+)}-\mu_{xm}E^{(+)}\langle B_{2q}\rangle, \tag{49}$$

$$G_{m,2q}^0(t)^{-1}\langle B_{2q}\rangle=-\mu_{xm}E^{(-)}\langle a_q\rangle+\Sigma_{m,2q}^{\mathrm{r}}\langle B_{2q}\rangle, \tag{50}$$

where $\Sigma_{m2q}^{\mathrm{r}}$ is the retarded self-energy of the biexciton due to the coupling to the bath. The stationary solutions of (49) and (50) are

$$\varepsilon_{xq}\langle a_q\rangle=(\mu_x+\mu_{xm}\langle B_{2q}\rangle)E^{(+)}, \tag{51}$$

$$(\varepsilon_{m,2q}+\Sigma_{m,2q}^{\mathrm{r}})\langle B_{2q}\rangle=\mu_{xm}\langle a_q\rangle E^{(-)}, \tag{52}$$

where

$$\Sigma^{\mathrm{r}}_{m,2q}=\sum_{\lambda}|g_{\lambda k}|^2/(\varepsilon_{m,2q}-\Omega_\lambda+\mathrm{i}\delta)\simeq-\mathrm{i}\gamma_m. \tag{53}$$

The solutions of these equations are

$$\langle a_q\rangle=\frac{\mu_x E^{(+)}(\varepsilon_{m,2q}-\mathrm{i}\gamma_m)}{\varepsilon_{x,q}(\varepsilon_{m,2q}-\mathrm{i}\gamma_m)-\mu_{xm}^2|E|^2}, \tag{54}$$

$$\langle B_{2q}\rangle=\frac{\mu_{xm}\mu_x^2|E|^2}{\varepsilon_{x,q}(\varepsilon_{m,2q}-\mathrm{i}\gamma_m)-\mu_{xm}^2|E|^2}. \tag{55}$$

Therefore, the first two terms of the polarization yield the following susceptibility:

$$\chi_1=P^{(+)}/E^+=-\mu_x^2G_x^{\mathrm{r}}(q,\omega_0), \tag{56}$$

$$\chi_2=P^{(+)}/E^+=-\mu_x^2\Sigma_x^{\mathrm{r}}(q,\omega_0)G_x^{\mathrm{a}}(q,\omega_0)G_x^{\mathrm{r}}(q,\omega_0), \tag{57}$$

where

$$G_x^{\mathrm{r}}(q,\omega_0)=1/(\omega_0-\omega_x-\Sigma_x^{\mathrm{r}}(q,\omega_0)), \tag{58}$$

and

$$\Sigma_x^{\mathrm{r}}(q,\omega_0)=\mu_{xm}^2|E|^2/(2\omega_0-\omega_m+\mathrm{i}\gamma_m). \tag{59}$$

The same results can alternatively be derived by nonequilibrium two-point Green's functions if an incoherent excitation is assumed. The corresponding nonlinear dielectric function is given by

$$\varepsilon(\omega_0)=\varepsilon_\infty[1-4\pi\mu_x^2(G_x^{\mathrm{r}}+\Sigma_x^{\mathrm{r}}G_x^{\mathrm{a}}G_x^{\mathrm{r}})]. \tag{60}$$

The first term $\propto G_x^{\mathrm{r}}$ is the result of the Frankfurt group [16] which gives a reasonable renormalized polariton spectrum. The full result (60) has first been derived by Abram and Maruani [21] by unitary transformation techniques (without damping terms). The appearance of the advanced Green's function in (60) makes it obvious, that $\varepsilon(\omega)$ has a pole in the upper half of the complex plane. If one calculates the renormalized eigenmodes of the polariton from

$$\varepsilon(\omega)\omega^2=c^2q^2, \tag{61}$$

it becomes again apparent that the result (60) is not acceptable. For simplicity, we neglect the damping. Then (61) can be written as

$$\omega^2-\frac{c^2q^2}{\varepsilon_\infty}=\frac{4\pi\mu_x^2\omega^2}{\omega-\omega_x}\left[\frac{(\omega-\omega_x)(2\omega-\omega_m)}{(\omega-\omega_x)(2\omega-\omega_m)-\mu_m^2|E|^2}\right]^2, \tag{62}$$

or

$$\omega-\frac{cq}{\sqrt{\varepsilon_\infty}}\simeq 2\pi\mu_x^2\omega\left[\frac{(\omega-\omega_x)(2\omega-\omega_m)^2}{(\omega-\omega_1)(\omega-\omega_2)}\right], \tag{63}$$

where

$$\omega_{1,2}=\tfrac{1}{2}((\omega_x+\omega_m/2)\pm[(\omega_x-\omega_m/2)+\mu_{xm}^2|E|^2]^{1/2}. \tag{64}$$

Neglecting spatial dispersion it is easy to calculate $cq/\sqrt{\varepsilon_\infty}$ as a function of ω. There are two zeros of the square bracket in (63), a simple one at $\omega=\omega_x$ and a quadratic one at $\omega=\omega_m/2$. ω_1 and ω_2, on the other hand, are poles slightly higher than the exciton level and slightly lower than half the molecule level.

After this discussion, the spectrum in fig. 3 is easily understood. Obviously it is for a number of reasons not acceptable. Sung and Bowden [21] argued that these unphysical results are due to the factorization of the second term in the polarization (47). However, this argument does not hold as will be shown below. The dispersion relation of Sung and Bowden [21] does look strange too (fig. 4). They get another resonance with anomalous dispersion at the frequency

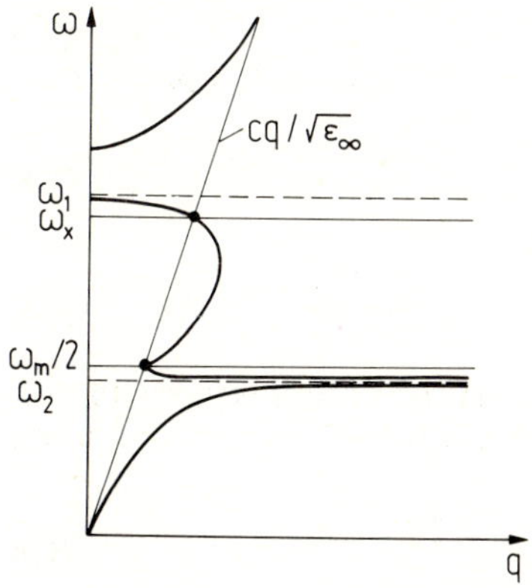

Fig. 3. Schematic plot of the dispersion (eq. (63)).

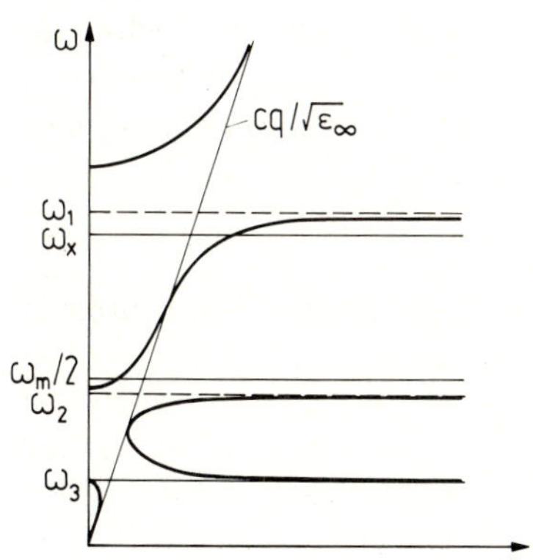

Fig. 4. Schematic plot of the dispersion due to Sung and Bowden [21].

$\omega = \omega_m - \omega_x$, which corresponds to induced emission. This resonance is present even with $N_m = N_x = 0$, which is strange.

We will now calculate the third nonfactorized contribution to the polarization (48).

$$P_3^{(+)} - \mu_{xm} \sum_q (\langle a_q^\dagger B_{q+q'} \rangle - \langle a_q^\dagger \rangle \langle B_{2q} \rangle) = \mu_{xm} \sum_q \bar{P}_{q,q'}^{(+)<}(t, t'). \tag{65}$$

For that purpose we introduce Green's functions

$$\begin{aligned} \bar{G}_{x,q'}(t, t') &= \langle T a_{q'}(t) a_q^\dagger(t') \rangle - \langle a_{q'}(t) \rangle \langle a_{q'}(t') \rangle, \\ \bar{G}_{x,q+q'}(t, t') &= \langle T B_{q+q'}(t) B_{q+q'}^\dagger(t') \rangle - \langle B_{q+q'}(t) \rangle \langle B_{q+q'}^\dagger(t') \rangle. \end{aligned} \tag{66}$$

Introducing again relative and central time arguments, one finds

$$\bar{P}_{q,q'}^{(+)}(\tau) = \int \frac{d\omega}{2\pi} \bar{P}_{q,q'}^{(+)}(\omega, \tau).$$

Following the same derivations as those given in section 3, the polarization function obeys the equation

$$\begin{aligned} &\left(i \frac{\partial}{\partial \tau} - \varepsilon_{m,q+q'} + \varepsilon_{x,q'} - \bar{\Sigma}_{q+q'}^{r}(\omega, \tau) \right) \bar{P}_{q,q'}^{(+)}(\omega, \tau) \\ &\qquad = -\mu_{xm} E^{(+)} (\bar{G}_{x,q'}^{<}(\omega, \tau) - \bar{G}_{m,q+q'}^{<}(\omega, \tau)) + \bar{\Sigma}_{m,q+q'}^{<}(\omega, \tau) \bar{P}_{q,q'}^{(+)a}(\omega, \tau), \end{aligned} \tag{67}$$

where $\bar{\Sigma}_{m,q+q'}^{<}$ is due to the interaction with the bath. The equations for the molecule and exciton particle propagators are

$$i \frac{\partial}{\partial \tau} \bar{G}_{m,q+q'}^{<} = i\mu_{xm} E \operatorname{Im} \bar{P}_{q,q'} + i(\Sigma_{m,q+q'}^{>} \bar{G}_{m,q+q'}^{<}(\omega, \tau) - \Sigma_{m,q+q'}^{<} \bar{G}_{m,q+q'}^{>}) \tag{68}$$

and

$$i \frac{\partial}{\partial \tau} \bar{G}_{x,q'}^{<} = -i\mu_{xm} E \operatorname{Im} \bar{P}_{q,q'}. \tag{69}$$

The corresponding rate equations for the total numbers of biexcitons and excitons are

$$\frac{\partial}{\partial \tau} \sum_{q'} \bar{n}_{m,q'} = \mu_{xm} E \sum_{q'} \operatorname{Im} \bar{P}_{q,q'} - \gamma_{\parallel m} \sum_{q'} \bar{n}_{m,q'}, \tag{70}$$

$$\frac{\partial}{\partial \tau} \sum_{q'} \bar{n}_{x,q'} = -\mu_{xm} E \sum_{q'} \operatorname{Im} \bar{P}_{q,q'} + \gamma_{\parallel m} \sum_{q'} \bar{n}_{m,q'}, \tag{71}$$

where, e.g., $\bar{n}_{x,q} = n_{x,q} - \langle a_q^\dagger \rangle \langle a_q \rangle$ is only the noncoherent part of the exciton population. The generation rate of the excitons is linked to the decay rate of the molecules, if other decay mechanisms of the molecules (longitudinal excitons, e.g.,) are disregarded. Under stationary, quasithermal conditions, where

$\mu_x - \omega_0 = \mu_m - 2\omega_0 = \bar{\mu}$, one gets

$$P_3^{(+)} = \sum_{q'} \int \frac{d\omega}{2\pi} \mu_{xm}^2 E^+ g(\omega) \left(\frac{A_x(q', \omega) - A_m(q+q', \omega)}{\varepsilon_{x,q'} - \varepsilon_{q+q'} - i\gamma_{\perp m}} \right) \tag{72}$$

with $g(\omega) = 1/[\exp(\beta(\omega - \bar{\mu})) - 1]$. The poles of the retarded exciton and molecule Green's functions are at

$$\omega_{1,2} = \tfrac{1}{2}((\varepsilon_{x,q'} + \varepsilon_{m,q+q'}) \pm \text{Sq.R.}),$$

where

$$\text{Sq.R.} = [(\varepsilon_{x,q'} - \varepsilon_{m,q+q'})^2 + 4\mu_{xm}^2 E^2]^{1/2}. \tag{73}$$

Decomposing the exciton and biexciton Green's functions, we find the spectral functions

$$A_x(q', \omega) = \tfrac{1}{2}(1 - a_{q,q'})\delta(\omega - \omega_1) + \tfrac{1}{2}(1 + a_{q,q'})\delta(\omega - \omega_2),$$
$$A_m(q+q', \omega) = \tfrac{1}{2}(1 + a_{q,q'})\delta(\omega - \omega_1) + \tfrac{1}{2}(1 - a_{q,q'})\delta(\omega - \omega_2), \tag{74}$$

where $a_{q,q'} = (\varepsilon_{m,q+q'} - \varepsilon_{x,q})/\text{Sq.R.}$ Using these relations, one gets

$$P_3^{(+)}/E^{(+)} = \sum_{q'} \mu_{xm}^2 [g(\omega_1) - g(\omega_2)]/\text{Sq.R.}, \tag{75}$$

which is exactly the third term of Abram and Maruani [21].

It is thus very unsatisfactory, that the improved treatments calculated with various techniques give the same, but unfortunately unacceptable results. This points to some basic inconsistent assumption common to all discussed approaches.

In order to find the origin of these difficulties, we investigated the problem using Green's function techniques by avoiding dangerous diagrams such as the one yielding the last term in (60). We used only renormalized Green's functions for the photons, excitons and molecules of the form

$$G_l^r = \sum_{j=1}^{N_l} \frac{f_{lj}}{\omega - \omega_{lj} + i\delta}, \tag{76}$$

where the eigenfrequencies ω_{lj}, the weight functions f_{lj} and the number of poles N_l have to be determined selfconsistently. We only found solutions around the resonance with a continuous spectrum. Continuous spectra can arise if the corresponding classical equations have chaotic solutions. In order to investigate this possibility, we are presently investigating the numerical solutions of the equations of motion of the classical amplitudes of three nonlinearly coupled oscillators, which above a critical intensity have chaotic solutions.

Acknowledgement

The author is grateful to L.L. Chase and G.D. Mahan for their hospitality and for the stimulating discussions. Furthermore, the author thanks M. Inoue for a cooperation in the problems described in section 4. The work done in Bloomington was supported by the National Science Foundation under the Grants No. DMR 81–05 005 and No. H 83–11765. Partial support is also acknowledged from the Commission of the European Community and from the Deutsche Forschungsgemeinschaft through the Sonderforschungsbereich Frankfurt/Darmstadt.

References

[1] For a review see: C. Klingshirn and H. Haug, Phys. Rep. 70 (1981) 315.

[2] K. Bohnert, M. Anselment, G. Kobbe, C. Klingshirn, H. Haug, S.W. Koch, S. Schmitt-Rink and F.F. Abraham, Z. Phys. B 42, (1981) 1;
A. Forchel, B. Laurich, W. Schmid, G. Maier and G. Mahler, J. Lumin. 24/25 (1981) 565.

[3] J.P. Löwenau, S. Schmitt-Rink and H. Haug, Phys. Rev. Lett. 49 (1982) 1511.
For a recent review see H. Haug and S. Schmitt-Rink, Progress in Quantum Electronics, in press.

[4] H. Schweizer, A. Forchel, A. Hangleiter, S. Schmitt-Rink, J.P. Löwenau and H. Haug, Phys. Rev. Lett. 51 (1983) 698.
D. Hulin, A. Antonetti, L.L. Chase, J.L. Martin, A. Migus, A. Mysyrowicz, J.P. Löwenau, S. Schmitt-Rink and H. Haug, Phys. Rev. Lett. 52 (1984) 779.

[5] M.E. Prise, M.R. Taghizadek, B.S. Wherrett and S.D. Smith, J. Lumin. 30 (1985) 369.
A. Miller, D.A.B. Miller and S.D. Smith, Adv. Phys. 30 (1981) 697.

[6] D.S. Chemla, J. Lumin. 30 (1985) 502.
D.S. Chemla, D.A.B. Miller, P.W. Smith, A.C. Gossard and W. Wiegmann, IEEE-QE 20 (1984) 265.

[7] L.V. Keldysh, Sov. Phys. JETP 20 (1965) 1018;
see also Landau–Lifshitz, Theoretical Physics, Vol. 10, Kinetics (Pergamon Press, New York, 1965).

[8] W. Schäfer, J. Treusch, private communication.

[9] A.L. Iwanov and L.V. Keldysh, Sov. Phys. JETP 57 (1983) 234.

[10] L.P. Kadanoff and G. Baym, Quantum Statistical Mechanics (W.A. Benjamin, New York, 1974).

[11] See, e.g., W. Hänsch and G.D. Mahan, Phys. Rev. B 28 (1983) 1902.

[12] This chapter is based on work done in collaboration with S. Schmitt-Rink and S.W. Koch.

[13] V.F. Elesin, Sov. Phys. JETP 32 (1971) 328.

[14] T. Harbich and G. Mahler, Phys. Status Solidi (b) 104 (1981) 185.

[15] This chapter is based on work done in collaboration with S. Ullmann, and during my stay in Bloomington with M. Inoue and G.D. Mahan.

[16] H. Haug, R. März and S. Schmitt-Rink, Phys. Lett. 77A (1980) 287;
R. März, S. Schmitt-Rink and H. Haug, Z. Phys. B 40 (1980) 9;
H. Haug, S.W. Koch, R. März and S. Schmitt-Rink, J. Lumin. 24/25 (1981) 621;
S. Schmitt-Rink and H. Haug, Phys. Status Solidi (b)108 (1981) 377.

[17] T. Itoh and T. Suzuki, J. Phys. Japan 45 (1978) 1939;
V.G. Lyssenko, K. Kempf, K. Bohnert, G. Schmieder, C. Klingshirn and S. Schmitt-Rink, Solid State Commun. 42 (1982) 401;
B. Hönerlage, J.Y. Bigot, R. Levy, F. Tomasini and J.B. Grun, Solid State Commun. 48 (1983) 803.
[18] Y. Masumoto and S. Shionoya, Solid State Commun. 38 (1981) 865.
[19] M. Kutawa, T. Mita and N. Nagasawa, Solid State Commun. 40 (1981) 911.
[20] H. Haug, Festkörperprobleme (Advances in Solid State Physics) Vol. XXII, 149, P. Grosse, ed. (Vieweg, Braunschweig, 1982).
[21] I. Abram and A. Maruani, Phys. Rev. B 26 (1982) 4759;
I. Abram, Phys. Rev. B 28 (1983) 4433; ibid B 29 (1984) 4480;
C.C. Sung and C.M. Bowden, Phys. Rev. A 29 (1984) 1957.

Journal of Luminescence 30 (1985) 188–203
North-Holland, Amsterdam

NONLINEAR OPTICS AND PLASMA-INDUCED BISTABILITY IN CdS

C. KLINGSHIRN, K. BOHNERT, H. KALT, V.G. LYSSENKO* and K. KEMPF

Physikalisches Institut der Universität, Robert-Mayer-Straße 2–4, D-6000 Frankfurt am Main, FRG

The investigation of highly excited semiconductors and of nonlinear optical properties recently tends to merge into one field of semiconductor physics. This is demonstrated here for the case of CdS. Two experimental methods are presented, which allow us both to detect optical nonlinearities and to understand their physical origins. These are the creation of laser-induced gratings and the excite- and- probe technique. The reasons for the nonlinearities are a broadening of the exciton resonances, transitions to the biexciton and the formation of an electron–hole plasma. The latter point is discussed in some detail and it is demonstrated, how the nonlinearities connected with this phase transition may be exploited to produce a dispersive and an intrinsic, absorptive bistability.

1. Introduction

In the frame of Maxwell's theory the relation between an electric field $\boldsymbol{E}$ applied to matter, the polarization of the medium $\boldsymbol{P}$ and the electric displacement $\boldsymbol{D}$ reads

$$\boldsymbol{D}=\varepsilon_0\boldsymbol{E}+\boldsymbol{P}, \tag{1}$$

with ε_0 known as the permittivity of vacuum ($\varepsilon_0=8.85\times10^{-12}$ As/Vm). In many cases one finds proportionality between $\boldsymbol{P}$ and $\boldsymbol{E}$. For (plane) harmonic waves with (angular) frequency ω this can be expressed by

$$\boldsymbol{P}=\varepsilon_0\chi(\omega)\boldsymbol{E}, \tag{2}$$

or

$$\boldsymbol{D}=\varepsilon_0(\chi(\omega)+1)\boldsymbol{E}=\varepsilon_0\varepsilon(\omega)\boldsymbol{E}, \tag{3}$$

* Permanent address: Institute of Problems of Microelectronic Technology, Academy of Sciences of the USSR, 142432 Chernogolovka, USSR.

0022-2313/85/$03.30

where $\varepsilon(\omega)$ is the dielectric function and $\chi(\omega)$ the susceptibility. The relation between the complex dielectric function $\varepsilon(\omega)=\varepsilon_1(\omega)+i\varepsilon_2(\omega)$ and the complex index of refraction $\tilde{n}(\omega)=n(\omega)+i\kappa(\omega)$ is given for nonmagnetic materials (permeability $\mu\equiv 1$) by

$$\varepsilon(\omega)=\tilde{n}^2(\omega). \tag{4}$$

It is obvious that the optical properties of a piece of matter like the absorption or transmission spectra are known, once $\varepsilon(\omega)$ or $n(\omega)$ are determined.

$\varepsilon(\omega)$ can be written as a sum over discrete or continuous resonances with eigenfrequencies ω_{0j} having an oscillator strength f_j and damping Γ_j.

$$\varepsilon(\omega)=1+\sum_j f_j/(\omega_{0j}^2-\omega^2-i\omega\Gamma_j). \tag{5}$$

In the vicinity of an isolated resonance ω_0 one can replace eq. (5) by

$$\varepsilon(\omega)=\varepsilon_b[1+f/(\omega_0^2-\omega^2-i\omega\Gamma), \tag{6}$$

where the background dielectric "constant" ε_b describes the contribution of all resonances with $\omega_{0j}>\omega_0$ to $\varepsilon(\omega)$ in the vicinity of ω_0 while the contribution of resonances with $\omega_{0j}<\omega_0$ are negligible.

The resonances which will be considered in this context are the exciton resonances. In simplest approximation, free excitons can be regarded as electron–hole pair states interacting via the screened Coulomb potential and forming a hydrogen-like series of states below the gap. In CdS we are dealing with three of these series in the vicinity of the absorption edge due to the threefold splitting of the valence band. They are labeled by the valence band (A, B or C), their total symmetry (Γ_1, Γ_5, etc.), and the main quantum number of the envelope function ($n=1, 2, \ldots$). The data for the A and B $n=1$, exciton states in CdS and the corresponding band gaps at liquid He temperature are the following according to [1,2], where the data in [1] have been corrected for the refractive index of air: $A\Gamma_6$: 2.5521 eV; $A\Gamma_5^T$: 2.5523 eV, $\Delta_{LT}(1.9\pm0.1)$ meV; $E_{gA}\approx$ 2.581 eV; $B\Gamma_{1/2}$: 2.5673 eV; $B\Gamma_5^T$: 2.5674 eV; $\Delta_{LT}=(1.4\pm1)$ meV; $E_{gB}\approx$2.5963 eV. A more detailed discussion of the exciton problem, e.g., in terms of the polariton concept including spatial dispersion, i.e., the dependence of ω_0 on the wavevector $\boldsymbol{k}$ or additional boundary conditions is beyond the scope of this contribution. The reader is referred to the numerous review articles (e.g. [3] and the references therein). It should be mentioned, that excitons can be bound and localized at impurities such as neutral acceptors or donors forming so-called bound excitons. In CdS I_1, I_2 and I_3 mean excitons bound to neutral acceptors, neutral donors and ionized donors, respectively.

As long as $\varepsilon(\omega)$ or $\tilde{n}(\omega)$ are independent of the field amplitude, one has a linear response of the polarization to the electric field. This is the range of linear optics. If $\varepsilon(\omega)$ or $\tilde{n}(\omega)$ depend on the amplitude or intensity of the light field one enters consequently the range of nonlinear optics. For a phenomenological description one often uses an expansion of $\chi(\omega)$ or $\tilde{n}(\omega)$ in a power series of the electric fields of the interacting light beams $E_j(\omega)$ or the intensities I [4]

$$P_i=\varepsilon_0\left\{\sum_j \chi_{ij}^{(1)}E_j(\omega_j)+\sum_{j,k}\chi_{ijk}^{(2)}E_j(\omega_j)E_k(\omega_k)+\sum_{j,k,l}\chi_{ijkl}^{(3)}E_j(\omega_i)E_k(\omega_k)E_1(\omega_1)+\cdots\right\}, \tag{7}$$

or

$$\tilde{n}(\omega, I_{exc})=\tilde{n}_0+\Delta\tilde{n}(I_{exc})=\tilde{n}_0+\tilde{n}_1 I_{exc}+\cdots. \tag{8}$$

The task is then to connect the coefficients $\chi^{(\nu)}$ to microscopic processes, e.g., the well-known $\chi^{(3)}$ effects. The above given expansion is adequate only for the description of instantaneous effects, i.e., effects where the polarization at a certain time t depends only on the field amplitude or light intensity at the same time. There are more complex cases, where the optical properties depend on the density, e.g., of electron–hole pairs n_p present at time t. Since these particles have a finite lifetime, they can have been created at $t'<t$. Thus we get $\chi=\chi(\omega, n_p)$. n_p itself may depend on the excitation conditions and the optical properties, i.e., $n_p=n_p$ (Excit., χ). The absorptive bisability described in section 4 is an example of this type of phenomena.

In semiconductors two ways towards the understanding of the phenomena of nonlinear optics have been followed by various research groups almost independently. One was the investigation of typical nonlinear optical phenomena like harmonic generation, frequency mixing, d.c. effect, phase conjugate reflection, etc., however often spectrally far away from the electronic resonances, e.g., [4,5]. The other is known as properties of highly excited semiconductors. In this field the interaction processes between various quasiparticles have been focussed like inelastic scattering between excitons, the formation and decay of biexcitons or the transition to the electron–hole plasma (EHP). References of this field are found, e.g., in [6]. Gradually one became aware that the above mentioned interaction processes result in optical nonlinearities and vice versa [7].

The aim of this contribution is to present two of the numerous experimental techniques, which allow us to detect and to interpret excitation induced variations (or renormalization processes) of $\varepsilon(\omega)$ namely the formation and diffraction from laser induced gratings (LIG) and the excite-and-probe technique. Various mechanisms will be presented with special emphasis on the transition from a low density gas of excitons to the electron–hole plasma. It will be demonstrated how the latter phase transition allows for the realization of various types of optical bistability. The contribution will be restricted to CdS though the phenomena are common to many other semiconductors, too. It is further written from an experimental point of view, and so for detailed theoretical investigations, the reader is referred, e.g., to the contributions to this volume by H. Haug, F. Henneberger, S.W. Koch or to [6,8] and the literature cited therein.

2. Laser-induced gratings

In this chapter we shall give a few introductory remarks about LIG and then present new data for CdS (see also, e.g., [9]). If two coherent, monochromatic and equally polarized plane waves (generally laser beams) coincide under a small angle θ they produce an interference pattern, fig. 1. There are planes of constructive interference with large field amplitudes (shown as dashed lines in fig. 1) with planes of destructive interference in between. If a sample is brought in this interference pattern and if the light intensities are sufficiently high as to produce a $\Delta\tilde{n}(I_{exc})$ a grating will be formed in the sample. We are dealing with an amplitude or a phase grating, depending whether the imaginary or the real part of $\Delta\tilde{n}$ is predominant. If the intensities I of beams a and b are equal the intensity in the interference pattern will vary periodically from 0 to $4I$. If we have, e.g., $I_b \ll I_a$ the intensity will only be modulated around I_a and the efficiency of the grating η will depend on

$$\eta \sim \left.\frac{\partial\tilde{n}}{\partial I}\right|_{I_a}. \tag{9}$$

In this case LIG may be regarded as a type of modulation spectroscopy. The spacing of the grating Λ is the same in vacuum and in the medium and is given by (λ = wavelength)

$$\Lambda = \lambda/2\,\sin(\theta/2). \tag{10}$$

Depending on the thickness d of the sample, two different types of gratings can be formed. Thin gratings for

$$d \ll 2\Lambda^2/\lambda. \tag{11}$$

This grating produces several diffracted orders as usual two-dimensional gratings do.

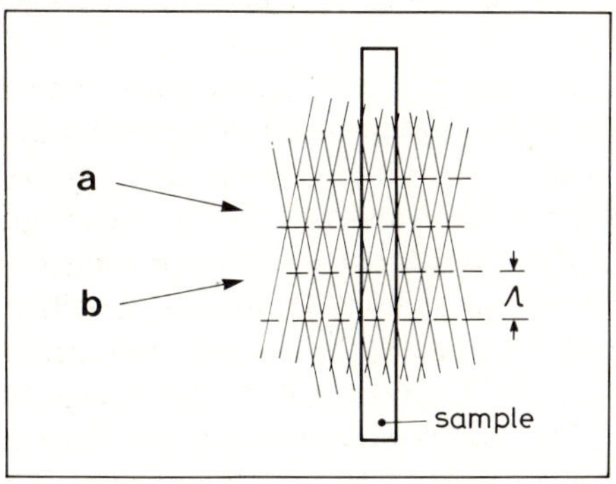

Fig. 1. Schematic drawing of the interference of two coherent laser beams, producing a grating in a nonlinear optical medium.

For

$$d \gg 2\Lambda^2/\lambda \tag{12}$$

we have a thick grating. In this case the diffracted beams have to fulfill the Bragg condition. If the beams which are forming the grating are simultaneously used as test beams, beam a will produce only one diffracted order (apart from zero order) which goes into the direction of beam b and vice versa.

If the two beams a and b have equal frequencies the diffraction grating is stationary in space. Consequently the diffracted beams will have the same frequency as the incident ones. If the two beams have different frequencies, a grating will be produced which moves through the sample. Due to the Doppler shift, the diffracted orders (except 0 order) will have a frequency shift with respect to the incident beam.

LIG have been investigated in CdS mainly at room temperature and far from the electronic resonances [9,10]. Correspondingly vague are partly the ideas about the physical origines mainly "free carries or temperature gratings" (see e.g., [9]). [11] is to our knowledge the only work, where LIG have been investigated in CdS at low temperature and in the vicinity of the exciton resonance. As can be deduced from the data in [11] the intensity of the main beam was about 3 MW/cm^2. This is the regime where EHP phenomena dominate in CdS in contrast to the assumption in [11] where the results have been interpreted in close analogy to CuCl where indeed biexciton effects are responsible for the formation of LIG. Here we have investigated the intensities of the diffracted orders (in other words the efficiency) of LIG in dependence of $\hbar\omega_{\mathrm{exc}}$ and I_{a} and I_{b}. I_{a} and I_{b} come from one excimer laser pumped dye laser of 3 ns and 0.03 meV temporal and spectral halfwidth, respectively. In some experiments I_{a} and I_{b} have been varied simultaneously, keeping their ratio constant. In other experiments I_{b} has been varied alone with fixed I_{a}. The geometry is shown schematically in the insert of fig. 2. The labelling of the various incident and diffracted beams is also indicated there. The two incident beams are polarized with $\boldsymbol{E} \perp \boldsymbol{c}$, $\boldsymbol{c}$ being the crystallographic axis. Samples of about 7 μm thickness have been used. The angle between beam a and b was 1.6° resulting in $\Lambda = 17.9$ μm, i.e., we are dealing with a thin grating. The sample temperature is about 7 K. The transmitted and diffracted beams are focussed on the vidicon target of an optical multichannel analyzer being used for an evaluation of the data. The lowest intensity at which we were able to detect diffracted orders from LIG was about 100 W/cm^2. From this value we increased the intensity up to 1 MW/cm^2. At higher intensities it was easy to observe orders up to 5. The intensity is decreasing rapidly with increasing order. In agreement with this, a direct imaging of the grating in the sample on the vidicon reveals only minor deviations from a sinusoidal structure. If some of the orders were filtered out the image changed according to Abbé's laws.

From the dependence of the LIG efficiency on $\hbar\omega_{\mathrm{exc}}$ and $I_{\mathrm{a,b}}$ three main

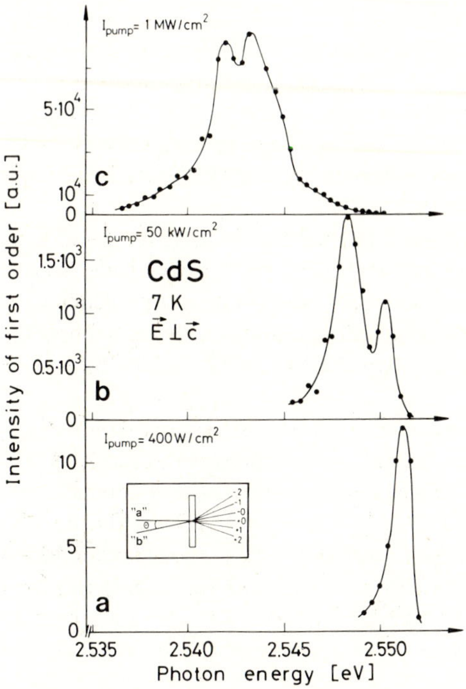

Fig. 2. The intensity of the first diffracted order of a laser-induced grating in CdS at 7 K as a function of $\hbar\omega_{exc}$ for three different excitation intensities. The insert shows schematically the geometrical arrangement.

processes have been found contributing to the formation of LIG. Typical spectra are shown in fig. 2. At intensities from 100 W/cm^2 up to about 1 kW/cm^2 the spectral width of the $A\Gamma_5$-exciton absorption band increases due to collision broadening. Consequently diffraction from LIG is observed at the flank of the absorption edge (fig. 2a). The intensity and thus the absorption coefficient are larger in the region of constructive interference. The LIG is therefore an amplitude structure.

At intensities from about 5 to 50 kW/cm^2 an additional process arises. The LIG is now mainly due to two-photon transitions to the biexciton. If the sample is illuminated with intense light of energy $\hbar\omega_{exc}$, photons with energy $\hbar\omega_{abs} = E_{biex} - \hbar\omega_{exc}$ are absorbed in a two-photon transition to the biexciton. The biexciton is a bound state of two excitons with eigenergy E_{biex}. The oscillator strength of this new excitation induced resonance at $\hbar\omega_{abs}$ is proportional to the polariton density or light intensity at $\hbar\omega_{exc}$. The excitation-induced peak in the imaginary part of $\Delta\tilde{n}$ (I_{exc}, $\hbar\omega_{exc}$) is connected with a resonance-like structure in the real part (according to Kramers–Kronig relations). For experimental and quantitative theoretical investigation of this additional resonance see, e.g., [3,6,8,12,13] and the literature cited therein. If $\hbar\omega_{exc}$ is tuned around $\frac{1}{2}E_{biex}$, the laser beam scans this additional resonance, resulting in the characteristic resonance structure seen in fig. 2b. The reabsorption dip situated in the center of the resonance corresponds to $\hbar\omega_{exc}=\frac{1}{2}E_{biex}$. From this one obtains for CdS $E_{biex}=5.0994$ eV. This value is in exact agreement with data deduced independently from two-photon Raman spectroscopy (TPRS) or luminescence-assisted two-photon spectroscopy (LATS) [13].

The appearance of the diffracted orders which are discussed as results of LIG in the wave picture can be equally well described in the quasiparticle picture [11, 14]. Two quanta of the beams a and/or b virtually excite a biexciton which then decays stimulated by beam a or b under energy and momentum conservation. More details about this aspect and the relation to TPRS will be published elsewhere [14].

At the highest excitation intensities around 1 MW/cm^2 the excitation spectrum of LIG is further shifting to the red and it becomes very broad (fig. 2c). At these excitation conditions the exciton concept breaks down and a new collective phase is formed the so-called electron–hole plasma (EHP). For more details about the EHP in CdS see the next section or [15,16] and the literature cited therein. The comparison of fig. 2c with fig. 4 shows that the LIG is due to an amplitude grating at photon energies above 2.54 eV and to a phase grating below. As usual this phase grating has lower efficiency as compared to the amplitude grating.

3. Electron–hole plasma

If the density of electron–hole pairs in a semiconductor is increased to sufficiently high values, the Coulomb interaction between electron and holes is screened to an extent that excitons cease to exist as individual bound pair states. Instead a new collective phase is formed, the EHP. This phase transition has consequences to the band structure. The forbidden gap decreases monotonously with increasing plasma density n_p mainly due to exchange and correlation energy. Generally a rigid shift of the bands is assumed. If the plasma temperature T_p is not too high, the population of the carriers in the bands becomes degenerate. In direct gap materials with dipole-allowed band-to-band transition one then observes optical amplification (i.e. gain) in the spectral region between the reduced gap $E_g'(n_p)$ and the chemical potential $\mu(n_p, T_p)$. The chemical potential is identical to the energetic distance between the quasi-Fermi levels of electrons and holes. The samples are transparent for $\hbar\omega < E_g'$ and absorbing for $\hbar\omega > \mu$. For details of the theoretical description of the EHP in CdS see, e.g., [8,17].

The most reliable experimental method to collect information about the EHP is the excite and probe technique. A spectrally broad, weak test beam is sent on the sample and its reflection or transmission is measured without and with an additional, intense, spectrally narrow pump beam. Care has to be taken about the spatial and temporal coincidence of both beams. Figure 3 gives transmission spectra of a probe beam (30 μm and 3 ns spatial and temporal half-widths, respectively) through an unexcited sample (solid line) and spatially and temporally centered on an excitation pulse of 50 μm diameter and a temporal FWHM of 5 ns (dashed line).

The transmission spectrum through the unexcited sample exhibits a modulation due to Fabry–Perot interferences in the platelet (thickness $d \approx 2.5\,\mu m$) and the absorption band of the $A\Gamma_5$ exciton. The excited sample shows gain for low photon energies, the absorption above the chemical potential and a blue shift of the mode structure indicating a variation of the real part of $\tilde{n}$ [16]. By fitting the gain and absorption spectra with theoretical models the plasma parameter n_p, T_p, $E'_g(n_p)$ and $\mu(n_p, T_p)$ may be determined. The fit has to take into account many-particle effects like an energy dependent damping, the "excitonic enhancement" [8,17], and – depending on the experimental conditions – also a drift of the plasma in case of nonhomogeneous excitation. From this type of investigation, from gain and reflection excitation spectroscopy [15] and from spatially resolved transmission and reflection spectra [18] a set of the plasma parameters has been deduced which is given in the following [15,18,19]: plasma density n_p increasing with excitation intensity from 10^{18} cm^{-3} to 4×10^{18} cm^{-3}; chemical potential of the plasma $\mu \approx 2.54$ eV slightly decreasing with n_p; drift length under inhomogeneous excitation conditions $l_D = (3 \pm 2)\,\mu m$; drift

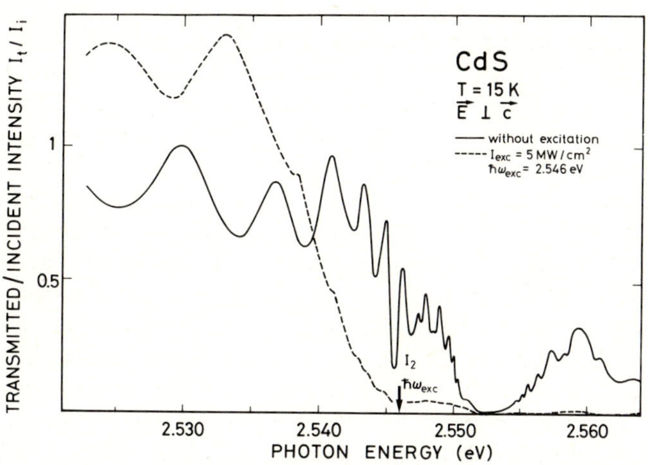

Fig. 3. Transmission spectra of a weak probe beam through a 2.5 μm thick CdS platelet at 5 K. Solid line: without additional excitation. Dashed line: with additional excitation ($I_{exc} = 5$ MW/cm^2; $\hbar\omega_{exc} = 2.546$ eV, $\boldsymbol{E} \perp \boldsymbol{c}$).

velocity $v_D \lesssim 2\times10^6$ cm/s. From the latter two data an effective lifetime τ_p of (150 ± 50) ps results.

The above mentioned set of data fulfills the necessary conditions

$$n_p = I_{exc}\tau_p/\hbar\omega_{exc}a. \tag{13}$$

Where a is given by l_D or the inverse of the absorption coefficient α^{-1}, whatever is longer. Larger deviations from the above mentioned set of data lead to internal contradictions. A few comments shall be made: The plasma does not reach the liquid like quasiequilibrium state since n_p increases with I_{exc} at constant T_p [15,19]. The attempt to fit the gain spectra with recombination under k-conservation and drift alone, neglecting many-particle effects is inadequate. It overestimates v_D and requires values of n_p which go considerably beyond those compatible with eq. (13) [18,19]. The lifetime determined here from l_D and v_D is in agreement with values determined in a completely different way [20]. For the relation between drift and Fermi velocities of electrons and holes the two inequalities hold:

$$v_D^e \ll v_F^e; \qquad v_D^h < v_F^h \tag{14}$$

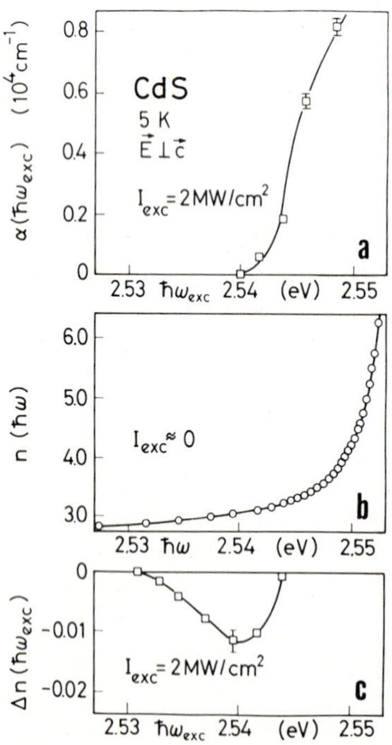

Fig. 4. The coefficient of absorption under high excitation at the position of the exciting laser itself. The absorption edge raises without excitation around 2.55 eV (see fig. 3) (a), the refractive index of the unexcited sample (b) and its variation at the position of the exciting laser itself (c).

indicating a "weak" drift [18,19]. The variation of Δn in the transition from a low density gas of excitons to the plasma, deduced from the shift of the Fabry–Perot modes is in agreement with theoretical calculations [8, 16]. Unfortunately this technique could be used only in the transparent spectral region. Recently the resonance region has been investigated by this technique in much thinner samples [21], however these experiments have been carried out only at $I_{exc} \leqslant 0.3$ MW/cm^2. This is just the onset of EHP phenomena. Furthermore the authors cannot exclude inhomogeneities which have been discussed already in [22].

The experimental data of fig. 3 can be exploited to determine $\Delta\tilde{n}(\hbar\omega, \hbar\omega_{exc}, I_{exc})$ i.e., the variation of $\tilde{n}$ at a certain energy $\hbar\omega$ produced by a laser situated at $\hbar\omega_{exc}$ (see also [16]). By extrapolating $\hbar\omega \rightarrow \hbar\omega_{exc}$ it is possible to deduce $\Delta\tilde{n}$ at the position of the exciting laser itself. Corresponding results are shown in fig. 4 for conditions at which a plasma is formed. The unexcited sample is transparent below the $A\Gamma_5$ exciton. At $I_{exc} \approx 1$ MW/cm^2 strong one-photon absorption occurs for $\hbar\omega_{exc} \gtrsim 2.54$ eV, i.e., above the chemical potential. Below this potential there is no efficient possibility of one-photon absorption, but, e.g., due to two-photon absorption a sufficient density of electron–hole pairs may be created to produce still a considerable variation $\Delta n(\hbar\omega_{exc})$. This point and the transition from low absorption at low excitation to high absorption at high excitation can be exploited to get optically bistable behaviour as will be discussed in the next section.

4. Plasma-induced bistability

An optically bistable device has under the same excitation conditions two reversible stable states of low and high transmission or reflection. The occurrence of one or the other state depends on the history. The presently most widely discussed optical bistabilities (OB) need the feedback from a Fabry–Perot resonator filled with an optically nonlinear medium. Bistability may occur either if n changes with the intensity or if the absorption bleaches with I_{exc}. The hysteresis loop in a plot of transmitted intensity versus incident intensity is revolved counterclockwise. A bistability of this type which involves $\Delta n(I_{exc})$ and partly also $\Delta\kappa(I_{exc}) < 0$ has been found, e.g., in GaAs and InSb (see [4,8,23] and the literature cited therein). A purely dispersive OB connected with bi-exciton transitions has been predicted by [8,24] and observed in CuCl [25]. Some indications in CdS in the spectral region just below 2.54 eV (see fig. 4) have been reported in [26]. More recent experiments with improved temporal resolution confirmed this phenomenon.

In this section we shall concentrate on an absorptive type of optical bistability which is caused by an increase of the absorption with increasing excitation $\Delta\kappa(I_{exc}) > 0$. The mechanism which will be described below has been presented, to our knowledge, for the first time at the spring meeting of the

German Physical Society in March 1983 by two of the authors (K.B. and C.K.) and has subsequently been published in [26,27]. Only a qualitative approach will be given here. For quantitative investigations the reader is referred to [28,29,30] and to the contributions of S.W. Koch and F. Henneberger to this volume.

The photon energy of a monochromatic laser pulse is selected in a spectral region, where the sample is transparent at low excitation and becomes strongly absorbing at high excitation. According to fig. 4 this is for $\boldsymbol{E}\perp\boldsymbol{c}$ the region between 2.54 eV and the free exciton absorption band around 2.552 eV or the transparent window between the $A\Gamma_5$ and $B\Gamma_5$ excitons. When the pulse rises, the sample is almost transparent. Due to two-photon band-to-band transition, or due to the weak one-photon absorption in the absorption tail, a small fraction of the incident light is absorbed, producing free carriers. If the density of free carriers is sufficiently high as to produce an EHP, the absorption edge shifts to lower values, and may be situated below $\hbar\omega_{exc}$. In this case the sample becomes opaque, the transmitted intensity switches to rather low values and the plasma is pumped very effectively by strong one-photon absorption. Due to this fact, the intensity may now be lowered to values considerably below those necessary to produce the switch down, and the sample remains in the absorbing state. At still lower values of I_{exc} even the efficient one-photon absorption is no more able to sustain the plasma, the carriers recombine and the sample becomes transparent again. As a consequence the sample has in a certain range of I_{exc} two stable states, one of high and one of low transmission. We call this type of absorptive OB an intrinsic one, for two reasons: it needs no optical feedback from an external cavity since the feedback occurs in the electronic system itself, and the excitation which is involved, namely the formation of an EHP, is also an intrinsic one. Figure 5 gives experimental results, confirming the above developed ideas. Figure 5a shows the incident and transmitted laser pulse shape recorded with a streak camera and an optical multichannel analyzer. The switch down is directly obvious, the switching up occurs at intensities which almost coincide with the noise level. The resulting hysteresis loop is shown in

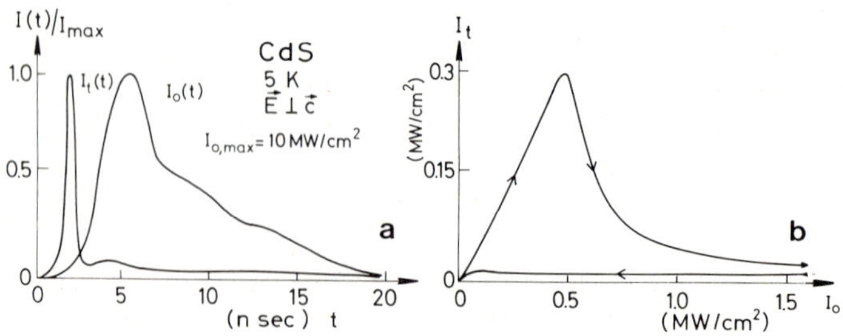

Fig. 5. The temporal shape of an incident and transmitted pulse $I_0(t)$ and $I_t(t)$ (a), and the resulting hysteresis loop $I_t = f(I_0)$ (b); $\hbar\omega_{exc} = 2.548$ eV.

fig. 5b. In contrast to the dispersive OB it is revolved clockwise. These data agree with those given in [26,27], however now with considerably improved time resolution. The main argument to attribute this OB to the formation of an EHP is the close coincidence of the spectral regions and intensities where this bistability can be observed [26] with data from plasma gain and reflection excitation spectroscopy [15] and with laser-induced grating spectroscopy (see above and [14,31]). Especially the transmission spectra of fig. 3 prove that in the maximum of the excitation pulse a plasma has been formed. A temporal analysis of the transmitted excitation pulse itself shows the absorptive OB. A hysteresis loop similar to fig. 5b has already been reported in [32], however without indication of its physical origin. Some details of the process shall be discussed below. The time resolution in fig. 5 is limited by the resolution of the OMA and the jitter of the system to about 200 ps. This is just the apparent switch down time τ_d (from 90% to 10%) in fig. 5a. The actual value of τ_d has been deduced from single-shot experiments to be $\tau_d = 70$ ps measured with optimized time resolution. From the physical point of view, there are two limiting factors for τ_d. One is the formation time of the EHP. If free carriers are created directly as in our experiment this value is supposed to be of the order of (or below) 10 ps. If excitons are excited resonantly the value may be somewhat longer (about 30 ps in GaAs [33].) The other constant is the time τ_g it takes the laser to increase the number of carriers from a value where the sample is still transparent to a value where it becomes opaque.

In the following we shall try to give some rough numbers. The absorbed excitation energy per time and volume is given by

$$dI_{exc}/dx = K_1 I_{exc} + K_2 I_{exc}^2, \tag{15}$$

if both one- and two-photon absorption contribute. For a photon energy of 2.548 eV K_1 has values at low excitation ranging from about 200 cm^{-1} to 10^3 cm^{-1} depending on the sample (see, e.g., [34]) and increasing with I_{exc} [35]. With $I_{exc} = 0.2$ MW/cm^2 and $K_1 = 500$ cm we find $K_1 I_{exc} = 10^8$ Wcm^{-3}. K_2 has, to our knowledge, not yet been measured at 5 K under the above conditions. Assuming that the matrix elements are rather energy and temperature independent, the value of $K_2 \approx 0.5 \times 10^{-7}$ cm/W [36] for 300 K and $\hbar\omega_{exc} = 1.78$ eV is scaled only by the influence of the resonance denominator to $K_2 \approx 2 \times 10^{-3}$ cm/W for 5 K and $\hbar\omega_{exc} = 2.548$ eV yielding $K_2 I_{exc}^2 = 0.8 \times 10^8$ W/cm^3. The generation rates G are proportional to $K_1 I_{exc} (\hbar\omega_{exc})^{-1}$ and $K_2 I_{exc}^2 (2\hbar\omega_{exc})^{-1}$, respectively. So the above conservative guess gives

$$G \gtrsim (K_1 I_{exc} + \tfrac{1}{2} K_2 I_{exc}^2)(\hbar\omega_{exc})^{-1} = 0.4 \times 10^{27}\ cm^{-3}\ s^{-1} \tag{16}$$

and $n_p = G\tau_p \gtrsim 0.5 \times 10^{17}$ cm^{-3}. This lower limit of n_p meets or even exceeds all values given in the literature for the Mott criterium in CdS [19], indicating again clearly that an EHP is formed. In order to shift the reduced gap $E'_g(n_p)$ from just above $\hbar\omega_{exc} = 2.548$ eV to below, i.e., to come from weak absorption to strong absorption, the plasma density has to be raised according to [17] from about 10^{17} cm^{-3} to 2×10^{17} cm^{-3}, i.e., $\Delta n_p \approx 10^{17}$ cm^{-3}. With eq. (13) limits for τ_g are

$$50 \text{ ps} \approx \Delta n_p \hbar\omega_{exc} a I_{exc}^{-1} < \tau_g < \Delta n_p G^{-1} \approx 250 \text{ ps}. \tag{17}$$

So this guess is in good agreement with the experimental value given above.

The actual intensity, where the switch-down condition is reached, depends on the carrier density and thus on the absorbed energy weighted with the carrier lifetime and the pulse rise time. By this aspect, parameters of the pulse enter, and an enhancement of the hysteresis may appear in fig. 5b in addition to the pure bistability. Similar arguments hold for the switch-up time τ_u. It has a lower limit given by the carrier lifetime which is of the order of 150 ps.

For the ratio between sample thickness d and the diffusion length l_D of the carriers there are two limiting cases. For $l_D > d$ the carrier-density and thus the optical properties are homogeneous through the sample. In the opposite case some oscillatory behaviour may appear (see the contribution by S. Koch). The experiments carried out here correspond to $l_D \lesssim d$.

If the density of the plasma is raised to a critical value n_p^c given by $\mu(n_p^c) = \hbar\omega_{exc}$ the sample becomes transparent again. Consequently the generation rate decreases, so does n_p and absorption sets in. This may lead to self-pulsation in the transmitted beam, but there are up to now no experimental results.

If the sample is illuminated with a laser beam, which has laterally, e.g., a Gaussian profile, it happens that the crystal switches in the central parts to low transmission, while the incident intensity is still transmitted in a ring-shaped structure around the centre. This has been observed by eye in our experiments looking on the sample through a microscope. Such a ring-shaped intensity distribution produces in the far-field an interference pattern. Thus great care has to be taken, if interference patterns which occur in connection with absorptive OB are evaluated to deduce variations of the real part of the refractive index [37,38].

A similar model as discussed here has been investigated theoretically in [30], however it is assumed that the increase of absorption comes only from an increasing damping of the excitons, e.g., due to collision broadening. It has been found, that only a hysteresis but no OB occurs, if one assumes that the exciton-damping Γ increases linearly with excitation, while a superlinear dependence of Γ on I_{exc} will lead to multivalued solutions and an OB. Such nonlinearity is likely to appear. The experimental conditions chosen in [30] $I_{exc} \gtrsim 6$ MW/cm^2; $\Delta E = E_{B\Gamma_1} - \hbar\omega_{exc} = 7$ meV for $\boldsymbol{E} \| \boldsymbol{c}$ will lead definitely to the formation of a plasma as can be judged from the above numbers and all transmission measurements [15,16,18,35] including plasma-gain excitation spectroscopy. For the transition to the plasma, the theory predicts true OB [28] in agreement with our experimental findings. The increase of the temporal resolution from [26] to the present results also shows a significant difference between the measured and calculated hysteresis loops involving an EHP as compared to those calculated with the exciton-damping model. Hysteresis loops which may be explained by the model of [30] can be observed at much lower excitation intensities around 50 kW/cm^2. Figure 6 gives an example.

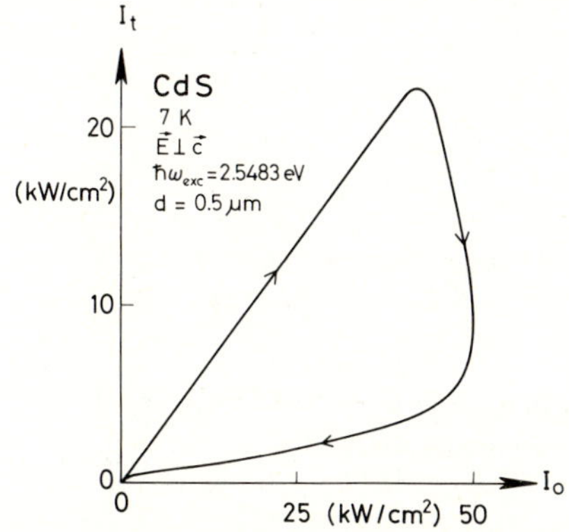

Fig. 6. Hysteresis loop observed in a 0.5 μm thick CdS sample at excitation conditions, where the formation of a EHP is still rather unlikely ($\hbar\omega_{exc} = 2.5483$ eV; $I_{exc} = 50$ kW/cm^2).

The type of resonatorless optical bistability making use of an increase of the absorption with increasing excitation seems to be of rather general nature as discussed, e.g., in [39]. There an absorptive bistability has been reported in multiple quantum wells of GaAs/Ga$_x$Al$_{1-x}$As which is also connected with an excitation induced red shift of the absorption edge, connected in this case, however with lattice heating. The transverse bistability, which has been observed under cw excitation in CdS at photon energies just below the I_2 bound exciton absorption peak [37,38] is explained meanwhile also along the same lines [38] involving again some lattice heating. The laser is tuned on the low energy wing of the I_2 absorption. Under d.c. excitation of the order of 1 kW/cm^2 this results in some heating of the sample, which causes a decrease of the gap and thus of E_{I_2}. This in turn shifts the center of the I_2 absorption towards $\hbar\omega_{exc}$. The resulting increase of the absorption allows to sustain the temperature rise with lower input intensity, thus resulting in OB. If I_{exc} is raised to values which cause the I_2 absorption to shift below $\hbar\omega_{exc}$ a self-pulsation may occur in the transmitted intensity. Finally the bistability observed in CdS under degenerate mixing of ruby-laser light at room temperature should be mentioned [10]. The types of absorptive bistability discussed above are much simpler than the dispersive ones concerning, e.g., the demands to the optical quality of the sample surfaces, their thickness and the coherence or monochromastism of the light. Thus they may have considerable technical interest in the future.

Acknowledgements

The work reported here is a project of the Sonderforschungsbereich Festkörperspektroskopie and is financed by the Deutsche Forschungsgemeinschaft. The high quality CdS samples are grown in the Kristall-und Materiallabor der Universität Karlsruhe.

References

[1] I. Broser, R. Broser and M. Rosenzweig, in: Landolt–Börnstein, New Series, Group III, Vol. 17b, O. Madelung, M. Schulz and H. Weiss, eds. (Springer, Berlin, 1982) p. 166;
I. Broser and M. Rosenzweig, Lecture Notes in Physics, Vol. 177, G. Landwehr, ed. (Springer, Berlin, 1983) p. 229.

[2] G. Blattner, G. Kurtze, G. Schmieder and C. Klingshirn, Phys. Rev. B 25 (1982) 7413.

[3] Topics in Current Physics, Vol. 19, K. Cho, ed. (Springer, Berlin, 1979).
Modern Problems in Condensed Matter Sciences, Vol. 2, Excitons, E.I. Rashba and M.D. Sturge, eds. (North-Holland, Amsterdam, 1982).
B. Hönerlage, R. Levy, J.B. Grun, K. Bohnert and C. Klingshirn, to be published in Phys. Rep.

[4] A. Müller, D.A.B. Miller and S.D. Smith, Advances in Physics, 30 (1981) 697.

[5] P.A. Franken and J.F. Ward, Reviews of Modern Physics 35, (1963) 23.
N. Bloembergen, Nonlinear Optics (W.A. Benjamin Inc., New York, 1965).

[6] C. Klingshirn and H. Haug, Phys. Rep. 70 (1981) 315.

[7] One of the first occasions at which this aspect has been stressed was the "Intern. Conf. on Excited States and Multiresonant Nonlinear Optical Processes in Solids", Aussois, France (1981).

[8] H. Haug, Advances in Solid State Physics, Vol. XXII, 149 (1982).
H. Haug and S. Schmitt-Rink, Electron theory of the optical properties of highly excited semiconductors, to be published in Progress in Quantum Electronics.

[9] H.J. Eichler, Advances in Solid State Physics, Vol. XVIII, 241 (1978).

[10] A. Borshch, M. Brodin, V. Volkov and N. Kukhtarev, Optics Commun. 41 (1982) 213;
M.S. Brodin, A. Borshch, V.I. Volkov, N.V. Kukhtarev and V.N. Sharkov, in: Proc. Intern. Conf. Lasers (1982).

[11] A. Maruani, D.S. Chemla and E. Batifol, Solid State Commun. 33 (1980) 805;
A. Maruani and D.S. Chemla, Phys. Rev. B 23 (1981) 841.

[12] K. Kempf, G. Schmieder, G. Kurtze and C. Klingshirn, Phys. Statatus Solidi (b)107 (1981) 297.

[13] V.G. Lyssenko, K. Kempf, K. Bohnert, G. Schmieder, C. Klingshirn and S. Schmitt-Rink, Solid State Commun. 42, (1982) 401.

[14] H. Kalt, Ph.D. Thesis, Frankfurt (1985) to be published.

[15] K. Bohnert, A. Anselment, G. Kobbe, C. Klingshirn, H. Haug, S.W. Koch, S. Schmitt-Rink and F. Abraham, Z. Phys. B 42 (1981) 1.

[16] A. Kreissl, K. Bohnert, V.G. Lyssenko and C. Klingshirn, Phys. Status Solidi (b)114 (1982) 537.

[17] M. Rösler and R. Zimmermann, Phys. Status Solidi (b)83 (1977) 85.

[18] K. Kempf and C. Klingshirn, Solid State Commun. 49 (1984) 23.

[19] K. Kempf, Ph.D. Thesis, Frankfurt (1984) to be published.

[20] H. Yoshida, H. Saito, S. Shionoya and V.B. Timofeev, Solid State Commun. 33 (1980) 161;
H. Yoshida, H. Saito and S. Shionoya, J. Phys. Soc. Japan 50 (1981) 881.

[21] F. Henneberger, J. Puls, H. Roßmann and I.N. Uraltsev, Phys. Status Solidi (b) 121 (1984) K191.
[22] K. Bohnert, G. Schmieder and C. Klingshirn, Phys. Status Solidi (b)98 (1980) 175.
[23] H.M. Gibbs, S.L. McCall, T.N.C. Venkatesan, A.C. Gossard, A. Passner and W. Wiegman, Appl. Phys. Lett. 35 (1979) 451.
[24] S.W. Koch and H. Haug, Phys. Rev. Lett. 46 (1981) 450;
E. Hanamura, Solid State Commun. 38 (1981) 939.
[25] B. Hönerlage, J.Y. Bigot and R. Levy, Proc. Conf. on Optical Bistability, Rochester (1983);
N. Peyghambarian, D. Sarid and H.M. Gibbs, ibid.;
R. Levy, J.Y. Bigot, B. Hönerlage and J.B. Grun, Solid State Commun. 48 (1983) 705;
N. Peyghambarian, H.M. Gibbs, M.C. Rushford and D.A. Weinberger, Phys. Rev. Lett. 51 (1983) 1692.
[26] K. Bohnert, H. Kalt and C. Klingshirn, Appl. Phys. Lett. 43 (1983) 1088.
[27] K. Bonhert, in: Proc. Summerschool on "Energy Transfer Processes in Condensed Matter", Erice, Sicily, June 1983), to be published.
[28] H.E. Schmidt, H. Haug and S.W. Koch, Appl. Phys. Lett. 44, (1984) 787.
[29] F. Henneberger and H. Roßmann, Phys. Status Solidi (b)121 (1984) 685.
[30] Hoang Xuam Nguyen and R. Zimmermann, Phys. Status Solidi (b) 124 (1984) 191.
[31] C. Klingshirn, K. Bohnert, K. Kempf and H. Kalt, to be published in Transactions of the Royal Society (1984).
[32] H. Rossmann, F. Henneberger and J. Voigt, Phys. Status Solidi (b) 115 (1983) K63.
[33] G.W. Fehrenbach, W. Schäfer, J. Treusch and R.G. Ulbrich, Physica 117B + 118B (1983) 303.
[34] M. Rosenzweig, Ph.D. Thesis, Berlin (1982).
[35] K. Bohnert, F. Fidorra and C. Klingshirn, Z. Phys., to be published.
[36] G. Kobbe and C. Klingshirn, Z. Phys. B 37 (1980) 9.
[37] M. Dagenais and H.G. Winful, Appl. Phys. Lett. 44 (1984) 574.
[38] M. Dagenais, to be published in Transactions of the Royal Society (1984);
M. Dagenais and W.F. Sharfin, to be published in Appl. Phys. Lett.
[39] D.A.B. Miller, to be published in Transactions of the Royal Society (1984);
D.A.B. Miller, A.C. Gossard and W. Wiegmann, to be published in Opt. Lett. (1984).

Journal of Luminescence 30 (1985) 204–216
North-Holland, Amsterdam

OPTICAL NONLINEARITY AND BISTABILITY OF CdS

F. HENNEBERGER, J. PULS and H. ROSSMANN
Humboldt-Universität zu Berlin, Sektion Physik, Bereich Halbleiteroptik, GDR

Experimental data concerning the nonlinear optical behaviour of CdS are reported. At cryo-temperatures an increasing nonlinear absorption at the exciton resonance is observed, with the aid of which optical hysteresis is demonstrated. At room-temperature saturation of the band-edge absorption is found at intensities as low as 10 kW/cm^2. The switch-on time is determined to be smaller than 6 ps.

1. Introduction

Nowadays the nonlinear optical properties of semiconductors are being extensively investigated with regard to an application for high-speed optical signal processing and laser design. Especially, new phenomena, as for example optical bistability, have attracted much interest. We report here experimental data for CdS at both cryo- and room-temperature, which should apply to most wide-gap II–VI compounds. The findings emphasize the potential applications of these materials in all optical devices.

2. Experimental

The samples investigated were platelet-like CdS-crystals grown from the vapour phase with thicknesses ranging from 0.5 to 5 μm. They were selected with respect to a high perfection of the surfaces. Excite-and-probe transmission experiments were performed using two dye lasers with a pulse duration of about 4 ns (narrow-band excitation with $\Delta\lambda = 0.03$ nm, broad-band probing with dye continuum). The transmitted radiation of the probing beam was analyzed by an OMA-2 system. In resonant single-beam experiments the nonlinear behaviour of the transmission under continuous variation of the intensity was deduced from pulse shape deformation measurements. Fast photodiodes and boxcar integrators (time resolution 600 ps) were used for simultaneous detection of both incident and transmitted pulses. Response times were studied using

0022-2313/85/$03.30

frequency-doubled single 6 ps pulses of a Nd:glass laser system for both excitation and probing. By a multi-probe beam arrangement the temporal behaviour of the transmission could be determined from a single laser shot.

3. Cryo-temperatures: increasing nonlinear absorption and resonatorless optical bistability

3.1. Exciton absorption under optical excitation

At low temperatures ($T = 1.8$–77 K) the transmission spectrum of CdS below the band edge is governed by pronounced exciton lines. The effect of high carrier concentrations on the spectrum was studied extensively in a number of papers, e.g., [1–3] and references therein, by excite-and-probe experiments. Main attention was paid to the occurrence of optical gain at the low-energy side of the spectrum as an evidence of an electron–hole plasma. On the other hand, an increasing absorption was observed with growing excitation at higher energies, being most pronounced in the spectral range of the lowest exciton line ($A_{n=1}$). Changes of the refractive index at this position were found to be comparatively small [3]. This increasing absorption dominating clearly refractive effects seems to be a pecularity of wide-gap semiconductors with large exciton binding energies (CdS: 28 meV) compared to materials with smaller energy gap as, e.g., GaAs [4], where the exciton line saturates like a homogeneously broadened resonance accompanied with strong refractive changes.

Figure 1 shows typical results of excite-and-probe measurements at $T = 77$ K for two, not specially selected spectral positions slightly below the $A_{n=1}$ exciton

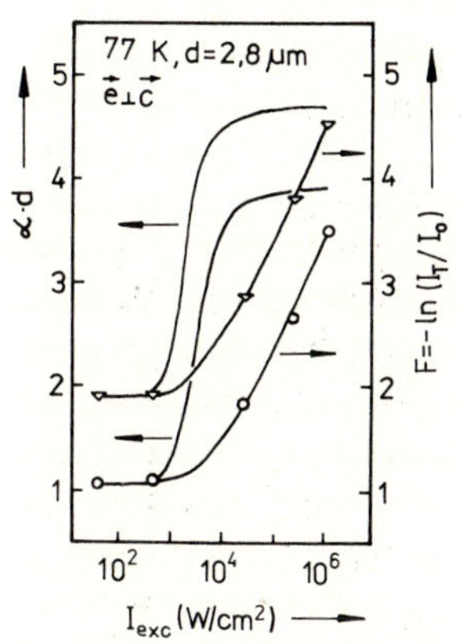

Fig. 1. Effective optical thickness F and calculated αd of the probing beam versus excitation intensity (energies of probing photons are 2.5333 eV (○) and 2.5362 eV (▽): probe beam polarization $\boldsymbol{e}$ is perpendicular to the crystal axis $\boldsymbol{c}$).

resonance ($E_{ex}^{A} = 2.545$ eV). The frequency of the excitation beam is adjusted to the $B_{n=1}$ resonance ($E_{ex}^{B} = 2.560$ eV). A strong decrease of the probe beam transmittivity with growing excitation is observed. The relation required for the physical understanding is the dependence of absorption coefficient α on the density n of photoexcited species or as a first step, on the excitation intensity. However, the quantity measured, namely, $F(I_{exc}) = -\ln(I_T/I_0)$, where I_{exc} is the excitation intensity at the front side of the sample and I_0 and I_T the in- and outcoming probe intensity, respectively, may not be directly identified with αd, because the sample thickness d is certainly larger than the penetration depth of the exciting light ($\alpha_{exc}^{-1} < 1\ \mu$m) leading to an inhomogeneous excitation profile. This fact seems to be disregarded sometimes. At the spectral positions investigated multiple-beam interferences are of minor importance and excitation-induced changes of the reflectivity are small [2]. Then we have, apart from the constant $-\ln(1-R)^2$,

$$F(I_{exc}) = \int_0^d \mathrm{d}x\, \alpha[I_{exc}(x)]. \tag{1}$$

Assuming $I_{exc}(x) = (1 - R_{exc}) I_{exc} \exp(-\alpha_{exc} x)$, that is, linear absorption of the excitation beam, we find

$$\alpha(I_{exc}) = \alpha[I_{exc} \exp(-\alpha_{exc} d)] + I_{exc} \alpha_{exc} \frac{\mathrm{d}F(I_{exc})}{\mathrm{d}I_{exc}}, \qquad (1 - R_{exc} \approx 1), \tag{2}$$

when (1) is derived with respect to I_{exc}. This relation allows for the construction of $\alpha(I_{exc})$ from $F(I_{exc})$. Characteristically, F itself does not enter but $\mathrm{d}F/\mathrm{d}I_{exc}$. Often the inhomogenity is accounted for by a two-layer model. The front layer having the thickness α_{exc}^{-1} is supposed to be homogeneously "filled" with excitation ($\alpha(I_{exc})$), while the second is considered as "empty" ($\alpha(0)$). This leads to $\alpha(I_{exc}) = \alpha(0) + \alpha_{exc}[F(I_{exc}) - \alpha(0)d]$, which deviates considerably from (2). Namely, excitation-induced changes of α are not pinned to a fixed layer, but affect a growing part of the sample with increasing I_{exc}. The analysis of our experimental data is complicated by nonlinear absorption of the excitation beam at the $B_{n=1}$ exciton, where saturable absorption is observed [5]. Therefore, a procedure to construct α in such a situation was also developed. Because the features of α already follows from (2) we refer to [6] for the details and present here only the final results (fig. 1). α increases rather abruptly in a relatively narrow intensity range. The plateau-like behaviour for larger I_{exc} results from the nearly constant derivative of F. Neglecting diffusion, which seems to be justified in view of recent findings [7] for CdS also at high carrier densities under excite-and-probe conditions, $\alpha(n)$ follows from $I_{exc} \sim n^{r}$, where r denotes the type of carrier recombination. $\alpha(n)$ deduced for the opposite case of a diffusion-controlled $n(x)$ is almost identical with the former apart from a rescaling.

Most probably, the behaviour found for α is due to a gap-shrinking caused by the dense gas of photoexcited excitons (excitonic screening [8]) leading to a transition from excitonic to band-to-band absorption. At the beginning an increase of exciton damping is contributing too. Slow thermal effects are ruled out by the results of measurements with optical delay between excite-and-probe pulses [6].

3.2. Optical hysteresis

The afore-described increasing absorption acts as a strong optical nonlinearity in a single-beam experiment, where pumping and probing are realized by the same beam. In 1982 we carried out the first experimental investigation [9] of this nonlinearity by studying the deformation of high-intensity ns laser pulses passing through the CdS sample resonantly in the A-line region at $T = 77$ K. Based on the short carrier lifetime of CdS in the ps range [10], steady-state conditions were expected to be fulfilled during the pulse. In the meantime similar experiments have been made by different groups at $T \approx 4$ K [11,12] and $T = 27$ K near the $B_{n=1}$ resonance [13].

Our findings are summarized in fig. 2. Above a critical input peak intensity I_{max} being about 40 kW/cm^2 in the example shown in fig. 2(a), a characteristic deformation appears at the trailing edges of the pulses. The temporal behaviour of the transmission for the different pulses is presented in fig. 2(b) demonstrating the transition from a lower to a higher absorbing state with a switching moment

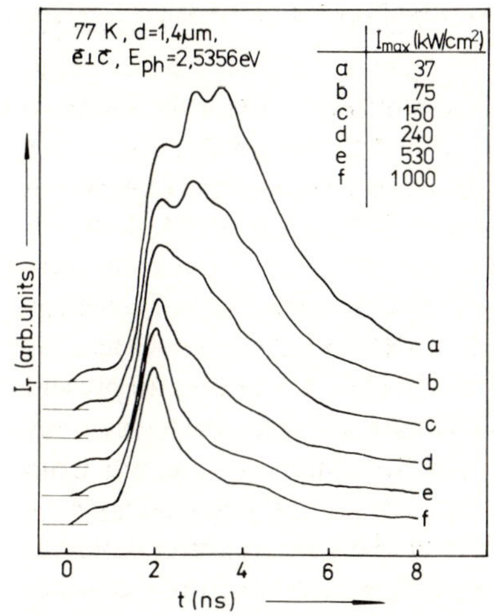

Fig. 2. Results of single-beam experiments on CdS at photon energies E_{ph} below the $A_{n=1}$ exciton. (a) Shapes of transmitted pulses in dependence on the input peak intensity I_{max}. The uppermost pulse (a) is identical with the incident pulse apart from a constant factor.

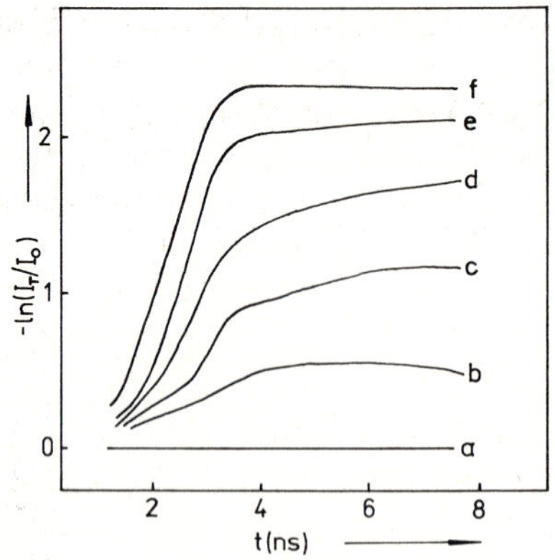

Fig. 2. (b) Temporal behaviour of the effective optical thickness according to fig. 2(a). The zero level is defined with respect to pulse (a).

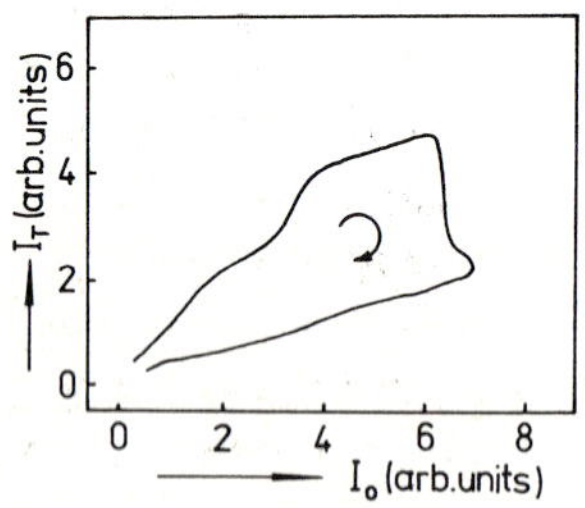

Fig. 2. (c) Plot of momentary incident versus momentary transmitted intensity for pulse (f) in 2(a).

of about 3 ns independent of I_{max}. In fig. 2(c) the momentary transmitted intensity is plotted versus the momentary input intensity for the largest I_{max}. A pronounced hysteresis cycle appears, which is, however, passed through in an opposite manner compared to the hysteresis of a FP-resonator filled with a saturable absorber.

For pulsed excitation the observation of hysteresis does not necessarily imply an optical bistability. Hysteresis appears also as transient effect in a monostable system, if the pulse duration is of the order of its relaxation time. The latter case was treated quantitatively in [13] for a linear density-dependence of the exciton damping, which excludes bistability. Enhanced by the nonlinearity of the system, distinct hysteresis appears even if the pulse duration is four times larger than the exciton lifetime. From the experimental point of view the decisive argument for a transient effect is, however, that the switching moment has to depend on the pulse area. This was indeed observed in [11] and [13] and, consequently, the hysteresis found was explained as transient phenomenon in [13]. We find, however, the switching moment to be independent of the pulse area, but fixed to the time moment, where the input pulse reaches its maximum (fig. 2(b)). An explanation in terms of a saturation level fails, because the absorption in the final state grows with I_{max}. At present, the origin for the differences between our results and those of [11,13] is not clear. In [13] pulses being distinctly shorter were used. In what follows, our findings are discussed with regard to bistability.

3.3. Resonatorless bistability

The method of achieving optical bistability (OB) most investigated experimentally involves combining a nonlinear optical medium with a FP-cavity providing the feedback [14]. Recently, growing interest has been paid to new variants of OB, where the cavity is replaced by other feedback mechanisms. As we have pointed out [15], resonatorless OB can be achieved in case of increasing absorption. Here, an intrinsic feedback based on the balance of optical generation and recombination of carriers is acting. The capability of this OB for comparatively broad-band, wavelength-insensitive operation is particularly attractive. The propagation of light with intensity I through such a medium is described by the equations

$$\frac{\partial I}{\partial x} = -\alpha(n)I, \qquad \frac{\partial n}{\partial t} = \frac{\partial}{\partial x} D(n) \frac{\partial}{\partial x} n + \left[\frac{\alpha(n)}{\hbar\omega} I - R(n)\right], \tag{3}$$

with $R(n)$ as recombination term. Under the condition $\mathrm{d}(\ln \alpha(n))/\mathrm{d}n > \mathrm{d}(\ln R(n))/\mathrm{d}n$, the balance of generation and recombination (term in brackets equal to zero) has two stable values of n for a given I in a certain intensity range. As seen from fig. 1, $\alpha(n)$ found at the A-exciton of CdS is well suited to achieve OB.

What is the transmission characteristic of such a system in the stationary case ($\partial n/\partial t = 0$)? For simplicity $\alpha(n)$ is modelled by a step-function, that is, $\alpha = \alpha_1$ for $n < n_c$ and $\alpha = \alpha_2$ for $n > n_c$ with some critical density n_c, and $R = n/\tau$ is used. As a first step, we ignore particle diffusion ($D = 0$). Then, n passes the hysteresis shown in fig. 3(a) at any point x of the sample ($0 \leqslant x \leqslant d$). Consequently, if the input intensity I_0 exceeds $I_\uparrow$, the sample is divided in two parts with $n > n_c$ and $n < n_c$, respectively, the boundary of which, x_c, migrates to larger x with growing I_0. Because the light intensity is given by $I(x) = I_0 \exp(-\alpha_2 x)$ for $x < x_c$ or $I(x) = I_0 \exp(\alpha_1 x_c - \alpha_2 x_c - \alpha_1 x)$ for $x > x_c$, it holds $I_\uparrow = I_0 \exp(-\alpha_2 x_c)$. When, however, I_0 is lowered, x_c remains fixed at the position x_c^{max} reached for the maximum input intensity I_0^{max} as long as $I(x_c^{max})$ equals $I_\downarrow$. After that, x_c returns to $x = 0$ according to $I_\downarrow = I_0 \exp(-\alpha_2 x_c)$. The corresponding hysteresis of x_c, and by this, of transmission, is shown in fig. 3(b). Characteristically, its shape depends on the maximum input. Only if I_0 exceeds $I_\uparrow \exp(-\alpha_2 d)$ a unique picture is obtained.

In the above treatment the particle density makes a jump at the boundary x_c from n_c to $n_\uparrow$ or $n_\downarrow$. However, diffusion acts very effectively at the boundary,

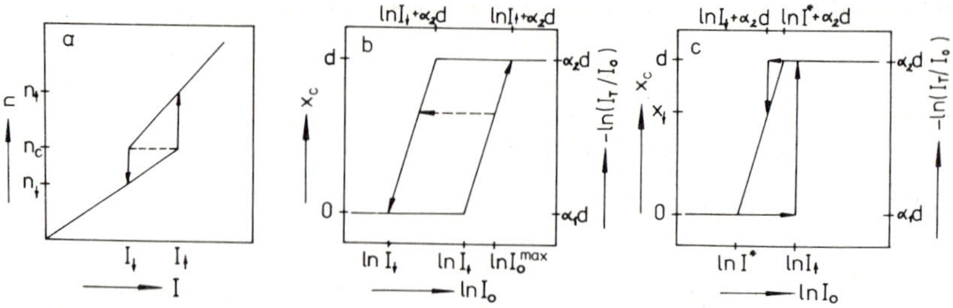

Fig. 3. Resonatorless optical bistability for the model described in the text. (a) Local hysteresis of n versus $I(I_\uparrow=\hbar\omega n_c/\tau\alpha_1,\ I_\downarrow=\hbar\omega n_c/\tau\alpha_2,\ n_\uparrow=(\alpha_2/\alpha_1)n_c,\ n_\downarrow=(\alpha_1/\alpha_2)n_c)$ as it follows from $\alpha(n)I/\hbar\omega=n/\tau$. (b) Position of the boundary x_c as well as $-\ln(I_T/I_0)=\alpha_1 d+(\alpha_2-\alpha_1)x_c$ versus incident intensity I_0, if diffusion is ignored. (c) The same as in 3(b) with diffusion in the limit $\alpha L\to 0$ and for $x_\uparrow<d(x_\uparrow=(1/\alpha_2)\ln(1/2+\alpha_2/2\alpha_1),\ x_\downarrow=d+(1/\alpha_2)\ln(1/2+\alpha_1/2\alpha_2))$. For finite αL the Z-like part of the characteristic becomes Ƨ-shaped.

even if the diffusion length $L=\sqrt{\tau D}$ is small compared to α^{-1}. As a consequence both n and dn/dx behave continuously at x_c. With these matching conditions (3) is simply solved for $D=$const. [15]. The result without surface recombination ($dn/dx=0$ at $x=0, d$) is shown in fig. 3(c), deviating appreciable from when diffusion is ignored from the beginning. Except for $x_c=0, d$ the relation $I^*=I_0\exp(-\alpha_2 x_c)$ holds, with $I^*=2I_\downarrow I_\uparrow/(I_\uparrow+I_\downarrow)$ being the intensity at which both possible n values in fig. 3(a) are stable in the sense of global stability. Hysteresis results only from the jumps of x_c at $x_c=0$ and $x_c=d$ to $x_\uparrow$ and $x_\downarrow$, respectively. If I_0 reaches $I_\uparrow$, the density front with $n=n_c$ propagates in the sample like a shock wave. At the beginning ($x_c=0$) its propagation velocity $dx_c/dt=-(\partial n/\partial t)/(\partial n/\partial x)|_{x=x_c}$ is infinite, while it tends to zero when x_c approaches $x_\uparrow$. The jump from $x_c=d$ to $x_\downarrow$ takes place in the same manner.

In the measurements pulses were used with a duration of about $\Delta\tau\approx 4$ ns, being about ten times larger than the carrier lifetime τ in CdS [10]. The diffusion length $\sqrt{D\tau}$ is of the order of 0.1 μm [16] in low-excited CdS. A possible increase of D at higher carrier densities should be compensated by a shortening of τ. Estimating the mean velocity of the shock wave by $\sqrt{D/\tau}$ and using the experimental data $\alpha_2\approx 1\ \mu m^{-1}$ and $\alpha_2/\alpha_1\approx 5$, we find the time $x_\uparrow/\sqrt{D/\tau}$, in which x_c approaches $x_\uparrow$, to be about 10τ. Thus, the steady-state with respect to the diffusion process is not reached during the pulse. The experimental results in such a situation should be intermediate between the predictions of fig. 3(b) and (c). Indeed, the switch-on behaviour shown in fig. 2(b) agrees almost perfectly with that expected from fig. 3(b). On the other hand, the recovery to the initial absorption state could not be observed indicating nonstationary.

Summarizing, our findings contain considerable evidence of OB. While transient effects due to carrier lifetime can be essentially excluded, the diffusion processes being characteristical of the bistable system considered, however, do not reach the steady-state. Much longer pulses have to be used for conclusive experiments.

4. Room-temperature: saturable band-edge absorption

4.1. Absorption edge under optical excitation

Large optical nonlinearities at room-temperature are of particular interest. We report here on saturable absorption at the band edge of CdS, which can be achieved at comparatively low intensities [17,18]. Typical low-intensity transmission spectra for a very thin CdS platelet crystal are presented in fig. 4. In the high-transmittive region below 2.45 eV ($e \perp c$) or 2.47 eV ($e \parallel c$), where an exponential decrease of the absorption coefficient (Urbach tail) is observed for thicker crystals [19], broad interference structures are seen. The differences in the spectra for $e \perp c$ and $e \parallel c$ result from the valence-band structure of CdS – the uppermost A-band is only dipole-allowed for $e \perp c$ – and the slightly different refractive indices.

Under optical excitation (fig. 5) a pronounced absorption bleaching is found at the high-energy side of the spectra affecting a range of about 50 meV at an intensity of only 30 kW/cm^2. At higher intensities the effect saturates. The maximum modulation degree of transmission is observed around 2.470 eV for $e \perp c$ ($T/T_0 \approx 7$) and 2.485 eV for $e \parallel c$ ($T/T_0 \approx 4$), respectively. Similar results were also obtained for CdS_xSe_{1-x} mixed crystals [17,18]. A decrease of the refractive index for both $e \perp c$ and $e \parallel c$ of about 0.25% follows from the shift

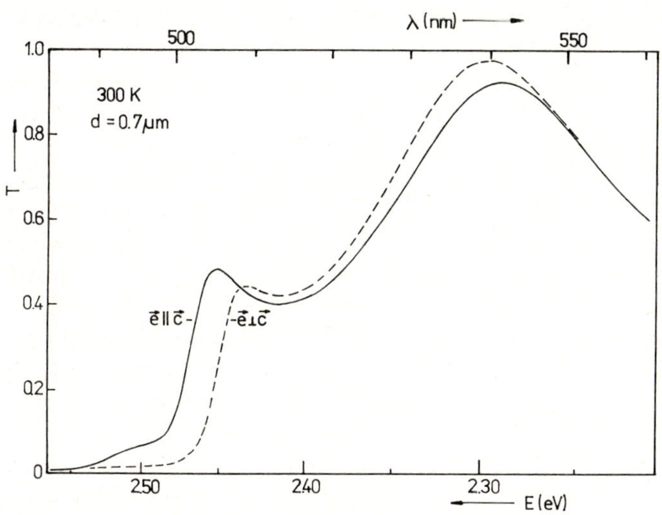

Fig. 4. Low-intensity transmission spectra in the band edge region of CdS.

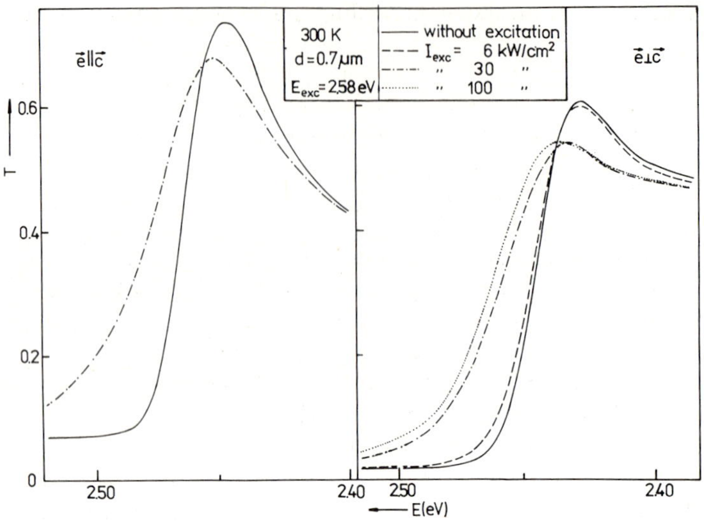

Fig. 5. Probe-beam transmission in dependence on the excitation intensity (energy E_{exc} of exciting photons is above the band edge).

of the respective interference maximum. During the shift the transmittivity at the maximum drops, which is simply connected with increasing absorption in the direction of the shift. Evidently, there is no bleaching in the spectral region below the maximum. Therefore, the absorption bleaching described here has nothing in common with that found at the Urbach tail of CdSe at very high excitation intensities of some ten MW/cm^2 [20].

In the spectral range slightly above the interference maximum the bleaching is a combined effect of absorption and reflection. This offers the possibility to reach much larger degrees of modulation by coating the platelets or matching their thickness.

Usually, absorption bleaching at the band edge of semiconductors is explained in terms of band filling. This requires, however, degeneracy of the excited carrier system, that is, $n\Lambda > 1$ where $\Lambda = (2\pi\hbar^2/kTm)^{3/2}$ is the thermal wavelength. However, one finds $n\Lambda \approx 0.01$ for the band masses m of CdS at a carrier density n of 10^{17} cm^{-3} and $T = 300$ K. Densities considerably larger than 10^{17} cm^{-3} should be hardly attainable under excitation in the kW range for carrier lifetimes smaller than 1 ns. But even if we assume much larger densities, say $n = 10^{19}$ cm^{-3}, the argument against band filling arises from the reduction of the band gap (shrinking) expected simultaneously with increasing density of the electron–hole plasma. According to [21] this reduction is about 100 meV at $n = 10^{19}$ cm^{-3} for CdS. On the other hand, the spectral range where the absorption bleaching appears is just the region of exciton resonances [19]. They are strongly broadened at room-temperature and, thus, not observable as discrete absorption lines. Therefore, we favour an explanation of the bleaching by a screening process of exciton states. According to the simple Mott criterion bound states do no longer exist, if the screening radius $r_s = (kT/8\pi e^{*2}n)^{1/2}$ (nondegenerated case) becomes smaller than $0.84a_B$ with a_B as the Bohr radius. Inserting CdS parameters we find a Mott density of about $10^{18}\ cm^{-3}$ for the A-exciton. This value becomes still smaller if the results of improved calculations accounting for dynamical screening [22] are used. Of course, this estimation is somewhat fictitious, because, strictly speaking, the bound states disappear due to the gap shrinking. Nevertheless, screening should act more effectively than band filling at room-temperature in CdS. The completely different behaviour of the band edge at cryo- and room-temperatures is left for further investigations. Our room-temperature findings point out once more, that the understanding of highly-excited wide-gap II–VI compounds is far from being satisfactory.

4.2. Saturable absorption in single-beam experiments

Similar to the cryo-temperature experiments the optical nonlinearity originating from the absorption bleaching described above was investigated by studying the behaviour of ns laser pulses passing the sample resonantly in the bleaching region. In fig. 6 the plot of transmission versus input intensity taken at the maxima of both transmitted and incident pulses is shown at a photon energy, where the largest bleaching effect is observed (fig. 5(a)). Almost the same characteristic is found at the trailing edges of the pulses indicating steady-state conditions.

The results in fig. 6 demonstrate directly that the band edge absorption of

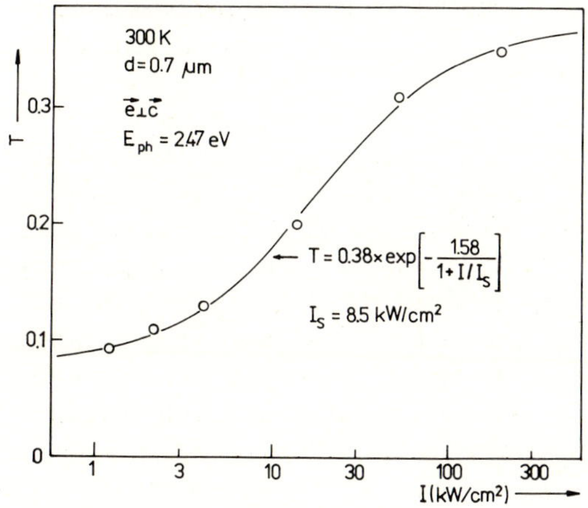

Fig. 6. Single-beam transmission T versus incident intensity I: (○) experimental results; full line: fit with Lorentzian model).

CdS may be saturated with milliwatt powers at room-temperature. The findings are surprisingly well fitted by the saturation of a homogeneously broadened resonance (full line in fig. 6), which hints again at the excitonic origin of the effect. Multiple-beam interferences are neglected because of the relatively large linear absorption at low intensities. However, the deviation of the prefactor 0.38 in fig. 6 from the pure reflection losses $(1-R)^2$ ($R \approx 0.25$) can be caused by Fabry–Perot effects in the saturated state.

Considering Gaussian beam saturation, the effective mean intensity is 1/3 of the intensity in beam center [23]. In this case, the saturation intensity of CdS becomes as low as that reported recently for a GaAlAs MQW structure [23]. The Lorentzian saturation model allows for an estimate of the nonlinear refraction coefficient n_2 defined through $n=n_0+n_2I$, where n_0 is the linear refractive index. As for the GaAlAs structure [23], we find $n_2 \approx 10^{-5}$ cm^2/W. This value is, of course, much larger than the nonresonant changes at the interference maxima in fig. 5.

4.3. *Temporal behaviour under picosecond excitation*

The response times of the bleaching effect were studied on a $CdS_{0.9}Se_{01}$ mixed crystal [17]. Here, the spectral position of the bleaching is matched to the second harmonic of the Nd:glass laser. A moderate pulse energy ($\approx$0.3 mJ/cm^2) was used to avoid saturation of the effect. As seen from fig. 7 the bleaching appears immediately with the excitation pulse of 6 ps. Assuming an exponential law $\Delta T = T - T_0 \sim \exp(-t/\tau)$, a recovery time of $\tau=(220+80)$ps was determined by regression (full line in fig. 7). However, there is a remarkable deviation from the simple exponential fit for the time just after excitation.

The longer time of about 200 ps is clearly connected with the band-to-band recombination of the excited carriers. Carrier diffusion was suppressed experi-

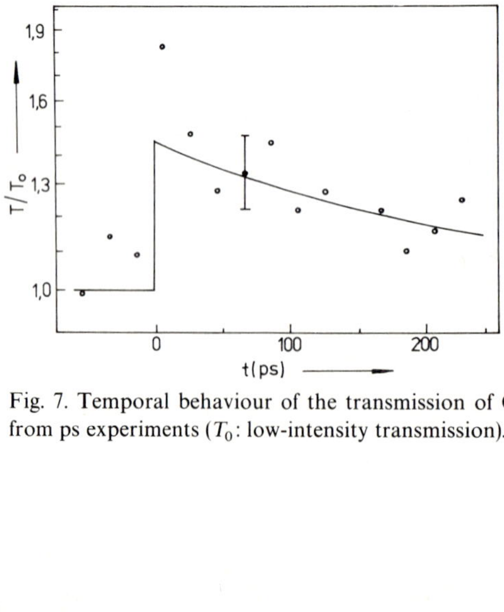

Fig. 7. Temporal behaviour of the transmission of $CdS_{0.9}Se_{0.1}$ at $E_{ph}=2.34$ eV for $e \perp c$ derived from ps experiments (T_0: low-intensity transmission).

mentally by using very thin platelets and a large excitation spot. The spike at about $t=0$ could be a hint to state filling. An alternative explanation would be coherent interaction between excite and probe pulse [24].

Summarizing, the saturable absorption demonstrated for CdSSe should be attractive for the following reasons:

- room-temperature operation at relatively low intensities, which should be achievable also under cw excitation;
- large tuning range of the operation wavelength by crystal composition;
- high transmission in the on-switched state;
- fast switching;
- combination with dispersive effects (e.g. optical bistability). Corresponding experiments are in progress.

Acknowledgements

The authors are very much indebted to Prof. E. Gutsche and Dr. J. Voigt for their remarks and interest in this work.

Note added in proof

Almost simultaneously and independently the general operation principle of the OB discussed in section 3.3 has been introduced by D.A.B. Miller et al. (Opt. Lett. 9 (1984) 162; J. Opt. Soc. Am. B Dec. (1984)). But only the situation of an infinite diffusion length has been treated.

References

[1] G.O. Müller, H.H. Weber and I. Höricke, Phys. Status Solidi (b) 91 (1979) 531.
R.F. Leheny and J. Shah, Phys. Rev. Lett. 37 (1976) 871.
[2] K. Bohnert, G. Schmieder and C. Klingshirn, Phys. Status Solidi (b) 98 (1982) 175.
I. Rückmann, V. May, F. Henneberger and J. Voigt, J. Lumin. 24/25 (1981) 593.
[3] F. Henneberger, J. Puls, H. Rossmann, and I.N. Uraltsev, Phys. Status Solidi (b) 121 (1984) K191.
[4] H.M. Gibbs, A.C. Gossard, S.L. McCall, A. Passner and W. Wiegmann, Phys. Rev. B 16 (1977) 1577.
[5] V.D. Egorov, G.O. Müller, H.H. Weber, R. Zimmermann, A.F. Dite, V.G. Lysenko and V.B. Timofeev, Nouvo Cimento B 39 (1977) 628.
[6] H. Rossmann, F. Henneberger and J. Voigt, to be published.
[7] K. Kempf and C. Klingshirn, Solid State Commun. 49 (1984) 23.
[8] G.W. Fehrenbach, W. Schäfer, J. Treusch and R.G. Ulbrich, Phys. Rev. Lett. 49 (1982) 1282.
R. Zimmermann and H. Stolz, Phys. Status Solidi (b) 124 (1984) 201.

[9] F. Henneberger, H. Rossmann and J. Voigt, Conf. Excitons in Semiconductors, Leningrad, 1982.
H. Rossmann, F. Henneberger and J. Voigt, Phys. Status Solidi (b) 115 (1983) K63.
[10] H. Yoshida, H. Saito and S. Shionoya, J. Phys. Soc. Japan 50 (1981) 881.
[11] K. Bohnert, H. Kalt and C. Klingshirn, Appl. Phys. Lett. 43 (1983) 1088.
[12] N.A. Vidmont, A.A. Maksimov, I.I. Tartakovski and V.M. Edelstein, Pis'ma Zh. Tekh. Fiz. 9 (1983) 1527.
[13] Hoang Xuan Nguyen and R. Zimmermann, Phys. Status Solidi (b) 124 (1984) 191.
[14] e.g.: H.M. Gibbs, Conf. Optical Bistability, Dynamic Nonlinearity, and Photonic Logic, in: Roy. Soc. (GB) March 1984.
[15] F. Henneberger and H. Rossmann, Phys. Status Solidi (b) 121 (1984) 685.
[16] G.V. Novikov, A.V. Ilinskij, K.F. Lider and N.S. Sokolov, Phys. Status Solidi (b) 48 (1971) 473.
[17] J. Puls, F. Fink and F. Henneberger, Proc. III. Int. Symp. Ultrafast Phenomena Spectroscopy, Minsk 1983, p. 339.
[18] J. Puls and F. Henneberger, Phys. Status Solidi (b) 121 (1984) K187.
[19] F. Spiegelberg, E. Gutsche and J. Voigt, Phys. Status Solidi (b) 77 (1976) 233.
[20] M.P. Lisitsa, N.R. Kulin and A.F. Marsnitschenko, Fiz. Tekh. Poluprov. 17 (1983) 73.
[21] R. Zimmermann and M. Rösler, Phys. Status Solidi (b) 75 (1976) 633; 83 (1977) 85.
[22] R. Zimmermann, K. Kilimann, W.D. Kraft, D. Kremp and G. Röpke, Phys. Status Solidi (b) 90 (1978) 175.
[23] D.A.B. Miller, D.S. Chemla, D.J. Eilenberger, P.W. Smith, A.C. Gossard and W.T. Tsang, Appl. Phys. Lett. 41 (1982) 679.
[24] C.V. Shank and D.H. Auston, Phys. Rev. Lett. 34 (1975) 479.

Journal of Luminescence 30 (1985) 217–231
North-Holland, Amsterdam

OPTICAL BISTABILITY IN CuCl

J.B. GRUN, B. HÖNERLAGE and R. LEVY

Laboratoire de Spectroscopie et d'Optique du Corps Solide, Associé au C.N.R.S. n° 232, Université Louis Pasteur, 5, rue de l'Université, 67000 Strasbourg, France

In CuCl, the optical properties of biexcitons imply an intensity dependence of the dielectric function. It is calculated within a density matrix formalism. The dielectric function is then used to discuss qualitatively experimental results for CuCl placed in a Fabry–Perot cavity. It turns out that the dispersive optical bistability is strongly limited by two-photon absorption.

1. Introduction

As it is well known, optical bistability in a Fabry–Perot may be obtained if the refractive index or the absorption coefficient of a material is intensity dependent and if optical feedback is provided [1]. Experimentally, it has first been demonstrated by Gibbs et al. [2] using a Fabry–Perot etalon filled with Na vapor. Since the optical non-linearities in vapors are quite small, the dimensions of the Fabry–Perot resonators had to be important. In order to diminish these dimensions, optically denser materials with strong non-linearities had to be employed. In semiconductors, the optical non-linearities may be very important, especially if one works near electronic resonances. This has first been done in GaAs [3] and InSb [4]. In both cases, the switch-on time of the device is limited by the round-trip time of the cavity, which is in the picosecond range. The switch-off times, however, are longer since the induced non-linearities of the dielectric function depend on the radiative lifetimes of the quasiparticles. These are excitons in the case of GaAs and free electron–hole pairs in InSb, and the resulting switching-off times are about 20 ns and 100 ns, respectively. They may be reduced by doping, but it was also interesting to look for materials, in which non-linearities are no more functions of the lifetime of quasiparticles. Such a system is CuCl, in which the existence of biexcitons [5] gives rise to an intensity dependence of the dielectric function [6]. In this material, biexcitons may be created by resonant two-photon absorption [5]. If, however, the photon energy of the exciting laser beam is detuned from the biexciton resonance,

0022-2313/85/$03.30

biexcitons are excited only virtually. They recombine radiatively creating two excitonic polaritons. This process is resonantly enhanced on exciton and biexciton states and has a giant oscillator strength. It gives rise to an intensity dependence of the absorption coefficient which is related to the imaginary part of the dielectric function. Since its real and imaginary part are connected to each other by a Kramers–Kronig relation, the refractive index becomes intensity dependent. It has been shown theoretically [7], that it should be possible to obtain optical bistability due to this process. In addition, the time constants which govern this non-linearity depend on the dephasing time of the exciting beam [8] which is in the picosecond range. Therefore, one should be able to obtain switch-off times limited by the round-trip time of the cavity, if biexcitons are excited off-resonance. At the resonance, however, the radiative lifetime of excitons will come into play and the switching time will be longer.

Experimentally, optical bistability has recently been observed using CuCl samples sandwiched between two external mirrors to provide optical feedback [9,10]. Switch-off times have been shown to depend on the photon energy [11,12], and optical bistability is resonant at half the biexciton energy. In order to discuss our experimental results, we will first calculate the dielectric function in the framework of the density matrix formalism. Then we will give resulting hysteresis loops using the mean field approximation. We will compare them qualitatively to our experimental results and give an outlook on experimental and theoretical work which should be pursued.

2. The dielectric function in the density matrix formalism

Let us consider a three-level system, corresponding to the crystal ground state $|1\rangle$, the exciton state $|2\rangle$, and the biexciton state $|3\rangle$, having the energies zero, E_{ex} and E_{Bi}, respectively. A light field $A(\omega)$ induces transitions between the states $|1\rangle$ and $|2\rangle$ or $|2\rangle$ and $|3\rangle$. The transition between $|1\rangle$ and $|3\rangle$ is forbidden for one-photon processes, but allowed, if two photons are involved. The Hamiltonian H then reads

$$H = H_0 - \mu A(\omega). \tag{1}$$

We assume for the electric field $A(\omega)$ a harmonic variation with a single frequency component

$$A(\omega) = A_0 \cos \omega t. \tag{2}$$

H_0 is the Hamiltonian of the non-interacting system and μ is the dipole operator. If ρ denotes the density matrix [13,14], the polarization function P is given by

$$P = N \operatorname{tr}(\rho\mu) = N\langle\bar{\mu}\rangle, \tag{3}$$

where N is the density of molecules in the crystal. $\langle\bar{\mu}\rangle$ denotes the ensemble average over the dipole matrix, which has in our case the form [13]

$$\mu=\begin{pmatrix} 0 & \mu_{\mathrm{ex}} & 0 \\ \mu_{\mathrm{ex}} & 0 & \mu_{\mathrm{Bi}} \\ 0 & \mu_{\mathrm{Bi}} & 0 \end{pmatrix}. \tag{4}$$

The dipole matrix elements μ_{ex} and μ_{Bi} are given by

$$\mu_{\mathrm{ex}}=\langle 2|\mu|1\rangle, \qquad \mu_{\mathrm{Bi}}=\langle 3|\mu|2\rangle. \tag{5}$$

The time dependence of the elements of the density matrix follows from the Schrödinger equation to [14]

$$\frac{\partial}{\partial t}\rho_{ij}=\frac{\mathrm{i}}{\hbar}[\rho, H]_{ij}-\rho_{ij}/\tau_{ij}, \tag{6}$$

where the coherence relaxation times τ_{ij} account for damping ($i \neq j$) and for the lifetime of the population of the excited states ($i=j$). The situation of eq. (1) has been discussed by different theoretical approaches [6,15–17]. Since we have noticed that population effects may have a considerable influence on the dielectric function [13], we reexamine this problem, using the density matrix formalism, where it is easy to include these effects.

Equation (6) gives rise to a set of coupled differential equations. We obtain for the off-diagonal elements:

$$\frac{\partial\rho_{12}}{\partial t}=\frac{\mathrm{i}}{\hbar}[-\mu_{\mathrm{ex}}A(\omega)(\rho_{11}-\rho_{22})+\rho_{12}E_{\mathrm{ex}}-\rho_{13}\mu_{\mathrm{Bi}}A(\omega)]-\rho_{12}/\tau_{12}, \tag{7}$$

$$\frac{\partial\rho_{13}}{\partial t}=\frac{\mathrm{i}}{\hbar}[A(\omega)(\mu_{\mathrm{ex}}\rho_{23}-\mu_{\mathrm{Bi}}\rho_{12})+\rho_{13}E_{\mathrm{Bi}}]-\rho_{13}/\tau_{13}, \tag{8}$$

$$\frac{\partial\rho_{23}}{\partial t}=\frac{\mathrm{i}}{\hbar}[-\mu_{\mathrm{Bi}}A(\omega)(\rho_{22}-\rho_{33})+\rho_{23}(E_{\mathrm{Bi}}-E_{\mathrm{ex}})+\mu_{\mathrm{ex}}A(\omega)\rho_{13}]-\rho_{23}/\tau_{23}. \tag{9}$$

The difference of population between two states i and j and their corresponding time evolution are derived from eq. (6) to

$$\frac{\partial(\rho_{11}-\rho_{22})}{\partial t}=-\frac{\mathrm{i}}{\hbar}A(\omega)[2\mu_{\mathrm{ex}}(\rho_{12}-\rho_{21})-\mu_{\mathrm{Bi}}(\rho_{23}-\rho_{32})] -\left[\frac{(\rho_{11}-\rho_{22})-(\rho_{11}-\rho_{22})_0}{\tau_1}\right], \tag{10}$$

$$\frac{\partial(\rho_{22}-\rho_{33})}{\partial t}=\frac{\mathrm{i}}{\hbar}A(\omega)[\mu_{\mathrm{ex}}(\rho_{12}-\rho_{21})-2\mu_{\mathrm{Bi}}(\rho_{23}-\rho_{32})] -\left[\frac{(\rho_{22}-\rho_{33})-(\rho_{22}-\rho_{33})_0}{\tau_2}\right], \tag{11}$$

where $(\rho_{ii}-\rho_{jj})_0$ give the equilibrium distribution of the population in the absence of an external field. We now develop the density matrix elements ρ_{ij} into a Fourier series

$$\rho_{ij}=\sum_n \rho_{ij}^n \, e^{in\omega t}. \tag{12}$$

Inserting eqs. (2) and (12) into eqs. (7) to (11), we obtain a set of recurrence relations involving the different Fourier coefficients ρ_{ij}^n, which have to be solved numerically:

$$\rho_{12}^n=\frac{1}{\Delta_{12}^n}\left\{\frac{\mu_{\text{ex}}A_0}{2}\left[(\rho_{11}-\rho_{22})^{n-1}+(\rho_{11}-\rho_{22})^{n+1}\right]+\frac{\mu_{\text{Bi}}A_0}{2}(\rho_{13}^{n-1}+\rho_{13}^{n+1})\right\}, \tag{13}$$

$$\rho_{13}^n=\frac{1}{\Delta_{13}^n}\left\{\frac{\mu_{\text{Bi}}A_0}{2}(\rho_{12}^{n-1}+\rho_{12}^{n+1})-\frac{\mu_{\text{ex}}A_0}{2}(\rho_{23}^{n-1}+\rho_{23}^{n+1})\right\}, \tag{14}$$

$$\rho_{23}^n=\frac{1}{\Delta_{23}^n}\left\{\frac{\mu_{\text{Bi}}A_0}{2}\left[(\rho_{22}-\rho_{33})^{n-1}+(\rho_{22}-\rho_{33})^{n+1}\right]-\frac{\mu_{\text{ex}}A_0}{2}(\rho_{13}^{n-1}+\rho_{13}^{n+1})\right\}, \tag{15}$$

$$(\rho_{11}-\rho_{22})^n=\delta_{n,0}+\frac{1}{\Delta_1^n}\left\{\frac{\mu_{\text{Bi}}A_0}{2}(\rho_{23}^{n-1}-\rho_{32}^{n-1}+\rho_{23}^{n+1}-\rho_{32}^{n+1})\right.$$
$$\left.-\mu_{\text{ex}}A_0(\rho_{12}^{n-1}-\rho_{21}^{n-1}+\rho_{12}^{n+1}-\rho_{21}^{n+1})\right\}, \tag{16}$$

$$(\rho_{22}-\rho_{33})^n=\frac{1}{\Delta_2^n}\left\{\frac{\mu_{\text{ex}}A_0}{2}(\rho_{12}^{n-1}-\rho_{21}^{n-1}+\rho_{12}^{n+1}-\rho_{21}^{n+1})\right.$$
$$\left.-\mu_{\text{Bi}}A_0(\rho_{23}^{n-1}-\rho_{32}^{n-1}+\rho_{23}^{n+1}-\rho_{32}^{n+1})\right\}, \tag{17}$$

with

$$\Delta_{ij}^n=E_j-E_i-n\hbar\omega+\mathrm{i}\hbar/\tau_{ij}, \tag{18}$$

$$\Delta_i^n=n\hbar\omega-\mathrm{i}\hbar/\tau_i, \tag{19}$$

These equations correspond to eqs. (15) to (19) of ref. [13] if the frequencies ω_t and ω_p are degenerated.

Since ρ is a Hermitian matrix, we find from eq. (12)

$$\rho_{ij}^n=(\rho_{ji}^{-n})^*. \tag{20}$$

In eqs. (16) and (17), we have assumed that the system is in its ground state for vanishing external fields; $\delta_{n,0}$ denotes the Kronecker symbol. Within this formalism, the dielectric susceptibility is given by [13]

$$\chi_\omega=\frac{2N}{\varepsilon_0 A_0}\{\mu_{\text{ex}}(\rho_{12}^1+\rho_{21}^1)+\mu_{\text{Bi}}(\rho_{23}^1+\rho_{32}^1)\}, \tag{21}$$

where ε_0 is the vacuum permitivity.

Equations (13) to (17) lead to an infinite system of coupled linear equations.

We restrict this system, taking into account all terms in which the denominators Δ_{ij}^n and Δ_i^n in eqs. (18) and (19) may become small (resonant terms) and their antiresonant terms, i.e., terms in which n is replaced by $(-n)$. To include these terms is necessary because then the correct pole structure is obtained [13], and our method goes beyond the rotating wave approximation.

It was shown in ref. [18] by continuous matrix inversion, that the approximation discussed above is quite good in the energy range of interest. The Fourier coefficients of interest are given by $(\rho_{11}-\rho_{22})^0$, $(\rho_{22}-\rho_{33})^0$, ρ_{12}^1, ρ_{12}^{-1}, ρ_{13}^2, ρ_{13}^{-2}, ρ_{23}^1 and ρ_{23}^{-1} and their complex conjugated terms. The system of coupled equations is now solved numerically, and eq. (21) is used to determine the dielectric susceptibility. It is applied to CuCl, where the relevant exciton and biexciton parameters are well-known [19,20]. As discussed in more details in ref. [13], the dipole matrix elements μ_{ex} and μ_{Bi} are given by

$$N\mu_{\mathrm{ex}}^2=\tfrac{1}{2}\varepsilon_0\varepsilon_{\mathrm{b}}(E_{\mathrm{L}}^2-E_{\mathrm{ex}}^2)/E_{\mathrm{ex}}, \tag{22}$$

and

$$\mu_{\mathrm{Bi}}^2A_0^2=4M^2n_{\mathrm{p}}, \tag{23}$$

respectively. $\varepsilon_{\mathrm{b}}=5$ is the background dielectric constant, $E_{\mathrm{L}}=3.208$ eV and $E_{\mathrm{ex}}=3.2025$ eV are the energies of the longitudinal and transverse exciton [19], respectively. $M^2=2.1\times10^{-22}\,\mathrm{eV^2cm^3}$ is the exciton–biexciton transition matrix element [20] and n_{p} the density of polaritons. The biexciton energy $E_{\mathrm{Bi}}=6.372$ eV is well-known from two-photon absorption [20] and the relaxation times are assumed to be such that $\hbar/\tau_2=\hbar/\tau_1=1.5\times10^{-6}$ eV and $\hbar/\tau_{12}=\hbar/\tau_{23}=\hbar/2\tau_{13}=2\times10^{-4}$ eV.

Using these parameters, we may now calculate the dielectric function, given by

$$\varepsilon/\varepsilon_0=\varepsilon_{\mathrm{b}}+\chi.$$

From this, we obtain the polariton wavevector Q and the absorption coefficient α by the relations

$$Q=\frac{\omega}{\sqrt{2}c}\{\mathrm{Re}\varepsilon+[(\mathrm{Re}\varepsilon)^2+(\mathrm{Im}\varepsilon)^2]^{1/2}\}^{1/2}, \tag{24}$$

and

$$\alpha=\frac{\omega}{\sqrt{2}c}\{-\mathrm{Re}\varepsilon+[(\mathrm{Re}\varepsilon)^2+(\mathrm{Im}\varepsilon)^2]^{1/2}\}^{1/2}. \tag{25}$$

They are related to the real (n') and imaginary part (n'') of the refractive index by

$$n'=cQ/\omega, \qquad n''=c\alpha/\omega. \tag{26}$$

Figure 1 gives the resulting polariton disperson $E(Q)$ ($E=\hbar\omega$ being the

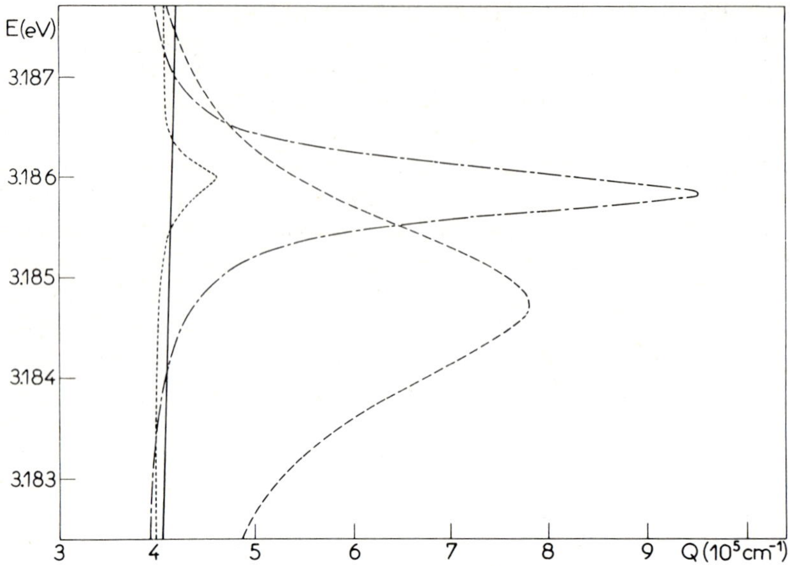

Fig. 1. Dispersion $E(Q)$ of polaritons near half the biexciton energy for different intensities (full line: $n_p=0$; dotted line: $n_p=10^{15}\ cm^{-3}$; dashed–dotted line: $n_p=10^{16}\ cm^{-3}$; dashed line: $n_p=10^{17}\ cm^{-3}$).

photon energy), and fig. 2 the absorption coefficient $\alpha(E)$ for different polariton densities n_p in the vicinity of half the biexciton energy at 3.186 eV. We find, in both figures, that an anomaly arises at half the biexciton energy, which is due to the two-photon absorption which was discussed in the introduction. At low intensities, it has essentially the same form as was found in ref. [16]. It increases

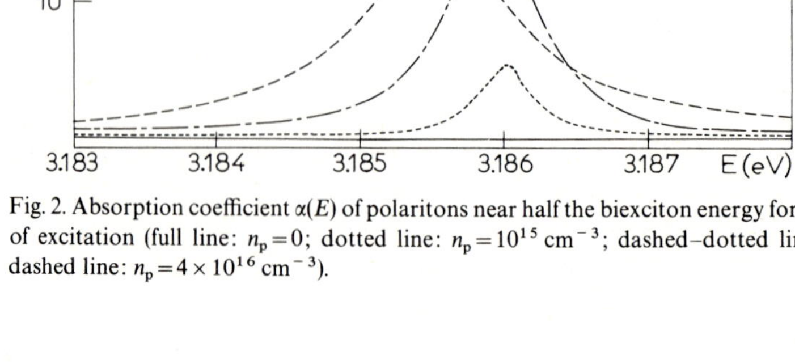

Fig. 2. Absorption coefficient $\alpha(E)$ of polaritons near half the biexciton energy for different intensities of excitation (full line: $n_p=0$; dotted line: $n_p=10^{15}\ cm^{-3}$; dashed–dotted line: $n_p=10^{16}\ cm^{-3}$; dashed line: $n_p=4\times10^{16}\ cm^{-3}$).

with increasing intensities. At $n_p \approx 10^{16}\,\mathrm{cm}^{-3}$, a Stark-shift of the biexciton resonance is observed. At even higher intensities, the resonance broadens and the anomaly in the dispersion and absorption diminishes in amplitude.

For $n_p \approx 6\times10^{16}\,\mathrm{cm}^{-3}$, strong fluctuations of the absorption constant as function of energy are observed, while the wavevector Q remains continuous up to $n_p \approx 3\times10^{17}\,\mathrm{cm}^{-3}$. It turns out, that these fluctuations are due to rapid variations of $\mathrm{Im}\,\rho^1_{23}$ with energy and intensity. They are of the order of several percent at $n_p = 3\times10^{16}\,\mathrm{cm}^{-3}$ and increase drastically for higher intensities. Since $\mathrm{Im}\,\rho^1_{12}$ is more important for the determination of α and remains continuous, these fluctuations do not show up in α at $n_p = 3\times10^{16}\,\mathrm{cm}^{-3}$ but only at higher intensities. They result from instabilities of the system of coupled equations with respect to small variations of E and n_p. This is due to the fact that the system of linear equations in ρ^n_{ij} (eqs. (13) to (17)) is strongly nonlinear with respect to E and n_p. It is interesting to notice that the system may be stabilized if $\hbar/\tau_i$ is increased or if the populations are blocked to fixed values.

Let us now discuss what happens if such a nonlinear medium is introduced into a Fabry–Perot etalon (FP) which consist of two parallel mirrors with a reflection coefficient R. The distance between the two surfaces is called l. We will discuss here only a very simple model in which we assume that the amplitude A_0 of the light field is constant throughout the crystal. The mean intensity inside the cavity I_0 is related to A_0 by

$$I_0 = A_0^2 c\varepsilon_0 n'/2, \tag{27}$$

and to the photon density n_p to I_0 by

$$I_0 = n_p \hbar\omega V_E, \tag{28}$$

where V_E is the energy velocity. For simplicity, we identify V_E with the group velocity $V_g = \partial\omega/\partial k$ for vanishing renormalization. V_E is therefore taken as a constant, independent of n_p. Then, the transmitted intensity I_t of the FP is assumed to be proportional to n_p and one obtains

$$I_t = (1-R) n_p \hbar\omega V_E. \tag{29}$$

On the other hand, the transmission τ of the FP is related to the incident intensity by

$$\tau = I_t/I_i = (1-R)^2/\{[\exp(k_0 n'' l) - R\exp(-k_0 n'' l)]^2 + 4R\sin^2 n' k_0 l\} \tag{30}$$

for a FP with absorption, excited under normal incidence. In eq. (30), k_0 is the photon wavevector in the vacuum. Introducing eq. (29) into (30), we obtain I_i as a function of n_p, which is now the average photon density inside the sample. Using eq. (26), n_p gives also n' and n'', and is used as a parameter to calculate I_t and I_i with eqs. (29) and (30). We can now plot I_t as a function of I_i and we obtain the usual hysteresis loops of the optical bistability. The validity of this approximation is discussed in ref. [17]. One should point out, however, that

$n''k_0l \leqslant 1$ should be respected so that the FP can provide sufficient feedback and our approximation works.

We will now give some experimental results and discuss them in the framework of our model for the dielectric function and optical bistability.

3. Experimental setup for optical bistability

Our experimental setup is shown in fig. 3. As it is discussed in more detail in refs. [9] and [11], we use an excimer laser (XeCl) to pump a dye laser, containing a diluted solution of αNND in ethanol or of BiBuQ in toluen. The laser emission is amplified in a second dye cell, pumped by the same excimer laser. Great care is taken to keep the superradiant emission in the beam small ($\leqslant 1\%$) compared to the laser emission. The shape of the pulses is kept well-defined in time with a width of about 3 ns (FWHM). The laser emission has a spectral width of 0.05 meV. After passing through a diaphragm, neutral density filters and a glass polarizer, the beam is split into two parts by a glass plate.

One part of the beam is focused onto the Fabry–Perot etalon containing the

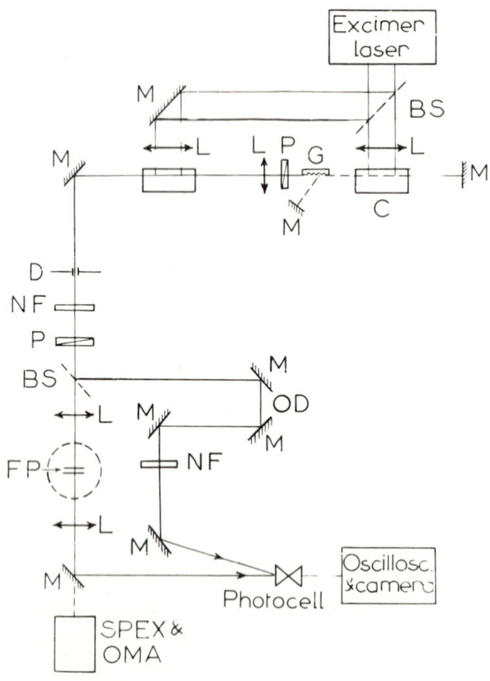

Fig. 3. Experimental setup. M: mirror, BS: beam splitter, L: lens, C: dye cell, P: polarizer, D: diaphragm, FN: neutral density filter, FP: Fabry–Perot, G: grating, OD: optical delay.

crystal in a spot of 100 μm diameter. Power density can be varied up to 60 MW/cm^2. The transmission of the laser pulses through the FP is detected by a fast photocell and is analyzed in time by a GHz oscilloscope. The overall time resolution is inferior to 500 ps. (The transmission spectrum may as well be analyzed by a spectrograph and an OMA system.) The other part of the beam passes, after an optical delay δ of 5.9 ns, through neutral density filters and is then detected by the same photocell. This beam is used as a reference for the shape of the pulse. The signal, visualized on the oscilloscope, is then photographed with a camera, so that single shots may be analyzed.

The FP consists of two thin glass plates coated with platinum films having a reflectivity of about 90%. A CuCl sample of 10 to 30 μm thickness is sandwiched between the two mirrors. Since the sample is in direct contact with the mirrors, its thickness is also the length of the FP which is excited perpendicular to its surface. Although the mirrors are highly reflective, only a maximum intensity variation of 40% is obtained. This is due to the absorption of the crystal near the 1S exciton line and to a lack of parallelism of the crystal surfaces. The FP and the sample are cooled down to low temperatures (1.9 K) in a dewar.

Figure 4 shows the transmitted and reference laser pulses analyzed by the oscilloscope when the FP is excited with a photon energy close to the biexciton energy. At this energy and intensity of excitation, the transmitted pulse is clearly deformed compared to the reference pulse. The intensity of the transmitted pulse is now analysed in detail.

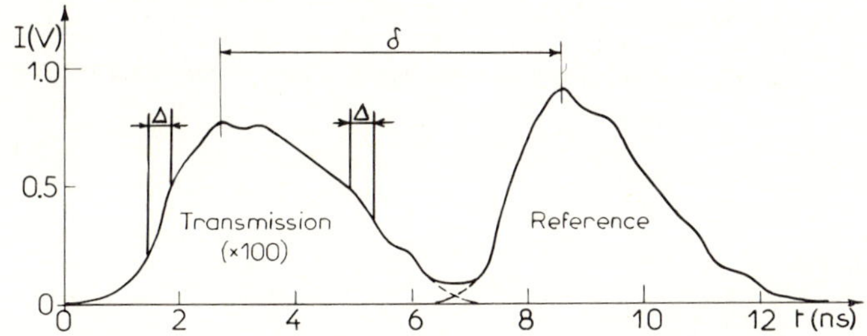

Fig. 4. Picture of the transmitted and incident dye laser pulses (the transmitted pulse is 1000 times attenuated by the FP). Δ = time resolution ($\sim$500 ps), δ = optical delay between the two pulses.

4. Experimental results and discussion

Figure 5 shows the output versus input intensity curves for different maximum intensities of excitation. Open circles correspond to increasing intensities and crosses to decreasing intensities. Figure 5(a) corresponds to fig. 4 where the bistability is clearly observed with a "switch-on" intensity of 15 MW/cm^2 and a "switch-off" intensity of 5 MW/cm^2. The switching point stays at a fixed intensity when the maximum intensity is decreased (fig. 5(b)). In fig. 5(d) no

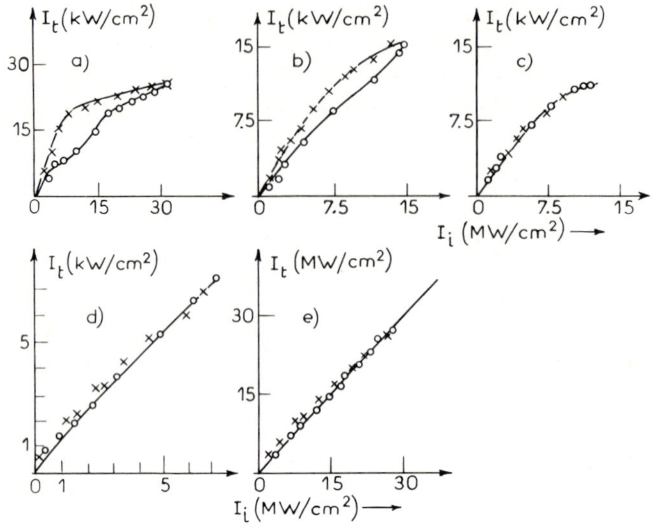

Fig. 5. Output (I_t) versus input (I_i) intensity curves for different maximum intensities (a)–(d). (e): no sample inside the cavity. (∘) for increasing I_i, (×) for decreasing I_i.

bistability is observed since $I_{max} < 15\ \mathrm{MW/cm^2}$ and, in fig. 5(e) we had no sample inside the cavity. Both corresponding curves are linear.

Let us now investigate the dependence of the hysteresis loop on the photon energy of the exciting beam. In fig. 6, we have plotted the areas of the bistability loop as a function of the photon energy of the exciting beam. This quantity is

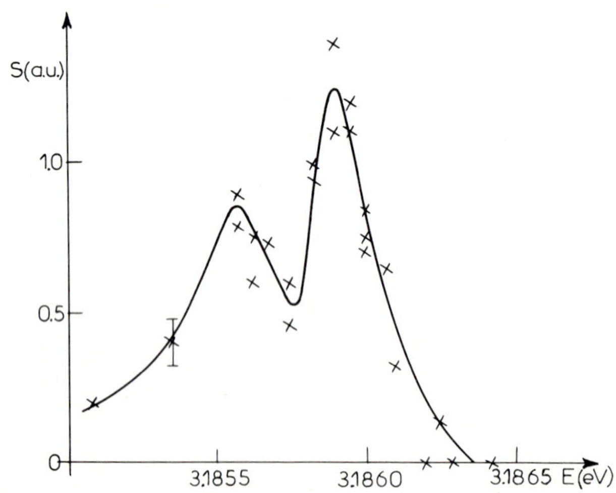

Fig. 6. Areas of the bistability loop as a function of the photon energy of the exciting beam.

proportional to the energy stocked inside the FP cavity. We obtain in this case a structure with two maxima. The first maximum is situated at an energy corresponding to a maximum of transmission of the FP at 3.1855 eV. The second maximum is situated near $E_{Bi}/2 = 3.186$ eV. This doublet structure depends on the position of the maximum of transmission of the FP relatively to $E_{Bi}/2$. In this case, they are well separated but we can also obtain excitation spectra with one maximum. In any case, we have shown that the switching behavior is different if the crystal is excited at the biexciton resonance or not.

Figure 7(a) shows a hysteresis cycle which is typically observed when we excite near the first maximum. In this case, the "switch-on" and "switch-off" times are not resolved by the setup and they are both below 500 ps. This can be concluded from the fact that increasing and decreasing pulses follow the same straight line near the origin.

If, however, the energy is near the second maximum, the hysteresis cycle (fig. 7(b)) has a completely different shape as the one shown in fig. 7(a). As our pulses are not long enough, we estimate the commutation time from the "on" to the "off" state to be of the order of 3 ns. Let us now discuss these results qualitatively using our theoretical model. In fig. 8, we give the transmission coefficient τ of a FP with a length of 10 μm as a function of the absorption coefficient α. The reflection coefficient R is taken equal to 0.95 and the refractive index $n' = 2.58$ is assumed to be constant. We are working at a maximum of transmission of the FP, i.e., $\tau = 1$ for $\alpha = 0$. For $\alpha > 1500\ \text{cm}^{-1}$, τ diminishes exponentially, indicating that the FP does no longer supply any feedback. The linear absorption constant α_L at this wavelength is equal to $200\ \text{cm}^{-1}$ [20]. In order to obtain optical bistability, the different damping constants in our model have to be chosen so that $\alpha < 1500\ \text{cm}^{-1}$ remains valid when the photon energy is tuned through the biexciton resonance. In addition, we see that $\alpha_L l = 0.2$ at low excitation intensities implies already a tiny value for the FP transmission rate; therefore our model is not applicable for thicker crystals. One should point out, however, that neglecting propagation effects leads to an overestimation of the influence of the absorption. Therefore, our qualitative discussion

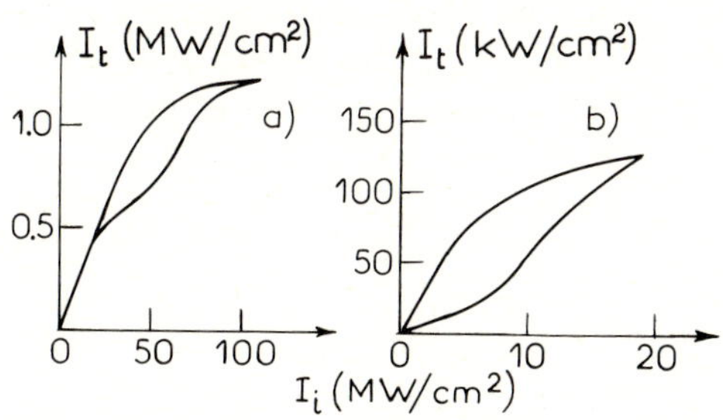

Fig. 7. (a) A typical hysteresis loop when exciting near the first maximum of figure 6, i.e., 3.1855 eV. (b) Hysteresis loop when exciting near the second maximum of fig. 6, i.e., 3.186 eV.

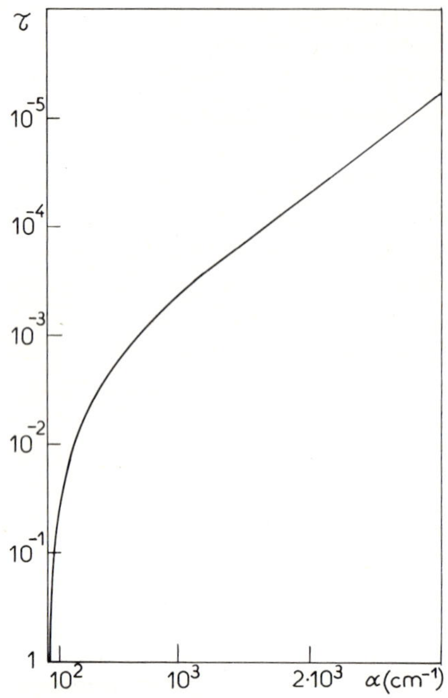

Fig. 8. Variation of the FP transmission τ as a function of the absorption coefficient α for a fixed cavity length $l=10\ \mu m$.

should even hold for the physical properties of thicker crystals. The damping constants τ_{12}, τ_{23}, τ_1 and τ_2 do not influence essentially n'' when varied by an order of magnitude around the values known approximately [21,22], and when n_p is increased up to $5\times10^{14}\ cm^{-3}$, which is the maximum value obtained in our experiments. Therefore, we have fixed the following damping constants: $\hbar/\tau_{12}=10^{-4}$ eV, $\hbar/\tau_{23}=2.5\times10^{-4}$ eV, $\hbar/\tau_1=10^{-6}$ eV and $\hbar/\tau_2=10^{-7}$ eV. There remains $\hbar/\tau_{13}$, which influences critically n'' around half the biexciton energy. It turned out that we were not able to reproduce an excitation spectrum as given in fig. 6 with a fixed value of τ_{13}, i.e., we could not obtain optical bistability at half the biexciton energy and apart from this value. In order to reproduce it qualitatively, we had to assume that τ depends on the excitation frequency as

$$\hbar/\tau_{13}=\Gamma_{13}^{0}+\Gamma_{13}^{1}\,\frac{(\Delta/2)^2}{(E_{Bi}-2\hbar\omega)^2+(\Delta/2)^2}, \tag{31}$$

where we have chosen $\hbar\Gamma_{13}^{0}=10^{-4}$, $\hbar\Gamma_{13}^{1}=5\times10^{-4}$ eV and $\Delta=5\times10^{-4}$ eV. This may be justified by the fact that, at the resonance, biexcitons are really created and that collisions between them broaden the level. It has the same physical origin as discussed in refs. [23,24]. With this choice of parameters, the calculated hysteresis loops of fig. 9 are obtained. We find in figs. 9(a), (b) that,

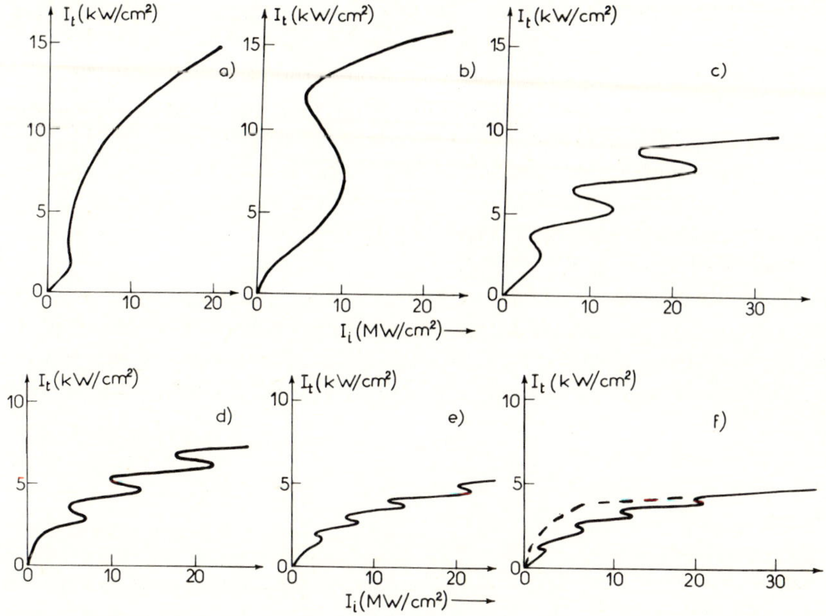

Fig. 9. Calculated transmission cycles for different photon energies $\hbar\omega_l$: (a) $\hbar\omega_l = 3.1845$ eV; (b) 3.185 eV; (c) 3.1855 eV; (d) 3.1856 eV; (e) 3.1857 eV; (f) 3.186 eV.

outside the biexciton resonance, a single bistability loop can be expected when I_i is increased up to 30 MW/cm^2. We suggest indeed that in fig. 5 one single hysteresis loop has been observed experimentally, when working slightly outside the biexciton resonance.

Closer to the biexciton resonance, as shown in figs. 9(c) to 9(e), several consecutive hysteresis loops are covered within the same intensity range of I_i. If the switching times are sufficiently rapid, one can expect to resolve them experimentally. This is shown in the curves of fig. 10, where several hysteresis loops are obtained experimentally, the frequency of the exciting laser being

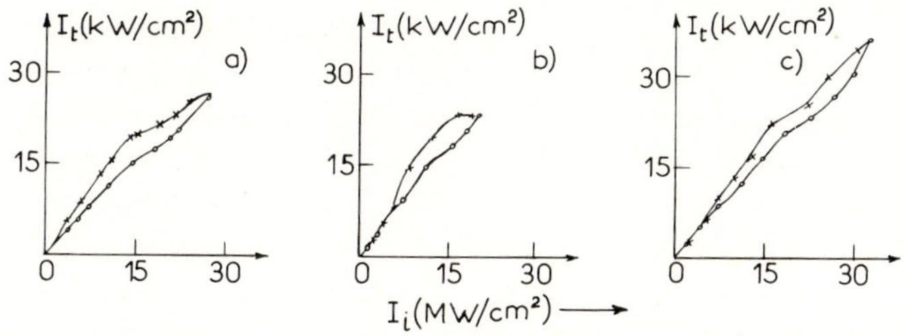

Fig. 10. Measured transmission cycles for different photon energies. $\hbar\omega_l$: (∘) increasing incident intensities, (×) decreasing incident intensities. (a) $\hbar\omega_l = 3.1861$ eV; (b) 3.1862 eV; (c) 3.1863 eV.

just outside the biexciton resonance. The form of the hysteresis loops is that of fig. 7(a).

At the biexciton resonance, fig. 7(b) suggests that the switching time from the "on" to the "off" state is in the nanosecond range. This implies in fig. 9, that for decreasing I_i, the FP stays on its high value of transmission and that the hysteresis loop would follow the dotted line, I_t being saturated. This analysis suggests that the second maximum in the excitation spectrum given in fig. 7 is artificial and should vanish in experiments, in which the variation of I_i in time is so slow that the device can follow.

Our theoretical analysis finally shows that more pronounced hysteresis loops with smaller switch-on and off intensities as shown here will be obtained, if thinner crystals are employed. This paradoxical situation is due to the importance of two-photon absorption in the case of optical bistability due to biexcitons.

Conclusion

In conclusion, we have given experimental evidence for optical bistability due to biexcitons in CuCl. We have shown that the phenomenon is strongly resonant at half the biexciton energy, which allows to state that the non-linearity is due to biexciton two-photon absorption. Working with virtual creation of biexcitons, the switching times are shorter than the experimental time resolution which is 500 ps. At the biexciton resonance, the switching times are about 3 ns long. The general features of the bistability curves can be explained by a dielectric function calculated for a three-level system. It turns out that damping constants in the model depend on the excitation frequency and that the hysteresis loops observed are mainly limited by two-photon absorption. Better results would probably be obtained if thinner samples were used.

Acknowledgements

The authors are grateful to J.Y. Bigot and F. Tomasini for many stimulating discussions. This work was supported by a contract with the "Ministère des P.T.T." of France – "Direction Générale des Télécommunications" – "Direction des Affaires Industrielles et Internationales". It has been carried out in the framework of an operation launched by the Commission of the European Communities under the experimental phase of the European Community Stimulation Action (1983–1985).

References

[1] For a recent review on optical bistability, see E. Abraham and S.D. Smith, Rep. Prog. Phys. 45 (1982) 815 and references cited therein.

[2] H.M. Gibbs, S.L. McCall and T.N.C. Venkatesan, Phys. Rev. Lett. 36 (1976) 1135.

[3] H.M. Gibbs, S.L. McCall, T.N.C. Venkatesan, A.C. Gross, A. Passner and W. Wiegmann, Appl. Phys. Lett. 35 (1979) 451.

[4] D.A.B. Miller, S.D. Smith and A. Johnston, Appl. Phys. Lett. 35 (1979) 658.

[5] For a recent review on biexciton properties, see: J.B. Grun, B. Hönerlage and R. Levy, in: Excitons, E.I. Rashba and M.D. Sturge eds. (North-Holland, Amsterdam, 1982).

[6] R. März, S. Schmitt-Rink and H. Haug, Z. Phys. B 40 (1980) 9.

[7] S.W. Koch and H. Haug, Phys. Rev. Lett. 46 (1981) 450.

[8] E. Hanamura, Solid State Commun. 38 (1981) 939.

[9] R. Levy, J.Y. Bigot, B. Hönerlage, F. Tomasini and J.B. Grun, Solid State Commun. 48 (1983) 705.

[10] N. Peyghambarian, H.M. Gibbs, M.C. Rashford and D.A. Weinberger, Phys. Rev. Lett. 51 (1983) 1692.

[11] R. Levy, B. Hönerlage and J.B. Grun, Phil. Trans., to be published.

[12] B. Hönerlage, J.Y. Bigot and R. Levy, Proc. of the Topical Meeting on Optical Bistability, Rochester (1983).

[13] J.Y. Bigot and B. Hönerlage, Phys. Status Solidi (b) 121 (1984) 649.
B. Hönerlage and J.Y. Bigot, Phys. Status Solidi (b) 123 (1984) 201.

[14] A. Yarif, Quantum Electronics, second edition (J. Wiley and Sons, New York, 1975).

[15] V. May, K. Henneberger and F. Henneberger, Phys. Status Solidi (b) 94 (1979) 911.

[16] I. Abram and A. Maruani, Phys. Rev. B 26 (1982) 4759;
I. Abram, Phys. Rev. B 28 (1983) 4433; ibid, 29 (1984) 4480.

[17] C.C. Sung and C.M. Bowden, Topical Meeting on Optical Bistability, Rochester, U.S.A., 1983.

[18] B. Hönerlage and J.Y. Bigot, Phys. Status Solidi (b) 124 (1984) 221.

[19] B. Hönerlage, A. Bivas and Vu Duy Phach, Phys. Rev. Lett. 41 (1978) 49.

[20] Vu Duy Phach, A. Bivas, B. Hönerlage and J.B. Grun, Phys. Status Solidi (b) 84 (1977) 731.

[21] R. Levy, B. Hönerlage and J.B. Grun, Phys. Rev. B 19 (1979) 2326.

[22] Y. Matsumoto, S. Shionoya and T. Takagaharn, Phys. Rev. Lett. 51 (1983) 923.

[23] N. Peyghambarian, L.L. Chase and A. Mysyrowicz, Opt. Commun. 42 (1982) 51.

[24] T. Itoh, T. Katohno and M. Ueta, J. Phys. Soc. Japan 53 (1984) 844; ibid. 53 (1984) 854.

Journal of Luminescence 30 (1985) 232–242
North-Holland, Amsterdam

THEORY OF ABSORPTIVE BISTABILITY

S.W. KOCH, H.E. SCHMIDT and H. HAUG

Institut für Theoretische Physik der Universität Frankfurt, Robert-Mayer-Str. 8, D-6000 Frankfurt-Main, FRG

A model for a resonatorless absorptive optical bistability in laser-excited semiconductors is presented. The intensity-dependent absorption coefficient is calculated microscopically taking into account band-gap renormalization and band filling due to formation of an electron–hole plasma. The spatial and temporal variations of the carrier density and the light intensity in the crystal are calculated from solutions of their coupled transport equations. An intrinsically bistable propagation mechanism for a density kink is found, which gives rise to additional structures in the bistable hysteresis loop. The influence of different carrier diffusion coefficients on the results is calculated.

1. Introduction

The intensity of light transmitted through a medium may show a nonlinear dependence on the intensity of the incident light. Under certain conditions one observes optical bistability, i.e., discontinuous switching between states of high and low transmission. This effect is a typical example of a first-order non-equilibrium phase transition in a physical system far away from thermal equilibrium.

Optical bistability is interesting for possible device applications since, e.g., bistable elements built of tiny semiconducting crystals may be used as optical switches in integrated optics. Up to now, optical bistability in semiconductors has been demonstrated experimentally, e.g., in GaAs [1], InSb [2] and CuCl [3]. In all these cases mainly dispersive nonlinearities have been utilized so that an additional nonlinear feedback for the light in the medium had to be provided by using a resonator geometry.

No such external feedback is necessary for a new type of optical bistability, which has been reported recently for highly excited CdS [4]. Here, the mechanism of induced absorption has been used to obtain the intensity-dependent nonlinear changes of the optical properties of the medium. The crystal has been excited in the frequency region below the A-exciton, where it is almost transparent for low incident light intensities. Only a small density of electron–hole

00-2313/85/$03.30

(e–h) pairs is created by two-photon absorption or by the weak one-photon absorption in the Urbach tail. However, when the incident intensity becomes sufficiently high, many-body effects in the e–h system yield an increasing reduction of the semiconductor band-gap. This can result in a situation in which the renormalized band-edge shifts below the laser frequency so that large one-photon absorption sets in. This mechanism of induced absorption, causes a switching from a state of weak absorption, i.e., high transmission, for low incident intensities, to a state of strong absorption, i.e., low transmission, for high incident light intensities. As mentioned above, no resonator geometry is necessary for this effect.

Especially this latter feature of the discussed new type of optical bistability necessitates the investigation of spatial effects. Only a single pass of the light beam through the nonlinear crystal takes place so that the light intensity decreases in the beam propagation direction in a nonexponential way. Thus, for this absorptive optical bistability longitudinal inhomogeneities occur. On the other hand, in situations *with* an external resonator and weak absorption a superposition of forward and backward travelling partial light waves causes an almost homogeneous longitudinal intensity distribution in the nonlinear medium. Spatial inhomogeneities can only build up in the direction perpendicular to the beam [5]. These transverse inhomogeneities are the only way by which spatial coexistence between regions of low and high transmission can be realized in this situation. If the spatial inhomogeneities are suppressed, e.g., by strong carrier diffusion, the system realizes the two-phase coexistence by temporal fluctuations [6]. In any case, the two-phase coexistence is a typical indication of a first-order phase transition.

In the present paper we investigate the longitudinal inhomogeneities for the absorptive bistability without resonator feedback. Under certain conditions we find a kink for the e–h density distribution, such that a region of high e–h density, i.e., low transmission, is followed by a region of low e–h density with high transmission. In this way, the longitudinal two-phase coexistence is realized. With increasing incident intensity the kink propagates through the crystal by an intrinsically bistable mechanism which is due to successive switching in consecutive sections of the medium. This leads to additional structures in the bistable hysteresis loop if the crystal is excited with sufficiently long pulses to avoid those hysteresis phenomena which are caused solely by the inertia of the system.

Before we go into further details of the results we describe in the following section 2 our theoretical model. In section 3 we investigate the static and dynamic properties of the absorptive bistability for the diffusion dominated limit. The solutions for temporally and spatially varying fields are discussed in section 4 for different diffusion coefficients.

2. Theoretical model

In this section we derive a simple model absorption coefficient and a set of transport equations for the e–h pair density N and the laser photons in the crystal. This model has been briefly introduced in ref. [7] and we will report its main ingredients here.

The absorption coefficient $\alpha(\omega)$ is given by

$$\alpha(\omega)=\omega \operatorname{Im} \varepsilon(\omega)/cn(\omega), \tag{1}$$

where $\varepsilon(\omega)$ is the optical dielectric function at frequency ω and $n(\omega)$ is the index of refraction. $\varepsilon(\omega)$ is expressed through the e–h polarization function P_{eh} using

$$\varepsilon(\omega)=\varepsilon_{\infty}-\frac{2\pi e^2 E_{\mathrm{g}}'}{\omega^2 mV}\sum_{\boldsymbol{k}} P_{\mathrm{eh}}(\boldsymbol{k},\omega). \tag{2}$$

Here, the optical matrix element has been evaluated in $\boldsymbol{k}\cdot\boldsymbol{p}$ perturbation theory [7]. m is the reduced e–h mass and E_{g}' is the renormalized band-gap. For the intensities at which the absorptive bistability is observed experimentally [4], excitons are already ionized. Therefore, we approximate the polarization function by the pair of renormalized electron (e) and hole (h) propagators P^0_{eh}: $(\hbar=1)$

$$P^0_{\mathrm{eh}}(\boldsymbol{k},\omega)=\frac{1-f_{\mathrm{e}}(\varepsilon_{\mathrm{e}k})-f_{\mathrm{h}}(\varepsilon_{\mathrm{h}k})}{\omega-E_{\mathrm{g}}'-k^2/2m+\mathrm{i}\gamma}, \tag{3}$$

and

$$E_{\mathrm{g}}'=E_{\mathrm{g}}+\Delta(N). \tag{4}$$

Here, the exchange and correlation energies have been approximated by a rigid band shift $\Delta(N)$, which has, e.g., been calculated by Rösler and Zimmermann [9].

$$f_{\mathrm{i}}(\varepsilon_{\mathrm{i}k})=\frac{1}{\exp\left[\beta(\varepsilon_{\mathrm{i}k}-\mu_{0\mathrm{i}})\right]+1},\qquad \mathrm{i}=\mathrm{e},\mathrm{h}, \tag{5}$$

are the Fermi distributions with $\varepsilon_{\mathrm{i}k}=k^2/2m_{\mathrm{i}}$ and the quasichemical potentials $\mu_{0\mathrm{i}}$. The damping constant γ in (3) takes into account the weak band-tail absorption. Inserting these expressions into (1) we find the one-photon absorption coefficient

$$\alpha(\omega,N)=\frac{2e^2E_{\mathrm{g}}\gamma}{\pi c\omega}\sqrt{\frac{2m}{\varepsilon_{\infty}}}\int_0^{\infty}\mathrm{d}x\sqrt{x}\,\frac{(1-f_{\mathrm{e}}-f_{\mathrm{h}})}{(\omega-E_{\mathrm{g}}'-x)^2+\gamma^2}. \tag{6}$$

The resulting absorption is plotted in fig. 1 for the example of CdS at $T=30$ K as a function of the e–h density N for a fixed frequency below the A-exciton resonance. For low densities $N\lesssim 10^{17}\,\mathrm{cm}^{-3}$ there is only weak band-tail absorption. The following strong increase in $\alpha(\omega,N)$ is caused by the shift of

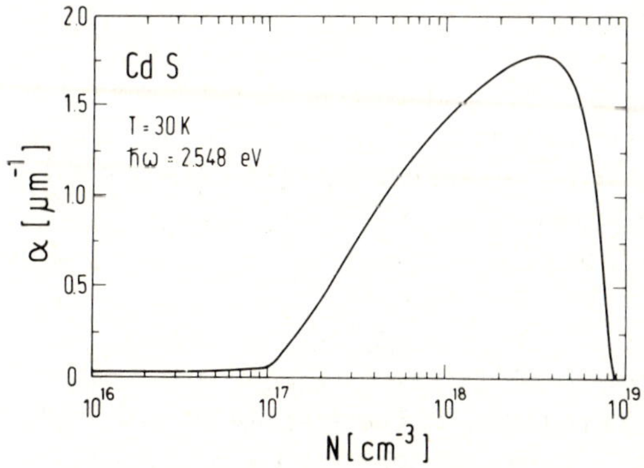

Fig. 1. Absorption coefficient $\alpha(\omega, N)$ for a fixed frequency as a function of e–h pair density N.

E_g' below the frequency ω of the exciting laser where the effective band-edge absorption sets in. For $\mu(N)\to\omega$ the crystal becomes transparent due to the bleaching by the intense light. Regions of negative α indicate optical gain, which however cannot be reached in single-beam experiments.

The rate equation for the e–h density $N(z, t)$ has the form (see also ref. [10])

$$\partial_t N - \partial_z D(N)\partial_z N = -N/\tau + \alpha(\omega, N)I/\omega, \tag{7}$$

where $\partial_t = \partial/\partial t$, $\partial_z = \partial/\partial z$ and $D(N)$ is the diffusion coefficient, τ the lifetime of the e–h pairs and I the intensity of the light in the crystal. For $I(z, t)$ the following transport equation is assumed

$$\partial_t I = -\frac{c}{n(\omega)}(\alpha(\omega, N)I + \partial_z I). \tag{8}$$

In eqs. (7) and (8) only spatial variations in z-direction are considered. These equations may be derived from the optical Maxwell–Bloch equations in the slowly-varying amplitude approximation, where the polarization field is eliminated adiabatically and the energy transport velocity is approximated by the phase velocity $c/n(\omega)$.

3. Diffusion dominated limit

In ref. [7] we discussed the so-called "mean-field" approximation of eqs. (7) and (8). In this case one assumes a spatially homogeneous density distribution

$$N(z, t) = \bar{N}(t), \tag{9}$$

and a simple exponential decay of the light intensity into the crystal

$$I(z, t) = I_0(t)\exp[-\alpha(\omega, \bar{N})z], \tag{10}$$

yielding the spatially averaged intensity

$$\bar{I}(t)=\frac{1}{L}\int_0^L \mathrm{d}z\, I(z,t)=\frac{I_0}{\alpha L}(1-\mathrm{e}^{-\alpha L}), \tag{11}$$

where L is the crystal length. This limiting case is appropriate in a diffusion dominated situation. The length on which $N(z)$ changes appreciably is given by $\sqrt{D\tau}$. Thus, $N(z)$ becomes approximately constant for $D\tau/L^2 \gg 1$, resulting in a simple exponentially decaying intensity. For very thin crystals with a high density of the e–h pairs this description becomes experimentally relevant if the temporal changes of the exciting laser pulse are slow compared to the characteristic time constants for e–h pair propagation. The appearance of intrinsic optical bistability can be determined by the stationary solutions of the rate equations. The results are shown in fig. 2. In fig. 2(b) we plot the transmitted intensity

$$I_t = T\bar{I}.$$

(T = transmission coefficient of the crystal surface) and in fig. 2(a) we plot the e–h pair density N as functions of the incident intensity I_0. The calculation have been done for CdS using the following material parameters: $E_g = 2.583$ eV, $\varepsilon_\infty = 8$, $m_e = 0.205\ m_0$, $m_h = 1.1\ m_0$, $\tau = 0.1$ ns, $T = 0.77$ and $\gamma = 0.6$ meV. Figure 2 shows in the range of I_0 between 0.3 MW/cm^2 and 0.85 MW/cm^2 intrinsic optical bistability, indicated by the multivalued behaviour of N and I_t. The lower branch of the density corresponds to small absorption and to a high transmitted intensity, whereas the upper branch of the density corresponds to low transmitted intensity. A stability analysis in the framework of a Fokker–Planck equation [7] shows that the intermediate branch is unstable since it

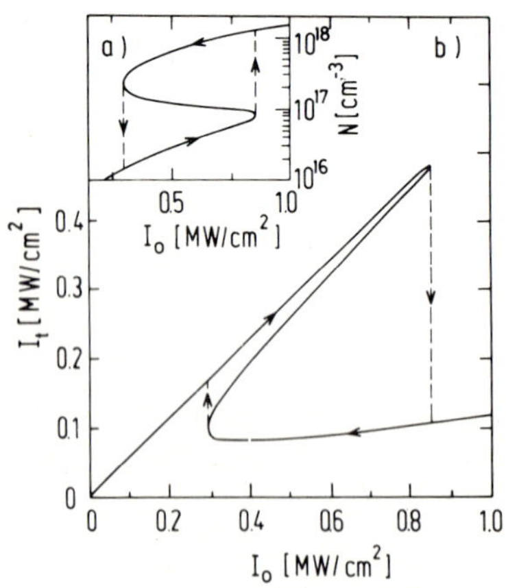

Fig. 2. Electron–hole pair density N (a) and transmitted intensity I_t (b) versus the intensity of the incident light I_0 for stationary conditions.

corresponds to a maximum in the generalized Ginzburg–Landau potential. It is interesting to note, that the intensity loop is traversed in a clockwise sense, whereas dispersive loops or absorptive loops due to saturation of a resonance are traversed counter-clockwise.

In most laboratory experiments bistable hysteresis loops are constructed comparing the shapes of a laser pulse before and after passing the sample. The parameter of such loops is the time and details of their shape depend on the temporal width and the risetime of the exciting pulses. We will refer to these constructions as dynamic hysteresis loops. To study details of these loops, we assume an incident pulse of the form

$$I_0(t)=I_0^{\max}\sin^2 \pi t/T_m, \qquad 0\leqslant t\leqslant T_m, \tag{12}$$

where $I_0^{\max}$ is the peak intensity and T_m is the pulse width. Our numerical calculations are simplified considerably by utilizing the fact, that in CdS the lifetime of the e–h plasma is of the order of 0.1 ns, whereas the relaxation time $n(\omega)L/c$ of the intensity is almost 100 times smaller. Therefore, temporal changes of the intensity will follow the temporal changes of the e–h density almost instantaneously. Under these conditions it is justified to eliminate the time derivative of the intensity adiabatically.

With this approximation we calculate numerically the dynamic hysteresis loop using incident pulses (eq. (12)) with different T_m. In fig. 3 we show the results for a CdS platelet crystal of 1 μm thickness for T_m=5.5 ns, 25 ns and 50 ns, respectively. The arrows indicate the region of I_0 for which the stationary solutions show intrinsic optical bistability. We find, that the shortest pulse yields the largest hysteresis loop. This loop is decreased systematically with increasing pulse width, until for T_m=50 ns it has almost reached the intrinsic

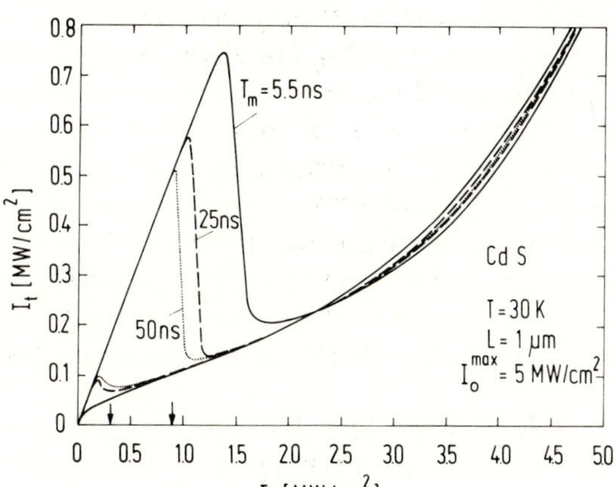

Fig. 3. Transmitted intensity I_t versus incident intensity I_0 for pulse excitation with $I_0^{\max}=5$ MW/cm^2 and different temporal pulse width. The arrows indicate the range of intrinsic bistability.

shape. The apparent increase in the dynamic hysteresis loop for short pulses is caused by "overshooting" effects, i.e., by the internally delayed nonlinear response of the e–h system. In order to eliminate these additional dynamic hysteresis phenomena, which are superimposed to the intrinsic bistable hysteresis loop, we use for the following calculations long pulses ($T_m = 50$ ns). In test calculations with short pulses (which will not be presented in detail) we found that many features of the structures were hidden. We obtained, e.g., practically identical dynamic loops for situations with very large and very small carrier diffusion, which however show pronounced intrinsic differences for long pulses.

4. Spatial effects

To study the influence of spatially inhomogeneous carrier-density and light-intensity distributions inside the crystal, we solve numerically the full transport equations (7) and (8), using only the adiabatic elimination of $\partial_t I$. The initial condition $N(z, t=0)=0$ determines the solutions uniquely for any given incident intensity $I(z=0, t)=I_0(t)$. For the numerical calculations we use a finite-difference method [11]. We investigate the influence of different carrier diffusion constants D, but since we find a somewhat unexpected additional structure in the dynamic bistable hysteresis loops for small diffusion, we discuss the limiting case $D=0$ first.

In fig. 4 we show as function of time the incident laser pulse $I_0(t)$ (eq. (12)) for $I_0^{\max} = 5\ \mathrm{MW/cm^2}$ and $T_m = 50$ ns, which has been used for all following calculations. Also plotted is the pulse after passing through a crystal with $D=0$ and a length of $L=1\ \mu\mathrm{m}$. For short times the transmitted pulse follows the incident pulse, since for those low intensities the absorption in the crystal is very low and only a small density of e–h pairs is created (see fig. 4(b)). For increasing incident intensity the density at the front of the crystal increases continuously until it reaches the value corresponding to the end of the lower branch of the density hysteresis loop (Fig. 2(a)). The still increasing $I_0(t)$ yields a jump of $N(z, t)$ to the upper bistable branch for small z values in the region of the front face of the crystal. The density distribution $N(z)$ in the crystal assumes the shape of density kink, as is shown in the insert to fig. 4(b) for different times. The appearance of such an inhomogeneous density distribution is a manifestation of the effect of longitudinal two-phase coexistence mentioned in the introduction to this paper.

Since a part of the crystal is now filled with a high e–h density, it is strongly absorbing and the intensity of the light transmitted through the whole crystal drops. This gives rise to the first peak in the temporal shape of the transmitted pulse (fig. 4(a)). However, the incident intensity is still increasing causing a further increase in the local e–h density and the transmitted intensity. In the

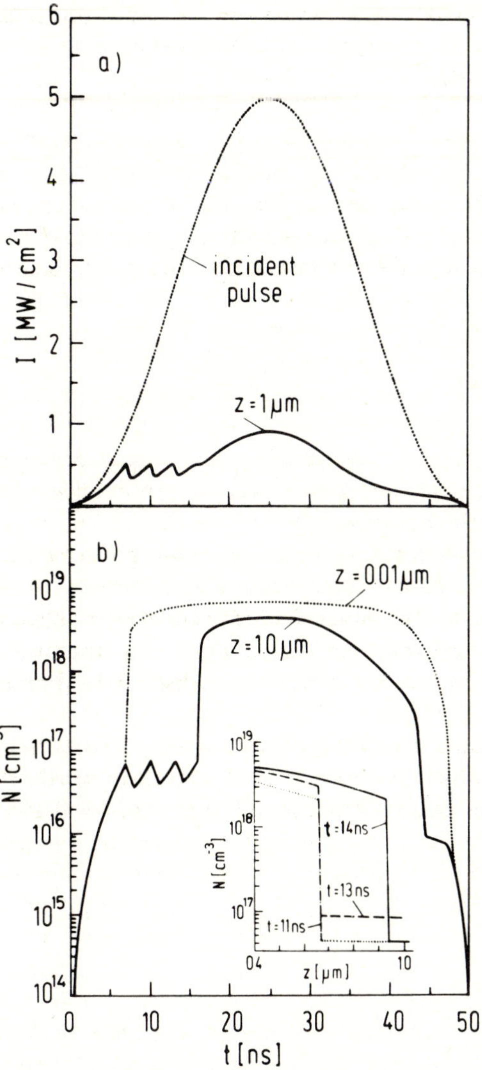

Fig. 4. (a) Incident pulse and pulse transmitted through a CdS sample of 1 μm thickness as function of time. (b) Versus time are plotted the local e–h density immediately after the front face of the crystal ($z = 0.01$ μm) and at $z = 1$ μm. The insert shows the spatial density distribution for different times.

crystal regions with high and low density, the local values of $N(z, t)$ follow the two different branches of the density hysteresis loop. Especially in the area with low $N(z, t)$ a continuous increase takes place, until $N(z)$ – for z immediately after the kink – reaches the value corresponding to the end of the lower branch and a jump occurs onto the upper branch. By this mechanism the kink moves stepwise with a steplength Δl deeper into the crystal (see the inset to fig. 4(b)). An increasingly larger crystal fraction goes into the state of high absorption resulting in a smaller total transmission.

This intrinsically bistable kink propagation causes the successive peaks in the transmitted light pulse. They stop only after the kink has passed the complete crystal, which is then totally filled with a high density of e–h pairs. These somewhat complex spatio-temporal effects manifest themselves in substructures of the dynamic bistable hysteresis loop. For $D=0$ we show in fig. 5 transmitted versus incident intensity for crystals of different length. For the smallest sample ($L=0.9$ μm, fig. 5(a)) the hysteresis loop ends around $I_0 \simeq 3$ MW/cm^2 indicating that for this incident intensity the density kink has reached the crystal end face. For the larger samples (figs. 5(b)–(d)) this never happens for the chosen pulse.

The extension of the dynamic loop is only limited by the peak intensity of the incident pulse, which is in our case $I_0^{max}=5$ MW/cm^2. The figures show, that the absolute values of I_t are only weakly reduced for increasing sample size. This is due to the fact, that under the present conditions the additional fraction of the crystal is always in the state of low absorption.

This whole picture changes somewhat if we introduce a nonzero finite diffusion coefficient D. Electron–hole pair diffusion will tend to smear out the local density distribution until $N(z)$ becomes constant in the diffusion dominated situation discussed in the preceding paragraph of this paper. With increasing density-independent diffusion constants, we find a continuously decreasing slope of the density kink. This is shown in fig. 6, where we plot N versus z at fixed time $t=25$ ns, i.e., at the time of highest incident intensity. Constructing the dynamic hysteresis loops for the assumed diffusion constants yields the results shown in figs. 7(a)–(c). With increasing carrier diffusion we find a reduc-

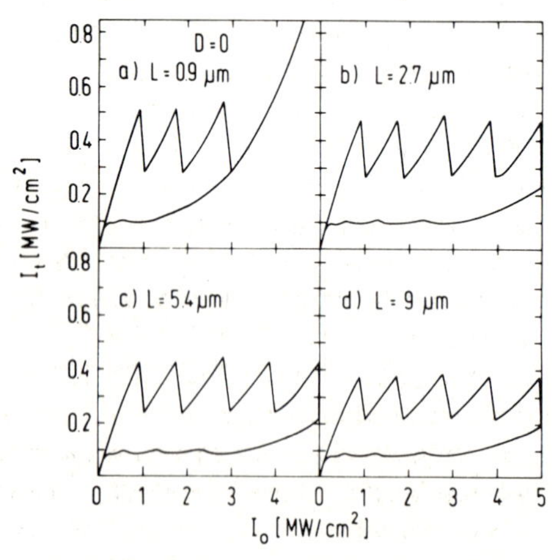

Fig. 5. Transmitted versus incident intensity for pulse excitation with $I_0^{max}=5$ MW/cm^2 and a pulse width of $T_m=50$ ns. The figures are for a CdS crystal without e–h pair diffusion ($D=0$) and different crystal length: (a) $L=0.9$ μm, (b) $L=2.7$ μm, (c) $L=5.5$ μm, and (d) $L=9$ μm.

tion of the additional structures. Diffusion constants larger than $D = 10\,\mathrm{cm}^2/\mathrm{s}$ lead practically to no additional changes, only the lower branch of the dynamic loop is reduced a little bit.

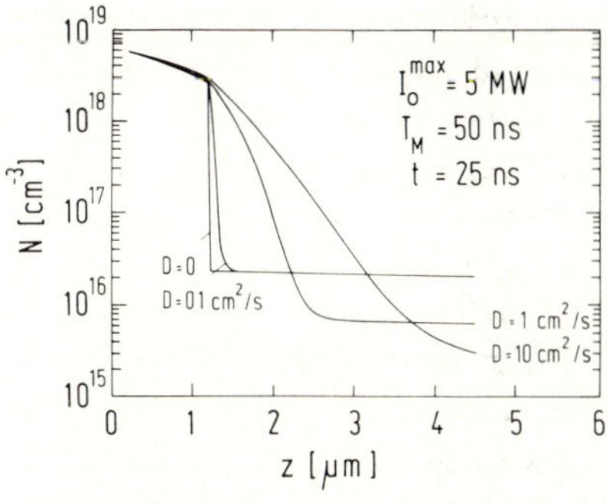

Fig. 6. Spatial density distribution $N(z)$ for different diffusion constants D at $t = 25$ ns after onset of the exciting pulse.

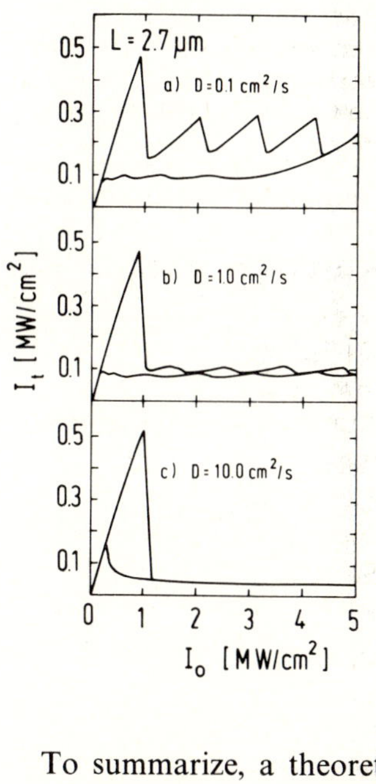

Fig. 7. Transmitted versus incident intensity for pulse excitation with $I_0^{max} = 5\,\mathrm{MW/cm}^2$ and $T_m = 50$ ns using different diffusion coefficients: (a) $D = 0.1\ \mathrm{cm}^2/\mathrm{s}$, (b) $D = 1.0\ \mathrm{cm}^2/\mathrm{s}$, and (c) $D = 10.0\ \mathrm{cm}^2/\mathrm{s}$ and a CdS crystal with length $L = 2.7$ μm.

To summarize, a theoretical model for optical bistability due to induced absorption has been developed. Dynamical hysteresis loops are calculated for pulse excitation of CdS samples. For small diffusion spatial inhomogeneities lead to additional structures which can be observed if the diffusion is not too strong and if the crystal length L is considerably larger than the length Δl of the stepwise kink progression.

Acknowledgements

This work has partly been supported by the Deutsche Forschungsgemeinschaft through the Sonderforschungsbereich 65, Frankfurt/Darmstadt and by the Commission of the European Communities (under the experimental phase of the European Stimulation Action). One of the authors (SWK) wants to thank the F. Thyssen-Stiftung, Köln, for financial support.

Note added in proof

In the meantime the predicted modulation of the transmitted intensity due to the kink motion has been found experimentally in CdS_xSe_{1-x} edge filters, where the induced absorption is due to a thermal shift of the band edge (H.M. Gibbs, G.R. Olbright, N. Peyghambarian and H. Haug, submitted to Opt. Lett.).

References

[1] H.M. Gibbs, S.L. McCall, T.N.C. Venkatesan, A.C. Gossard, A. Passner, and W. Wiegmann, Apll. Phys. Lett. 35 (1979) 451.

[2] A. Miller, D.A.B. Miller and S.D. Smith, Adv. Phys. 30 (1981) 697 and the given references to the author's work, partly in collaboration with other authors.

[3] R. Levy, J.Y. Bigot, B. Hönerlage, F. Tomasini and J.B. Grun, Solid State Comm. 48 (1983) 705; N. Peyghambarian, H.M. Gibbs, M.C. Rushford and D.A. Weinberger, Phys. Rev. Lett. 51 (1983) 1692.

[4] K. Bohnert, H. Kalt and C. Klingshirn, Appl. Phys. Lett. 43 (1983) 1088.

[5] J.V. Moloney, R.M. Belic and H.M. Gibbs, Opt. Commun. 41 (1982) 379.

[6] H.E. Schmidt, S.W. Koch and H. Haug, Z. Phys. B 51 (1983) 85.

[7] H.E. Schmidt, H. Haug and S.W. Koch, Appl. Phys. Lett. 44 (1984) 787; S.W. Koch, H.E. Schmidt and H. Haug, Appl. Phys. Lett. (Nov. 15, 1984).

[8] H. Haug and S. Schmitt-Rink, Progress in Quantum Electronics (to be published) (1984).

[9] M. Rösler and R. Zimmermann, Phys. Status Solidi (b) 83 (1977) 85.

[10] Y. Toyozawa, Solid State Commun. 28 (1978) 533 and ibid. 32 (1979) 13.

[11] G.E. Forsythe and W.R. Wasow, Finite-Difference Methods for Partial Differential Equations (Wiley, New York, 1964).

Journal of Luminescence 30 (1985) 243–247
North-Holland, Amsterdam

PROGRESS IN FEMTOSECOND MEASUREMENT TECHNIQUES

C.V. SHANK

AT&T Bell Laboratories, Holmdel, New Jersey 07733, USA

This paper describes the recent advances in femtosecond measurement techniques and the application of these methods to the study of highly excited semiconductors.

Rapid advances have taken place in the generation of optical pulses for making measurements in the femtosecond time domain. Dye lasers can now be made to produce optical pulses as short as 60 femtoseconds [1] and amplified to gigawatt optical intensities [2]. Progress continues. Two years ago, optical pulses as short as 30 femtoseconds were reported [3], and most recently, 16 femtosecond optical pulses have been generated. [4]

As we move into the femtosecond realm, a whole new range of problems becomes accessible. If we consider that a 30 femtosecond pulse corresponds in the frequency domain to 1000 cm^{-1}, it is clear that we can study the properties of solids and liquids in a time less than a phonon collision. Experiments in this time domain present the possibility of exciting coherent or correlated states in matter that evolve to some statistical observable. The transient evolution of such processes provides a means of studying fundamental interactions.

An important application of femtosecond techniques to the solid state physics has been the investigation of phase transitions induced by intense optical pulses on semiconductor surfaces. This problem has been the subject of intense interest and controversy. Time resolved reflectivity techniques performed with nanosecond [5] and picosecond [6] time resolution were able to establish that intense optical pulses modified the reflectivity of a Si surface to that of molten Si in a time too short to measure. Measurements of time resolved reflectivity using femtosecond optical pulses [7] clearly resolved the process of energy transfer from an optically excited electron hole plasma to the crystal lattice, followed by a subsequent phase change to a "melted state".

To further elucidate the melting process, a measure of the transition from order to disorder is useful. Since reflectivity is a scalar quantity, it is not possible to obtain information concerning crystalline order. However, the process of second-harmonic generation is governed by a tensor quantity that contains elements of crystal symmetry. For the case of the ⟨111⟩ surface of Si, the

0022-2313/85/$03.30 © Elsevier Science Publishers B.V.
(North-Holland Physics Publishing Division)

projection of the bulk symmetry onto the surface is three-fold symmetric. The second-harmonic radiation from this surface with p-polarized fundamental and second-harmonic radiation is given by

$$I(2w)=kI^2(w)[\cos 3\theta+A(\theta)]^2,$$

where θ is the angle of rotation about the surface normal, K is a constant and $A(\theta)$ is the isotropic contribution to nonlinear polarization which is a function of the angle of incidence, θ. For this surface $A(\theta)\approx 1$ resulting in a three-fold symmetry.

Second-harmonic radiation from a weak 2.0 eV pulse was measured [8] following excitation with an intense 90 femtosecond optical pulse. The measured second-harmonic radiation is plotted as a function of crystal rotation and is plotted in polar form in fig. 1. Before excitation, a three-fold symmetric pattern is observed. By 240 femtoseconds the second harmonic crystal has become significantly disordered and by 3.0 ps an isotropic variation of the second harmonic with angle of rotation is observed indicating a loss of crystalline order and melting.

At intensities significantly above the melting threshold, material is ejected from the crystalline surface leaving permanent damage. To investigate this process an imaging technique has been developed which can be used to produce images with femtosecond time resolution. Such images provide a detailed study of the physics of a highly excited semiconductor surface as a function of time, spatial position, and wavelength.

The images were obtained using a variation of the pump and probe technique. An intense 80 fs pulse was used to excite a silicon surface. A weak probing pulse, generated using white light continuum generation, [9] was directed to the point of excitation at a fixed time delay. The specularly reflected probe radiation was imaged onto a screen or photographic film with a magnification of approximately 100×. Either still photographs or motion pictures could be taken by synchronizing the camera shutter with the 12 Hz repetition rate

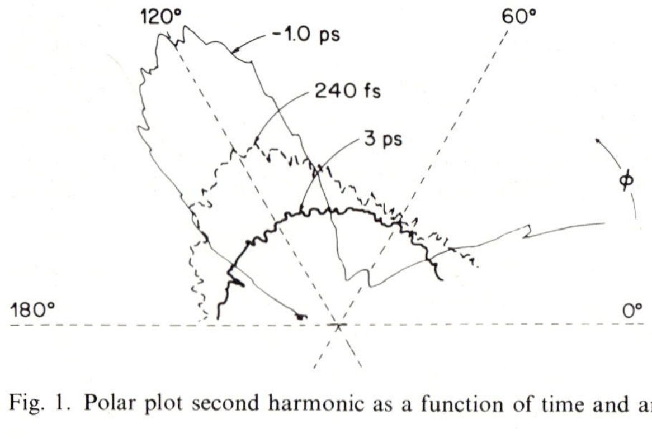

Fig. 1. Polar plot second harmonic as a function of time and angle θ for pump energy 2.0 E_{TH}.

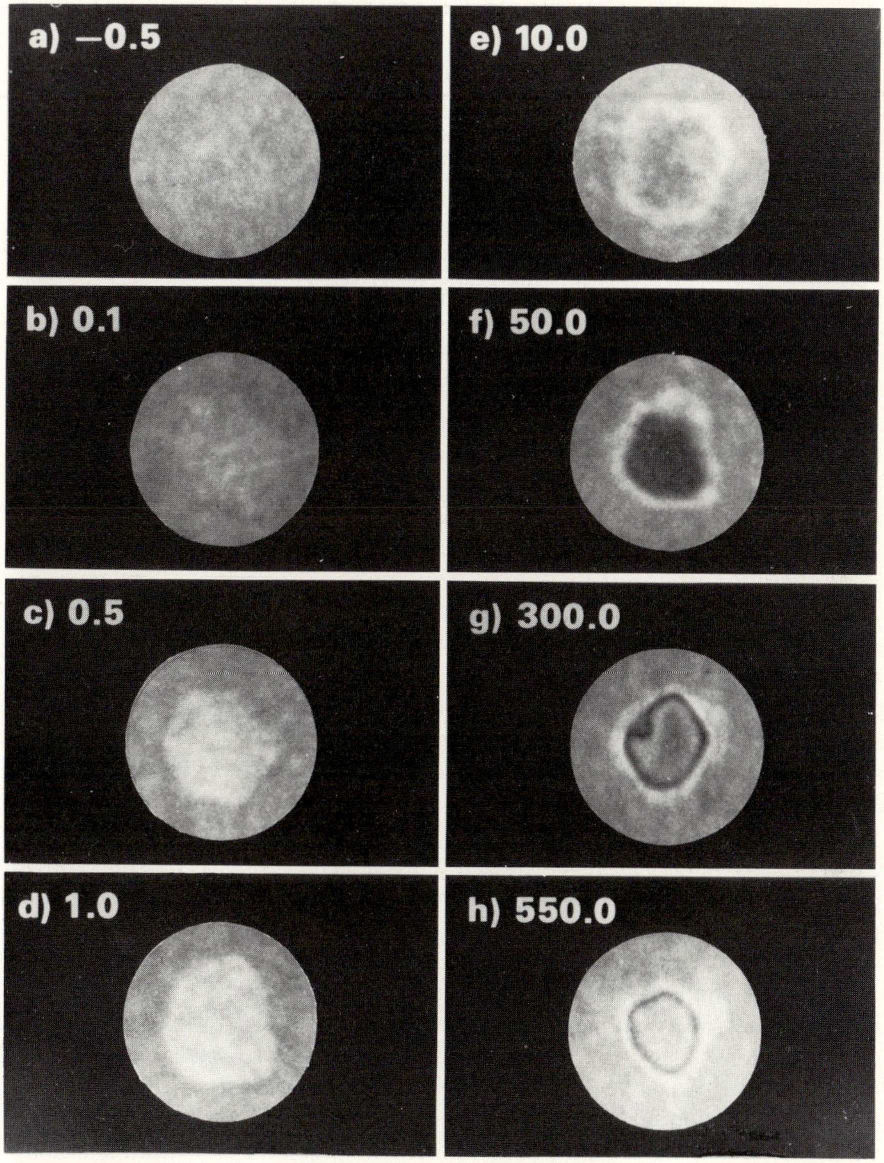

Fig. 2. Time-resolved photos of a silicon [111] surface following photoexcitation by an 80 fs optical pulse of 0.5 J/cm^2. Numbers indicate pump-probe optical delay in ps. Note the rapid appearance of the highly reflective molten silicon ((b) through (d)), followed by the ejection and dissipation of evaporated material (dark central spot in (e) through (h)).

of the laser. The time delay between the pump and the probe was varied by a controllable optical path delay.

The images of the excited silicon surface at time delays ranging from $\Delta =$ -0.5 to $+600$ ps. are shown in figs. 2(a) to (h). Here the pump fluence was approximately 5 times threshold for melting. Before arrival of the excitation pulse, only a uniformly illuminated region of the surface is seen (see fig. 2(a). After arrival of the excitation pulse, the reflectivity of the excited region is selectively increased because significant melting has occurred in this region (see fig. 2(b)–(d)). At $\Delta t = +0.1$ ps, this region is faint (see fig. 2(b)) for two reasons. First, melting is not complete by this time and second, there is a frequency sweep on the probe pulse because of dispersive optics in the path of the probe beam.

The appearance of the highly reflective molten spot at later times ($\Delta t > 1.0$ ps.) depends dramatically on the pump fluence. For fluence between the melting threshold and about 2.5 melting threshold, its appearance remains essentially unchanged from $\Delta t = 1.0$ ps out to $\Delta t = 600$ ps. This is consistent with earlier results [4] showing that for these fluence levels the reflectivity levels off after about 1.0 ps and remains unchanged until later than a nanosecond, when re-solidification beings.

At higher fluence levels, a dark region begins to appear in the center of the molten spot at $\Delta t = 5$ to 10 ps, as shown in fig. 2(e). This central region continues to darken, becoming darkest between $\Delta t = 50$ and 100 ps, although the edge remains bright (see fig. 2(f)). At still later times, the center of the dark spot begins to become transparent again (fig. 2(g)) and by $\Delta t = 600$ ps, it has substantially dissipated (fig. 2(h)), except for a narrow dark ring at the outer edge of the original dark spot.

We believe that this dark region originates from a gradually thickening cloud of material ejected from the hot molten silicon surface for as long as 100 ps following excitation. Maximum particle emission occurs in the hot, central portion of the molten silicon, while the edge remains cool enough that a substantial absorptive cloud never develops, thus explaining the bright outer ring of unobscured molten silicon which persists throughout the time of observation. Earlier studies have demonstrated substantial emission above melting threshold of both charged particles and neutral silicon atoms.

In summary, this brief discussion reviews recent progress in femtosecond optical pulse techniques to make measurements of dynamic processes in solids on this extremely short time scale. These new tools give us a chance to peer into the as yet unexplored world of processes that take place in solids on an ultrafast time scale.

References

[1] C.V. Shank, Science, 219 (1983) 1027.

[2] R.L. Fork, C.V. Shank and R.T. Yen, Appl. Phys. Lett. 45 (1982) 223.

[3] C.V. Shank, R.L. Fork, R.T. Yen, R.H. Stolen and W.J. Tomlinson, Appl. Phys. Lett. 40 (1982).

[4] J.G. Fujimoto, A.M. Weiner and E.P. Ippen, Appl. Phys. Lett. 44 (1984) 832.

[5] D.H. Auston, J.A. Golovchenko, A.L. Simons, C.M. Surko and T.N.C. Venkatesan, Appl. Phys. Lett. 33 (1979) 539.

[6] R. Yen, J.M. Liu, H. Kurz and N. Bloembergen, Appl. Phys. A27 (1982) 153.

[7] C.V. Shank, R. Yen and C. Hirlimann, Phys. Rev. Lett. 50 (1983) 454.

[8] C.V. Shank, R. Yen and C. Hirlimann, Phys. Rev. Lett. 51 (1983) 900.

[9] R.L. Fork, C.V. Shank, C. Hirlimann, R. Yen and W.J. Tomlinson, Opt. Lett. 8 (1983) 1.

Journal of Luminescence 30 (1985) 248–261
North-Holland, Amsterdam

CHIRPED AND CHIRP-FREE FEMTOSECOND PULSES IN PASSIVELY MODE-LOCKED DYE LASERS

V. BRÜCKNER, W. DIETEL, E. DÖPEL, W. RUDOLPH and B. WILHELMI

Friedrich-Schiller University Jena, Department of Physics, GDR–6900 Jena, Max-Wien-Platz 1, GDR

Femtosecond light pulse generation in a ring laser and its propagation properties are described. The basic principles and some theoretical aspects of the pulse generation process in dye lasers are summarized taking into account the influence of the amplifier, of the saturable absorber, of the glass path and of mirrors. Phase modulation arising from multilayer dielectric mirrors has been calculated and measured by determination of an equivalent intracavity glass path length. The main source of phase modulation (chirp) seems to be the saturable absorber.

1. Introduction

Using ultrashort light pulses it is possible to study extremely rapid physical, chemical and biological phenomena on the picosecond and even femtosecond time scale [1,2]. The investigation of these processes stimulates an enormous progress in the whole field of femtosecond lasers and measuring technique. At present the shortest pulses are achieved in passively mode-locked dye lasers, where the pulse duration has been shortened down to about 50 fs. These pulses have been obtained in ring lasers with counter-propagating pulses [3] in which the absorber induces a certain frequency chirp due to saturation. Using additional dispersive elements outside or inside the laser cavity [4,5] the pulses can be compressed. Figure 1 shows a typical ring configuration containing Rh 6 G as amplifier and DODCl as saturable absorber. The wavelength tuning and the internal pulse compression were achieved by use of a glass prism. Note that the duration of such pulses have been further reduced by high-power amplification, nonlinear optical chirp formation and linear optical pulse compression [6]. In the present paper the light pulse generation in ring lasers and the properties of such pulses are described. The basic principles and the theory of the pulse generation process in dye lasers are summarized including the influence of the glass pathlength, of the cavity mirrors and of semiconductors on the properties of the light pulses generated.

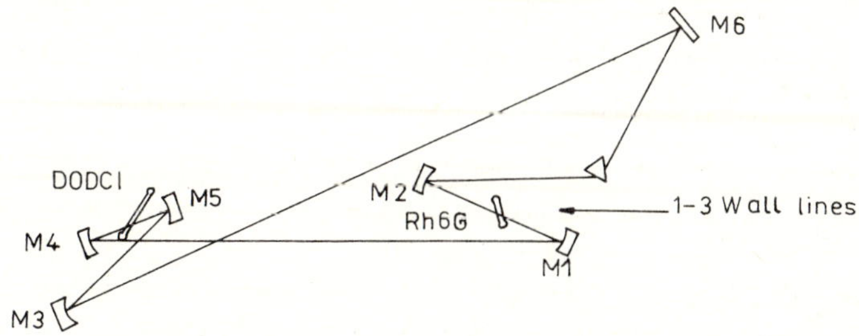

Fig. 1. Ring laser configuration.

2. Pulse formation process – basic principles

2.1. Action of amplifier and absorber

Although the rate equation approximation (REA) for the amplifier and the absorber fails in explaining some experimental details it is possible to use it to clarify some basic features of the pulse formation in passively mode-locked dye lasers [1,7]. When the pump radiation exceeds the threshold for a certain mode the field of this mode is amplified in an active medium starting from quantum noise. The multimode operation leads to a large number of fluctuation peaks in the initial laser intensity. The interaction cross section σ_{21}^{a} of the active medium is large. Therefore, a value of the laser intensity being sufficient for saturation of the absorber is reached already within a time of about 20 cavity round trips. The absorber favours fluctuation peaks or groups of fluctuation peaks of high energy. The average duration of the fluctuation peaks as well as the duration of the pulse formed later on are small compared with the fluorescence life times T_{21}^{a} and T_{21}^{b} of the amplifier (a) and of the saturable absorber (b). The saturation of both media is controlled by the pulse energy. The absorption saturation and the gain depletion become remarkable when the radiation energy per unit area achieves the order of magnitude of the saturation energy density $\varepsilon_s = \hbar\omega_L/\sigma_{21}$, where ω_L is the center frequency. Such an absorber attenuates the leading edge of the pulse or of the fluctuation peak until the energy passed the medium attains a value which is sufficient to modify the absorption considerably (see fig. 2a). Hence the pulse becomes asymmetric with the leading edge being steeper. In order to generate ultrashort (fs) light pulses it is necessary to suppress not only the leading edge of the pulse but also the trailing one. It can be achieved by gain depletion. The gain decreases because the occupation number inversion decreases during the interaction with the short light pulse. Under suitable conditions the gain at the trailing edge is smaller than the gain at threshold and, thus, the pulse finally experiences losses (fig. 2b). In this way the combined action of absorber and amplifier yields conditions where a certain

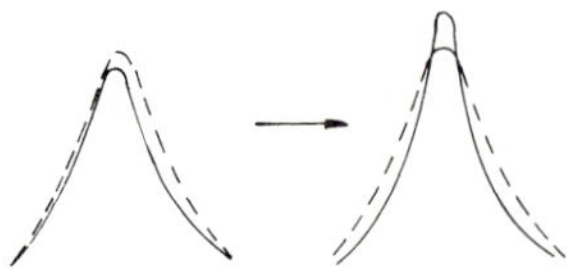

Fig. 2. Pulse shaping in the absorber and the amplifier (dotted line – initial pulse shape).

group of fluctuations of comparatively high energy is favourably amplified and shortened by suppressing both leading and trailing parts of the pulse in every round trip until a single ultrashort light pulse is formed. During this stability range the pulses become shorter and shorter by the combined action of the amplifier and the absorber down to the pulse durations where the frequency filters counterbalance this pulse shortening. If all additional frequency selective optical components are carefully removed from the resonator the spectral profiles of gain and absorption remain as filters only. In this way the laser can operate in a steady-state regime emitting a train of pulses that are equal in energy and temporal shape. Using such arrangement pulses of about 0.4 ps were achieved [8].

2.2. *Colliding pulse mode-locking* (*CPM*)

In agreement with the used arrangement (ring resonator) we have to take into account that the counterpropagating pulses collide in the saturable absorber. The regime of CPM can be obtained by using ring cavities [3] where the distance between absorber and amplifier is one quarter of the ring perimeter [6,9]. The influence of these special positions originates from the coherent superposition of the pulses counterpropagating in the absorber. Therefore, the absorber can be saturated at smaller pulse energies due to spatial hole burning, which is favourable with respect to stable mode-locking. In agreement with calculations [10] it was shown experimentally [9] that the pulse duration can be decreased from about 400 fs to less than 100 fs using a CPM arrangement.

2.3. *Nonlinear optical chirp formation in the saturable absorber and internal pulse compression*

In [11] the fs pulses generated in passively mode-locked cw dye lasers have been found to exhibit down chirp, i.e., the carrier frequency decreases from the leading to the trailing edge of the pulses. This down chirp mainly arises from the action of the saturable absorber of transition frequency ω_{21} on the light pulse of frequency ω_L, where $\omega_L < \omega_{21}$. Approximately, this action can be described

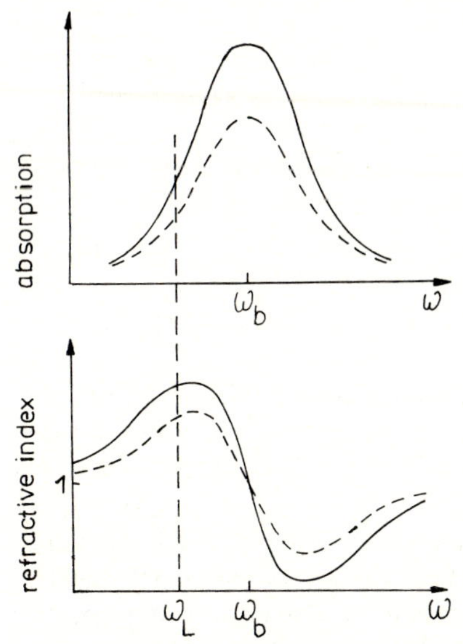

Fig. 3. Idealized absorption profile and corresponding dispersion curve before (——) and after (- - - -) the passage of the pulse.

in the rate equation approximation using only the saturation of the absorber as the source of the chirp (fig. 3). Over the main part of the pulse the frequency has been calculated to vary linearly with time [11] (see fig. 4). These down chirped pulses can be compressed by passing them through a medium with positive curvature of the dispersion curve ($d^2n/d\lambda^2 > 0$), e.g., through optical glasses in the visible region and also in semiconductors in the spectral region of small losses. After passage through an appropriate pathlength L^{min} the pulse duration has a minimum and the pulses are almost bandwidth-limited. For pathlengths $L < L^{min}$ the pulses have uncompensated down chirp, in the case

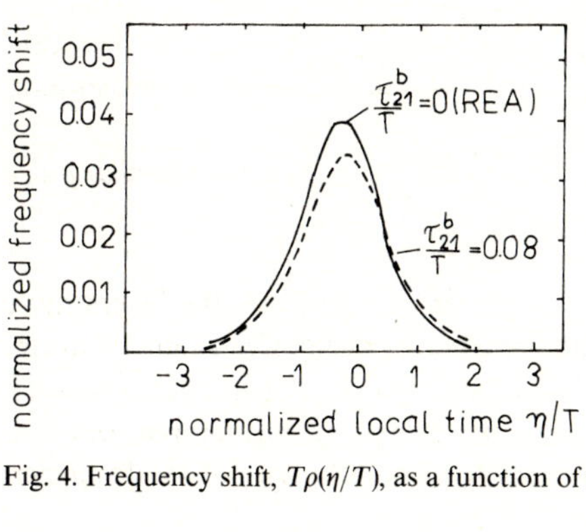

Fig. 4. Frequency shift, $T\rho(\eta/T)$, as a function of time.

of $L > L^{min}$ the dispersion yields up chirp and, hence, an increase of the pulse duration [12]. The down chirp arising from the action of the saturable absorber most advantageously can be compressed by use of intracavity glass elements. In the arrangement shown in fig. 1 the prism serves as such an element whose pathlength can be changed by translation [5]. Thus an internal pulse compression is going on, and the pulses extracted from the cavity are chirp-free and almost bandwidth-limited. The intercavity pathlength in glass has to be chosen very carefully with an accuracy of about 10 μm to achieve the broadest spectrum and the shortest pulses [13].

3. Theoretical concept

In the present paper we restrict our considerations to a steady-state pulse regime and establish an approximate round trip equation taking into account modulus and phase of the pulses like the framework of [7,10,14]. Furthermore, a certain phase memory of the absorber and amplifier was taken into consideration. The results allow us to complete the insight into such laser regimes in which up to now the shortest pulses have been observed.

3.1. Light–matter interaction – basic equations

The amplifier (a) and the absorber (b) are supposed to consist of two-level systems with negligible energy relaxation during the pulse passage. These assumptions seem to be justified in common used absorbing (DODCI) and lasing (Rh 6 G) dyes with vibronic relaxation times $>10^{-13}$ s and pulse durations $<10^{-13}$ s. After the rather long round-trip time u the influence of all relaxation transitions can approximately be described by the effective life times, T_{21}^{a} and T_{21}^{b}, of the two-level systems where $u > T_{21}^{a}$ and $u \gg T_{21}^{b}$ hold true. For the case of plane waves travelling in the z-direction the following system of partial differential equations has to be solved [15]:

$$\left(\frac{\partial}{\partial t} - \mathrm{i}\omega_j\right)\rho_{12}^{j} = \frac{\mathrm{i}}{\hbar}\mu_{12}^{j}\Delta^{j}E - \rho_{12}^{j}/\tau_{21}^{j}, \tag{1}$$

$$\frac{\partial}{\partial t}\Delta^{j} = -\frac{2\mathrm{i}}{\hbar}(\mu_{12}^{j}\rho_{21}^{j} - \mu_{21}^{j}\rho_{12}^{j})E - \frac{\Delta^{j} - \Delta_{\mathrm{e}}^{j}}{T_{21}^{j}}, \tag{2}$$

$$-\frac{\partial^2}{\partial z^2}E + \frac{\varepsilon_0\varepsilon_j}{c^2}\frac{\partial^2}{\partial t^2}E = -\gamma^{j}\mu_0\left(\mu_{21}^{j}\frac{\partial^2}{\partial t^2}\rho_{12}^{j} + \mu_{12}^{j}\frac{\partial^2}{\partial t^2}\rho_{21}^{j}\right), \qquad j = a, b, \tag{3}$$

where ω_j is the mid-frequency of the transition, τ_{21}^{j} the phase decay time, $\Delta^{j} = \rho_{22}^{j} - \rho_{11}^{j}$ the inversion probability, Δ_{e}^{j} is the equilibrium value of Δ_j, and μ_{12}^{j} the transition moment. First let us consider the pulse propagation in one direction. With $E(z, t) = \frac{1}{2}\bar{E}(z, t)\exp[\mathrm{i}(\omega_{\mathrm{L}}t - k_{\mathrm{L}}^{j}z)] + \mathrm{c.c.}$ ($\bar{E} = A(z, t)\exp[\mathrm{i}\varphi(z, t)]$

is the complex pulse envelope, ω_L the center frequency of the pulse, k_L^j the wave number at ω_L) eq. (1) yields for the off-diagonal element ρ_{12}^j [16]

$$\rho_{12}^j(t) = \frac{\mathrm{i}}{2\hbar} \mu_{12}^j \tau_{21}^j \exp[\mathrm{i}(\omega_L t - k_L^j z)] \mathscr{L}^j \left[\sum_{n=0}^{\infty} (-\mathscr{L}^j \tau_{21}^j)^n \frac{\partial^n}{\partial t^n} (\bar{E} \Delta^j) \right], \tag{4}$$

where $\mathscr{L}^j = [1 + \mathrm{i}\tau_{21}^j(\omega_L - \omega_j)]^{-1}$ is the complex line shape factor. Note that the term with $n=0$ leads to the well-known rate equation approximation (REA). For the sake of convergence a negligible bandwidth of the pulses compared with the bandwidth of the transitions is required. To approach the situation in passively mode-locked dye lasers (pulse duration $T <$ or $\ll \tau_{21}^j$) we consider the phase memory by using terms in eq. (4) up to $n=2$. After simplifying (2) and (3) by the slowly varying envelope and rotating wave approximations [1] we employ an iteration procedure similar to [14] to calculate modulus and phase of the pulses after a single pass of the absorber and amplifier. Assuming small amplification and absorption coefficients κ^j, small pulse energies ε_j normalized to the saturation energy and $|\tau_{21}^j/\bar{E}\cdot(\partial\bar{E}/\partial t)| < 1$ we find for the complex amplitudes $\bar{E}(\eta)$ after passing the medium [15],

$$\begin{aligned} \bar{E}(\eta) = \Bigg\{ 1 - \tfrac{1}{2}\kappa^j \mathscr{L}^j \Bigg[1 - \varepsilon_j + \tfrac{1}{2}\varepsilon_j^2 + \tau_{21}^j \mathscr{L}^j F^j + \beta^j \tau_{21}^j \int_{-\infty}^{\eta} \mathrm{d}\eta' \, \mathrm{Re}\left(\mathscr{L}^{j^2} \bar{E}_0^* \frac{\partial}{\partial \eta'} \bar{E}_0 \right) \Bigg] \\ + \tfrac{1}{2}\kappa^j \tau_{21}^j \mathscr{L}^{j^2} (1 - \varepsilon_j) \frac{\partial}{\partial \eta} - \tfrac{1}{2}\kappa^j \tau_{21}^{j^2} \mathscr{L}^{j^3} \frac{\partial^2}{\partial \eta^2} \Bigg\} \bar{E}_0(\eta), \end{aligned} \tag{5}$$

where η is the local time and

$$\varepsilon_j = \beta^j \, \mathrm{Re}(\mathscr{L}^j) \int_{-\infty}^{\eta} \mathrm{d}\eta' |\bar{E}_0(\eta')|^2,$$

$$F^j = \beta^j \, \mathrm{Re}(\mathscr{L}^j) |\bar{E}_0(\eta)|^2, \qquad \beta^j = \tau_{21}^j |\mu_{21}^j|^2 / \hbar^2.$$

Note that in the case of $\tau_{21}^j \to 0$ (no phase memory) we obtain the analogous REA equation. From eq. (5) it can be seen that the phase memory modifies the saturation and gives rise to terms describing approximately the action of a frequency filter (i.e., phase modulation already in the low signal case $\varepsilon^j \to 0$).

3.2. Pulse shaping in the saturable absorber

To discuss the action of the absorber we provide an unchirped $\mathrm{cosech}^2(\eta/T)$ input pulse, where T represents a scaling factor with $T \simeq 0.57$ FWHM pulse duration. From eq. (5) we can conclude that both the frequency extent of the light pulse and the frequency dependent absorption have to be taken into account. At resonance ($\omega_L = \omega_b$) it results in a decreasing absorption with increasing τ_{21}^b/T (fig. 5). Outside resonance ($\omega_L \gtrless \omega_b$), higher (lower) frequencies in the pulse spectrum are more heavily absorbed if $\tau_{21}^b(\omega_L - \omega_b) < 0$ [$\tau_{21}^b(\omega_L - \omega_b) > 0$]. It results in a shift of the mean carrier frequency $\dot{\varphi}$,

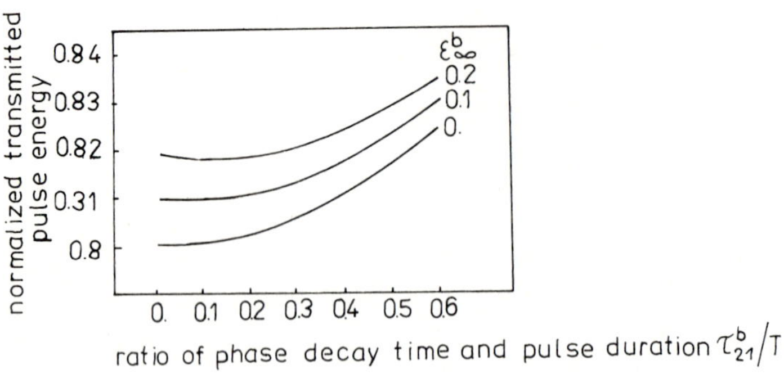

Fig. 5. Transmitted pulse energy E_{out}/E_{in} as a function of τ^b_{21}/T at different saturation.

$$\dot{\varphi} = \int_{-\infty}^{+\infty} \left(\frac{\partial \varphi}{\partial \eta} |\bar{E}(\eta)|^2 \right) \mathrm{d}\eta \Big/ (\varepsilon^b_\infty / \beta^b \operatorname{Re} \mathscr{L}^b),$$

with $\varepsilon^b_\infty = \varepsilon_b(\infty)$ towards lower (higher) frequencies as shown in fig. 6. It was demonstrated experimentally [4,5,17], that the combined action of the saturation and the phase memory outside the resonance results in phase-modulated (chirped) pulses. The resulting mean chirp

$$\ddot{\varphi} = \int_{-\infty}^{+\infty} \frac{\partial^2 \varphi}{\partial \eta^2} |\bar{E}(\eta)|^2 \, \mathrm{d}\eta \Big| (\varepsilon^b_\infty / \beta^b \operatorname{Re} \mathscr{L}^b)$$

strongly depends on the detuning $\tau^b_{21}(\omega_L - \omega_b)$, the pulse energy ε^b_∞ and τ^b_{21}/T (fig. 7). It can be shown from these calculations that REA should be valid if $\tau^b_{21}/T <$ (or $\ll$) ε^b_∞, i.e., as long as saturation dominates in the pulse and pulse

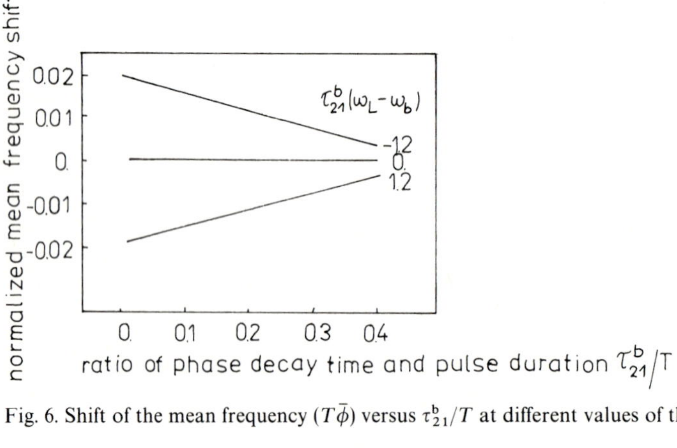

Fig. 6. Shift of the mean frequency ($T\bar{\dot{\varphi}}$) versus τ^b_{21}/T at different values of the detuning.

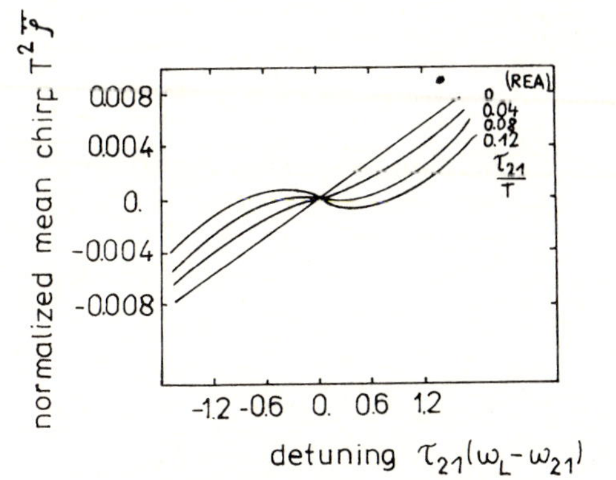

Fig. 7. Mean chirp versus detuning at different values of τ_{21}^b/T ($\tau_{21}^b/T = 0$ corresponds to the REA).

shaping process [15,18]. As it is seen from fig. 7 for red shifted pulses, i.e. $\tau_{21}^b(\omega_L - \omega_b) < 0$, at $\tau_{21}^b/T < \varepsilon_\infty^b$ a frequency decreasing within the pulse duration can be expected (down chirp), where this down chirp is smaller than obtained by REA [15]. Furthermore, a vanishing phase modulation at the pulse edges and an approximately linear frequency shift near the pulse maximum may be expected. With respect to the shortest achievable pulses the interaction of light pulses with passive absorbers is more favourable below or above resonance than in resonance.

3.3. Passive mode-locking and chirp generation by laser mirrors

We start our treatment with the resonator configuration shown in fig. 8. Equation (5) already contains the bandwidth limiting properties of the media, additional frequency selective elements (prisms . . .) are lumped in the effective frequency filter. As it was shown in section 3.2. the pulses generally pass the media outside resonance and become phase modulated. As it was demonstrated experimentally [5,12,19] this phase modulation can be controlled and changed using linear optical elements (glass path, mirror coatings and so on). Their

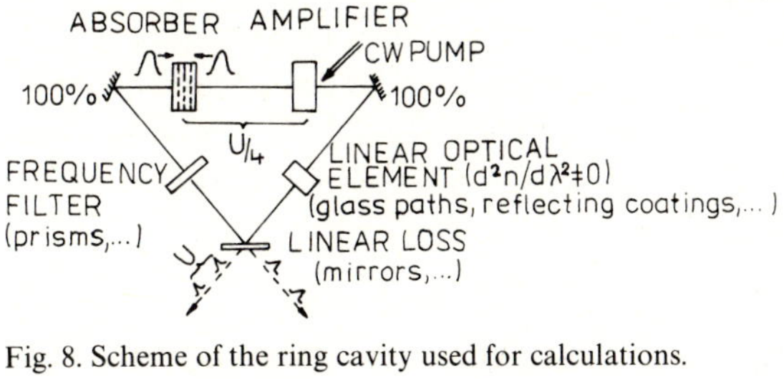

Fig. 8. Scheme of the ring cavity used for calculations.

dispersive properties ($\partial^2 n/\partial\lambda^2 \neq 0$) are summarized in an effective glass path L and we define a dispersion parameter of the resonator

$$R = (4\pi^2 Lc/\omega_L^3)(d^2n/d\lambda^2).$$

The steady-state round trip equation for the complex amplitude $\bar{E}$ can be written in a simplified manner as [15]

$$\left\{g(\bar{E}) + f(\bar{E})\frac{d}{d\eta} + (d - iR)\frac{d^2}{d\eta^2}\right\}\bar{E}(\eta) = 0, \tag{6}$$

where

$$\bar{E}(\eta) = A(\eta)\exp[i\varphi(\eta)].$$

In eq. (6) g, f, d are complex functions of the parameters of the absorber, the amplifier and possible additional frequency filters. To calculate the regime of the shortest pulses we have to find solutions of eq. (6) without phase modulation,

$$d^2\varphi/d\eta^2 = d\varphi/d\eta = 0, \tag{7}$$

i.e., the value of R must be determined from the requirement of compensation of the chirp caused by the two media. Equation (6) can be satisfied by the ansatz

$$A(\eta) = A_0/\cosh(\eta/T), \tag{8}$$

and yields algebraic equations for the unknown pulse duration T, pulse energy ε_∞^b normalized to the saturation energy of the absorber and the dispersion parameter R. Numerical calculations were performed using parameters of Rh 6G and DODCI as amplifier and absorber, respectively.

As it is seen in fig. 9 the glass path that is necessary for chirp compensation has a maximum value at a certain small signal absorption and laser frequency in agreement with the experimental results [19]. The position of this maximum value corresponds roughly with the minimum pulse duration (fig. 10). Comparing the cases of resonance and off resonance (fig. 10) we can conclude that the pulse duration is substantially smaller for off-resonance interaction which can be explained by a favourable combined action of the chirp generation and compression mechanisms. Assuming for an estimation $\tau_{21} = 2/\Delta\omega_a$ with $\Delta\omega_a = 3.6 \times 10^{14}\,s^{-1}$ in Rh 6G [10] we obtain for $T/\tau_{21}^a \simeq 4$–6 (fig. 10) a FWHM pulse duration of about 40–60 fs in agreement with the experiments [5,12,19]. In order to obtain minimum pulse durations it is necessary to take into account all resonator elements of fig. 8 influencing the chirp. Let us discuss the influence of multilayer dielectric cavity mirrors on the pulse chirp [13]. Note that this influence was also estimated in [20]. The electromagnetic wave reflected from the mirror (E_r) is connected to the incident wave E_i by

$$E_r(\omega) = E_i(\omega)|r(\omega)|\exp[i\varphi(\omega)],$$

where $|r|^2 = R$ is the reflectivity, and φ is the phase shift. Typically the mirror

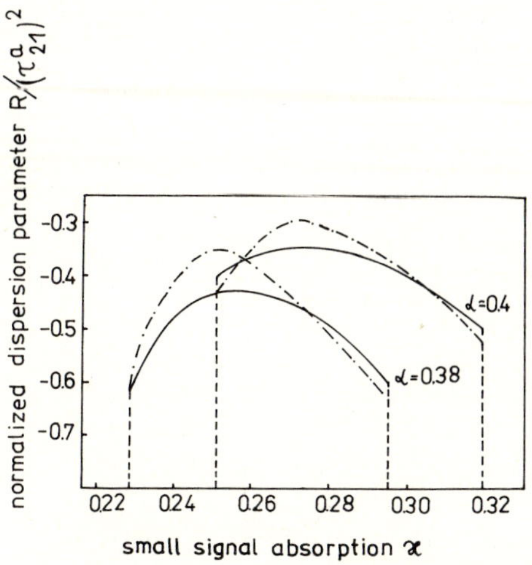

Fig. 9. Dispersion parameter normalized to the square of the phase relaxation time of the amplifier versus small signal absorption for different amplification coefficients.

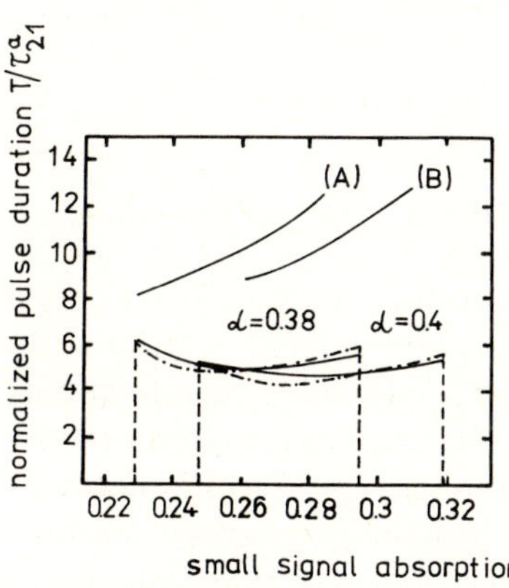

Fig. 10. Pulse duration versus small signal absorption at different amplification coefficients (curves A and B correspond to unchirped pulses, i.e., $\omega_L = \omega_a = \omega_b$).

consists of a dielectric multilayer system with an optical layer thickness of $\lambda_0/4$ (λ_0 is the resonance wavelength of the mirror). The reflectivity $R(\lambda)$ and the phase shift $\varphi(\lambda)$ was calculated for several numbers of layers and angles of incidence neglecting any absorption and dispersion. For example, in figs. 11(a) and (b) the results of two different layer systems are shown (type A: 19 layers of $\lambda_0/4$ thickness, type B: $18 \times \lambda_0/4$ and a $\lambda_0/2$ layer as the first one). The phase shift depends nonlinearly on the wavelength and gives rise to a chirp in the reflected pulses because of $d^n\varphi/d\omega^n \neq 0$ (for $n \geqslant 2$). At high reflectivity ($R \simeq 1$) the chirp of the lowest order ($n=2$) is equivalent to the chirp caused by a passage

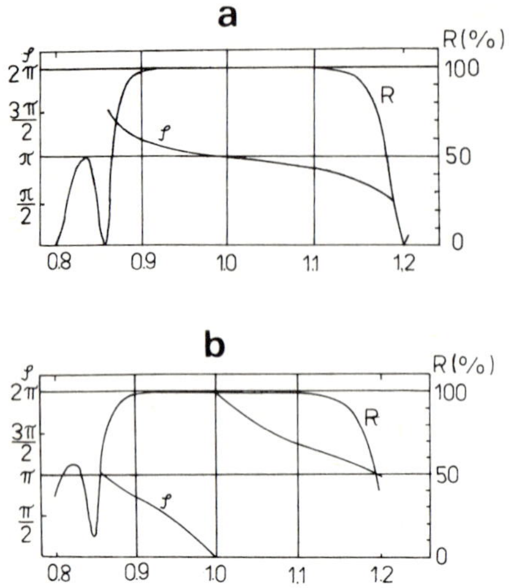

Fig. 11. Reflectivity R and phase shift φ versus normalized wavelength λ/λ_0 for two types of mirrors (type A: (a), type B: (b)).

through a dispersive material of length l,

$$d^2\varphi/d\omega^2 = l\, d^2k/d\omega^2, \tag{9}$$

where $k = 2\pi n/\lambda$. In the wavelength scale this equivalent glass path length l is

$$l = \frac{1}{2\pi}\left[2\frac{d\varphi}{d\lambda} + \lambda\frac{d^2\varphi}{d\lambda^2}\right]\left(\frac{d^2n}{d\lambda^2}\right)^{-1}. \tag{10}$$

Using the dispersion parameters of fused quartz (SQ 1) and the phase shift in the mirror (figs. 11(a) and (b)) we can calculate the equivalent path length l (figs. 12(a) and (b)). Thus, the laser mirrors can produce up chirp ($l > 0$) as well as down chirp ($l < 0$) in dependence on the layer system used and on the ratio λ/λ_0. For mirrors of type A the chirp is small near the resonance ($\lambda/\lambda_0 \simeq 1$), it increases rapidly at the edges of the reflectivity band. Therefore, at the edge of the reflectivity band the optimum glass pathlength (for generation of the shortest pulses) may be reduced significantly because a considerable part of the down chirp is already compensated by these mirrors. This may be the reason why some groups did not observe any chirp of the pulses without any glass path in the laser cavity [3]. In a recent work the same authors also use glass prisms to compensate the chirp inside the laser cavity [3]. Independent of the number of layers the blue side of the reflectivity band leads to up chirp, at the red side to down chirp (fig. 12(a)). The value of the chirp increases with increasing number of layers [13]. Laser action at one of the two edges of the reflectivity (for output coupling) is connected with a strong variation of losses and, therefore, with limitations of the oscillating bandwidth. More advantageously output coupling

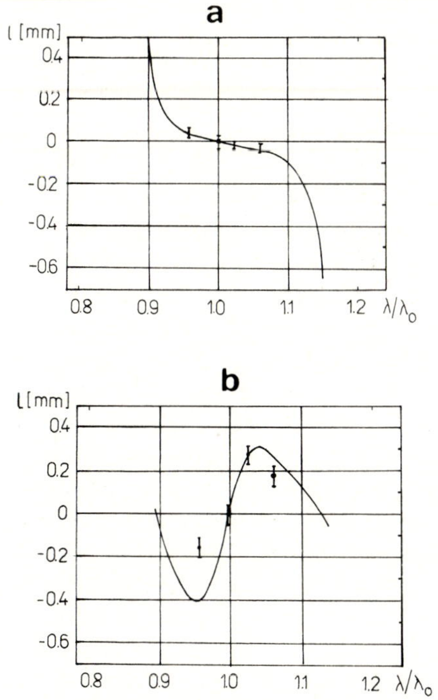

Fig. 12. The calculated equivalent glass pathlength l in SQ1 corresponding to the phase shift $\varphi(\lambda)$ in mirrors of type A: (a)) and type B: (b).

can be achieved using a type B mirror (fig. 12b). These calculations were checked experimentally using two sets of high-quality mirrors of different λ_0, the experimental results (figs. 12(a) and (b)) are in a good qualitative agreement with the calculations. Note that any deviations from the $\lambda_0/4$ thickness leads to a shift in the phase curve $\varphi(\lambda)$ of fig. 11(a) to the blue (red) edge if the thickness is smaller (higher) than $\lambda_0/4$.

Finally, it should be noted that higher orders of $d^n\varphi/d\omega^n$ (i.e. $n>2$) also can influence the pulse compression process. In this case the current pulse shape plays an important role, further details are now under investigation.

3.4. Pulse compression and semiconductor applications

The role of glass dispersion was demonstrated experimentally by intracavity as well as extracavity pulse compression [19]. Intracavity compression was achieved by translating the intracavity prism, thus adjusting the pathlength of glass inside the cavity. At all wavelengths a minimum in the dependence of pulse duration versus glass thickness was obtained. At 617 nm the shortest pulses (53 ps) were obtained using a 2 mm BK7 glass pathlength [5]. The measured pulse durations after transmission through various thicknesses of the glass are at least in a good qualitative agreement with the theoretical concept developed by us. It should be mentioned that the contribution of the frequency selective elements fixed already in eq. (5) is now under consideration in order to check its influence on the pulse compression process. Outside the resonator the chirp compensation needs far longer glass paths (about two orders of magnitude higher than intracavity). Furthermore, note that it is not possible to obtain the ultimate pulse duration by extracavity compression, i.e., in the extracavity pulse compression certain phase modulations remain uncompensated.

Finally, we want to discuss some applications of semiconductors to the generation process. First, instead of glass semiconductors can be used. Because in semiconductors $d^2k/d\omega^2$ is about two orders of magnitude more than in fused silica (e.g., in CdS $d^2k/d\omega^2 \simeq 3.9 \times 10^{-27}$ s^2/mm at 615 nm can be calculated from the dispersion data [21]) shorter samples can be used for chirp compensation. Experimentally it was found that an external CdS pathlength of 3 mm leads to an increase in the pulse duration of about a factor 2. This is in rough correspondence with the calculations using the dispersion parameters of CdS. Note that the same duration prolongation of bandwidth-limited pulses can be achieved using a pathlength of about 250 mm in fused silica. The second favourable application of semiconductor is the external or internal creation of a chirp by nonlinear optical means, which leads to an increase of the bandwidth of the pulse due to the large value of the third order susceptibility (e.g., in CdS in the visible range we have $\chi^{(3)} \simeq 2.2 \times 10^{-20}$ m^2/V^2 [22], i.e., more than two orders of magnitude larger than in fused silica, which is widely used in pulse compression in fibers [23]). Substantial nonlinearities can already be achieved at moderate light intensities, which correspond to a field strength of about 10^8 V/m. Some experiments in this field are in preparation. Of course, it is also possible to use semiconductors in the absorption range. Assuming the pulse duration to be less than the intra- and interband relaxation processes [24] a large number of free carriers can be generated. Strong changes in the carrier density also can lead to a chirp [25], which can be compensated in the manner described above.

References

[1] J. Hermann and B. Wilhelmi, Lasers for ultrashort light pulses (Akademie-Verlag, Berlin, 1984).

[2] Picosecond phenomena III, K.B. Eisenthal, R.M. Hochstrasser, W. Kaiser and A. Laubereau, eds. (Springer-Verlag, Berlin, 1982).

[3] R.L. Fork, B.I. Greene and C.V. Shank, Appl. Phys. Lett. 38 (1981) 671.
J.P. Gordon, R.L. Fork and D.E. Martinez, Paper presented at IQEC '84, Anaheim, 1984.

[4] W. Dietel, E. Döpel, D. Kühlke and B. Wilhelmi, Opt. Commun. 43 (1982) 433.

[5] W. Dietel, J.J. Fontain and J.-C. Diels, Opt. Lett. 8 (1983) 4.

[6] R.L. Fork, C.V. Shank, R. Yen, C. Hirlimann, IEEE QE-19 (1983) 500.
J.G. Fujimoto, A.M. Weiner and E.P. Ippen, Paper presented at IQEC '84, Anaheim, 1984.
[7] H.A. Haus, IEEE QE-11 (1975) 736.
[8] C.V. Shank and E.P. Ippen, Appl. Phys. Lett. 24 (1974) 373.
[9] W. Dietel, Opt. Commun. 43 (1982) 64.
[10] D. Kühlke, W. Rudolph and B. Wilhelmi, IEEE QE-19 (1983) 526.
[11] W. Rudolph and B. Wilhelmi, Preprint FSU Jena 1983.
[12] W. Dietel, W. Rudolph, B. Wilhelmi, J.-C. Diels and J.J. Fontaine, Proc. 3rd Conf. Ultrashort Phenomena in Spectroscopy (UPS) Minsk, 1983.
[13] W. Dietel, E. Döpel, K. Hehl, W. Rudolph and E. Schmidt, Opt. Commun. 50 (1984) 179.
[14] J. Hermann, F. Weidner and B. Wilhelmi, Appl. Phys. B26 (1981) 197.
[15] W. Rudolph, B. Wilhelmi. Appl. Phys. B 34 (1984) 2741.
[16] M.D. Crisp, Phys. Rev. A 8 (1973) 2128.
[17] E.P. Ippen and C.V. Shank, Appl. Phys. Lett. 27 (1975) 488.
[18] W. Rudolph and B. Wilhelmi, Opt. Commun. (submitted).
[19] J.J. Fontaine, W. Dietel and J.-C. Diels, IEEE QE-19 (1983) 1467.
[20] S. De Silvestri, P. Laporta and O. Svelto, IEEE QE-20 (1984) 533.
[21] S.J. Czyzak, W.M. Baker, R.C. Crane and J.B. Howe, JOSA 47 (1957) 240.
[22] Laser Handbook, Vol. 2 (Sovietskoje Radio, Moscow, 1978).
[23] T. Damm, M. Kaschke, F. Noack and B. Wilhelmi, Opt. Commun. (in press) Preprint FSU 1984.
[24] H. Bergner, V. Brückner, F. Kerstan and W. Nowick, Phys. Status Solidi (b) 117 (1983) 603.
[25] D. Marcuse and J.M. Wiesenfeld, Appl. Opt. 23 (1984) 74.

Journal of Luminescence 30 (1985) 262–271
North-Holland, Amsterdam

HIGH DENSITY ELECTRON–HOLE PLASMA IN Si INDUCED BY FEMTOSECOND PULSES

D. HULIN*, A. MIGUS, C. TANGUY* and A. ANTONETTI

Laboratoire d'Optique Appliquée, E.N.S.T.A., Ecole Polytechnique, Batterie de l'Yvette, 91120 Palaiseau, France

High electron and hole (e–h) densities of about 10^{22} cm^{-3} have been produced in silicon, using 620 nm wavelength laser pulses of 100 fs duration. These density values are determined by measuring the dependence of the pulse self-reflectivity on its energy. By comparison with a fully non-linear model of light propagation, we show that dissipation processes inside the plasma are dominated by e–h collisions, with a characteristic time of 3×10^{-16} s. The onset of melting within 100 fs and its nature, considering the high plasma density, are also discussed in view of scattered light measurements.

Introduction

In recent years, melting of semiconductors by laser pulse has received a great deal of interest in the perspective of pulsed laser annealing. The fundamental mechanism of this phase transition is not completely understood and particularly its exact nature has been intensely controversial. However, it is now well established that, within the duration of nanoseconds or even picoseconds pulses, melting can be described as a thermal process [1]. The problem is still open in the case where the laser energy is absorbed by the crystal in a time shorter than the electron–phonon interaction time and if the produced carrier concentration is close from the total number of valence band electrons. Advances in the generation of intense ultrashort light pulses make such studies possible now.

Let us review schematically the fundamental mechanisms which may cause a phase transition. Melting can be considered as occurring when the root mean square of the atom displacement is a fraction of the lattice interspace [2]. This displacement can be caused by the thermal vibrations (phonons) of the heated lattice: this is the classical mechanism, referred to as "thermal melting".

* Also at Groupe de Physique des Solides de l'E.N.S., Tour 23, 2 place Jussieu, Paris 5°, France.

0022-2313/85/$03.30 © Elsevier Science Publishers B.V.
(North-Holland Physics Publishing Division)

One other recently proposed mechanism [3], the "plasma-assisted melting", is related to the fact that the energy is coupled into the semiconductor by the optical excitation. The principle is the following: a photon of high enough energy excites an electron from the valence band towards the conduction band, i.e., more or less, from a bonding orbital to an antibonding one. If the number of excited electrons is a sizeable fraction of the total number of valence band electrons ($N=2\times10^{23}/\text{cm}^3$ in Si), the bonds between atoms are no more effective and the ions are now free to move with respect to each other: the material becomes a fluid. If the properties of this liquid are not known, we may however expect it to be cold because no thermal energy transfer to the ions is needed to obtain this phase transition; we will herefrom refer to this fluid as the "cold liquid".

In order to determine the mechanism involved in the melting process, it is then of great importance to know the exact nature of the obtained liquid. If it can be shown that the resulting melted phase is not the well-known usual hot liquid state, this would demonstrate that melting occurs through the plasma-assisted process. However, in addition to the difficulty for characterizing the cold liquid, the major experimental problems are due to its very short lifetime. Indeed, during the excitation by the laser pulse, the absorbed input energy is initially stored by the carriers, raising their effective temperature far above the equilibrium value. The time during which the liquid can remain cold is then limited by the transfer of the heat from the electrons towards the ions. In the crystal, this corresponds to the usual electron–phonon interaction time, and there is no reason to assume a very different value for the liquid. Since this time is of the order of one picosecond or less, only laser pulses of subpicosecond duration are able to bring evidences of a plasma-assisted melting, if any.

Reflectivity measurements

We have performed reflectivity experiments on a bulk crystalline silicon of orientation [100] with a laser pulse of 100 fs duration at 620 nm, energetic enough to melt (or even to evaporate) the crystal at each shot. The penetration depth of this light in the unexcited silicon is 2 μm [4] but can be reduced to less than 300 Å if a large number of electrons and holes are already present, created by the previous incident photons in the pulse leading edge. The energy deposited per volume unit (and then the corresponding heating) will be very different in these two extreme configurations and the knowledge of the energy deposition profile in the crystal is then of prime importance. For that purpose, we have measured the self-reflectivity of the laser pulse to obtain a time averaged information on the events inside this pulse, and consequently the mechanisms of the laser energy transfer towards the silicon. This type of experiment may give less information than a two-beam pump–probe experiment where the test beam probes the system at adjustable delays after the pump pulse. However, in our case where the pump beam is its own test beam, the advantage is the lack of error on the determination of the absolute delay between the pump and the probe, an important aspect for the study of events faster than the pulse itself.

The self-reflectivity of the excitation beam has been measured with the experimental scheme described on fig. 1. Our set-up consists of a CPM dye laser producing pulses with a 80 fs duration at 620 nm. These pulses are further amplified in a four-stage dye amplifier. A double spatial filtering, obtained through the use of diaphragms D_1 and D_2 is needed to insure a spatial homogeneity of the beam, at the expense of some energy losses. A small part of the incident beam is collected by the photodiode PH2. The excitation (around 0.1 mJ) is then focused onto a spot of 40 μm diameter on the silicon surface. A fresh surface is obtained for each pulse by moving the silicon wafer parallel to its plane. The reflected signal is then detected by the photodiode PH1, through a diaphragm D_3 eliminating any light scattered by the surface. A computerized data acquisition system collects *for each shot* the ratio between the signals (1) and (2) as a function of (2), i.e., the reflectivity in function of the input energy. In order to get a large range of variation for the excitation, we use a set of calibrated optical densities in front of the sample and increase the laser fluctuations. To prevent any non-linearity of the photodiode response over a too large range of intensity, we manage to stay around the same value of the detected signal by inserting on the reflected beam the optical densities removed from the input beam. The computer output is then a semi-log plot of the changes in reflectivity over more than 4 decades of the excitation amplitude.

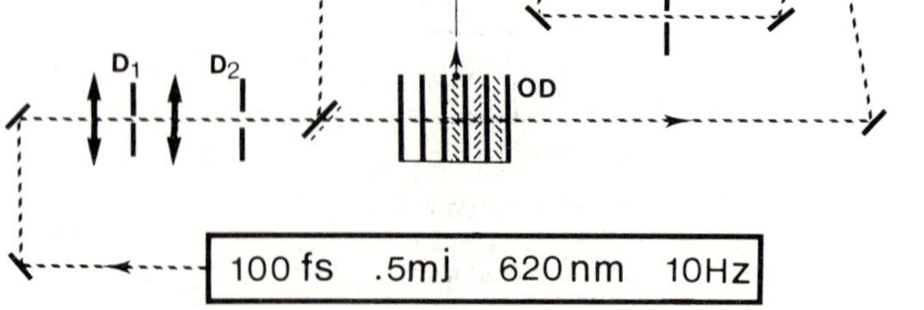

Fig. 1. Experimental setup; D_1, D_2, and D_3 are diaphragms; PH1 and PH2 are photodiodes. The optical densities (neutral filters) are placed either in front of the sample or on the detected beam (see text).

Figure 2 shows the experimental results obtained on a crystalline silicon surface with an excitation at a 620 nm wavelength. The same type of results has been obtained for a germanium crystal. The reflectivity variations are displayed for a two orders of magnitude variation range of the pump intensity, bracketing the region where a dip is observed for the ratio R/R_0. The maximum of the excitation intensity corresponds to the maximum of available energy density when focused onto a 40 μm diameter spot. The shallow dip amplitude is of the order of 5 % and the reflectivity maximum is far from the liquid value which would be approximately larger by a factor 2 than the unperturbed value for crystalline silicon. At high excitation intensities, the increase of reflectivity levels off without reaching a plateau.

The changes in reflectivity observed during the pulse can be explained either by the presence of an electron–hole plasma excited by the light pulse or by the onset of melting. In the first case, the dip is related to the fact that the carrier density value obtained has reached the plasma frequency density. In the latter case, the onset of melting prevents the achievement of this density value and causes the increase of reflectivity before reaching the true minimum. Let us first examine the possibility of thermal melting during the pulse.

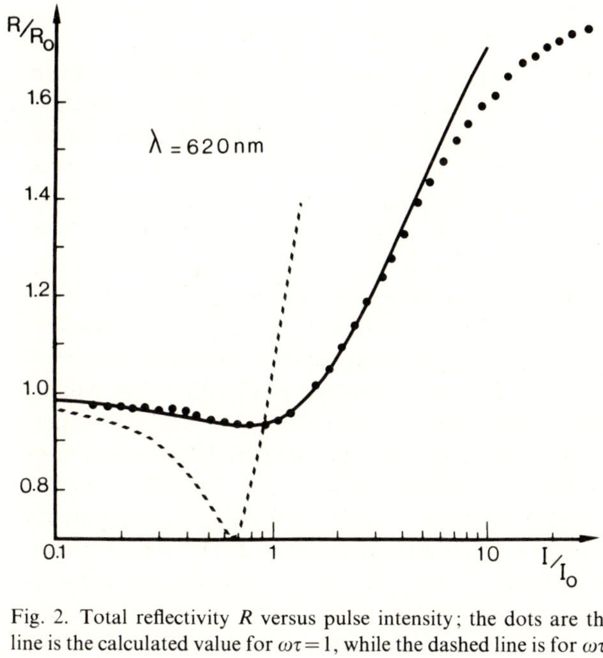

Fig. 2. Total reflectivity R versus pulse intensity; the dots are the experimental data. The solid line is the calculated value for $\omega\tau = 1$, while the dashed line is for $\omega\tau = 10$.

Can thermal melting occur in 100 fs?

The actual possibility of thermal melting in 10^{-13} s can be seriously challenged when considering the orders of magnitude for the different energies involved in the experiment. Knowing the specific heat of silicon at room temperature and its variation with temperature [5], we deduce that about 3000 J/cm^3 are needed to raise the crystal temperature up to the melting point (T=1410°C). When adding the latent heat for melting, a total energy density of $\approx$7000 J/cm^3 is found necessary to bring the crystalline silicon to the ordinary liquid state. In our experiment, typically, an energy of 0.1 mJ per pulse is focused onto a 40 μm diameter spot producing an excitation energy density of 6 J/cm^3, half of it being effective, the other half being reflected. In the most unfavorable case, this energy is absorbed inside a 2 μm thick penetration layer so that at least 15000 J/cm^3 are available, well above what is needed to melt silicon. The fundamental remaining question is how fast can this heat be transferred to the lattice.

Whatever the interactions between the electrons and holes inside the plasma, the excess energy of the excited carriers is ultimately transferred to the lattice through an emission of phonons. The rate at which this energy transfer per unit volume takes place is proportional to the number of carriers able to trigger the event, to the energy transferred at each event (phonon energy) and to the inverse of a time t_{e-ph} characteristic of the electron–phonon interaction:

$$dE/dt = E_{phonon} \times Ne/t_{e-ph}.$$

The phonon energy is close to 50 meV; for the density, we can take in a first step a value close to the plasma frequency density: 10^{22} cm^{-3}. With the usually quoted range 10^{-13}–10^{-12} s for t_{e-ph}, this leads in 100 fs to a maximum energy transfer of 80 J/cm^3 towards the lattice, an amount greatly insufficient to induce the melting. The value of the electron–phonon coupling constant can be discussed because some enhancement may appear due to the high effective temperature of the carriers (in each scattering of an electron emitting a phonon, the final density of states is larger than it would be at the bottom of the conduction band [6]). On the other hand, with such a high carrier density, the screening effect for the electron–phonon interaction has also to be taken into account. For $N=10^{22}$ cm^{-3}, this interaction has been found to be reduced by almost two orders of magnitude from its unperturbed value [7]. Furthermore, due to the high density of phonons, absorption should also occur concurrently to emission of phonons. In these conditions, it is difficult to predict the trend for the increase or the decrease of t_{e-ph}. Anyway, the comparison between the 7000 J/cm^3 needed to melt and the computed 80 J/cm^3 supports the idea that thermal melting cannot occur during the 100 fs time lapse. If we had taken the hypothesis of a smaller density value (for the case where the plasma frequency density cannot be reached), the discrepancy between the two energies would have been more striking. If a thermal melting during the pulse appears unlikely, we cannot rule out, at this stage, the possibility of the other type of melting. Even though, let us first analyze the reflectivity change during the laser pulse assuming the presence of a high density e–h plasma and come back later on this point. Since our data corresponds to the silicon modifications averaged during the excitation pulse, their interpretation is not straightforward and needs some modeling.

A non-linear model

The response of a material to an applied electromagnetic field is fully characterized by the knowledge of the dielectric constant $\varepsilon(\omega)$ at every location in the material at any time. On the other hand, ε is modified by the presence of the created charged carriers:

$$\varepsilon=\varepsilon_0+i\varepsilon_0'-Ne^2/\mu(\omega^2+i\omega/\tau),$$

where $\varepsilon_0+i\varepsilon_0'$ is the unperturbed crystal dielectric constant at frequency ω, N the density of e–h pairs, μ their reduced effective mass and τ a dissipation time inside the plasma [9]. The plasma influence is negligible as long as the carrier density is smaller than N_p defined by:

$$N_p=\mu\varepsilon_0\omega^2(1+1/\omega^2\tau^2)/4\pi e^2.$$

In the usual case where ε is slowly varying (i.e., its variation are very small for length and time increments of the order of the wavelength and the period of the incident light) the reflectivity for normal incidence is related in a very simple way to $\varepsilon=(n+ik)^2$ by a well-known formula. These expressions are no more valid in our experimental conditions where the carrier density becomes of the order of, or larger than, N_p. The dielectric constant is now varying spatially on short distances due to the presence of the electrons and holes created on a thin layer close to the surface, with an exponential initial profile. This profile will, of course, be modified progressively during the pulse due to the continuous modification of the absorption. To deal with this highly non-linear problem, one has to go back to the fundamental Maxwell equations and solve them without using the usual simplifications.

The reflected electric field is the difference between the incident field $E_0(t)$ and the amplitude $E(0, t)$ at $z=0$ of the field penetrating into the crystal. The reflectivity can be defined as

$$R(t)=|E_0(t)-E(0, t)|^2/|E_0(t)|^2.$$

The electric field at each time $E(z,t)$ follows the Maxwell equations. These equations are becoming non-linear due to the fact that $\varepsilon(\omega,t)$ depends on $|E|^2$ through the number of carriers whose creation rate is proportionnal to $|E|^2$. They can be solved [9] taking into account the two boundary conditions:

(1) $\varepsilon = \varepsilon_{\text{non-perturbed}}$ for $z \to \infty$,
(2) continuity of $\boldsymbol{E}$ and $\boldsymbol{B}$ at the surface.

This set of equations have been numerically solved to determine $E(z, t)$ at every time and location and particularly $E(0,t)$, from which the instantaneous reflectivity $R(t)$ can be derived. Since the dielectric constant depends on N, the number of already created carriers, the temporal variation of the electric field in the pump pulse has to be introduced. In our case, we take a biexponential shape for the laser pulse with a FHWM of 100 fs. For the self-reflectivity, the only useful quantity is

$$R = \int_T R(t)|E_0(t)|^2 \, \mathrm{d}t \bigg/ \int_T |E_0(t)|^2 \, \mathrm{d}t.$$

T is a time lapse long enough to include the whole pulse. Note that, although this quantity is defined only with the values of the electric fields at the surface, the knowledge of these fields implies the complete knowledge of what happens inside the crystal and particularly of the carrier density profile.

The variation of R with the density N exhibits first a decrease as N get closer to the plasma frequency N_p and then an increase when N gets above this value. The detailed shape of the curve depends on the parameter τ [9], the smaller the value of τ the smaller the dip in the reflectivity. In order to fit the calculated curve with the experiment (fig. 2), we choose an appropriate value for τ and adjust the density scale on the excitation intensity scale. In the density

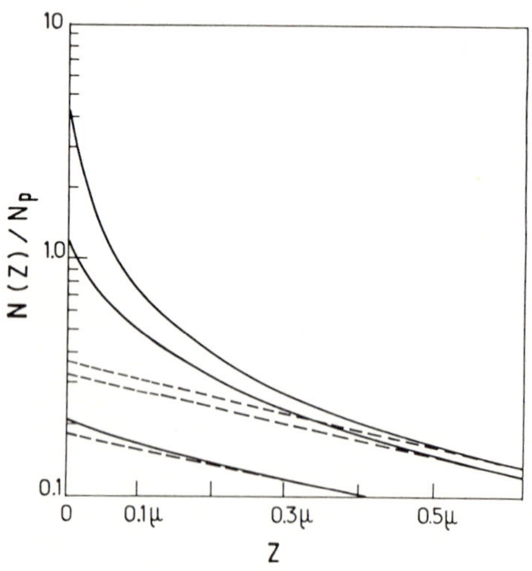

Fig. 3. Calculated electron–hole density profile $N(z)$ at the end of the pump pulse for $I/I_0 = 0.2$, 1 and 5. The dotted lines show the behavior for a profile decreasing exponentially over the virgin Si penetration depth.

range around N_p, we can reasonably assume that N at the surface varies linearly with the excitation intensity [8]. The location of the reflectivity minimum indicates the pump energy value for which the plasma density N_p is obtained [10], i.e., a density close to $10^{22}\,\text{cm}^{-3}$ (see discussion in [9] for the estimation of N_p). An important information comes from the value of τ needed to fit the experimental curve: $\omega\tau \approx 1$ ($\omega = 2\pi\nu = 3 \times 10^{15}\ \text{s}^{-1}$) so that $\tau \approx 3 \times 10^{-16}$ s, a surprisingly small value. It is difficult to explain this dissipation time inside the plasma by the electron–phonon processes whose interaction times range around 10^{-13} s. More probably, it corresponds to the collision time between charged carriers and more specifically between electrons and holes, this last process being the only one to produce a damping for the motion of electrons and holes in presence of an electric field [8]. The value found here is in good agreement with a theoretical estimate using a screened Coulomb interaction and the Rutherford scattering cross section [9]. Note that this short collision time between carriers allows one to use the principle of an effective temperature to describe the electronic states.

Light scattering measurements

Figure 4 shows the light scattering from the surface (diffusion) collected in the same experimental conditions as the reflectivity (above) except that the diaphragm D_3 is replaced by a small mask. The directly reflected beam is then

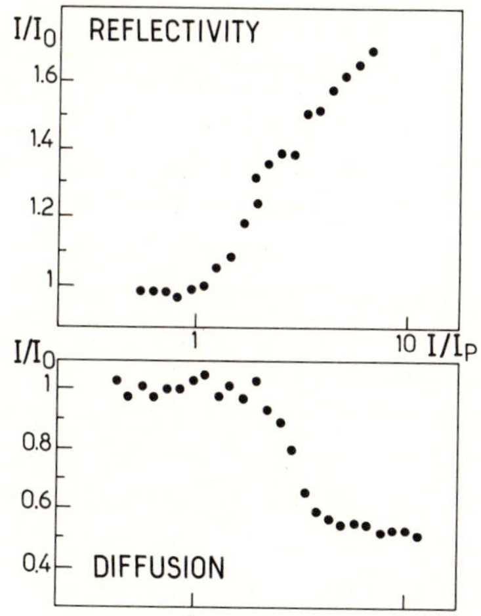

Fig. 4. The lower part of the figure shows the light scattering signal as a function of the pulse intensity recorded in the same experimental conditions as the reflectivity signal shown in the upper part of the figure.

suppressed and only the light of the pump pulse scattered by the silicon surface is collected on the photodiode PH1. The ratio of scattered and incident light is almost constant up to an excitation value 2 or 3 times larger than I_0 (the laser energy for the reflectivity minimum), and then drops rapidly to about 0.6 times its initial value, decreasing only very slowly as the excitation further increases. The amount of light scattering by a surface is related to deviations from a perfect flatness by a fraction of the light wavelength. The observed decrease of the scattered light may be due to a decrease of either the number of diffracting centers or their efficiency. It is most unlikely that a spatial rearrangement of the crystal surface over distances of the order of 1000 Å could occur in 100 fs and therefore the number of diffracting centers should stay constant. On the other hand when N reaches the plasma frequency N_p, the penetration depth of the light inside the crystal decreases strongly and can be reduced to a few hundred angstroms (fig. 3). The "landscape" of the surface seen by the light when the penetration depth is 300 Å has a small probability to be the same as when it was 2 μm. Then, the decrease in the scattering rate seems to be related to a change of the penetration depth, occurring for an excitation intensity around three times larger than the one required at the reflectivity minimum. It is in this e–h density range that the optical properties of the semiconductor are becoming similar to the ones of a metal. The plasma effects then are most likely responsible for the changes in the scattering efficiency occurring at $3I_0$. However, we cannot rule out the onset of a non-thermal melting also to be the origin of the abrupt diminution of the penetration depth.

Discussion and conclusion

In conclusion, we have measured the self-reflectivity of a 100 fs pump laser pulse at 620 nm to determine the laser energy absorption profile inside the silicon crystal. When increasing the laser intensity, a minimum in the reflectivity is seen, followed by a further increase of the signal. This upwards trend does not seem to be related to the onset of any melting: (i) the amount of heat needed to be transferred towards the lattice to obtain the thermal melting is too large for such a small period of time (10^{-13} s); (ii) the change of light scattering occurring at a higher excitation intensity supports the idea that the plasma is still present at $3I_0$ (or that melting, if any, occurs only at $3I_0$). Then, the reflectivity variations can be related to the presence of a high density electron–hole plasma, at least for the excitation energies not too far above the reflectivity minimum. Since the carriers are created in a small layer close to the surface, the dielectric constant varies rapidly over a short distance and the usual reflectivity formula are not valid any more. Going back to the Maxwell equations, this non-linear problem has been solved and the instantaneous reflectivity has been predicted. The signal obtained in the experiment corresponds to an average of the events during the pump pulse, and more precisely, to a correlation of the crystal modifications with the laser pulse (in a pump–probe experiment, it is also a correlation of the "perturbation" with the probe pulse). Introducing the time dependence of the laser pulse amplitude, we have compared the computed averaged reflectivity curve with the experiment. The other two noticeable informations are first that, using 100 fs pulses, an e–h density of 10^{22} cm^{-3} can be obtained, a value close to the one required for the "plasma-assisted melting". Secondly, a very small value of 3×10^{-16} s is obtained for the relaxation time inside the plasma which cannot be attributed to the electron–phonon interaction but rather to the electron–hole collisions. It is the first time that such a process has been observed since it requires e–h densities only obtainable with femtosecond lasers.

Acknowledgement

We gratefully acknowledge the theoretical support of M. Combescot and J. Bok of the Ecole Normale Superieure, Paris, and the help of J.Y. Vinet of Laboratoire d'Optique Appliquée, Palaiseau, for the computations.

References

[1] H. Kurz, J.M. Liu and N. Bloembergen, in: Cohesive Properties of Semiconductors under Laser Irradiation, ed., L.D. Laude), NATO AS, n°69 Martinus Nijhof, The Hague (1983) p. 419.
[2] J.M. Ziman, Principles of Theory of solids (Cambridge University Press, London (1965) p. 63.
[3] M. Combescot and J. Bok, Phys. Rev. Lett. 48 (1982) 1413.
[4] D.E. Aspnes and A.A. Studna, Phys. Rev. B 27 (1983) 985.
[5] See for example, W.R. Runyan, Silicon Semiconductor Technology (McGraw-Hill Book Company, New York, 1965).
[6] J.R. Chelikowsky and M.L. Cohen, Phys. Rev. B 14 (1976) 556.
[7] E.J. Yoffa, Phys. Rev. B 23 (1981) 1909.
[8] D. Hulin, M. Combescot, J. Bok, A. Migus, J.Y. Vinet and A. Antonetti, Phys. Rev. Lett. 52 (1984) 1998.
[9] M. Combescot and J. Bok, J. Lumin. 30 (1985) 1.
[10] This statement is not absolutely correct: the final surface density at the end of the pump pulse is already above N_p because what is measured is the averaged value R, and I_0 corresponds in fact to the minimum of R. However, this is true within less than a factor of 2 and can be retained for the sake of simplicity.

Journal of Luminescence 30 (1985) 272–289
North-Holland, Amsterdam

PULSEWIDTH-DEPENDENCE OF NONLINEAR ENERGY DEPOSITION AND REDISTRIBUTION IN Si, GaAs AND Ge DURING 1 μm PICOSECOND IRRADIATION

Arthur L. SMIRL, Thomas F. BOGGESS, Steven C. MOSS and Ian W. BOYD

Center for Applied Quantum Electronics, Department of Physics, North Texas State University, Denton, Texas 76203, USA

We have used 1 μm pulses ranging in duration from 4–260 ps to measure the pulsewidth dependence of the nonlinear absorption, melting threshold, and resolidification morphologies of Si, GaAs, and Ge. With these materials, we have been able to quantify a variety of nonlinear absorption processes with a single excitation wavelength. We find that the fluence required to melt Si and GaAs is roughly proportional to the square root of the pulsewidth while that required for Ge is nearly pulsewidth independent. A crystalline-to-amorphous transition is observed in Si for pulses less than 10 ps and in GaAs for all pulsewidths, but no such transition is observed in Ge. These observations are shown to be consistent with the various energy deposition and redistribution mechanisms present in each material. Finally, we have used the active nonlinearities in Si and GaAs to construct optical limiters designed to protect sensitive optical components from intense 1 μm radiation.

Si, GaAs and Ge are the most commonly used materials for fabrication of electronic components. Within the optics industry, they are used as detector and infrared transmissive optical materials. They (particularly GaAs) have also been investigated for nonlinear optical device applications such as bistability and phase conjugation. In addition, they (particularly Si) are the semiconductors most often studied in laser processing applications. Here, we summarize preliminary measurements in which we contrast the pulsewidth-dependence of the nonlinear absorption, melting thresholds, phase transitions, and resolidification morphologies of these materials induced by 1 μm pulses ranging from 4–260 ps in duration. We then describe the use of resonant nonlinear optical self-action mechanisms in Si and GaAs to construct passive optical limiting switches or regulators.

Experimental apparatus:

The measurements reported here were conducted with three separate, well-characterized, picosecond laser systems [1]: (1) an actively/passively

0022-2313/85/$03.30

modelocked Nd:YAG laser, (2) a passively modelocked Nd:YAG laser and (3) a passively modelocked Nd:glass laser. All three lasers operated in the TEM_{00} transverse mode. The Nd:YAG lasers produced pulses at 1.064 μm that could be varied in temporal duration from 35–260 ps by selecting etalon output couplers of varying thicknesses. The glass laser produced pulses at 1.054 μm that had an average temporal duration of 7 ± 2 ps. For each laser firing, a single pulse was selected from each modelocked train, and a portion of that pulse was delivered to a calibrated pulsewidth monitor so that the temporal extent could be determined on a shot-to-shot basis. The majority of the pulse was amplified and then passed through a continuously variable attenuator consisting of a half-wave plate placed between a pair of crossed polarizers. Finally, the pulse was focused onto the sample at near normal incidence, and the transmitted and reflected energies were measured. Near threshold the measurements were conducted in a one-on-one fashion. That is, the sample was mechanically translated before each laser firing so that virgin material was exposed. Each irradiated spot was then inspected with a high-contrast Nomarski microscope. The threshold for damage or melting was defined as the fluence at which any change in the self-reflectivity or any permanent change in the sample surface was observed.

To ensure that the transverse spatial profiles were Gaussian and to determine the focused spot size, the beams were scanned in both the vertical and horizontal dimensions by a small diameter pinhole placed at various positions before, near and following focus. The spatial uniformity was checked by melting the surface of a highly-polished wafer of crystalline Si and by examining the recrystallized spot with a Nomarski microscope. In addition, vidicon scans indicated that there were no significant shot-to-shot variations in the spatial profile.

The Si, GaAs, and Ge samples were high-purity, high-resistivity, optically-polished, crystalline wafers which were 1 mm, 1.75 mm and 5.7 μm in thickness, respectively. The Si wafers were antireflection coated on the second surface to avoid Fabry–Perot effects.

Nonlinear absorption

Silicon, GaAs and Ge were chosen as candidates for study because they are relatively well-characterized and because they allow us to use a single excitation wavelength to investigate and contrast a variety of nonlinear absorption mechanisms. For our purposes, the relevant features of the band structures of these three semiconductors are schemed in fig. 1. In Si at 1 μm, the excitation energy (~1.17 eV) is very near the bandgap and only indirect single-photon transitions are initially allowed. By contrast, direct two-photon and direct single-photon absorption are the dominant energetically-allowed processes in GaAs and Ge, respectively.

The transmission of Si at 1 μm was measured as the excitation level was increased from the linear regime to well above the melting threshold. The results for pulsewidths (FWHM) of 48 ± 10 ps and 180 ± 37 ps are shown in fig. 2. Below 30 mJ/cm^2, the Si is relatively transparent. The transmission in this region is linear and is consistent with an indirect absorption coefficient of

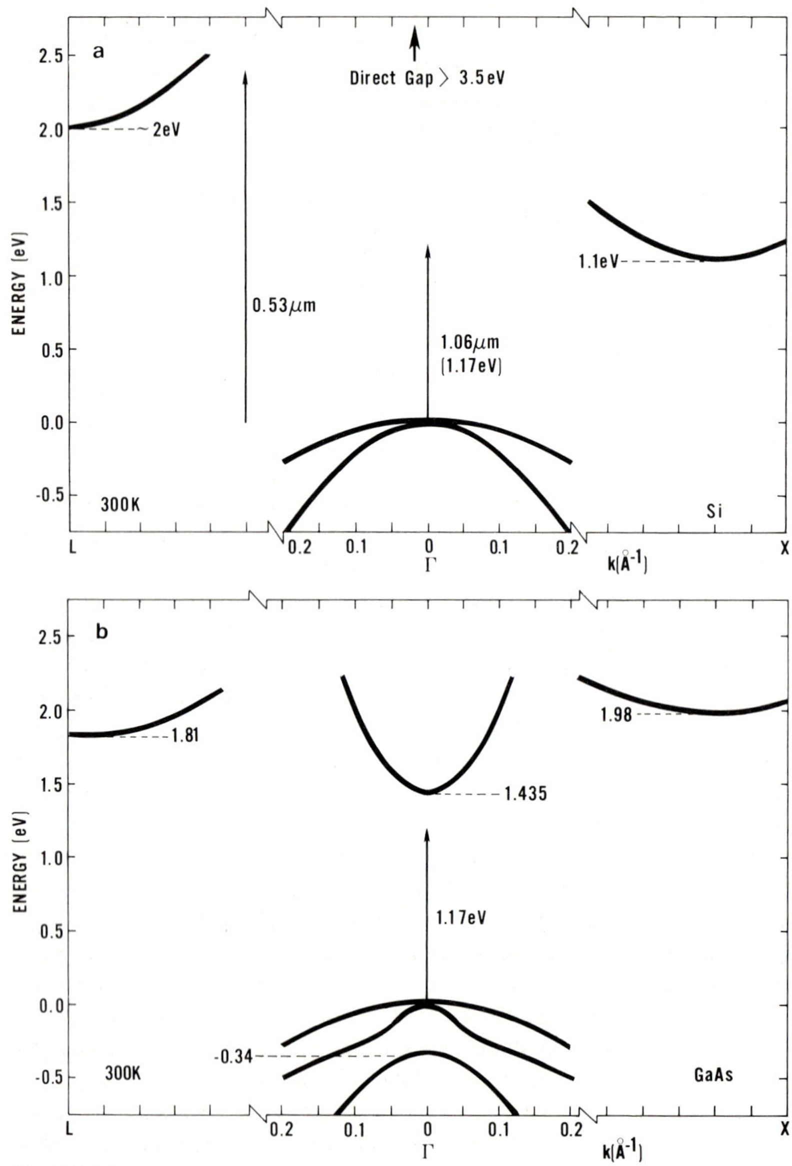

Fig. 1(a), (b).

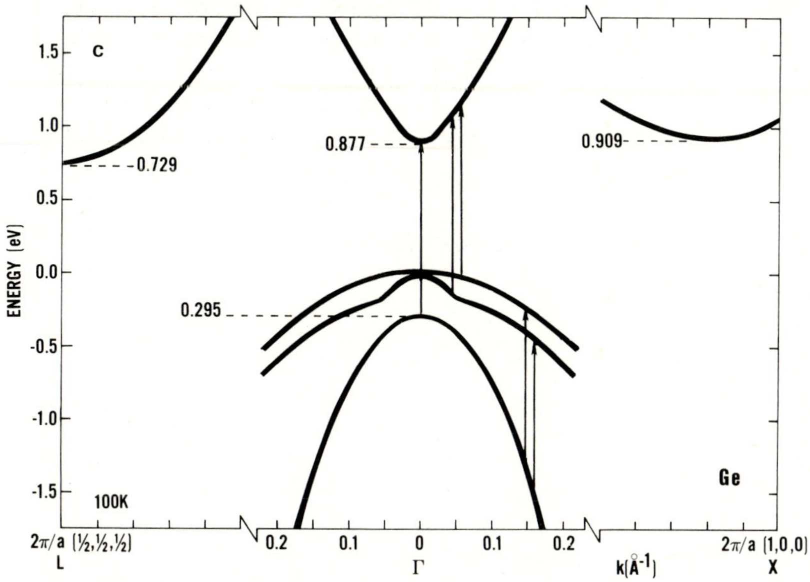

Fig. 1. Relevant features of the band structures of (a) Si, (b) GaAs, and (c) Ge.

approximately 10 cm^{-1}. At higher incident fluences, the absorption is increasingly dominated by free-carrier absorption, as well as by increased indirect absorption that is caused by bandgap narrowing with increasing lattice temperature. The latter effect results in a larger number of photogenerated free carriers that in turn provides further lattice heating, and near the melt threshold, a thermal runaway process develops. The solid curve in fig. 2 is the result of a calculation of the Si transmission that includes only indirect and free-carrier absorption, using previously reported values for the temperature-dependence of the free-carrier cross section [2], bandgap [3], and indirect absorption coefficient [4]. No adjustable parameters were used. The theoretical curve can be made to pass through the data points by decreasing the cross section slightly. Because of the highly nonlinear and rapid nature of the thermal runaway, lattice heating is significant only for fluences very near the melting threshold. The data in fig. 2 show that, below threshold, the transmission of Si does not measurably depend on pulsewidth, but depends strictly on the incident fluence. This suggests that processes such as Auger recombination and two-photon absorption are not significant during the optical pulsewidth. Carrier diffusion is expected to be significant but will not alter the transmission. Since free-carrier absorption depends only on the total integrated number of carriers between the front and back surfaces, it will be unaffected by longitudinal carrier position and insensitive to carrier diffusion. The absence of two photon effects in our measurements is in contrast to recent reports [5] that indirect two-photon

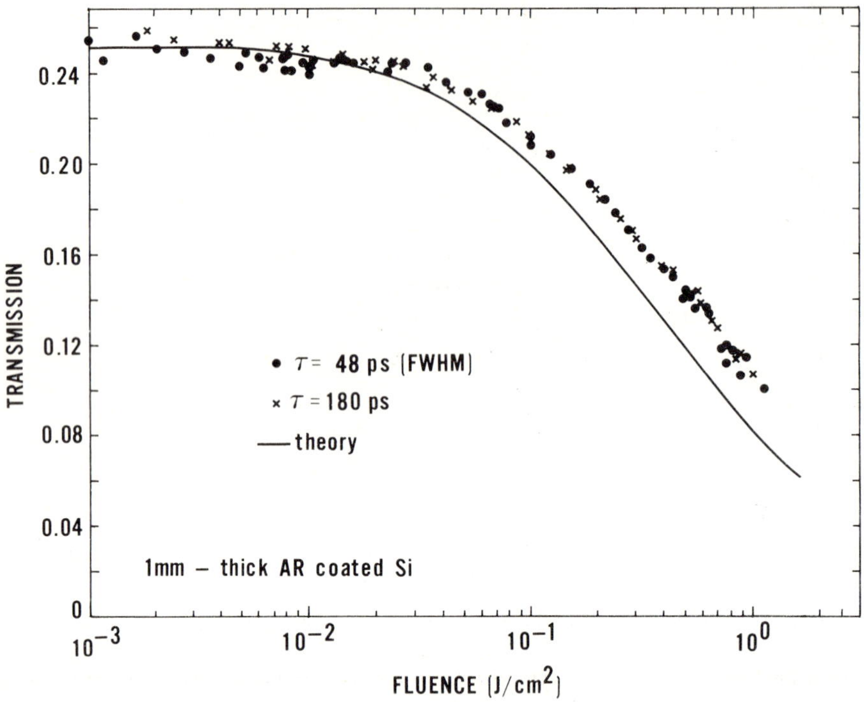

Fig. 2. Nonlinear transmission of a 1 mm thick Si wafer for pulsewidths of 48 (dots) and 180 ps (crosses). The solid curve is the result of a calculation using only indirect and free-carrier absorption, with values for the temperature dependence of the free-carrier cross section, bandgap, and indirect absorption coefficient obtained from the literature.

absorption dominates the energy deposition in Si at 1 μm. In the latter measurements, a two-photon absorption coefficient β of 52 cm/GW was extracted by plotting the inverse of the transmission versus the intensity, by noting that such a plot is very nearly a straight line, and by extracting β from the slope. We comment that the differential equation relating the intensity I to the two-photon absorption coefficient β is identical in form to that relating the fluence $\mathscr{F}$ to the free-carrier cross section σ. Consequently, when measurements at a single pulsewidth are performed, inverse transmission plots for free-carrier absorption are difficult to distinguish from those for two-photon absorption.

When the data shown in fig. 2 for two pulsewidths are plotted as the inverse of the transmission versus fluence (rather than intensity), the plots are independent of pulsewidth, confirming that free-carrier absorption not two-photon absorption is the dominant absorption mechanism. Moreover, a free-carrier cross section of approximately 3.5×10^{-18} cm^2 can be extracted from the slope of such a plot [6]. When this cross section is used, the theoretical curve of fig. 2 is coincident with the data.

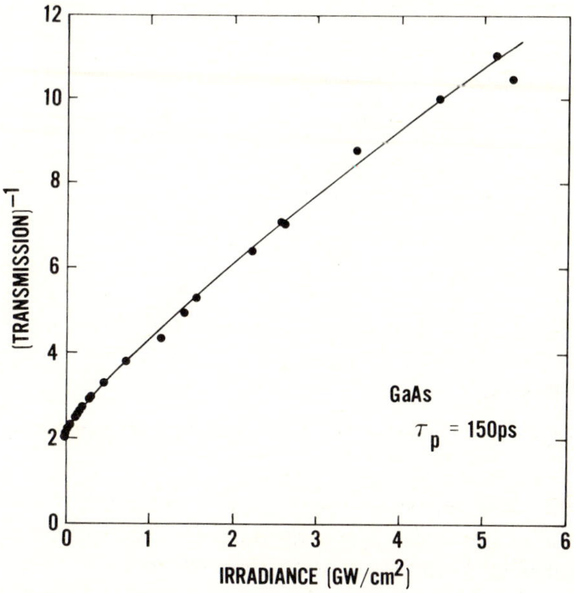

Fig. 3. GaAs transmission as a function of incident irradiance for 150 ps pulses. The solid curve is the result of a numerical calculation using a linear absorption coefficient of 0.55 cm^{-1}, a free carrier cross section section of 3×10^{-18} cm^2, and a two-photon absorption coefficient of 28 cm/GW.

The nonlinear transmission of GaAs was measured on five occasions [1], using three separate laser systems and pulsewidths of 40, 45, 90, 108, and 150 ps (FWHM) at 1.064 μm and 7 ps at 1.054 μm. One measurement of the inverse transmission as a function of incident irradiance for the 150 ps pulses from the actively/passively modelocked Nd: YAG laser is shown in fig. 3. The solid curve is a numerical fit with a linear absorption coefficient α of 0.55 cm^{-1}, a free-carrier cross section σ of 3×10^{-18} cm^2, and a two-photon absorption coefficient $\beta = 28$ cm/GW. Similar data were obtained and similar curves generated for the other pulsewidths. Since reported values for the free-carrier cross section range from 2×10^{-18} to 3×10^{-17} cm^2, we treated both σ and β as fit parameters. At low irradiances, the choice of σ used in our analysis did not strongly influence the extracted value for β. Conversely, these measurements do not allow a determination of σ with any precision. By conservatively estimating our knowledge of the pulse energy, temporal width, spatial width and absolute transmission and by considering the relative error associated with the numerous measurements, we estimate a total root-mean-square error of approximately 32% or a β of 26 ± 8 cm/GW. Although there is a wide spread in reported values for the two-photon coefficient, this value is in good agreement with the smallest values reported in the literature [7–11].

The transmission of Ge has been extensively measured and analyzed as a

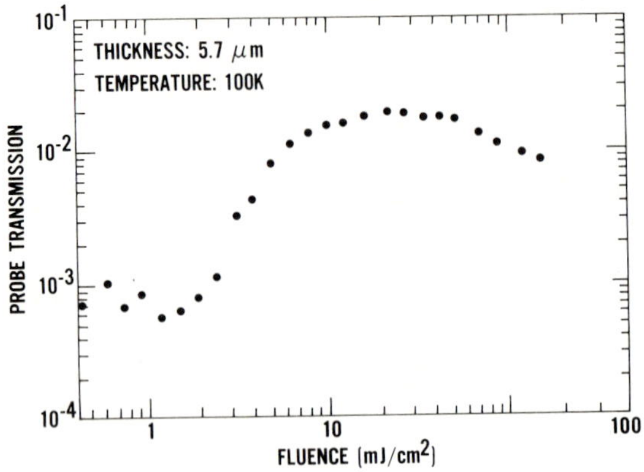

Fig. 4. Ge transmission as a function of incident fluence for a fixed probe delay of 250 ps following excitation.

function of optical pulsewidth, excitation level, time delay following excitation, sample thickness, temperature, and surface preparation [12]. The tendencies in the nonlinear transmission of a 5.7 μm thick sample are shown in fig. 4 for a fixed time delay of 250 ps following excitation. In contrast to Si and GaAs, the Ge transmission initially increases with excitation level and, for longer delays, peaks at approximately 1% (far below complete saturation) and eventually decreases with increasing excitation level. These tendencies can be grossly understood as follows. Because the optical radiation is coupled relatively near the valence band edge, the Ge transmission is easily saturated and is extremely sensitive to carrier temperature and spatial location. Consequently, saturation effects dominate thermal effects in determining the nonlinear transmission, as is evident from the data shown in fig. 4. The Ge transmission initially increases with increasing fluence as the valence states needed for direct absorption are depleted for hole densities above $\sim 10^{19}$ cm^{-3}. At higher excitation levels, the transmission peaks and decreases as direct intervalence band transitions (see fig. 1(c)) become allowed when carrier densities reach approximately 10^{20} cm^{-3}. Analysis of the tendencies in the data indicate that the carriers have cooled to near lattice temperature by the end of the excitation pulse. The analysis also suggests that carrier diffusion and acoustic phonon-assisted cooling determine the transmission for the first few hundred picoseconds following excitation – although surface recombination will dominate for samples thinner than 5 μm [12].

Melting threshold

We have also determined that the fluence $\mathscr{F}_{TH}$ required for the single shot melting of Si depends strongly on the pulsewidth [13]. The data in fig. 5 clearly

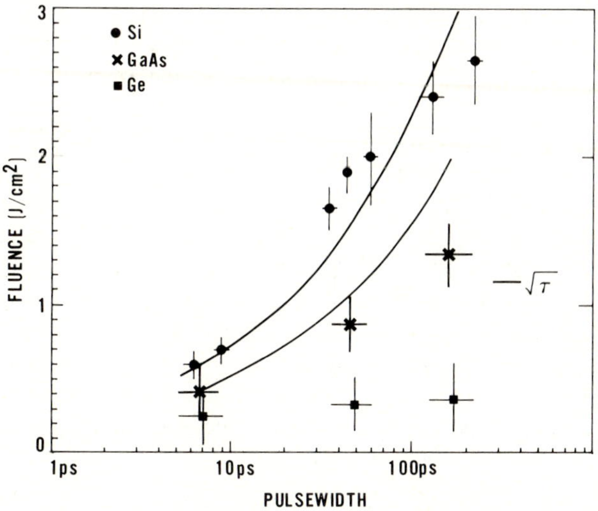

Fig. 5. Pulsewidth dependence of the melt threshold of Si (circles), GaAs (crosses), and Ge (squares).

show that a smaller fluence is required to melt c-Si with the shorter pulses. The threshold is found to range from 0.6 ± 0.15 J/cm^2 for 6 ps pulses to 2.7 ± 0.3 J/cm^2 for 225 ps pulses. Measurements were performed on several separate optically polished samples (some with the back surface antireflection coated and some not) and on one etched sample. No significant sample-to-sample variations in threshold were observed. Including the heat of fusion, an energy density greater than 7000 J/cm^3 is needed to melt Si initially at 300 K. If only simple indirect absorption and the accompanying free-carrier absorption are included, the energy density deposited at the surface of Si can be calculated in closed form and is given by

$$\rho_s = \alpha \mathscr{F}_0 [1 + (\sigma / 2\hbar\omega) \mathscr{F}_0], \qquad (1)$$

where $\mathscr{F}_0$ is the fluence at the first surface (taking the reflection into account), σ is the free-carrier cross section and $\hbar\omega$ is the photon energy. Such a simple model predicts that there is no pulsewidth dependence to the threshold and, at threshold, that ~ 30 J/cm^3 is deposited at the surface. Clearly, the dependence of the free-carrier cross section and indirect absorption coefficient on lattice temperature must be included. At fluences near threshold, the inclusion of lattice heating results in a thermal runaway process, whereby the free-carrier absorption heats the lattice, the increased lattice temperature in turn causes bandgap narrowing, the bandgap narrowing results in increased indirect absorption, the indirect absorption creates more carriers, more carriers increase the free-carrier absorption and so on. Such thermal effects must be included numerically. A simple thermal runaway model that includes only temperature-dependent indirect and free-carrier absorption predicts a threshold for 7 ps pulses of *roughly* the correct order of magnitude but predicts no pulsewidth-dependence to the threshold. Clearly other processes must be included. Notice that the melting threshold for Si exhibits roughly a square-root dependence on pulsewidth (see fig. 5), as would be expected for a transport-dominated process (where the diffusion length is proportional to the square root of time). This suggests that carrier or thermal diffusion (or both) are important. However, the thermal diffusion length in Si for 10 ps is only 40 nm at 300 K and less at higher temperatures. Consequently, we would not expect thermal diffusion to contribute significantly for pulses of this width. By contrast, the carrier diffusion length is 5 times this magnitude. The possible role of carrier diffusion in determining the melting threshold can be readily understood by remembering that, prior to thermal runaway, free-carrier absorption is the dominant mechanism in Si for transferring energy from the pulse to the lattice. The amount of energy taken from the pulse by free-carrier absorption depends only on the total number of carriers contained between the front and back surface, independent of longitudinal location. The lattice temperature at a given position, on the other hand, depends on how many carriers are located at that position when the energy is transferred from the pulse to the carriers and, for our purposes, instantaneously from the carriers to the lattice. Thus, carrier diffusion is expected to affect the overall transmission only weakly and indirectly, while strongly influencing the lattice temperature at the surface. These expectations are consistent with the observed fluence-dependence of the transmission measurements and the pulsewidth-dependence of the melt threshold.

The dependence of the melting threshold of GaAs on pulsewidth has also been measured and is also shown in fig. 5. Again the threshold is roughly (though not exactly) proportional to the square root of the pulsewidth. This pulsewidth-dependence is not only consistent with diffusion, as discussed in the preceding paragraph, but is consistent with two-photon absorption as the primary absorption mechanism. The energy per unit volume deposited at the front surface, if only two-photon absorption is included, is given by

$$\rho_s = (2\ln 2/\pi)^{1/2} \beta \mathscr{F}_0^2/\tau. \tag{2}$$

Under these circumstances, the threshold fluence is proportional to the $\sqrt{\tau}$. At threshold, two-photon absorption alone would result in the transfer of less than 400 J/cm^3 to the crystal near the surface (>6000 J/cm^3 are required to melt) and cannot solely account for melting. Possibly either a two-photon-assisted avalanche or surface absorption is the dominant damage mechanism.

By contrast, little variation in the melt threshold with pulsewidth was observed for Ge (see fig. 5), indicating that energy transport during the pulse is ineffective in determining the threshold. Again, as with Si, thermal diffusion lengths (<30 nm) are expected to be small for our pulsewidths, but carrier diffusion lengths (~ 15 times larger) are not. This lack of pulsewidth-dependence to the threshold would be expected if direct absorption dominated the energy deposition and if there were no saturation of the direct absorption coeficient. Under the latter circumstances and assuming that the electrons transfer their energy to the lattice rapidly compared to the pulsewidth, the energy would be deposited in a layer approximately one absorption depth thick, and diffusion would play no role. Given that the nonlinear absorption measurements in Ge at this wavelength (fig. 4) are dominated by saturation effects (that is, they are sensitive to carrier position), this result is at first somewhat surprising. However, it can be understood by recognizing that, at excitation levels near threshold, the direct absorption coefficient is bleached and the carrier density is uniform for several microns into the sample. Since approximately 100 ps are required for carriers to diffuse 1 μm (at 300 K), carrier diffusion will be negligible over these distances for our pulsewidths.

Phase transitions and surface morphology

These pulsewidth-dependences are also reflected in the surface morphologies. Close inspection of the Si morphology shows that, for pulses longer than 30 ps, bulk melting is followed by recrystallization, but with some point or line defects incorporated into the structure. By contrast, for pulses shorter than 10 ps, we consistently observe [14] a crystalline-to-amorphous transition (fig. 6a), contrary to published expectations at 1 μm. Laser-induced amorphization of c-Si has been reported previously only using nanosecond ultraviolet [15, 16] and picosecond visible or uv radiation [17]. It had been generally assumed that the absorption depth associated with 1 μm radiation was not small enough to allow the formation of a-Si from the melt [18]. Figures 6a – 6d show Nomarski micrographs of the surface morphology of c-Si after irradiation by 7 ps pulses at 1 μm. At $\mathscr{F}_{TH}$ (fig. 6a), the c-Si is transformed to an amorphous disc. At slightly higher fluences (fig. 6b), only the perimeter of the irradiated area undergoes an amorphous transition while the central region actually recrystallizes. At higher energy densities, the recrystallized region becomes the dominant feature while the amorphous area appears only as a very thin surrounding ring (fig. 6c). At still higher fluences (fig. 6d), a narrow heavily-rippled ring always appears within the recrystallized material. In grossly over-simplified terms, a critical melt thickness exists such that the time required for thermal diffusion or cooling is exactly equal to the time required for recrystallization. For melt thicknesses less than this critical value, the melt at the surface will freeze in the amorphous state before recrystallization can occur. For melt depths thicker than this critical value, the regrowth time is less than the time required for quenching and the melt recrystallizes. Well above threshold, the melt depth in the center of the irradiated area is driven beyond this critical depth, while remaining below that thickness in the wings. Thus, c-Si that is

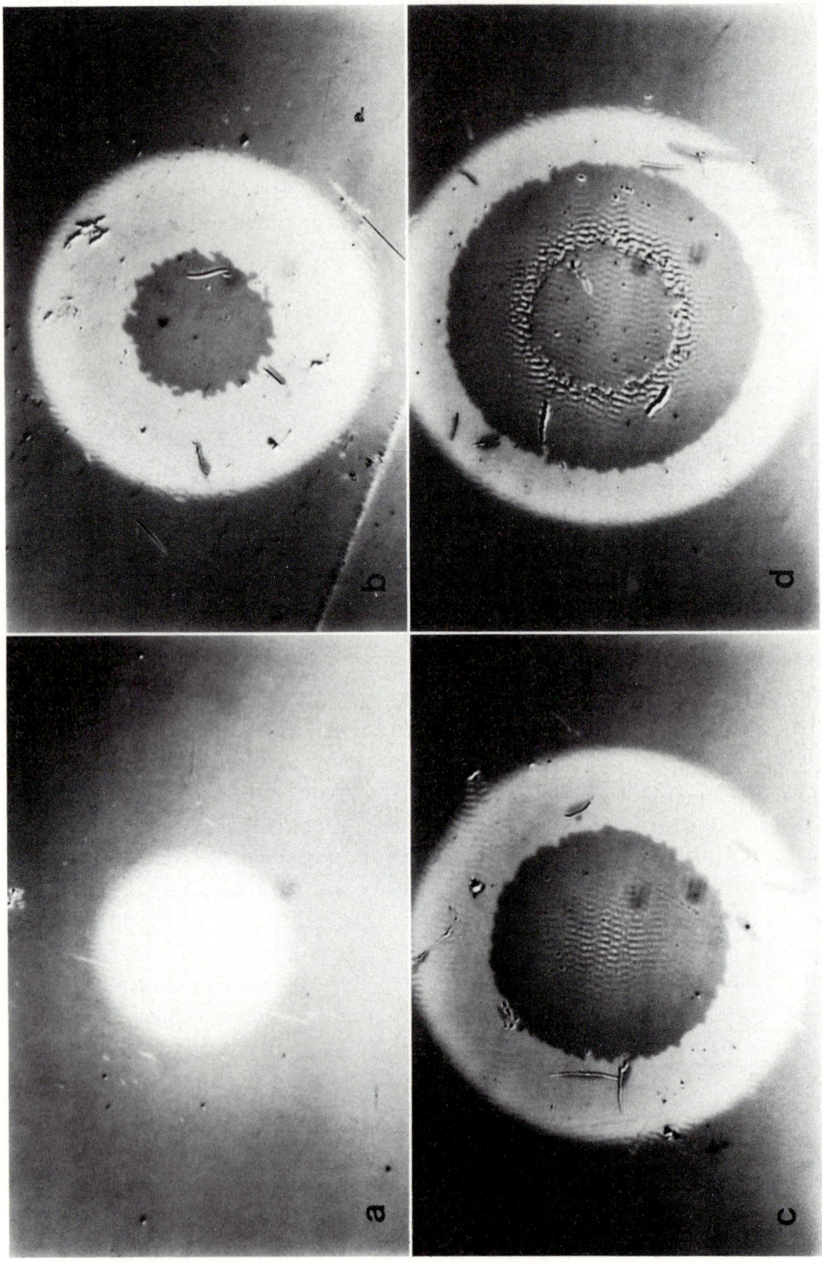

Fig. 6. Surface morphology of c-Si after irradiation with nominally 7 ps pulses, with the pulse energy increasing from (a) to (d).

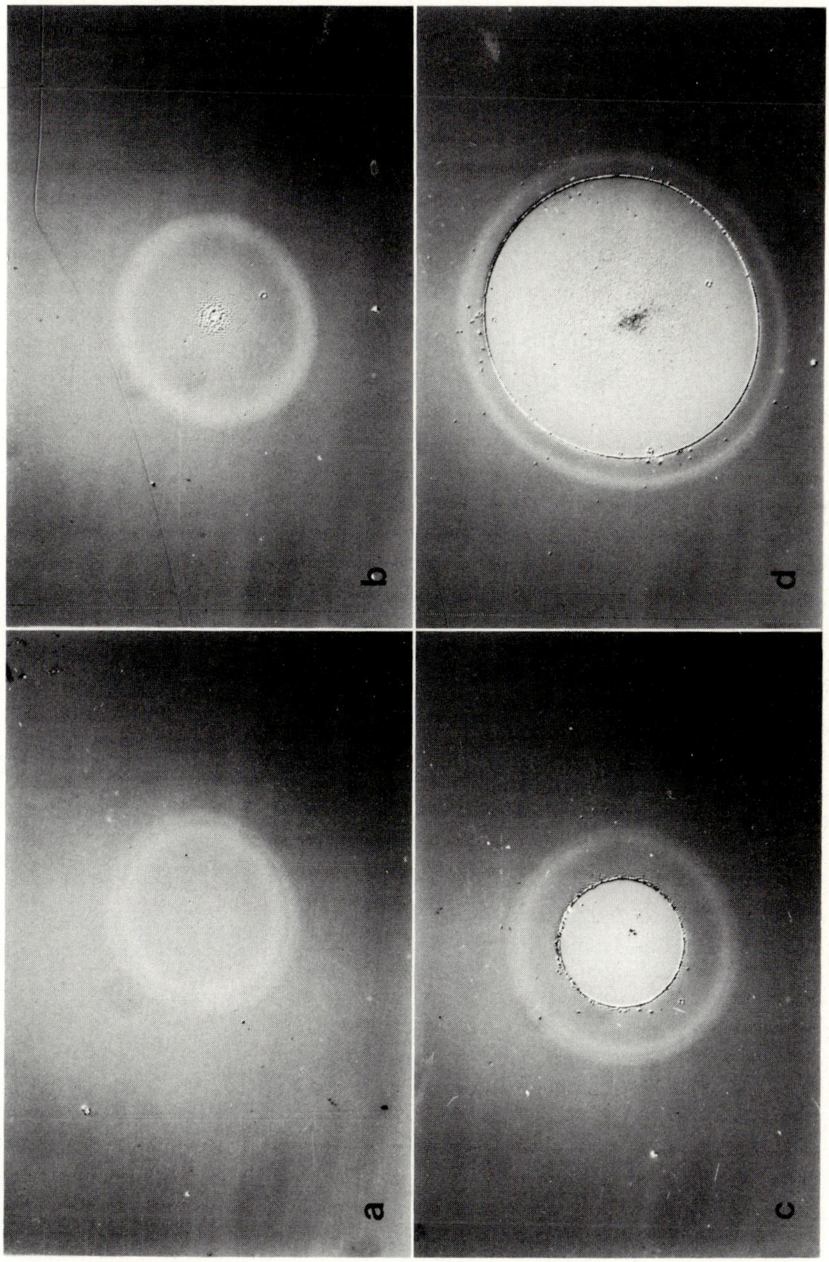

Fig. 7. Surface morphology of c-GaAs after picosecond irradiation, with the pulse energy increasing from (a) to (d).

surrounded by an amorphous ring would be expected to form from the melt – as observed. For longer pulsewidths (> 30 ps), carrier diffusion and the accompanying free-carrier absorption will prevent the formation of gradients sufficiently steep or melt thicknesses sufficiently thin for amorphous material to form. Thus, the observed phase transitions and surface morphologies are consistent with the preceding discussions of the nonlinear energy deposition and pulsewidth-dependence of the threshold.

Amorphous regions or rings are also observed in the resolidified melt morphologies of GaAs, although they are faint and more difficult to observe or photograph. Four Nomarski micrographs of the GaAs surface following melting are shown in fig. 7. No amorphous regions were observed for Ge. The latter observation is consistent with thicker melt depths and more gradual temperature profiles that result from saturation effects.

Optical limiting and regulation

The fluence-dependent nonlinear absorption, the accompanying self-refraction, and laser-induced melting and subsequent crystalline regrowth in Si can be used together with spatial filtering techniques to construct a simple effective energy limiter or regulator. This completely passive device has a high transmission for low input energies, but it clamps the output at a low value for high input energies. Applications for such a device include the restriction of pulse energy incident upon sensitive optical components and the regulation of fluences for use in materials processing.

Optical limiting by nonlinear absorption in semiconductors was proposed and demonstrated as early as 1967 [19]. Moreover, the spatial filtering technique that we have employed has been used previously with liquid and gas filled cells to demonstrate optical limiting [20, 21] and bistability [22]. What we have done differently here is to use nonlinear refraction together with spatial filtering, in addition to the nonlinear absorption in semiconductors, to demonstrate a more effective limiter. In fact, in the present geometry, nonlinear refraction is the dominant limiting mechanism in the device.

The device geometry is shown in fig. 8. Here, a single 48 ps (FWHM) pulse

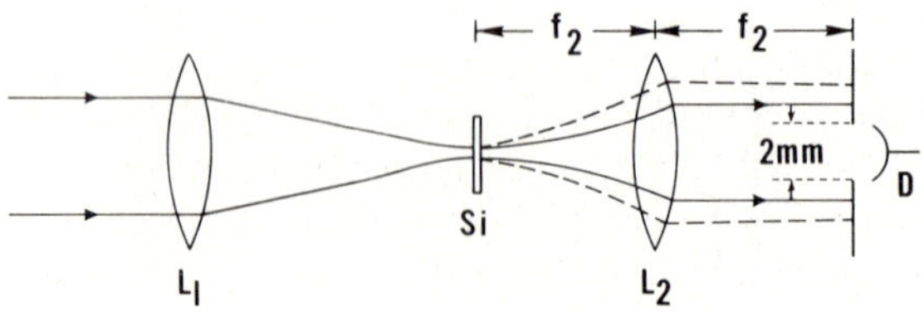

Fig. 8. Geometry of the Si optical limiter.

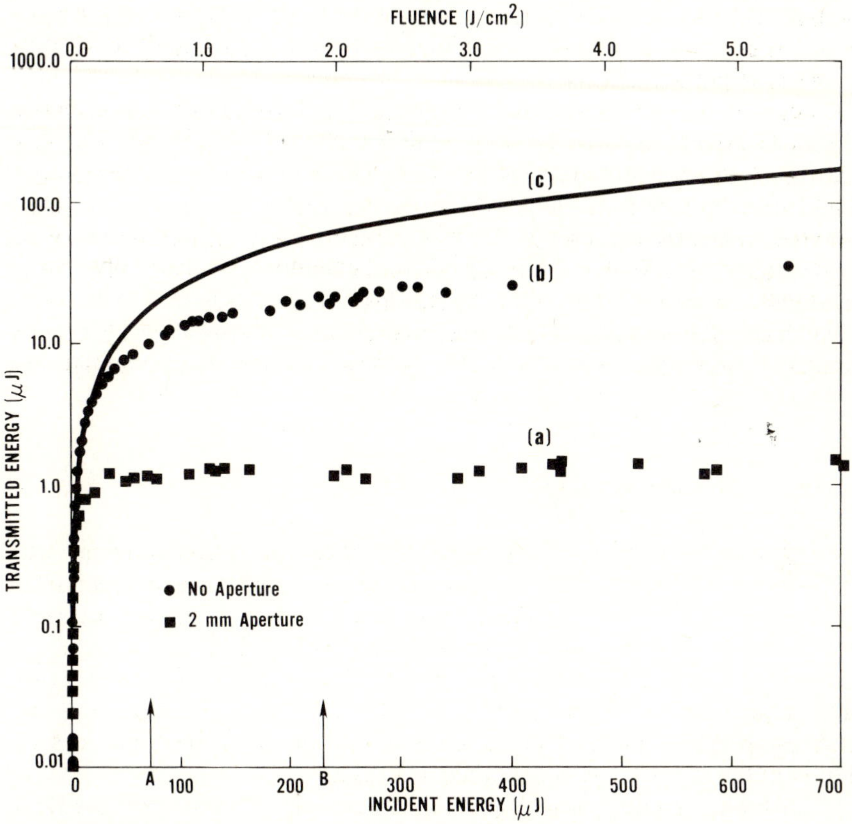

Fig. 9. Si optical limiter response (squares), nonlinear absorption of Si as measured with aperture removed (circles), and Beer's law absorption of Si (solid curve).

at 1 μm was focused onto the Si wafer by a lens L_1. The transmitted pulse was then recollimated by lens L_2 onto a 2 mm aperture placed in front of a photodiode. The energy restricting capabilities of the device are shown by the squares in fig. 9. At low energies (< 1 μJ), the device response is linear and is consistent with the 26% linear transmission of the Si and the 72% transmission of the pinhole. At higher input energies, the device clamps the output energy at 1 μJ for more than 2 orders of magnitude.

Both nonlinear absorption and self-defocusing are responsible for the energy-restricting nature of this device. The contribution of nonlinear absorption can be measured by carefully collecting all of the transmitted energy with the pinhole removed. As seen from the circles in fig. 9, the nonlinear absorption that was discussed in a preceding section (see fig. 2), when acting alone, reduces the transmitted energy by a factor of 5 over that expected from Beer's law. Self-defocusing reduces it by another factor of 40. The latter contribution is investigated by placing a vidicon in the plane of the pinhole. Typical beam profiles at energies of 0.2 and 100 μJ are shown in fig. 10, clearly illustrating the effects of self-defocusing.

Notice that the switch continues to regulate the transmitted energy even at fluences above that required for single-shot melting in Si. Since Si recrystallizes following melting at this pulsewidth, *single* shot irradiations at fluences well above the single-shot threshold produced negligible change in the linear response. Repeated exposure above threshold, however, caused cratering on the Si surface and a noticible degradation of the linear transmission. Details of device performance are (and will be) published elsewhere [23–25].

We have also constructed a similar device using GaAs [1]. The device geometry is identical to that used for Si. The initial response, though not

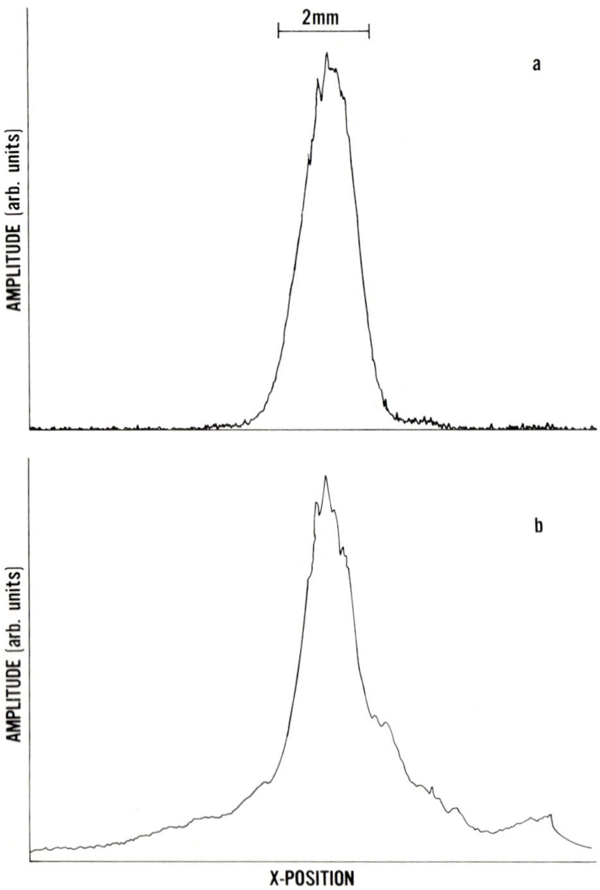

Fig. 10. Beam profiles taken in the plane of the aperture for (a) 0.2 μJ and (b) 100 μJ of energy incident on the Si.

resolved in fig. 11, is linear. For incident energies above 10 μJ, the output energy is clamped at 1 μJ. Additional data (not shown in fig. 11 for clarity) indicate that the clamping action continues even for input energies as high as 1.2 mJ. Over the full range of operation the system transmission is reduced from 33% to 0.1%. As with Si, the contribution of nonlinear absorption was determined by measuring the transmitted energy with the pinhole removed. At an input energy of 160 μJ nonlinear absorption reduces the device transmission by a factor of ~7, while defocusing associated with the free carriers reduces it by an additional factor of ~9. Although the free carriers contribute very little to the nonlinear absorption in GaAs compared to two-photon transitions, they clearly affect the device response. The contribution of self-defocusing was again investigated by placing a vidicon detector in the plane of the pinhole. At the higher irradiances, the beam is highly distorted with a large amount of energy redistributed into the wings. As with Si, regulation continues for energies well above the single-shot threshold. Again details of the device performance will be published elsewhere [1].

The distorted beam profile shown in fig. 10 illustrates why care must be taken in designing collection optics for nonlinear transmission measurements. Nonlinear self-refraction usually accompanies significant nonlinear absorption. This nonlinear refraction coupled with an incorrectly designed collection system

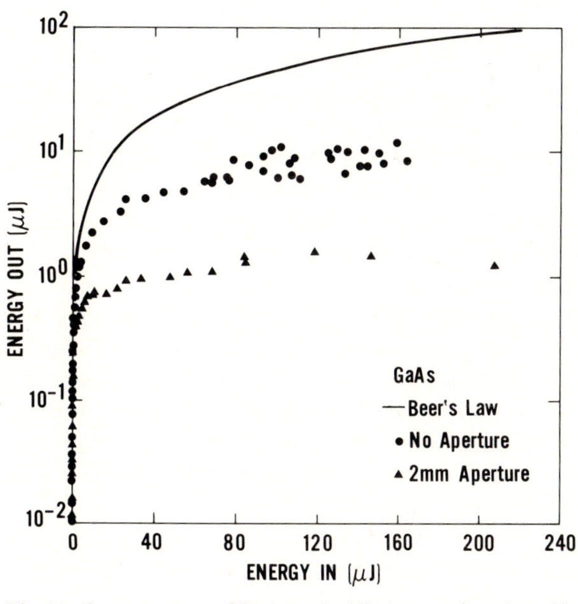

Fig. 11. Output energy of GaAs optical limiter as a function of input energy with the 2 mm aperture in place (triangles) and with it removed (circles). The solid curve shows the linear absorption of the GaAs.

and a small area detector can lead to a drastic overestimation of the nonlinear absorption.

Summary

We have measured the pulsewidth-dependence of the nonlinear absorption, damage threshold, phase transitions and resolidification morphologies of Si, GaAs and Ge at 1 μm.

In Si, the nonlinear absorption is found to be strictly fluence-dependent, the principal nonlinearity is caused by optically-generated free carriers, and, contrary to some reports, there is no measurable contribution from two-photon absorption at room temperature at this wavelength. From these measurements, a free-carrier cross section of 3.5×10^{-18} cm^2 is extracted without requiring a knowledge of the carrier number. Near the melting threshold, a thermal run-away process develops that leads to melting of the Si surface. The fluence required for single shot melting of crystalline Si is found to be proportional to the square root of the pulsewidth, as would be expected for significant carrier transport during our 4–260 ps pulsewidths. In addition, a crystalline-to-amorphous transition is observed for pulses shorter than 10 ps (contrary to published expectations [18]) but not for pulses longer than 30 ps (in agreement with previous observations [18]).

A two-photon absorption coefficient of 26 ± 8 cm/GW is extracted from the nonlinear absorption measurements in GaAs. The single shot melting threshold for GaAs exhibits a square-root dependence on pulsewidth similar to that of Si. Although this is expected from, and is consistent with, two-photon absorption as the dominant energy deposition mechanism or with energy transport, these two processes cannot account for the magnitude of the melting threshold. We believe that a two-photon-assisted avalanche or surface absorption may cause melting at the surface. Amorphous rings are observed in the morphology at 1 μm for all of the pulsewidths used here.

By contrast to Si and GaAs, the Ge absorption coefficient decreases with increasing excitation level. The nonlinearity is associated with a saturation of the direct absorption coefficient and the onset of direct intervalence band absorption. There is little or no pulsewidth-dependence to the melting threshold, and no amorphous structure is observed in the morphology. These are consistent with the relatively gradual temperature and carrier gradients that one expects to be associated with these saturation effects.

Finally, we have used the nonlinear absorption and the self-refraction associated with the generation of free carriers in Si and GaAs, together with spatial filtering, to construct simple passive optical limiters.

Acknowledgements

This research was supported by the U.S. Office of Naval Research, The Robert A. Welch Foundation, and the North Texas State University Faculty Research Fund.

References

[1] T.F. Boggess, A.L. Smirl, S.C. Moss, I.W. Boyd and E.W. Van Stryland (1984) unpublished.
[2] K.G. Svantesson and N.G. Nilsson, J. Phys. C (Solid State Phys.) 12 (1979) 3837.
[3] C.D. Thurmond, J. Electrochem. Soc. 12 (1975) 1133.
[4] G.E. Jellison, Jr. and D.H. Lowndes, Appl. Phys. Lett. 41 (1982) 594.
[5] M.F. Becker, R.M. Walser, Y.-K. Jhee and D.Y. Sheng, in: Picosecond Lasers and Applications, SPIE 322 (1982) 93.
[6] T.F.Boggess, A.L. Smirl, S.C. Moss and I.W. Boyd (1984) unpublished.
[7] J.M. Ralston and R.K. Chang, Appl. Phys. Lett. 15 (1969) 164.
[8] D.A. Kleinman, R.C. Miller and W.A. Nordland, Appl. Phys. Lett. 23 (1973) 243.
[9] J.H. Bechtel and W.L. Smith, Phys. Rev. B13 (1976) 3515.
[10] B. Bosacchi, J.S. Bessey, and F.C. Jain, J.Appl. Phys. 49 (1978) 4609.
[11] A.F. Stewart and M.Bass, Appl. Phys. Lett. 37 (1980) 1040.
[12] A.L. Smirl, in: Semiconductor Processes Probed by Ultrafast Laser Spectroscopy, ed., R.R. Alfano (Academic Press, New York, 1983) in press.
[13] I.W.Boyd, S.C. Moss, T.F. Boggess, and A.L. Smirl, Appl. Phys. Lett. 45 (1984) 80.
[14] I.W.Boyd, S.C. Moss, T.F. Boggess and A.L. Smirl, in: Energy Beam–Solid Interactions and Transient Thermal Processing, eds., J.C.C. Fan and N.M. Johnson (Elsevier Science, New York, 1984) p. 203.
[15] R.Tsu, R.T. Hodgson, T.Y. Fan and J.E.E. Baglin, Phys. Rev. Lett. 42 (1979) 1356.
[16] A.G. Cullis, H.C. Webber, N.G. Chew, J.M. Poate and P. Baeri, Phys. Rev. Lett. 49 (1982) 219.
[17] P.L. Liu, R. Yen, N. Bloembergen and R.T. Hodgson, Appl. Phys. Lett. 34 (1979) 864.
[18] P.L. Liu, R. Yen, N. Bloembergen and R.T. Hodgson, in: Laser and Electron Beam Processing of Materials, eds. C.W. White and P.S. Peercy (Academic Press, New York, 1980) p. 156.
[19] J.E. Geusic, S. Singh, D.W. Tipping and T.C. Rich, Phys. Rev. Lett. 19 (1967) 1126.
[20] R.C.C. Leite, S.P.S. Porto and T.C. Damen, Appl. Phys. Lett. 10 (1967) 100.
[21] M.J. Soileau, W.E. Williams and E.W. Van Stryland, IEEE J. Quantum Electron. QE-19 (1983) 731.
[22] J.E. Bjorkholm. P.W. Smith, W.J. Tomlinson and A.E. Kaplan, Opt. Lett. 6 (1981) 345.
[23] T.F. Boggess, S.C. Moss, I.W. Boyd, and A.L. Smirl, Opt. Lett. 9 (1984) 291.
[24] T.F. Boggess, S.C. Moss, I.W. Boyd and A.L. Smirl, in: Proc. Ultrafast Phenomena IV, eds., D.H. Auston and K.B. Eisenthal (Springer-Verlag, New York, 1984) in press p. 202.
[25] S.C. Moss, A.L. Smirl, T.F. Boggess and I.W. Boyd (1984) unpublished.

Journal of Luminescence 30 (1985) 290–302
North-Holland, Amsterdam

SUBPICOSECOND TIME-RESOLVED MOTT TRANSITION IN CuCl

D. HULIN*, A. MYSYROWICZ*, A. MIGUS and A. ANTONETTI

Laboratoire d'Optique Appliquée, Ecole Polytechnique E.N.S.T.A Batterie de l'Yvette 91120-PALAISEAU, France

The Mott transition in CuCl from a metallic phase of free electrons and holes towards an insulating phase of bound particles (excitons or biexcitons) has been studied by time-resolved luminescence with a resolution slightly better than 1 ps. The phase change takes place in a very short but finite time (about 3 ps) at a carrier density $N \simeq 10^{19}\ \mathrm{cm}^{-3}$.

Recent progress in the technology of short laser pulse generation [1] has opened new perspectives in the study of solid state matter. By the use of such lasers and associated optical detection methods, it becomes possible to reach a temporal resolution in the picosecond and even femtosecond regime. This allows one to unravel mechanisms which could only be apprehended globally before. Another new aspect is the fact that such ultrashort lasers provide the only tool for injection of extremely high densities of electron–hole (e–h) pairs in semiconductors. The electronic excitation efficiency is reduced if the pulse duration exceeds the characteristic time constants for recombination mechanisms of the generated carriers. Since the total energy per pulse is limited by the onset of irreversible damage in the crystal, the yield of created pairs at the end of the pulse is strongly dependent on the pulse duration for the same given number of photons. With femtosecond lasers, one can readily achieve excited particle densities up to some $10^{21}\ \mathrm{cm}^{-3}$ without destroying the crystalline structure.

Electrons and holes created by photons absorption in a semiconductor interact with each other via the Coulomb potential. At small densities and temperatures, the lowest electronically excited state of the crystal is the exciton, a bound electron–hole pair (or the biexciton, a molecular complex of two bound excitons). This situation has already been explored in great detail in a large number of materials. Interesting new problems arise when the density

* Permanent address: Groupe de Physique des Solides de l'E.N.S., Tour 23, 2 Place Jussieu, Paris 5°, France.

0022-2313/85/$03.30

increases, because interactions are no more limited to paired particles and the N-particle system must be considered as a whole. The various correlations between carriers are responsible for modifications such as decreases in the band gap energy (renormalization of the band gap), changes in the dielectric constant (screening of the Coulomb interaction), etc. At high densities and low temperature, the free energy of the system may be minimized by the appearance of the so-called e–h liquid, as first observed in Si and Ge [2]. In that case, a spatial separation takes place between a condensed metallic phase of unbound electrons and holes (liquid) and a dilute phase of excitons (gas). One speaks of a liquid phase because the density of carriers is practically constant, any increase in the number of carriers being merely translated into an increase of the spatial dimension of the condensed phase. The occurrence of this electron–hole liquid phase is favoured by a many-valley structure such as in indirect gap materials.

The situation is very different if one deals with a system of bound particles converting into an e–h plasma of variable density with no spatial separation. This Mott-like phase transition [3] is typical of direct band gap materials. At small density, the exciton is well identified, each electron interacting with only one hole, except for infrequent collisions. Upon increase of the density N, the average distance between the particles becomes comparable to the diameter of the bound particle. The excitons overlap so much that they loose their identity completely. The mutual interactions lead to a screening of the Coulombic binding between the electron and the hole and for $N^{1/3}a_x \simeq 0.4$ (a_x being the excitonic Bohr radius) [4] the system undergoes a transition from the insulating phase of excitonic particles into the conducting phase of free electrons and holes.

There are still a number of unsolved problems concerning the exact nature of the Mott transition in semiconductors, particularly regarding its dynamical aspect. From what precedes, it is clear that femtosecond lasers are very well suited for such a study, because of the very large attainable range in the density of electron–hole pairs and of course because of the very high temporal resolution. CuCl is a good candidate for the study of the metal–insulator transition, since the exciton binding energy is large $E_x = 212$ meV [5]. As a consequence, the excitonic states are well separated spectrally from band to band transitions. Also, the large binding energy excludes a dissociation of excitons arising from a trivial thermal effect (as it is well known, at high excitations, the temperature of carriers may be raised well above the temperature of the coolant). CuCl is a direct gap semiconductor ($E_g = 3.146$ eV) with two series of excitons from the lowest conduction band, and the two upper valence bands. The $n=1$ terms of the two series are located at 3870 Å and 3780 Å, respectively, at 4.2 K. In what follows, we will be concerned mainly with the lowest dipole allowed $n=1$ exciton at 3870 Å. At low temperature and intermediate electron–hole pair densities, the system may further reduce the energy per pair by forming excitonic molecules, also called biexcitons. The presence of biexcitons is signaled by a characteristic emission from their decay into excitons and photons. The predominant biexciton emission is at 3910 Å at $T \leqslant 15$ K, and is usually referred to as M_L line in the literature [6].

The experiments were performed starting from a passively mode-locked ring dye laser. The output pulse from the oscillator is amplified through a four stage amplifying system pumped by the second harmonic of a Nd–Yag laser. Pulse characteristics after amplification are the following: wavelength $\lambda = 620$ nm, energy per pulse ~ 0.4 mJ; pulse duration: 120 fs or less, repetition rate: 10 Hz. In order for the pulse wavelength to fall in the fundamental absorption region of CuCl, the second harmonic frequency is generated in a 1.5 mm thick KDP crystal, yielding an energy per pulse at 310 nm of 10 μJ and similar pulse duration. The UV pulse is focused onto a spot of dimension $(300\ \mu m)^2$ on the sample cooled at $T = 15$ K. The sample itself consists of a 1000 Å thin crystalline layer of CuCl deposited on a fused silica substrate. Since less than half the incident UV pulse energy is absorbed over the sample thickness, it is considered that a nearly homogeneous volume excitation is achieved.

Time-resolved absorption experiments are performed using the classical pump and probe technique. For the femtosecond probe, light continuum is generated in a 20 mm long water cell. Time-resolved luminescence experiments use an optical Kerr shutter as a fast optical gate (fig. 1). A variable delay line allows to adjust the aperture of the Kerr cell with respect to the onset of the signal to be monitored. Instead of CS_2 as an active medium in the Kerr shutter, as is usually the case, we use benzene in a 2 mm thick cell. Benzene has the advantage of a shorter response time with slightly less efficiency [7]. After passage through the Kerr shutter, the signal is spectrally analyzed with a $f = 20$ cm spectrograph and recorded with an optical multichannel analyzer (OSA, BM- Spektronik). Since the spectral width of interest extends over 200 Å, the

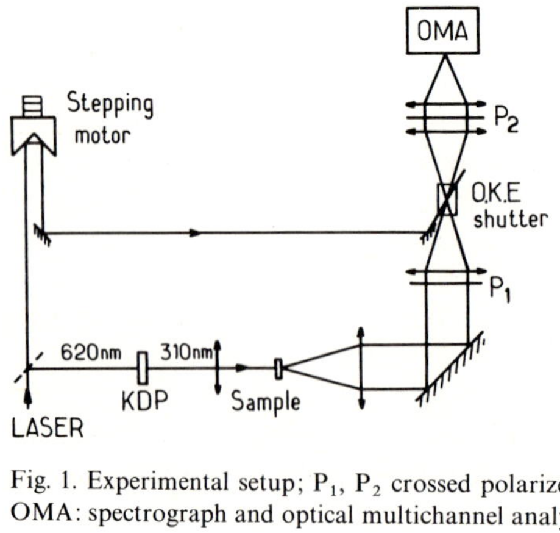

Fig. 1. Experimental setup; P_1, P_2 crossed polarizers; OKE shutter: Optical Kerr effect shutter; OMA: spectrograph and optical multichannel analyzer.

difference in the group velocities between the various frequency components as they pass through optical elements (window of the dewar, lenses, polarizer, benzene cell itself) deteriorate the temporal resolution of the shutter from ~200 fs to slightly less than 1 ps. In some cases where accurate subpicosecond measurements are required at specific wavelengths, the OMA detector is replaced by a slit followed by a photodiode. The output from the diode is fed to an HP 1000 computer which samples the signal only for a fixed, predetermined pump pulse intensity. The computer also controls the variable delay line so that repetitive scanning can be performed to improve the signal to noise ratio.

Figure 2 shows time-resolved absorption spectra recorded with the OMA at different delays from the pump. The continuous curve corresponds to the test beam preceding the pump beam and is therefore representative of the unexcited crystal. One recognizes the $n=1$ terms of the two exciton series from the two upper valence bands and the lowest conduction band. The band edge is not apparent in the figure since it starts around $\lambda=363$ nm. For a delay $\Delta t=1$ ps after carrier injection, the absorption spectrum of the sample is greatly modified. The two main changes are the strong decreases of the exciton oscillator strength, and the simultaneous appearance of a new absorption continuum extending to longer wavelengths. This reflects the presence in the medium of a very high density of free carriers which screen the Coulomb interaction of the exciton states, and lead to a reduction of the band gap energy (band gap renormalization). The value of the renormalized gap depends upon the density of injected carriers, but its study is not straightforward in absorption, as we will see later.

Figure 3 shows the dynamics of exciton recovery, measured with the diode and repetitive scanning. The exciton reappears in less than 20 ps, this value

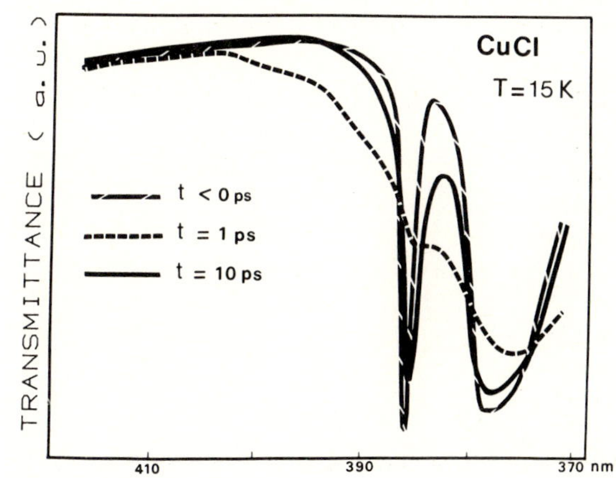

Fig. 2. Transmittance spectra of CuCl (at $T=15$ K) taken at different time delays between excitation and probing.

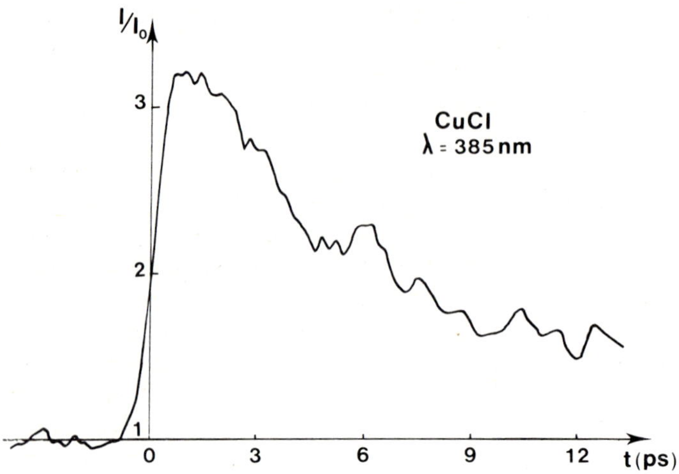

Fig. 3. Induced transmission and relaxation of the $n=1$ exciton resonance (at $T=15$ K) showing the recovery of $n=1$ exciton bleaching.

depending upon the initial excitation intensity. Note that, as it recovers, the exciton does not retrieve its initial position, but is displaced towards shorter wavelengths; this effect is partly responsible for the remaining induced increased transmission at wavelengths of observation (see also fig. 2).

The changes in absorption occurring below the exciton around 400 nm, are weaker and strongly dependent of excitation intensity. In all cases of fig. 4 an induced absorption is observed initially, the amplitude of which depends

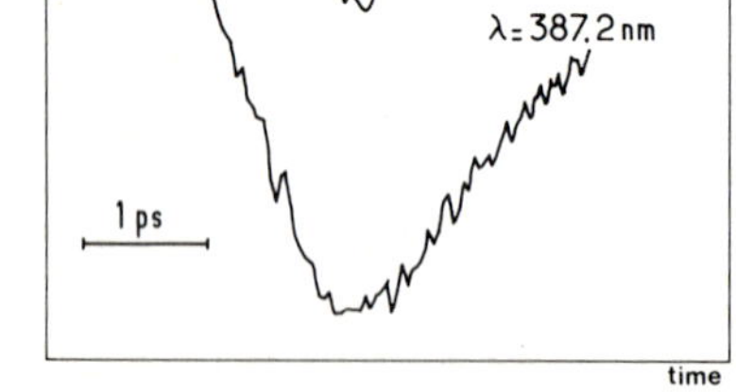

Fig. 4. Kinetics of the induced transmission at different wavelengths below the $n=1$ exciton, in the region of the gap renormalization. The onset of the signal is not the same for the three curves due to the group velocity dispersion in the continuum (not corrected here). The vertical amplitude scale is the same for the three curves and corresponds to 60% for the maximum at 387.2 nm.

upon the wavelength. This induced absorption decreases rapidly with time, converting into optical gain at λ equal or larger than 400 nm. These results are well understood if one remembers that the injected carriers have an excess energy of 600 meV with respect to the bottom of the band gap. At carriers densities of 10^{20} cm^{-3} or more, interparticle collisions are extremely rapid compared to other relaxation processes, so that in first approximation, it is reasonable to assume that the carriers have thermalized between themselves in less than 10^{-13} s. At the end of the pump pulse, the injected carrier have a very high effective temperature $T \sim 2500$ K. As a consequence, the bottom of the renormalized gap around 400 nm is initially little occupied, and a net absorption is observed. As cooling of carriers proceeds via LO phonon emission, filling of low energy band states takes place, which reduces the induced absorption and can lead to optical gain when population inversion is realized. The thermal cooling of carriers is initially very fast as long as kT_{eff} exceeds the energy of LO phonons ($E = 26$ meV) with a typical relaxation time $\tau \sim 10^{-14}$ s for an unscreened electron – LO phonon interaction, but evolves at a much slower rate for $T_{\text{eff}} \sim 100$ K.

On the other hand, the density of carriers in CuCl diminishes rapidly, even on a time scale of a few picoseconds, leading to a high energy shift of the renormalized gap, and therefore a reduction of the induced absorption below the exciton. As a consequence, it is difficult to extract from time-resolved transmission experiments alone the contributions from band filling and change in band gap renormalization. In this respect time-resolved luminescence experiments, although they provide less time resolution (1 ps instead of 100 ps) bring important complementary informations.

Figure 5 shows time-integrated luminescence spectra recorded for different excitation intensities I_0. At low I_0 ($\sim 10^{-3}$ J/cm^2) only the M_L line at 392 nm from the biexciton decay is apparent. This line is slightly shifted and broadened to low energies if compared with lower intensity, nanosecond pulsed excitation ($\lambda = 391$ nm). As I_0 is increased, M_L shows a saturation behaviour, whereas the plasma emission band becomes apparent on the lower energy side.

Figure 6 shows time-integrated luminescence intensities at different wavelengths, either in the plasma or biexciton region (log–log representation) in function of pump intensity. The biexciton luminescence follows first a quadratic dependence, as expected, and evolves towards a sublinear dependence in the presence of the plasma. It is probable that a near complete saturation of the biexciton would be reached for a spatially homogeneous excitation spot. The signal from the plasma appears at a threshold intensity, which depends upon the observation wavelength. This reflects the increase of gap renormalization with injected carrier density.

Time-resolved luminescence spectra have been recorded at $T = 15$ K (figs. 7 and 8). For an excitation intensity just below the threshold for plasma appear-

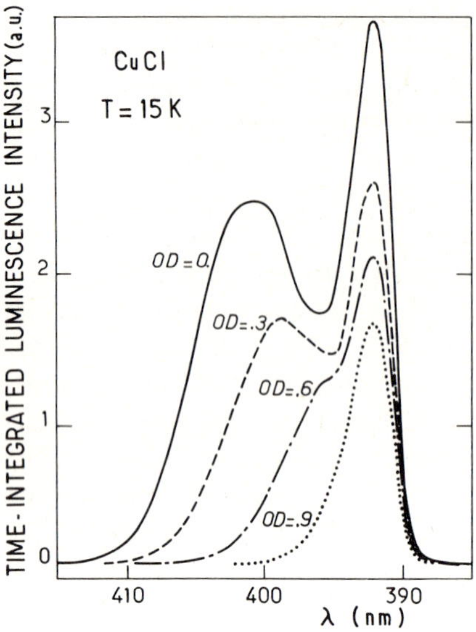

Fig. 5. Time-integrated luminescence for different excitation intensities.

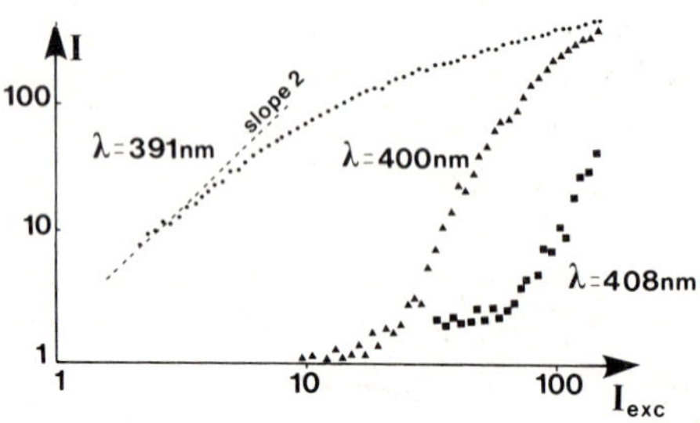

Fig. 6. Time-integrated luminescence intensity at different wavelengths as a function of the excitation intensity (log–log plot). $\lambda = 391$ nm corresponds to the position of the biexciton. $\lambda = 400$ nm and $\lambda = 408$ nm are inside the plasma luminescence.

ance (fig. 7), the biexciton emission appears immediately with a build-up time limited by the response of the optical Kerr shutter. One concludes therefore that the free carriers initially injected well above the minimum gap energy form biexcitons in less than 1 ps. It can also be noticed that during the first few picoseconds, the spectrum of the biexciton luminescence is broadened and then narrows as time increases, probably because of the reduction of the particle density with time.

The most striking features of our data appear when the excitation is increased by a factor 10 with respect to the preceding case: at early times after excitation *only* a broad emission characteristic of the plasma recombination is present

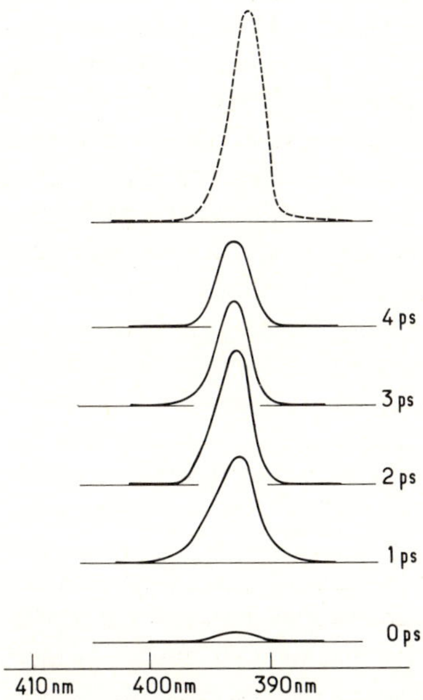

Fig. 7. Time-resolved spectra of CuCl at 15 K at different delays from the UV pump beam ($\lambda = 310$ nm). The estimated initial electron–hole pairs density is $n \sim 10^{19}\,\mathrm{cm}^{-3}$. The dashed curve is a time-integrated emission spectrum.

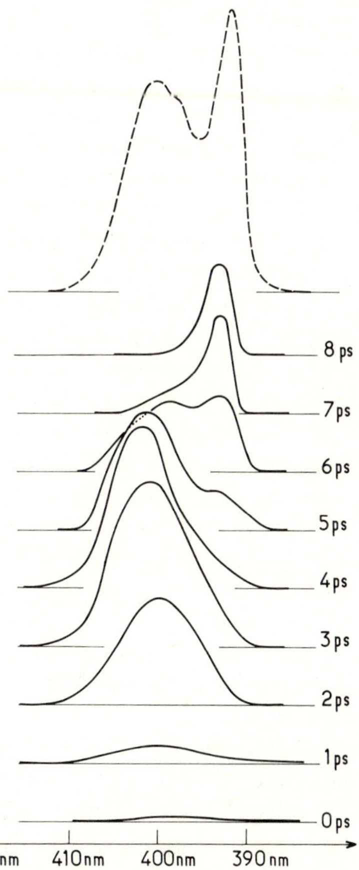

Fig. 8. Time-resolved luminescence of CuCl at 15 K. The UV pump intensity is 10 times that of fig. 7. The dashed curve is a time-integrated emission spectrum.

decaying in about 7 ps. This evidences an initial electron–hole pair density above the Mott initial value. It is only after a *true delay of 4 to 5 ps* that the biexciton emission rises. We note that the process of conversion of the plasma into biexcitons is very fast and takes place in less than three picoseconds. This estimation is only an upper limit because we average different initial e–h densities inside the pump spot. More precisely and even if a Gaussian pulse profile is assumed, the spatially integrated detection collects emission from parts of the sample with different initial excitation conditions, so that, for instance, the regions with lower carrier density reach the transition in advance to other regions. This effect may lengthen artificially the detection of the conversion time. However it should also be noticed that the biexcitonic lineshape undergoes an evolution during the transition, being broader at 5 ps than at 7 ps. This suggests that the measured 3 ps conversion time can be mostly attributed to the switch from the free electron and hole regime to the excitonic one with no spatial separation [8].

An important aspect of this study of the Mott transition is the characterization of the plasma at the time of conversion. In principle, the effective temperature, density and chemical potential may be obtained at any delay from the lineshape analysis of the time resolved luminescence [9]. However, such an analysis will be difficult to perform, during the first picoseconds, because the temperature changes rapidly during the measurement interval of 1 ps, or, during the transition from plasma to biexcitons because the plasma and biexciton luminescence spectra are mixed up. In the intermediate time domain 2–5 ps, we have fitted the experimental results with calculated lineshapes, assuming a non-k conservation, an assumption known to give a good approximation in this high density regime [9].

The following formula was used:

$$I(h\nu)=A\iint_0^{\Delta E} f_e f_h \sqrt{E_e}\sqrt{E_h}\,\delta(E_G+E_e+E_h-h\nu)\,\mathrm{d}E_e\mathrm{d}E_h$$

$$=A'\int_0^{\Delta E} f_e f_h \sqrt{E_e}\sqrt{\Delta E-E_e}\,\mathrm{d}E_e,$$

where $\Delta E=h\nu-\mathrm{E}_{\mathrm{gap}}=E_e+E_h$, E_e and E_h are the excess energies of the electron and the hole with respect to the bottom of the conduction band or the top of the valence band; f_e and f_h are the respective Fermi distributions which depend on N and T. Due to the non-k conservation, the calculated lineshape is rather sensitive to the temperature, as well as to the chemical potential. For a given lineshape, one obtains a set of coupled parameters (N, T) that give a line profile in reasonable agreement with the experiment, a larger N corresponding to a smaller T.

In order to choose amongst the different sets of coupled parameters obtained from the luminescence lineshape analysis, transmission spectra provide useful complementary information. For instance, calculated absorption spectra predict large optical gain around 4000 Å for high densities N and low temperatures T. On the other hand, no gain at all is predicted for a low value of N and high temperatures T. Experimentally, neither of the two situations is observed, but only a small gain is apparent after a delay of 3 ps. This allows one to deduce a value of $N\cong 3\times 10^{19}$ cm^{-3} and $T\sim 150$ K for the plasma at $\Delta T=3$ ps. By taking into account that there is a factor 10 in particle density between time delay $\Delta t=0$ and $\Delta t=8$ ps, where the luminescence is identical to the low excitation regime of fig. 7, one can infer that the initially injected particle density in the case of fig. 8 is $N\sim 10^{20}$ cm^{-3}, and more important, that at the time of conversion the plasma has a carrier density of 10^{19} cm^{-3} with an effective temperature of ~100 K. This small value (in comparison with the other characteristic values of the system like E_x/k or the initial temperature) allows to assess that in our experiment the Mott transition is observed at low temperature.

Taking advantage of the knowledge of the density, one can now plot the renormalization experimental curve as a function of the carrier density (fig. 9). The theoretical curve is deduced from a general expression [11,12] which for a given material depends only on the Rydberg energy and the Bohr radius of the exciton. In the case of CuCl, the binding energy is large so that the exciton radius a_x is small and consequently not well defined. The comparison between the experiment and theory is made for $a_x = 7.3$ Å [13]. The agreement could be improved for larger but still reasonable values of a_x (~10 Å).

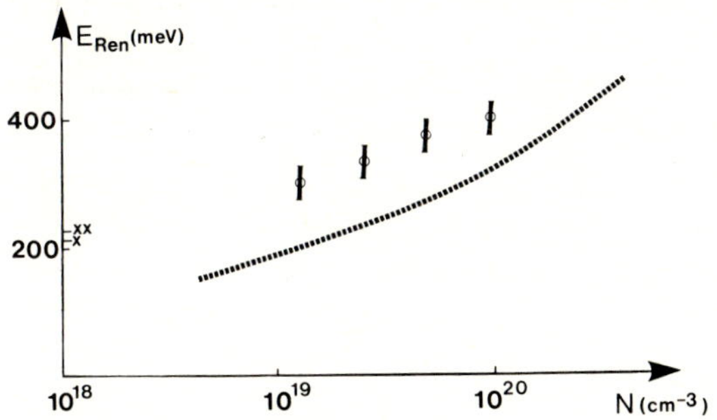

Fig. 9. Energy renormalization as a function of carrier density. The curve is a theoretical plot (see text) with $a_x = 7.3$ Å, the points are experimental values.

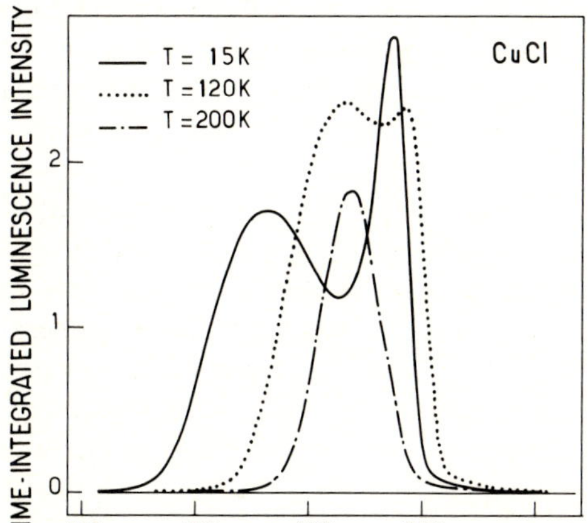

Fig. 10. Time-integrated luminescence intensity at different temperatures for the same excitation. Note the high energy shift of the biexciton between 15 K and 120 K due to the gap increase with temperature in CuCl.

Finally, it can be noticed that the time during which the plasma emission is observable is only limited by the conversion of the plasma into biexcitons. In a situation where no bound particle exist, the Mott transition is not expected and the *total* lifetime of the plasma can be longer. This situation is reached at higher temperature ($T \gtrsim 200$ K). At 200 K, the biexciton line should appear at 386 nm due to the high energy shift of the gap in CuCl with increasing temperature: no such line is observable in our time-integrated luminescence spectra (fig. 10). Only the plasma is present. In the time-resolved luminescence at 200 *K* (fig. 11) the main features are similar to the one observed at low temperature except for the important point of the Mott transition. No plasma conversion is observed and contrary to the case of $T = 15$ K, the plasma luminescence is still present at 8 ps.

In conclusion we have studied the Mott transition between a metallic phase of free electrons and holes towards an insulating phase of bound particles (excitons or biexcitons) in CuCl. This study has been done for the first time by time-resolved luminescence at the subpicosecond scale. Compared to

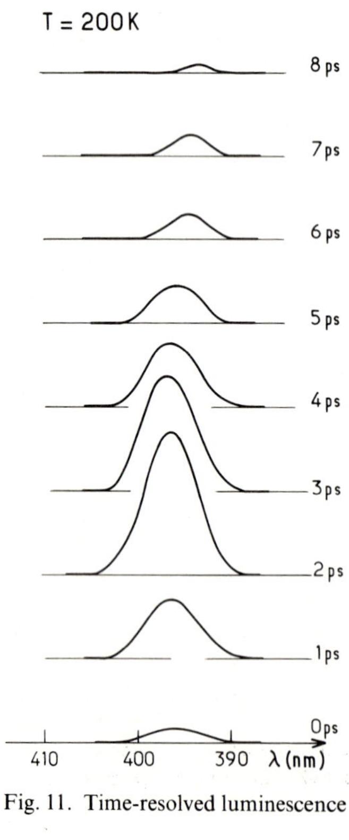

Fig. 11. Time-resolved luminescence of CuCl at 200 K in the same conditions as fig. 8.

time-resolved absorption, this technique has the advantage of a direct identification of each transient species through their luminescence line. At moderate excitation intensity, corresponding to a created carrier density below the critical Mott density N_c, the formation of biexcitons takes place in a time within the experimental resolution (1 $\lesssim$ps). On the other hand, for a carrier density higher than N_c, there is no evidence of biexciton emission in the first picoseconds. Only electron–hole plasma luminescence can be observed. After a few picoseconds, the luminescence spectra indicate a transformation of the plasma into bound particles, a phase change occurring in a very short but finite time (of about 3 ps). The spectral lineshape analysis indicates a density $N \simeq 10^{19}\,\mathrm{cm}^{-3}$ at the transition. The corresponding effective carrier temperature is around 100 K. This sufficiently small value insures that the Mott transition is occurring at a low temperature so that the phase change does not get smeared by heating effect, giving its full significance to this study.

Acknowledgement

We gratefully acknowledge the assistance of Dr. J. Etchepare and G. Grillon in the Kerr shutter experiment and Pr. L.L. Chase for the samples. This work was partly supported by D.R.E.T.

References

[1] R.L. Fork, B.I. Greene and C.V. Shank, Appl. Phys. Lett. 38 (1981) 671; A. Migus, J.L. Martin, R. Astier, A. Antonetti and A. Orszag, in: Picosecond Phenomena, Springer series in Chemical Physicals Vol. 23 (1982) pp. 6–9.

[2] T.M. Rice, J.C. Henoel, T.G. Phillips and G.A. Thomas, in: Solid State Physics, Vol 32, eds., H. Ehrenreich, F. Seitz and D. Turnbull (Academic, New-York, 1977).

[3] N.F. Mott, Proc. Phys. Soc., London A 62 (1949) p. 416.

[4] V.E. Bisti, and A.P. Silin, Sov. Phys. Solid State 90 (1978) 1068;
V.E. Bisti, Sov. Phys. Solid State 23 (1981) 855.

[5] See for instance, A. Goldmann, Phys. Status Solidi b 81 (1977) 9.

[6] See for instance, J.B. Grun, B. Hoverlage and R. Lévy, in: Excitons, eds., I.E. Rashba and M.D. Sturge (North-Holland, Amsterdam, 1982) p. 505.

[7] D. Hulin, J. Etchepare, A. Antonetti, L.L. Chase, G. Grillon, A. Migus and A. Mysyrowicz, Appl. Phys. Lett. 45 (1984) 993.

[8] Theoretical considerations predict a smooth change from one limit to the other, even at $T=0$; see for example: C. Comte and P. Nozières, J. Physique 43 (1982) p. 1069.

[9] The gain in the direction of observation (perpendicular to the layer) is less than 5% and consequently does not affect the emission lineshape. The delay ($\sim$2 ps) in the establishment of the plasma luminescence corresponds, as seen in absorption, to the time necessary for the cooled carriers to fill the bottom of the band. As shown previously the screening of the electron–phonon interaction at high carrier densities is responsible of this delay (see D.

Hulin, A. Antonetti, L.L. Chase, J.L. Martin, A. Migus, A. Mysyrowicz, J.P. Löwenau, S. Schmitt-Rank and H. Haug, Phys. Rev. Lett. 52 (1984) pp. 779–782).

[10] Y. Yoshikuni, H. Saito and S. Shionoya, Solid State. Commun 32 (1979) p. 665 and references therein; M. Capizzi, A. Frova, F. Martelli, S. Modesti, L.G. Quagliano, J.L. Staehli, M. Guzzi, R.A. Logan, G. Chiaretti, Physica B&C (Netherlands) Vol. 117/118 (1983) pp. 333–335.

[11] C. Benoit à la Guillaume, in: Collective Excitation in Solids, ed., di Bartolo (Plenum Press, New York, 1983) p. 633.

[12] P. Vashita and R.K. Kalia, Phys. Rev. B 25 (1982) 6492.

[13] We take the value calculated by M. Ojima et al., J. Phys. Soc. Jap. 14 (1978) 1294 after the effective mass obtained in their reference [20] (C.I. Mu, T. Goto and M. Veta, J. Phys. Soc. Jap. 34 (1973) 693).

Journal of Luminescence 30 (1985) 303–317
North-Holland, Amsterdam

PICOSECOND STUDIES OF HIGHLY EXCITED CdS

Hiroshi SAITO,

Department of Applied Physics, Okayama University of Science, Ridai-cho, Okayama 700, Japan

Results on picosecond luminescence and excite-and-probe transmission as well as transient grating measurements for highly excited CdS measured at a bath temperature of 5 K will be presented. The luminescence and optical gain both due to electron–hole plasma and excitonic molecule recombination are observed. The electron–hole plasma decays very fast by bimolecular recombination of electrons and holes in the plasma and diffusion of the carrier toward the low density regions, and transforms into excitons and excitonic molecules within 100–200 ps. The possibility of electron–hole liquid formation is definitely excluded. The exciton and excitonic molecule decay rather slowly and govern the optical properties for times longer than ~ 200 ps.

1. Introduction

It is well known that recombination of electron–hole plasma (EHP) and of excitonic molecules are important carrier loss processes in highly excited CdS at low temperature. Luminescence as well as optical gain due to these processes under quasi-stationary conditions has been studied extensively within the last years [1]. Interrelation between the EHP and excitonic system (excitons and excitonic molecules), especially transformation of the EHP into the excitonic system, has not been discussed in detail yet. The recombination of the EHP and the excitonic molecule nearly coincides in their energetic position in CdS. This makes it difficult to discriminate which process contributes to luminescence or gain spectrum under quasi-stationary condition. It has been discussed theoretically that formation of electron–hole liquid [2,3] or plasma clusters [4,5] is likely for CdS because of relatively high critical temperature for liquid formation predicted theoretically [2]. Experimentally, however, this problem remains still an open question. Investigation of relaxation and recombination behavior in picosecond time regime is essential to discuss these problems in detail. It is the aim of this paper to present experimental data with respect to

* The work was performed in collaboration with E.O. Göbel at Max-Planck-Institut für Festkörperforschung, Stuttgart, FRG.

0022-2313/85/$03.30

these problems. Definite conclusions can be drawn only if different experimental techniques are applied.

In this paper recent results on time-resolved luminescence studies and two kinds of excite-and-probe studies, i.e., transmission measurements under high intensity excitation and picosecond transient grating measurements for CdS will be reported. The results show that for high excitation ($N_{e-h} > 10^{18}$ cm^{-3}) the optical properties within the first 100–200 ps are governed by the EHP. The decay of the EHP is both due to carrier recombination and carrier diffusion and the EHP cannot be maintained for times longer than 200 ps, because it transforms into an excitonic system. For times longer than 200 ps the optical properties are thus determined by the excitonic system. We further show that the EHP does not undergo a phase transition to electron–hole liquid during its lifetime. We will confine ourselves here to discussing the interplay between the EHP and the excitonic system on the base of these results. Detailed discussion on the relaxation behavior of the EHP will appear in a separate paper.

This paper will be organized as follows: In section 2 of this paper we report the results on time-resolved luminescence experiments. Section 3 concerns with the time-resolved transmission spectroscopy under high intensity excitation. Excitation was provided by means of two different ways, that is, one is the excitation by one-photon absorption at $\lambda = 355$ nm and the other is the excitation by two-photon absorption at $\lambda = 532$ nm (CdS is transparent at 532 nm). For $\lambda = 355$ nm CdS has high absorption coefficient of about 2×10^5 cm^{-1}. The excitation by this wavelength, therefore, makes it possible to generate carrier density high enough to create the EHP, but, on the other hand, a steep gradient of the carrier density normal to the excited surface cannot be avoided. Homogeneous volume excitation can only be provided by two-photon absorption, because of its low absorption coefficient.

The steep density gradient brought about by one-photon excitation results in carrier diffusion to the low density direction, affecting the dynamical behavior of the excited carriers [6]. Transient grating technique is the direct way to measure the diffusion constant of free carriers as well as of an excitonic system. In section 4 we report the picosecond dynamics of laser-induced transient phase grating. The diffusion coefficient of the photo-generated carriers and of the excitonic system is obtained by measuring the decay of diffraction intensity for different grating spacings and excitation intensities.

Experiments were performed at a bath temperature of 5 K. Optical excitation was provided by one-photon absorption of the third harmonic pulses (355 nm) or by two-photon absorption of the second harmonic pulses (532 nm) of a passively mode-locked Nd–YAG laser (pulse width ~25 ps). The excitation energy of 1 μJ corresponds to a power density of about 10 MW/cm^2 throughout the experiments for one-photon absorption.

2. Luminescence experiments

A CS_2 optical Kerr shutter with time resolution of ~25 ps was used in the time-resolved luminescence experiments for a picosecond light gate. Crystals with a size of about $5 \times 5 \times 1\ mm^3$ were used in order to suppress stimulation effects, and to obtain spontaneous luminescence [7].

Figure 1 shows time-resolved luminescence spectra for excitation with $\lambda = 355$ nm. A very broad luminescence band is observed just after the excitation. The broad band becomes sharper with increasing time. For times later than ~150 ps the change of the spectral shape with time delay becomes inconspicuous. It is noticed that the position of the sharp line in the later time regime coincides with the luminescence line due to excitonic molecule recombination [8].

Figure 2 shows semi-logarithmic plots of the spectrally integrated luminescence intensity versus time delay for two different excitation levels. For an excitation intensity (I_{exc}) of 0.28 μJ the luminescence intensity shows an exponential decay with a time constant of about 300 ps. For $I_{exc} = 1.5$ μJ a com-

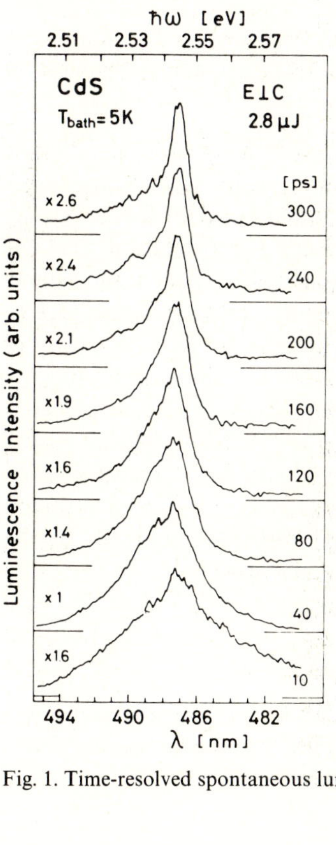

Fig. 1. Time-resolved spontaneous luminescence spectra.

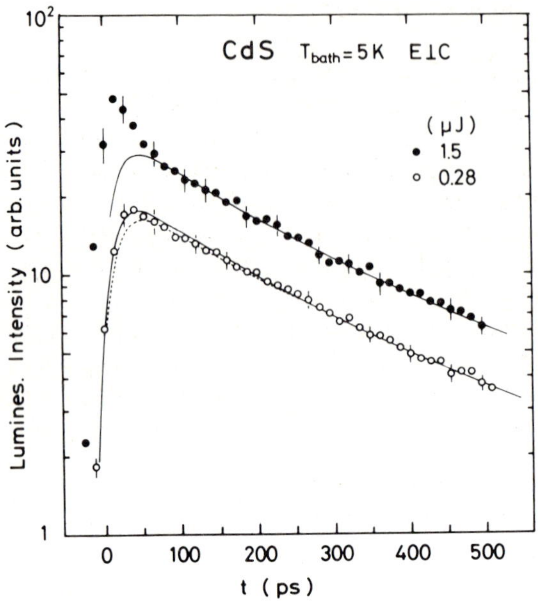

Fig. 2. Spectrally integrated spontaneous luminescence intensity as a function of delay time for two excitation levels. Solid and dashed curves are calculated time behavior of the excitonic molecule density as described in the text.

ponent with a fast decay comes into appearance just after the excitation, followed by the exponential decay with the same time constant as for $I_{exc} = 0.28\ \mu J$.

Comparing the results in figs. 1 and 2, the luminescence with the fast decay corresponds to the broad band observed just after the excitation and the luminescence with the slow decay to the sharp line in the later time regime. These facts indicate that two different recombination processes contribute to the luminescence decay, that is, the sharp line is due to the recombination of excitonic molecules because of its energetic position and line width, and the broad band arises from the recombination of the EHP. The luminescence spectra in fig. 1 can be understood as follows: the EHP created by the strong laser excitation decays rapidly and within 200 ps the EHP transforms to excitons and excitonic molecules. The present interpretation is in contradiction to that proposed by Yoshida et al. [7,9], but will be further supported by the transmission and the transient grating data (sections 3 and 4).

3. Transmission experiments

Transmission experiments were performed using output pulses from a dye laser (Coumarin 152A) synchronously pumped by the third harmonics (355 nm) of the Nd–YAG laser as probe pulses.

3.1. Excitation by one-photon absorption for $\lambda = 355$ nm

Figure 3 depicts time-resolved optical density (OD) spectra for two excitation levels. Negative OD values correspond to optical gain. For $I_{exc} = 1.1$ μJ (fig. 3(b)) the OD spectrum at 47 ps delay time has a peak at 2.528 eV accompanied by a shoulder at 2.52–2.49 eV. The peak value of the gain coefficient is about 5600 cm^{-1}, if we assume a depth of the excited volume to be 0.5 μm. The crossover position from absorption to gain (2.540–2.536 eV) shifts to lower energies with time. For $I_{exc} = 0.08$ μJ (fig. 3(a)), the spectral width is narrow in the entire time regime, and the peak position is situated at about 2.535 eV. The crossover position of about 2.542 eV is slightly higher than for $I_{exc} = 1.1$ μJ, and remains almost unchanged with time delay. Furthermore the gain at the peak position for $I_{exc} = 1.1$ μJ decays very fast and no gain is observed for times longer than 200 ps, whereas the gain at the peak position for $I_{exc} = 0.08$ μJ decays rather slowly.

The recombination process both due to the EHP and the excitonic molecule should contribute to optical gain. The two processes can be distinguished by their different time behavior [10] as shown already. The optical gain observed

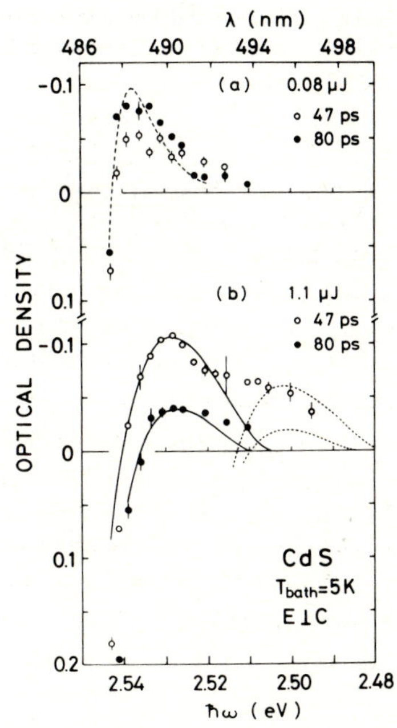

Fig. 3. Time-resolved optical density spectra for the excitation by one-photon absorption at 355 nm. Figure 3(a) is for 0.08 μJ and 3(b) is for 1.1 μJ. Negative OD value corresponds to optical gain. Solid and dotted curves in 3(b) are calculated gain spectra due to the EHP, and the dashed curve in 3(a) is the calculated gain spectrum due to the excitonic molecule recombination as described in the text.

for $I_{exc}=0.08\ \mu J$ is therefore attributed to excitonic molecule recombination. We further conclude that the broad gain band with faster decay time observed for $I_{exc}=1.1\ \mu J$ is due to the recombination of the EHP. The peak gain value of 5600 cm^{-1} is also consistent with this interpretation.

The effective temperature of the EHP (T_{eff}) is determined from time-resolved luminescence spectra, because the high energy tail of the luminescence spectrum reflects T_{eff} via the respective distribution functions of the electrons and the holes [11]. The initial high T_{eff} decreases steeply down to 60–90 K within the first 50 ps, and thereafter the decrease becomes slower. This decrease in T_{eff} results in a high energy shift of chemical potential of the EHP if the electron–hole density (N_{e-h}) in the EHP remains constant. The experimental results in fig. 3(b), on the contrary, show a low energy shift of the crossover, which corresponds to the chemical potential of the EHP [12]. This fact definitely rules out the possibility of electron–hole liquid formation and therefore N_{e-h} decreases continuously because of recombination and diffusion of the carrier (see also section 4).

N_{e-h} is obtained by analyzing the gain spectra in fig. 3(b). Calculations are performed with the simplified assumption that *k*-selection rule for the recombination of electrons and holes in the EHP is neglected [13]. T_{eff} used for the calculation is 80 K and 65 K for $t=47$ ps and $t=80$ ps, respectively. Results are shown by the solid curves in fig. 3(b). N_{e-h} used for the fitting is $2.8\times10^{18}\ cm^{-3}$ for $t=47$ ps and $1.9\times10^{18}\ cm^{-3}$ for $t=80$ ps. Dotted curves in the low energy part are ascribed to a coupled plasmon–phonon sideband of the main EHP band [14,15]. The sideband is shifted to the low energy side with respect to the main EHP band by 27 meV for $N_{e-h}=2.8\times10^{18}\ cm^{-3}$ and by 26 meV for $N_{e-h}=1.9\times10^{18}\ cm^{-3}$. Good agreement between the experimental points and the calculated gain spectra is obtained. We therefore conclude that the rather broad gain spectrum at 47 ps and 80 ps can be attributed to the occurrence of a plasmon–LO phonon coupled mode sideband of the main EHP recombination band.

The recombination of the electron and hole in the EHP represents a bimolecular recombination process. The recombination rate thus is given by $R_r=\beta(N_{e-h})^2$, where β is the bimolecular recombination coefficient. Using N_{e-h} at 47 and 80 ps, we can estimate β to be $5–7\times10^{-9}\ cm^3/s$. N_{e-h} at 47 ps and this β value yields N_{e-h} at 200 ps to be $7–9\times10^{17}\ cm^{-3}$. The EHP with these carrier densities and $T_{eff}=50$ K is still degenerate, that is, the sum of Fermi energy of the electron and the hole is positive. Experimentally, however, almost no gain due to the EHP is observed for $t\geqslant200$ ps. Therefore it is concluded that the EHP cannot be maintained in CdS for $N_{e-h}=7–9\times10^{17}\ cm^{-3}$, although these densities are far above the Mott density of $\sim1\times10^{17}\ cm^{-3}$ which is calculated using Debye–Hückel screening length and a temperature of 50 K.

The observed gain for $I_{exc}=0.08\ \mu J$ is ascribed to excitonic molecule recombination. If we assume that the excitonic molecule and the free exciton are at the same temperature (T), their distribution can be described by Boltzmann statistics. The optical gain spectrum due to the recombination of the excitonic molecule is obtained by a simple calculation in which the k-dependence of the matrix element is neglected and is given by

$$G(\hbar\omega)\propto T^{-3/2}\sqrt{E_m-\hbar\omega}\{2^{-3/2}(N_m/N_{ex})\cdot\exp[-(E_m-\hbar\omega)/kT] \\ -\exp[-2(E_m-\hbar\omega)/kT]\}, \qquad (1)$$

where N_m and N_{ex} are the density of the excitonic molecule and the exciton, respectively. $E_m=E_{ex}-E^b_m$, where E_{ex} is the energy of the free exciton and E^b_m is the binding energy of the excitonic molecule. One finds that the crossover position from absorption to gain, which is given by solving $G(\hbar\omega)=0$ except for the case of $\hbar\omega=E_m$, is a function of the ratio N_m/N_{ex} and T. The experimental result in fig. 3(a) gives the crossover to be at $\hbar\omega=2.5424$ eV. If we use $E_m=2.5473$ eV [16] and $T=60$ K, eq. (1) yields the ratio N_m/N_{ex} to be about 1.1. The calculated gain spectrum is depicted in fig. 3(a) by the dashed curve, where the only fitting parameter is the absolute OD value. Good agreement is found, giving confirmation of the present interpretation.

The time variation of N_{ex} and N_m is obtained by solving the rate equation for the exciton and excitonic molecule densities given by

$$\mathrm{d}N_m/\mathrm{d}t=-N_m/\tau_m+B(N_{ex})^2, \qquad \mathrm{d}N_{ex}/\mathrm{d}t=G(t)+N_m/\tau_m-2B(N_{ex})^2-N_{ex}/\tau_{ex}, \qquad (2)$$

where τ_m is the lifetime of the excitonic molecule, τ_{ex} the lifetime of the free exciton and B the formation rate of the excitonic molecule from the exciton. Here we assumed that only excitons, and no excitonic molecules, are photogenerated because we employed the excitation via band-to-band absorption. $G(t)$ is the generation rate of the exciton given by

$$G(t)=G_0/(\cosh(2.634t/\tau_p)); \qquad (\tau_p=25\text{ ps}).$$

In solving eq. (2) N_m and N_{ex} were normalized to the exciton density N^0_{ex} which corresponds to the density if the excitation is given by the δ function at $t=0$. Then B is expressed by the formation time τ_f as $\tau_f=(BN^0_{ex})^{-1}$, and also $N_m=N^0_{ex}m$ and $N_{ex}=N^0_{ex}n$, where m and n are dimensionless densities of the excitonic molecule and the exciton, respectively. The rate equations are solved numerically using $\tau_m=17$ ps[17], and τ_{ex} and τ_f as parameters. The fit to the experimental values of $N_m(t)$ does not depend critically on the values for τ_{ex} and τ_f.

In fig. 2 two examples of the calculated N_m are shown by the solid and dashed curves. The solid curves represent the case that $\tau_f=400$ ps and $\tau_{ex}=2200$ ps, and the dashed curves represent the case that $\tau_f=800$ ps and $\tau_{ex}=1000$ ps. Good agreement between the experimental data and the calculated curves is found in each case except only for the very early time stage. An important point obtained from the calculation is that N_m decays nearly exponentially with a time constant much longer than τ_m, and that N_{ex} also decays nearly exponentially with the time constant of about 600 ps.

Under the resonant excitation of the excitonic molecule, in which almost no excitons are excited directly, the excitonic molecule luminescence shows an extremely short lifetime [17]. In the present case, however, the exciton and the excitonic molecule are generated via low density carriers excited by band-to-band absorption. Because of the extreme short lifetime of the excitonic molecule, N_m is determined by the longer τ_f and τ_{ex}, resulting in a slower decay of N_m. This causes the slow decay of the luminescence as well as the optical gain due to the excitonic molecule recombination.

3.2. Excitation by two-photon absorption for $\lambda = 532$ nm

Figure 4 depicts time-resolved optical density spectra observed for the excitation by two-photon absorption of 532 nm laser pulses. The excitation density amounts to about 700 MW/cm^2. Again negative OD values correspond to optical gain. The OD spectrum at 140 ps delay time has a peak at 2.532 eV accompanied by a low energy tail. The peak gain value is about 23 cm^{-1}, which is evaluated using the sample thickness of ~150 μm, because the entire volume of the crystal is occupied by almost homogeneous carriers. At 840 ps the OD value itself decreases. Simultaneously two peaks at 2.537 eV and

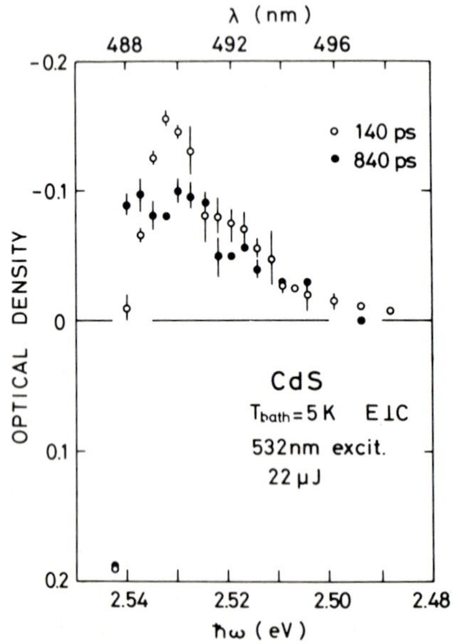

Fig. 4. Time-resolved optical density spectra for the excitation by two-photon absorption at $\lambda =$ 532 nm. Negative OD value again corresponds to optical gain.

2.53 eV seem to appear, but unfortunately the experimental data are not so reliable to discuss in detail.

In fig. 5 is shown the time behavior of the negative OD value at 2.537 eV, which corresponds to the high energy tail of the OD peak at 140 ps spectrum. Absorption takes place just after the excitation (not shown in fig. 5). This absorption changes to gain at ~40 ps. The gain decays with time constant of about 370 ps. Although not shown here, optical gain in the low energy tail also decays with nearly the same time constant as mentioned above (c.f. the case of 355 nm excitation).

Remarkable differences of the present results (figs. 4 and 5) compared with those obtained under one-photon excitation shown in fig. 3 are as follows: The peak gain value is two orders of magnitude smaller. The decay time of the gain is roughly the same for the entire spectral range measured here, and is much slower than the decay time of the gain due to the EHP recombination but is nearly twice the decay time due to the excitonic molecule recombination observed under 355 nm one-photon excitation. Furthermore the peak position and the crossover position at 140 ps are close to those due to the excitonic molecule recombination (see fig. 3(a)). These facts suggest that the gain observed for the present excitation can be ascribed to the excitonic molecule recombination. The accurate density of the excited electron–hole pairs has to be known for further information. The accurate density was determined from absorption measurements for 532 nm laser pulses. Net absorption of ~23% for the same sample used for the transmission measurements is observed under the incident laser power of 21 μJ, which is nearly the same as for the transmission measurements. This absorption amounts to an absorption coefficient of $\sim 18\ cm^{-1}$.

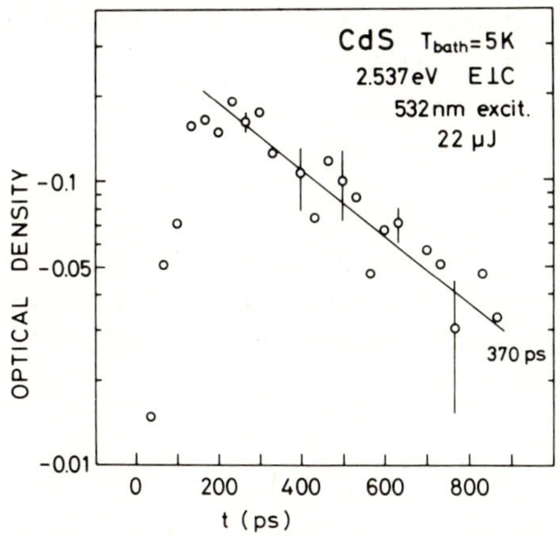

Fig. 5. Optical gain at $\hbar\omega = 2.537$ eV as a function of delay time between the excitation pulse at $\lambda = 532$ nm and the probe pulse at $\lambda = 488.5$ nm (2.537 eV). The solid line indicates the decay time of 370 ps.

If we make a correction for the reflection loss of $\sim 18\%$ for the incident intensity and if we take the laser pulse as a δ function for simplicity, a carrier density of $\sim 1 \times 10^{18}\ \mathrm{cm}^{-3}$ is generated. It is noted that the actual carrier density should be less than this value if we take into account the actual shape of the incident laser pulse.

This density is found to be nearly equal to the carrier density around 200 ps delay time for one-photon excitation by 355 nm, where almost no EHP is present as mentioned already. If we increase the incident laser intensity up to twice the present intensity, the crystals often suffer damage. We therefore conclude that it is very difficult to excite carrier density high enough to generate the EHP in CdS by means of two-photon absorption because of its low absorption coefficient. However, homogeneous volume excitation can be achieved by this excitation. We further conclude that the optical gain spectra as well as the decay of the gain shown in figs. 4 and 5 are ascribed to the excitonic molecule recombination.

4. Transient grating experiments

The experimental configuration used in the transient grating measurements was nearly the same as described by Smirl et al. [18]. The interference between the two pump beams at 355 nm produces a modulation of the optically created carrier density to form a grating. The grating decay is monitored by measuring the first-order diffracted light of a weak probe pulse at 532 nm (which is transparent for CdS). The diffraction of the probe beam is due to index modulation of the sample induced by free carriers or excitonic system.

The kinetics of the gratings can be understood in terms of the continuity equation. The time change of the density of the free carrier or the excitonic system, $N(x, z, t)$, is given by

$$\frac{\partial N(x, z, t)}{\partial t} = G(x, z, t) + D\nabla^2 N(x, z, t) - \alpha N(x, z, t) - \beta N^2(x, z, t), \tag{3}$$

where D is ambipolar or excitonic diffusion constant, and α and β are linear and bimolecular recombination rates, respectively, and where x is the direction along the sample surface and z is the direction normal to the surface. We assumed that D is density (and therefore position) independent. The spatially modulated generation rate is given by

$$G(x, z, t) = G_0(t)[1 + \cos(2\pi x/\Lambda)] \exp(-az),$$

where Λ is the grating spacing, a is the absorption coefficient and $G_0(t)$ is the time-dependent generation rate given by

$$G_0(t) = G_0/(\cosh(2.634t/\tau_p)); \qquad (\tau_p = 25\ \mathrm{ps}).$$

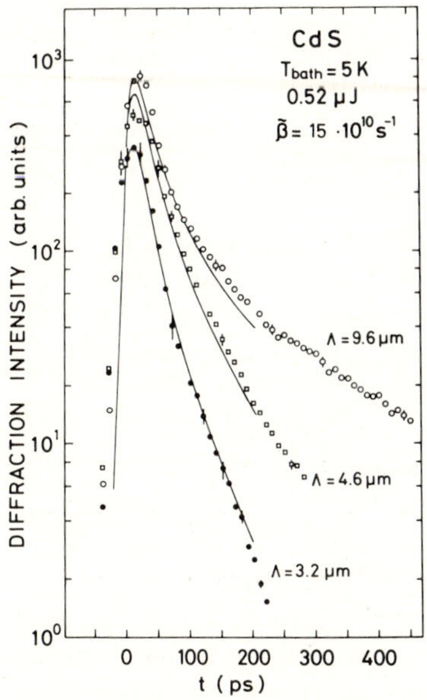

Fig. 6. Diffraction intensity as a function of delay time between the excitation pulse at $\lambda=355$ nm and the probe pulse at $\lambda=532$ nm for three grating spacings for $I_{exc}=0.52$ μJ. Solid curves are the calculated diffraction intensity due to the EHP as described in the text.

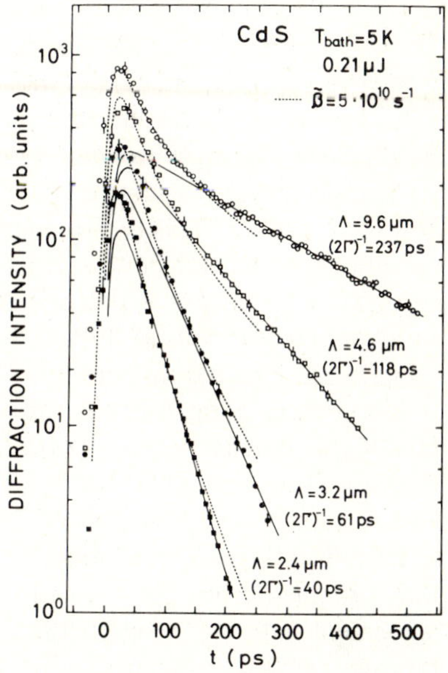

Fig. 7. Diffraction intensity as a function of delay time between the excitation pulse at $\lambda=355$ nm and the probe pulse at $\lambda=532$ nm for four grating spacings for $I_{exc}=0.21$ μJ. Solid curves are the calculated diffraction intensities due to the excitonic grating and dotted curves are due to the plasma grating as described in the text.

Figures 6 and 7 depict semi-logarithmic plots of the diffraction intensity against time delay for three and four grating spacings, respectively. Excitation intensity used in fig. 6 is about half of that for the highest intensity (1.1 μJ) in the transmission measurements in order to generate the same peak carrier density, because the carrier density at the grating peak becomes twice the mean carrier density, which corresponds to the case of the transmission experiments. The decay of the diffraction intensity in fig. 6 is non-exponential and furthermore gets faster with decreasing grating spacings. The fact that the decay is non-exponential can be attributed to the bimolecular recombination of the EHP. The fact that the decay of the grating depends on the grating spacing Λ clearly reveals that diffusion of the carrier contributes to the decay and hence to the decrease of the EHP density.

In solving eq. (3) we can neglect the contribution of a linear recombination process in this case. $N(x, z, t)$ was normalized to electron–hole density (N_0) at $x=0$, $z=0$ and $t=0$, which corresponds to the density if the excitation is given by the δ function at $t=0$. Then $N(x, z, t)=N_0 \cdot n(x, z, t)$, where $n(x, z, t)$ is a dimensionless density. The bimolecular recombination rate β is reduced to $\hat{\beta}=\beta N_0$. For the absorption coefficient we used $a=5\times10^4\ \text{cm}^{-1}$ for the present calculation, instead of using the real value of $2\times10^5\ \text{cm}^{-1}$ because of the following reason. Diffusion normal to the sample surface will also be present in the transmission experiments. This means that the value for β as obtained from the transmission measurements (5–$7\times10^{-9}\ \text{cm}^3/\text{s}$) accounts for the decay of the electron–hole density not only due to the recombination but also due to the diffusion. Therefore it seems rather reasonable to use an appropriate absorption coefficient which should be smaller than the real value. Then the nonlinear recombination-diffusion equation for $n(x, z, t)$ was solved numerically using D and $\hat{\beta}$ as parameters.

Solid curves in fig. 6 depict the result for $D=10\ \text{cm}^2/\text{s}$ and $\hat{\beta}=15\times10^{10}\ \text{s}^{-1}$. This $\hat{\beta}$ value yields $N_0=2.5\times10^{19}\ \text{cm}^{-3}$ with using $\beta=6\times10^{-9}\ \text{cm}^3/\text{s}$. Using the peak value for $n\approx0.31$ obtained at 8 ps, we can determine the peak electron–hole density to be $N(0,0,8)=7.7\times10^{18}\ \text{cm}^{-3}$, and also the density at 47 ps to be $N(0,0,47)=3.2\times10^{18}\ \text{cm}^{-3}$. The density at 47 ps is nearly the same as that obtained from the transmission measurement ($2.8\times10^{18}\ \text{cm}^{-3}$). Good agreement is found between the observed and calculated results. As expected for this excitation intensity, the plasma grating decays by the bimolecular recombination and diffusion, and disappears within 200 ps.

Deviation for $\Lambda=9.8\ \mu\text{m}$ and $t>200$ ps seen in fig. 6 can be ascribed to the fact that the EHP transforms to the excitonic system in this time regime resulting in the change of D and the lifetime. Transformation of the EHP to the excitonic system is clearly seen in the data for $I_{exc}=0.21\ \mu\text{J}$ shown in fig. 7. The time behavior of the diffraction intensity shown in fig. 7 changes from a faster decay to a slower decay at 100–150 ps. This behavior is clearly seen especially for larger grating spacings. It is noticed that the diffraction intensity for $t\geqslant150$ ps (the slow component) decays exponentially. This time behavior is in accordance with the results that the EHP decays faster, disappearing within 150–200 ps, and thereafter optical properties are determined by the excitonic system which has longer lifetime as already stated in the former sections. We therefore ascribe the faster component to the decay of the plasma grating and the slower component to the decay of the excitonic grating.

As mentioned in section 3, N_{ex} and N_m decay nearly exponentially. Then the kinetics of the excitonic grating can be expressed by eq. (3) in which bimolecular recombination process is neglected, that is, only the linear recombination process and the diffusion take part in the decay of the grating. $N(x, z, t)$ represents the density of the exciton and the excitonic molecule in this case. The linear recombination-diffusion equation can be solved analytically [18]. The diffraction intensity is proportional to

$$\{\phi(t)\}^2 \propto \left\{ \int_{-\infty}^{t} G_0(t')\exp\left[-\Gamma(t-t')\right]\mathrm{d}t' \right\}^2, \tag{4}$$

where the grating decay rate Γ is $\Gamma = 4\pi^2 D/\Lambda^2 + 1/\tau_R$, and where τ_R is the linear recombination lifetime. When the grating lifetime is long enough compared with the excitation pulse width, the diffraction intensity is expressed by

$$\{\phi(t)\}^2 \propto \exp(-2\Gamma t).$$

The diffraction intensity thus decays exponentially with a time constant of $(2\Gamma)^{-1}$. Calculated time evolution of the diffraction intensity using eq. (4) is shown in fig. 7 by solid curves for each grating spacing. The used $(2\Gamma)^{-1}$ values are shown in fig. 7. Good agreement is found in the later time regime. The D value obtained from the fitting is about 16 cm^2/s and τ_R is about 600 ps. This τ_R value is just the same as the decay time of N_{ex} as already shown in section 3.

Deviations from the calculated curves for $t < 100$ ps clearly show the occurrence of the EHP. The time evolution of the grating decay for the fast component is also calculated in the same manner as used for $I_{exc} = 0.52$ μJ. Results are shown by dotted curves in fig. 7. The parameters used are $D = 10$ cm^2/s and $\hat{\beta} = 5 \times 10^{10}$ s^{-1}. No β value or no N_0 value is known for $I_{exc} = 0.21$ μJ. If we assume that N_0 is proportional to the excitation intensity, we get $N_0 = 1 \times 10^{19}$ cm^{-3} and therefore $\beta = 5 \times 10^{-9}$ cm^3/s. Using the peak value for $n \sim 0.4$ at 8 ps delay time, we can determine the peak density to be $N(0, 0, 8) = 4 \times 10^{18}$ cm^{-3}. The fitting is good within the first 150 ps for $\Lambda = 9.6$ μm and $\Lambda = 4.6$ μm cases. For $\Lambda = 3.2$ and 2.4 μm cases, however, the fitting is good for the entire time regime measured here. This indicates as if the EHP (plasma grating) is persistent until ~200 ps. The calculated density at $x = 0$, $z = 0$ and $t = 100$ ps is $N(0, 0, 100) = 9 \times 10^{17}$ cm^{-3} for $\Lambda = 2.4$ μm. This density is nearly equal to the density at ~200 ps in the transmission measurement for $I_{exc} = 1.1$ μJ (see section 3). It is sure, therefore, that the transformation of the plasma grating to the excitonic grating takes place some time after 100 ps. Thus the grating decay is explained in terms of the recombination and the diffusion of the EHP within the first 100–150 ps, and thereafter in terms of the recombination and the diffusion of the excitonic system.

It is concluded that the electron–hole density in the EHP decreases continuously because of the recombination and the diffusion of the carrier, and the EHP transforms to the exciton and the excitonic molecule within 100–200 ps. This conclusion also excludes definitely the possibility of electron–hole liquid formation during its lifetime.

6. Summary

The time behavior of highly excited CdS is studied with picosecond time resolution by luminescence, transmission and transient grating experiments. Luminescence and optical gain due to both recombination of electron–hole plasma and excitonic molecule are definitely identified. In contrast to quasi-stationary experiments these two recombination processes can be distinguished by their different time behavior.

The EHP is created only at the highest excitation levels for one-photon surface excitation. The EHP decays very fast by recombination and diffusion, and transforms into excitons and excitonic molecules within 100–200 ps. The time-dependence of the spectral shape of the gain is well explained in terms of the EHP recombination taking account of plasmon–phonon sideband emission. The bimolecular recombination rate for the electron–hole recombination within the EHP is determined to be $5\text{–}7 \times 10^{-9}\ \mathrm{cm^3/s}$.

Transient grating experiments point out that diffusion of the optically excited carriers into the low density regions also contributes to the decrease of the EHP density. The initial plasma grating also transforms to the excitonic grating within 100–200 ps. The density of the EHP decreases continuously with time and the transformation to electron–hole liquid is definitely excluded.

No EHP is generated by two-photon volume excitation at excitation levels up to the damage threshold because of the low two-photon absorption coefficient. Optical properties are therefore determined by the recombination of the exciton and the excitonic molecule.

The time variation of the optical properties at low excitation levels for one-photon excitation is also determined by the exciton and the excitonic molecule. The position and spectral shape of the gain observed at these excitation levels can be well explained by the excitonic molecule recombination.

Acknowledgements

The author gratefully expresses thanks to H.J. Queisser for giving him the possibility to perform the present experiments at Max-Planck-Institut für Festköperforschung and for helpful discussions. He is also indepted to S. Shionoya for valuable discussions and to H. Sawami for his efforts in the numerical calculations.

References

[1] C. Klingshirn and H. Haug, Phys. Rep. 70 (1981) 315.
[2] M. Rösler and R. Zimmermann, Phys. Status Solidi (b) 83 (1977) 85.
[3] G. Beni and T.M. Rice, Phys. Rev. B 18 (1978) 768.
[4] S.W. Koch, Solid State Commun. 35 (1980) 419.
[5] K. Bohnert, M. Anselment, G. Kobbe, C. Klingshirn, H. Haug, S.W. Koch, S. Schmitt-Rink and F.F. Abraham, Z. Phys. B 42 (1981) 1; see also ref. [1].
[6] A. Forchel, H. Schweizer and G. Mahler, Phys. Rev. Lett. 51 (1983) 501.
[7] H. Yoshida, H. Saito, S. Shionoya and V.B. Timofeev, Solid State Commun. 33 (1980) 161.
[8] S. Shionoya, H. Saito, E. Hanamura and O. Akimoto, Solid State Commun. 12 (1973) 223.

[9] H. Yoshida, H. Saito and S. Shionoya, J. Phys. Soc. Japan 50 (1981) 881.
[10] For resonant excitation of the excitonic molecule, the lifetime of the excitonic molecule is very fast, ~20 ps [17].
[11] W. Graudszus and E.O. Göbel, Physica B 117/118 (1983) 555 (Proceeding of the 16th International Conference on the Physics of Semiconductors).
[12] If the diffusion of the carrier in the EHP is important, the shift of the chemical potential is also related to the velocity of the carrier, see ref. [6]. For the maximum carrier velocity of $\sim 10^5$ cm/s estimated from the obtained diffusion constant of $D = 10\,\mathrm{cm}^2/\mathrm{s}$ (see section 4), the gain spectra are almost unaffected.
[13] Though this assumption cannot be physically motivated, it provides good agreement with exact calculations. H. Haug and D.B. Tran Thoai, Phys. Status Solidi (b) 98 (1980) 581.
[14] E.O. Göbel, P.H. Liang and D. v.d.Linde, Solid State Comm. 37 (1981) 609.
[15] H. Saito, Solid State Commun. 39 (1981) 71.
[16] E. Hanamura, J. Phys. Soc. Japan 39 (1975) 1516; see also ref. [1].
[17] R. Baumert, Phys. Rev. in press.
[18] A.L. Smirl, S.C. Moss and J.R. Lindle, Phys. Rev. B 25 (1982) 2645.

Journal of Luminescence 30 (1985) 318–332
North-Holland, Amsterdam

ULTRAFAST RELAXATION PROCESSES IN SEMICONDUCTORS

Roberto LUZZI

Departamento de Física do Estado Sólido e Ciência dos Materiais, Instituto de Física, Universidade Estadual de Campinas, 13100 – Campinas, SP, Brasil

A brief review of theoretical studies concerning ultrafast relaxation processes in direct-gap polar semiconductors is presented. Comparison with measurement of time-resolved optical spectra in GaAs is shown.

1. Introduction

Studies of the optical properties of semiconductors under high-excitation conditions has shown novel and quite interesting features in optical spectroscopy. For systems slightly deviated from equilibrium exact close expressions for their response functions to mechanical perturbations can be obtained in the form of correlation functions in equilibrium [1,2]. A practical way to calculate such correlation functions is the double-time thermodynamic Green function formalism of Bogoliubov and Tyablikov, described in the already classic paper by Zubarev [3,4]. The actual calculations may be difficult for the case of interacting many-body systems but they are formally closed at this level. However, measurements can be performed on systems strongly departed from equilibrium, when the responses of the system depend on its instantaneous and local nonequilibrium state. This is the case of highly excited semiconductors (HEPS) when spectra obtained by means of ultrafast laser spectroscopy (UFLS) depend on the characteristics of the nonequilibrium distributions of elementary excitations during the lapse of instrumental time resolution.

Hence the usual response function theory needs be replaced by another capable to incorporate these features. Among the actual techniques to deal with nonequilibrium many-body systems the nonequilibrium statistical operator (NSO) method seems to offer an elegant and concise closed-off formalism in the theory of irreversible processes adequate to deal with a large class of experimental situations [5,6,7]. It is considered a far reaching generalization of the Chapman–Enskog approach for the derivation of transport equations in the kinetic theory of gases [5], providing a macroscopic description

0022-2313/85/$03.30 © Elsevier Science Publishers B.V.
(North-Holland Physics Publishing Division)

for systems arbitrarily away from equilibrium whose evolution includes nonlinear, nonlocal, and retardation effects. The NSO method allows for the derivation of the nonlinear transport equations (NLTE), i.e., the set of generally complex equations relating the time average behavior of some physical observables of the system to the detailed motion of the particles of which it is composed. The solution of these equations is necessary in order to obtain the information on the time-dependent macrostate of the system to evaluate the response function to external stimuli [8]. In a low order of approximation (the so-called linear approximation in the relaxation processes [9]) the NLTE take the form of Boltzmann-like equations. This limit has been used in calculations by several authors [10,11,12] to interpret measurements performed using UFLS [13–18]. These are pulse-probe experiments which permit to follow the temporal evolution of the nonequilibrium macroscopic state of the HEPS.

This theoretical treatment has its shortcomings. First, it is well known that to go from the microscopic to the macroscopic description of a many-body system, a contraction of the information required to describe its states has to be performed: the macrostate of the system is characterized by a small number of thermodynamic extensive variables, say $Q_j(t)=\mathrm{Tr}\{P_j\rho(t)\}$, $j=1,2,\ldots,s$, which are the average values over the nonequilibrium ensemble, characterized by the NSO $\rho(t)$, of the dynamical quantities P_j. According to Bogoliubov [7] a contracted description is possible if there exists a relaxation time for microinformation, τ_μ, such that for $t\gg\tau_\mu$ correlations with lifetime smaller than τ_μ can be ignored and the $Q_j(t)$ form a set of variables enough for the description of the nonequilibrium macroscopic state of the system. At present there seems to be no wholly satisfactory theory to generate this information allowing to make a unique decision [5,19]. The ever increasing time-resolution being obtained in UFLS laboratories, where pulses of femtoseconds duration are becoming available poses increasing limitations to the application of methods based on the NSO.

In this communication we briefly describe theoretical results, based on the quasilinear approximation in the NSO method, to discuss pulse-probe UFLS experiments in polar direct-gap semiconductors.

2. Ultrafast relaxation processes in semiconductors

In a typical pump-probe experiment a concentration n of electron–hole pairs is generated by a pulse of intense laser light. This can be done either by direct interband absorption of one photon or by means of nonlinear effects such as two-photon absorption or second-harmonic generation. The carriers (electrons and holes) acquire an excess kinetic energy $\hbar\Omega_L - E_G$ per pair, where E_G is the gap energy and Ω_L is equal to one or twice the laser frequency ω_L

in one- or two-photon absorption processes respectively. An optical response is recorded while the HEPS is relaxing the excess energy through different channels. The relaxation processes occur via different types of interactions between the several subsystems of the semiconductor sample: carriers and lattice vibrations interacting through deformation potential and Fröhlich interaction, anharmonic effects between lattice modes, luminescent recombination, non-radiative recombination, Auger effect, etc. It is assumed that the excitation conditions are such that the carriers remain on the metallic side of Mott transition.

The free subsystems are described by the usual Hamiltonians for phonon and photon fields, and carriers are treated as Landau's quasiparticles in a two parabolic bands Hamiltonian in the electron–hole representation [20]. Coulomb interaction, incorporated in a self-consistent way in the single-particle energy dispersion relation, is responsible for the very rapid internal thermalization of carriers in the subpicosecond time scale [10,21] and for the polarization of the carrier system that should screen Fröhlich interaction. The main effect of carrier–carrier interaction is to ensure the very rapid thermalization of carriers making it possible to describe their macroscopic (thermodynamic state) in terms of an instantaneous internal energy, $E_c(t)$, and the reciprocal of an effective temperature, $\beta_c(t)=1/kT_c^*(t)$, the instantaneous concentration, $n(t)$, and quasichemical potentials, $\mu_e(t)$ for electrons and $\mu_h(t)$ for holes [11]. Secondly the screening effect of Fröhlich interaction is taken into account by means of the introduction of a dielectric constant calculated in the random phrase approximation (RPA).

The excess energy the carriers have received from the laser pump is dissipated exciting lattice vibrations while cascading down the band states. In the early stages of the relaxation processes, following the initial departure from equilibrium the rate of production of LO phonons is very large. This production of LO phonons is not uniform throughout the Brillouin zone, but because of the wavevector dependence of Fröhlich interaction LO phonons in certain regions of the Brillouin zone are preferentially warmed up [22,23]. Therefore, differently to the case of the carrier system a macroscopic description of the LO-phonon system should not be possible in terms of solely its internal energy, $E_{LO}(t)$, and a unique instantaneous effective temperature, $\beta_{LO}(t)=1/kT_{LO}^*(t)$, for very short time intervals after optical pumping. Similar arguments apply to the acoustic phonon system, which receives energy directly from the carriers via deformation potential, but at a small rate, and preferentially from the warmed up LO phonons via anharmonic interactions. For the first kinetic stage in the relaxation processes a description of the thermodynamic state of the LO phonons requires the use of the whole set of occupation numbers functions $\nu_{q\mu}(t)$, with q running over the entire Brillouin zone. All calculations at present avoid this cumbersome difficulty, which is enhanced by the fact that one would need information on the initial nonequilibrium values of $\nu_{q\mu}$ at some time t_0 after the lapse τ_μ of microrandomization of the initial distribution has occurred. Here we approximately describe the nonequilibrium thermodynamic state and kinetic evolution of the HEPS in terms of the internal energies of the carriers and the lattice, and the concentration of carriers. If we denote these variables by $Q_j(t)$, and the corresponding mechanical quantities by P_j, in compact form the nonlinear transport equations in Zubarev's method are [24]

$$\frac{dQ_j(t)}{dt} = \sum_{l=0}^{\infty} J_j^{(l)}(t), \tag{1}$$

where on the right-hand side we have a series expansion in scattering operators of increasing order in the interaction strengths. See the article by Luczka [25] for their detailed general form. For the case of the HEPS $J^{(0)}$ and $J^{(1)}$ vanish, and the lowest order surviving collision operator is $J_j^{(2)}(t)$. This is the only one to be retained in the quasilinear approximation in the relaxation processes, and the set of NLTE becomes closed when the usual time-dependent Hartree–Fock approximation is used in the calculation of averages over the coarse-grained nonequilibrium ensemble; then eqs. (1) reduce to a set of coupled nonlinear Boltzmann-like equations.

Details of the calculations are given elsewhere [11,12], and next we present here a summarized review of the phenomena of relaxation processes in the HEPS.

To solve the set of NLTE initial values for the carriers concentration and effective temperature were obtained from the experimental data, and the effective temperature of LO phonons was obtained from considerations of conservation [11,12]. It was assumed that A phonons do not change appreciably their temperature and remain at near the thermal bath temperature T_0.

Before proceeding with the description of the relaxation processes in the HEPS we will comment on the approximations involved in the calculations.

(a) *The LO-phonon distribution functions*

We have already noted that the use of a unique effective temperature for the whole LO-phonon system disregards the expected inhomogeneous distribution of excess energy over the different modes. We have calculated the emission time of LO phonons due to scattering by carriers for different carriers effective temperatures [23]. Our results show that the rate of energy transfer is large in regions of the reciprocal space around a given LO-phonon wavenumber, as shown in fig. 1.

(b) *Laser-induced lattice heating of A phonons*

The acoustic phonons are heated by transfer of excess energy from the LO phonons via anharmonic effects, and next dissipate it via heat diffusion processes

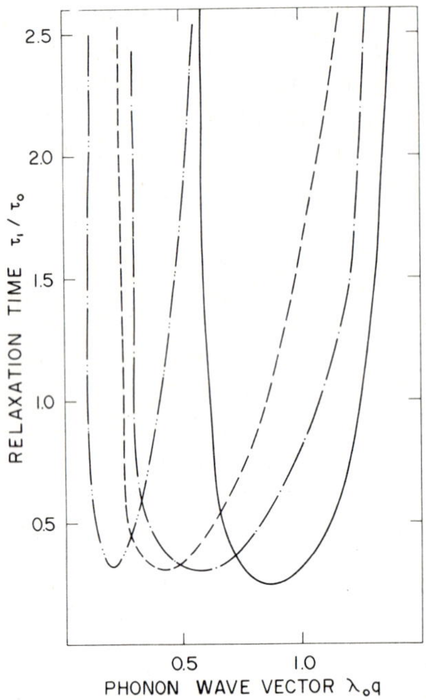

Fig. 1. Emission time of LO phonons due to scattering with carriers as a function of LO-phonon wavevector. τ_0 and λ_0 are scaling units (see ref. [23a]) which for conduction electrons in GaAs are $\tau_0 \simeq 0.5$ ps, $\lambda_0 \simeq 38$ Å. Each curve corresponds to a given carriers effective temperature $\theta = T_c^*/\theta_0$, where θ_0 is Einstein temperature of LO phonons ($\theta_0 = 430$ K in GaAs): $\theta = 0.1$ for full the line, $\theta = 0.5$ for the dash–dot line, $\theta = 1$ for the dashed line, and $\theta = 5$ for dash–double dot line.

from the active volume of the sample. We have calculated the rates of energy transfer to and from the A-phonon system in polar semiconductors under high levels of photoexcitation [26]. It is verified that for very short pulses (picosecond duration), except at very high levels of excitation (say laser energies of several eV and excitation by short pulses of *M* watts), the A phonons show negligible to moderate heating. In conditions resembling laser annealing experiments, under illumination by large pulses (nanosecond duration and up) or continuous illumination, the lattice temperature increases smoothly until it approached a critical value of laser photon density when the steady-state A-phonon temperature increases steeply and melting should follow [26]. See fig. 2.

(c) *Time-dependent Hartree–Fock approximation*

Higher order correlation functions that appear in the calculation of $J_j^{(2)}(t)$ were factorized using the time-dependent Hartree–Fock approximation in its normal form (i.e. retaining only diagonal terms). This allowed us to close

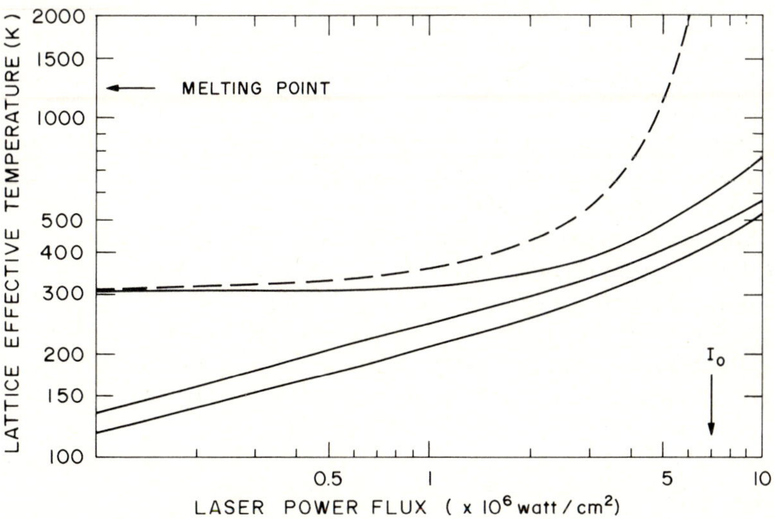

Fig. 2. The steady-state effective temperature of A phonons in GaAs as a function of the power flux of a c.w. laser [26]. The thermal bath temperature is 300 K, and the heat diffusion coefficient was calculated in the nondegenerate statistical limit.

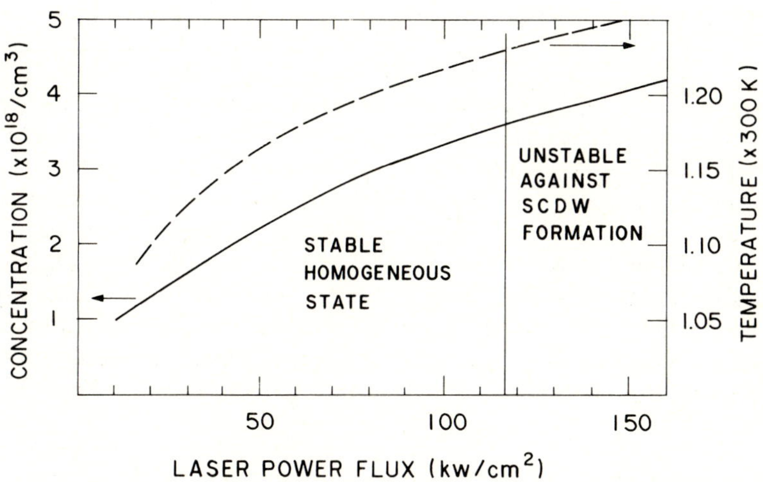

Fig. 3. Dependence of the steady-state values of concentration and carriers effective temperature on laser power for a GaAs sample under c.w. laser illumination. The stationary uniform inverted population of electrons becomes unstable against the formation of a steady-state charge density wave in conditions of strong photoexcitation.

the set of NLTE without necessity to resort to the calculation of additional correlation functions, as sometimes required by the method [19]. The solution corresponds to an homogeneous inverted population of electrons and holes. Hydrodynamic effects [27] that may be present in the HEPS have then been neglected. However, it is worth mentioning that the uniform steady state that results in the HEPS under continuous illumination may become unstable against the formation of a steady-state charge density wave (SCDW) [28]. Such disorder–order condensation phenomena may take place before melting [29], and this transformation of the excess energy into an ordered structure may only occur in far-from-equilibrium conditions governed by NLTE like eqs. (1) [30]. Description of the new nonequilibrium ordered structure implies the use of a generalized time-dependent Hartree–Fock approximation which allows for the inclusion of the amplitude of the SCDW. Figure 3 shows the

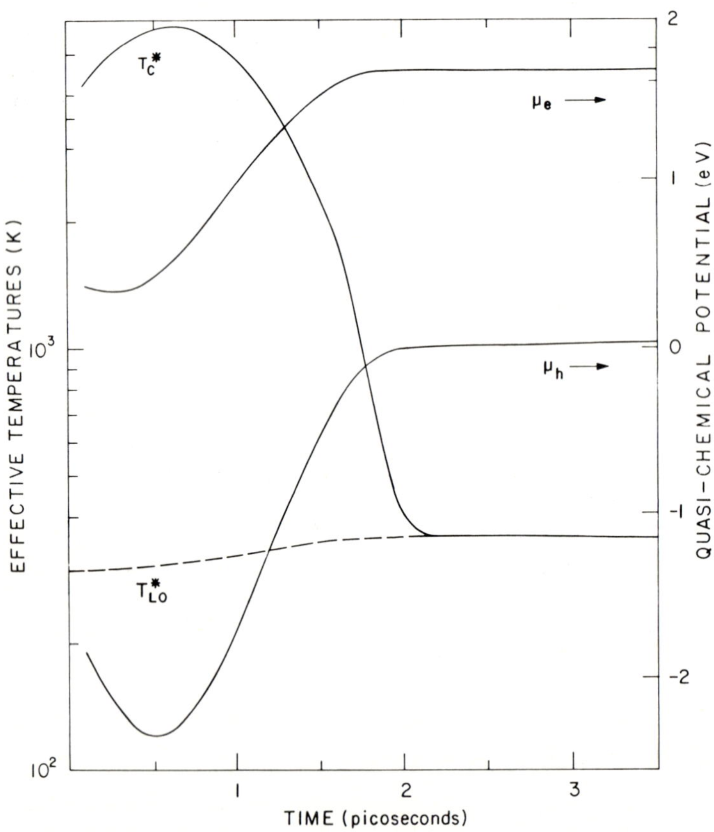

Fig. 4. Time-evolution of the carriers and LO-phonons effective temperatures calculated in the conditions of the experiment of ref. [13] (after ref. [32]).

stationary uniform values of n and T_c^* as a function of laser power, and the point of instability of the homogeneous electron system with population inversion.

We consider next the time evolution of relaxation processes in the HEPS. Our theoretical calculations have closely followed the conditions of several experiments reported in the literature on the subject [13–15], and details are given in refs. [11] and [12]. The LO-phonon system was described by an Einstein model, with the value of the phonon frequency being the one obtained in Raman scattering experiments. A Debye model was used for the A-phonon system. The relaxation time associated with anharmonic processes was estimated from Raman linewidth measurements [31]. Induced recombination, self-absorption, and Auger effect were neglected since they are not relevant in the early stages of the relaxation processes.

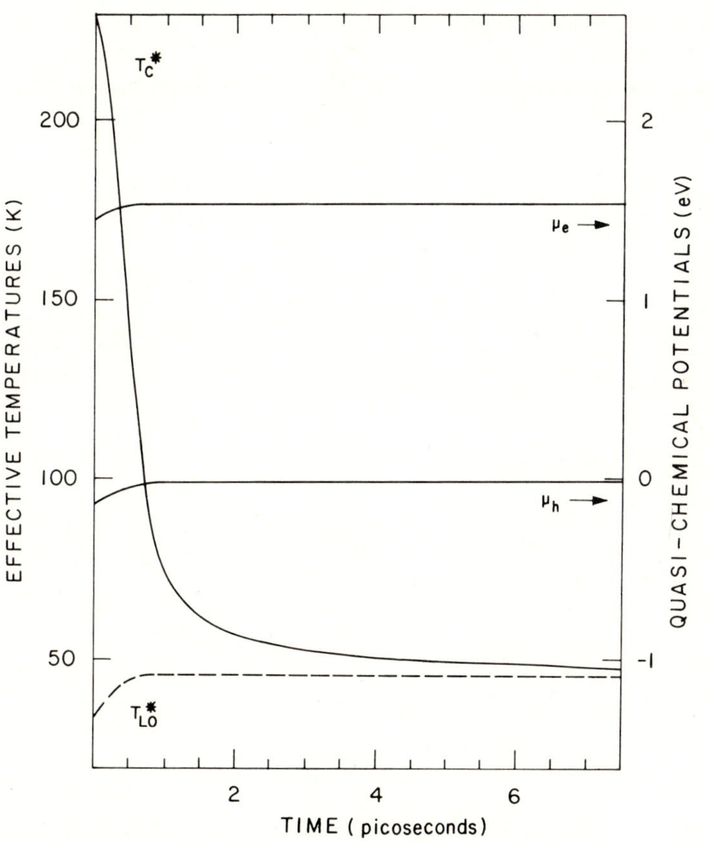

Fig. 5. Time-evolution of the carriers and LO-phons effective temperatures in the conditions of the experiment of ref. [14] (after ref. [32]).

We compared the curves for the time evolution of the carriers effective temperatures with the experimental data obtained by best fitting of optical spectra of GaAs reported in refs. [14] and [15], and the agreement is very good within experimental error. To illustrate the results we show in figs. 4 and 5 two limiting cases: (a) reflectivity experiments [13] performed at room temperature ($T_0 = 300$ K), excess energy per pair was 2.6 eV ($\lambda_L = 3100$ Å) and laser power flux was roughly 660 MW cm^{-2}, and (b) transmission measurements [14] performed at $T_0 = 10$ K, with an excess kinetic energy of 0.120 eV, ($\lambda_L = 1.64$ eV), and laser power flux of roughly 20 MW cm^{-2}. The calculation have been performed using an exciting laser pulse of Gaussian time profile, and the beginning of the curves corresponds to the point when carriers concentration is of the order of 10^{16} cm^{-3}, in order to ensure that the electron–hole system is on the metallic side of Mott transition. Figures 6 and 7 show the rates of energy transfer through different channels [32].

A general picture of the behavior of ultrafast relaxation processes in polar semiconductors can be drawn from these results:

(1) In all cases there occurs a very rapid transference of a large part of the carriers excess energy to the LO-phonon system. Both systems become mutually

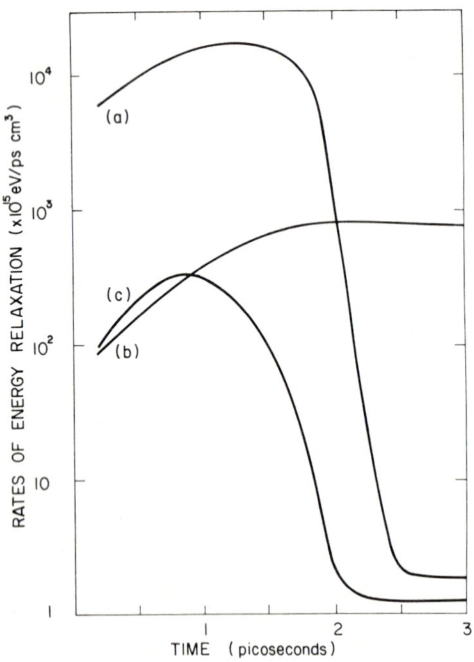

Fig. 6. Evolution of the rates of energy transfer between the different subsystems in the HEPS in GaAs: (a) from carriers to LO phonons, (b) from carriers to A phonons, (c) from LO phonons to A phonons (anharmonic effect), for the case of ref. [13].

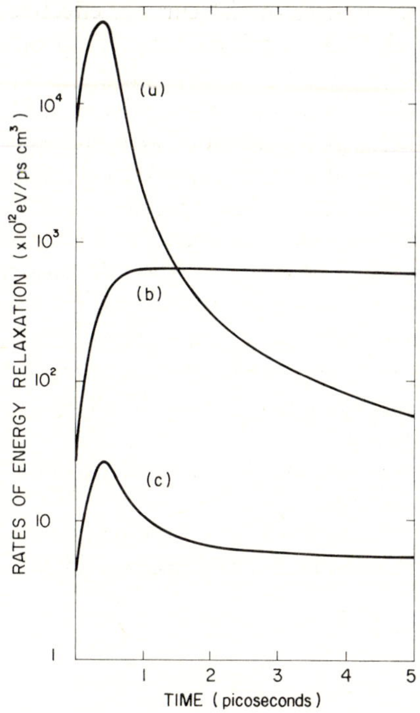

Fig. 7. Evolution of the rates of energy transfer between the different subsystems in the HEPS in GaAs: (a) from carriers to LO phonons, (b) from carriers to A phonons, (c) from LO phonons to A phonons, for the case of ref. [14].

thermalized at a common effective temperature, $T_c^* = T_{LO}^*$, in times of the order of a few picoseconds.

(2) For intermediate to high reservoir temperatures (say $T_0 \lesssim 77$ K), T_{LO}^* increases only slightly above T_0. This is a result of the fact that the concentration phonons in equilibrium is larger than the excess created for pair concentrations of up to 10^{19} cm^{-3}. A noticeable warming up of LO phonons would follow for values of n of the order of 10^{20} cm^{-3} which is difficult to obtain in the materials we considered because of optical saturation.

(3) When reservoir temperature is low (say $T_0 \lesssim 20$ K), under high levels of excitation the number of LO phonons in excess of equilibrium is usually large. T_{LO}^* increases well above T_0 while T_c^* decreases until both effective temperatures become rapidly equalized. From then on the common temperature of both subsystems decreases towards final equilibrium at a slow pace. This explains the long near plateau in the curve T_c^* vs. time observed in several experiments. It has been argued that other possible reason for this effect may be the screening of Fröhlich interaction. We have compared results obtained

using a bare and a screened Fröhlich interaction (using Debye–Hückel and Lindhardt RPA dielectric constants). No noticeable differences of behavior have been detected in all cases; see figs. 8 and 9 [32].

(4) The rate of energy transfer to the A-phonon system through the channel due to anharmonic effects is roughly an order of magnitude larger than the rate of energy transfer through the channel resulting from carriers colliding with A-phonons, in accord with Van Driel conclusions [33].

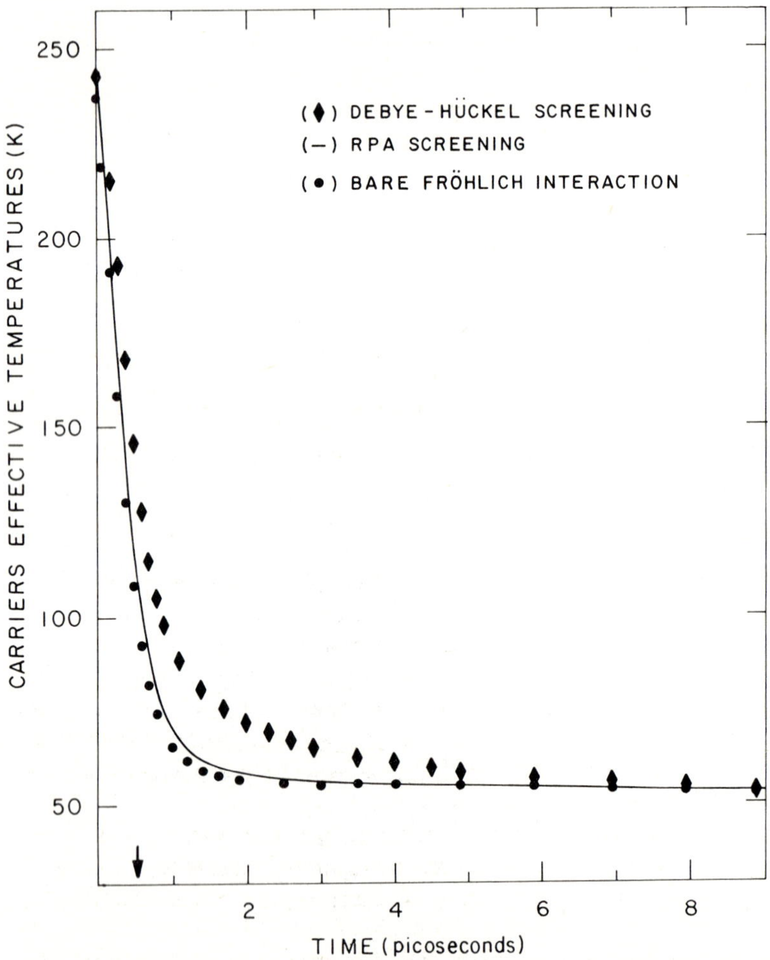

Fig. 8. Evolution of the carriers effective temperature of GaAs in the conditions of the experiment of ref. [13] for the cases of (a) a bare Fröhlich interaction (FI), (b) a screened FI using Lindhardt RPA dielectric function, and (c) a screened FI using a Debye–Hückel dielectric function (after ref. [32]).

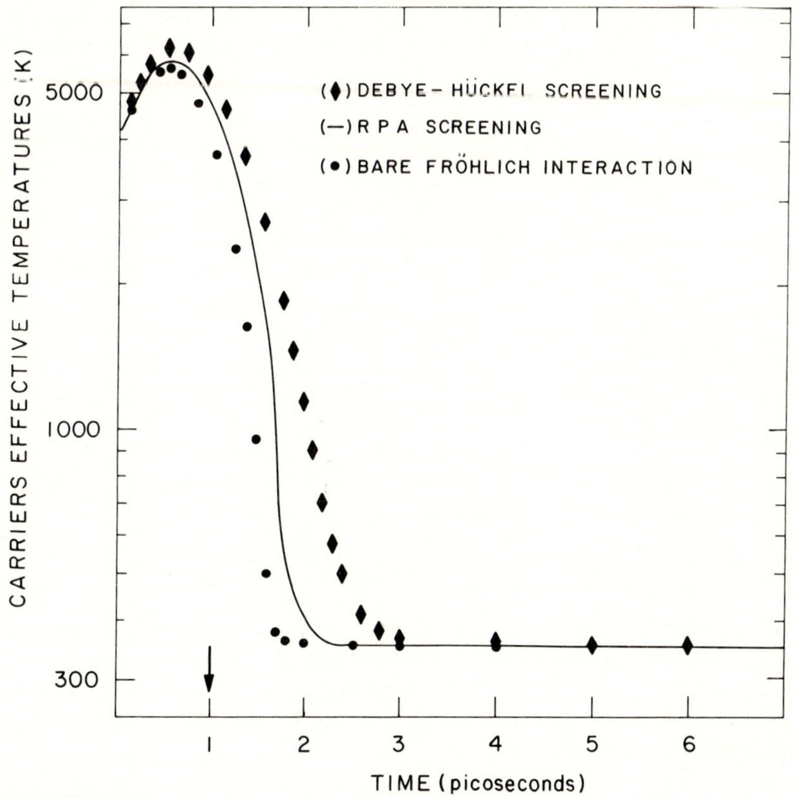

Fig. 9. Evolution of the carriers effective temperature of GaAs in the conditions of the experiment of ref. [14] for the cases of (a) a bare Fröhlich interaction (FI), (b) a screened FI using Lindhardt RPA dielectric function, and (c) a screened FI using a Debye–Hückel dielectric function (after ref. [32]).

This picture of ultrafast relaxation processes in semiconductors derived from an analysis of optical spectra is also consistent with the one that can be obtained from the study of ultrafast transient transport [34]. The mobility and diffusion coefficients of carriers in the central valley of polar direct-gap semiconductors show two different regimes of behavior, one involving "overshoot" effects in the conditions of item 2 above, or near smooth increase towards the steady-state value in the conditions of item 3 above. Figure 10 shows the dependence of the mobility of carriers in a parabolic central valley of a polar semiconductor as a function of the carriers effective temperature and for different values of the LO-phonons effective temperature.

The conclusions we have drawn (items 1 to 4) apply to any polar direct-gap semiconductor under excitations of picosecond duration and energy transfer of the order of up to a few Joules per cubic centimeter. Optical saturation does

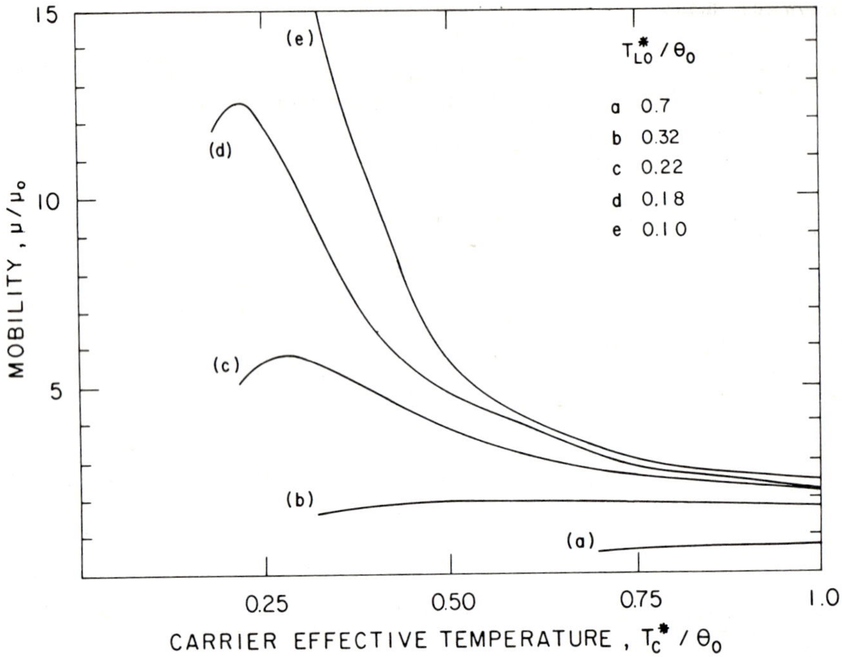

Fig. 10. Dependence of the mobility of carriers in a parabolic central valley of a polar semiconductor on carriers effective temperature T_c^*/θ_0. θ_0 is the Einstein temperature of LO phonons (430 K in GaAs), and μ_0 is defined in ref. [36].

not allow to pump larger amounts of energy to obtain carrier concentrations above roughly 10^{20} carriers cm^{-3}. This situation is altered in the case of indirect-gap semiconductors, e.g., Si, Ge; then laser-annealing processes are possible and an intense heating of the whole lattice occurs [35]. The case of polar direct-gap semiconductors under illumination by long pulses or c.w. laser has been previously discussed in this note.

Finally we may say that techniques of ultrafast laser light spectroscopy in conjunction with statistical mechanics methods, which yield nonlinear transport equations for the description of the kinetic and relaxation processes in non-equilibrium physical systems, provide a powerful tool for the study of picosecond responses in highly excited semiconductors. Besides the interest in the comprehension of the basic physical problems involved, there exists a parallel technological interest since performance of semiconductor electronic devices is ultimately determined by how fast they operate. Development of high-speed technology requires one to explore semiconductor behavior in picosecond and sub-picosecond time scales.

Acknowledgements

The author gratefully acknowledges a grant from IBM of Brasil, which made it possible for him to participate in the 3rd IUPAP Semiconductor Symposium where the results summarized in the present article were presented.

References

[1] H.B. Callen and T. Welton, Phys. Rev. 82 (1951) 34.
[2] R. Kubo, J. Phys. Soc. Japan 12 (1957) 570.
[3] D.N. Zubarev, Usp. Fiz. Nauk. 71 (1960) 71 [Soviet Phys. Usp. 3 (1960) 320].
[4] V.L. Tyablicov, Methods in the Quantum Theory of Magnetism (Plenum, New York, 1967).
[5] R. Zwanzig, in: Perspectives in Statistical Physics, H.J. Raveché, ed. (North-Holland, Amsterdam, 1981).
[6] B. Robertson, Phys. Rev. 114 (1966) 151; 160 (1968) 175.
[7] D.N. Zubarev, Neravnovesnaia i Staticheskaya Termodinamika (Nauka, Moscow, 1971) [English Translation: Nonequilibrium Statistical Thermodynamics, in: Studies in Soviet Physics Sciences Series, P. Gray and P.J. Shepherd, eds. (Consultants Bureau, New York, [1974)].
[8] R. Luzzi and A.R. Vasconcellos, J. Stat. Phys. 23 (1980) 539.
[9] V.P. Kalashnikov, Teor. Mat. Fiz. 34 (1978) 412 [Theor. Mat. Phys. (USSR) 34 (1978) 263].
[10] A. Elci, M.O. Scully, A.L. Smirl and J.C. Matter, Phys. Rev. B 16 (1977) 191.
[11] A.C. Algarte, Ph.D. Thesis (UNICAMP, 1983);
A.C. Algarte and R. Luzzi, Phys. Rev. B 27 (1983) 7563.
[12] R. Luzzi and A.R. Vasconcellos, in: Semiconductor Processes Probed by Ultrafast Laser Spectroscopy (Academic, New York, forthcoming).
[13] C.V. Shank, D.H. Auston, E.P. Ippen and O. Teschke, Solid State Comun. 26 (1978) 567.
[14] R.F. Leheny, J. Shah, R.L. Fork, C.V. Shank and A. Migus, Solid State Commun. 31 (1979) 809.
[15] D. von der Linde and R. Lambrich, Phys. Rev. Lett. 42 (1979) 1090.
[16] H. Yoskida, H. Saito and S. Shionoya, Phys. Status Solidi (b) 104 (1981) 331.
[17] M. Pugnet, A. Cornet, J. Collet, M. Brousseau, B.S. Razkirin and G.M. Michailov, Solid State Commun. 36 (1980) 85.
[18] R.J. Seymour, M.R. Junnakar and R.R. Alfano, Solid State Commun. 41 (1982) 657.
[19] V.P. Kalashnikov, Teor. Mat. Fiz. 35 (1978) 127 [Theor. Math. Phys. (USSR) 35 (1978) 362].
[20] J.M. Ziman, Electrons and Phonons (Clarendon, Oxford, 1960).
[21] J. Collet, T. Armand, and M. Pugnet, Phys. Lett. 96A (1983) 368.
[22] E.J. Yoffa, Phys. Rev. B 23 (1981) 1909.
[23] (a) A.R. Vasconcellos and R. Luzzi, Solid State Commun. 44, (1982) 1019;
(b) ibid 49 (1984) 587.
[24] D.N. Zubarev, in ref. [7], Section 25, Chapter IV.
[25] J. Luczka, Phys. Lett. A 69 (1979) 393.
[26] A.R. Vasconcellos and R. Luzzi, to be published.
[27] K. Bohnert, M. Anselment, G. Kobbe, G. Klinshirn, H. Haug, S.W. Koch and S. Schmitt-Rink, Z. Phys. B 42 (1981) 1.
[28] A.R. Vasconcellos and R. Luzzi, Phys. Rev. B 29 (1984) 1052.
[29] J.A. van Vechten, Solid State Commun. 39 (1981) 1825; in: Laser and Electron–Beam Processing of Materials, C.W. White and P.S. Percy, eds. (Academic, New York, 1980).

[30] I. Prigogine, Nature (London) 246 (1973) 67;
G. Nicolis and I. Prigogine, Self-Organization in Nonequilibrium Systems (Interscience, New York, 1977).
[31] R.K. Chang, J.M. Ralston and D.E. Keating, in: Light Scattering Spectra of Solids I, G.B. Wright, ed. (Springer, New York, 1969).
[32] A.C. Algarte, private communication. See also refs. [11] for calculations using a rectangular laser pulse time profile.
[33] H.M. van Driel, Phys. Rev. B 19 (1979) 5928.
[34] A.R. Vasconcellos and R. Luzzi, Phys. Rev. B 27 (1984) 3874.
[35] N. Bloembergen, in: Laser–Solid Interactions and Laser Processing, S.D. Ferris, H.J. Leamy and J.M. Poate, eds. (AIP, New York, 1979).
[36] A.R. Vasconcellos and R. Luzzi, to be published.

Journal of Luminescence 30 (1985) 333–339
North-Holland, Amsterdam

HOT ELECTRON RELAXATION IN $In_{0.53}Ga_{0.47}As$

Kathleen KASH and Jagdeep SHAH

AT&T Bell Laboratories, Holmdel, N.J. 07733, USA

We have determined the hot carrier energy loss rate to the lattice by measuring the cooling curve of a photoexcited hot plasma in $In_{0.53}Ga_{0.47}As$. These measurements were made by using a sensitive upconversion technique to measure the infrared ($\sim$1.2–1.6 μm) luminescence with 10 ps time resolution. We find the carrier energy loss rate to be about an order of magnitude smaller than predicted by a simple model, and surprisingly insensitive to carrier density at high densities.

An understanding of the dynamics of carrier energy loss processes in highly excited semiconductors is of fundamental interest as well as of importance in the design of high speed devices. Time-resolved measurements of luminescence and absorption of hot carriers provide direct information about the evolution of the carrier energy distribution. Picosecond luminescence experiments using a streak camera or an optical Kerr shutter [1,2,3] and subpicosecond pump-and-probe absorption experiments [4,5] have been used to study carrier dynamics in GaAs. From these measurements it was found that the energy relaxation of the electron–hole plasma was slower at higher carrier densities. This trend was attributed [1–5] to screening of the carrier–polar optic phonon interaction, which is the primary energy loss mechanism in GaAs at carrier temperatures above $\sim$40 K. Subsequent work [6,7] has pointed out the importance of non-equilibrium phonons and of using the frequency and wave vector dependent dielectric constant in determining carrier energy loss rates to the lattice.

The focus of the present work is twofold. First, by measuring luminescence spectra with picosecond time resolution, we have determined the carrier energy loss rate to the lattice for the infrared semiconductor $In_{0.53}Ga_{0.47}As$ ($E_g \simeq 0.83$ eV at 2 K). We find that the energy loss rate at high density is smaller than expected for a simple model, just as in the case of GaAs [4,5]. Second, we have investigated the dependence of this rate on carrier density and find that the energy loss rate is surprisingly insensitive to carrier density at high carrier densities.

The measurements reported here were made on a 0.5 μm thick layer of mode-

00-2313/85/$03.30

rate purity ($<10^{16}$ cm^{-3}) $In_{0.53}Ga_{0.47}As$ grown lattice matched to InP by liquid phase epitaxy. The sample was mounted on a cold finger and maintained at 10 K. The experimental setup was described earlier [8] and is shown in fig. 1. Pulses from a synchronously pumped R6G dye laser, of 8 ps duration, 610 nm wavelength, and 4 MHz frequency were split into two pulse trains. One pulse train was focused to an 80 μm diameter spot onto the sample, while the other was sent through a variable delay path. The luminescence from the epilayer and the delayed pulse train were then focused collinearly onto a 1 cm long crystal of $LiIO_3$, cut with the optic axis 40° from the normal, to generate the sum frequency. The time resolution of the up-conversion gating of the luminescence was $\leqslant 10$ ps. The luminescence spectrum over a large energy range at a particular time delay was obtained by scanning the angle of the crystal and the spectrometer simultaneously.

We show in figs. 2 and 3 the luminescence spectra at several different times after an excitation of 2 nJ/pulse and 6 nJ/pulse, respectively. At the shortest time, the spectra are broad, extending over 200 meV above the bandgap energy

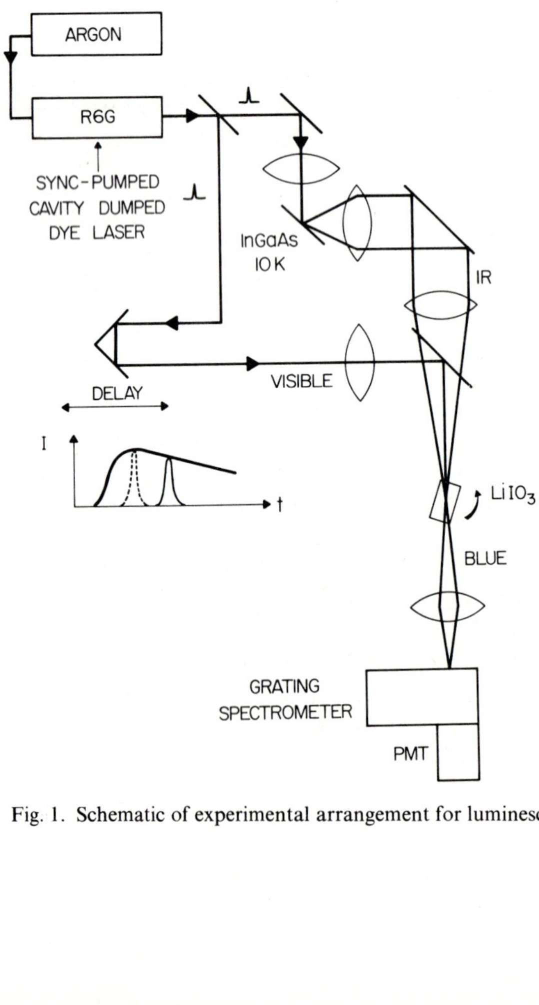

Fig. 1. Schematic of experimental arrangement for luminescence up-conversion.

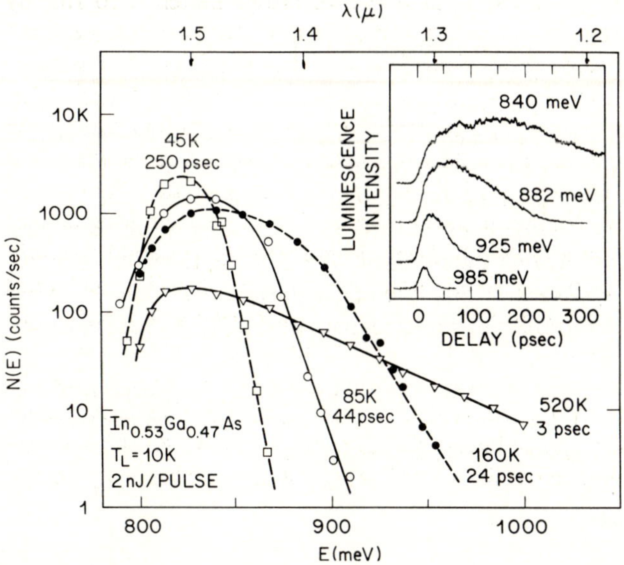

Fig. 2. Luminescence spectra of $In_{0.53}Ga_{0.47}As$ at four time delays, measured after the center of the 8 ps excitation pulse; excitation energy 2 nJ/pulse. The carrier temperatures determined from the spectra are shown. The inset shows time evolution at various photon energies.

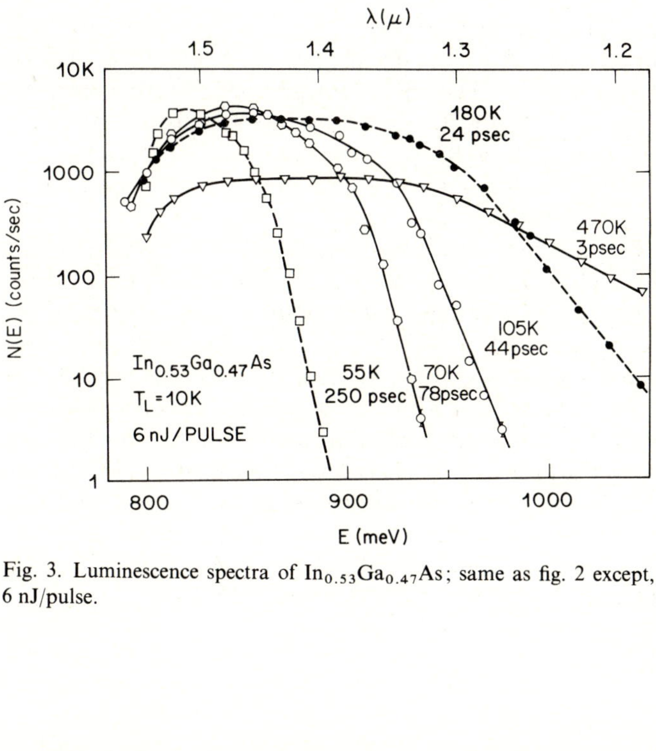

Fig. 3. Luminescence spectra of $In_{0.53}Ga_{0.47}As$; same as fig. 2 except, the excitation energy is 6 nJ/pulse.

(~820 meV at 10 K). The long high energy tails decay exponentially with energy. At longer time delays the spectra become narrower, the exponential tails steeper, and the intensity of the luminescence near the bandgap increases. These results are consistent with those obtained by fixing the luminescence wavelength and varying the time delay. The typical time dependence of the luminescence is shown for several photon energies in the inset to fig. 2.

In order to obtain information about the carrier energy loss rate to the lattice from these data, we need to obtain the carrier temperature from the luminescence lineshape at various time delays. We then compare the experimental "cooling curve" obtained in this manner with the predictions of theoretical models to obtain information about carrier energy loss processes and rates [9].

In order to obtain carrier temperatures from the spectra, a lineshape analysis is required. The spectra, neglecting the effect of the energy dependence of reabsorption through the epilayer, are described by

$$N(E)\mathrm{d}E \sim E^2|M|^2 f_e f_h \rho_e \rho_h \mathrm{d}E,$$

where $N(E)\mathrm{d}E$ is the number of photons observed within $\mathrm{d}E$ of energy E, f_e and f_h are the electron and hole occupation numbers, ρ_e and ρ_h are the densities of states, and $|M|^2$ is the recombination matrix element. The low energy portion of the spectra, where the effects of bandgap renormalization, plasma screening of the exciton enhancement of the matrix element, and band filling are most important, depends in a complicated way on the carrier density. At energies well above the quasi-Fermi energy, where these considerations are less important, the spectra are well described by the factor $E^2\exp(-E/kT_c)$ containing the carrier temperature T_c. The effect of neglecting reabsorption is to lower the apparent temperature slightly (in this case by no more than 10 percent). We have also neglected the slight energy dependence of the factor $|M|^2\rho_e\rho_h$ at high energies. This approximation raises the apparent carrier temperature, again by approximately 10 percent.

We have deduced the carrier temperatures from the spectra at various time delays by fitting the high energy tails to $E^2\exp(-E/kT_c)$ and have plotted these T_c versus the time delay in fig. 4 for the two values of the incident energy/pulse. There are several interesting aspects of the data. We note that temperatures as high as 500 K have been observed. Also, the carriers remain at temperatures considerably higher than the lattice temperature for the longest time delay we have investigated. Finally, within our experimental accuracy, the carrier cooling curves are approximately the same for the two excitation densities investigated.

The cooling curve for a hot plasma is determined by the average rate (per electron–hole pair) at which the plasma loses energy to the lattice, $\langle \mathrm{d}E/\mathrm{d}t \rangle$, and the specific heat per electron–hole pair, $\mathrm{d}\langle E \rangle/\mathrm{d}T$, and can be calculated by integrating the equation

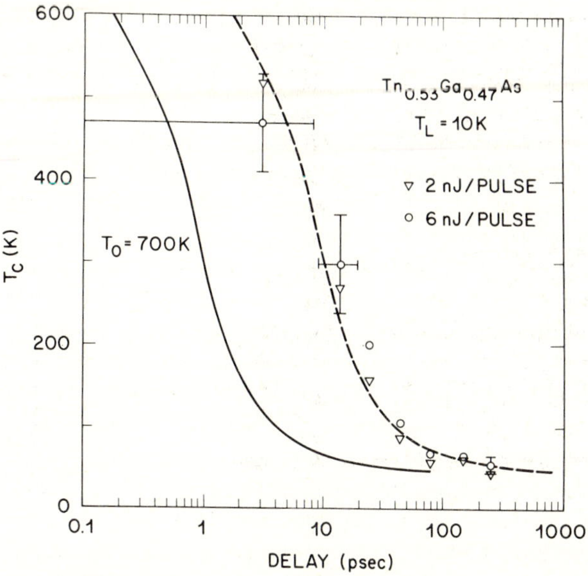

Fig. 4. Carrier temperature versus time. The solid line is calculated using the simple model discussed in the text, with an initial temperature of 700 K. The dashed line is this calculation shifted by a factor of 10 to longer times.

$$\frac{dT}{dt} \times \frac{d\langle E \rangle}{dT} = \left\langle \frac{dE}{dt} \right\rangle.$$

The simplest model one can consider is that of a non-degenerate plasma losing energy to the lattice by optical phonons [9,10]. The cooling curve for InGaAs, calculated by taking into account the reduction in carrier–LO phonon interaction by a factor of three compared to GaAs [11] is shown in fig. 4 as a solid line. The observed cooling curve is approximately parallel to the calculated curve but displaced by approximately a factor of 10 to longer times.

The actual situation in the semiconductor following intense photoexcitation is considerably more complicated than the simple model considered above. It is, therefore, not surprising to find a large difference between the model calculation and the experimental results. There are several factors that need to be included in a realistic model. The effect of degeneracy on the energy loss rates and the specific heat must be considered [12]. The effect of frequency and wave vector dependent dielectric constant must also be considered [6]. An approximate method of taking this second factor into account is to use Fermi–Thomas screening in the static approximation. Finally, the effect of non-equilibrium phonons on the cooling curve must be considered. Pötz and Kocevar [7] have recently reported that this third point is a major factor in the slowed cooling of the plasma. It is important to realize that these effects are interdependent and must all be taken into account simultaneously in a realistic calculation. In addition to these effects, which are intrinsic to a high density plasma, there may be experiment-dependent effects such as cooling due to plasma expansion [13] and a time dependent carrier density because of plasma expansion, stimulated emission and Auger recombination. Thus, a realistic calculation is quite complex. We note, however, that most of the factors mentioned above lead to a slowing of the cooling curve compared to the simple model, in better agreement with the experimental results.

Finally, we comment on the insensitivity of the cooling curve to a change in excitation density. This result is in contrast to the case of GaAs [5] where, at a given time delay, higher temperatures were observed with increasing density. One might also expect that increased screening and hot phonon effects might continue to slow down the cooling curve with increasing density. We propose the following as a possible explanation of the observed insensitivity to excitation density. With increasing excitation density the carrier–polar optical phonon interaction becomes less effective because of screening and hot phonon effects. When this happens, the dominant coupling mechanism is the hole–non-polar optical phonon interaction. The short range non-polar interaction is less sensitive to screening; it is also less sensitive to hot phonon effects because of wave-vector-independent matrix elements and the large wave-vector phonons involved in hole–phonon scattering. As a result, the energy loss rate due to non-polar interaction and, hence, the cooling curve, is less sensitive to changes in photoexcitation density. Since carrier–polar phonon coupling is reduced in InGaAs compared to GaAs [11], and the non-polar interaction is expected to be about the same [14], the insensitivity of the cooling curve to excitation density is likely to occur at a lower density in InGaAs than in GaAs. Further experimental and theoretical work is necessary to check this hypothesis.

In summary, our measurement of the infrared luminescence spectra of the ternary semiconductor $In_{0.53}Ga_{0.47}As$ with 10 ps time resolution clearly shows the existence of a very hot carrier distribution soon after the excitation pulse. These hot carriers subsequently cool by losing energy to the lattice at a rate which is approximately an order of magnitude smaller than that calculated from a simple model and which is insensitive to the photoexcited carrier density at high densities.

Acknowledgment

We wish to thank A.E. DiGiovanni for technical assistance.

References

[1] S. Tanaka, H. Kobayashi, H. Saito and H. Shionaya, J. Phys. Soc. Japan 49 (1980) 1051.
[2] E.O. Goebel, W. Graudzus and P.H. Liang, J. Lumin. 24 (1981) 573; W. Graudzus and E.O. Goebel, Proc. 16th Int. Conf. Physics of Semiconductors I, 555 (Montpelier, 1982).
[3] R.J. Seymour, M.R. Junnarkar and R.R. Alfano, Solid State Commun. 41 (1982) 657.
[4] C.V. Shank, R.L. Fork, R.F. Leheny and J. Shah, Phys. Rev. Lett. 42 (1979) 112.
[5] R.F. Leheny, Jagdeep Shah, R.L. Fork, C.V. Shank and A. Migus, Solid State Commun. 31 (1979) 809.
[6] M. Pugnet, J. Collet and A. Cornet, Solid State Commun. 38 (1981) 531 ; J. Collet, A. Cornet, M. Pugnet and T. Amand, Solid State Commun. 42 (1982) 883.
[7] W. Pötz and P. Kocevar, Phys. Rev. B 28 (1983) 7040.
[8] Kathleen Kash and Jagdeep Shah, Appl. Phys. Lett. 45 (1984) 401.
[9] Jagdeep Shah, J. de Phys. C 7 (1981) 445.
[10] E.M. Conwell, High Field Transport in Semiconductors (Academic Press, New York, 1967).
[11] Jagdeep Shah, R.F. Leheny, R.E. Nahory and M.A. Pollack, Appl. Phys. Lett. 37 (1980) 475.
[12] We have calculated the effect of degeneracy on energy loss rate and specific heat and shown that increasing degeneracy is not responsible for the observed insensitivity of the cooling curves to the excitation density.
[13] S. Modesti, L.G. Quagliano and A. Frova, J. Lumin. 24 (1981) 581.
[14] J.D. Wiley, in: Semiconductors and Semimetals, eds., R.K. Willardson and A.C. Beer (Academic Press, New York, 1975) Vol. 10, p. 101.

Journal of Luminescence 30 (1985) 340–354
North-Holland, Amsterdam

TRANSIENT ANISOTROPY EFFECTS IN THE ABSORPTION SATURATION OF GaAs

J.L. OUDAR and I. ABRAM

Laboratoire de Bagneux – CNET, 196 rue de Paris, 92220 Bagneux, France

A. MIGUS, D. HULIN and J. ETCHEPARE

Laboratoire d'Optique Appliquée, Ecole Polytechnique – ENSTA, 91120 Palaiseau, France

The absorption of polarized light imparts a transient anisotropy on GaAs due to the preferential orientational excitation of electron–hole pairs. We present a series of experiments designed to measure the relaxation of the wavevector orientation of these photoexcited electron–hole pairs.

1. Introduction

The interaction of polarized light with a material system is inherently anisotropic, since the polarization of the light beam imposes a preferential direction in the orientation of the excited states produced through light absorption. On the other hand, the material system is subject to random and sudden processes (such as thermal fluctuations, or interactions and "collisions" among the excitations present in the material) which tend to randomize the light-excited states [1] and thus destroy their anisotropy. A study of the light-induced anisotropy in a material, therefore, provides information on the relaxation processes which its excited states may undergo.

At low light intensities the anisotropy induced by a "pump" light beam is very weak, and thus the propagation of a "test" beam in the material is not appreciably affected. In other words, at low intensities the superposition principle familiar from linear optics is valid, and the two beams propagate inside the material without interacting on each other. At higher intensities however a large number of excitations may accumulate and a test beam will be affected by the presence or the passage of a pump pulse. Thus the principle of superposition is no longer valid, and the interaction of light with the material is now nonlinear. This fact permits us to use the whole experimental and theoretical arsenal of nonlinear optics in the study of relaxation phenomena [2].

A case in point is the study of the relaxation of photoexcited carriers in semiconductors [3,4]. In this paper we shall describe a series of nonlinear optical experiments on the relaxation of hot electrons in GaAs. Throughout the paper we shall refer, therefore, to the properties of direct gap semiconductors, but similar phenomena are also observable when indirect gap electronic transitions are involved [5]. The paper is organized as follows: In section 2, we review the nonlinear optical response that results from the accumulation of photoexcited electron–hole pairs. In section 3 we discuss the anisotropic absorption saturation which is induced in a semiconductor by linearly polarized light. In section 4 we describe four-wave mixing-type experiments which bring into play the light-induced anisotropy, and in section 5 we present a summary of our results and our conclusions.

2. Absorption saturation

A photon of frequency ω incident on a semiconductor may be absorbed and give rise to the excitation of an electron–hole pair, if its energy is higher than the bandgap of the semiconductor E_g

$$\hbar\omega \geqslant E_g. \tag{2.1}$$

According to the optical selection rules, since the photon wavevector is very small, the electron and the hole have equal and opposite wavevectors

$$\boldsymbol{k}_e = -\boldsymbol{k}_h \tag{2.2}$$

whose magnitudes satisfy

$$\hbar k_e^2/2m_e + \hbar k_h^2/2m_h = \hbar\omega - E_g, \tag{2.3}$$

where m_e and m_h are the electron and hole effective masses respectively. Once a pair state is created, it cannot be excited a second time, because of the Pauli exclusion principle. Thus the transition is optically bleached (that is, it cannot contribute to the absorption of a second photon) unless the electron–hole pair is scattered out of that state through a change of the electron (or hole) wavevector either in direction or in magnitude. A change in the magnitude of the wavevector corresponds to a change in the kinetic energy of the corresponding particle.

When using unpolarized light successive photons populate all possible directions of the wavevectors which satisfy eqs. (2.2) and (2.3) following the cylindrical symmetry imposed by the propagating beam, until all allowed transitions of energy $\hbar\omega$ are bleached. The optical transmission of the sample thus progressively passes from a low value (because of absorption at $\hbar\omega$) to a high value (because of the saturation of that absorption). This can be described mathematically by a formalism similar to the one used for absorption saturation in a statistical ensemble of two level systems. That is the absorption coefficient of the material at frequency ω which is given by

$$\alpha(\omega) = \alpha_0(\omega)(1 - f^{e}(\omega) - f^{h}(\omega)), \tag{2.4}$$

where α_0 is the low-intensity value of α, while $f^{e}(\omega)$ and $f^{h}(\omega)$ are respectively the fractions of electrons and holes, optically acessible at frequency ω, that are populated by light. Clearly as the fractional occupation increases, the absorption coefficient decreases and drops to zero when

$$f^{e}(\omega) + f^{h}(\omega) = 1. \tag{2.5}$$

From that point on, the material ceases to absorb and no more electron–hole pairs can be created.

In a preliminary experiment [6], the bleaching induced on a thin sample of GaAs by an intense beam was observed by monitoring the transmission of that beam as a function of its intensity. The incident beam consisted of pulses obtained from an optical parametric generator driven by the frequency-doubled output of a mode-locked YAG laser. The output of the parametric generator was tunable between 0.7 and 2 μm, with 10 μJ pulse energy, 7 ps pulse duration and 3 Hz repetition rate. It was generally used at wavelengths slightly above the GaAs bandgap (0.77 to 0.83 μm). The sample was a 1.5 μm thick GaAs layer sandwiched between two $Al_xGa_{1-x}As$ layers ($x = 0.35$) grown by molecular beam epitaxy, and was kept at 77 K in all experiments. The resulting curves (see fig. 1) are typical of the saturation of absorption through the accumulation of optically excited electron–hole pairs (eq. (2.4)). They are compatible with

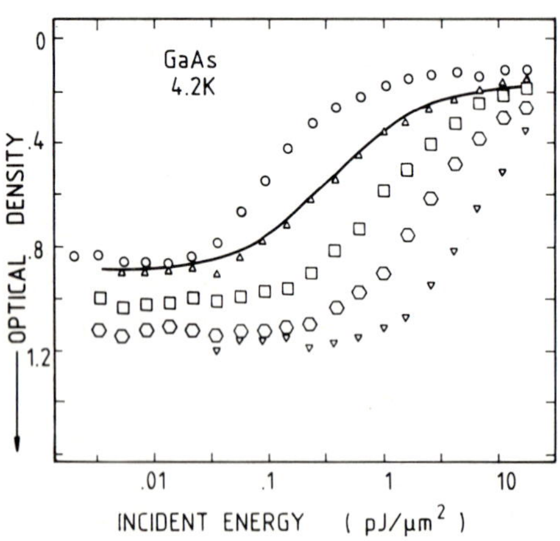

Fig. 1. Absorption saturation in GaAs at different wavelengths. Excess above the bandgap is: (○) 1 meV; (△) 10 meV; (□) 40 meV; (⬡)100 meV; (▽) 270 meV. The curve at 10 meV was fitted using a thermalized distribution at a temperature $T = 90$ K.

several simple models for the functional dependencc of $f^{e,h}$ on the incident pulse energy. However, these curves contain little information which could permit one to distinguish between these models and thus understand the dynamics of the photo excited hot clectron–holc pairs.

While the optical absorption process causes the accumulation of electron–hole pairs, energy relaxation removes some pairs from the states accessible by light (i.e., satisfying eq. (2.3)) and populates electron and hole states at different energies, bringing about a relaxation-induced bleaching at those energies. In other words $f^{e,h}(\omega')$ at frequencies ω' other than that of the incident beam ω can become nonzero, while $f^{e,h}(\omega)$ diminishes. We note that since the creation of pairs at $\omega' \neq \omega$ does not proceed by direct absorption of light, it does not cease when condition (2.5) is met at ω'. It is possible to overpass the value of 1 (one) and obtain thus negative absorption coefficients in eq. (2.4) implying the possibility of amplification at those frequencies.

By studying the temporal and spectral characteristics of the bleaching process, we may get a handle on the electronic energy relaxation processes that occur in the semiconductor. This can be accomplished in a pump-and-test type configuration. A short intense light pulse ("pump") of a given frequency, excites the material and induces a bleaching. A weaker second pulse ("test"), which can be scanned spectrally or temporally with respect to the pump, monitors the transmission of the sample and thus yields information on the extent of the bleaching.

In a series of experiments [7] performed in our laboratory the pump and probe pulses were derived from two independent optical parametric generators, however both pumped simultaneously and in parallel by the same frequency-doubled, mode-locked YAG laser pulse. Thus the two generators were independently tunable, but were synchronized. The pump pulse could be advanced or retarded with respect to the probe pulse by an optical delay line which could scan 250 ps at a time. In order to facilitate the interpretation of the experimental results, the probe pulse was attenuated by a factor of 100 to 1000 so that its interaction with the semiconductor was linear, that is, it did not induce any measurable amount of bleaching by itself.

Figure 2 presents the transmission of an excited sample as a function of the probe beam frequency. The pump beam is fixed at 1.53 eV, i.e., approximately 15 meV above the bandgap. The pump and probe pulses arrive simultaneously on the sample. We note that the bleaching is not confined to the pump frequency. In fact it is much more pronounced at energy values lower than 1.53 eV, but is also present at higher energies. The energy distribution of the electron–hole pairs corresponding to this bleaching can be fitted by a thermal distribution at a temperature of about 120 K, under the assumption that the electron–hole correlations which give rise to excitonic effects (e.g., the transmission dip at 1.508 eV for $I_{pump}=0$) are completely quenched (screened) at the electron densities of these experiments ($N=10^{17}\text{cm}^{-3}$).

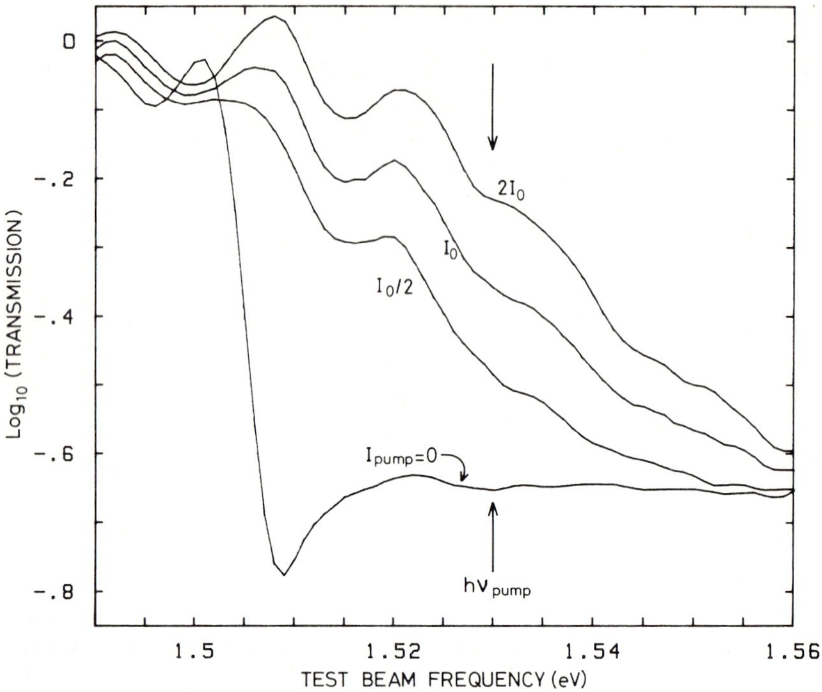

Fig. 2. Bleached transmission spectrum of GaAs. For the pump $I_0=100\,\mu\text{J/cm}^2$.

That is

$$f^{i}(E)=\left[1+\exp\left(\frac{E-\mu_i}{kT}\right)\right]^{-1}, \tag{2.6}$$

where μ_i is the chemical potential of particle i, i = e, h. In most calculations it is sufficient to consider a simplified band structure for GaAs consisting of one valence and one conduction band. The reason is that even though there are two valence bands (heavy and light holes) there is a very large difference in their effective masses, implying a much larger density of states for the heavy holes than for light holes of the same energy.

The observation of a quasi-thermalized distribution indicates that within the pulse duration of 7 ps there is an extensive energy redistribution of the photo-excited electrons and holes, which closely resembles the thermalization process. Clearly, if the effective temperature of the electron–hole plasma were 0 K, absorption saturation at the pump frequency (measured by 7 ps pulses) would occur only after the conduction and valence bands were completely filled with electrons and holes respectively, up to the energy levels which directly interact with light. However, since the plasma is found at a finite temperature, the quasi-thermalized distribution has an exponential tail towards higher energies. Thus bleaching sets in progressively as the incident pulse energy is increased. This nonlinear optical mechanism has been termed "absorption saturation by band filling" [2].

A temporal study with 7 ps pulses clearly cannot give much information on the electron and hole relaxation dynamics by which the quasi-thermalized distribution is attained. It can elucidate however the subsequent evolution of that distribution. Figure 3 present the results of experiments in which the relative time delay between the pump and test pulses (both at the same frequency) is scanned. The decay of the absorption saturation at each energy is compatible with the decay of the quasi-thermal distribution, essentially because of the radiative recombination of the electron–hole pairs. Cooling of the distribution to the bath temperature (77 K) occurs within a few picoseconds and is thus at the limit of our experimental resolution [7].

Thus in order to study the relaxation dynamics of the electrons and holes, femtosecond temporal resolution would be necessary. To this end, a series of experiments was performed [8], in which two sub-picosecond pulses were used as pump and test. Their cross-correlation (fig. 4(b)) was fitted to a Gaussian

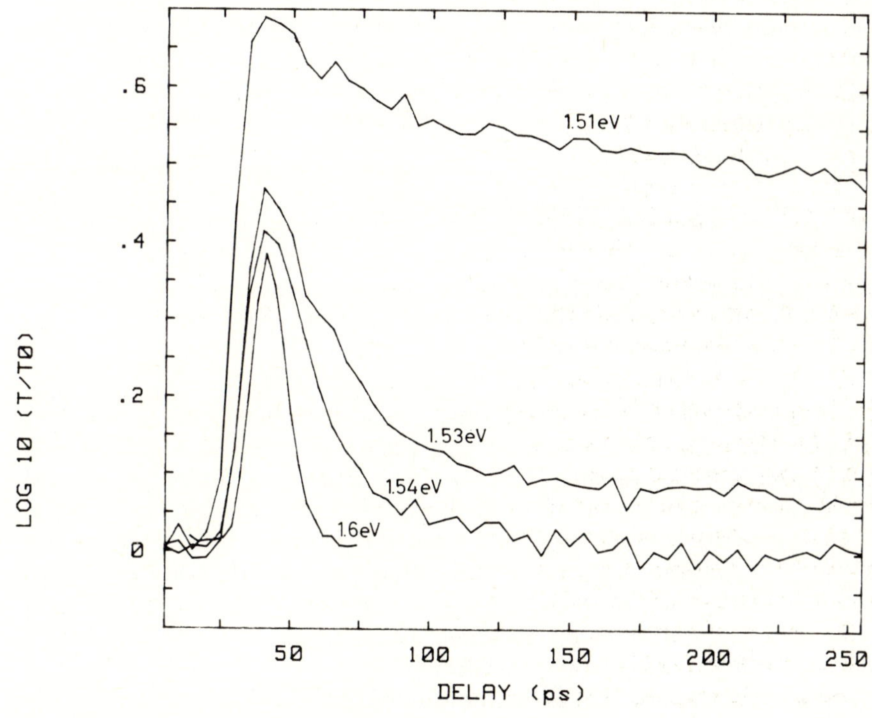

Fig. 3. Bleached transmission of GaAs vs. test delay, at different wavelengths.

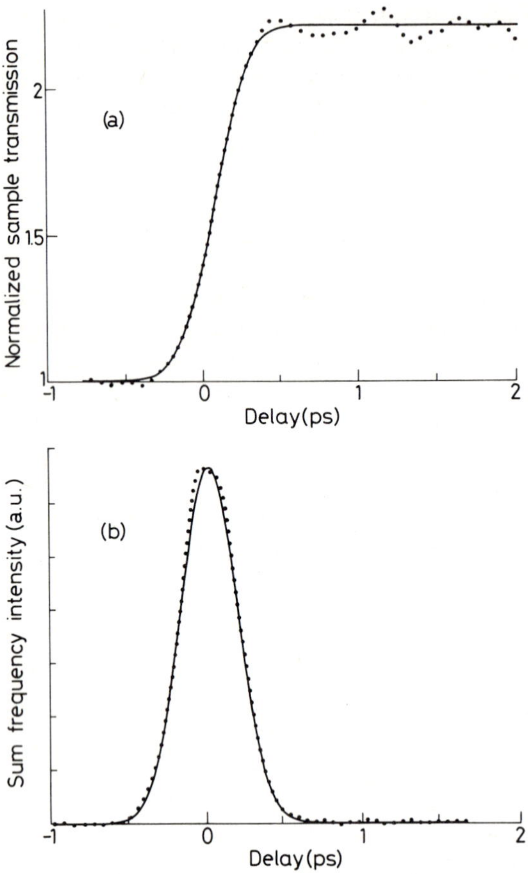

Fig. 4. (a) Bleached transmission of GaAs vs. test delay. Solid line is integral of curve 4(b). (b) Pump–test intensity cross-correlation measured by sum-frequency mixing in KDP. Solid curve is $G(\delta)=\exp-(\delta^2/\delta_0^2)$ with $\delta_0=0.25$ ps.

$G(\delta)=\exp(-\delta^2/\delta_0^2)$ with $\delta_0=0.25$ psec, which represents the instrument response. The pump pulse had an energy density of $100\,\mu J/cm^2$ exciting thus an estimated 6×10^{17} electrons/cm^3, and was centered at 806 nm ($\sim$1.54 eV) that is about 25 meV above the GaAs bandgap. Its spectral width was about 12 nm ($\sim$25 meV). The test pulse had a broad spectrum, but its detection was done at the pump wavelength. The absorption saturation observed is given in fig. 4(a). We note that the experimental points are fitted well by the time integral of the pulse cross-correlation function. This suggests that the absorption changes are roughly proportional to the number of electron–hole pairs created, and that relaxation of the distribution established in 250 fs by our relatively broad-band pump is very slow near the bandgap.

Amelioration of the temporal resolution to study the electron–hole relaxation, is not yet technically possible. We can take advantage however of the anisotropic nature of the interaction between the semiconductor and light to examine redistribution in the photoexcited pairs, within the temporal resolution available.

3. Anisotropic absorption saturation

When linearly polarized light is incident on a cubic III–V semiconductor at a frequency above the direct bandgap, it can excite an electron–hole pair only to the extent that the polarization vector has a projection on the transition matrix element of the pair. Since the orientation of this vectorial matrix element depends on the electron and hole wavevectors, polarized light will excite preferentially pairs whose electron (or hole) wavevector points in a given direction [9]. Using for the valence band the wavefunctions that originate from the p-type atomic orbitals quantized along the wavevector, and for the conduction band the s-type wavefunctions, it can be shown that the probability of exciting an electron–(heavy) hole pair of wavevectors $\pm \boldsymbol{k}$ with linearly polarized light, is proportional to

$$|\hat{\boldsymbol{e}} \cdot \boldsymbol{p}_{\mathrm{ehh}}|^2 \propto \sin^2 \theta, \tag{3.1}$$

where θ is the angle between the wavevector $\boldsymbol{k}$ and the electric field vector $\hat{\boldsymbol{e}}$. That is, wavevectors perpendicular to the polarization vector are preferentially excited. This probability distribution is a condition which is in addition to the momentum and energy conservation conditions expressed by eqs. (2.2) and (2.3).

The electron and hole distributions $f^{\mathrm{e,h}}$ have cylindrical symmetry along the polarization vector $\hat{\boldsymbol{e}}$; therefore each distribution can be expanded in a series of Legendre polynomials of $\cos \theta$. That is

$$f(\boldsymbol{k}) = f_0(\omega) + f_2(\omega) P_2(\cos \theta) + f_4(\omega) P_4(\cos \theta) + \cdots, \tag{3.2}$$

where P_i is the ith order Legendre polynomial and the odd orders are discarded by symmetry. Since the transition probability is proportional to

$$\sin^2 \theta = \tfrac{2}{3}(1 - P_2(\cos \theta)) \tag{3.3}$$

the interaction of light with the semiconductor excites mainly the angular distributions f_0 and f_2, because of the orthogonality of the Legendre polynomials. In particular, the probability of exciting a pair with wavevectors in the direction θ, is given by

$$\frac{\mathrm{d}}{\mathrm{d}t} f(\theta) = \tfrac{3}{2}\alpha_0 I \sin^2 \theta (1 - f(\theta)), \tag{3.4}$$

where $f(\theta) = f^{\mathrm{e}}(\theta) + f^{\mathrm{h}}(\theta)$. Substituting eq. (3.2) and performing the angular averages, we obtain

$$\frac{\mathrm{d}}{\mathrm{d}t} f_0 = \alpha_0 I(1 - f_0 + \tfrac{1}{5} f_2), \tag{3.5a}$$

$$\frac{\mathrm{d}}{\mathrm{d}t} f_2 = -\alpha_0 I(1 - f_0 + \tfrac{5}{7} f_2 - \tfrac{2}{7} f_4), \tag{3.5b}$$

and so on.

An electron–hole pair excited by an incident pump beam contributes to the bleaching measured by a test beam, since its contribution to the absorption coefficient of the test beam is

$$\alpha = \alpha_0(1 - f(\theta)) \sin^2 \theta', \tag{3.6}$$

where θ and θ' are the angles between the electron (or hole) wavevector $\boldsymbol{k}$ and the polarization vectors of the pump and test beam respectively. Summing the contributions of all pairs and using the properties of the Legendre polynomials, we may obtain the absorption coefficient for a test beam whose polarization forms an angle χ with the pump polarization as

$$\alpha(\chi) = \alpha_0(1 - f_0 + \tfrac{1}{5} f_2 p_2(\cos \chi)). \tag{3.7}$$

Clearly, the saturation of the semiconductor absorption by a polarized beam imparts an anisotropy on the semiconductor, since the propagation of a test beam through the material is different depending on its polarization. This anisotropy persists, until scattering processes randomize the directions of the wavevectors of the excited electron–hole pairs, relaxing them into an isotropic distribution. Such a time evolution corresponds to the decay of the second-order angular distribution f_2 and the persistence of only the zeroth-order (isotropic) term f_0.

The absorption anisotropy corresponding to eq. (3.7) is given by

$$\alpha_\parallel - \alpha_\perp = \tfrac{3}{10} \alpha_0 f_2, \tag{3.8}$$

and is thus directly proportional to the second-order angular distribution, f_2. The existence of this anisotropy implies that the propagation of a test beam through the excited material, brings about a rotation of its polarization in addition to the attenuation of its intensity. For a small rotation angle ε, we have

$$\varepsilon = \tfrac{1}{4} \sin 2\chi \int_0^d (\alpha_\parallel - \alpha_\perp)\, \mathrm{d}z, \tag{3.9}$$

where the integral is over the sample thickness d. Thus, a time-resolved measurement of the polarization rotation monitors the time-evolution of the second-order electron–hole angular distribution, which in turn gives information on the relaxation processes that bring about a randomization of the electron and hole wavevectors. Such an experiment is much more sensitive than those described in section 2 since it examines directly only the part of the distribution that may decay, subtracting automatically the nondecaying "background bleaching" due to the isotropic part of the distribution.

Such an experiment was performed [8] using two sub-picosecond pulses with a Gaussian cross-correlation of 0.25 ps width (as described in section 2), whose polarizations formed an angle of $\chi \approx 30^\circ$. The polarization of the test pulse transmitted through the sample was analyzed into two mutually perpendicular components, and the ratio of their intensities (which is measure of the angle of polarization rotation) was plotted as a function of the relative time delay between the pump and test pulses. The curve obtained is slightly wider than the cross-correlation Gaussian at the two pulses and is asymmetric towards the positive delay times (i.e., test after pump), indicating that the anisotropy decays at time scales comparable to the duration of the pulses.

In order to extract the information on the decay of the anisotropy from these curves we must consider explicitly the excitation and relaxation kinetics through equations such as (3.5). If we assume that f_2 is small compared to 1 and to $1-f_0$, and that it decays exponentially to zero, eqs. (3.5) can be re-written as

$$\frac{\mathrm{d}}{\mathrm{d}t} f_2 = -\frac{\mathrm{d}}{\mathrm{d}t} f_0 - \frac{f_2}{T_2}, \tag{3.10}$$

where T_2 is the exponential decay time of f_2. Since the time derivative of the isotropic bleaching (i.e. $(\mathrm{d}/\mathrm{d}t) f_0$) was determined experimentally to be fitted well by the cross-correlation of the two pulses, the time-evolution of f_2 can be calculated as the convolution of the corresponding Gaussian with a decaying exponential of characteristic time T_2. A fit of the experimental points gives a value of $T_2 \approx 0.19$ ps. This value may be somewhat modified when the model of eq. (3.10) is further refined. It represents however the first direct measurement

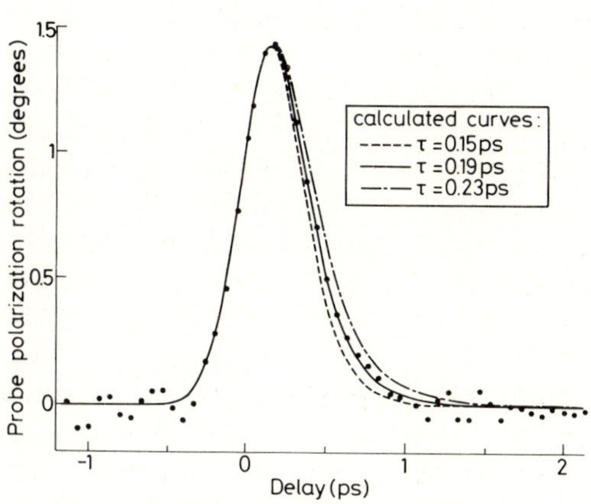

Fig. 5. Amplitude of polarization rotation of transmitted test pulse vs. pump–test relative delay.

of the orientational relaxation of the photon-excited pairs, probably due mainly to carrier–carrier collisions.

4. Four-wave mixing

The accumulation of photoexcited electron–hole pairs, can give rise also to multi-beam nonlinear interactions in addition to absorption saturation. In particular, when two noncolinear beams (pump and test) are incident on the material, a third beam (signal) is generated, such that its wavevector $\boldsymbol{k}_s$ obeys

$$\boldsymbol{k}_s = 2\boldsymbol{k}_p - \boldsymbol{k}_t, \tag{4.1}$$

where $\boldsymbol{k}_p$ and $\boldsymbol{k}_t$ are the pump and test wavevectors respectively. The generation efficiency of the signal beam is directly related to the efficiency of the absorption saturation (bleaching) process, and thus, four-wave mixing experiments can provide information on the electron–hole relaxation dynamics. Such experiments involve generally the detection of the signal beam propagating in a background-free direction, and are thus much more sensitive than classic bleaching experiments which consist essentially of measuring a small intensity variation of a relatively intense test beam.

One way of describing four-wave mixing is in terms of transient gratings [2]: The interference of the pump and test beams is imprinted on the semiconductor as a sinusoidal variation in the density of excited pairs. This constitutes a bleaching grating of wavevector $k_p - k_t$, which can subsequently diffract the incident pump photons in the direction k_s. Two types of bleaching gratings can be distinguished, which bring into play the isotropic and the anisotropic absorption saturation respectively. In the former case, the pump and the test beam polarizations are parallel, and their interference gives rise to a sinusoidal spatial variation of the light intensity. The excitation of the semiconductor in the region of the interference, thus, involves a spatial periodic variation in the concentration of photoexcited electron–hole pairs. The diffracted signal beam carries thus information on the overall electron–hole population, which is given mainly by the isotropic angular distribution f_0. In the latter case, a pump and test beam with perpendicular polarizations interfere, giving rise to a region of spatially uniform intensity but with a periodic variation in the direction of the polarization. This interference pattern imprints on the semiconductor a spatially uniform electron–hole population, however with a variation in the orientation of their wavevectors, since in the excited region the electron–hole pairs obey locally eq. (3.1). This grating involves therefore a periodic variation of the absorptive anisotropy induced on the semiconductor by the polarized light. The signal beam can be diffracted only as long as this anisotropy persists. It carries therefore information on the wavevector re-orientation of the electron–hole pairs described by the decay of the second-order angular distribution f_2.

Some of the properties of four-wave mixing originating from isotropic and anisotropic absorption saturation were investigated in two series of experiments using 7 ps long pump and test pulses [10]. In time-resolved experiments, the test was delayed optically with respect to the pump and the intensity of the signal generated was plotted as a function of the relative time delay (fig. 6). When the polarizations of the two pulses were parallel, the curve obtained was close to the coherence profile of the two pulses, and was thus significantly narrower than in the perpendicular configuration; in this latter case the curve was close to the intensity cross-correlation of the two pulses.

The difference between the two types of curves can be understood if we consider the temporal characteristics of the pump–test interference pattern.

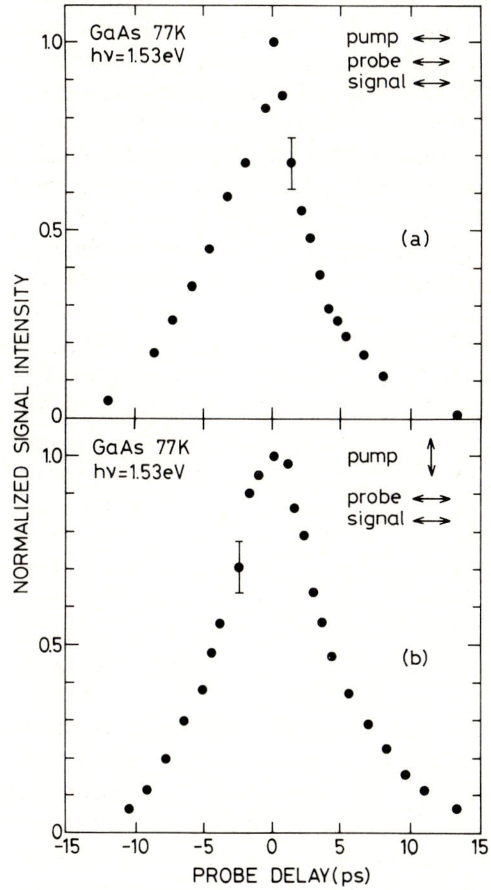

Fig. 6. Four-wave-mixing signal intensity vs. pump–test relative delay: (a) pump and test polarizations parallel, (b) pump and test polarizations perpendicular.

When the two pulses overlap within their coherence time, the pattern persists throughout their duration, and an efficient saturation grating can be constituted both in the perpendicular and in the parallel configuration. When the two pulses overlap outside their coherence time, the interference fringes shift rapidly. A grating with a slow decay time integrates all positions of the fringes and thus quickly becomes diffuse and ceases to diffract. On the other hand, a grating that decays faster than the fringe shift is constantly re-constituted (albeit with a phase shift) and thus continues to diffract. Thus, the fact that no signal is observed in the parallel configuration outside the coherence profile of the pulses, confirms that the decay time of f_0 is long compared to the picosecond time scale. Similarly the larger width of the curve in the perpendicular configuration confirms that the re-orientation time for the electron and hole wavevectors is in the sub-picosecond regime. However the 7 ps duration of the pump and test pulses in these experiments does not permit a quantitative measurement of this relaxation times.

In a second series of experiments, the efficiency of the signal beam generation $\eta = E_{\text{signal}}/E_{\text{test}}^{\text{in}}$ was measured as a function of pump energy density F_{pump} at various photon energies (fig. 7). The conversion efficiency is significantly smaller for perpendicular polarizations ($\eta_\perp$) than for parallel ones ($\eta_\parallel$). This difference reflects essentially the much faster decay of the anisotropic absorption saturation, even though geometric factors account also for part of the difference. An interesting point on the curves in fig. 7, is that they cannot be described by a

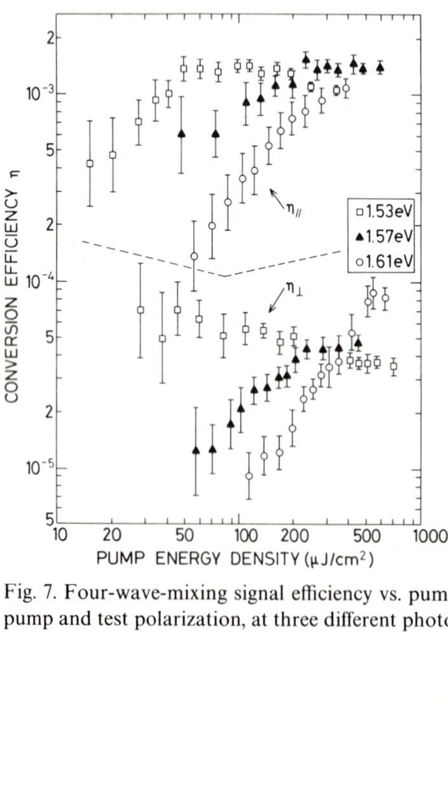

Fig. 7. Four-wave-mixing signal efficiency vs. pump energy density for parallel and perpendicular pump and test polarization, at three different photon energies.

third-order susceptibility $\chi^{(3)}$ often used in describing four-wave mixing. In the $\chi^{(3)}$ regime one would expect η to increase as $(F_{\text{pump}})^{\gamma}$ with $\gamma = 2$, while these data are described by γ ranging from 1.2 to -0.25. These deviations are due essentially to the accumulation of an electron–hole population which, because of saturation, reduces the depth of the periodic modulation which constitutes the grating, reducing thus its efficiency. Other effects could also contribute to this deviation, such as the power broadening of the valence-to-conduction band transitions, or the variation of the scattering rates due to the increased electron–hole pair density. Further experimentation is however necessary to sort out the different contributions.

5. Conclusions

We have presented in this paper the broad lines of absorption saturation in semiconductors, underlining especially the anisotropic saturation which is induced by linearly polarized light. This anisotropy results from the fact that electron–hole pairs with wavevectors perpendicular to the polarization are preferentially excited, and thus the anisotropy persists until the electron–hole wavevectors are randomized. Different types of experiments have been devised, which probe directly the anisotropic part of absorption saturation, thus giving information on the scattering processes which randomize the electron–hole wavevectors.

One such experiment consists of the time-resolved measurement of the polarization rotation a test beam undergoes as it propagates through a GaAs sample excited by a polarized pump pulse. In a simple model for this experiment, the angle for the rotation is directly proportional to the anisotropic part of the electron–hole distribution; applying this model to the experimental results a value of $T \approx 0.2$ ps is obtained for the decay time constant of the anisotropic wavevector distribution, at carrier densities of about $6 \times 10^{17}/\text{cm}^3$.

Another experiment involves the observation of a four-wave mixing signal, when the two incident pulses (pump and test) have perpendicular polarizations. In this configuration the interference pattern of the two pulses consists of a periodic variation of the polarization, thus the spatial distribution of electron–hole pairs excited by this pattern corresponds to a grating composed of the anisotropic part of the induced absorption saturation. The fact that a four-wave mixing signal could be observed outside the overlap of the coherence times of the two pulses confirms that the decay time of the anisotropy is an order of magnitude shorter than the pulse coherence time of 5 ps.

References

[1] J. Shah, J. de Phys. 42 (1981) C7-445.

[2] B.S. Wherett, A.L. Smirl and T.F. Bogess, IEEE J. Quantum Electron. QE-19 (1983) 680.

[3] C.V. Shank, R.L. Fork, R.F. Leheny and J. Shah, Phys. Rev. Lett. 42 (1979) 112;
D. von der Linde and R. Lambrich, Phys. Rev. Lett. 42 (1979) 1090.

[4] C.L. Tang and D.J. Erskine, Phys. Rev. Lett. 51 (1983) 840.

[5] A.L. Smirl, T.F. Boggess, B.S. Wherett, G.P. Perryman and A. Miller, Phys. Rev. Lett. 49 (1982) 933; IEEE J. Quantum Electron. QE-19, (1983) 690;
Z. Vardeny and J. Tauc, Phys. Rev. Lett. 46 (1981) 1223; 47 (1981) 700.

[6] C. Minot, J. Chavignon, H. Le Person and J.L. Oudar Solid State Commun. 49 (1984) 141.

[7] J.L. Oudar, J. Dubard, R. Raj, I. Abram (to be published).

[8] J.L. Oudar, A. Migus, D. Hulin, G. Grillon, J. Etchepare and A. Antonetti, Phys. Rev. Lett. 53 (1984) 384.

[9] B.P. Zakharchenya, D.N. Mirlin, V.I. Perel' and I.I. Reshina, Sov. Phys. Usp. 25 (1982) 143.

[10] J.L. Oudar, I. Abram and C. Minot, Appl. Phys. Lett. 44 (1984) 689.

Journal of Luminescence 30 (1985) 355–368
North-Holland, Amsterdam

BOUND EXCITON FORMATION AND EXCITONIC LOCALIZATION IN SEMIMAGNETIC SEMICONDUCTORS

A.V. NURMIKKO

Brown University, Division of Engineering Providence, RI 02912, USA

We review recent experimental work in which kinetics of exciton localization at impurities and alloy potential fluctuations are studied by picosecond laser spectroscopy, specifically in the mixed crystals $Cd_{1-x}Mn_xSe$ and $Cd_{1-x}Mn_xTe$.

1. Introduction

Following recent developments in ultrashort pulse laser technology, there are now many examples of applications to transient spectroscopy of semiconductors where distinctly new physical information has been obtained. In this article we wish to focus on picosecond spectroscopy of excitons while discussing selected aspects of exciton kinetics in the so-called 'semimagnetic' semiconductors, specifically $Cd_{1-x}Mn_xSe$ and $Cd_{1-x}Mn_xTe$. These II–VI materials, while possessing many of the conventional features of their parent compound semiconductors, have additional spin–spin interactions which lead to a number of interesting and useful properties [1]. The emphasis below, however, is to show more generally that the experimental techniques in question can yield real time kinetic information of excitonic energy relaxation processes with considerable detail. Since the article is not intended as a general review of exciton phenomena, only specifically relevant references will be made to the vast exciton literature [2].

The organization of the paper is as follows. In section 2 we review some of the technical aspects of a two-laser excite–probe approach, termed here as photomodulation spectroscopy. In section 3 we show by examples of recent work how the competition for free exciton confinement by impurities and alloy potential fluctuations may, in some instances, be separately identified from time-resolved spectra. Some concluding remarks follow in section 4.

* This work was supported by AFOSR.

00-2313/85/$03.30

2. Photomodulation spectroscopy – instrumentation

The experimental apparatus used throughout the work described here employs a pair of synchronously pumped cw modelocked dye lasers in an excite–probe configuration as shown in fig. 1. The apparatus has been previously used by us to examine free electron and impurity bound exciton kinetics in several compound semiconductors [3]. The two independently wavelength tunable lasers provide the excitation and direct optical interrogation of semiconductor samples at approximately 80 MHz repetition rate, usually with a small angle separating the two beams. The individual, approximately 6 ps long pulses from each laser can usually be cross-correlated to better than 10 ps with good consistency, and their optical polarization readily controlled. The photon energies can be independently varied with a spectral resolution typically close to 0.3 meV. The arrival of a pulse of excitation at $t=0$ injects, for example, an initially monoenergetic exciton gas into a sample. The presence of this excitation is measured as a weak optically induced modulation of the transmitted (or reflected) probe pulse intensities at a variable delay time t_d. All of the transient information is obtained optically by varying the delay time. In fact, our photodetectors are conventional photodiodes with no particular high speed performance. The excitation beam is electro-optically chopped at a frequency of 10 MHz to take advantage of a 'quiet' region in the dye lasers noise spectrum. With lock-in detection and digital signal averaging, our equipment permits the measurement of a fractional change in the probe beam transmission (or reflection) coefficient to better than 10^{-7} with a one-second integration time. This

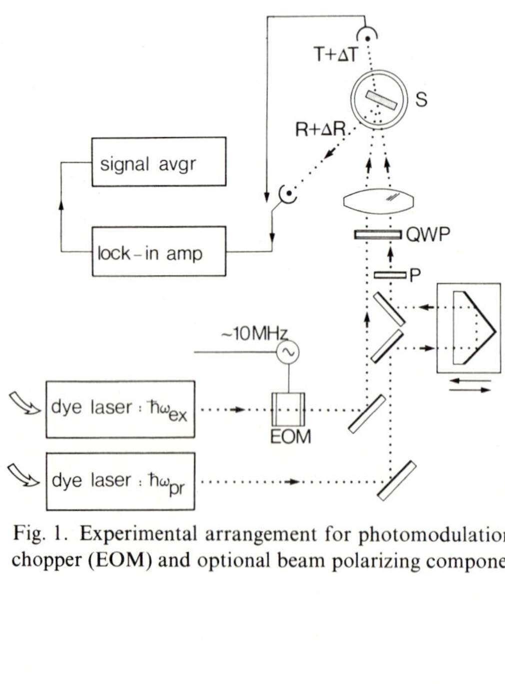

Fig. 1. Experimental arrangement for photomodulation spectroscopy including the electro-optic chopper (EOM) and optional beam polarizing components (QWP and P).

value is near the shot noise limit for such fractional change δ which can be readily calculated from

$$\delta=(4\hbar\omega\Delta\nu/P_0\eta)^{1/2}, \tag{1}$$

where P_0 refers to the average probe power (typically not exceeding few mW), η the detector quantum efficiency, and $\Delta\nu$ the detection bandwidth.

This arrangement differs from measurements by transient photoluminescence by being essentially a modulation technique with added sensitivity to interactions amongst the excited state populations. Further, instead relying on spontaneous emission, the use of a weak probe beam permits the information be collected by induced emission in a well collimated direction of the probe beam. Thus, for example, there are usually no difficulties in having the excitation and probe beams at comparable photon energies. In those instances where sample surfaces scatter significant amount of pump light, addition of a low frequency chopper into the probe beam (with a serially connected second lock-in amplifier) readily eliminates the problem. The overall data acquisition is most conveniently obtained through microcomputers, including continuous monitoring of the laser output powers. In this way a change in the transmitted probe beam intensity (in transmission mode) is measured with and without the effect of excitation

$$\Delta I_{\mathrm{T}}(\hbar\omega_{\mathrm{p}},\tau)=I_{\mathrm{T}}(\hbar\omega_{\mathrm{p}},\tau)\big|_{\text{with exc.}}-I_{\mathrm{r}}(\hbar\omega_{\mathrm{p}},\tau)\big|_{\text{no exc.}}\,. \tag{2}$$

The data acquisition apparatus usually directly normalizes the measurement yielding

$$\Delta I_{\mathrm{T}}/I_{\mathrm{T}}\equiv\Delta T/T=-\Delta\alpha\cdot\ell, \tag{3}$$

where T is the transmission coefficient and $\Delta\alpha$ the induced change in the absorption coefficient at the probe photon energy. In this simplified discussion we neglect corrections to $\Delta T/T$ by index of refraction changes. Thus the spectral lineshape information about the electronic excitation to which the probe is tuned is contained in the quantity $\Delta\alpha$. The details of the lineshape depend on the mechanism of photomodulation, whether the electronic resonance of interest is homogeneously or inhomogeneously broadened, spatial uniformity factors, and so on, and are usually easiest to consider separately for specific instances.

3. Bound and localized excitons in semimagnetic semiconductors

In this section we discuss results of experiments where exciton kinetics have been studied in two semimagnetic semiconductors, $Cd_{1-x}Mn_xSe$ and $Cd_{1-x}Mn_x$. Many of the details which are focussed on below are likely to be similar also in other direct gap semiconductor alloys without the magnetic constituent.

We start by posing the question of experimental determination of localization of free excitons in mixed crystals by two primary mechanisms, namely, their capture by suitable impurities and localization in random potential wells, the latter originating from statistical alloy compositional fluctuations. These two capture mechanisms are likely to compete in those circumstances where the corresponding impurity densities are low or moderate. The impurity bound excitons in CdSe and CdTe have been extensively studied through photoluminescence and are dominantly those associated with neutral impurities. The formation of such a bound exciton from the free state is driven by electrostatic forces and the relevant capture cross sections can indirectly be inferred to be large. The energy gain in the process equals the additional binding energy, and may be further altered by dynamic self-energy changes (polaron effects) as discussed in a specific instance below. The question of localization in alloy potential fluctuations is usually more difficult to gauge apriori because of uncertainties about the appropriate density-of-states distribution, but roughly, involves the center-of-mass motion of an exciton and the availability of a distribution of suitably sized local potential wells. Figure 2 shows for schematic reference the two paths of energy relaxation, which will be discussed now in specific experimental context. The emphasis will be on qualitative presentation of the main physical ideas involved.

3.1. Donor bound exciton in n-$Cd_{1-x}Mn_xSe$

Studies of photoluminescence in weakly n-type $Cd_{1-x}Mn_xSe$ have shown that a neutral donor bound exciton dominates the near bandedge emission spectra [4]. This conclusion is inferred by monitoring the spectra in samples of increasing Mn concentration (up to $x = 0.10$) and comparing these with the well-known bound exciton spectrum of CdSe. Of interest here is the correlation between such cw luminescence spectra and the transient spectra gene-

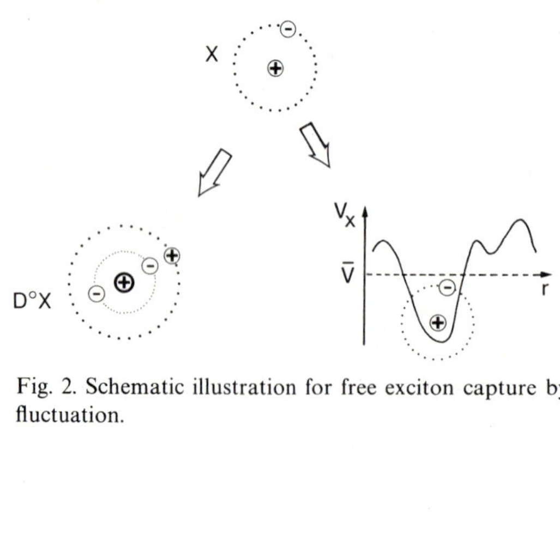

Fig. 2. Schematic illustration for free exciton capture by a neutral donor and an alloy potential fluctuation.

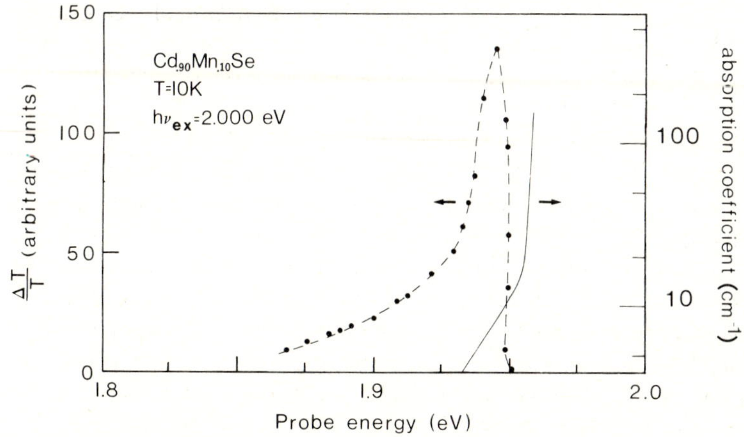

Fig. 3. Photomodulation spectrum of n-$Cd_{0.90}Mn_{0.10}Se$ at 10 K, 60 ps following excitation at 2.00 eV. Portion of absorption edge tail is also shown (right).

rated by picosecond photomodulation techniques described above. Figure 3 shows an example of modulation spectra in a sample of $Cd_{1-x}Mn_xSe$ ($x=0.10$, donor density $N_D = 3\times10^{16}$ cm^{-3}, and sample temperature 10 K), obtained 60 ps following the excitation at 2.00 eV [5]. This energy is some 50 meV above the bound exciton resonance. The excitation is kept weak so that the absorbed photon density of approximately 1×10^{14} cm^{-3} is much smaller than the donor density. A portion of the absorption edge tail is also shown. The distinct peak is identified as the donor bound exciton from its spectral position which coincides with that of the cw luminescence peak at 10 K. The finite linewidth is interpreted to originate from alloy compositional fluctuations within the effective Bohr volume of the exciton. This is expected to cause inhomogeneous broadening because of (i) the dependence of the energy gap on x (assuming a constant binding energy) and (ii) because of the influence of the spin exchange on the exciton state by the localized Mn-ion magnetic moments (the so-called thermodynamical magnetic fluctuation limit [8]). A transient excitation spectrum (probe energy within the exciton line) is displayed in fig. 4 after 20 ps delay, and also shows a distinct peak at and about the bound exciton spectral position.

For a single, homogeneously broadened bound exciton resonance, a photomodulation spectrum is expected to show a spectral lineshape which is similar to that obtained from a time-resoved luminescence experiment. This follows immediately if the absorption coefficient for the exciton can be written as

$$\alpha_h(\omega) = (N_{D^0} - N_{D^0X})\sigma(\omega), \tag{4}$$

where N_{D^0} is the total density of neutral donors, N_{D^0X} the bound exciton density, and $\sigma(\omega)$ the resonance lineshape function. Inhomogeneous broadening

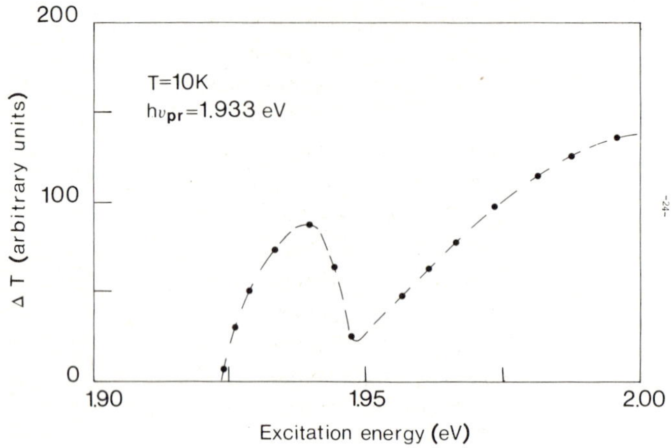

Fig. 4. Excitation spectrum for the sample of fig. 3 recorded at 20 ps time delay and probe energy of 1.933 eV.

can be included trivially, and leads again to analogy with time-resolved luminescence as long as interactions between the (localized) excitons can be neglected. This is in contrast to the case of free excitons where collisional broadening from exciton–exciton scattering gives a distinctly different photomodulation spectra [6]. The spectral lineshape obtained in fig. 3 shows some asymmetry with a rather pronounced low energy tail. In the simple model, this reflects primarily the presence of a corresponding tail in the density of states, either associated with the bound exciton itself or with other near bandedge contributions.

The rates of formation and decay of the donor bound excitons can be obtained from time-dependent study of spectra such as in fig. 3. Figure 5 shows the growth of the modulated probe signal in the same sample at $\hbar\omega_p = 1.933$ eV for two different photon energies of excitation at $T = 10$ K (also shown is the cross correlation signal, obtained through frequency summation in a nonlinear crystal). A good fit to the data is obtained by using a Gaussian source term for the laser pulses with a material response function represented as a sum of two exponentials. One of the time constants thus obtained is relatively independent of the photon energy of excitation (equal to approximately 7 ps). The other shows a distinct dependence on the photon energy of excitation, becoming as short as 2–3 ps once the photon energy of excitation increases to above the bound exciton energy. Such results have been interpreted in terms of an inhomogeneous bound exciton where, in addition to the initial capture of a free exciton, subsequent spectral diffusion within the localized bound states need also to be considered [5]. For the free exciton capture step, the data yields a cross section of approximately 3×10^{-12} cm^{-2} at the measurement temperature of 10 K. This is a very large value and shows directly the strength of the electrostatic forces which underlie the trapping process.

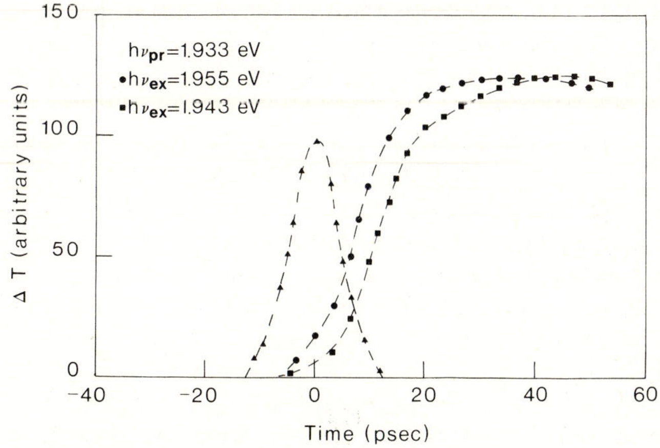

Fig. 5. Formation of the bound exciton signal at 1.933 eV at two different energies of excitation (also shown is the pulse cross correlation).

The D^0X peak of the cw photoluminescence spectrum for the n-type $Cd_{0.90}Mn_{0.10}$ samples in question had also shown a striking Stokes shift and line broadening effect when the temperature was lowered from 10 K to 2 K [4]. A tentative interpretation of this behavior was made by evoking the formation of the so-called bound magnetic polaron (BMP), in analogy to first such interpretation by Golnik et al. of observed temperature dependent binding energy of the neutral acceptor bound exciton in $Cd_{1-x}Mn_xTe$ [7]. The concept of the BMP in semimagnetic semiconductors supposes the presence of a sufficiently large exchange interaction between localized carrier(s) and the spins of the Mn ions within the former's Bohr volume. A fully developed BMP represents a locally magnetically ordered spin system within the effective Bohr volume and can occur in the absence of externally applied fields. The energy gain to the complex from this ordering (ferro- or antiferromagnetic depending on the sign of the exchange interaction) is the polaron energy. Equivalently, the large internal magnetic fields which accompany the BMP split the electronic levels of the carrier(s) thus providing a spectroscopic means to study the effect. In the instance of a single hydrogenically bound carrier, good theoretical description has been developed to treat the BMP in semimagnetic semiconductors [8]. On the other hand, multiparticle complexes such as D^0X or A^0X have until recently been discussed only through semiquantitative arguments in connection with the BMP effect [19,20].

While the single electron associated with the neutral donor in n-$Cd_{1-x}Mn_xSe$ does not, in fact, show a particularly strong BMP effect, the possibility of an enhanced exchange interaction in the presence of a bound exciton stems mainly from the substantially larger exchange constant of the valence hole. (The square of the exchange coefficient enters into the one-particle BMP theory; this gives a factor of 20 advantage for the hole over the electron.) In connection with the low temperature (2 K) D^0X luminescence spectrum of $Cd_{0.90}Mn_{0.10}$, the possible formation of a BMP would then be a dynamic event triggered by the arrival of a hole–electron pair at the neutral donor site. This argument offers a good opportunity for testing in a direct time-resolved experiment where the bound exciton energy might be dynamically measured.

Figure 6 shows the modulated transmission spectra obtained at $T=2$ K on the same 100 μm thick platelets of $Cd_{0.90}Mn_{0.10}Se$ whose transient spectra at a higher temperature was discussed above. The spectra are recorded at different time delays following the picosecond excitation at photon energy of 1.968 eV at $t=0$ ps [9]. The initial formation of the bound exciton was again confirmed to take place rapidly (few ps) but the subsequent spectra show a distinct Stokes shift, leading to an overall energy shift of the exciton peak of approximately 15 meV during its existence. This dynamic shift is in strong contrast with time-resolved spectra observed at 10 K (and above) which exhibited no measurable shifts but merely a time-dependent amplitude, consistent with the formation and decay of an energetically stationary bound exciton. Study of the excitation spectrum at the low temperature showed a rapid amplitude decrease for photon energies below 1.945 eV. This further confirms that the time-dependent behavior in fig. 6 is one of dynamic self-energy relaxation

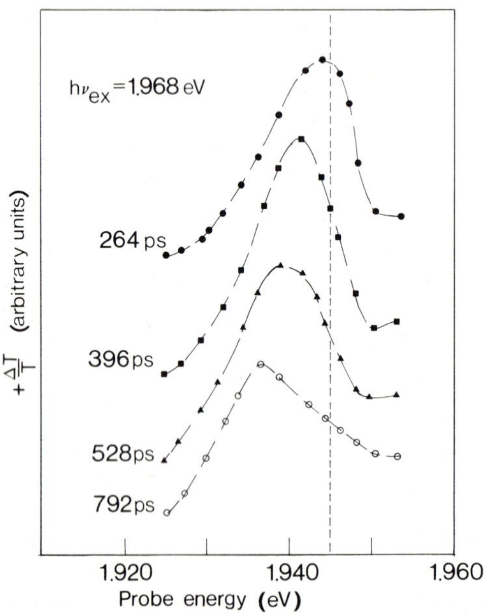

Fig. 6. Time resolved spectra of the donor bound exciton at $T=2$ K. Vertical line indicates the energy of excitation below which rapid loss of signal amplitude occurred.

of a bound exciton with a well-defined higher energy initial state only which is accessible for direct excitation. Further details of arguments that these results bear witness to the dynamical formation of a BMP from an initially 'bare' exciton are given in ref. [9]. A rough estimate can be made from the observed 15 meV shift with the one carrier BMP model of Dietl and Spalek [8] but this leads to an effective Bohr volume of approximately 25 Å (for the outer hole). Such a small radius is not expected from a normally configured D^0X complex [10] and this suggests that significant distortion of the Bohr orbit in the BMP–D^0X complex may have taken place in the very large (megagauss) internal effective magnetic fields. The precise details of the lineshapes seen in fig. 6 are also not easily interpreted but they clearly involve inhomogeneous broadening from alloy compositional fluctuations, as well as some probable contribution to the high energy side by an effect inherent to our photomodulation technique [11]. Finally, the measured time constant for the formation time of an BMP is at present open to theoretical interpretation although some initial recent calculations have been made [12].

3.2. *Exciton localization by alloy compositional fluctuations*

In any mixed crystal (e.g. $A_{1-x}B_xC$) there will be local fluctuations in the composition $x(r)$ on the scale of the lattice constant about a mean value x expressed for the macroscopic crystal. In the effective mass approximation such statistical compositional variations lead to fluctuations in the energy gap $E_g(r)$, and thus to bandedge broadening as measured in optical experiments. From the viewpoint of an exciton, these fluctuations present a system of random potential wells and a possibility for localization at a sufficiently low temperature. Experimentally, there have been reports in several mixed crystals of evidence for such localization, obtained through analysis of photoluminescence spectra [13], including recent transient spectroscopy [14]. The probability for localizing depends on the available density of exciton states below an appropriate (though difficult to define) mobility edge, below which the center-of-mass motion of the exciton can be represented by a non-extended envelope function. In a real crystal there is concurrent competition from the capture of excitons by impurities, this often being a dominant process as discussed above. Further, in a semimagnetic crystal, there will be an expected contribution from magnetic fluctuations as well [20]. The localization in random potential fluctuations is not driven by Coulombic forces but rather a gradient in the exciton potential and does not necessarily have a large cross section for exciton capture. Therefore, a judicious material choice is useful in localization studies in an effort to minimize the interfering influence by impurity capture.

We discuss here recent results [15] of time-resolved studies on subnanosecond timescale of excitonic spectra in $Cd_{1-x}Mn_xTe$ ($x=0.15$) where picosecond laser techniques have been applied to look for evidence of exciton localization. Golnik and Lavallard have also performed similar work recently [16]. Since $Cd_{1-x}Mn_xTe$ usually grows slightly p-type but with substantial compensation, we concentrated selectively on $Cd_{1-x}Mn_xTe$ samples ($x=0.15$) where cw photoluminescence spectra did not show the strong emission characteristic of the neutral acceptor bound exciton (A^0X). (In samples where such A^0X emission dominates, a large temperature-induced shift in the emission energy occurs, associated earlier with the BMP effect by Golnik et al. [7].) When excited above the free exciton energy, the cw photoluminescence spectra showed a well-defined peak below the free exciton and typically 10 meV or more above the known A^0X energies. A distinct blue shift of the peak energies with temperature and the narrowing of the spectra was measured in the range from 2 to 20 K (the shift was, however, considerably less than that observed with the A^0X emission). By $T=30$ K the high energy side of the spectrum had a width comparable to kT, compatible with a free exciton. (In contrast, there was a substantial red shift in the tail of the absorption edge with increasing temperature. For example, at a value 100 cm^{-1} for the absorption coefficient, the red shift of the edge had an average value of 1.5 meV/K in the range of 2 to 20 K). This behavior suggests a (direct gap) material where close compensation has sufficiently reduced the density of neutral acceptor sites so that low temperature free exciton localization to near bandedge states may have become possible. To verify such a hypothesis more directly, the techniques of time-resolved spectroscopy are of particular value. Specifically, one looks for the details of spectral diffusion in the exciton spectra immediately below the free exciton energy (as determined, e.g., from reflectance spectra).

Figure 7 shows transient photomodulated spectra below the free exciton energy, initiated by picosecond laser excitation in the same samples at $T=2$ K. The absorbed photon density per pulse was maintained below 10^{14} cm^{-3} to avoid intensity dependent (saturation) effects. The photon energy of excitation in the left panel corresponds approximately to that of the free exciton (at 1.835 eV) and that in the right panel lies below it. The spectral lineshapes do not have features which can be associated with free excitons [6] and we associate them with nonextended states. There is clearly a pronounced influence on the spectra by excitation energy with a rather narrow and distinct peak appearing at this energy when it is below the free exciton energy. For energies of excitation varying over a range of nearly 10 meV below the free exciton such a resonant-like peak was evident, tracking well the energy of excitation. Additionally, at delay times $t>0$ a distinct low energy shoulder began to grow, with a subsequent time-varying Stokes shift and eventual decay of amplitude. A time-dependent shift is also seen in the broader 'nonresonantly' excited spectrum in the left panel of fig. 7. In all cases we also saw a long lived spectrally stationary 'background' contribution which became increasingly apparent with the decay of the exciton (~ 1 ns), as is evident in the longest graphed delay times in fig. 7.

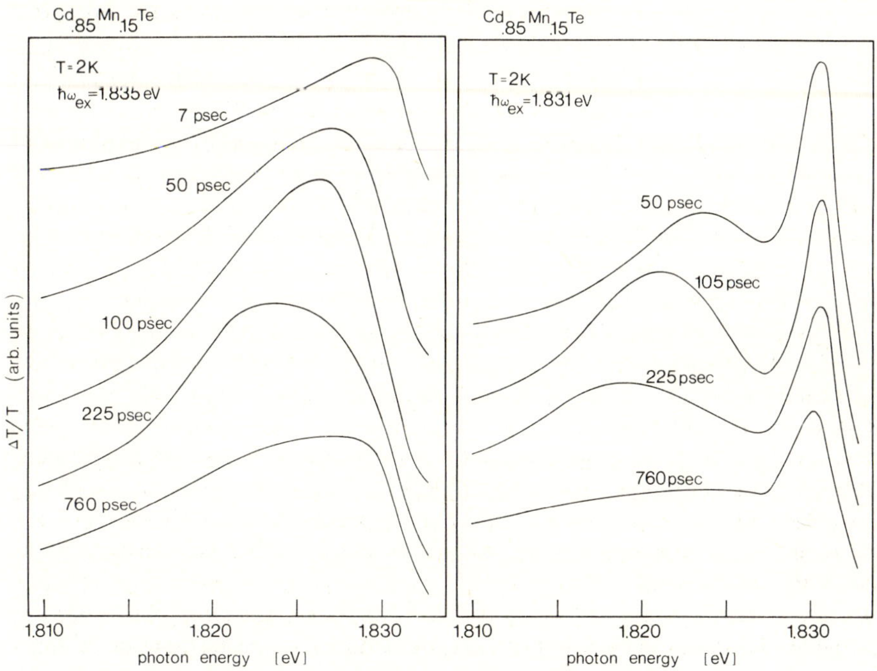

Fig. 7. Photomodulation spectra of $Cd_{0.85}Mn_{0.15}Te$ at $T=2$ K for two energies of excitation: at the free exciton reflectance resonance (left) and below it (right).

The resonantly excited peaks and their associated shoulders lost their distinctive shapes as well as decreased in amplitude when sample temperatures reached $T=10$ K. This change was also clearly seen through an increasingly rapid decay of the peaks (e.g., about 100 ps at $T=5$ K for the peak in fig. 7). At the same time, at higher photon energies spectral features were seen which showed a sign change in the experimentally measured quantity dT/T, as expected for a collision broadened free exciton line [6].

A qualitative interpretation of these results can be made by using some of the concepts already developed for describing free exciton localization by alloy potential fluctuations [17]. At the low temperatures, data such as in fig. 7 are taken as strong evidence of direct observation of the localization phenomena. We can crudely estimate [18] the range of energies for such localized states below the free exciton edge to be approximately 10 meV, if the Mn ion distribution is statistical (although there is uncertainty at present about the degree of clustering which may take place in this alloy). The resonant peaks which can be excited at low temperature show the presence of localized states from which finite spectral diffusion can be seen to lower energy states. The absence of the lower energy shoulder at $t=0$ and its appearance only at

$t>0$ provides the argument for such energy relaxation, originating from acoustic phonon assisted tunneling within the continuum distribution of localized states (the details of such density-of-states are difficult to address from our data, but the simplest and most common assumption is that of an exponential form). An energy dependent tunneling cross section together with an exponential density-of-states can give rise to the appearance of a distinct shoulder as calculated in more detail by Cohen and Sturge [17]. The disappearance of the resonant peaks with increasing temperature is due to an increasing rate of spectral diffusion as well as thermal excitation to extended exciton states.

When the initial excitation occurs to the extended states, free exciton capture takes place with comparable probability to most of the localized states, followed by subsequent energy relaxation within these states. From the time-dependent data we infer capture rates typically on the order of 10^{11} s^{-1} at $T=2$ K, slower than those seen earlier with impurity bound exciton formation discussed above. At this low lattice temperature it is unlikely that a significant thermal ionization is present and a multiple trapping like relaxation process should be less probable than the phonon assisted tunneling in energy relaxation within the localized states. From the observed spectral diffusion tunneling rates can be estimated to be on the order of 10^9 s^{-1} at $T=2$ K, i.e., comparable to the exciton lifetime (among the approximations made in this estimate is that of a constant optical matrix element within the localized states). Thus, a complete thermalization is not achieved. At higher temperatures, the role of multiple trapping is expected to increase the thermal equilibriation rate within the localized states as is qualitatively seen in our data.

The evidence that free exciton localization is being directly seen in these experiments cannot be completely separated from the unavoidable effects of impuritites and other defects in the presently available $Cd_{1-x}Mn_xTe$. There is also a general difficulty in distinguishing a localized exciton from an inhomogeneously broadened bound exciton or a trapped single carrier. Apart from an A^0X, there is not much spectroscopic evidence for other, more weakly bound impurity related excitons in $Cd_{1-x}Mn_xTe$. In a nearly compensated p-type (high resistivity) material the position of the equilibrium Fermi level lies in the vicinity of the valence bandedge at low temperatures possibly overlapping the disorder induced valence band tails. With increasing temperature some thermal ionization of electrons into the unoccupied localized states takes place thus shifting E_F deeper into the gap and giving rise to the observed shift in the tail of the absorption edge. This also makes possible single-electron (nonexcitonic) contributions in the photomodulated spectra, which follow the excitation of an electron from a defect state into the conduction band tail states. Any possible subsequent single carrier localization would be expected to result in longer lifetimes and is likely to be associated with the 'background' signal spectra observed by us. Finally, while we have not explicitly considered the role of magnetic fluctuations on exciton localization, they clearly are present and effect a finite contribution. Intriguing questions of their influence remain and require careful additional experimental work.

4. Summary

In this article we have demonstrated the versatility of picosecond time-resolved techniques in examining two principal processes which provide initial energy relaxation mechanism in mixed direct gap crystals $Cd_{1-x}Mn_xSe$ and $Cd_{1-x}Mn_xTe$. The experimental results discussed above have been focussed on attempts to isolate the formation of impurity bound excitons from the localization phenomena associated with alloy potential fluctuations. In the former case, there is evidence that the magnetic (exchange) interactions in n-Cd_{1-x}-Mn_xSe lead to further self-energy adjustment through the BMP effect. The observation of localization in the random potential wells, on the other hand, has been inferred from a characteristic behavior of spectral diffusion. In this instance the contribution by the exchange effects is yet to be separated from the nonmagnetic fluctuations.

Acknowledgement

The author wishes to thank Dr. Jonathan Harris (now at SOHIO Corporate Research Center), and Mr. Xi-Cheng Zhang, who performed most of the experimental work reviewed here at Brown University. Thanks are also due to Dr. W. Giriat at IVIC, Venezuela, Prof. J. Furdyna and Prof. A.K. Ramdas at Purdue University, Dr. R.L. Aggarwal at MIT, and Prof. A. Wold at Brown University for their generosity with samples.

References

[1] J. Furdyna, J. Appl. Phys. 53 (1982) 7637.
[2] 'Excitons', E.I. Rashba and M.D. Sturge, eds. (North-Holland, Amsterdam, 1982).
[3] J.H. Harris, S. Sugai and A.V. Nurmikko, Appl. Phys. Lett. 40 (1982) 885; S. Sugai, J.H. Harris and A.V. Nurmikko, Physica 117A&B (1983) 368.
[4] C. Huber, A.V. Nurmikko, M. Gal and A. Wold, Solid State Commun. 46 (1983) 41.
[5] J.H. Harris and A.V. Nurmikko, Phys. Rev. B 28 (1983) 1181.
[6] Y. Hefetz, X.-C. Zhang and A.V. Nurmikko, Bull. Am. Phys. Soc. 29 (1984) 476, and Phys. Rev. B, in print.
[7] A. Golnik, J. Gaj, M. Nawrocki, R. Planel and C. Benoit à la Guillaume, Proc. 15th Int. Conf. Phys. of Semicon., Kyoto 1980, p. 819.
[8] T. Dietel and J. Spalek, Phys. Rev. Lett. 48 (1982) 355.
[9] J.H. Harris and A.V. Nurmikko, Phys. Rev. Lett. 51 (1983) 1472.
[10] B. Stebe and G. Munschy, Solid State Commun. 35 (1980) 557.
[11] The author is grateful to Prof. Benoit à la Guillaume for pointing out an additional spectral contribution which is relevant to the photomodulation technique in the case of dynamical spectral shifts.

[12] J. Spalek, private communication.

[13] S. Lai and M.V. Klein, Phys. Rev. Lett. 44 (1980) 1087;
S. Permogorov, A. Reznitsky, V. Travnikov, S. Verbin, G.O. Muller, P. Flogel and M. Nikoforova, Phys. Status Solidi B 106, (1981) K 83;
E. Cohen and M.D. Sturge, Phys. Rev. B 25 (1982) 3828;
Y. Sasaki, H. Serizawa and Y. Nishina, J. Non-Cryst. Solids 59&60 (1983) 1003.

[14] J.A. Kash, A. Ron and E. Cohen, Phys. Rev. B 28 (1983) 6147.

[15] X.-C. Zhang and A.V. Nurmikko, Proc. 17th Int. Conf. Semiconductor Physics, San Francisco (1984).

[16] A. Golnik and P. Lavallard, Proc. Conf. Physics V. 6, eds., R.R. Galazka and J. Rauluszkiewicz (Polish Acad. Sciences, 1983) p. 360.

[17] E. Cohen and M.D. Sturge, ibid.

[18] O. Goede, L. John and D. Henning, Phys. Status Solidi B 89 (1978) K 183;
S.D. Baranovskii and A.L. Efros, Sov. Phys. Semicon. 12 (1978) 1328.

[19] Advances in such calculations have recently been reported by R. Planel, J. Cernogora, J. Gaj and C. Benoit à la Guillaume, Proc. 17th. Int. Conf. Semiconductor Physics, San Francisco (1984).

[20] A. Golnik, J. Ginter and J.A. Gaj, J. Phys. C 16 (1983) 6073.

Journal of Luminescence 30 (1985) 369–374
North-Holland, Amsterdam

CARRIER RELAXATION IN InGaAs

M.E. PRISE, M.R. TAGHIZADEH, B.S. WHERRETT
and S.D. SMITH

Physics Dept., Heriot–Watt University, Edinburgh, UK

A modelocked Nd:YAG pumped optical parametric amplifier providing 35 ps pulses in the wavelength range $\lambda = 1.45–2.1$ μm has been used to study photoexcited carrier recombination in InGaAs. At carrier densities in excess of 10^{18} cm^{-3} Auger processes are found to dominate the carrier recombination. An Auger rate of $2.5 \pm 0.5 \times 10^{-28}$ cm^{6} s^{-1} is determined.

Introduction

InGaAsP, epitaxially grown on InP has been used widely in the construction of such optoelectronic devices as semiconductor lasers and detectors for fibre optic communications systems operating at 1.3 μm and 1.55 μm. Whilst GaAs has been studied extensively using tunable near bandedge optical excitation, very few alloy investigations have been reported. Our development of a Nd:YAG-pumped optical parametric amplifier [1] enables us to use a number of excite wavelengths in the vicinity of the bandedge and the 35 ps pulse length of this system allows us to use the excite–probe technique to investigate phenomena on a picosecond timescale.

The investigation reported here is of carrier dynamics in the alloy $In_{0.53}Ga_{0.47}As$, which can be grown epitaxially on InP and is the narrowest bandgap alloy in the InGaAsP–InP system, with the gap at $\lambda = 1.67$ μm. A highly temperature-dependent threshold current has been measured in InGaAsP–InP injection lasers, for which several mechanisms have been put forward. Strong non-radiative recombination, carrier leakage, and hot carrier effects have been suggested as possible causes [2]. The recombination rates and their variation with band parameter is hence of considerable interest. Here we concentrate on Auger recombination, which is the strongest non-radiative recombination mechanism at the high carrier densities reached in lasers [3–6]. Our results are compared with measurements of the Auger coefficient in 1.3 μm and 1.55 μm InGaAsP, the latter obtained using photoluminescence [7–9] and excite–

0022-2313/85/$03.30

probe techniques with a modelocked Nd:YAG laser operating at the fixed wavelength of 1.06 μm [10].

Experimental method

The samples used were 4.4 μm thick InGaAs layers (n-type, 10^{16} cm^{-3}) grown by vapour phase epitaxy on InP (n-type, 2×10^{18} cm^{-3}) 300 μm thick. The composition of these samples was determined using double crystal X-ray diffraction.

Measurements are reported for two wavelengths corresponding to 775 and 806 meV, which are respectively 25 meV and 56 meV above the initial room temperature bandedge.

The temporal behaviour of the sample absorption under optical excitation was examined using an excite–probe technique. The configuration is shown in fig. 1.

A single pulse from our optical parametric amplifier was split to provide one intense excite pulse and a second probe pulse one hundred times weaker. A corner-cube reflector on a translator produced relative delays of up to 4 ns. The external angle of the excite beam and the probe beam was 5°, to enable probe detection. The input energy, the excite-pulse transmission and the probe pulse transmission, were monitored using PbS detectors calibrated using

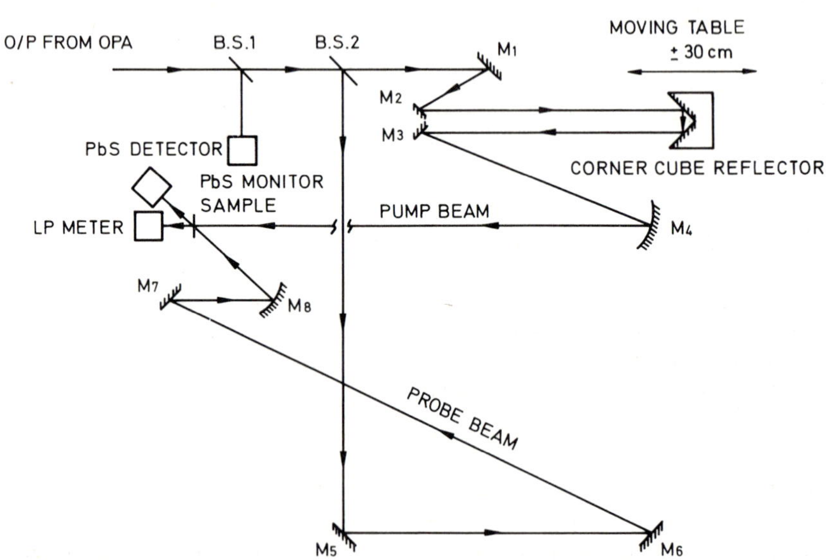

Fig. 1. Experimental layout.

neutral density filters. A minicomputer controlled the delay and collected the data.

The transmission of the excite pulse itself increased as a function of the pulse input energy due to absorption bleaching. A limiting transmission of 28% was obtained, caused by the sample reflectivity and substrate absorption. In this high energy limit (with peak energy densities 1 mJ/cm^3) the InGaAs absorption is completely bleached, and a uniform carrier density is induced throughout the sample.

In order to reduce complications due to Gaussian spatial profile and non-uniform carrier distribution through the sample, the probe beam was focussed to half of the 700 μm excite beam ($1/e^2$) diameter and measurements were made under high excitation conditions. This high excite energy limit provides an ideal regime for measurements since the probe beam then traverses a sample region in which the photoexcited carrier density generated by the excite pulse is completely uniform. We can also obtain the magnitude of the initial carrier density, since the quasi-Fermi levels in the conduction and valence bands are equal under completely bleached conditions.

Figure (2) shows the resulting probe energy transmission as a function of time delay. A rapid transition recovery of the order of a few hundred picoseconds is apparent.

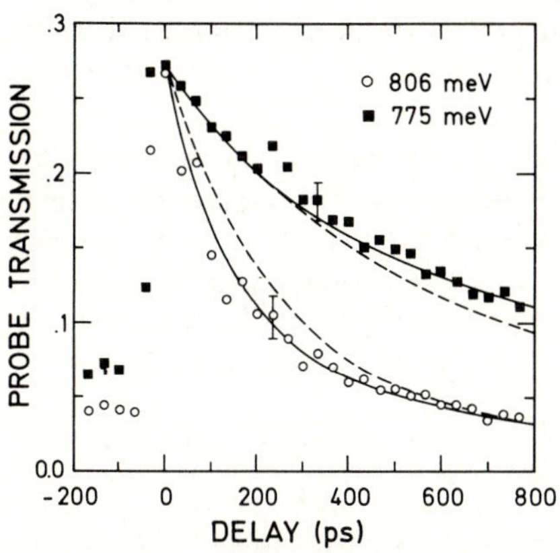

Fig. 2. Probe transmission as a function of time-delay and wavelength. The solid lines are the best theoretical fit including Auger recombination. The dashed line, for comparison is a fit accounting for only extrinsic and radiative recombination.

Interpretation

The rapid transmission recovery is due to recombination of the photoexcited carriers. In order to obtain the carrier decay rates from our data we first determine the transmission as a function of carrier density and then curve-fit directly to our results. A number of assumptions must be made.

Our principle assumption is that the carriers thermalize very rapidly within the conduction and the valence bands, so the carrier distribution can be characterized by a constant representative temperature, throughout the experiment. Since carrier thermalization times are typically less than 1 ps and our pulse length is 35 ps this assumption is valid. From the resulting Fermi distributions of the photoexcited carriers one obtains the absorption [2] as a function of excited carrier density N and temperature at any given frequency ω,

$$\alpha(\hbar\omega, N) = \alpha_0(\hbar\omega)[f_v(E_v, N, T) - f_c(E_c, N, T)]. \tag{1}$$

Here $\alpha_0(\hbar\omega)$ is the linear absorption in the absence of excitation, E_c and E_v are the energies of the optically coupled states, T is temperature, and f_v and f_c are the electron Fermi occupancies.

In order to determine the Fermi energies the bandedge effective masses of the conduction and heavy hole band were taken as $m_e = 0.041 m_e$ and $m_h = -0.50 m_e$ [11]. Conduction band non-parabolicity was estimated using the Kane four-band model, setting the spin-orbit splitting as zero [12]. The carrier-density dependence of the bandgap due to exchange and correlation effects was assumed to be the same as that of GaAs [13]. The Joyce–Dixon approximation [14], which is valid under these experimental conditions, was used to obtain the Fermi-energies directly as a function of carrier density. Hence we could evaluate the transmission change as a function of carrier density through eq. (1).

We now curve-fit by estimating the induced carrier density at zero time delay, N_0, and then by modelling the time dependence of the carrier density N. Since we are working under initial bleaching conditions N_0 is determined as that density for which $f_c = f_v$. Thus if the carrier temperature is taken as 300 K, this implies $N_0 = 1.8 \times 10^{18}$ cm^3 and 2.3×10^{18} cm^3 for $\hbar\omega = 775$ eV and 806 meV respectively.

The recombination rate can be written as

$$dN/dt = -AN^3 - BN^2 - CN. \tag{2}$$

Here A, B, and C are the Auger, radiative and extrinsic rates, respectively. Using the momentum matrix element ($2m_e P^2/\hbar^2 = 23.2$ eV [4]) one calculates the radiative rate, obtaining $B = 2.3 \times 10^{-10}$ cm^3 s^{-1} [15]. The solid lines in fig. 2 indicate the results of fitting to the experimental data at the two frequencies, allowing the same temperature and Auger rate to vary. For a temperature of 300 K fits to both sets of data are obtained for $A = 2.5 \pm 0.5 \times 10^{-28}$ cm^6 s^{-1}. An upper limit on the extrinsic recombination rate of 2×10^8 s^{-1} is obtained.

It has been suggested that the temperature of the carriers may be different from that of the lattice (the excitation of electrons high into the conduction band as part of the Auger recombination process may cause a heating). Experimental measurements of carrier heating in 1.3 μm InGaAsP indicate an increase in carrier temperature by much less than 100 K [16]. However, if in our model we take a carrier temperature of 400 K this yields an Auger coefficient of $1.8 \pm 0.3 \times 10^{-28}$ cm^6 s^{-1}.

Limitations due to the uncertainty of the curve fitting are shown by the dashed lines on the figure. If Auger recombination is ignored completely the best fit is considerably less acceptable than that chosen.

Conclusions

We are generating carrier densities of $1.5–2.5 \times 10^{-18}$ cm^{-3}, similar to those at threshold in a semiconductor laser. We deduce an Auger coefficient in $In_{0.53}Ga_{0.47}As$ that lies in the range $2–3 \times 10^{-28}$ cm^6 s^{-1} implying that Auger recombination is dominant mechanism at carrier densities 10^{18} cm^{-3}. Our result can be compared to that measured in other composition InGaAsP alloys. Measurements in 1.3 μm InGaAsP alloys yield a rate of $2.3 \pm 1 \times 10^{-29}$ cm^6 s^{-1} [10]. Scaling arguments suggest that the Auger rate should increase by approximately an order of magnitude as we move to the narrower bandgap material, but the exact scaling depends on the effective masses of the Auger excited electrons and holes, which are somewhat uncertain. An increase in Auger rate has also been found from measurements of luminescence decay times in going from 1.3 μm to 1.67 μm [8,9].

Experimental results, including our own are lower than those predicted theoretically by several authors [4,5]. A possible reason is contained in the suggestion that the overlap integrals in the theoretical calculations may be too large [17].

Acknowledgements

We wish to acknowledge partial financial support from the Rome Air Development Centre at Griffiss Air Force Base, on Grant No. AFOSR-83-151, and the provision of samples by British Telecom Research Laboratories, England.

References

[1] M.E. Prise, Ph.D. Thesis, Heriot-Watt University, Edinburgh (1983).

[2] Y. Horokoshi, in: GaInAsP Alloy Semiconductors, ed., T.P. Pearsall (John Wiley and Sons, New York, 1982) p. 379.

[3] A. Haug, Appl. Phys. Lett. 42 (1983) 512.

[4] A. Sugimura, IEEE J. of Quantum Electron 18 (1982) 352.
[5] R.J. Nelson and N.K. Dutta, J. Appl. Phys. 54 (1983) 2923.
[6] L.C. Chiu, P.C. Chen and A. Yariv, IEEE J. Quantum Electron. QE18 (1982) 934.
[7] A. Mozer, K.M. Romanek, W. Schmid, M.H. Pilkuhn and E. Schlosser, Appl. Phys. Lett. 41 (1982) 964.
[8] C.H. Henry, B.F. Levine, R.A. Logan and C.G. Bethea, IEEE J. Quantum Electron. QE 19 (1983) 905.
[9] C.H. Henry, B.F. Levine, R.A. Logan and C.G. Bethea, Electron. Lett. 20 (9) (1984) 354.
[10] B. Sermage, H.J. Eichler, J.P. Heritage, R.J. Nelson and N.K. Dutta, Appl. Phys. Lett. 42 (1983) 259.
[11] Appendix in: GaInAsP Alloy Semiconductors, ed., T.P. Pearsall (John Wiley and Sons, New York, 1982) 456.
[12] E.O. Kane, J. Phys. Chem. Solids 1 (1957) 249.
[13] N.K. Dutta, J. Appl. Phys. 52 (1982) 55.
[14] W.B. Joyce and R.W. Dixon, Appl. Phys. Lett. 31 (1977) 354.
[15] See for instance, A. Yariv, Quantum Electronics (John Wiley and Sons, New York, 1975) p. 160.
[16] B. Etienne, J. Shah, R.F. Leheny and R.E. Nahory, Appl. Phys. Lett. 41 (1982) 1019.
[17] M.G. Burt and C.S. Smith, J. Phys. C., Solid State Phys. 17 (1984) L47.

Journal of Luminescence 30 (1985) 375–394
North-Holland, Amsterdam

RAMAN PROBE STUDIES OF Nd:YAlG LASER GENERATED NON-EQUILIBRIUM EXCITATIONS IN GaAs

Ralph BRAY and K. WAN

Physics Department, Purdue University, West Lafayette, Indiana 47907, USA

A survey is presented of the non-equilibrium excitations in GaAs that can be probed by Raman scattering. Intense Q-switched Nd:YAlG laser pulses are used both for generating the non-equilibrium excitations at low temperature, and for in-situ Raman scattering probe measurements. Particular emphasis is placed on the search for non-equilibrium, zone-edge slow TA phonons. We discovered that the difference frequency combination spectrum involving slow TA phonons is surprisingly coincident with the acceptor excitation line spectrum for the Zn impurity. The discrimination between the non-equilibrium phonon population and a non-equilibrium hole population on minority acceptors in n-GaAs is discussed.

1. Introduction

The Raman scattering spectrum is capable of providing a very comprehensive probe of non-equilibrium excitations in semiconductors. The excitations that can be probed include (i) the free carriers in both the single particle and collective mode (plasmon) regimes, (ii) the bound carriers through the excitation spectra of the impurities, (iii) the zone-center optical phonons, and (iv) both the acoustical and optical zone-edge phonons in the second-order spectrum. The Raman spectrum can thus show the correlated changes in the various excitations in the material as generated, for example, by intense laser pulses.

In this paper we shall describe a study of the non-equilibrium excitations generated in n-GaAs by intense, Q-switched Nd:YAlG laser pulses, with the same laser used as the Raman probe. Earlier Raman probe studies were mainly concerned with the analysis of changes in zone-center optical phonon population [1], and more recently with the changes in the free carrier population [2]. In the present work, a much more comprehensive analysis of the Raman spectrum was undertaken. A prime objective was to search for evidence of the generation of a non-equilibrium population of long-lived, large wavevector,

0022-2313/85/$03.30

slow TA (transverse acoustic) phonons. The stimulus for this study was the report of anomalous results of heat pulse measurements at low temperatures in GaAs [3]. An apparent bottleneck was found in the decay of high frequency (>1 THz) slow TA phonons. The phonon frequencies were deduced indirectly from the dispersion-reduced group velocities in the transit of the heat pulses from the laser excitation point to the detector, a Pb superconducting bolometer with a low-frequency cutoff of 0.7 THz. Phonon lifetimes exceeding 1 μs were deduced from the apparent ballistic nature of the phonon transit over several millimeters. Such long lifetimes are not surprising at very low temperature for the case of purely anharmonic decay. The strong dispersion of the high frequency, slow TA phonon branch prohibits the spontaneous decay of the phonons into two or more lower-frequency modes [4]. The anomalous feature is the apparent weakness of the elastic isotope scattering. Two Ga isotopes are present in a 60:40 ratio. This high isotopic content is expected to provide a much smaller ballistic mean free path [3,5]. Since the frequency of isotope scattering varies at least as the fourth power of the phonon frequency, the isotope scattering time can be expected to be quite short, e.g., only a few nanoseconds for zone-edge TA phonons (~ 2 THz in GaAs).

These results motivated our attempt to verify the existence of long-lived populations of very large wavevector TA phonons, using the Raman spectrum [6,7]. The latter should be capable of providing a direct identification of the frequency of non-equilibrium phonons, especially near the zone-edge, and also their wavevectors or positions in the Brillouin zone. A critical evaluation of the viability of the Raman spectrum for this purpose occupies a major part of this paper. Large wavevector, zone-edge phonons can be seen only in the second-order of the Raman spectrum. Although broad, the spectrum contains strong structural features, including shoulders and sharp peaks, which reflect primarily the pattern of the joint density of states [8]. The structure can be analyzed with the aid of the phonon dispersion curves obtained from neutron scattering data, and the various features can be associated with high symmetry points (e.g., at X, K and L) in the Brillouin zone. The second-order spectrum contains contributions from overtones (two phonons in the same branch, with equal and opposite wavevector), and from sum and difference frequency combinations (two phonons in different branches with opposite wavevectors). The same phonon can thus participate in several different combinations, thus providing redundancy in the detection process and hence a means for checking the reliability of the identification of the signals. On the other hand, the very versatility of the Raman spectrum permits it to detect many other competing excitations, which may obscure or simulate the desired signals. As we shall see, analysis of the non-equilibrium excitations in the Raman spectrum is not a trivial process.

2. Scheme of generation and detection of non-equilibrium excitations

Generation of non-equilibrium excitations in GaAs was accomplished with weakly absorbed, Q-switched Nd:YAlG laser pulses at 1.064 μm, in order to achieve in-depth, bulk excitation. The same laser beam was used for the Raman scattering analysis, thus providing in-situ detection of non-equilibrium excitations of all wavevectors [6,7]. We did not resort to physical separation of the generation and detection processes in the sample, as is done in heat pulse measurements where transport of phonons is an essential feature for focussing on excitations with specific wavevectors or group velocities. That approach would have unduly restricted the scope of the investigations which were aimed at simultaneous observation of excitations of all possible wavevectors. Also, we opted for in-depth, bulk excitation, in spite of much more intense surface excitation possible with above-band-gap radiation, in order to minimize intrusion of surface scattering and decay mechanisms for the phonons, and also to avoid dealing with the problem of very rapid escape of the non-equilibrium population from the generation and detection volume.

The use of sub-band-gap radiation for excitation was based on the expectation that there existed sufficiently strong extrinsic excitation mechanisms, and that these could be supplemented at sufficiently high laser intensity by intrinsic non-linear excitation by two-photon or second-harmonic processes. Although the absorption coefficient in GaAs at 1.064 μm is relatively weak (a few cm^{-1}), it is known that this wavelength is at the center of an absorption peak. The latter has been attributed recently to EL2 defects [9], i.e., As atoms at Ga vacancies. Both substantial photoionization and intra-center absorption have been reported for these defects. The photoionization is responsible for increases in free electron concentration of the order of $10^{16}/cm^3$, which have been detected both by photoconductivity measurements [9,10] and by single particle Raman scattering [2]. The intra-center absorption has been shown to be associated with step-like, multiple zone-edge TA phonon emission processes [9]. Luminescence in GaAs also reveals strong contributions by phonon-assisted processes, with phonon structure very dependent on the type of impurity present [11]. One can therefore expect several processes to be active for extrinsic generation of non-equilibrium excitations, and results to be strongly sample-dependent, reflecting the variation in the nature and density of the defects or defect complexes present in the individual GaAs samples.

3. Experimental arrangement

The basic excitation and Raman scattering equipment was described earlier [2] for a study of single-particle (electron) scattering in GaAs at 300 K. A difference in the present work was the use of a closed-cycle cryogenic refrigerator which facilitated long runs, greater than 24 h if necessary, at a cold finger temperature of ~12 K. Other useful details, relevant to the search for and analysis of the second-order spectrum, and to low-temperature operation, deserve mention. The Nd:YAlG laser was used either in CW or Q-switched modes to provide respectively, weak or strong intensities. The difference between the two spectra, after normalization to the intensity of the TO line (assumed unaffected by laser intensity), provided the measure of the generation of non-equilibrium excitations. The laser beam was focussed to a diameter of 100 to 150 μm in the sample, furnishing CW intensities of $\sim 10^3$ W/cm^2 and Q-switched intensities $\sim 10^6$ W/cm^2. The Q-switched pulses were ~100 ns in half-width when applied at a repetition rate of 1000 Hz. At higher repetition rates, the pulses become broader and less intense. This feature was used sometimes to achieve lower pulse intensities, close to the CW range. This was important for certain samples where strong luminescence in the CW regime made it difficult to establish a baseline for the Raman spectrum. The use of pulsed excitation, combined with gating of the photon counting system, minimized the contribution from the delayed luminescence. The gating procedure also minimized sample luminescence contributions in the intense Q-switched regime.

The 90° Raman scattered light was collected and analyzed by a Spex double grating spectrometer. Operation with slit widths of 1 mm gave resolution of 4 Å. Measurements were made at intervals of 2 Å. The data were collected in a computer, corrected for spectral response of the gratings and the S1 photomultiplier and for variations in transmitted laser intensity, and finally binomially averaged. The computer storage and analysis of the data permitted detailed determination of the variation of the spectrum with exciting laser intensity, even though a combined excitation and probe beam technique was used.

The sample temperature can be an important factor in determining the phonon lifetime. The minimum temperature in the present setup is the cold finger temperature of 12 K. The laser radiation undoubtedly further heats the sample locally, in the path of the beam. Upper limits of the local heating were determined from the temperature of the excited electrons as observed in the Rayleigh wing of the laser line. The Stokes to anti-Stokes ratio was used for this purpose. From the absence of anti-Stokes zone-center TO contributions in both the CW and Q-switched spectra, we could ascertain that the local lattice temperatures remained below 40 K.

The symmetry of the Raman spectra was always specified. For this purpose, the samples were cut to provide specific crystallographic orientations, and the polarization of the incident and scattered light was carefully controlled. This is important for analyzing the second-order Raman spectrum and distinguishing contributions from overtones and sum and difference frequency combinations.

Our studies were limited to bulk GaAs samples, $\geqslant 2 \times 2 \times 2$ mm^3, to accommodate the focussed laser beam in a scattering volume, 100 μm in diameter and nearly 2 mm in length. This minimized contributions of surfaces to the scattered light. Thus it was required to use samples cut from grown ingots, which do not generally have the high quality available in epitaxial samples. Measurements were restricted to samples with low equilibrium carrier concentrations, ranging from semi-insulating samples to n-type sample with electron concentrations below $10^{15}/cm^3$ at 300 K. Low carrier concentrations were necessary to keep single-particle and plasmon scattering contributions from swamping the second-order phonon contributions in the Raman spectrum, and to minimize luminescence effects at low temperatures. In most cases, characterization of the samples could best be made from the Raman spectrum itself.

4. Overview of the non-equilibrium Raman spectra

In fig. 1, we show a series of Raman spectra for sample G2-1336, covering the frequency shift range ± 320 cm^{-1}. The spectra are shown for various exciting laser intensities, from CW through increasing Q-switched pulse intensities. The relative peak power per pulse, obtained from the TO lines, is given in the inset in terms of the CW power. The laser intensity was varied by changing the Q-switch repetition rate. The various spectra are shown normalized to each

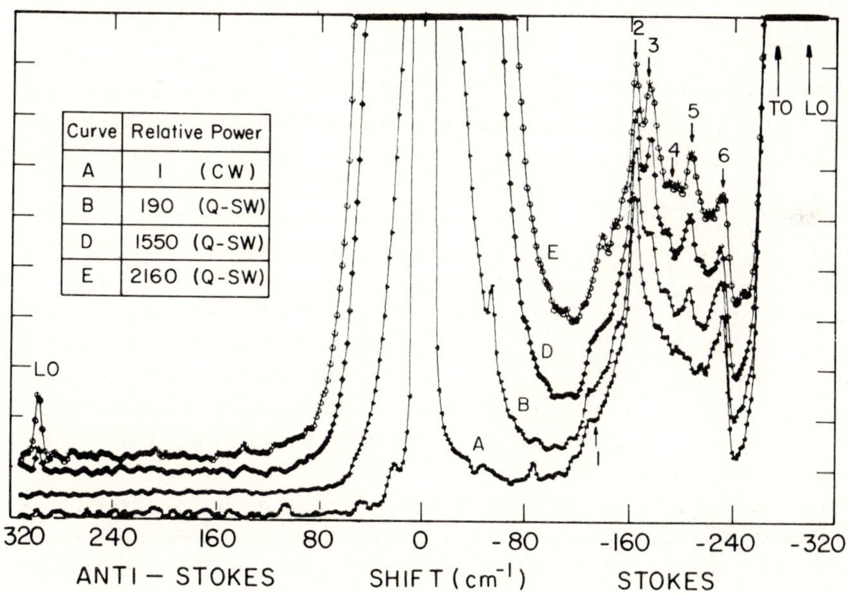

Fig. 1. Low temperature ($\sim$20 K) Raman spectra for n-GaAs, sample G2-1336, as a function of Nd:YAlG laser intensity. The inset shows peak power of Q-switched pulses, relative to the CW power (case A). The symmetry for the scattering configuration is $\Gamma_1 + \Gamma_{15}$, with scattering strength $(a^2 + \frac{4}{3}d^2)$.

other through the integrated TO Stokes line. Thus, any departure from the base CW spectrum, represents enhancement in signal from change in population of excitations.

The strongest contributions to the spectra, far off-scale in height, in fig. 1, come from the zone-center TO and LO peaks on the Stokes side. In terms of the normalized TO peaks, the LO Stokes contributions are independent of exciting laser intensity.

The rapid growth and broadening of the scattered light spectra in the wings of the Rayleigh line (i.e., within $\pm 80\ \mathrm{cm}^{-1}$), is due to the increase in the free electron concentration. This is shown on a more appropriate scale in fig. 2. The data were taken for polarization of incident and scattered light perpendicular to the scattering plane, and correspond to the charge density scattering mechanism [2]. The increase in area of the wing represents increase in electron concentration. The sample had an electron concentration of $\sim 10^{13}/\mathrm{cm}^3$ at 300 K, too low to be visible in the CW measurements. The increase in carrier concentration observed here, with increasing laser intensity, occurred in all n-type samples tested, both at low temperatures and at 300 K. This is in accord with the photoionization of the ubiquitous EL2 centers. The extrinsic nature of the generation is evident from the relatively low intensities at which excitation begins. The increase in electron temperature, obtained from Stokes to anti-Stokes ratios, is tabulated in the inset. A feature that is not understood is the

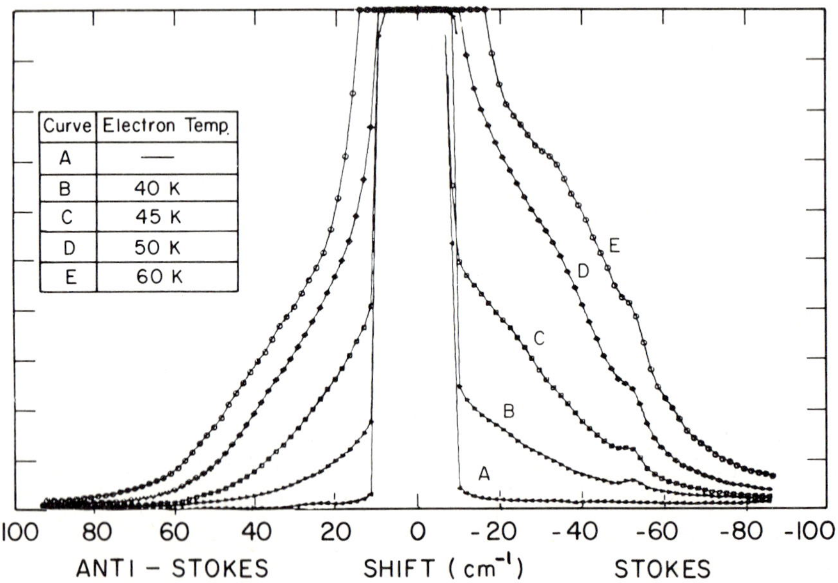

Fig. 2. Raman spectra of the Rayleigh wing as a function of laser intensity (from fig. 1), showing the increase in free electron concentration (related to area of wing), and electron temperature (from Stokes/anti-Stokes ratios).

sharp rise in the Rayleigh wing very near the laser line, occurring at the very highest laser intensities. It may be related to the onset of intrinsic generation of electron–hole pairs by non-linear processes. At these same intensities, we also see the onset of an anti-Stokes LO signal at 296 cm^{-1}. This LO excitation may be due to defect-associated relaxation processes, or thermalization of hot carriers, possible following intrinsic excitation high in the conduction band. The highest occupation number $\tilde{n}$ measured for the LO phonon line was $\sim 10^{-3}$.

Not shown in fig. 1 is a sharp peak at 310 cm^{-1} or 38.7 meV, on the Stokes side, which decreases with increasing laser intensity. It is perhaps an electronic Raman signal, associated with depopulation of electrons on a defect, but it has not been otherwise identified.

In the anti-Stokes spectrum, there appears a flat uplift of the spectrum, which increases with laser intensity. This up-conversion is matched by a stronger uplift of the spectrum on the Stokes side; an extended measurement shows it to be a long, decreasing tail of the Rayleigh wing. The uplift, or enhancement in the baseline signal, which can interfere seriously with analysis of the enhanced phonon signals, may be due to a weak radiative recombination of the electrons excited from the defects.

The most interesting and striking effect is the increase in structure on the Stokes side, between 100 and 240 cm^{-1}. Analysis of this structure is the central problem of this paper.

5. Non-equilibrium structure in the Stokes spectrum

5.1. Two-phonon overtone structure

Between 100 and 240 cm^{-1} on the Stokes side of the equilibrium CW spectrum, is the basic two-phonon structure. At low temperature, it is due solely to overtones of acoustic phonons. Its temperature dependence is determined by the phonon emission factor $(1+\tilde{n})^2$, where $\tilde{n}$ is the phonon occupation number. At low temperature, $\tilde{n} \ll 1$. The overtone contributions are absent on the anti-Stokes side, where their strength is proportional to $\tilde{n}^2$ for phonon absorption. The identification of the structural features is based on the phonon spectrum derived from neutron scattering data [12]. The shoulder (1) at 126 cm^{-1} represents zone-edge TA phonons near the L point in the $\langle 111 \rangle$ direction; the very sharp peak (2) at 162 cm^{-1} is due to slow TA phonons near the X and K points in the $\langle 100 \rangle$ and $\langle 110 \rangle$ directions; the smaller peak (6) at 227 cm^{-1} is due to the fast TA phonons near the K point. The overtone structure is determined primarily by the large density of states associated with the strong dispersion of the phonon spectrum in the large wavevector regions of the Brillouin zone.

In the Q-switch regime, there is a progressive uplift of the overtone spectra with increasing laser intensity, and the onset of additional structural features.

Before analyzing the new structure, we examine critically whether the uplift of the overtone Stokes spectra could be due to an increase in the overtone contributions, indicative of a non-equilibrium increase in zone-edge phonon population.

To analyze this question, we took the difference between each of the several Q-switched spectra and the base CW spectrum. The difference spectra are plotted in fig. 3(b), with the enlarged relevant portion of the two-phonon structure repeated in fig. 3(a). If the uplift is due to an increase in phonon population distributed in zone-edge states according to the density of states, then the difference spectra would have to show the same structure as in the original CW spectra. Conceivably, phonon enhancement might be restricted to say, slow TA phonons, in which case only the structure associated with shoulder (1) and peak (2) would be seen. We find however that there is no trace of the overtone structure in the difference spectra, although the new peaks labelled (3), (4) and (5) persist in exaggerated form, and even some additional peaks become evident, e.g., (7) and (8). There is only a smooth background, within the noise of the data, where the overtone structure should be present. From this analysis we conclude that the uplift of the overtone structure provides no evidence of phonon en-

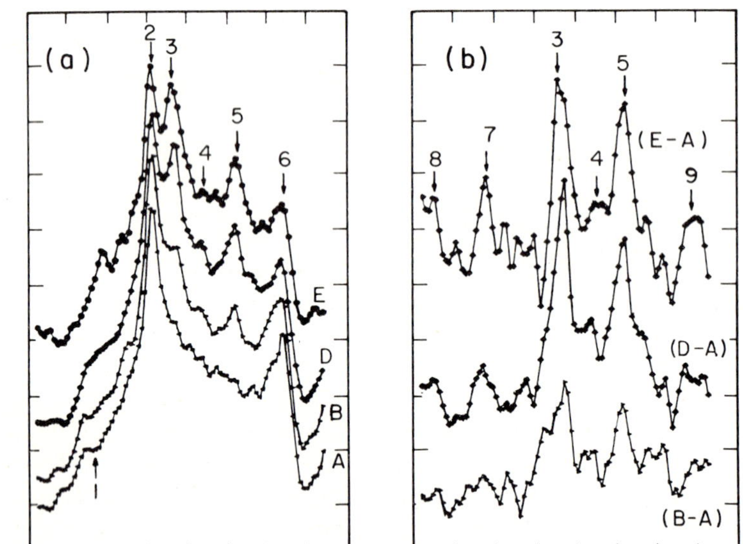

Fig. 3. (a) Stokes shift data from fig. 1, emphasizing the second-order scattering regime for acoustic overtones, and difference frequency combinations. (b) The difference spectra between the high intensity, Q-switched pulsed spectra B, D, E and the low intensity CW spectrum A, revealing additional non-equilibrium structure in the spectra.

hancement, at least in the regime of high density of states. This conclusion is in contradiction with the one reached earlier [6,7], that there was a nearly uniform, 20% increase in the overtone structure, and hence an increase in phonon occupation number of about 0.1 for both slow and fast zone-edge TA phonons. The new conclusion is based on the more detailed data taken over a larger range of laser intensity, the greater uplift achieved, and in particular, the use of the difference spectral analysis.

5.2. The new peaks: difference frequency combinations

The most striking result of the Q-switched spectra is the appearance of the new peaks (3), (4) and (5) in figs. 1 and 3. These appear to be augmented by additional, but less certain structure at (7) and (8) and perhaps (9). The latter peaks are not reproducible in other measurements on the same sample, and we give them less weight.

We searched for an identification of the sharp, non-equilibrium peaks in the phonon dispersion curves of Waugh and Dolling (hereafter WD) [12], obtained from neutron scattering data, and reproduced here as fig. 4. Table 1 contains the measured positions of all the peaks from fig. 3 and the closest phonon combinations from the neutron data. Rather amazingly, most of the new peaks [except (9)], can be represented as phonon difference frequency combinations, in every case involving the absorption of a slow TA phonon at the X, K or L

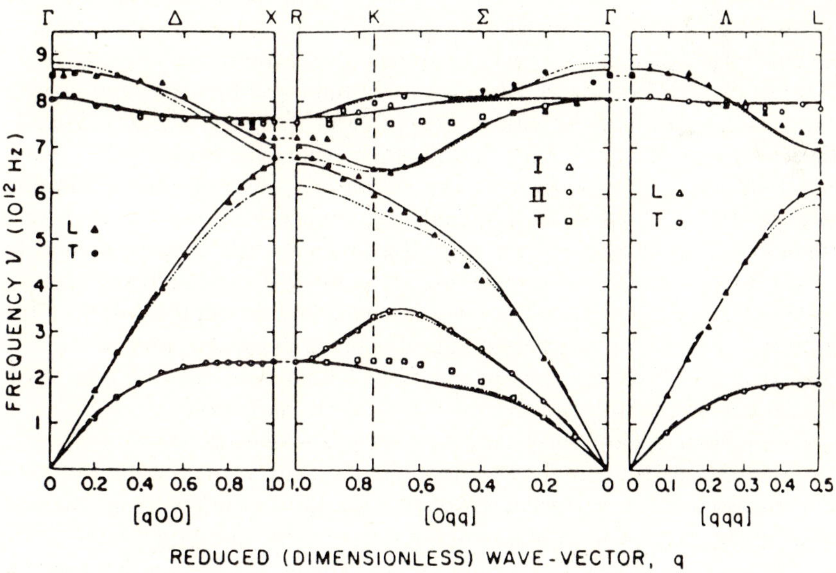

Fig. 4. The phonon dispersion curves of GaAs at 300 K. From neutron scattering data (Waugh and Dolling, ref. [12]).

Table 1
Difference frequency combinations attributed to sharp peak structure

Raman peak no.	Phonon differences	Measured Raman shifts (cm^{-1})	Calculated from neutron data (300 K)	Calculated from Raman data (low T)
3	TO – TA (X, K)	173	172	176.5
4	IO – TA (K)	184	184	–
5	TO – TA (L)	203	199	201
7	IIO – TA (K)	134	135	–
8	IIA – TA (K)	109	109	–

points, along with the emission of a higher frequency mode. Thus TO–TA (X, K) represents the emission of a zone-edge TO phonon at (X) or (K), and the absorption of a TA phonon with opposite wavevector at the corresponding symmetry point. There are not even any close competitors for most of the combinations chosen from the neutron scattering data. (The high-frequency modes labelled IO, IIO and IIA (see fig. 4), represent optical and acoustical modes of mixed transverse and longitudinal character in the ⟨110⟩ direction.) Since the neutron and Raman data are at very different temperature, we used the Raman data at low temperature for a self-consistency check, where possible. From 2TO and 2TA overtone peaks in the low-temperature Raman data, we can deduce TO–TA (X, K) and TO–TA (L) as given in the last column of table 1. These fall reasonably close to the data in the other columns.

The results presented above, demonstrating the reproducibility and growth of the peaks (3), (4) and (5) with increasing laser intensity, and the possible presence of supplementary peaks (7) and (8), provide what appears to be compelling evidence for the success of our attempt to observe non-equilibrium zone-edge, slow TA phonons. Although the temperature at which we were presumably seeing these phonons (>15 K), was relatively high compared to that used for heat pulse measurements [3], it should be remembered that we were looking, in-situ, on a time scale of less than 100 ns, the half-width of the laser pulses, and much shorter than that of the heat pulse measurements.

This conclusion posed a host of questions which needed to be addressed, both to assess its validity, and to exploit it to learn more about the specifics of the non-equilibrium phonons generated. One puzzling aspect was the absence of confirmatory evidence of TA phonon enhancement in the Stokes and anti-Stokes overtone spectrum. Analysis of this question requires a quantitative estimate of the increase in phonon population. Another question related to the significance of the sharpness of the difference frequency peaks, and its compatibility with the spectrum for the joint density of states. The sample dependence of the non-equilibrium spectrum was still another aspect requiring investigation, in view of the expectation that the phonon generation process depends on the particular defects present. We turn to the discussion of these questions.

5.3. Significance of sharp peaks in the phonon difference frequency spectrum

Although the non-equilibrium set of sharp peaks fit well to difference frequency combinations, there is no a priori reason to expect the spectrum for these combinations to have that particular form. The only points of reference for making a prediction on the form of the spectrum, are in the room tempera-

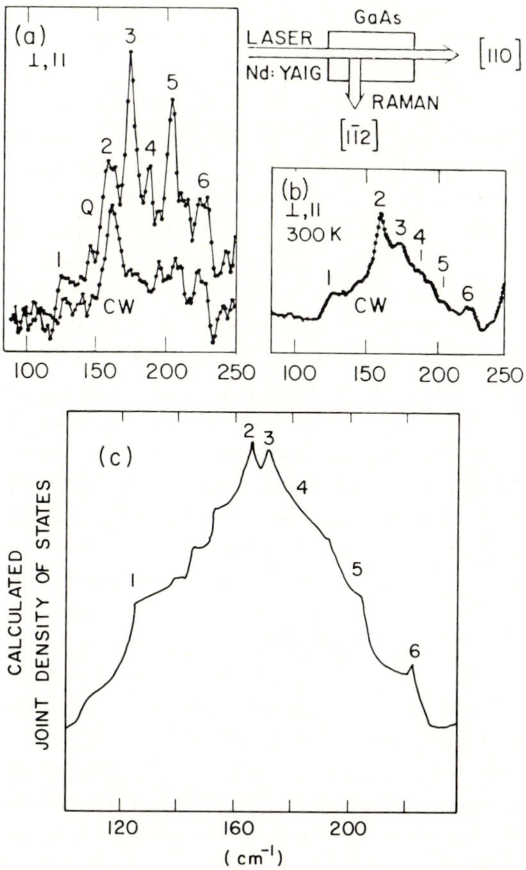

Fig. 5. (a) Low-temperature Q-switched and CW Raman scattering data for n-GaAs G2–1336, in symmetry configuration $\Gamma_{12}+\Gamma_{15}$, with scattering strength $(2b^2+\frac{1}{3}d^2)$. The overtone structure (1), (2) and (6) is strongly suppressed in this symmetry, leaving the non-equilibrium sharp peaks (3), (4) and (5) better defined. (b) Thermal equilibrium spectrum (300 K) shows absence of the sharp peak structure (relative to the overtone base). (c) Joint density of states (300 K) calculated by M. Lax (ref. [8]), showing good agreement with the data of (b), and absence of sharp peaks (4) and (5).

ture data for an equilibrium population of phonons and in the calculated spectrum for the joint density of states.

In fig. 5, we show a comparison of (a) the relevant portion of the non-equilibrium spectrum at low temperature, (b) the equilibrium (CW), room temperature spectrum and (c) a joint density of states spectrum, calculated by Lax [13]. We chose to present here a case for a different symmetry, one in which the overtone contributions (1), (2) and (6), are strongly suppressed, so that we could focus better on the difference frequency contributions. In this symmetry, the peaks (3), (4) and (5) in the low temperature data in fig. 5(a), stand out much more sharply. The symmetry-induced transformation that is obtained, is a necessary but not sufficient condition for the justification of the phonon difference frequencies model. The change observed is not necessarily unique to our analysis.

We note that in fig. 5(b) and (c), the equilibrium and joint density of states spectra do not show a series of sharp peaks, but rather a continuum, starting at the position of peak (3). To rationalize this result in terms of the non-equilibrium phonon model, we would have to assume that the low temperature, non-equilibrium phonon distribution is very different from that of thermal equilibrium phonons, which is not unreasonable. Two consequences follow from this deduction: (1) there is a selectivity to the generation of phonons, reflecting an anisotropic, phonon generation pattern, peculiar to the specific defects or defect complexes, and not simply reflecting the density of phonon states, and (2) it is necessary that the non-equilibrium phonons cannot be efficiently thermalized, and their momentum cannot be very rapidly randomized, during the 100 ns duration of the laser pulse. Although the evidence is rather indirect, the analysis has provided just the kind of information we hoped to obtain when we undertook this investigation. While we certainly can not claim that this analysis has established the validity of the non-equilibrium phonon picture, it has provided compelling grounds for further investigations.

5.4. Significance of absence of enhancement of TA overtone structure

One of the troubling aspects of the phonon enhancement interpretation was that only the difference frequency combinations seemed affected; there was no corroboratory evidence of enhancement of the Stokes and anti-Stokes overtones for the same slow TA phonons. The analysis of this question requires an estimate of the increase in phonon population, and provides a criterion for judging the validity of the interpretation.

It can be shown that the difference frequency combinations are the most sensitive to small changes in the TA phonon population. The strength of these combinations, say TO–TA, is proportional to the factor $(1+\tilde{n}_0+\Delta\tilde{n})_{TO}(\tilde{n}_0+\Delta\tilde{n})_{TA}$ where $\tilde{n}_0$ represents the thermal equilibrium populations and $\Delta\tilde{n}$ is the enhancement in population. At low temperature, $\tilde{n}_0 \ll 1$, and if significant enhancement occurs only in the TA mode, the whole factor reduces simply to $\Delta\tilde{n}_{TA}$ itself, as compared to $(\tilde{n}_0)_{TA}$ in the equilibrium case. For Stokes overtones, say 2TA, the population factor reduces to $(1+\Delta\tilde{n})^2 \approx (1+2\Delta\tilde{n}_{TA})$. For $\Delta\tilde{n}_{TA}<0.1$, the increase in Stokes overtone signal is below 20%, and barely distinguishable due to the uplift in the baseline described earlier, whereas the increase in the Stokes difference frequency combination is by a factor >100. For the anti-Stokes overtones the corresponding factor is $\Delta\tilde{n}^2_{TA}$, which is <0.01, and this signal is not detectable within the noise level. However, should $\Delta\tilde{n}$ reach a level of ~0.3, then the enhancement should become quite strong in both Stokes and anti-Stokes overtone signals, and even in the sum frequency combination, say TO+TA. Thus, where we fail to observe enhanced overtone signals, we must conclude that $\Delta\tilde{n}_{TA}$ is <0.1, or that the phonon interpretation may not be valid.

Unfortunately, there is no straightforward means of quantitatively measuring $\Delta\tilde{n}_{TA}$. It can not be done from a direct comparison of non-equilibrium and equilibrium signals at low temperature, because the latter are too small there to be distinguishable from background. The equilibrium difference frequency spectrum at room temperature is useful as a calibration base only within broad limits, because it is not sufficiently distinguishable from other, overlapping contributions. Also, we tried to use the Stokes sum frequency combination peak, TO+TA (X,K) for calibration; it is most easily identified and remains strong at low temperature because it involves emission processes only, and consequently is governed by the factor $(1+\tilde{n}_0)_{TO}(1+\tilde{n}_0)_{TA} \approx 1$. Since they involve the same modes, the ratio of scattering cross sections for TO+TA and TO−TA is easily calculated, if the joint density of states can be assumed to be the same (which is probably not grossly off). From this, we can compare heights of non-equilibrium TO−TA (X,K) and equilibrium TO+TA (X,K) peaks, and deduce $\Delta\tilde{n}_{TA}$.

We have attempted all the above analyses for the data presented in fig. 3. The results were borderline. We found $\Delta\tilde{n}_{TA}$ to be in the vicinity of 0.1; it could not be established with any degree of certainty that it was sufficiently larger than 0.1 as to bring into question the validity of our phonon picture.

There is a hidden assumption in the validity test described above that deserves some comment. That assumption is that the excess phonon populations are proportional to the joint density of states. If the states are filled more selectively (a possibility which was discussed earlier), then it is conceivable that the states determining the difference frequency peaks and those determining the overtones are not filled equally, in which case the arguments discussed above, and the validity test derived, are not applicable. However, we should still expect a close relation between difference and sum frequency enhancement, because these two combinations are more likely to depend on similar sets of states in the Brillouin zone.

5.5. Sample dependence of the non-equilibrium Raman structure

Since samples with different defect-complexes are expected to generate different distributions of phonons, we examined a variety of samples to look for possible changes in the non-equilibrium spectrum. Results for four samples are compared in fig. 6. In the upper half of each figure, we compare Q-switched and CW spectra, and in the lower half, the difference between these two spectra. The data in fig. 6(a) is a repeat for the sample of fig. 3. In fig. 6(b) we show data for sample G2–1345, with a higher equilibrium electron concentration, $\sim 5 \times 10^{14}/\mathrm{cm}^3$ at 300 K. Here the peaks (3), (4) and (5) are similar to that in (a), although somewhat weaker. Also, there is either more noise or some subsidiary structure in the spectrum for this sample. The spectra in (c) are for a sample used at Bell Laboratories for the heat pulse measurements which demonstrated the existence of the slow TA decay bottleneck. The non-equilibrium effects are quite different here from those shown in (a) and (b). The edge of the non-

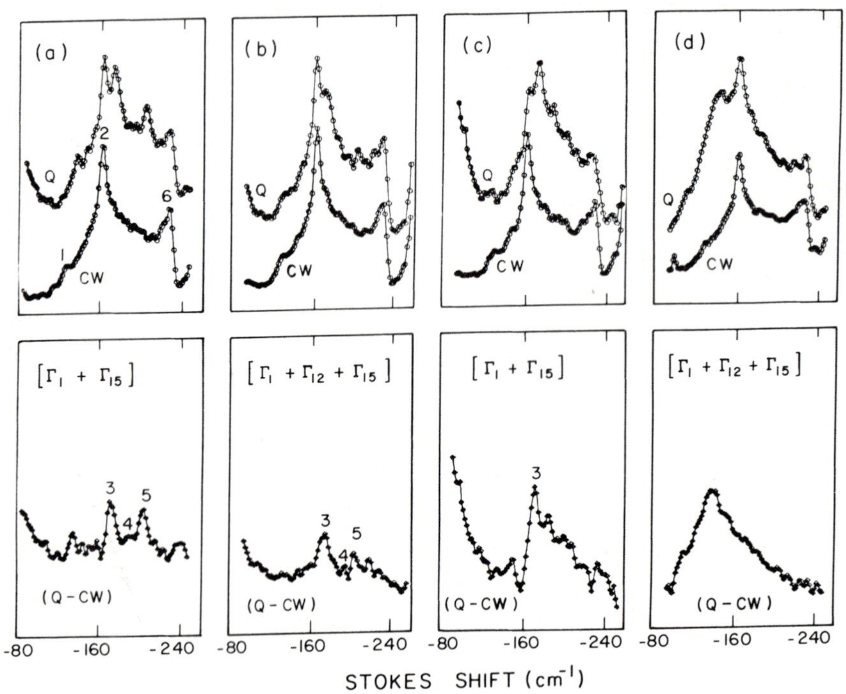

Fig. 6. Comparison of spectra for four different samples of n-GaAs, with symmetry specified. The Q-switched (Q) and CW spectra are shown above, and the difference spectra (Q-CW) below: (a) sample G2–1336; (b) sample G2–1345; (c) Bell Laboratory sample used for heat pulse measurements (loaned by V. Narayanamurti); (d) Cr-doped, semi-insulating sample (from P. Yu, with identical results for sample from V. Narayanamurti).

equilibrium structure at peak (3) is much stronger, and it is followed by more of a continuum than a sharp line structure, with the rest of the structure [peaks (4) and (5)], barely discernible. By the tests we established in section 5.4, if the non-equilibrium structure here is due to difference frequency combinations, it is strong enough that it should also manifest itself in other portions of the spectrum. However, we found that it does not do so This argues against the validity of the difference frequency phonon model, at least for explaining the whole of the non-equilibrium spectrum for this sample.

Finally in fig. 6(d), we show spectra for a Cr-doped, semi-insulating sample. Although the equilibrium spectrum is normal, the non-equilibrium spectrum shows none of the sharp line structure observed in the other samples, but instead a very broad structure, peaking near 140 cm^{-1}, below the large 2TA overtone peak. Another Cr-doped sample, from a different source gave identical results. The very broad peak is unrelated to any recognizable structure, phonon or otherwise, and may be due to the highly complex Cr level system. It is interesting to note that Cr is regarded as a strong phonon scatterer [14] and it is tempting to argue that it is fulfilling its role here and suppressing the observation of excess TA phonons.

The variety of results presented above is consistent with the view that non-equilibrium phonon structure should be strongly sample-dependent, but it does not prove that we are indeed seeing such phonons. In fact, the spectra in (c) and (d), demonstrate contributions to the non-equilibrium spectra from other than phonon difference frequency combinations.

5.6. Could there be an alternative mechanism?

At no stage in this work could we conceive of another interpretation, a different mechanism, that could account for the series of non-equilibrium peaks. The fit to the difference frequency combinations, and the relationship to the original objectives of the research, were too beguiling. It would require an amazing and shocking coincidence for an alternative mechanism to provide an equally good fit to the several non-equilibrium peaks. Nevertheless, we encountered just such a situation near the completion of the work described above. Indirectly and by chance, we discovered in a recent journal article [15], that the neutral acceptor Zn in GaAs has transitions between the ground state and the excited states that just fit the three major peaks (3), (4) and (5). This mechanism also has its surprises and unexpected features, as we shall see below, where we discuss electronic Raman scattering by Zn acceptors in n-type GaAs!

6. Electronic Raman scattering by Zn acceptors in n-type GaAs

The energy level scheme for holes on Zn acceptors in GaAs is given in fig. 7. In table 2, we list the energies of the so-called G, E, D and C transitions, as

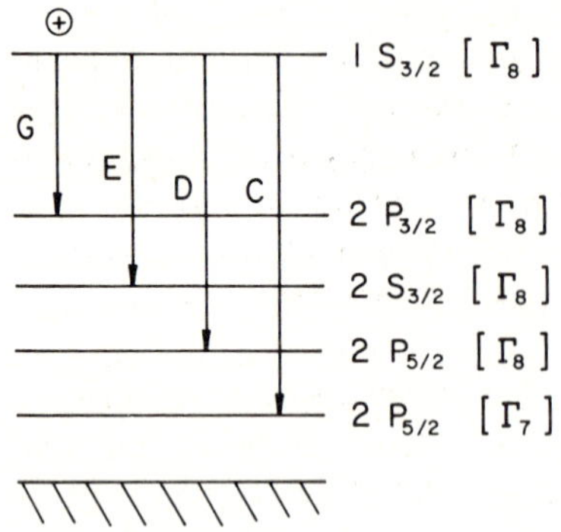

Fig. 7. Energy level scheme (not to scale) for shallow acceptors in GaAs, showing the dominant transitions G, E, D and C and states.

Table 2
Zn-acceptor excitation lines (cm^{-1})

Zn transitions	Far-IR photocond. [17]	Selective excitation lumin. [15]	Two-hole shift [18]	Theory (L+B) [16]	Raman shift (a)
G	156	154	–	154	–
E	–	172	174	185	173
D	187	185	–	188	184
C	202	200	–	203	203
Ionization	246	–	246	–	–

(a) Present work.

obtained from various types of measuring techniques [15], and fitted to the theory of Lipari and Baldereschi [16]. Our relevant Raman data are repeated for comparison. The best impurity excitation data are obtained from far-infrared, thermally-assisted, photoconductivity measurements [17], utilizing high quality, p-type epitaxial samples to obtain sharp line spectra. The selective-excitation luminescence peaks [15] were obtained in a bulk-grown (liquid-encapsulated Czochrolski), Cr-doped, semi-insulating sample. Here the lines are barely distinguishable, except for the strong E line ($1S_{3/2}$–$2S_{3/2}$ transition), which is allowed in this measurement, but not in infrared absorption. This technique involves donor–acceptor transitions, at specific pair separations selected by specific, sharp laser pump lines near the 1.5 eV absorption edge.

As shown in table 2, the agreement of our Raman peaks (3), (4) and (5) with the E, D and C lines, is excellent, although the G line is missing in the Raman spectrum. The strongest Raman line, the E line, is allowed in Raman scattering.

This agreement is not without its quirks! It must be realized that the excitation spectra described above are for holes in the ground state of neutral Zn acceptors in p-type material. But our Raman data are for n-type GaAs, where Zn, if present, would be a minority impurity and hence negatively charged by compensation by the majority donors. Thus to obtain the hole excitation spectrum, it is necessary to postulate that at high Q-switched laser intensities, a non-equilibrium hole population is generated and maintained in the ground state of Zn acceptors. This requires some very special conditions: (1) Zn must be accidentally present as the dominant, minority shallow acceptor in the sample. Other acceptors would give additional, distinctly different sets of spectral lines. (We need not worry about contributions from donors because they are too shallow to contribute to the part of the spectrum of interest here). (2) A mechanism is required, involving extrinsic, sub-band-gap Nd:YAlG laser excitation, for generating bound holes on the Zn impurities. This means that there must be a receptive trap state in the gap to which Nd:YAlG laser radiation can excite electrons from the valence band (or directly from the Zn acceptors), leaving holes behind which are frozen out in the ground state of the Zn. Furthermore, the free electrons, observed in the enhanced Rayleigh wing of the n-type sample, can recombine rapidly with the non-equilibrium holes in the Zn. Presumably, the high intensity laser pulses are necessary here to continuously replenish the non-equilibrium, trapped holes. The hole generation process outlined above may not contribute to photoconductivity if the holes are trapped sufficiently rapidly, but it would necessarily contribute to the absorption. Such an absorption contribution must be present in all compensated n-type samples, and it may be related to the absorption band always seen [9] in the vicinity of the Nd:YAlG laser line.

7. Discrimination between the alternative interpretations

We are faced with discriminating between two interpretations, the one based on a non-equilibrium population of zone-edge, slow TA phonons, and the other on a non-equilibrium population of holes, specifically on Zn acceptors. And, of course, there is the possibility that both are present.

Although the Zn acceptor interpretation seems the simpler one, this, of course, does not mean it is the correct one, and it certainly can not be excluded that both mechanisms are present, perhaps in different strengths in different samples. Also, we have not yet accounted for all the non-equilibrium signals observed. Some of them do not fit any impurity excitation scheme.

The varied nature of the results on different samples, as shown in fig. 6, can be rationalized for either model. Zn was not necessarily present in all samples, at any rate in equal concentrations. (If the Zn interpretation is correct, then it appears that it is at least a frequent, if hitherto hidden, contributor.) We have already noted that we expect strong sample dependence in the non-equilibrium phonon case. In terms of the experiments already carried out there seems no way of positively discriminating between the two mechanisms.

We consider then what future experiments may be used to obtain more definitive answers to the several problems posed here. Consider first the search for non-equilibrium phonons. This requires that we find samples that do not contain Zn, or in fact other shallow acceptors that can interfere with the search for non-equilibrium phonons. That is likely to prove impossible, in which case it may become necessary to separate excitation and probe sites – give up the in-situ type of measurement, and look for non-equilibrium phonons in the process of transport, as in heat pulse measurements.

The test for the presence of Zn acceptors poses an easier problem. It is necessary to compare the form and symmetry of the spectra observed here, with those in a p-type sample having known Zn doping. In fact, however. we have found no reports in the literature of electronic Raman scattering studies in GaAs which have yielded sharp line spectra for any acceptors. The only relevant study was for highly-doped Zn and Cd ($>10^{16}/\mathrm{cm}^3$), p-type material [19], and it showed very broad spectra, which did not conform at all to those observed here. If bulk samples are not available with sufficiently low Zn content to give sharp line spectra, it will be necessary to attempt to do the Raman scattering on epitaxial samples, such as were used in thermally-assisted, photoconductivity measurements [17].

Another test for the dominant role of impurities is to apply uniaxial strain to the samples, which should be capable of shifting the electronic Raman lines, but have negligible effects on phonon difference frequency spectra. Such experiments are in progress.

8. Summary and conclusions

We have shown that the Raman spectrum is capable of probing a rich variety of non-equilibrium excitations in GaAs, in addition to the zone-center optical phonons and free electrons. As mentioned in the introduction, the overlap of contributions from different excitations can provide a melange of data which may require much ingenuity to unravel. We have encountered in this work a particularly unexpected and difficult situation, where a set of non-equilibrium peaks fits very well to two very different mechanisms, the one involving a non-equilibrium population of zone-edge slow TA phonons, and the other a non-equilibrium population of holes in the ground state of minority acceptors. In both cases, the explanations of the experiments have been far from straightforward. The phonons are required to be distributed in the Brillouin zone in some selective manner not dictated by the density of states. On the other hand, the minority acceptor is required to be a very specific one (Zn), and furthermore it must be possible to create a non-equilibrium hole population in the ground state of this acceptor by virtue of the fortuitous presence of trap states in the forbidden gap to which electrons can be excited from the valence band by the laser radiation.

Although the minority acceptor interpretation appears more favorable, it will take many more experiments to reliably discriminate between these two interpretations, especially since both situations may be present. We have outlined some of the tests that can be made to provide this discrimination.

Finally we note that these experiments have been extremely stimulating for forcing both speculation and deeper insight into the possible properties of non-equilibrium excitations, both phonons and holes. As frequently happens, these speculations and insights may bear fruit elsewhere, even when they are not completely applicable to the present situation. In any case, we have reached a much better definition of the conditions needed to probe for non-equilibrium, large wavevector phonons by second-order Raman scattering.

9. Acknowledgements

We wish to express our gratitude to D. Abramsohn and K.T. Tsen who built much of the apparatus and participated in the initiation of the use of Raman spectroscopy as a probe of non-equilibrium excitations, and in the earlier stages of the search for non-equilibrium zone-edge phonons. The support of this work by the National Science Foundation, through Grant No. 82–17442, is also gratefully acknowledged.

References

[1] See e.g., D. von der Linde, J. Kuhl and H. Klingerberg, Phys. Rev. Lett. 44 (1980) 1505.
[2] D.A. Abramsohn, K.T. Tsen and Ralph Bray, Phys. Rev. B 26 (1982) 6571.
[3] R.G. Ulbrich, V. Narayanamurti and M.A. Chin, Phys. Rev. Lett. 45 (1980) 1432.
[4] R. Orbach and L. A. Vredevoe, Physics 1 (1964) 91.
[5] M. Lax, V. Narayanamurti, P. Hu and W. Weber, J. de Phys., Colloque C6, supplement No. 12, Vol. 42, 1981, p. C6-161.
[6] K.T. Tsen, D.A. Abramsohn and Ralph Bray, Phys. Rev. B 26 (1982) 4770.
[7] K.T. Tsen, D.A. Abramsohn and Ralph Bray, Physica 117B and 118B (1983) 543.
[8] M. Lax, V. Narayanamurti, R.C. Fulton, R. Bray, K.T. Tsen and K. Wan, in: Proc. Fourth Int. Conf. Phonon Scattering in Condensed Matter, Stuttgart, 1984, eds., W. Eisenmenger, K. Lassman and S. Dottinger (Springer-Verlag, Berlin, 1984) p. 133.
[9] M. Kaminska, M. Skowronski, J. Lagowski, J.M. Parsey and H.C. Gatos, Appl. Phys. Lett. 43 (1983) 302.
[10] G.K. Celler, S. Mishra and Ralph Bray, Appl. Phys. Lett. 27 (1975) 297;
G.K. Celler and Ralph Bray, Phys. Rev. B 13 (1976) 5397.
[11] H.J. Queisser, in: Festkorper Probleme XI, Advances in Solid State Physics, ed., O. Madelung (Pergamon, Oxford, 1971) p. 45.

[12] J.L.T. Waugh and G. Dolling, Phys. Rev. 132 (1963) 2410.
[13] M. Lax, private communication. The calculated spectrum shows both the combination modes and the partially suppressed overtones. It fits well the room temperature Raman spectrum in (b); see also ref. [8] for the isolated combination mode spectrum.
[14] V. Narayanamurti, M.A. Chin and R.A. Logan, Appl. Phys. Lett. 33 (1978) 481.
[15] A.T. Hunter and T.C. McGill, Appl. Phys. Lett. 40 (1982) 169.
[16] A. Baldereschi and N.O. Lipari, Phys. Rev. B 9 (1974) 1525.
[17] R.F. Kirkman, R.A. Stradling and P.J. Lin-Chung, J. Phys. C: Solid State Phys. 11 (1978) 419.
[18] D.J. Ashen, P.J. Dean, D.T.J. Hurle, J.B. Mullin, A.M. White and P.D. Greene, J. Phys. Chem. Solids 36 (1975) 1041.
[19] G.B. Wright and A. Mooradian, Proc. Int. Conf. on Phys. Semiconductors, Moscow, 1968 (Nauka, Leningrad, 1968) p. 1067.

Journal of Luminescence 30 (1985) 395–408
North-Holland, Amsterdam

STUDIES OF PULSED LASER MELTING AND RAPID SOLIDIFICATION USING AMORPHOUS SILICON*

D.H. LOWNDES AND R.F. WOOD

Solid State Division, Oak Ridge National Laboratory, Oak Ridge, Tennessee 37831, USA

Pulsed laser melting of ion implantation-amorphized silicon layers, and the subsequent solidification of undercooled liquid silicon, have been studied experimentally and theoretically. Measurements of the time of the onset of melting of amorphous silicon layers, during an incident laser pulse, have been combined with measurements of the duration of melting, and with modified melting model calculations to demonstrate that the thermal conductivity, K_a, of amorphous silicon is very low ($K_a \simeq 0.02$ W/cm K). K_a is also found to be the dominant parameter determining the dynamical response of amorphous silicon to pulsed laser radiation; the latent heat of fusion and melting temperature of amorphous silicon are relatively unimportant. Transmission electron microscopy indicates that bulk (volume) nucleation occurs directly from the highly undercooled liquid silicon that can be prepared by pulsed laser melting of amorphous silicon layers at low laser energy densities. A modified thermal melting model has been constructed to simulate this effect and is presented. Nucleation of crystalline silicon apparently occurs at a nucleation temperature, T_n, that is higher than the temperature, T_a, of the liquid-to-amorphous phase transition. The model calculations demonstrate that the release of latent heat by bulk nucleation occurring during the melt-in process is essential to obtaining agreement with experimentally observed depths of melting. These calculations also show that this release of latent heat accompanying bulk nucleation can result in the existence of buried molten layers of silicon in the interior of the sample after the surface has solidified. It is pointed out that the occurrence of bulk nucleation implies that the liquid-to-amorphous phase transition (produced using picosecond or ultraviolet nanosecond laser pulses) cannot be explained by purely thermodynamic considerations.

1. Introduction

In this paper we summarize the results of recent time-resolved reflectivity measurements [1–3] and model calculations [1–4] which, together with post-irradiation transmission electron microscopy (TEM) studies [3,5], have allowed us to study both the transformation of amorphous (a) silicon to a

*Prepared by the Solid State Division, Oak Ridge National Laboratory, Oak Ridge, Tennessee 37831. Operated by Martin Marietta Energy Systems, Inc. for the U.S. Department of Energy, under Contract No. DE-AC05-840R21400, June 1984.

0022-2313/85/$03.30

highly undercooled liquid (ℓ) phase and the subsequent rapid solidification process. Our work was motivated by the following considerations.

On the one hand, there have been a number of recent theoretical and experimental attempts [6–12] (the experiments using both pulsed and continuous heating) to determine thermodynamic properties of a-Si, such as its melting temperature, T_a, and latent heat of fusion, L_a. These efforts have resulted in estimates for the difference between T_a and T_c, the melting temperature of crystalline (c) Si, ranging from no experimentally observable difference ($T_a \approx T_c$) [6,7] to $(T_c - T_a) \sim 500$ K [11]. The two most recent studies, by Donovan et al. [8] and by Olson and co-workers [6,7], used the results of continuous heating of a-Si in the solid phase to estimate properties such as T_a and L_a; both studies resulted in estimates for T_a near the middle of this range [i.e., $(T_c - T_a) \sim 200$–300 K], though neither directly observed an a→ℓ transformation. We have been interested in whether an alternative approach, using time-resolved measurements and related model calculations, could provide additional information about the thermodynamic parameters or thermophysical properties of a-Si (such as its thermal conductivity, K_a), and help to illuminate the reasons for the wide range of values obtained in previous studies.

Our time-resolved measurements employ a newly developed technique [1,2] that allows us to determine the time, t_m, of the onset of melting of ion-implanted a-Si, during an incident laser pulse, as well as conventional measurements of the duration, τ_s, of the high reflectivity phase (HRP) associated with surface melting, both as functions of the laser energy density, E_ℓ. One of the principal results of the work reported here is a demonstration of the dominant role played by the low thermal conductivity, K_a, of a-Si, relative to that of c-Si, in determining both the melting threshold value of E_ℓ for a-Si layers of varying thicknesses, and the subsequent solidification behavior of partially molten layers of a-Si, at higher E_ℓ. The t_m measurements provide a particularly powerful constraint in fixing the value of K_a. We conclude that the dynamics of pulsed laser melting and resolidification are relatively insensitive to the values of the melting temperature, T_a, and the latent heat of fusion, L_a, of a-Si. This result seems particularly significant in view of the conflicting conclusions regarding T_a in earlier measurements.

The second factor motivating our work was the realization that, since a-Si apparently does melt at $T_a \ll T_c$, then pulsed laser melting of a-Si, using laser pulses of low energy density, E_ℓ, provides a means for preparing a melt of ℓ-Si that is dramatically undercooled, relative to its normal freezing point. Relatively well-characterized a-Si (free of voids and gaseous impurities) can be prepared by ion implantation into a c-Si substrate. By melting only partially through such an a-Si layer, the undercooled ℓ-Si remains isolated from the c-Si beneath and the mechanism for resolidification under conditions of large undercooling can then be studied. Does solidification proceed from the ℓ–a boundary (as in liquid phase epitaxy from a crystalline substrate)? Or, will some form of bulk (volume) nucleation and growth be observed in the absence of a crystalline template?

Cross-sectional TEM micrographs [3,5] provide important evidence that bulk nucleation does occur under these conditions. Furthermore, thermal melting model calculations of melt depth and melt duration can be accurately "calibrated," by comparison with time-resolved reflectivity measurements of the duration of the high reflectivity phase (surface melt duration). These model calculations then can be used to infer temperature profiles in the undercooled ℓ-Si, and the nucleation behavior of undercooled ℓ-Si can be determined by comparison with the solidification profiles observed in TEM and by using the theory of homogeneous nucleation [3,4]. Using this approach, we have been able to estimate a nucleation temperature, T_n, where $T_a < T_n < T_c$. Thus, bulk nucleation apparently occurs in undercooled ℓ-Si at temperatures above that of the $\ell \rightarrow$a transition[4]. This conclusion is important since the conventional thermodynamic interpretation of the laser-induced conversion of c-Si to a-Si apparently requires undercooling to T_a for this conversion to occur [13]. Our work, revealing bulk nucleation at $T_n > T_a$, seems to imply that the $\ell \rightarrow$a transition cannot be explained by purely thermodynamic considerations and that nonequilibrium and kinetic effects must play an important role. Thus, the combination of techniques described in this paper provides a powerful method for studying the dynamics of rapid solidification under conditions of large undercooling of the melt.

2. Time-resolved measurements

A frequency-doubled Nd laser followed by four dichroic mirrors was used to provide a harmonically pure 532 nm beam with (18 ± 1) ns pulse duration. The laser was operated at fixed amplifier gain to avoid slight gain-dependent changes of second-harmonic pulse width; E_ℓ was varied with lenses; a beam splitter was used for *in-situ* monitoring of the energy of every laser shot; a separate vacuum planar photodiode and storage oscilloscope were used to monitor the temporal shape of each pulse. Only laser pulse shapes that were temporally nearly Gaussian were used in the analyses. A cw HeNe probe laser beam, incident at 10° to the sample normal and focused to a 35 μm diameter ($1/e$) spot, a silicon avalanche photodiode detector, and a storage oscilloscope (time constant ~1 ns) were used to monitor the time-resolved reflectivity, R. The a-Si layers were produced by implantation of 40 keV Si ions at a dose of $6 \times 10^{15}/\text{cm}^2$ into (100) n-type silicon of 1–5 Ω cm resistivity; transmission electron microscopy (TEM) showed the amorphous layer to be slightly greater than 100 nm thick.

Figure 1 shows results for t_m as a function of E_ℓ [2,3]; the data for c-Si were taken from ref. [1]. These data were obtained by simultaneously monitor-

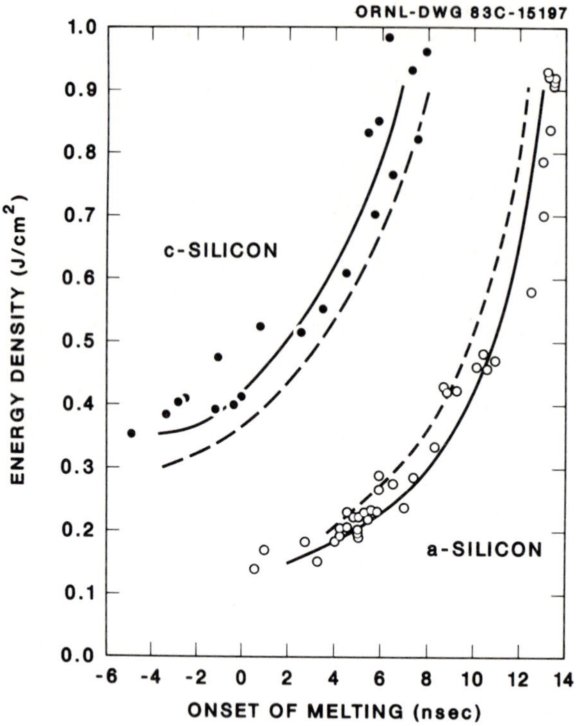

Fig. 1. Measured time of onset of melting (t_m) for a-Si (circles) and for c-Si (dots) vs. E_ℓ. Also shown are the results of model calculations of t_m for c- and a-Si (see text).

ing the time evolution of the R signal and of the incident 532 nm laser pulse and determining t_m during the incident laser pulse. To further reduce errors due to slight variations in pulse shape, the zero of time for these measurements was taken to be the center of the incident laser pulse; *positive* times correspond to melting *before* the peak of the pulse. The experimental t_m is defined as the time when the HRP is first reached. As is expected intuitively, c-Si melts only during the latter half of the laser pulse at low E_ℓ, with t_m moving to the front of the laser pulse with increasing E_ℓ (fig. 1). t_m for a-Si exhibits a similar qualitative dependence on E_ℓ but a-Si, in contrast to c-Si, is never observed to melt as late as the peak of the laser pulse. In fact, for $E_\ell > 0.6$ J/cm^2, a-Si melts within the first few nanoseconds of the laser pulse, some 12–13 ns prior to its peak.

Figure 2 shows the results of our measurements of $\tau_s$vs. E_ℓ for a-Si [2,3]. Also shown for comparison are results of recent τ_s measurements, under nearly identical conditions, for c-Si [1]. The most apparent differences between the two sets of results are that (1) a-Si melts at a significantly lower value of E_ℓ ($\sim$0.15 J/cm^2) than does c-Si (E_ℓ=0.35 J/cm^2) and (2) τ_s for a-Si displays a quasiplateau from $\sim$0.25 to $\sim$0.4 J/cm^2.

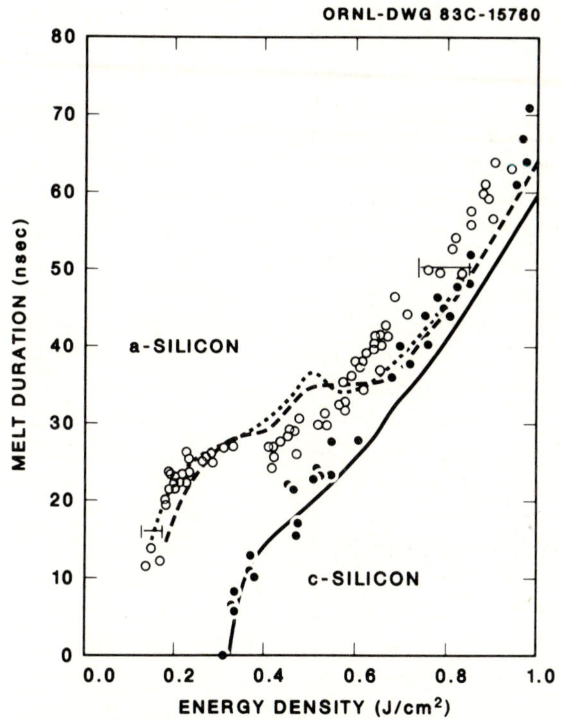

Fig. 2. Measured duration of the high-reflectivity phase (surface melt duration, τ_s) for a-Si (circles) and for c-Si (dots) vs. E_ℓ. The error bars are derived from the measured rms precision of the E_ℓ calibrations. Also shown are the results of model calculations of τ_s for a-Si and for c-Si (see text).

3. Thermal conductivity of amorphous silicon

For c-Si we recently found [1] that near the melting threshold, model calculations of τ_s and t_m are quite sensitive to the values used for the optical and thermal properties of c-Si. This result provides confidence that the process may be inverted, with the observed behavior of t_m vs. E_ℓ being used as a constraint in order to infer thermophysical properties of interest for a-Si [2,3]. In melting model calculations for a-Si, we used a reflectivity $R_a=0.42$ and absorption coefficient $\alpha=5\times10^5$ cm^{-1}, independent of T in the solid [14]. When the first cell (10 nm) in the finite difference calculations melts, α is increased to 10^6 cm^{-1} and R to 0.70. Exploratory calculations were made with many different pairs of values of T_a and L_a (see discussion below) but our baseline calculations assumed $L_a=1320$ J/gm and $T_a=1150$ °C, in close agreement with the estimates of Donovan et al. [8]. The thermal conductivity, K_c, of c-Si was taken from Glassbrenner and Slack [15] while K_ℓ for ℓ-Si was related to the electrical conductivity in ℓ-Si by the Wiedemann–Franz law [1].

We at first assumed a temperature-independent K_a for a-Si equal to the high temperature c-Si value of $\simeq 0.25$ W/cmK [15], but the t_m measurements showed that this could not be correct. K_a was then successively halved until it was in the range 0.01–0.03 W/cmK, when satisfactory agreement with experiment was finally obtained [2]. The value $K_a = 0.02$ W/cmK was used subsequently in the calculations, unless otherwise stated; this value represents an average over the expected weak T-dependence of K_a above room temperature.

The solid curves lying close to the measured data in fig. 1 summarize the agreement that can be obtained between experiment and calculations of t_m vs. E_ℓ for a- and c-Si. The dashed curve in the "c-Si" region illustrates the results of calculations in which we used $T_a = 1150$ °C, and $L_a = 1320$ J/gm but took $K_a = K_c$. It is clear that the reduction of melting temperature and latent heat from crystalline values ($L_c = 1799$ J/gm and $T_c = 1410$ °C) has little effect on t_m (E_ℓ): It is the value of K_c that defines the dynamical melting behavior as being that of c-Si. The dashed curve in the "a-Si" region of fig. 1 reinforces this conclusion, since it was obtained by putting $T_a = T_c = 1410$ °C, $L_a = 1320$ J/gm but keeping $K_a = 0.02$ W/cm K. These results provide convincing evidence for the dominant role of K_a in determining t_m [2].

The value $K_a \simeq 0.02$ W/cmK determined here for self-implantation-amorphized a-Si is quite reasonable. A lattice thermal conductivity calculation, using a Debye phonon spectrum for a-Si, gives K_a (300 K) = 0.011 W/cmK, while Goldsmid et al. [16] recently obtained K_a (293 K) = 0.026 W/cmK for a 1.15 μm thick deposited film of a-Si on sapphire. Webber et al. [17] have also estimated that $K_a \sim 10^{-2}$ W/cmK via computer modeling of pulsed laser melting of ion-implanted a-Si layers.

4. Transmission electron microscopy: evidence for bulk nucleation

Cross-sectional TEM micrographs have been published [3,5] illustrating the solidification morphologies that are observed following pulsed ruby laser (693 nm wavelength, 12 ns FWHM pulse duration) irradiation of a 500 nm thick a-Si layer produced by $^{30}Si^+$ ion implantation of (100) c-Si at liquid nitrogen temperatures. The published TEM results [3,5] are typical of the solidification morphologies that were observed in 100 to 500 nm thick a-Si layers, using both pulsed Nd:YAG and pulsed ruby lasers and will simply be summarized here for the case of a 500 nm thick a-Si layer.

No microstructural change was observed compared to the as-implanted state at $E_\ell = 0.1$ J/cm^2. However, for E_ℓ slightly less than 0.2 J/cm^2, the surface melts as indicated by a jump in reflectivity. At $E_\ell = 0.2$ J/cm^2, the top 183 nm of a-Si was transformed into a very fine-grained polycrystalline (FG) region upon solidification, with average grain size ~10 nm. At 0.3 J/cm^2, two distinct regions were formed: a surface region (80 nm deep) containing large columnar polycrystalline grains (known henceforth as the LG region) and a deeper layer (233 nm thick) containing FG material similar to that formed alone at 0.2 J/cm^2. With further increase of E_ℓ, the thickness of the LG region grows linearly with E_l, while the thickness of the FG region reaches a maximum of about 240 nm at $E_\ell = 0.5$ J/cm^2 and then decreases with further increase of E_ℓ. The LG region disappears completely only when complete epitaxy occurs, with melting through of the a–c interface at high E_ℓ. Thus, for lower E_ℓ (insufficient to melt through to the a–c interface), the sequence of materials observed is LG p-Si near the surface, followed in order of increasing depth by FG p-Si, the remaining unmelted a-Si, and the c-Si substrate; for sufficiently low E_ℓ the LG region is absent.

A microdiffraction pattern taken from the FG region showed that diffraction rings characteristic of amorphous Si were not observed; thus, the FG p-Si region marks an abrupt boundary with the a-Si beneath. The possibility that the FG region results from solid phase nucleation and growth (SPNG) can be ruled out because (1) there is insufficient time for SPNG to result in ~10 nm crystallites (the minimum time required even at the melting point of c-Si is ~1 μs) [6]; and (2) a sharp boundary is observed between the FG and underlying amorphous regions, rather than the diffuse boundary expected for SPNG [5]. The grains of FG p-Si were randomly oriented and equiaxed, indicating that there is no preferred direction for crystal growth during formation of the FG p-Si region, and certainly no microscopic evidence for the existence of a planar solidification front. This behavior is strongly suggestive of bulk nucleation occurring in highly undercooled ℓ-Si. Insight into its origins is obtained from the calculations described next.

5. Model calculations: solidification of highly undercooled ℓ-Si

To our knowledge, a formulation of the heat flow problem incorporating bulk nucleation, large undercoolings, and a pulsed heat source is not yet available. Nevertheless, we have found that finite-difference calculations [18] can be modified to *simulate* nucleation effects under such conditions. Calculations of the type discussed in refs. [1] and [18] should be adequate for determining the onset of melting, t_m, but it is not clear that they can be modified to describe subsequent solidification from a highly undercooled melt. More specifically, such calculations in the past have generally assumed that the material gives up latent heat at the same temperature at which it is absorbed. In the case of an a-Si layer with $T_a = 1150$ °C, the material would have to cool to 1150 °C before solidifying. This is an unnecessarily restrictive condition which, if valid, would seem to imply from thermodynamic considerations that the material could only resolidify as a-Si. However, the presence in typical transmission electron micrographs of randomly oriented FG p-Si suggests that the FG material marks a region in which homogeneous bulk nucleation has occurred at some $T<T_n$, the nucleation temperature, lying between T_a and T_c. The volumetric release of latent heat on bulk nucleation of the c phase is accompanied by either a very rapid solidification or the creation of a "slush" zone [18,19] from which solidification subsequently proceeds at a rapid rate; during this time it may be that no well-defined melt front exists. Nucleation does not occur (or occurs very slowly) in the molten Si with $T>T_n$, until a melt front is reestablished at the interface with the fine-grained material.

With the foregoing in mind, the calculations were modified from those of refs. [1] and [18] to include not only the presence of the a-Si layer but also a simulation of the effects of bulk nucleation. We emphasize that only one physical and mathematical model of an a-Si layer on a c-Si substrate was used in all of the calculations discussed in this paper. We discussed the results for t_m before considering the need to include undercooling and bulk nucleation in the model only because these two effects are not expected to have any significant influence on t_m, whereas they may greatly influence τ_s. While the melt front is still advancing into the a-Si layer, L_a, T_a, and K_a, K_c, or K_ℓ are used depending on the phase of the material in a given finite-difference cell. The latent heat is always switched from L_a to L_c as soon as an a-Si cell has melted; this is the simplest assumption consistent with the transmission electron microscopy observations. If the melt front does not penetrate to the a–c interface, a layer of a-Si with its low thermal conductivity remains. The ℓ-Si in a given cell is allowed to solidify at the temperature of that cell, provided that this T is less than T_n and that the melt front has begun to recede from its point of maximum penetration. For those molten cells in which T is initially $>T_n$ but $<T_c$, crystallization cannot occur until T drops below T_n or until the rapidly growing region of fine-grained Si from an adjacent cell can provide nucleation sites for the liquid with $T>T_n$. When the melt front initially contacts the a–c interface, the temperature of the interface, T_{ac}, will be T_a. The melt front cannot penetrate further until T_{ac} rises to T_c and enough latent heat is supplied to begin to melt the c-Si region. If E_ℓ is not great enough for T_{ac} to reach T_c, crystallization will occur at temperatures between T_a and T_c. A thin layer of fine-grained Si may be formed if $T_{ac}<T_n$ but otherwise only large-grained c-Si will be formed. Finally, when $T_{ac}\geqslant T_c$, epitaxial regrowth of single-crystal Si from the c-Si substrate can proceed just as it would for an entirely c-Si sample.

The results of calculations of τ_s for a 100 nm thick a-Si layer on a c-Si substrate are shown in fig. 2. The solid curve in the c-Si region was taken from ref. [1] and demonstrates that acceptable agreement can be obtained with the experimental τ_s for c-Si. The origin of the apparent displacement of the calculated curve from the experimental points at higher values of E_ℓ is not clear (however, note the experimental error bars). The short-dashed curve in the a-Si region was obtained from calculations with $T_a=1150°C$, $L_a=1320$ J/g, and $K_a=0.02$ W/cmK, while the long-dashed curve used the same values of

L_a and K_a and $T_a = T_c = 1410$°C. Both curves give reasonably good agreement with experiment, given the experimental error bars; thus, a choice between the two sets of parameters is not obvious. In fact, a series of calculations in which T_a was gradually increased to T_c simply gave a family of curves lying between the two shown. Since all of these pairs of T_a, L_a values also give satisfactory agreement with the t_m (E_ℓ) curves of fig. 1, we again see that it is K_a, not T_a or L_a, which strongly influences the results. The peak in the short-dashed curve at 0.5 J/cm^2 may be an artifact of the present modeling; however, both the calculations and TEM results show that this is approximately the value of E_ℓ for which the remaining a-Si layer has become so thin as to greatly increase heat flow to the c-Si substrate. Thus, reflectivity measurements and model calculations of surface melt duration are in satisfactory overall agreement for the 100 nm thick a-Si layer, lending confidence that the model calculations can be used to infer both the depth of melting into a-Si and temperature profiles in the undercooled ℓ-Si, prior to solidification.

Typical results of calculations of temperature vs. depth [3,4] are shown in fig. 3, while melt depth profiles vs. time [4] are shown in fig. 4, for a 100 nm thick a-Si layer and a 523 nm wavelength, 18 ns FWHM laser pulse. (The surface melt durations presented in fig. 2 were obtained from calculations similar to those shown in fig. 4.) In these calculations we again took $T_a = 1150$ °C and $L_a = 1320$ J/gm and $K_a = 0.02$ W/cm K, with other parameter values for

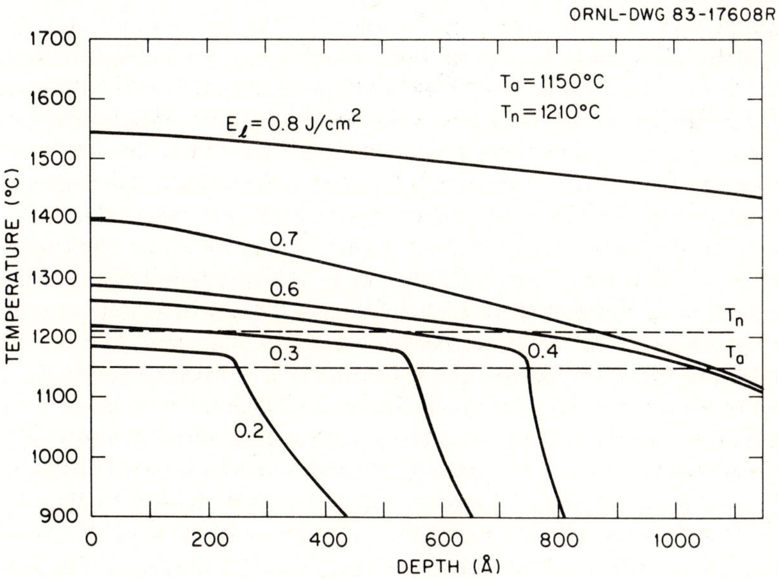

Fig. 3. Profiles of temperature vs. depth in undercooled liquid silicon. Each profile is shown at the time of maximum melt-front penetration for the given E_ℓ.

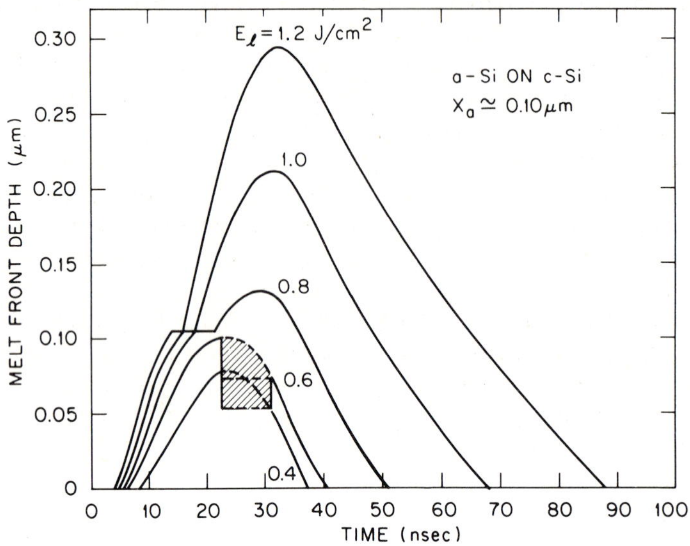

Fig. 4. Melt-front profiles vs. time for different laser E_ℓ. The cross-hatching indicates the region in which TEM indicates that a well-defined melt front does not exist (see text).

ℓ- and c-Si as described above. Figure 3 shows temperature profiles at the times when the melt front for each E_ℓ reaches its maximum penetration. T_n was varied to obtain approximate agreement with the TEM results for the widths of the FG, LG, and a-Si regions; fig. 3 shows $T_n \simeq 1210$ °C but this value depends on the assumed value of T_a. Figure 3 indicates that the temperature of the ℓ-Si for 0.2 J/cm^2 does not rise above T_n, so that only FG p-Si is expected at the surface (in agreement with TEM), followed by the remainder of the a-Si. For 0.3 J/cm^2, $T > T_n$ in the region nearest the surface, and a thin region of LG material is expected to grow from the underlying, bulk-nucleated FG material. As E_ℓ increases, the LG region grows in size while the FG region shrinks until at ~0.6 J/cm^2 only ~20 nm of FG p-Si remains; for $E_\ell \gtrsim 0.8$ J/cm^2 the melt front penetrates completely through the a-Si. The kink in each of the curves in fig. 4 at ~100 nm occurs because $T_a < T_c$ and $L_a < L_c$; the melt front pauses until enough energy is absorbed for it to penetrate the a–c interface. For $0.6 \lesssim E_\ell \lesssim 0.8$ J/cm^2, the front contacts the interface but does not penetrate it. For $E_\ell \lesssim 0.6$ J/cm^2, the temperature of part of all of the ℓ-Si is $< T_n$ and nucleation occurs in an extended region for which no well-defined melt front exists, as indicated schematically by the cross hatching on fig. 4.

Initial calculations for 400 nm thick a-Si layers gave melt-front penetrations much less than the measured combined thickness of the FG and LG regions. This seemed to imply that bulk nucleation and grain growth, accompanied by the release of latent heat, occurred before the melt front reached its maximum penetration and increased the depth of melting. We might expect (and the calculations confirmed) this to be an important effect for thick layers but not for 100 nm thick layers. Bulk nucleation *during* melt-in was simulated by allowing an increment ΔE_n of latent heat to be released *during* melt-front penetration. ΔE_n was estimated from the relationship $\Delta E_n = F\rho L_c X_m$, where X_m is the measured maximum penetration at a given E_ℓ, ρ is the density, and F is a factor interpreted as the fraction of ℓ-Si that solidifies before the melt front reaches X_m. To be physically meaningful, F must be $\leqslant 1$ and we found $F \simeq 0.5$–0.6 for most E_ℓ. Moreover, the temperature profiles when X_m was reached again indicated that the widths of the LG and FG p-Si regions extracted from the calculations were in good agreement with the TEM values if $T_n \simeq 1210$ °C. The model calculations clearly indicate that *molten material can exist in the interior of the sample after the surface has resolidified.*

The theory of homogeneous bulk nucleation [19] yields an expression for the nucleation rate I of the form

$$I = I_0 \exp(-\Delta G^*/kT). \tag{1}$$

I_0 is a factor having units of $\mathrm{cm^{-3}\,s^{-1}}$ and magnitude of approximately 10^{35}. ΔG^* is the free energy of a critical nucleus at temperature T. For spherical nuclei of radius r the free energy is given by $\Delta G = 4\pi r^2\sigma + 4\pi r^3 \Delta G_v/3$, in which σ is the surface free energy and ΔG_v the volume free energy. Setting $\mathrm{d}\Delta G/\mathrm{d}r = 0$ gives

$$r^* = -2\sigma/\Delta G_v, \qquad n^* = 4\pi r^{*3} N_v/3, \qquad \Delta G^* = 16\pi\sigma^3/3(\Delta G_v)^2. \tag{2}$$

r^* is the radius of the critical nucleus, n^* is the number of atoms in it, N_v is the number of atoms/unit volume. If the difference in the heat capacity C_p between the liquid and the solid is neglected, ΔG_v can be expressed in terms of the undercooling $\Delta T \equiv T_c - T$ as $\Delta G_v = -L_c \Delta T/T_c$.

If indeed the FG p-Si is due to bulk nucleation, σ can be calculated as follows. TEM [3,5] indicated an average grain size of ~10 nm or a volume of 10^{-18} $\mathrm{cm^3}$. Knowing the thickness x_p of the FG p-Si region from TEM and an approximate value of its formation time t_p from the model calculations, I can be determined under the assumption that each 10 nm grain resulted from a single nucleation event. For example, from the results leading to figs. 3 and 4 for $E_\ell \simeq 0.3$ J/cm^2, we estimate $x_p \simeq 50$ nm, $t_p \simeq 10$ ns, and find $I \simeq 10^{26}$ $\mathrm{cm^{-3}\,s^{-1}}$. With a value of I established, eqs. (1) and (2) and the expression for ΔG_v can be used to relate σ to ΔT. Following ref. [8], we took $\Delta T = 265$ K and calculated σ, r^*, and n^* with the results: $\sigma = 219$ erg/cm^2, $r^* = 0.66$ nm, and $n^* = 60$ atom. For the lower limit of ΔT (=390 K) estimated in ref. [8], we found $\sigma = 275$ erg/cm^2, $r^* = 0.57$ nm, and $n^* = 39$ atom. Having fixed σ, ΔT can be varied to determine its effects on I, R^*, and n^*, with typical results shown in table 1.

Table 1
I, r^*, and n^* vs. $\Delta T/T_c$ for $\sigma = 275$ erg/cm^2

$\Delta T/T_c$	I (K)	I (cm^{-3} s^{-1})	r^* (nm)	n^*
0.13	1464	7.9×10^6	1.01	214
0.15	1431	5.0×10^{15}	0.87	139
0.17	1397	4.1×10^{19}	0.77	96
0.19	1363	2.4×10^{22}	0.69	69
0.21	1330	2.6×10^{24}	0.62	51
0.23	1296	8.7×10^{25}	0.57	39
0.25	1262	1.3×10^{27}	0.52	30

We note that the values of I in table 1 are much greater than those estimated by Turnbull [20] but are not inconsistent with values found from molecular dynamics calculations [21].

We will summarize our view of the physical processes that take place. The 18 ns laser pulse causes the a-Si to melt at a $T \sim 300$ K less than T_c. Melting is either incomplete, in the sense that small clusters of atoms remain in the melt, or nucleation occurs so rapidly that the two situations are nearly indistinguishable on the time scale of the present experiments. In either case, a fraction of the latent heat is, in effect, released while the melt front is still advancing; this release is essential for forcing the melt front to its maximum penetration. In the FG p-Si region the temperature of the ℓ-Si is $\leqslant T_n$ long enough for nucleation to occur. Because nucleation will occur in those regions first melted,

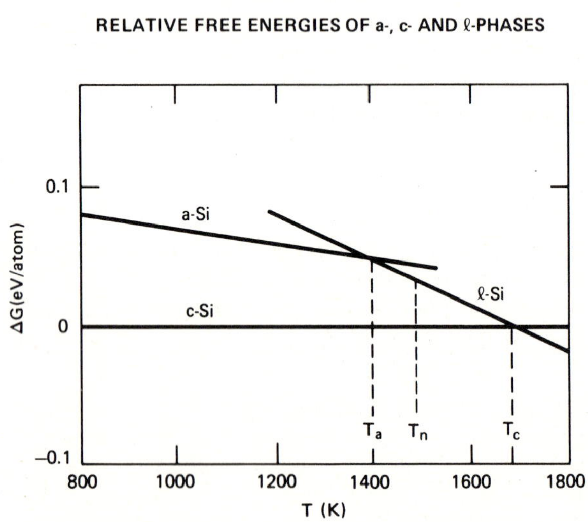

Fig. 5. Schematic representation of Gibbs free energy vs. temperature for a-Si and ℓ-Si relative to c-Si.

the surface may solidify before the deeper-lying regions. We have not yet determined the extent of the "slush zone" in which ℓ-Si and c-Si coexist at any particular time, but we note that *the physical properties of such a mixture are likely to be quite different from those of either pure* ℓ-Si *or c-Si.* Finally, we observe that T_n should appear on the free-energy diagram, as in fig. 5. From this diagram [4] it can be seen that slow cooling of ℓ-Si from T_c will result in nucleation of p-Si rather than formation of a-Si. Therefore, the ℓ→a transition that can be produced using picosecond or ultraviolet nanosecond laser pulses must occur either because the system is undercooled to T_a so quickly that nucleation cannot occur or because a large number of mistakes are made in the incorporation of atoms into the growing interface at high solidification velocities. (To the extent that the free-energy diagram of fig. 5 can be used, the latter situation corresponds roughly to a path from the ℓ-Si line to, and along, the superheated a-Si line at temperatures above T_n.) Thus, we see that the a-Si phase must be "trapped" by nonequilibrium thermodynamic and kinetic effects [23].

Acknowledgements

The authors thank F.W. Young, Jr. and G.E. Jellison, Jr. for several informative discussions. The research was sponsored by the Division of Materials Sciences, U.S. Department of Energy under Contract No. DE-AC05-840R21400 with Martin Marietta Energy Systems, Inc.

References

[1] D.H. Lowndes, R.F. Wood and R.D. Westbrook, Appl. Phys. Lett. 43 (1983) 258.
[2] D.H. Lowndes, R.F. Wood and J. Narayan, Phys. Rev. Lett. 52 (1984) 561.
[3] D.H. Lowndes, R.F. Wood and J. Narayan, Mat. Res. Soc. Symp. 23 (1984) 99.
[4] R.F. Wood, D.H. Lowndes and J. Narayan, Appl. Phys. Lett. 44 (1984) 770.
[5] J. Narayan and C. W. White, Appl. Phys. Lett. 44 (1984) 35.
[6] G.L. Olson, S.A. Kokorowski, J.A. Roth and L.D. Hess, Mat. Res. Soc. Symp. 13 (1983) 141 and references herein to earlier papers by these authors.
[7] G.L. Olson, J.A. Roth, L.D. Hess and J. Narayan, Proc. U.S.–Japan Seminar on Solid Phase Epitaxy and Interface Kinetics, Oiso, Japan (June 20–24, 1983).
[8] E.P. Donovan, F. Spaepen, D. Turnbull, J.M. Poate and D.C. Jacobson, Appl. Phys. Lett. 42 (1983) 698.
[9] B.G. Bagley and H.S. Chen, in: AIP Conf. Proc. 50 (1979) 97.
[10] F. Spaepen and D. Turnbull, ibid., p. 73.
[11] P. Baeri, G. Foti, J.M. Poate and A.G. Cullis, Phys. Rev. Lett. 45 (1980) 2036.
[12] J.A. Knapp and S.T. Picraux, Appl. Phys. Lett. 38 (1981) 873.
[13] See, e.g., P.L. Liu, R. Yen, N. Bloembergen and R.T. Hodgson Appl. Phys. Lett. 34 (1979) 864; R. Tsu, R.T. Hodgson, T.Y. Tan and J.E. Baglin, Phys. Rev. Lett. 42 (1979) 1356;

A.G. Cullis, H. C. Weber, N.G. Chew, J.M. Poate and P. Baeri, Phys. Rev. Lett. 49 (1982) 219.

[14] G.E. Jellison, Jr. (private communication).

[15] C.J. Glassbrenner and G.A. Slack, Phys. Rev. A 134 (1964) 1058.

[16] H.J. Goldsmid, M.M. Kaila and G.L. Paul, Phys. Status Solidi A 76 (1983) K31.

[17] H.C. Webber, A.G. Cullis and N.G. Chew, Appl. Phys. Lett. 43 (1983) 669.

[18] R.F. Wood and G.E. Giles, Phys. Rev. B 23 (1981) 2923.

[19] R.F. Wood, to be published; see J.H. Holloman and D. Turnbull, Progr. Metal Phys. IV (1953) 333.

[20] D. Turnbull, J. Appl. Phys. 21 (1950) 1022.

[21] C.S. Hus and A. Rahman, J. Chem. Phys. 70 (1979) 5234.

[22] The possibility that a-Si can be substantially superheated has been suggested previously; see, e.g., ref. [8].

[23] R.F. Wood, Mat. Res. Soc. Symp. 13 (1983) 83.

Journal of Luminescence 30 (1985) 409–424
North-Holland, Amsterdam

THERMAL DESCRIPTION OF LASER ANNEALING

P. BAERI

Istituto Dipartimentale di Fisica, Università di Catania,
Corso Italia, 57 I95129 Catania, Italy

Heating and cooling rates as well as melting and solidification velocity of surface layers of irradiated samples by laser pulses are semi-quantitatively described in terms of heat flow concepts based on the assumption that laser light is directly converted in lattice heating. The range of validity of this approach compared with a more complete scheme taking into account the free carrier plasma evolution is sketched and the importance of Auger effect in the plasma–lattice coupling mechanism is detailed. The most important consequences of the quencing rates achievable by short laser pulse irradiation on the structure modification of semiconductor surface layers are reviewed with more details on the liquid to amorphous silicon transition. This is in fact the more new and less understood fast solidification process induced by pulsed laser irradiation.

Introduction

The basis of the simplest description of laser annealing is the hypothesis that light energy $\mathrm{d}E$ absorbed in a volume $\mathrm{d}V$ of the irradiated sample in a time interval $\mathrm{d}t$ is converted into heat "locally" and "instantaneously" giving rise to a local positive variation of temperature $\mathrm{d}T = \mathrm{d}E/\rho C_\mathrm{p}\,\mathrm{d}V$ in the same time $\mathrm{d}t$, ρ being the mass density and C_p the specific heat of the sample material. Heat diffusion, controlled by the thermal conductivity K (or the thermal diffusivity $D = K/\rho C_\mathrm{p}$), provides a redistribution in temperature both during the heating (i.e., during the irradiation with a laser pulse) and during the cooling (after the laser switch-off).

A thermal diffusion equation with a "source" term which describes the heating of the sample is generally solved to obtain detailed information on the sample temperature evolution [1,2]

$$\frac{\partial T}{\partial t} = \frac{\partial}{\partial x}\left(\frac{K}{\rho C_\mathrm{p}}\frac{\partial T}{\partial x}\right) + \frac{\alpha I(x,\tau)}{\rho C_\mathrm{p}}, \qquad (1)$$

where α is the light absorption coefficient and I the light intensity at a depth x

0022-2313/85/$03.30

inside the sample and at time t. Numerical solution of the equation is often needed because of the involved parameters (α, C_p, K) with the temperature and with the structure of the sample which in many practical cases is a layered one (amorphous layer on crystal bulk or metal deposited on semiconductors, and so on). A numerical solution also allows one to take into account the local enthalpy aborption and release during phase transition and so it is possible to simulate the evolution of the sample melting and solidification [3].

Heat flow concepts

Without going into detail with the heat diffusion equation solutions, some heat flow concepts can be very easily predicted. Assuming a planar geometry, the sample is heated by a laser pulse of time duration τ_p through a thickness of the order of the light absorption length α^{-1} or the heat diffusion length $\sqrt{2D\tau_p}$ depending on which of these two parameters is larger. The energy density E_{th} required to reach the melting point at the sample surface (which is normally the hottest point) is proportional to the melting temperature and to the inverse of the heated thickness.

Pulses of energy density E greater than E_{th} will melt a layer whose thickness, within the first μm, will be proportional to $E - E_{th}$. In silicon the light penetration depth can range from 10^{-6} cm in the case of uv irradiation to 10 cm in the case of Neodymium 1.06 μm wavelength (in this last case it strongly varies with temperature approaching the value of 10^{-4} cm close to the melting point). Pulse time duration can be varied from hundred nanoseconds down to a couple of nanoseconds in Q-switched mode, or it can be reduced to some ten picoseconds; the corresponding heat diffusion length are 30 μm, 5000 Å and 300 Å respectively. We will here neglect any discussion about femtosecond laser pulses. The energy density threshold for sample melting is 0.2 J/cm^2 for 2.5 ns, 0.347 μm, pulse [4], it is also 0.2 J/cm^2 for a 20 ps [5] 0.53 μm wavelength. In the first case the heated thickness is 5000 Å due to heat diffusion, in the second case it is 10000 Å due to light absorption depth. This factor two is perfectly compensated by a factor two in the reflectivity coefficient for these two wavelengths. These examples are the case of the lowest energy density threshold for the melting of Si which up till now are experimentally verified. For a 30 ns ruby laser pulse crystal silicon the energy density threshold to melt the surface is 0.8 J/cm^2; it changes very slow with the pulse duration due to the fact the heated thickness is largely determined by the absorption length. For amorphous silicon instead, the absorption length being a few thousand angstroms, the melting threshold varies from 0.4 to 1.2 J/cm^2 changing the pulse duration (and then the heat diffusion) from 10 to 100 ns. E_{th} of 4.5 J/cm^2 is needed for a nanosecond pulse at 1.06 μm, due to the very low absorption coefficient of this wavelength in Si; but due to the increase of absorption coefficient with temperature, 3 J/cm^2 is enough if the sample is pre-heated at 150°C.

Heating of the sample surface happens in the same time scale of the pulse itself. Heating rates ranging from 10^{12} to 10^{10} K/s for pulses of 1 and 100 ns time duration respectively, can be reached with energy density close to E_{th}. Heating inside the sample is delayed by heat diffusion (about 100 ns is needed for heat to be diffused through 1 μm in hot silicon). So, even for very short pulses, the maximum heating rate before melting at 1 μm depth will be of the order of 10^{10} K/s. Cooling after the laser switch-off is faster the steeper the temperature gradient. This temperature gradient near the end of the laser pulse is of the order of $T_m/\sqrt{2D\tau}$ or $T_m \times \alpha$ [3], T_m being the melting temperature. It can reach several 10^7 K/cm for the extreme case of the 2.5 ns uv. For a 30 ns ruby or neodymium laser on amorphous silicon it is about 7×10^6 K/s, and is about 3×10^6 K/cm for ruby laser on crystal Si due to greater absorption depth in crystal compared with amorphous Si, whilst it is about 5×10^5 K/cm for neodymium laser on crystal Si for the same reason. With increasing time or increasing depth inside the sample the temperature gradient becomes smaller and smaller. For example in the previous cited case of ruby laser on crystal silicon, the temperature gradient at the surface becomes a factor five smaller 200 ns after the pulse switch off.

Surface cooling rate is proportional to the temperature gradient. Maximum cooling rates of several 10^9 K/s are easily achieved at the surface whilst at 1 μm depth they will be about one order of magnitude smaller. As an example, in the following the first direct time-resolved temperature measurement is reported [6].

A thin 2000 Å thick film thermocouple (Fe and constantan) has been in vacuum deposited onto a alumina or Si substrate. On top of it a 1 μm thick Ge film has been sequentially deposited without vacuum breaking. A front and lateral view of this device is showed in fig. 1a. The circular central part of the thermocouple has a diameter of 3 mm. Regions 1 and 2 are the cold points at room temperature and are connected to the oscilloscope by thin copper wires.

The Ge layer was shot by Nd glass laser pulses with 50 ns FWHM. The beam spot diameter was 3 mm and spatially uniform within 10%. The output voltage of the thermocouple during irradiation is directly measured by a fast storage oscilloscope (fig. 1b). Results for a 0.3 J/cm^2 shot are shown in fig. 2a and 2b for an alumina and Si substrate, respectively. In accordance with a previous calibration of the thermocouple of 16.6 μV/°C a maximum temperature of about 600°C is reached in both samples. The measured quenching rates are 3×10^8 and 10^9 K/s respectively. The different values are related to the different thermal conductivity of the two substrates. In Si it is about 1W/s°C while in Al_2O_3 it is at least one order of magnitude smaller. The time delay between the pulse peak (shown as a dashed line) and the temperature maximum is due to heat diffusion process because the sample is heated essentially at the surface whilst the temperature is recorded 1 μm inside the sample. For simplicity the calculation to be

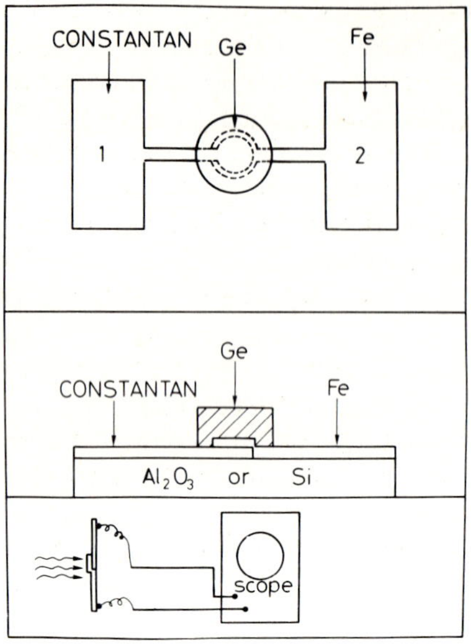

Fig. 1. Plan view (a) and lateral view (b) of the deposited thin film thermocouple. The insert shows the schematic electrical circuit.

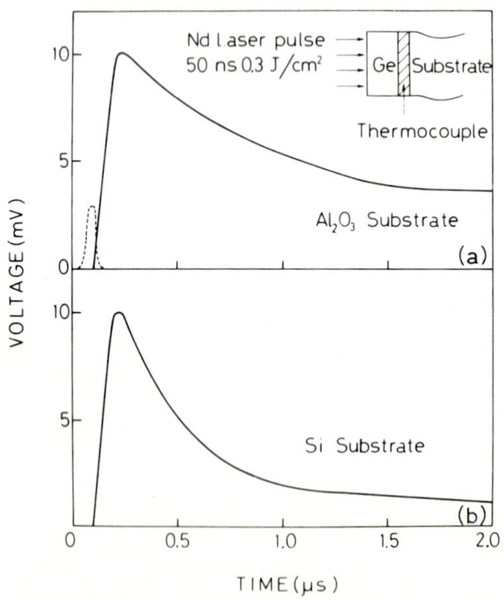

Fig. 2. Oscilloscope trace of the output voltage vs time obtained by irradiation of a 1.2 μm thick Ge layer deposited onto a thin iron–constantan thermocouple on Al_2O_3 (a) and on Si (b) substrate. The pulse shape is shown as a dashed line for comparison.

compared with the experimental data were made using a heat flow procedure assuming a bulk Ge sample of infinite thickness and irradiated with 50 ns 1.06 μm laser pulse. The temperature–time dependence at the sample surface (upper part) and at 1 μm depth (lower part) computed for 0.3 and 0.15 J/cm^2 energy density values are reported in fig. 3. The threshold for surface melting is 0.27 J/cm^2 and the maximum temperature computed at the surface and at 1 μm depth differs about 500 K for the 0.3 J/cm^2 irradiation. The difference becomes negligible 0.5 μs after the pulse. The average quencing rate is about 4×10^8 K/s which is comparable to the values extracted from the experimental results of fig. 2.

The liquid–solid interface velocity during the epigrowth, following the laser melting, plays an important role in many nonequilibrium thermodynamic questions, as impurity trapping and metastable phase formation. It is then important, within the framework of the heat flow computations, to understand how to control it.

At the liquid–solid interface the heat flow must be equal to the release of the latent heat, so

$$K\,\mathrm{d}T/\mathrm{d}x = H\rho v, \tag{2}$$

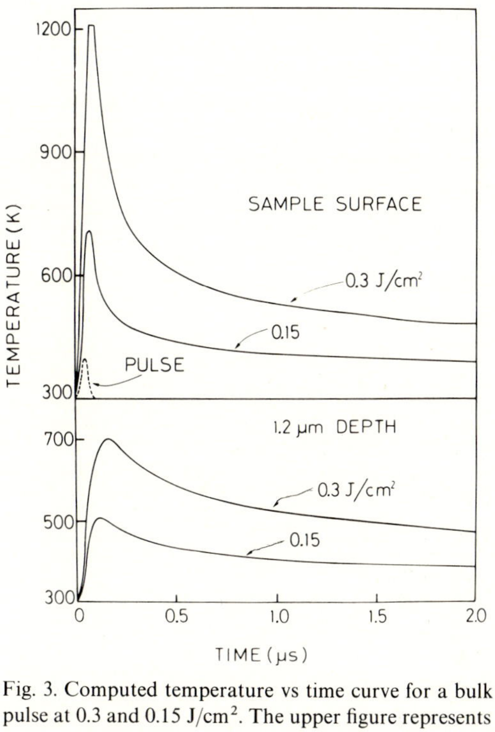

Fig. 3. Computed temperature vs time curve for a bulk Ge sample irradiated with 50 ns Nd laser pulse at 0.3 and 0.15 J/cm^2. The upper figure represents the temperature at 1 μm inside the sample.

where v is solidification velocity, H the melting enthalpy, $\mathrm{d}T/\mathrm{d}x$ the temperature gradient in the solid behind the liquid–solid interface.

Another term taking into account the temperature gradient in the liquid layer should be included in eq. (2) but it can be neglected due to the much lower value of $\mathrm{d}T/\mathrm{d}x$ in the liquid [10].

The temperature gradient just at the end of the pulse is strongly determined by the energy deposition depth as previously discussed. It turns out that the velocity, at least for energy densities close to the melting threshold, is inversely proportional to the square root of the pulse duration, or directly proportional to the absorption coefficient, depending on which of the two is more influencing the energy deposition length. Due to the large variation of α and τ_p it has been possible to control the liquid–solid interface velocity from less than 1 m/s to 20 m/s. As an example in fig. 4 are shown the calculated melt front position vs time for four different cases. Case (a) is a free running neodymium laser on amorphous silicon, case (b) is a Q-switched Nd 30 ns laser pulse on crystal silicon pre-heated at 300°C, case (c) is the same as case (b) but on an amorphous silicon and case (d) is an ultraviolet (ruby double frequency 2.5 ns pulse on crystal Si). The energy densities are 30, 3.2, 2.5 and 0.45 J/cm^2 for case (a), (b), (c) and (d) respectively. The velocity of the liquid–solid interface decreases also on increasing the energy density above the threshold.

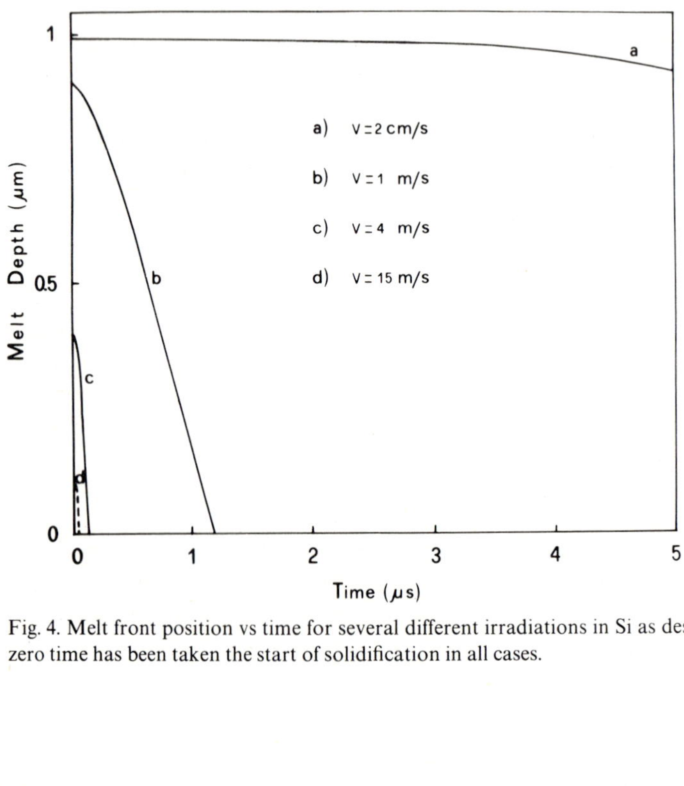

Fig. 4. Melt front position vs time for several different irradiations in Si as described in the text. As zero time has been taken the start of solidification in all cases.

Energy coupling

Till now we have neglected all the intermediate mechanisms of energy transformation from laser light to sample heating, but the previous scheme is valid in the following hypothesis:

(1) The diffusion length of free carriers before recombination must be shorter than the light penetration depth or the heat diffusion length during pulse duration.

(2) Free carrier absorption must be negligible compared with lattice (band-to-band) absorption.

(3) The lifetime of free carriers must be shorter than pulse duration.

If these conditions are not fulfilled a more complicated scheme of calculation which include space and time evolution of free carrier density and energy like the one proposed by Lietoila and Gibson [7] must be used. Two characteristic time constants are important in this scheme: the electron–phonon relaxation time τ_e and the Auger decay time τ_A. τ_e has been estimated to be of the order of 1–10 ps although Yoffa [8] has calculated that under conditions pertinent to laser annealing the rate of energy release of the carrier to the lattice is slowed down; recent experiments of time-resolved reflectivity in the picosecond regime by Liu, Kurtz and Bloembergen [9] have shown that 1 ps is a suitable value for τ_e so its influence on the simplest "thermal" scheme is negligible.

The Auger decay rate τ_A is related to the free carrier concentration N by the Auger decay constant γ_3: $\tau_A = 1/\gamma_3 N^2$. An evaluation of the value of γ_3 in silicon can be done by a transmission measurement with Nd 1.06 µm wavelength laser pulses. This wavelength can create electron–hole pairs because the energy associated with a photon is just above the silicon band gap; moreover free carrier absorption becomes comparable with lattice absorption at concentration levels exceeding some 10^{18} carriers/cm^3. The pulse which creates the plasma is then a suitable probe to determine its concentration through the measurement of transmittance. In fig. 5 is plotted the transmitted against the incident energy density measured in the range 0.1–3 J/cm^2 of a 25 ns laser pulse. The silicon sample was a 10 Ω cm, 100 µm thick single crystal (100) oriented with both faces optically polished. The trend is clearly not linear, the transmitted incident energy ratio being 0.25 at 1 J/cm^2 and 0.16 at 2.8 J/cm^2. Although lattice heating plays some role for this nonlinearity, mainly responsible for the absorption increase with energy density is the free carrier plasma. Calculations based on the previous quoted model [7] assuming a free absorption cross section of 5×10^{-18} cm^2 at this wavelength are reported in the same fig. 5 for several values of the Auger decay constant γ_3.

The best fit is found for $\gamma_3 = 8 \times 10^{-31}$ cm^6 s^{-1}. Free carrier diffusion is negligible because of the in-depth light penetration. (Lattice absorption ranges from 10 cm^{-1} to 110 cm^{-1} from R.T. to 450 K.) Lattice absorption temperature dependence was previously determined at low intensity incident light in the

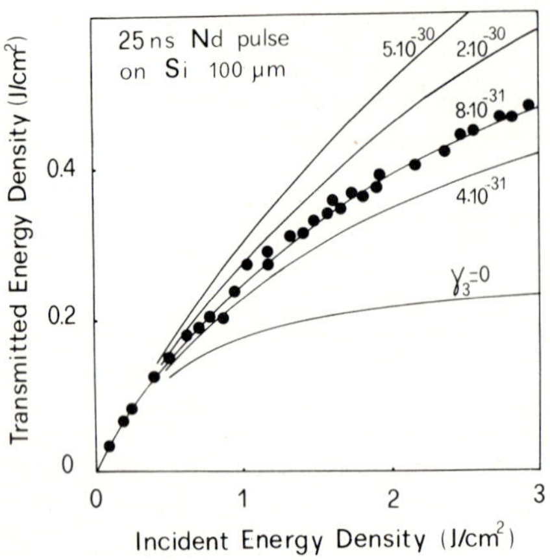

Fig. 5. Transmitted energy density through a 100 μm Si sample for a 25 ns 1.06 μm pulse vs incident energy density (black dots). Full lines refer to calculations assuming different values of the Auger decay constant γ_3.

range 300–500 K. In the case of $3 J/cm^2$ pulse a maximum carrier concentration of $3 \times 10^{19}/cm^3$ is reached 10 ns after the pulse peak and at this concentration the Auger lifetime is 1.38 ns. A maximum temperature of 450 K is reached 35 ns after the pulse peak. A temperature rise without taking into account free carrier absorption would have been, instead, much lower. This example shows that the Auger decay mechanism can be a limit in the energy exchange between electrons and lattice. In table 1, for different laser pulses on Si characterized by the first 4 columns (energy density, time duration, wavelength and lattice absorption coefficient) are reported the energy deposition depth L (the maximum between the absorption length and the heat diffusion length); the maximum free carrier concentration, N, reached during the pulse; the corresponding minimum Auger decay time, τ_A; the average diffusion length, L_D, of carriers during the time τ_A, assuming an ambipolar diffusion coefficient of $100 cm^2/s$; and the maximum free carrier absorption coefficient. Conditions 1, 2 and 3 stated at the beginning of this paragraph are fulfilled in all the ns pulse cases except the 1.06 μm pulse in

Table 1

E (J/cm^2)	τ_p (s)	λ (μm)	α_L (cm^{-1})	L (μm)	N (cm^{-3})	τ_A (s)	L_D (μm)	α_{FC} (cm^{-1})
3	3×10^{-8}	1.06	10	10^3	2×10^{19}	3×10^{-9}	5.4	100
0.8	3×10^{-8}	0.64	10^3	10	6.4×10^{19}	3×10^{-10}	1.7	320
0.2	2.5×10^{-9}	0.315	10^6	0.5	6.4×10^{20}	3×10^{-12}	0.17	3200
1	2×10^{-11}	1.06	10	10^3	1.4×10^{20}	6×10^{-11}	0.7	700
0.2	2×10^{-11}	0.53	10^4	1	5×10^{20}	5×10^{-12}	0.22	2500

which absorption by free carrier is greater than lattice absorption. In the picosecond regime the Auger decay time is comparable with the pulse duration itself; plasma evolution must be taken into account for a correct prediction of lattice heating. It must be pointed out however that in any case the final result is heating or melting of the lattice even if somewhat delayed in respect of the light pulse itself.

Structure modification of surface layers

The very high heating and cooling rates achievable by short laser pulses assume relevance because they allow the formation of particular structure in the irradiated surface layers. In fact by the fast heating, it is possible to reach the melting point preventing other phase changes in solid phase of any metastable structure present at the sample surface. An example of this effect is the melting of the amorphous silicon. Amorphous silicon is a metastable structure whose free energy is higher than that of crystal silicon [11]. It undergoes spontaneously the transition to poly or single crystal in solid phase. The maximum velocity for epygrowth for amorphous layers on a single crystal is about 10^3 Å/μs. To melt an amorphous layer of about 1000 Å without regrowing a single crystal in solid phase, a heating rate of at least 10^9 K/s must then be employed. The melting temperature of amorphous Si is some hundred degree below the crystal melting point, the liquid produced by the previous melting can then be at a temperature much lower than the crystal–liquid equilibrium temperature.

Recrystallization from such an undercooled liquid may produce very peculiar structures [12] and its behaviour is still under investigation. A quite good understanding of this kind of phenomena is presented by D.H. Lowndes in the Proceeding of this conference; for precision's sake reference must be made to a quite different interpretation of the same phenomena reported by Thompson et al. [13].

Another important aspect is related to the cooling rate and the solidification velocity. Solidification velocity of several m/s can result in some metastable structure formation; quencing of this structure from the melting point down to room temperature in some hundred nanoseconds is enough to prevent any change in a thermodynamically more stable structure. The diffusion coefficient of dopants in silicon has a maximum value of the order of 10^{-12} cm^2/s close to the melting temperature. The average diffusion length in the hundred nanosecond time scale is only 1 Å, which means that any long-range atomic reordering is prevented.

The first effect studied as a function of the liquid–solid interface velocity was the impurities surface accumulation. After the complete solidification of the irradiated sample implanted with some dopant, part of the impurity is rejected on the sample surface.

According to the classical theory of normal freezing this has been explained as follows: whilst the melt front comes toward the surface, the impurity profile is broadened by the diffusion in the liquid, a fraction of dopant is rejected by the advancing interface in the phase characterized by the higher solubility, i.e., in the liquid, and then accumulates the sample surface. At the interface the concentrations of impurities in liquid, c_1, and in the solid, c_s, will stay in a given ratio $K_0 = c_s/c_1$, K_0 is the equilibrium distribution coefficient defined as the ratio of the concentration in the solid and the liquid phase determined by the phase diagram at a fixed temperature close to the melting point. But far from thermodynamic equilibrium, at the very high speed reached in the case of laser irradiation, the process of surface rejection of impurity can be described still in terms of normal freezing but using instead of K_0 an interface distribution coefficient $K'(v)$ which is velocity dependent.

Usually K_0 is much less than unity and the very common result at high velocity reached by laser annealing is that K' is several orders of magnitude greater than K_0 and depend strongly on the velocity. With all the other parameters fixed, the amount of impurity rejected on the sample surface increases with decreasing interface segregation coefficient K'. The amount of impurity which is retained in the solid is located into the lattice sites of the silicon crystal at a concentration far above the solid solubility limit thus forming a supersaturated solid solution.

To determine $K'(v)$ one has to compare the measured dopant profile after laser irradiation with the results of a model calculation for dopant redistribution [3] using K' as a fitting parameter. The kinetic of the melt front remain fixed by the irradiation condition through the heat flow calculation. As an example of different redistributions due to different solidification rates fig. 6 reports the depth profiles of Te implanted in Si and irradiated with a Nd laser pulse at 1.06 μm. The lower part shows the profiles after irradiation with a Nd laser pulse at 1.06 μm. The lower part shows the profiles after irradiation with 2.0 J/cm^2 of the implanted sample. The amount of surface accumulation is ~20%, the calculated solidification velocity is about 3 m/s. If the implanted sample is first thermally annealed so that a single crystal is obtained before the Nd laser irradiation, the Te profile reported in the upper part of the figure shows a considerably larger surface accumulation.

The small absorption coefficient of the 1.06 μm wavelength in Si crystal gives rise to a small temperature gradient and the calculated solidification velocity is about 0.8 m/s. The corresponding K' fitting values are ~0.5 and 0.03 for the higher and the lower velocity, respectively. Several theories [14,15,16,17] have been developed to describe nonequilibrium dopant segregation at a fast moving liquid–solid interface. It is far from the purpose of this work to review them and the reader is referred to the original papers or to recent reviews [18].

The common feature to these models is a quantity D_i defined as the impurity diffusion coefficient in the near interface region. It has been shown [19,20] that

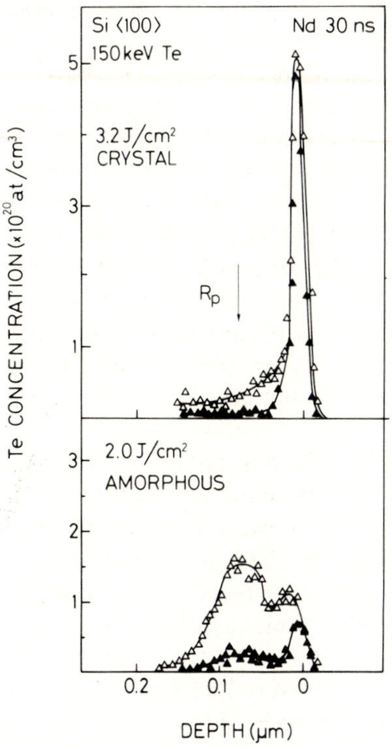

Fig. 6. The depth profile of a Si (100) sample irradiated with 150 keV Te after irradiation with 30 ns Nd laser pulse for a single crystal (up) and with 1500 Å amorphous layer (down). The R_p arrow represents the average penetration of Te before the irradiation.

assuming $D_i = (D_s D_1)^{1/2}$ D_s and D_1 being the diffusion coefficients in the solid and liquid phase respectively, both at the melting point, it is possible to predict if a given impurity will be trapped or rejected on the surface. If $v_c = D_i/\lambda \geqslant v$, λ being the interface thickness (i.e. few monolayers) the impurity is faster than the interface and will accumulate in the high solubility phase, i.e. the liquid, the final result will be a K' value very close to the equilibrium K_0 value. If $v_c = D_i/\lambda \leqslant v$ then the impurity will be frozen by the fast advancing solid phase and K' will reach a value very close to unity.

Quencing of the amorphous phase

Another example of a fast quenced metastable structure is the amorphization of silicon single crystals. It has been shown in fact that for solidification velocity approaching 15 m/s [21] amorphous silicon is obtained. The picture of this phenomenon is not yet completely understood as the one of the supersaturated solutions previously discussed.

In fig. 7 [22] the measured regrowth velocity is shown as a function of the maximum melt depth for a 2.5 ns 0.347 μm wavelength laser pulse oriented on a silicon single crystal (100) and for energy densities ranging from 0.2 to 0.8 J/cm^2. The melt front kinetics has been followed by means of time resolved conductivity measurements and closely agree with the one computed on the basis of the thermal model. The black dots refer to samples in which the amorphous layer was produced after irradiation. A maximum liquid–solid interface velocity of 15 m/s then exists for the regrowth of a single Si crystal from the melt in the (100) direction. A possible approach to the explanation of this effect comes from thermodynamic considerations. A finite solidification velocity can be obtained only by freezing the liquid at a temperature below the melting point.

Theoretical evaluation [23,24] predicted about 15 K undercooling for 1 m/s velocity. In fig. 8 is plotted the free energy vs temperature for crystal, amorphous and liquid crystal. The liquid curve intersects the solid curve at two different temperatures for the amorphous and the crystal phases.

Liquid to amorphous transition is possible only at a temperature about 300 K below the crystal melting point. Following the previous estimate, a liquid–solid interface velocity of 20 m/s will produce an undercooling of about 300 K that is enough to bring the liquid silicon at the energy level required to undergo the transition to the amorphous phase. But kinetic limitation concerning the nuclea-

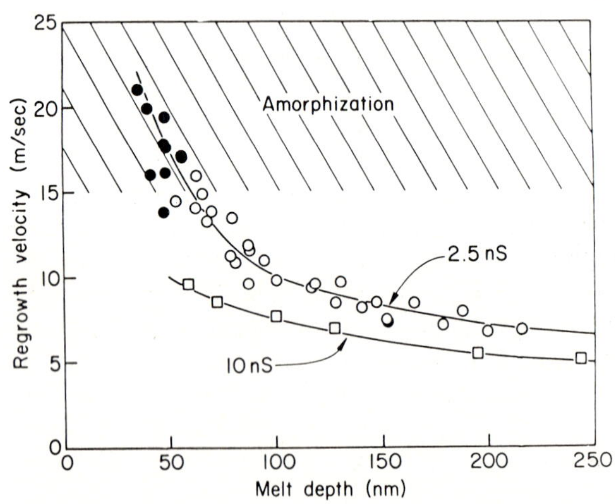

Fig. 7. Regrowth velocity vs maximum melt depth for uv pulse irradiation measured by transient conductance measurements. Solid dots indicate that amorphous Si forms from the melt (from ref. [22]).

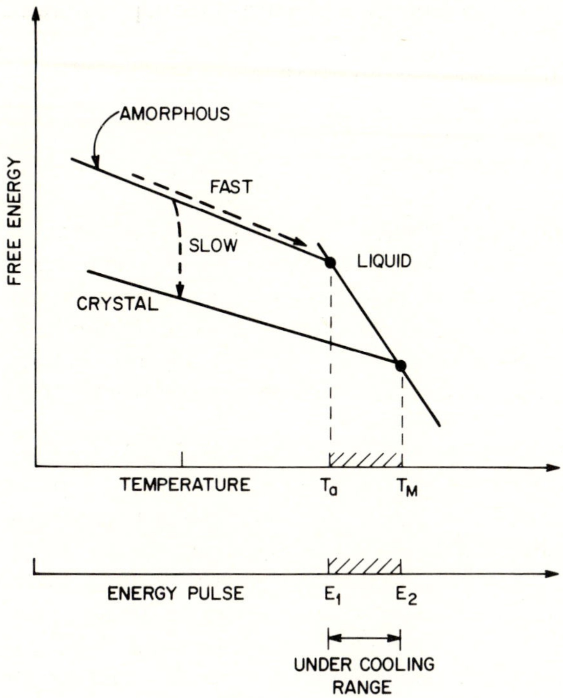

Fig. 8. Free energy vs temperature (schematic) for amorphous, crystal and liquid silicon. T_α and T_m differ by about 300 K.

tion of the amorphous phase must be also taken into account to explain this effect.

Further indications can be obtained from a comparison between the amorphous formation in a doped and undoped Si crystal.

In fig. 9 [25] are shown the aligned spectra for a 2 MeV He incident on Si (111) samples irradiated with 0.33 J/cm^2 2 ns 347 nm laser pulse doped (full dots) with 50 keV Te 2×10^{15} cm^{-2} and undoped (open dots).

The thickness of the amorphous layer is almost a factor 2.5 greater for the doped sample. It can be interpreted as an influence of the dopant atoms that act as heterogeneous nucleation centers for the amorphous phase.

The effect can also be explained with some difference in the undercooling–velocity relationship between a doped and an undoped crystal but kinetic considerations about nucleation rates are strongly proposed by this other following effect [26].

A sample was first implanted with 120 keV, 2×10^{15} at. cm^{-3} In ions and then annealed by a 2 J/cm^2 30 ns q-switch ruby laser pulse. The resulting structure was a single crystal free of defects with an average concentration of substitutional

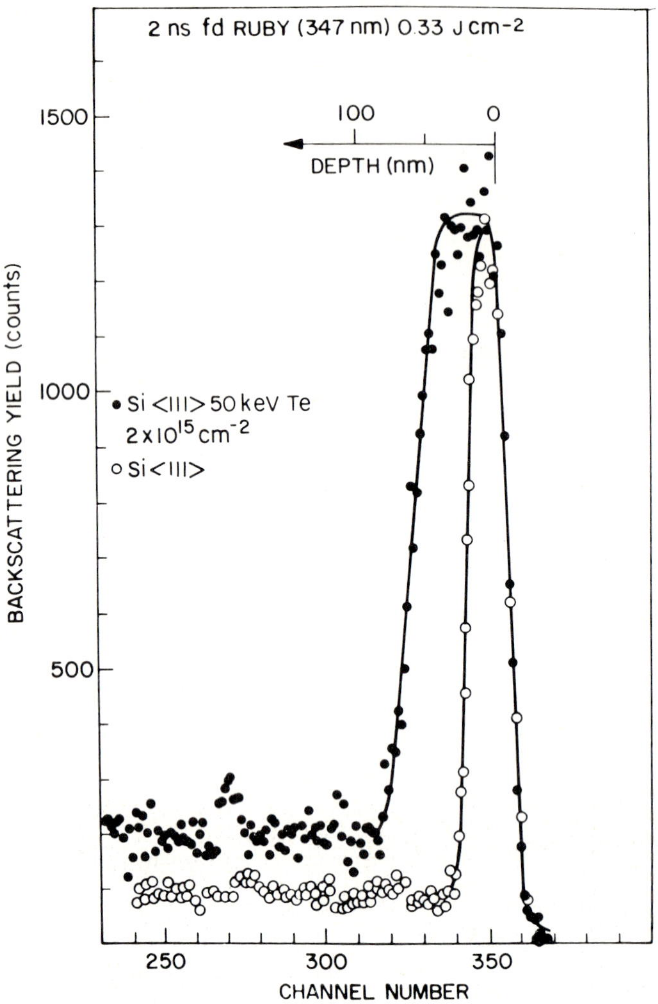

Fig. 9. ⟨111⟩ channeling analysis of a Si sample performed by 2 MeV He. Open dots refer to a pure sample and black dots to a sample implanted with 50 keV, 2×10^{15} Te atoms and then furnace annealed. Both samples are irradiated with 2 ns ultraviolet pulses at 0.33 J/cm^2 (from ref. [25]).

In atoms of 4×10^{19} at cm^{-3} in the outer 2000 Å and a strong peak of about 1.5×10^{15} cm^{-2} of In atoms rejected on the sample surface.

Backscattering spectrum of the indium signal is shown in fig. 10 (triangles). Following an irradiation with 0.36 J/cm^2 2 ns 0.347 nm laser pulses which produce a 500 thick amorphous layer it changes. In the same fig. 10 (open circle) the backscattering spectrum of this new distribution is shown: the surface peak disappeared and another peak appears at lower energy indicating a strong

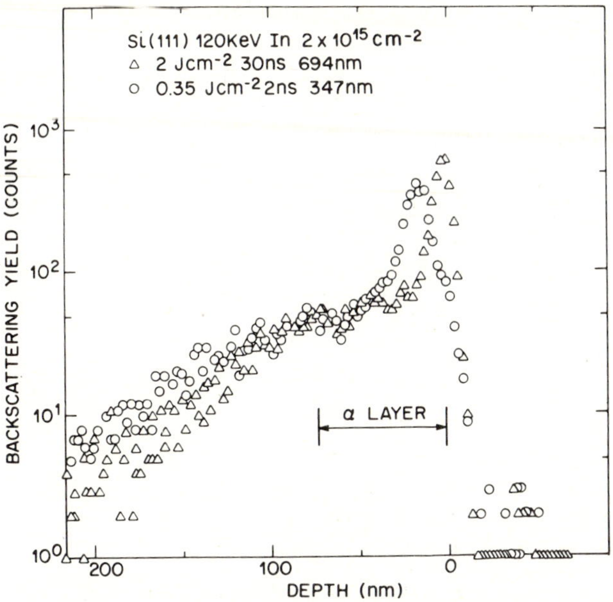

Fig. 10. 2 MeV He backscattering yield of the indium distribution obtained by laser annealing of an implanted sample with 30 ns ruby laser (△) and after another irradiation with 2 ns ultraviolet laser which produced an amorphous layer, 800 Å thick (≫) (from ref. [26]).

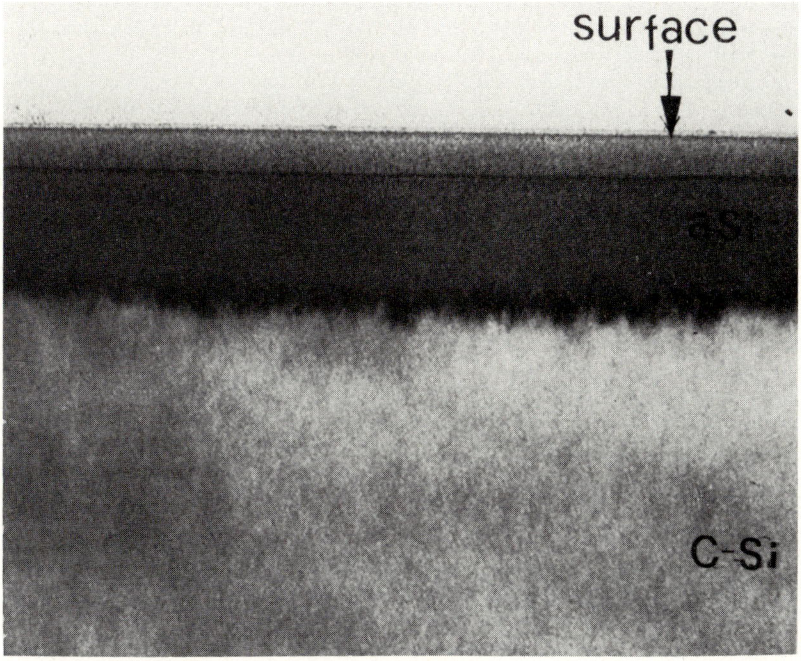

Fig. 11. TEM micrograph of the same sample of fig. 10 after the 2 ns pulse irradiation (from ref. [26]).

concentration of In located at 160 Å from the same surface. A cross sectional TEM of this sample is shown in the next fig. 11. The T_m peak results in a narrow (20 Å thick) buried layer of In imbedded in 500 Å thick amorphous silicon layer. This appears as a clear evidence that a solid amorphous–liquid interface, nucleated at the sample surface and segregated the In toward the inside, until a second liquid–solid interface starting from the interior of the sample is encountered at 160 Å from the surface. Nucleation processes of the amorphous phase is then important to determine the kinetics of the amorphous layer formation following laser pulse irradiation.

References

[1] P. Baeri, S.U. Campisano, G. Foti and E. Rimini, J. Appl. Phys. 50 (1979) 788.

[2] R.F. Wood and G.E. Giles, Phys. Rev. B 23 (1981) 2923.

[3] P. Baeri and S.U. Campisano, in: Laser Annealing of Semiconductors, eds., J.W. Mayer and J.M. Poate (Academic Press, New York, 1982) chap. 4.

[4] A.G. Cullis, H.C. Webber, N.G. Chew, J.M. Poate and P. Baeri, Phys. Rev. Lett. 49 (1982) 219.

[5] R. Yen, J.M. Liu, H. Kurz and N. Bloembergen, Appl. Phys. A 27 (1982) 153.

[6] P. Baeri, S.U. Campisano, E. Rimini and J.P. Zhang, Appl. Phys. Lett. 45 (1984) 398.

[7] Lietoila and J.F. Gibbons, J. Appl. Phys. A 27 (1982) 153.

[8] E. Yoffa, Phys. Rev. 21 (1980) 3415.

[9] L.A. Lompré, J.M. Liu, H. Kurtz and N. Bloembergen, Appl. Phys. Lett. 44 (1984) 3.

[10] P. Baeri, in: Laser and Electron Beam Interaction with Solids, eds., B. Appleton and G.K. Celler (North-Holland, Amsterdam, 1982) p. 151.

[11] E.P. Donovan, F. Spapen, D. Turnbull, J.M. Poate and D.C. Jacobson, Appl. Phys. Lett. 42 (1983) 698.

[12] P. Baeri, S.U. Campisano, M.G. Grimaldi and E. Rimini, J. Appl. Phys. 53 (1982) 8730.

[13] M.O. Thompson, G.J. Galvin, J.W. Mayer, P.S. Peercy, J.M. Poate, D.C. Jacobson, A.G. Cullis, N.G. Chew, Phys. Rev. Lett. (in press).

[14] R.F. Wood, Appl. Phys. Lett. 37 (1980) 302.

[15] M.J. Aziz, J. Appl. Phys. 53 (1982) 1158.

[16] R.N. Hall, J. Phys. Chem. 57 (1953) 836.

[17] J.C. Brice, The Growth of Crystals from the Melt (North-Holland, Amsterdam, 1975) p. 66.

[18] S.U. Campisano, Appl. Phys. A 30 (1983) 195.

[19] B.K. Jindall and W.A. Tiller, J. Chem. Phys. 49 (1968) 4632.

[20] S.U. Campisano and P. Baeri, Appl. Phys. Lett. 42 (1983) 1023.

[21] A.G. Cullis, M.C. Webber, N.G. Chew, J.M. Poate and P. Baeri, Phys. Rev. Lett. 49 (1982) 219.

[22] M.O. Thompson, J.W. Mayer, A.G. Cullis, M.C. Webber, N.G. Chew, J.M. Poate and P.C. Jacobson, Phys. Rev. Lett. 50 (1983) 896.

[23] M.C. Flemings, Solidification Processing (McGraw-Hill, New York, 1974) p. 306.

[24] K.A. Jackson, in: Surface Modification and Alloying, eds., J.M. Poate, G. Foti and D.C. Jacobson (Plenum, New York, 1983) p. 59.

[25] S.U. Campisano, D.C. Jacobson, J.M. Poate, A.G. Cullis and N.G. Chew, Laser quencing of amorphous Si from the melt containing dopants, Appl. Phys. Lett. (in press).

[26] S.U. Campisano, J.M. Poate, D.C. Jacobson, A.G. Cullis and N.G. Chew, Impurity redistribution at the liquid–amorphous Si interface, Phys. Rev. Lett. (in press).

Journal of Luminescence 30 (1985) 425–445
North-Holland, Amsterdam

PHONON POPULATIONS DURING PULSED LASER ANNEALING

A. COMPAAN

Kansas State University, Manhattan, KS 66506, USA

A summary is presented of pulsed Raman and several other time-resolved experiments and the information they provide on the dynamics of electron–phonon and phonon–phonon interactions under pulsed laser annelaing (PLA) conditions. A number of recent cw Raman studies provide a much improved basis for understanding the consequences on Raman spectra of temperature-dependent resonance effects, high carrier density effects, phonon anharmonicity, and strain effects. New pulsed Raman studies, under PLA conditions, of the shift and shape of the first-order line and of the full first- and second-order spectra are discussed. Emphasis is placed on explaining the results of time-resolved Raman experiments in a framework consistent with other types of experiments. The Raman data appear to be compatible with the existence of a fluid phase followed by a hot solid near the normal melting point of Si. The data, however, indicate the presence of additional broadening of the first-order Raman line probably due to carrier–phonon interactions, and cooling rates which appear to be higher than predicted by thermal models.

1. Introduction

When visible lasers are used for pulsed laser annealing (PLA) the optical energy in most semiconductors is initially absorbed into the electronic system. As the carriers cool and electron–hole recombination occurs, energy is fed into the phonon system as energy of atomic motion in the crystal lattice. Better understanding of the rate of energy transfer from electrons to phonons and of the rate of thermalization within the many branches of the phonon system have been the stimuli for a large number of experiments, many of which are discussed in these proceedings.

Already with some of the first PLA experiments [1], it was recognized that any assumptions regarding thermalization rates of the electrons and phonons should be made with great care. It was suggested early on that nonthermal mechanisms may be important with pulse durations of $\sim$10 ns and below. Van Vechten et al. [2] presented a number of observations which suggested that models were inadequate which assumed essentially complete thermalization of electron and phonon systems. Some of these objections have since been

0022-2313/85/$03.30

adequately answered. In fact, many subsequent experiments have given support to models which propose that melting occurs at or very near the normal 1412 °C transition in Si [3,4,5]. However it remains unsettling to many observers that thermal models require a process which begins with a tetrahedrally coordinated amorphous phase, passes through a 12-fold coordinated metallic phase (which is ~15 % more dense) [6], and arrives at a nearly perfect, tetrahedrally coordinated, crystalline phase – all in a few nanoseconds or less. Contrapuntally, the thermal models have the elegance of simple assumptions and fairly straightforward calculations to obtain cooling rates, melt depths, etc.

The question of whether melting occurs below the normal melting point is best answered with time-resolved measurements of phonon populations. Such measurements have been performed by two independent groups using Raman scattering with pulsed lasers [7,8]. Early results indicated that the phonon populations (at least in the optic branch) remain well below the normal melting point of Si. However it has also become clear that this type of measurement is a challenging one and must be interpreted with considerable care [9,10].

In the period of time since the first report of Raman data under PLA conditions [7], a number of cw laser Raman studies have been performed in Si which have greatly improved the base of knowledge of the Raman effect in Si upon which the pulsed Raman measurements depend. The cw measurements include studies of phonon frequency shifts and broadening as a function of crystal temperature using oven heating [11,12] and cw CO_2 laser heating [13]. Other studies have been performed on very heavily doped Si to elucidate the effect on phonons of high carrier densities [14,15,16], and to study the light scattering properties of the carriers themselves [17]. Finally the high temperature behavior of electronic resonance effects in phonon Raman scattering is now much more completely understood [18].

The major objective of this paper is to review critically the latest pulsed Raman data and its significance in terms of electron–phonon relaxation. The discussion will begin with a brief review of a variety of measurements other than Raman scattering which have yielded information on the dynamics of PLA. Throughout the discussion we shall emphasize results for Si which is the most thoroughly studied semiconductor under PLA conditions.

2. Review of time-resolved measurements

Simple but convincing evidence of the existence of a change of phase occurs in the the transient reflectivity rise [19] which is easily observed with a small He–Ne laser. The value of the reflectivity during this transient phase is within experimental error of that of normal molten Si. A number of other experiments give further evidence of the nature of this "high reflectivity" phase. Copious electron emission occurs [20], and thermionic emission of Si neutrals and Si^+

(Ga and As emission from GaAs samples) with velocity distributions near that expected of Si (GaAs) at the melting point [21]. Transient conductance measurements [22] indicate conductivities near the value for normal molten Si. Second harmonic generation from Si surfaces [23,24] similarly indicates a change of symmetry under PLA conditions, although the detailed nature of the required structural changes is still under discussion. Pulsed Raman measurements during the high reflectivity phase [25] indicate an absence of signal in the frequency shift range from 200–600 cm^{-1}. All of these are consistent with the existence of a fluid phase during this period.

The absorption spectrum during the "high reflectivity" phase has been measured directly using a pulsed dye laser as a transmission probe [26] (see fig. 1). (Actually the two points at lowest photon energy were obtained with the Nd:YAG laser fundamental, 1.17 eV, and with a beam generated by stimulated Raman scattering in H_2, 0.78 eV.) An additional measurement has been obtained by Lyon et al. [27] at $h\nu = 0.366$ eV which also shows strong attenuation. Figure 1 also shows the absorption coefficient for the normal molten phase of Si [28]. Clearly the absorption during the transient, laser annealing phase is quite consistent with that of normal molten Si. At a minimum, one must conclude that this phase has metallic properties (high carrier density, appropriate optical constants, etc.)

The conditions in the solid just before the onset of the high reflectivity phase, have been studied by reflectivity measurements on bulk Si with picosecond

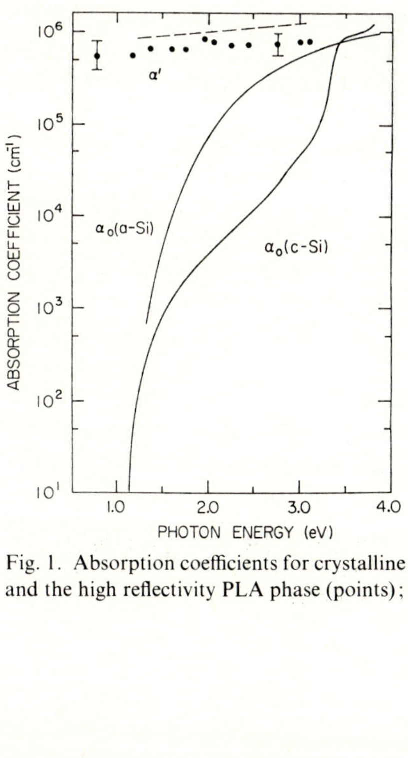

Fig. 1. Absorption coefficients for crystalline Si (c-Si) amorphous Si (c-Si), molten Si (dashed line) and the high reflectivity PLA phase (points); from ref. [26].

[29,30] and subpicosecond [31,32] pulses. The results show evidence of photoexcited carrier densities well into the $10^{22}/cm^3$ range. These appear to be well above the density needed for a plasma-induced phase change [33,34]. However, on a time scale of ~ 0.1 ns and greater, the carrier densities photogenerated in the solid phase appear to be only $\sim 10^{20}/cm^3$, well below that needed for a plasma-induced phase change.

The conditions in solid Si immediately following the high reflectivity phase are not as thoroughly explored, but some results are clear. Measurements of optical constants with nanosecond pulses and cw probe lasers have shown that the recrystallized Si is indeed hot after recovery from the "high reflectivity" phase. Murakami et al. [35] have shown that reflectivity oscillations occur in SOS samples due to multiple reflections between the air–Si and Si–sapphire surfaces. Similar measurements by Lee [36] are shown in fig. 2 for two excite wavelengths and two sample thicknesses. The oscillations arise from the temperature dependence of the index of refraction. Clear evidence of this was obtained by scanning across the SOS samples to find slight variations in Si thickness. Thus the oscillations from a point with a minimum reflectivity, R_{min}, before the pulse are complementary to those from a point with maximum reflectivity, R_{max}, before the pulse. From calibrations with oven-heated samples, we have inferred temperatures as high as 600–800 °C within a few tens of nanoseconds of the end of the high reflectivity phase. Murakami et al. [35] have inferred temperatures up to 860 °C.

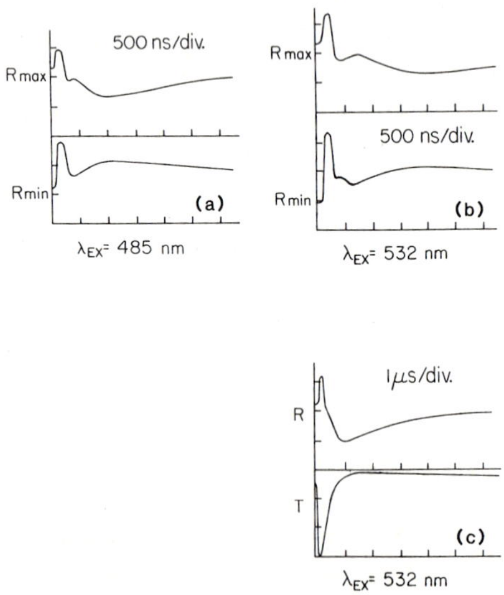

Fig. 2. Time-resolved reflectivity measured on 1.83 μm SOS, (a) and (b), and 0.6 μm SOS, (c), with two different wavelengths showing thin film interference oscillations; from ref. [36]. Probe wavelength 633 nm (note, the narrow peak is from high reflectivity phase).

With picosecond pulses and similar SOS samples Lompré et al. [37] obtain average temperatures to 700°C using 30 ps pulses. However, a difficulty with these R and T measurements on SOS is that one necessarily obtains a temperature averaged through the 0.5 to 2 μm sample thickness. In addition strong absorption sets in at high temperatures. The traces of fig. 2 show how this absorption damps the reflectivity oscillations in the time just after the high reflectivity phase. Finally, it must be recognized that measurements of optical constants of Si in the visible are really measurements of electronic properties and inferences of lattice temperature require assumptions regarding thermalization as a starting point. Furthermore, optical constants depend on a variety of factors including carrier densities and temperature and lattice dilation and strain as well as lattice temperature.

3. Energy flow into the phonon system: electron–phonon interactions

The rate of energy transfer from electrons and holes into phonons is a central issue in the discussion of PLA mechanisms. In some semiconductors high plasma densities slow down or even choke off this relaxation rate [38]. In polar semiconductors the electric field of the longitudinal optic (LO) phonon couples very strongly to the carriers producing a very rapid energy flow out of the carrier system. This Fröhlich mechanism couples electrons and holes with LO phonons, scattering predominantly to near-zone-center ($q \ll \pi/a$) phonons – just the region studied by first-order Raman scattering. Picosecond pulse Raman experiments by Von der Linde et al. [39] have shown directly the heating and cooling of LO phonons in, e.g., GaAs. On the other hand, similar experiments in Si (a nonpolar semiconductor) showed no evidence of heating on a picosecond time scale [39]. Deformation potential coupling is somewhat weaker and therefore electron–phonon relaxation rates slower; however, the lack of observable heating is most likely a consequence of energy being spread over a much wider region of the optic phonon branch of Si. Deformation potential coupling is not strongly q-dependent; thus a larger amount of laser-deposited energy may be required to observe an effect.

3.1. Energy and momentum conservation

It is useful to review the energy and crystal momentum conservation conditions for electron–phonon and hole–phonon scattering in Si. Figure 3(a) shows relevant parts of the electron energy band diagram for Si omitting spin–orbit splitting [40]. Figure 3(b) indicates the phonon dispersion curves [41]. For frequency-doubled Nd–YAG laser excitation ($h\nu = 2.33$ eV) band-to-band absorption is phonon-mediated. About 1.2 eV of excess energy is shared by the electron and hole. For parabolic bands and a dispersionless optic phonon,

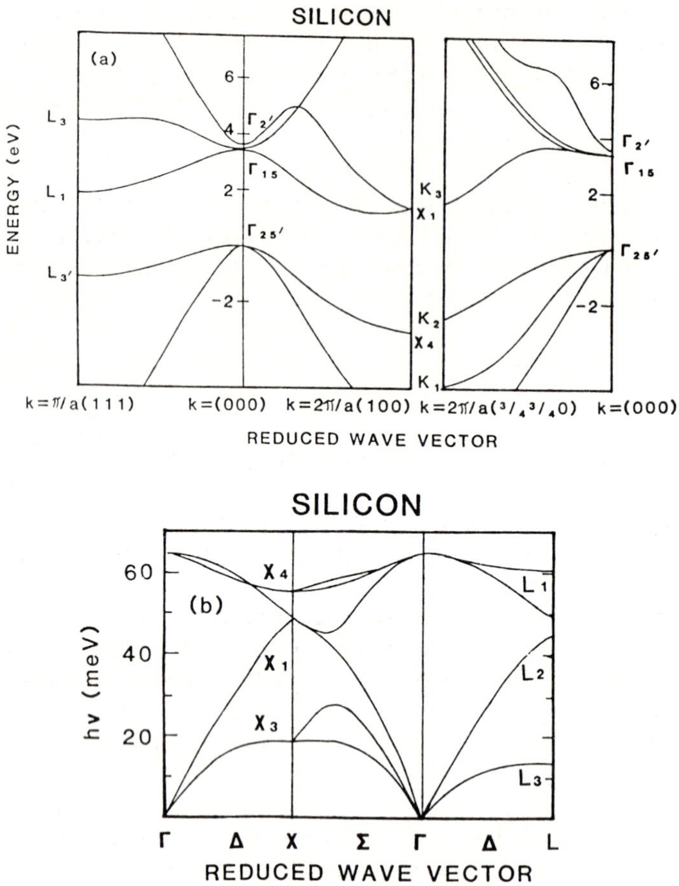

Fig. 3. (a) Electronic band structure and (b) phonon dispersion curves for Si from refs. [40] and [41] respectively.

maximum and minimum phonon momenta, q, can be obtained by considering energy and momentum conservation. This gives [42]

$$|q(^{\text{min}}_{\text{max}})| = |k|[1 \mp (1 - \hbar\omega_0/E)^{1/2}], \tag{1}$$

where $\hbar\omega_0$ is the optic phonon energy and E and k are the initial excess electron energy and momentum. For a 1.0 eV electron scattering in the [100] direction ($m^* = 0.91\ m_0$), the minimum momentum change would be $\Delta q_{\text{min}} \simeq 1.6 \times 10^6$ cm^{-1}. Similarly in the transverse direction (perpendicular to [100]), $m^* = 0.19\ m_0$ and $\Delta q_{\text{min}} \simeq 7.1 \times 10^5$ cm^{-1}. For light holes and heavy holes with ~ 1.0 eV a similar range of values is obtained. Maximum momentum changes are of the order of the distance to the zone boundary for 1 eV electrons and decrease in proportion to the square root of the excess energy. Clearly energy may be desposited over a large fraction of the optic phonon branch, but as the electrons and holes cool Δq_{min} increases and Δq_{max} decreases.

Anticipating the Raman studies described below, we find the crystal momentum transfer in a backscattering Raman experiment is $\Delta q = 4\pi n/\lambda = 1\times10^6$ cm^{-1} at $\lambda = 532$ nm and $\Delta q = 1.7\times10^6$ cm^{-1} at $\lambda = 405$ nm. Thus Raman experiments at these wavelengths do overlap with the range of phonon momenta initially produced in electron–phonon scattering. However as the electrons and holes cool, phonons are no longer directly generated at wave vectors probed by Raman scattering.

3.2. *Effects of high carrier densities*

In the presence of high densities of free carriers, electron–phonon relaxation may be modified. In polar semiconductors there exists evidence [38] that the Fröhlich coupling mechanism is screened by carrier densities in the range of 10^{17} cm^{-3}. However the deformation potential coupling responsible for scattering in Si is not likely to be modified until much higher densities [43] ($\sim10^{21}$ cm^{-3}).

An estimate of the density of photoexcited carriers produced by a 1 J/cm^2, 10 ns laser annealing pulse may be obtained from measured recombination rates. Three-body Auger recombination [44] should limit the peak carrier density. Equating the photogeneration rate at the surface with the Auger recombination rate yields,

$$n = (\alpha I/Ch\nu)^{1/3} \simeq 2\times10^{20}\ \mathrm{cm}^{-3}, \tag{2}$$

where α is the absorption coefficient (1×10^4/cm at 532 nm), C is the Auger rate ($C\simeq4\times10^{-31}$ cm^6/s), I is the laser intensity and $h\nu$ the photon energy. Recent ir reflectivity measurements give a similar number [29]. This quantity should be treated with care, however, since C and α may be carrier density and temperature-dependent. In fact femtosecond reflectivity measurements [31,32] have given indication that Auger recombination is inhibited in Si at carrier densities of $\sim10^{22}$/cm^3.

3.3. *cw Raman in doped Si*

At carrier densities of 10^{19}–10^{21} cm^{-3} electron–phonon scattering is significantly increased in Si. This evidence comes from cw Raman measurements on heavily doped Si [14–16]. As free electrons or holes are added to Si by doping, a spectrum of possible electronic excitations is added with excitation energies which overlap the phonon energies. The optic phonon Raman spectrum displays this coupling in the form of an asymmetric line (Breit–Wigner–Fano shape) which is broadened and shifted in energy [14].

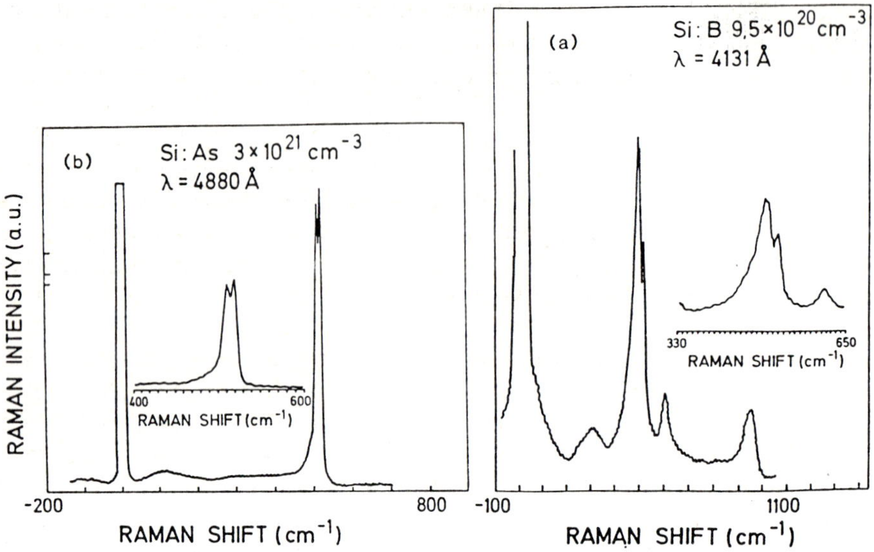

Fig. 4. Raman spectra from extremely heavily B-doped Si (a) and As-doped Si (b). Note softening and broadening of the optic phonon peak near 500 cm^{-1}; from ref. [15].

Figure 4(a) illustrates the effect observed in heavily B-doped Si, and fig. 4(b) illustrates the effect for heavily As-doped Si [15]. These samples were prepared by ion-implantation and pulsed excimer laser annealing with a heavily doped surface layer ~ 0.2 μm thick. Some penetration of laser light into the unimplanted substrate leads to the Raman peak which occurs at the unshifted, 520 cm^{-1} position for pure Si. Note that both n- and p-type doping produce a phonon softening, i.e., a reduction of phonon frequency. This softening is the real part of the phonon self-energy due to interaction with electronic excitations. The imaginary part leads to the increased broadening of the line which is also apparent in fig. 4.

The density dependence of the real part of the phonon self-energy (phonon softening) is shown in fig. 5 for n-type doping [15]. The imaginary contribution to the self-energy, which appears as broadening of the Raman line, has a similar dependence on carrier density. For a hole density of $p = 1 \times 10^{21}$ cm^{-3} we find a width of $\Delta\bar{\nu} \simeq 15$ cm^{-1} compared with the normal room temperature optic phonon width of $\Delta\bar{\nu} \simeq 2.5$ cm^{-1}. Thus the broadening is increased by nearly an order of magnitude due to interaction with single particle electronic excitations. In undoped Si the optic phonon line broadening is due only to anharmonic interactions with other phonon modes. In doped Si, however, the optic phonon can decay into electronic excitations, thus introducing an important new consideration into the phonon dynamics.

The origin of these single-particle electronic excitations is shown in fig. 6.

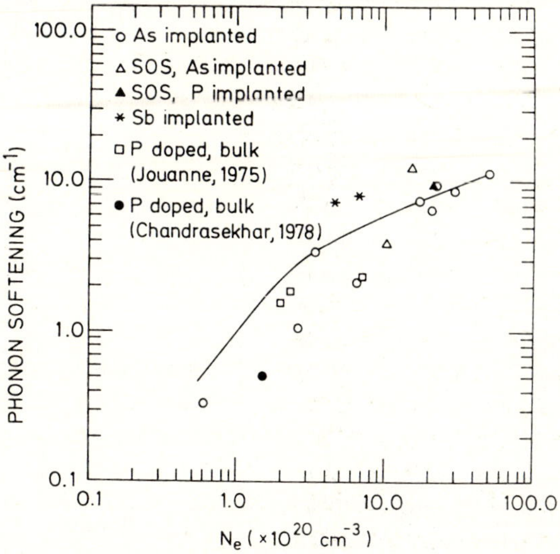

Fig. 5. Phonon softening as a function of carrier concentration for n-type dopants, from refs. [14 and 15].

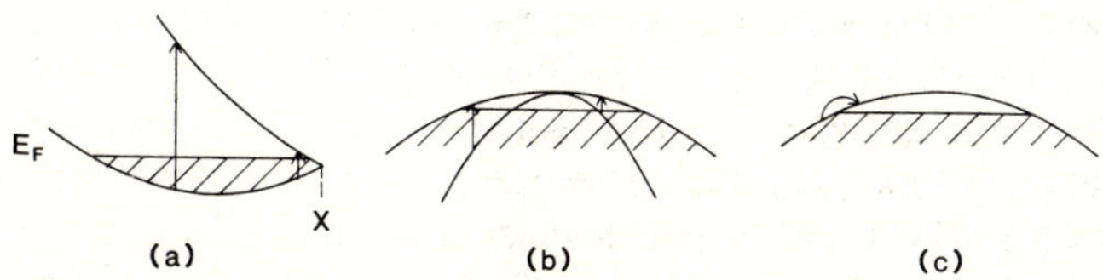

Fig. 6. Portions of the Si band structure showing the type of single-particle excitations which produce phonon softening and broadening.

Here the essential parts of the electronic band structure are qualitatively sketched with arrows indicating the type of transitions which may be produced by phonon decays. These excitations will affect not only near-zone-center optic phonons, cf. fig. 6(a) and (b), but via intervalley scattering, zone boundary optic and acoustic modes will also be affected. Finally low frequency acoustic modes [16] may be involved in the type of excitation sketched in fig. 6(c). Evidence of the validity of this interpretation of the phonon softening and broadening comes from the observation that in very heavily p-doped Ge ($n \sim 10^{21}$ cm^{-3}) almost no effects are produced [15]. This is understood from the fact that the L-point conduction band valley is separated from the next conduction band by more than 2 eV. Thus there are no electronic excitations which overlap the zone-center optic phonon spectrum and hence no additional scattering or broadening occurs.

A consequence of these additional electron–phonon couplings is a strong enhancement of the rate of thermalization of the electronic system with the phonon system. In fact the electrons and holes will be coupled strongly with many regions of the phonon system reducing the possibility of occurrence of phonon hot spots at isolated regions of the Brillouin zone.

3.4. Anharmonic phonon interactions

In pure Si the lifetime of optic phonons is determined by decay into two acoustic branch phonons. This decay rate has recently been studied as a function of lattice temperature by Hernandez and Tsu [11] and by Balkanski et al. [12] using cw Raman scattering. They find that the optic phonon lifetime decreases from 2 ps (2.5 cm^{-1}) at room temperature to 0.4 ps at 1140 K. Extrapolating the fit of ref. [12] using third- and fourth-order anharmonic terms yields an optic phonon lifetime of 0.2 ps at the Si melting point. Nemanich et al. [13] have recently measured the Raman line in Si heated to melting with a cw CO_2 laser. The width of the Raman line which they measure is consistent with the extrapolation discussed above.

The phonon–phonon interactions also renormalize the phonon frequencies significantly at high temperature. The cw Raman experiments described above show the general softening of frequencies due to these interactions and to the lattice expansion, which of course is also a consequence of the anharmonicity.

The above measurements of phonon–phonon interactions have been performed essentially under thermal equilibrium conditions. But the existence of these data now allow the possibility of interpreting similar shifts in pulsed Raman data as a further check for consistency with the temperature inferred from Stokes/anti-Stokes ratios.

4. Time-resolved Raman measurements

Pulsed Raman measurements under PLA conditions have attracted considerable attention for at least two reasons. First the technique is probably the best way of probing directly phonon populations with good time resolution at specific points in the Brillouin zone. This contrasts with other techniques such as ion emission [21], optical properties [29–31,35–39] and X-ray studies of lattice dilation [46], which respond to average properties of the lattice and the electronic structure. Second, the Raman measurements have yielded phonon populations which stand in strong disagreement with temperatures calculated on the basis of the thermal models.

However, it has also been emphasized, by both groups involved in the pulsed Raman studies, that the interpretation of the Raman Stokes/anti-Stokes ratios must be done with a considerable amount of caution. Thus in our first report we pointed out the need to include in the analysis electronic resonance effects and their potential temperature dependence. Unfortunately there existed at the time no studies of such changes with temperature. As an alternative we showed [9] how it is possible to obtain directly the resonance-dependent correction factors for obtaining phonon populations from Stokes/anti-Stokes ratios. The method relies on the time-reversal invariance of Raman scattering. However, the technique requires the use of a continuously tunable dye laser which limits its applicability. The temperature-dependent resonance effects have now been significantly clarified in a cw Raman study of oven-heated Si [18]. Consequently it is now possible to include resonance effects on the Stokes/anti-Stokes ratio and on the overall strength of the Raman signal, provided the system is not far from thermal equilibrium.

4.1. Resonant Raman effects

The lowest direct band-to-band transitions in Si occur at ~ 3.4 eV and arise from the E_0' and E_1 gaps. The E_1 gap occurs over an extended region in k-space along the Λ direction whereas the E_0' occurs at the Γ point (see fig. 3). Therefore, the E_1 transitions have a much higher density of states and dominate in the optical properties, including the resonant behavior of the Raman susceptibility. We have demonstrated [18] that the Raman susceptibility, χ_s'', over the range of 1.5 eV to 3.7 eV follows very closely the frequency derivative of the complex dielectric constant, $d\varepsilon/d\omega$. However the $d\varepsilon/d\omega$ spectrum [47,48] must be shifted to higher frequency by one-half the phonon frequency to account for both "incoming" and "outgoing" photon resonances.

The rate of Stokes photon scattering, R_s', is related to χ_s'' by [18]

$$R_s' = \frac{T_s T_L P_L' \omega_s^3}{(\alpha_s + \alpha_L)\eta_L \eta_s} \; \frac{\Delta\Omega'[n(\omega_0)+1]}{32\pi^2 c^4 \omega_0} \, |\varepsilon_L \varepsilon_s \chi_s''|^2, \tag{3}$$

where T_s and T_L are transmission coefficients of the Si surface, P_L' is the incident laser power, ω_s the scattered photon frequency, ω_0 the phonon frequency, $\alpha_s(\alpha_L)$ the absorption coefficient and $\eta_s(\eta_L)$ the index of refraction at the scattered (incident) frequency, $\Delta\Omega'$ the external collection solid angle, n the phonon occupation factor, and ε_L (ε_s) the polarization unit vectors. Note that the Cartesian indices have been suppressed on $\varepsilon_L, \varepsilon_s$ and χ_s'. Temperature dependence occurs in the factors T_s, T_L, α_s, α_L, η_L, η_s and χ_s'' in addition to the phonon population factor $n(\omega_0)$ [18].

By far the strongest temperature dependence other than for $n(\omega_0)$ occurs in the absorption coefficients [47], α_s and α_L, and in the Raman susceptibility, χ_s''. (We have found [18], however, that over much of the spectrum the temperature dependence of $(\alpha_s+\alpha_L)^{-1}$ nearly cancels the temperature dependence $|\chi_s''|^2$.) The temperature dependence of χ_s'' is shown in fig. 7. The shift and broad-

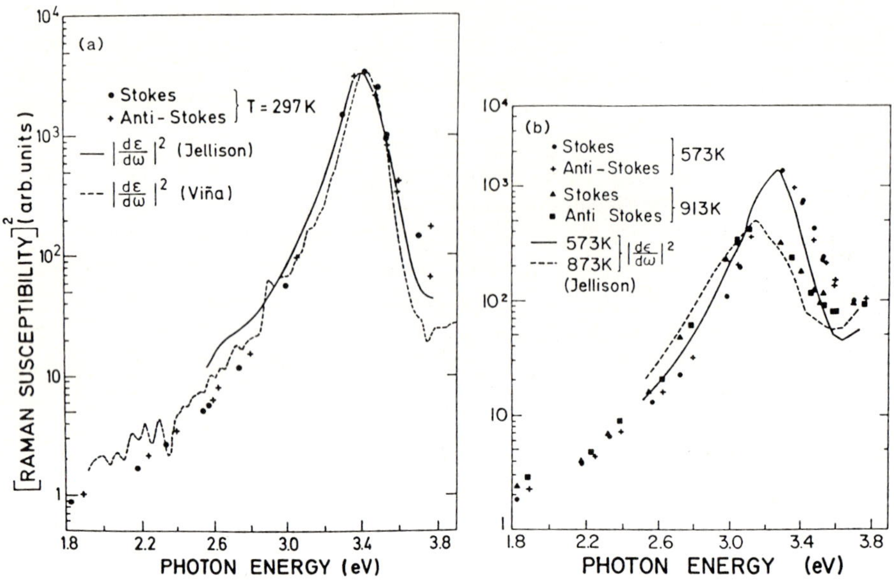

Fig. 7. Raman susceptibility, χ_s'', as a function of laser photon energy for room temperature (a) and for two elevated temperatures (b), from ref. [18].

ening of the resonance is clearly apparent in the curves of fig. 7(b). Note that we have compared our measurements of $|\chi_s''|^2$ with the measurements by Jellison et al. [47] and by Vina [48] of $|d\varepsilon/d\omega|^2$ after shifting upward in frequency by $\omega_0/2$. The Raman and ellipsometric data extend only to ~900 K. However the shift and broadening of the resonance are linear in temperature to 900 K (see fig. 8) and we believe a linear extrapolation to the melting point is reasonable. This extrapolation is included in the data fit described below.

The temperature-induced shift and broadening of the Raman resonance cause a gradual decrease in the strength of the optic phonon Raman signal. This is plotted in fig. 9 for several laser frequencies. Our original pulsed Raman experiments were done at 405 nm. Figure 9 shows this wavelength to be a good choice, i.e., the integrated area of the Raman peak decreases very slowly with temperature. Our most recent experiments were done at 532 nm (see below). Near this wavelength the drop in Raman strength is larger and this is included in our analysis.

4.2. Raman scattering with 10 ns pulses

The experimental configuration for these pulsed Raman measurements is illustrated in fig. 10. The main part of the amplified, frequency-doubled Nd:YAG pulse is used to pump a dye laser with two amplifier stages. The dye laser

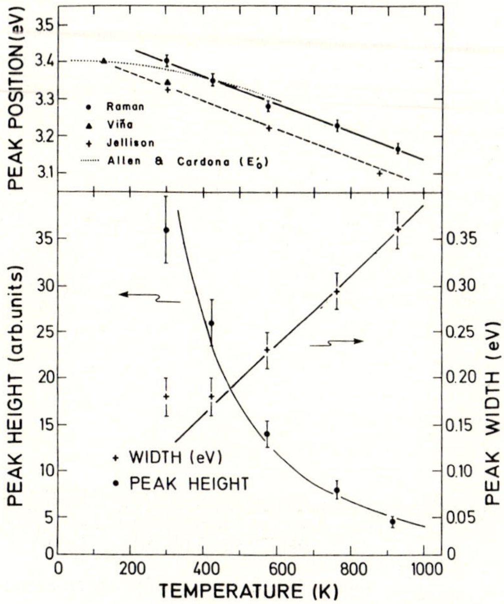

Fig. 8. Shift and broadening of the Raman susceptibility as a function of temperature; from ref. [18].

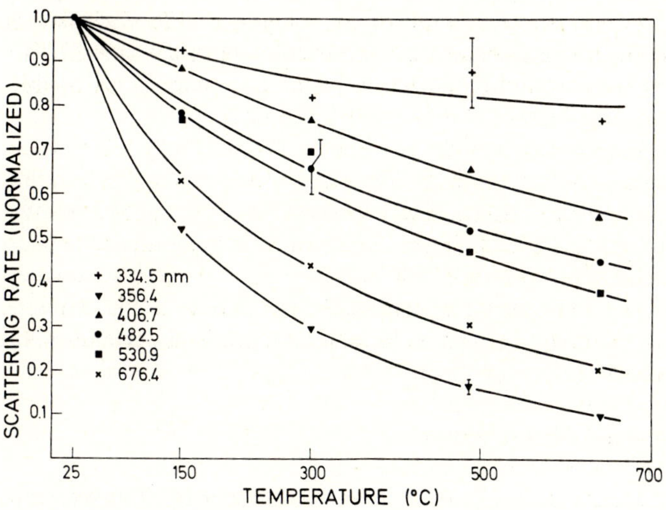

Fig. 9. Integrated area of the first-order Raman peak in Si as a function of temperature for several laser lines; from ref. [18].

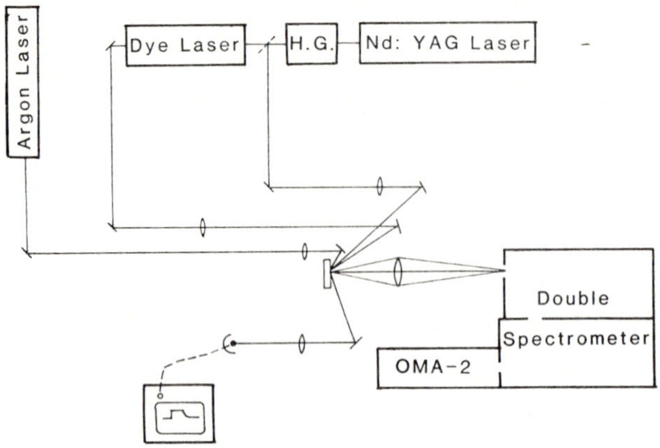

Fig. 10. Experimental set-up for the time-resolved measurements shown in fig. 11.

pulse at 565 nm has excellent homogeneity with a half-power width of 800 μm on the sample. A small fraction of the 532 nm beam is split off and delayed in an optical delay line to be used as the Raman probe. It is focussed to a diameter of ~120 μm on the sample. Pulse durations are 12 ns. Other aspects of the system are the same as described in earlier work [7,9].

Figure 11 shows the Stokes–Raman signals obtained with 0.18 J/cm^2 probe energy at 532 nm with probe delays of 8 ns and 90 ns. The heat pulse power at 565 nm was adjusted to give a full reflectivity rise lasting 2–5 ns for the 8 ns delay and an 80 ns flat-topped reflectivity rise for the 90 ns delay. The probe power is about 3 times greater than used previously and was chosen so that the probe produces significant heating which presumably keeps the sample temperature high during the probe pulse. We find, under these conditions, Stokes/anti-Stokes ratios of 1100 ± 300 °C.

The data of fig. 11a shows a significant asymmetry. We believe this asymmetry may arise from temperature variations in space and time since the Raman line shifts with temperature. The solid curve in fig. 11(a) is a fit using temperature weighting factors as sketched in the inset. This temperature profile probably arises from a very rapid cooling rate following the excitation pulse, coupled with the fact that the probe pulse has enough power density at its center to maintain the sample essentially at its melting point. However the fringes of the probe beam have too little power to prevent cooling of the Si surface. [The temperature inhomogeneity is not caused by the heat beam since its power variation over the probe size (1/8 power points) is less than 20 %.]

The fit of fig. 11(a) was made by using the measured shift of the phonon with temperature [12], the decrease in scattering strength shown in fig. 9, and correcting for an assumed two-dimensional compressive strain [49] for the pulsed case: i.e., we assume thermal expansion only normal to the surface. These strain

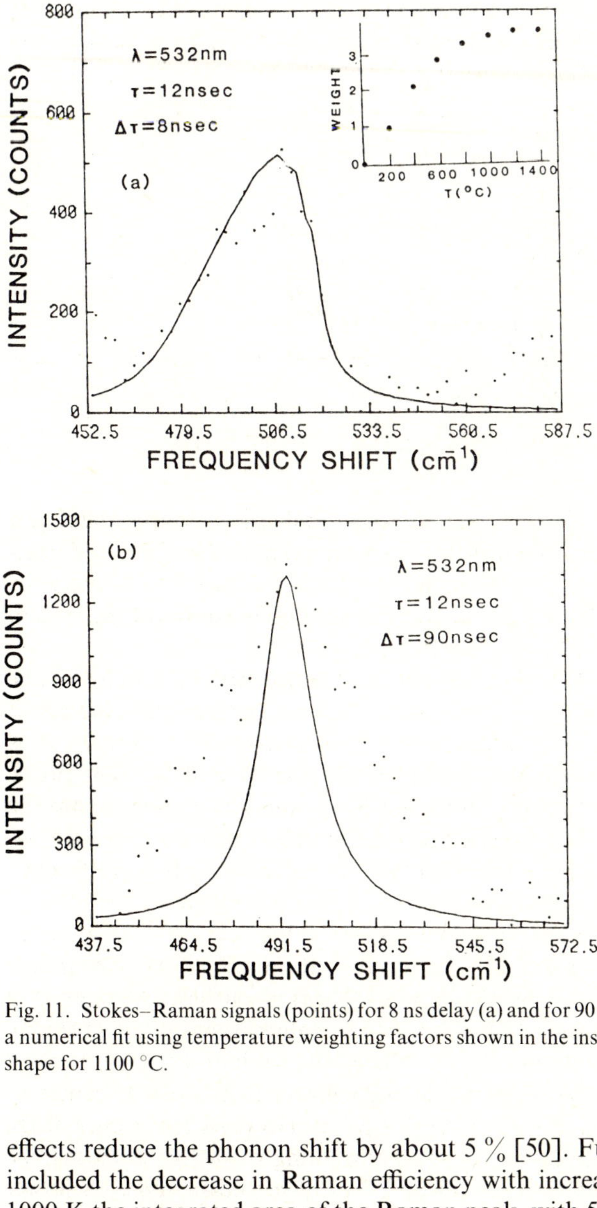

Fig. 11. Stokes–Raman signals (points) for 8 ns delay (a) and for 90 ns delay (b). Solid curve in (a) is a numerical fit using temperature weighting factors shown in the inset. Solid curve in (b) is expected shape for 1100 °C.

effects reduce the phonon shift by about 5 % [50]. Furthermore, we have also included the decrease in Raman efficiency with increasing temperature. Below 1000 K the integrated area of the Raman peak, with 532 nm excitation, follows the empirical form [18,50]

$$\text{Area} = A_0 \exp[-(T-293)/700], \tag{4}$$

where T is measured in Kelvins. The appropriate Raman width [12] at each temperature was also included in the fitting. Thus we are able to explain the asymmetry if about 5 % of the sample surface cools to 200 °C by 15 ns after the peak of the excitation pulse. The existence of additional broadening due to enhanced carrier–phonon scattering, as discussed below for the case of 90 ns delays, would raise this minimum temperature slightly.

The data points of fig. 11(b) show much less asymmetry, probably because the sample has spent ~80 ns in the high reflectivity phase. Temperature gradients should be smaller and cooling rates less. The peak position fits a curve expected for 1100 °C. However the width is much too large. The size of temperature variation required to fit this width is unreasonable and we believe another broadening mechanism must be involved.

Our conclusion regarding these data from 12 ns pulses is that they are consistent with the normal melting temperatures having been reached during the high reflectivity phase. However the data require extremely rapid cooling rates and a broadening mechanism in addition to the normal thermal shift and broadening.

4.3. Raman scattering with 150 ns pulses

First-order Raman scattering gives a direct sample of phonon populations at only one point on the phonon dispersion curves. Therefore we obtain no information on the degree of thermalization among the phonon branches. Second-order scattering, however, leads to peaks in the Raman spectrum which generally arise from points of high phonon density of states. In fact most second-order peaks in Si arise from overtones ($\omega_1+\omega_1$) rather than combinations ($\omega_1+\omega_2$) making identification simple [51]. Second-order scattering, unfortunately, has only ~1 % or less of the first-order intensity in Si at 300 K. Thus the second-order spectrum is only marginally observable with low repetition rate (~10 Hz) lasers. The 10 kHz frequency of a repetitively Q-switched "cw" Nd:YAG, however, permits second-order studies.

With intracavity frequency doubling, we obtained ~5 W of average power in ~150 ns pulses at 532 nm. Raman scattering was obtained [52] from the same beam by using a 20 ns gate on the photon counting electronics. This gate was adjusted to begin about 30 ns after the peak of the laser pulse. A He–Ne reflectivity probe indicated this as the time when recovery from the high reflectivity phase occurred for the central region of the laser spot. A 300 μm aperture was used at the entrance to the spectrometer so that only the central 100 μm region of the sample was imaged into the spectrometer. The sample was rotated at ~20 Hz to avoid damage from the laser.

Spectra obtained with this system are shown in fig. 12. The low power spectrum, fig. 12(a), was obtained with the laser beam attenuated by a factor of 10 and the gate switched off. The high power spectrum fig. 12(b), shows the large enhancement of anti-Stokes peaks relative to the Stokes peaks (both first and

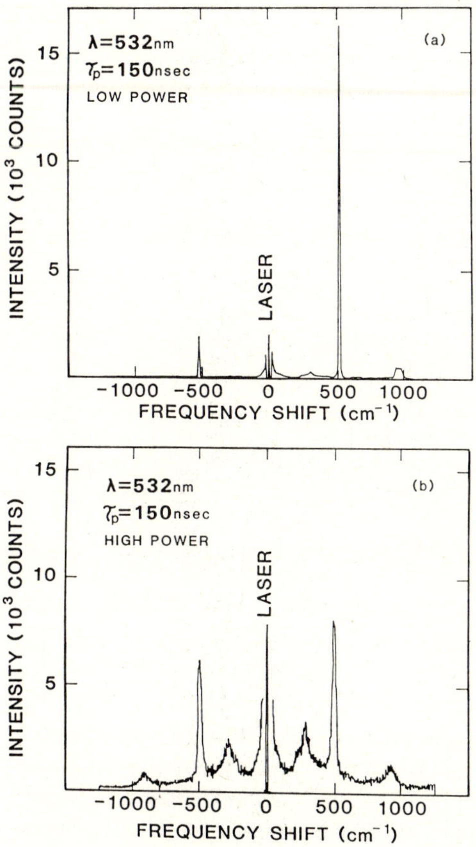

Fig. 12. Raman spectra showing both first- and second-order Stokes and anti-Stokes peaks for low power (a) and high power (b). Nd:YAG laser excitation ($\lambda = 532$ nm) as described in the text.

second order). Stokes/anti-Stokes ratios for the first-order line at $\sim 500\ \text{cm}^{-1}$, and for the second-order features at $\sim 950\ \text{cm}^{-1}$ and $280\ \text{cm}^{-1}$ indicate phonon "temperatures" of 1500 ± 300 K. Thus phonon populations in the optic branch at zone center and zone boundary (X and L points) appear within experimental error to be in equilibrium with each other and with the zone boundary transverse acoustic mode (X and L points).

One difficulty with inferring phonon populations from Stokes/anti-Stokes ratios is that although the ratio grows exponentially at low temperatures, it saturates to unity at high temperatures ($kT \gg \hbar\omega_0$). However, the ratio of a second-order peak at $2\omega_1$ to an allowed first-order Raman peak at ω_2 scales linearly with T at high temperature. Thus for Stokes scattering,

$$[n(\omega_1)+1]^2/[n(\omega_2)+1] \simeq \omega_2 kT/\hbar\omega_1^2, \tag{5}$$

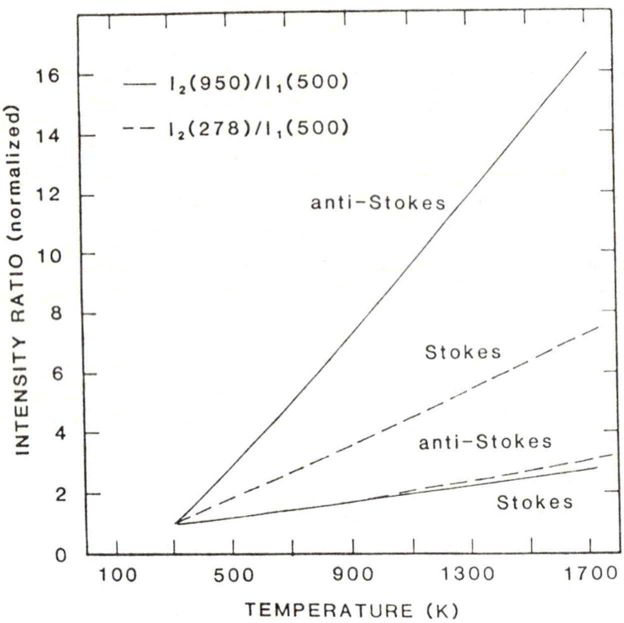

Fig. 13. (a) Anti-Stokes/Stokes ratios for Raman features at three different frequency shifts. (b) Predicted second-order Raman to first-order Raman ratios for two prominent second-order peaks in Si.

and similarly for the anti-Stokes ratio. In fig. 13 we illustrate the behavior of these ratios for some of the Raman features. The second-order to first-order ratios have been normalized to unity at $T=300$ K to remove the large differences in scattering strengths.

The spectrum of fig. 12(b) clearly shows a strong enhancement of the second-order features at high laser power. Taking second-order to first-order (500 cm^{-1}) ratios from figs. 12(a) and (b) and the curves of fig. 13 we find the phonon populations again to be consistent with a lattice temperature of $T=1500 \pm 300$ K.

For 150 ns pulses the data were acquired with a single beam serving for excitation and probing. Thus temperature gradients temporally and spatially are of some concern. The 20 ns gate late in the laser pulse was chosen to allow most of the laser energy to be deposited before signal acquisition. Furthermore, 50–100 ns should be sufficient time for temperature gradients normal to the surface to be reduced. (The Raman probe depth for 532 nm in Si above $T=1000$ K is less than ~ 0.2 μm.) Therefore the temperature estimates given above should apply to a reasonably homogeneous situation. The results indicate complete equilibration of the phonon modes at temperatures near the normal melting point of Si.

5. Summary

Using the excite/probe technique, with ~10 ns pulses at 565 nm and 532 nm respectively, we have found that the shape and shift of the Raman line as well as the Stokes/anti-Stokes ration to be indicative of phonon temperatures of 1100 ± 300 °C. This is much higher than previous studies with either a 532 nm/405 nm excite/probe combination or a 485 nm/405 nm combination where temperatures of 300–450 °C were found. We believe these differences are not strongly influenced by any temperature dependence of corrections to the Stokes/anti-Stokes ratio [9,52]. These corrections approach unity at high temperatures.

We find that it is important, particularly for fitting line shapes with the 532 nm probe, to include the fact that the Raman strength at the melting point decreases by a factor extrapolated to be 0.13 for 532 nm excitation and 0.26 for 405 nm excitation [18,50].

However the major difference in the present study is the use of much higher probe powers (~0.2 J/cm^2). In the present work the probe significantly affects the Si by slowing the cooling after the high reflectivity phase. In fact we observe a lengthening of the transient reflectivity signal when the probe is on. Thus the major reason for the observation of higher phonon populations is probe heating: i.e., with a probe power of 0.06 J/cm^2 rapid cooling could take place during the probe pulse. Thus, even with a probe delay of 10 ns or less, when the probe pulse duration of 8 or 12 ns is included there is time for some cooling to occur. The published calculations of Jellison et al. [35] indicate for 0.6 J/cm^2 pulses a cooling to ~1000 °C at 10 ns and to ~850 °C at 20 ns after the end of the molten phase. Our results could be explained with a cooling rate two to three times as fast. Undercooling of the melt could also be involved.

Experiments described here with 150 ns pulses have provided firm evidence regarding phonon thermalization on this time scale. The Raman data obtained with 150 ns pulses clearly indicate good equilibration among a number of regions of the dispersion curves. Thus we have found populations characteristic of $T = 1500 \pm 300$ K in the optic mode at the zone center and zone boundaries and for the TA mode at the X and L points. Second-order spectra have not yet been obtained with 10 ns pulses, but the fact that the phonon frequency shifts are consistent with Stokes/antiStokes ratios suggests that thermalization is largely complete on this time scale as well.

Finally, we find that there is significant additional broadening of the first-order Raman line which cannot be explained by temperature gradients. These can be plausibly explained by carrier–phonon interactions [14,15] at densities in the range of ~$10^{20}/cm^3$.

Acknowledgements

The author is indebted to M.C. Lee and G.J. Trott for their collaboration on the experiments with 10 ns pulses, and to M. Cardona, H. Vogt and the Max Planck Institut für Festkorperforschung–Stuttgart for making possible the work with 150 ns pulses. The support of the U.S. Office of Naval Research and A. von Humboldt Foundation is gratefully acknowledged.

References

[1] I.B. Khaibullin, B.I. Shtyrkov, M.M. Zaripov, R.M. Bayazitov and M.F. Galjautdinov, Radiation Effects 36 (1978) 225;
I.B. Khaibullin, E.I. Shtyrkov and M.M. Zaripov, Proc. 15th Int. Conf. Physics of Semiconductors, Kyoto, 1980 (J. Phys. Soc. Japan 49 (1980) Suppl. A) p. 1281.
[2] J.A. Van Vechten, R. Tsu, F.W. Saris and D. Hoonhout, Phys. Lett. 74A (1979) 417.
[3] C.M. Surko, A.L. Simons, D.H. Auston, J.A. Golovchenko, R.E. Slusher, and T.N.C. Venkatesan, Appl. Phys. Lett. 34 (1979) 635.
[4] P. Baeri and S.U. Campisano, in: Laser Annealing of Semiconductors, eds., J.W. Mayer and J.M. Poate (Academic Press, New York, 1982) p. 75.
[5] R.F. Wood and G.E. Giles, Phys. Rev. B 23 (1981) 2923;
A. Leitoila and J.F. Gibbons, J. Appl. Phys. 53 (1984) 3207.
[6] Liquid Semiconductors, V.M. Glazov, S.N. Chizhevskaya and N.N. Glagoleva, eds. (Plenum, New York, 1969).
[7] H.W. Lo and A. Compaan, Phys. Rev. Lett. 44 (1980) 1604.
[8] D. von der Linde and G. Wartman, Appl. Phys. Lett. 41 (1982) 700.
[9] A. Compaan, Mat. Res. Soc. Symp. Proc. 13 (1983) 23;
A. Compaan, H.W. Lo, M.C. Lee and A. Aydinli, Phys. Rev. B 26 (1982) 1079.
[10] D. von der Linde, G. Wartman and A. Ozols, Mat. Res. Soc. Symp. Proc. 13 (1983);
G. Wartman, D. von der Linde and A. Compaan, Appl. Phys. Lett. 43 (1983) 613.
[11] R. Tsu and J.G. Hernandez, Appl. Phys. Lett. 41 (1982) 1016.
[12] M. Balkanski, R.F. Wallis and E. Haro, Phys. Rev. B 28 (1983) 1928.
[13] R.J. Nemanich, D.K. Biegelson, R.A. Street and L.E. Fennell, Phys. Rev. B 29 (1984) 6005.
[14] See for example, M. Balkanski, K.P. Jain, R. Beserman and M. Jouanne, Phys. Rev. B 12 (1975) 4328 and M. Chandrasekhar, J.B. Renucci and M. Cardona, Phys. Rev. B 4 (1978) 1623.
[15] A. Compaan, G. Contreras, M. Cardona and A. Axmann, J. Phys. (Paris) 44 (1983) C5-197;
A. Compaan, G. Contreras and M. Cardona, Mat. Res. Soc. Symp. Proc. 23 (1984) 117;
G. Contreras, A.K. Sood, M. Cardona A. Compaan, Solid State Commun. 49 (1984) 303.
[16] A.K. Sood and M. Cardona, Solid State Commun. 49 (1984) 306.
[17] M. Cardona, in: Light Scattering in Solids IV, Topics in Applied Physics Vol. 54, eds., M. Cardona and G. Guntherodt (Springer-Verlag Berlin, 1984) p. 5.
[18] A. Compaan and H.J. Trodahl, Phys. Rev. B 29 (1984) 793.
[19] D.H. Auston, C.M. Surko, T.N.C. Venkatesan, R.E. Slusher and G.A. Golovchenko, Appl. Phys. Lett. 33 (1978) 437.
[20] J.M. Liu, R. Yen, H. Kurz and N. Bloembergen, Appl. Phys. Lett. 39 (1981) 755;
J.P. Long, R.T. Williams, T.R. Royt, J.C. Rife and M.H. Kabler, Mat. Res. Soc. Symp. Proc. 13 (1983) 89;
M. Bensoussan and J.M. Moisson, Physica 117B and 118B (1983) 404.

[21] A. Pospieszczyzk, M. Abdel Harith and B. Stritzker, J. Appl. Phys. 54 (1983) 3176.
[22] M.O. Thompson and G.J. Galvin, Mat. Res. Soc. Symp. Proc. 13 (1983) 57.
[23] D. Guidotti, T.A. Driscoll and H.J. Gerritsen, Solid State Commun. 46 (1983) 337; T.A. Driscoll and D. Guidotti, Phys. Rev. B 28 (1983) 1171.
[24] C.V. Shank, R. Yen and C. Hirlimann, Phys. Rev. Lett. 51 (1983) 900.
[25] A. Compaan, A. Aydinli, M.C. Lee and H.W. Lo, Mat. Res. Soc. Symp. Proc. 4 (1982) 43.
[26] A. Compaan, in: Cohesive Properties of Semiconductors Under Laser Irradiation, NATO ASI series E-69, ed., L. Laude (Martinus Nijhoff, The Hague, 1983) p. 391.
[27] S.A. Lyon, Y.H. Chen, J.F. Lin and J.M. Worlock, Appl. Phys. Lett. 42 (1983) 978.
[28] K.M. Shvarev, B.A. Baum and P.V. Gel'd, Sov. Phys. Solid State 16 (1975) 2111.
[29] J.M. Liu, H. Kurz and N. Bloembergen, Appl. Phys. Lett. 41 (1982) 643; H.M. van Driel, L.A. Lompré and N. Bloembergen, Appl. Phys. Lett. 44 (1984) 285.
[30] D. von der Linde and N. Fabricius, Appl. Phys. Lett. 41 (1982) 991.
[31] C.V. Shank, R. Yen, C. Hirliman, Phys. Rev. Lett. 50 (1983) 454.
[32] D. Hulin, M. Combescot, J. Bok, A. Migus, J.Y. Vinet and A. Antonetti, Phys. Rev. Lett. 52 (1984) 1998.
[33] V. Heine and J.A. Van Vechten, Phys. Rev. B 13 (1976) 1622; J.A. Van Vechten, in: Semiconductor Processes Probed by Ultrafast Spectroscopy, ed. R. R. Alfano (Academic, New York, 1983).
[34] J. Bok, Phys. Lett. 84A (1981) 448.
[35] K. Murakami, K. Takita and K. Masuda, Jpn. J. Appl. Phys. 20 (1981) L867.
[36] M.C. Lee, M.S. Thesis, Kansas State University, 1982 (unpublished).
[37] L.A. Lompré, J.M. Liu, H. Kurz and M. Bloembergen, Appl. Phys. Lett. 43 (1983) 168.
[38] E.O. Göbel and O. Hildebrand, Phys. Status Solidi (b) 88 (1978) 645; W. Graudszus and E.O. Göbel, Physica 117 and 118 B + C (1983) 555.
[39] D. von der Linde, J. Kuhl and E.H. Rosengart, J. Phys. Soc. Jpn. 49 Suppl. A (1980) 653.
[40] Landoldt-Bornstein, New Series Gp. III, Vol. 7a, eds., O. Madelung, M. Schulz and H. Weiss (Springer, Berlin, 1982).
[41] H. Bilz and W. Kress, Phonon Dispersion Relations in Insulators, Springer series in Solid-State Sciences, Vol. 10 (Springer-Verlag, Berlin, 1979).
[42] A. Battacharya, B.G. Streetman and K. Hess, J. Appl. Phys. 53 (1982) 1261; M.C. Lee, A. Aydinli, H.W. Lo and A. Compaan, J. Appl. Phys. 53 (1982) 1262.
[43] E.J. Yoffa, Appl. Phys. Lett. 36 (1980) 37; Phys. Rev. B 21 (1980) 2415.
[44] J. Dziewior and W. Schmid, Appl. Phys. Lett. 31 (1977) 346; K.G. Svantessen and N.G. Nilsson, Solid State Electron. 21 (1978) 1603.
[45] G.E. Jellison and F.A. Modine, Appl. Phys. Lett. 41 (1982) 180.
[46] B.C. Larsen, C.W. White, T.S. Noggle, J.F. Barhorst and D.M. Mills, Mat. Res. Soc. Symp. Proc. 13 (1983) 43.
[47] G.E. Jellison and F.A. Modine, Phys. Rev. B 27 (1983) 7466; D.E. Aspnes and A.A. Studna, Phys. Rev. B 27 (1983) 985.
[48] L. Vina and M. Cardona, Phys. Rev. B 29 (1984) 6739.
[49] I.I. Novak, V.V. Baptizmanskii and L.V. Zhoga, Opt. Spectrosc. (USSR) 43 (1977) 145; E. Anastassakis, A. Pinczuk, E. Burstein, F.H. Pollak and M. Cardona, Solid State Commun. 8 (1970) 133.
[50] M.C. Lee, G. Trott and A. Compaan (unpublished).
[51] P.A. Temple and C.E. Hathaway, Phys. Rev. B 7 (1973) 3685.
[52] G.E. Jellison, Jr., D.H. Lowndes and R.F. Wood, Mat. Res. Soc. Symp. Proc. 13 (1983) 35.

Journal of Luminescence 30 (1985) 446–471
North-Holland, Amsterdam

LASER-INDUCED COHERENT MODULATION OF SOLID AND LIQUID SURFACES

H.M. van DRIEL, J.E. SIPE and Jeff F. YOUNG*

Department of Physics and Erindale College, University of Toronto, Toronto, Canada, M5S 1A7

The interaction of a single spatially coherent, polarized laser beam with an interface possessing microscopic roughness leads to inhomogeneous energy deposition due to the interference between the incident beam and scattered interface fields. At high intensities this can lead to spatially periodic melting or vaporization of solid or liquid layers, and temporary or permanent modification of the interface. We review our experimental and theoretical work and discuss the detailed electrodynamics, spatial spectra, surface structure and temporal growth characteristics of this phenomenon.

1. Introduction

The interaction of high intensity laser radiation with semiconductors has led to the study of many processes in high density solid state plasmas and in the rapid thermal processing of materials. For most of the work performed to date, the only laser parameters that are considered important are the intensity, pulse width and wavelength. In modelling the interactions that take place, one is quite often safe in using a one-dimensional model to describe absorption and (heat or plasma) diffusion processes, with depth effects being of primary interest. The laser is used primarily as an ultrafast source of narrow bandwidth energy, and all of the processes can be understood in terms of bulk parameters. In this review we will describe the role of the surface, or interface between two bulk media, in the absorption of laser radiation. Because of the abrupt change in dielectric properties which occur at many interfaces and because no interface is atomically smooth, radiation can be scattered along the interface and interfere with the incident beam. This interference occurs because of the spatial *coherence* of the laser and the direction and strength of the scattered field depend on the *polarization* of the beam. This leads to inhomogeneous energy deposition at

*Present address: Division of Physics, National Research Council of Canada, Ottawa, Canada, K1A OR6.

0022-2313/85/$03.30 © Elsevier Science Publishers B.V.
(North-Holland Physics Publishing Division)

the interface which, at sufficiently high energy, manifests itself as inhomogeneous melting or vaporization, depending on the fluence level and the type of interface [1–3]. Figure 1, showing the structure formed on Ge using a linearly polarized, pulsed 1.06 μm beam at normal incident is typical of the regularity of the effect. Photodeposition processes also lead to periodic structure formation via this interference effect [4]. Whatever the mechanism or material the permanent or temporary structures formed as a result of this provide information about the electrodynamics of the interface and the nature of material redistribution in the presence of a spatially inhomogeneous phase transition. Of particular interest is the universality of this phenomenon [5] provided a sufficient amount of energy can be deposited near the interface: if the absorption coefficient is large enough at the incident wavelength, the same types of effects can occur in metals, semiconductors and insulators, solids or liquids, for laser pulses of arbitrary length, and wavelength.

In what follows we will outline the role of interfaces in high excitation laser phenomena. The basic interface electrodynamics will be discussed in the next section together with the spatial spectra that result from inhomogeneous energy deposition. This will be followed by material redistribution considera-

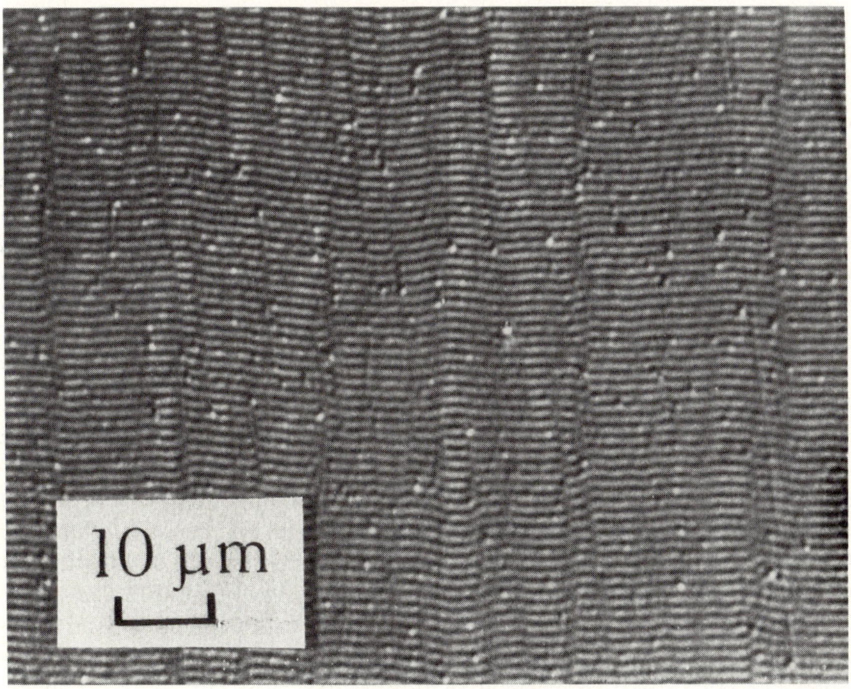

Fig. 1. Photograph of the surface structure produced on Ge with linearly polarized, 1.06 μm radiation incident at 0°.

tions and the formation of permanent structure. Finally, we will discuss the transient evolutionary processes that occur during and after the laser beam, as revealed by time-resolved diffraction studies. Overall, the emphasis will be on outlining the richness of the effects that can be observed and understanding them from a simple physical point of view. The reader who is interested in a rigorous justification of the physical picture presented here can find such in our series of articles in Physical Review B [2,3,6]. For simplicity of presentation and because this conference is based on semiconductor phenomena, we will mainly confine our attention to effects induced by pulsed 1.06 μm beams on germanium. Our earlier articles as well as those of others [7–14] describe similar effects in other materials at different excitation wavelengths.

2. Electrodynamics and spatial spectra

The appearance of laser-induced periodic surface structure evokes two fundamental questions: first, what is the nature of the interface electro dynamics which leads to interference and inhomogeneous energy deposition; second, even if one has inhomogeneous energy deposition, why is there either a permanent (in solids) or temporary (in liquids) redistribution of material leading to the periodic structure. We will consider the first question in this section and the second in subsequent sections.

Spatially inhomogeneous energy deposition is intimately related to how incident electromagnetic radiation scatters from a microscopically rough surface and the interference that occurs between the incident and scattered fields. Apart from any material parameters that influence this process, the beam parameters that are important are polarization type (linear, circular, etc.) and degree (characterized by a correlation time), wavelength and spatial coherence. The interaction of a plane electromagnetic wave with an interface is depicted in fig. 2. If the beam is incident at an angle θ_i and the component of the wavevector in the plane of the surface is $\boldsymbol{k}_i$, then for a textbook smooth surface there will be refracted and reflected waves, both of which will also have $\boldsymbol{k}_i$ as the parallel components of their wavevectors. If the interface is rough, either because of spatial variations of its profile or dielectric constant due to, e.g. composition, then the light will be scattered in general in all directions into the vacuum and the bulk. If, for example, the interface possesses a low amplitude grating of wavevector $\boldsymbol{q}$, in the so-called "selvedge" region separating the vacuum and bulk, then the scattered fields will have parallel wavevector components $\boldsymbol{k}_i \pm \boldsymbol{q}$; for large amplitude gratings higher-order scattered fields can exist with $\boldsymbol{k}_i \pm n\boldsymbol{q}$, where $n=1,2,3 \ldots$. A randomly rough surface can be considered to be composed of a spectrum of gratings of general Fourier components $\boldsymbol{q}$, in which case there then exists a spatial spectrum of scattered light fields. For the *particular* electromagnetic fields which are scattered *along* the interface,

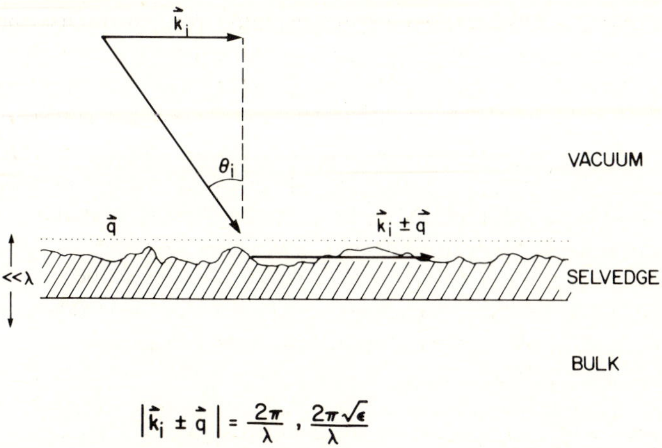

Fig. 2. The geometry of light incident from vacuum on a rough surface of bulk dielectric constant ε. The symbols are explained in the text.

the magnitude of the wavevector is, of course, equivalent to the magnitude of the parallel component.

If the incident side of the interface is characterized by a (real) dielectric constant ε_a, while the other side is characterized by a negative ε_b, then if $\varepsilon_a\varepsilon_b/(\varepsilon_a+\varepsilon_b)>1$, surface polaritons which propagate along the interface exist, and it is well known [15] that a scattered field couples resonantly with these interface excitations which then dominate the resulting spatial spectra of scattered light. These surface polaritons can exist as phonon polaritons in insulators or semiconductors in the far infrared, or as plasmon polaritons in metals. They represent true surface electromagnetic modes, and, in the absence of damping through the imaginary part of the dielectric constant, would have infinite lifetimes. These modes can be excited under the restrictions of conservation of momentum, which requires that the wavevector of the scattered light matches that of the surface polariton. From the dispersion relation of the surface polariton this implies

$$|\boldsymbol{k}_i \pm \boldsymbol{q}| = (2\pi/\lambda)[\varepsilon_a\varepsilon_b/(\varepsilon_a+\varepsilon_b)]^{1/2} \qquad (1)$$

where λ is the vacuum wavelength. Since surface polaritons have both transverse and longitudinal field components, they can always interfere with the incident beam. The interference of the refracted beam with the field of the surface polariton will lead to spatially inhomogeneous energy deposition at wavevector $\boldsymbol{q}$ (which must also satisfy eq. (1)); this dominates the spectrum of inhomogeneous energy deposition because of the strong scattering into surface polariton modes. Then, in $\boldsymbol{q}$ or Fourier space, the locus of points of inhomogeneous energy depostion will essentially lie on two intersecting circles.

Many surfaces, such as those of insulators and semiconductors, are not polariton active at incident wavelengths of interest. Nonetheless, it is still possible to induce non-radiative field structures at the interface which have characteristics somewhat similar to those of polariton fields, but which decay in amplitude faster than they do. Some characteristics of these fields were discussed by Sommerfeld [16] in connection with the propagation of radio waves along the surface of the earth, and similar phenomena exist in the propagation of acoustic waves near an interface [17]. These field structures, which are not true surface modes, have been referred to as lateral or ground waves. In the optical domain we refer to the corresponding field structures as "radiation remnants" and have discussed their characteristics in detail [2]. For dielectric interfaces there are actually two components to the field structure, one with wavevector $2\pi\sqrt{\varepsilon_a}/\lambda$ and the other $2\pi\sqrt{\varepsilon_b}/\lambda$. It is thus tempting, and often heuristically convenient, to think of these fields as radiation fields propagating parallel to the interface in the upper and lower medium respectively. Of course, although such fields would satisfy the wave equation in one medium, they would not satisfy it in the other medium, nor would they satisfy the Maxwell boundary conditions at the interface. Precisely because of this they must be more rigorously understood as non-radiative fields which do not propagate energy parallel to the surface, hence the term "radiation remnants". Nonetheless, these field structures dominate the scattered field spectrum in the presence of an interface, somewhat similarly to the way surface polaritons would dominate the spectrum if they were present. Thus, for non-polariton active media, eq. (1) is replaced by:

$$|\boldsymbol{k}_i \pm \boldsymbol{q}| = \begin{cases} 2\pi\sqrt{\varepsilon_a}/\lambda \\ 2\pi\sqrt{\varepsilon_b}/\lambda \end{cases} \tag{2}$$

For $\varepsilon_b \gtrsim \varepsilon_a$, the component associated with the bottom half of this equation can in general have a larger amplitude, while for $\varepsilon_b/\varepsilon_a \gg 1$ the other component completely dominates. This is in agreement with experimental observations [12] which indicate that the spacings of structures induced on NaCl and MgF_2 in air is dictated by ε_b while for high index materials like Ge ($\varepsilon_b \geqslant 16$) and Si ($\varepsilon_b \geqslant 12$) the structure is determined by the vacuum wavelength. It might be noted as well that for polariton active interfaces the radiation remnant fields at $2\pi\sqrt{\varepsilon_a}/\lambda$ will also occur if $\varepsilon_a > 0$ but will have an amplitude which is small compared to the polariton field structures, and so can be neglected. In any event if the incident medium is air or vacuum ($\varepsilon_a = 1$) the two types of field structures would be indistinguishable with respect to their induced structures for infrared or visible incident beams on metals, since for $\varepsilon_b \ll -1$ the polariton dispersion relation lies very close to the light line. It follows that for metals and semiconductors, the spacings of induced structures will be essentially the same, being almost independent of their dielectric constants. Only for the low index dielectrics or metals in the UV will the spacings reflect the material dielectric constant. This, in part, accounts for the universal characteristics of laser induced periodic surface structure. Figure 3 represents a graphical illustration of the locus of inhomogeneous energy deposition based on the top half of eq. (2) with $\varepsilon_a = 1$. For normal incidence where $\boldsymbol{k}_i \equiv 0$, the two circles will collapse into a single circle.

Equations (1) and (2) represent the most general locus of induced surface structures, particular cases of which have been widely reported in the literature over the past 20 years [7–14]. For example, if p-polarized light is incident on a surface, the dominant features in the induced structures have been noted at $q = (2\pi/\lambda)(1 \pm \sin\,\theta_i)$, with the ripples being perpendicular to the polarization vector as noted in fig. 3. The actual spectrum of structure is much richer than this, as will be seen below.

In general, a more detailed treatment [2] than that presented here indicates that the strength of the scattered fields for different propagation directions is a function of the angle of incidence and the polarization of light through such factors as the Fresnel coefficients and the initial surface topography. The

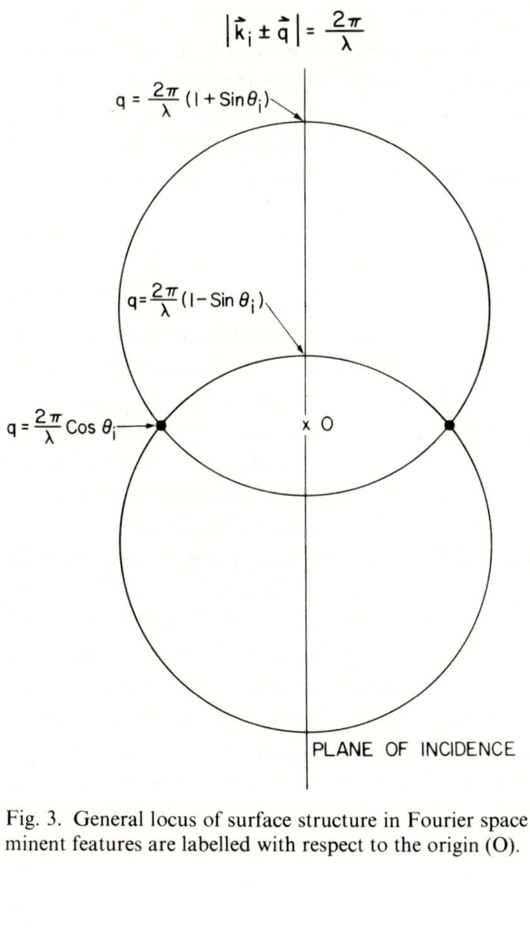

Fig. 3. General locus of surface structure in Fourier space given by $|\boldsymbol{k}_i \pm \boldsymbol{q}| = 2\pi/\lambda$. The most prominent features are labelled with respect to the origin (O).

amplitude spectrum along the two intersecting circles indicated in fig. 3 will vary. The observation of the structures in Fourier space therefore provides a map of the scattered field strengths, with the induced structures essentially yielding a hologram of the scattered fields. A convenient way of obtaining the hologram in Fourier space is to shine a cw laser beam on the surface following high intensity illumination and to view the far-field diffraction pattern in reflection or transmission as appropriate. This provides complimentary and in many cases, more detailed information than that obtained by viewing the sample in real space under an optical microscope.

From the above discussion, the dominant spacings are seen to be independent of material type in many cases and merely scale with the wavelength. Given the universal characteristics [5] which have been observed by us, in this paper we will confine the remainder of the discussion to structures induced on Ge using 1.06 μm, 25 ns pulses from a Q-switched Nd:YAG laser. For these conditions the threshold of melting is approximately 100 mJ/cm^2. Although in other materials the threshold for observing the effects discussed here and the details of material redistribution will depend on optical and thermodynamic parameters, germanium at least illustrates many of the different and salient features associated with the phenomenon. With respect to the detailed spectra, fig. 4 shows the far-field diffraction pattern obtained from a germanium surface using a 0.51 μm Ar^+ laser after it had been repeatedly illuminated at an angle of incidence of 60° by the p-polarized pulsed laser at a fluence of 150 mJ/cm^2. The distortion in the circles is due to the projections of the cones of diffracted light onto a plane surface for photography. Although evidence is seen for the $\lambda/(1 \pm \sin\theta)$ fringes, they do not appear as points, but rather are parts of arcs in the spectrum. At $\theta_i = 60°$ they do not correspond to the strongest features in the spectrum, which actually occur at the intersection points of the two circles, these according to fig. 3 have the spacing $\lambda/\cos\theta_i$, and correspond to ripples which are parallel to the incident polarization. The Fourier spectra observed for germanium for different θ_i and polarizations are shown schematically in fig. 5. Our theory is able to account for the amplitude of the Fourier spectra around the two circles for the different cases illustrated. One particularly interesting feature occurs when one considers incident circularly polarized light. As fig. 5 illustrates, right- and left-handed circularly polarized light yield diffraction patterns which are mirror images of each other in the plane of incidence. One such diffraction pattern is shown as fig. 6. On the one hand this symmetry is not unexpected since right and left circularly polarized light are mirror images of each other in the plane of incidence. However, the fact that the handedness of the polarization can be revealed by a process depending only on the linear optical properties of the material is intriguing. For normally incident circularly polarized light, the symmetry becomes complete and the diffraction pattern reveals a uniformly illustrated circle as indicated in fig. 7. This implies that light is scattered isotropically along the surface and, as shown

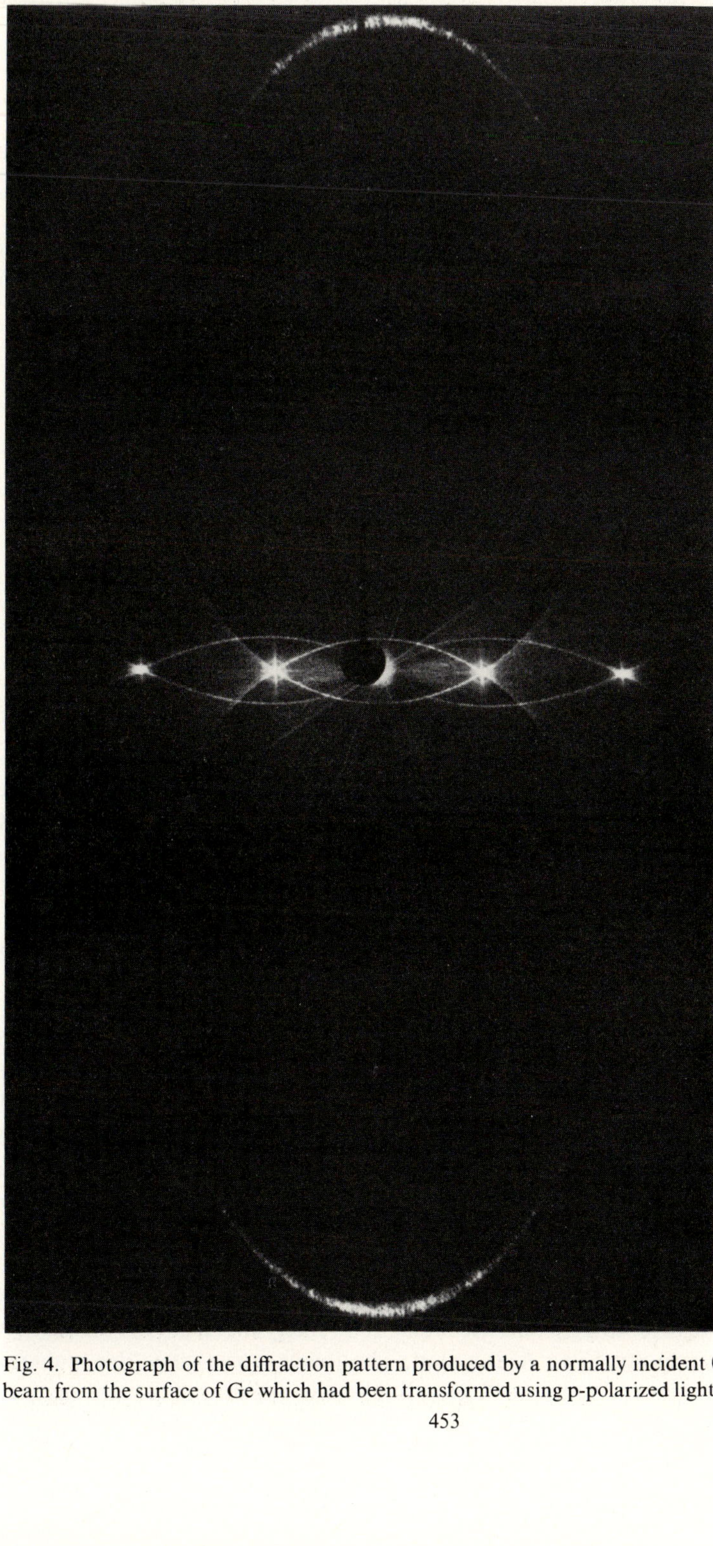

Fig. 4. Photograph of the diffraction pattern produced by a normally incident 0.51 μm Ar^+ laser beam from the surface of Ge which had been transformed using p-polarized light at $\theta_i = 60°$.

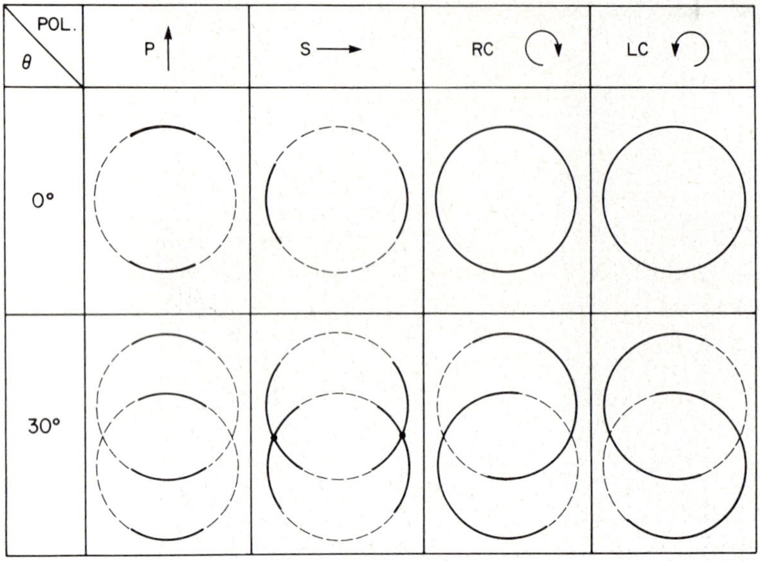

Fig. 5. Schematic summary of far-field diffraction patterns for Ge for various angles of incidence, θ, and polarization states of the impacting beam. Solid curves indicate high-intensity regions.

by the microscope picture in fig. 7, the resulting interference between all the ripple structures yields what looks like an isotropically rough surface.

If the inhomogeneous energy deposition is initiated from the solid phase, as it is if the solid is illuminated near the melting threshold, then the interference is between the incident beam and the radiation remnant. On the other hand, if the laser beam has a fluence of, say, 500 mJ/cm^2, which is far above the threshold then the solid can melt early in the pulse and inhomogeneous energy deposition can be initiated in the liquid, which is metallic. Inhomogeneous energy deposition then occurs because of the interference between the incident beam and a surface plasmon on the metallic liquid. Equation (1) will then dictate the spacings of the structure which form after the liquid resolidifies, rather than eq. (2). For liquid Ge at 1.06 μm, the real part of the dielectric constant is -32 [18] so that the difference in spacings between the solid and liquid induced structures is about 2% at normal incidence. This difference is increased for off-normal incidence and we have used the spacings of the cosine type ripples to test the two equations. Figure 8 shows the spacings observed as a function of the parallel wavevector component which was altered by varying the angle of incidence of the beam. Data points were obtained at the two extreme fluence levels from the intersection points of the two diffraction circles as illustrated in the inset to fig. 8. As the incident intensity increases the diffraction peaks shift from point D to A as noted. The lower curve gives the spacings extracted at the lower fluence corresponding to the radiation remnant regime,

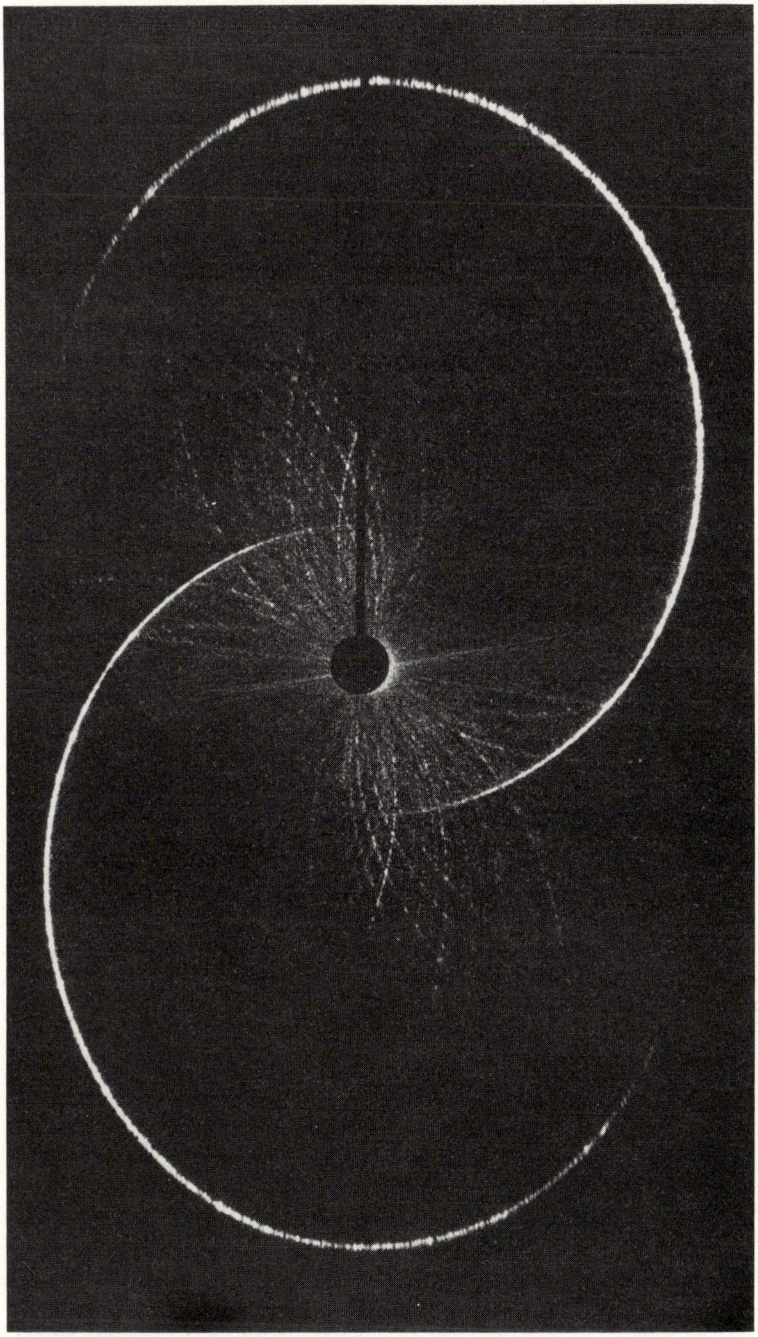

Fig. 6. Photograph of far-field diffraction pattern obtained in reflection from Ge surface with normally incident 0.51 μm Ar^+ laser; Ge surface was damaged by right circularly polarized beam at 30°.

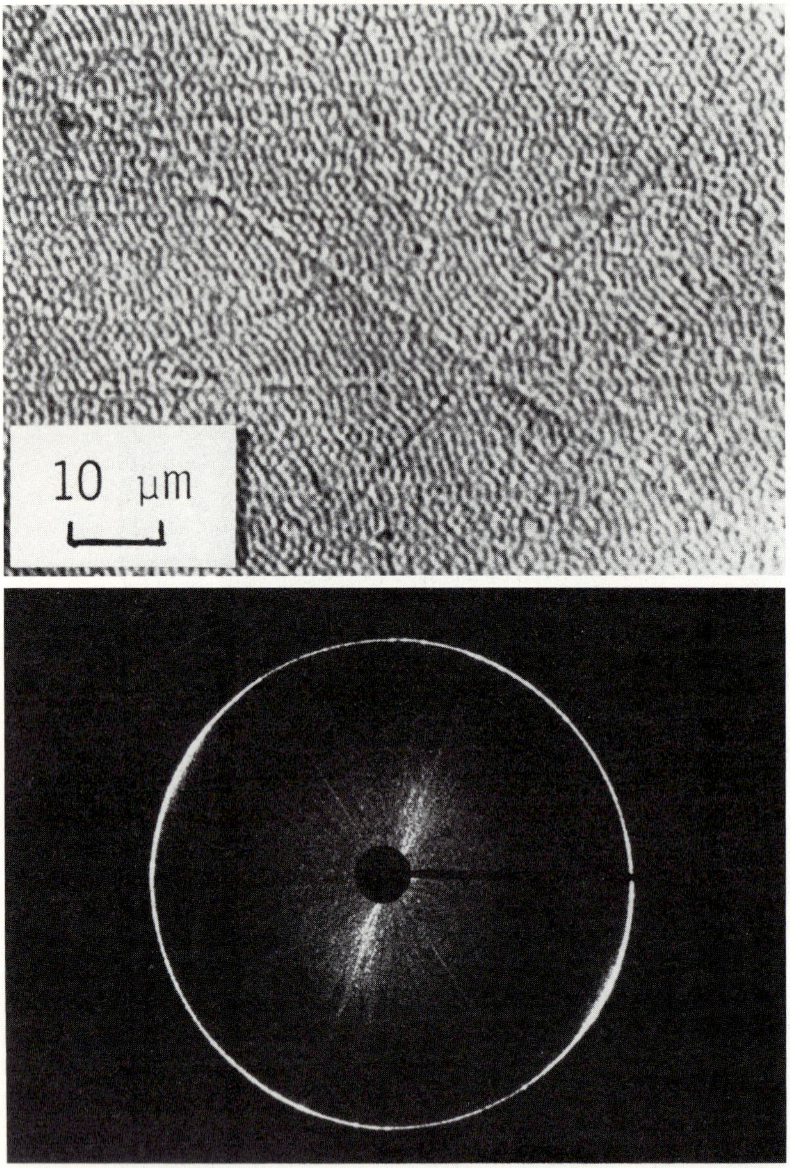

Fig. 7. (Top): Normarski microscope photograph of surface structure on Ge induced by 1.06 μm circularly polarized beam at normal incidence; (bottom): the corresponding far-field diffraction pattern.

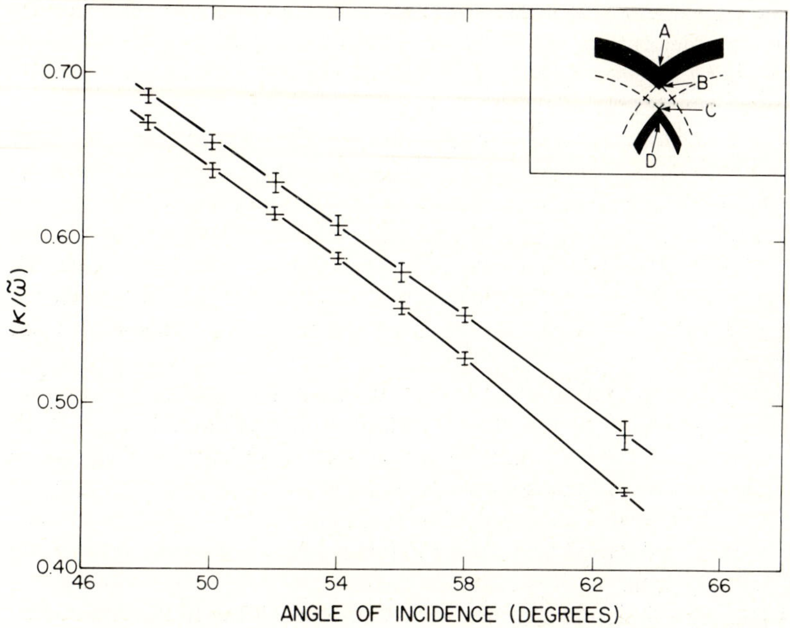

Fig. 8. A plot of the normalized wavevector ($\tilde{\omega} \equiv 2\pi/\lambda$) of induced structure as a function of the $\lambda = 1.06$ μm beam's angle of incidence. The vertical extremities of the data points along the top (bottom) curve correspond to the measured positions of points A and B (C and D) shown in the inset. The inset is a magnified view of the intersection of the two diffraction circles (see fig. 3). The solid lines represent points calculated using the radiation remnant (lower curve) and surface plasmon dispersion relations (upper curve).

while the upper curve is associated with the predicted surface-plasmon induced structures. Good agreement is obtained in both cases with the theory.

As noted above the induced structures primarily reflect the wavelength, angle of incidence, polarization, and spatial coherence of the incident beam. The interference conditions can be destroyed (as would be desired for material processing applications) by reducing the correlation time of the polarization or by destroying the spatial coherence. If the correlation time is reduced to much less than the time required for structure formation (see below), or if the spatial coherence distance is less than a few wavelengths, then obviously interference effects, if any, would be weak.

3. Material redistribution and feedback for growth

Whereas the surface electrodynamics, which determines the spatial inhomogeneous energy deposition, is primarily a function of incident beam parameters, the resultant permanent or temporary surface structure associated with material redistribution is determined by the material's physical parameters. In certain situations, such as laser photodeposition of metals from organometallic gases [4] the growth rate of ripples is determined by the dissociation rate of the gas, whereas for materials like germanium and silicon, which are forced to undergo a phase transition at high laser intensity, the main parameters are thermodynamic. The nature of material redistribution depends also on the laser fluence. In all we have identified four qualitatively different fluence regimes for periodic structure formation on germanium. We will outline the details of growth in the lowest fluence regime, which occurs near the threshold of melting and will briefly discuss the other regimes at the end of this section and in the next section.

The growth characteristics in the lowest fluence regime reflect the fact that inhomogeneous melting occurs at the surface of germanium. Figures 9–11 show a sequence of electron microscope pictures of an initially flat surface of germanium which was successively illuminated by s-polarized 150 mJ/cm^2 pulses at 30°. After the first shot (fig. 9) the surface develops what appears to be randomly distributed raised features or bubbles. This occurs because of the almost isotropic scattering of surface waves from the initially microscopically rough surface and their interference with the incident beam. An additional contributing factor is the instability of a thin melt layer with respect to spatial variations in its thickness [19]: Such a layer will tend to break up into smaller elements. The first shot generates a sufficient amount of roughness to cause large amplitude surface scattering on subsequent shots. On the following shots the preferential scattering of light along certain directions is more strongly determined by the beam polarization and so inhomogeneous energy deposition occurs principally along certain strips. These strips grow in amplitude and in width from shot to shot until after about 20 shots the process saturates as the amplitude reaches a small fraction of the incident wavelength. Note, however, that the bubble features do not change after the first shot, indicating the inhomogeneous melting is now primarily confined to just the strips. Feedback is obviously important in determining growth. As the amplitude of a particular Fourier component of the surface roughness, $\boldsymbol{q}$, begins to grow there is even more scattering leading to more interference and growth of surface features at $\boldsymbol{q}$. The growth characteristics of the ripples in many ways resemble those of a three-wave-parametric-mixing process. In this case the three waves are the incident and scattered light and the material ripple. The ripple is then seen to build up from initial "noise", which in this case represents surface roughness. Although there is a "gain" associated with the growth, it certainly is not independent of time even if the beam intensity is. During the interaction process the surface undergoes spatially inhomogeneous melting with the fraction of the surface which is molten in strips changing as a function of time. This effect alone causes the scattered light intensity to change rapidly with time

Fig. 9. An electron microscope picture of the surface of germanium after being irradiated at 30° by 1 pulse of s-polarized, 1.06 μm radiation in fluence regime A.

[6]. Also, as will be shown below, the surface develops raised features following resolidification with the amplitude and shape changing on a shot to shot basis. At best one can say that there is strong nonlinear growth and that both intra- and inter-pulse feedback determine the ripple "gain". A detailed feedback theory would be very complicated involving time and temperature dependent

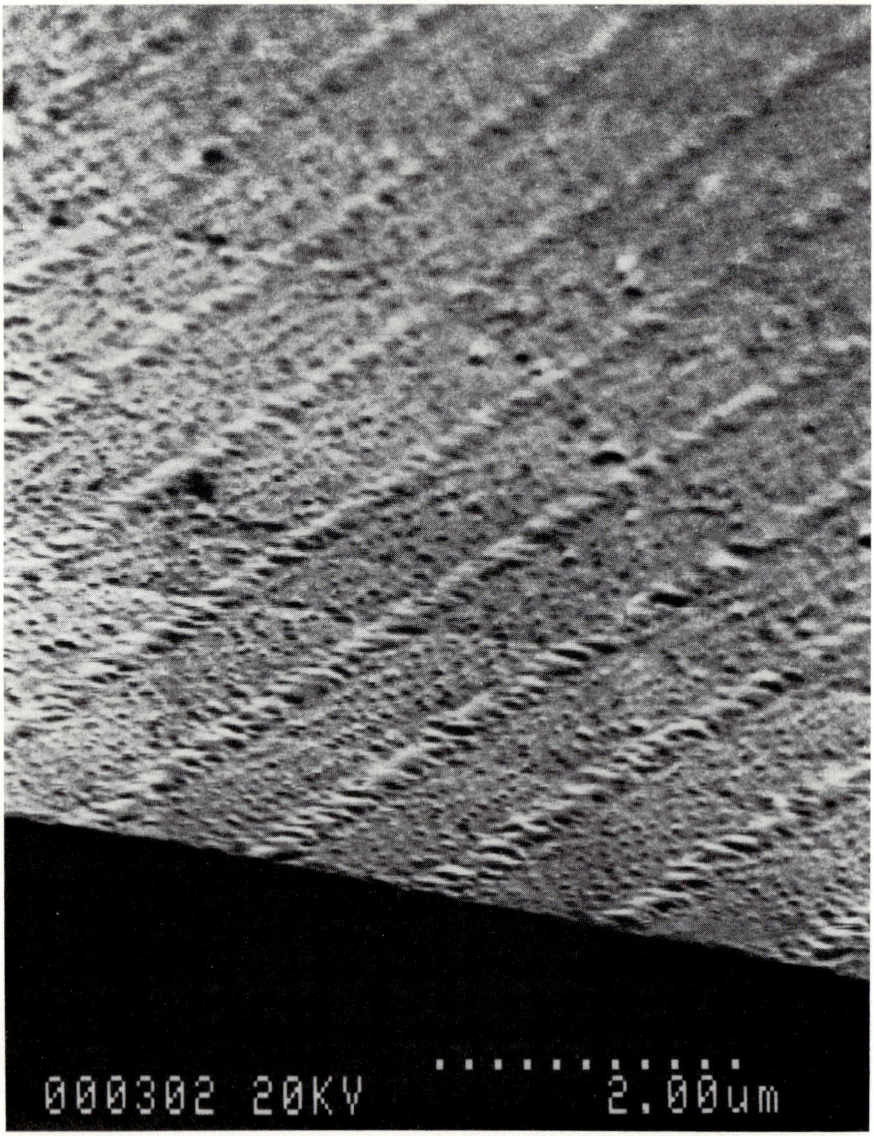

Fig. 10. An electron microscope picture of the surface of germanium after being irradiated at 30° by 2 pulses of s-polarized, 1.06 μm radiation in fluence regime A.

material parameters and diffraction calculations on a microscopic scale [6].

The basic question of why the material, once partially melted, does not yield a flat surface on resolidification is answered with the aid of the top of fig. 12. This shows the profile of a germanium surface which has undergone localized

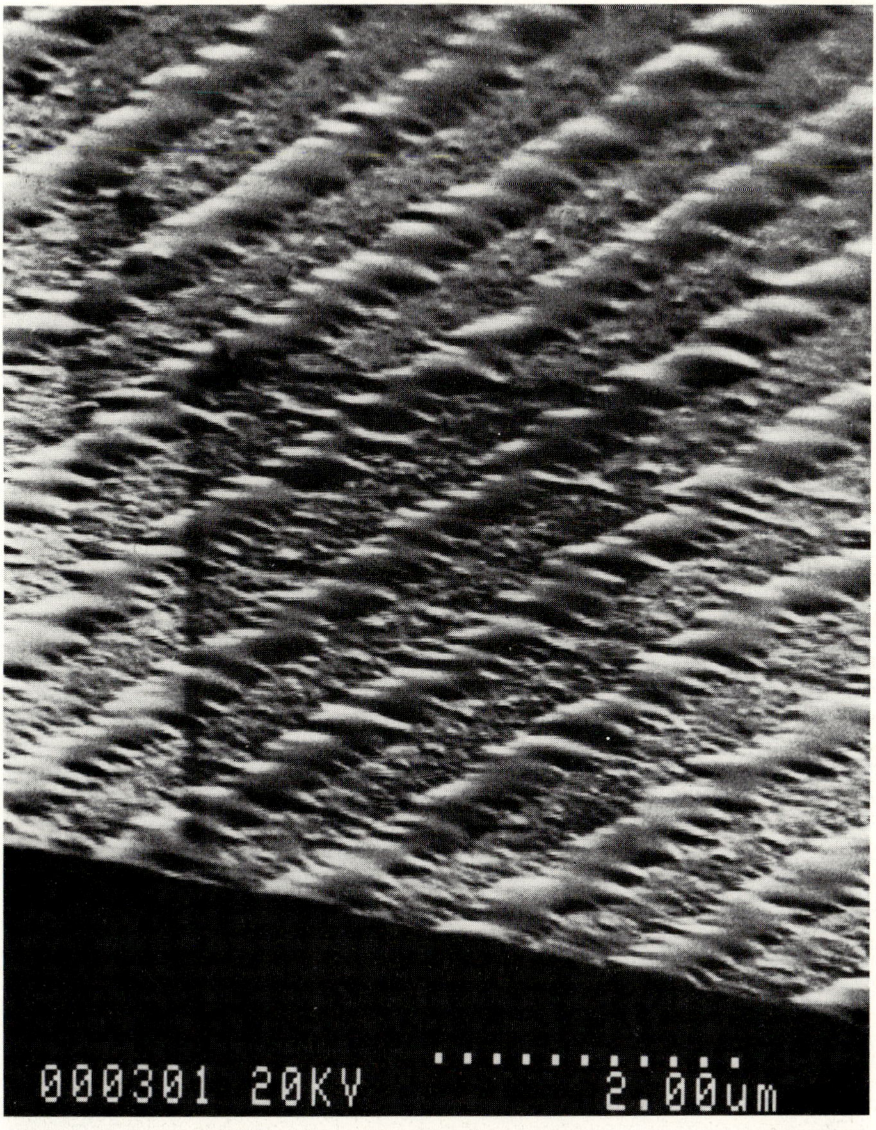

Fig. 11. An electron microscope picture of the surface of germanium after being irradiated at 30° by 3 pulses of s-polarized, 1.06 μm radiation in fluence regime A.

melting. Because there is a density increase on melting, the surface shows a miniscus shaped molten region. When cooling begins, it initiates at the sides of the miniscus where the melt thickness is smallest, causing the depression angle to be frozen in. As resolidification of the rest of the miniscus proceeds, the central portion *must* be raised above the ambient level due to conservation of

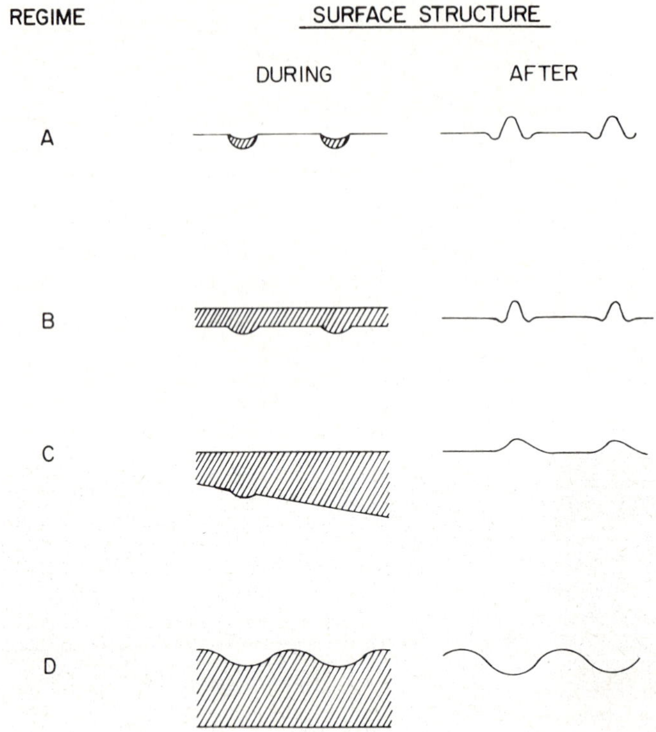

Fig. 12. Surface profiles of Ge which occur during and well after laser radiation for different laser fluences: A (100–150 mJ/cm^2); B (200–300 mJ/cm^2); C (300–400 mJ/cm^2); D (500–1000 mJ/cm^2).

mass. This causes a "sombrero-like" profile to be formed in place of the miniscus as indicated by the "after" illustration. This picture is borne out both by the experiments of Surek and Chalmers [20] who heated millimeter sized regions of Ge wafers over a time scale of seconds and observed this type of profile, and by more rigorous arguments for structure formation involving surface tension effects [21]. In our case the resolidification process takes place on a nanosecond time scale as will be seen in the following section. Figure 13 shows an electron microscope picture of the actual surface profile of a cleaved germanium wafer. The depressions on either side of the 200 Å high peaks is easily seen if the page is viewed on edge. This scenario of the development of the permanent structure also explains why it is the laser fluence and not the intensity which is the key energy parameter in influencing the growth. It is therefore not surprising that pulses as short as picoseconds [13] or as long as seconds [22] yield the same effects. Once the energy has been inhomogeneously deposited in the solid then the same effects will be noted *when the liquid regions resolidify*.

As mentioned above, there are four qualitatively different regimes for

Fig. 13. An electron microscope picture of the cross section of fringes formed by p-polarized, 1.06 μm pulses incident on germanium at 60° in regime A.

structure formation as the fluence is raised from 100 to 500 mJ/cm^2. Figure 12 illustrates the surface profiles of these regimes during and after the laser pulse. These regimes have, in large part, been identified with the aid of the transient diffraction experiments of the next section. Having dealt with the first regime above (henceforth referred to as regime A), let us consider the other three briefly now. In the second regime or regime B, which occurs for fluences near 200 mJ/cm^2, the fluence is sufficiently high so that a melt layer can form over the entire surface. This melt layer does not have uniform thickness, however, due to inhomogeneous energy deposition in the solid during the early stages of the pulse. As a result, when the spatially periodic melt front approaches the surface after the laser pulse, parts of the surface resolidify before others yielding liquid strips and the scenario developed for regime A unfolds with the characteristic sombrero profiles resulting. Regime C which occurs near the spatial edge of the pulse and for fluences of about 300 mJ/cm^2 is better understood following a discussion of regime D which occurs for fluences greater than 500 mJ/cm^2. In this regime the solid liquifies to a large depth quickly during the early part of the pulse. Once the liquid is optically thick interference can occur between the incident beam and an excited surface plasmon on the metallic liquid as mentioned in the previous section. This causes inhomogeneous energy deposition resulting in spatially varying boiling of the liquid. The ablation of the material from the liquid surface and the resulting back pressure on the liquid due to conservation of momentum cause the excitation of a surface capillary wave similar to that observed by Keilmann on liquid metals [23]. The surface capillary wave has a sinusoidal profile which will be permanently frozen into the surface if the liquid resolidifies in a time which is short compared to the capillary wave lifetime. Figure 14 shows a picture of the surface profile corresponding to regime D. When this figure is compared to fig. 13 it is evident that a more sinusoidal profile is formed in the higher fluence regime. Of course, if the fluence is too high such that the capillary wave dies out before the liquid resolidifies an essentially flat surface will result. For germanium this occurs in the vicinity of 1 J/cm^2. Finally the intermediate regime, C, can be understood as a combination of regimes B and D. Near the edge of the pulse in fluence regime B one can obtain inhomogeneous energy deposition as discussed above. Towards the center of the pulse the fluence is large enough to generate a thick molten layer but not intense enough of itself to cause excitation of capillary waves. However, a sufficient amount of scattered light can be launched from the rippled surface in regime B to interfere with the incident beam and cause inhomogeneous energy deposition in a melt front which is returning to the surface. The ripples which form from the liquid in this regime, unlike in the other three regimes, are sensitive to the lateral gradient of beam fluence.

4. Transient diffraction and reflectivity experiments

Observations of the time-resolved diffraction and reflectivity of a cw Ar^+ laser during ripple formation corroborate the ideas discussed above and provide

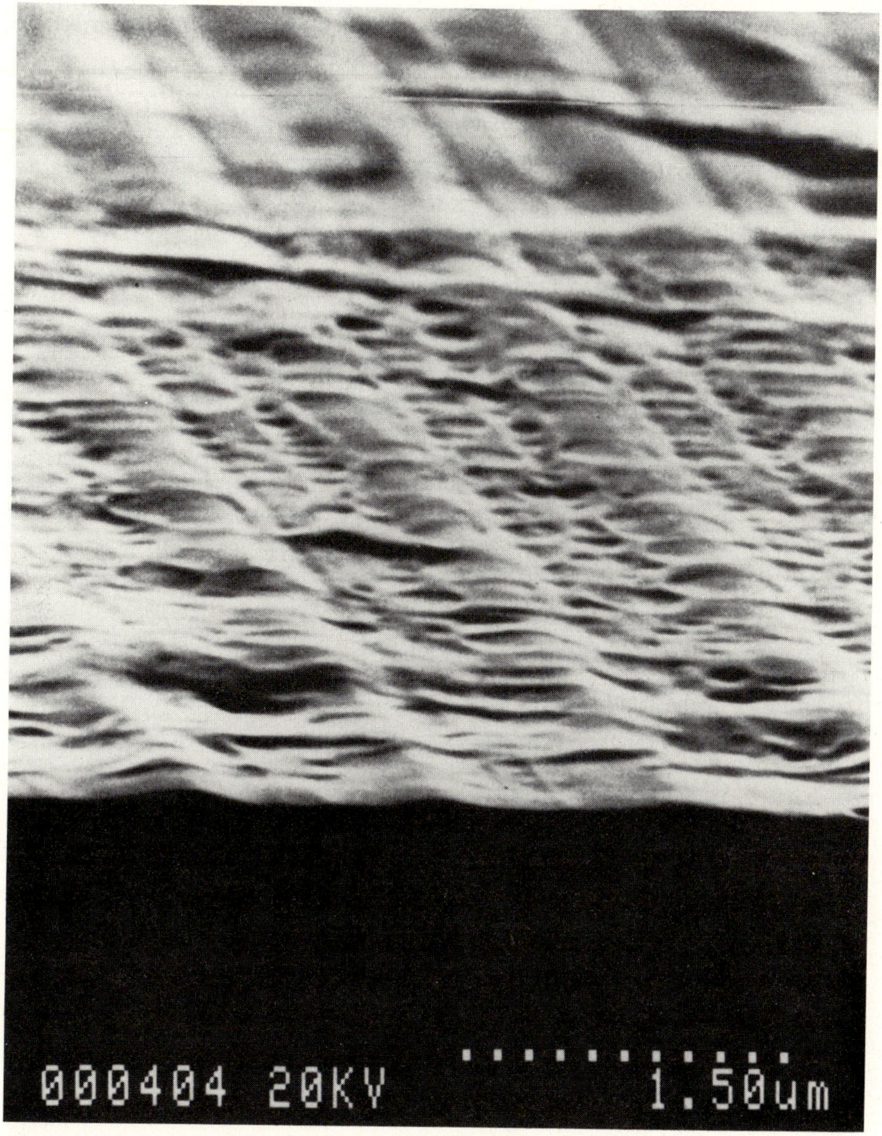

Fig. 14. An electron microscope picture of the smoothly varying ripples formed by 3 s-polarized, 1.06 μm pulses incident at 45° with fluences in regime D.

interesting new insights into the dynamics of the phenomenon. Figure 15 shows the transient reflectivity and diffraction from germanium during and after illumination by a succession of laser pulses in regime A. The detector which monitors the diffraction signal was placed in a position where ripple

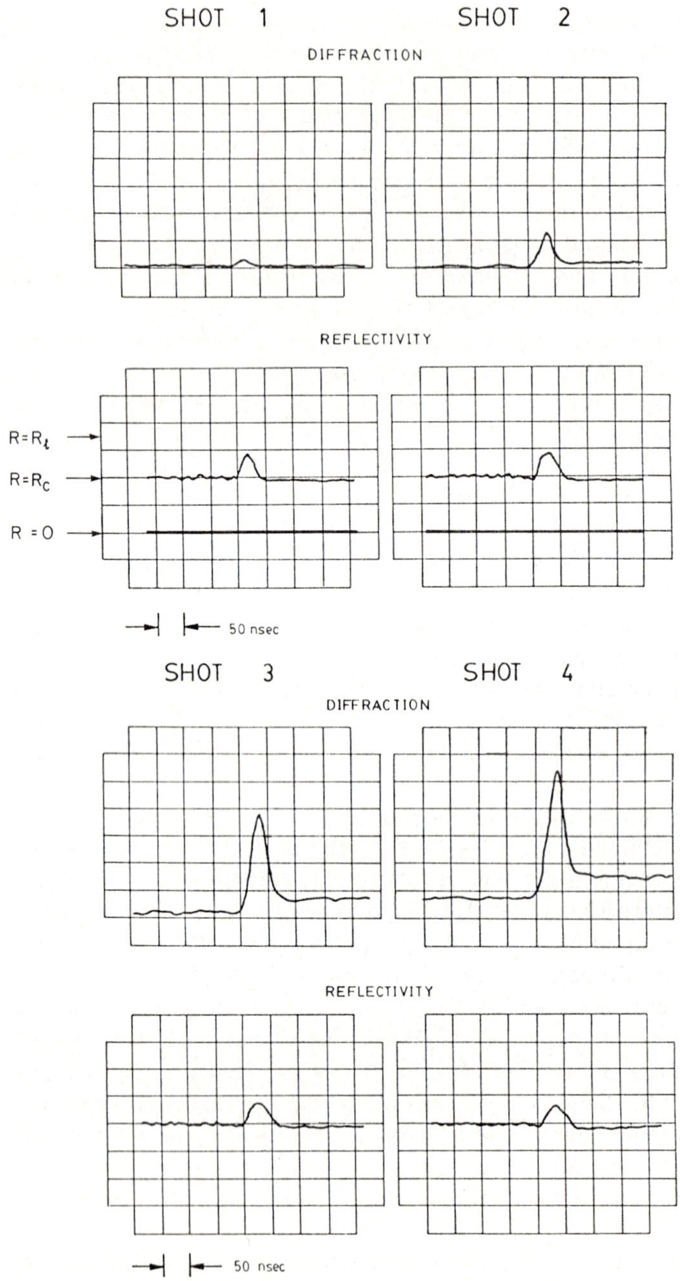

Fig. 15. Diffraction and reflectivity oscilloscope traces of a cw Ar^+ laser beam from a Ge surface while illuminated by a 25 ns, 1.06 μm laser beam in fluence regime A. The transient signals were obtained by appropriate placements of two separate photomultiplier tubes. $R_c R_\ell$ indicate bulk crystal and bulk liquid reflectivities.

structure, once developed, would scatter the light as dictated by geometry. During each shot the reflectivity rises to a value which is intermediate to that between the bulk liquid and solid values. This indicates that the same fraction of the solid is melting on each shot, although as figs. 9–11 show the laser energy deposition certainly has different characteristics. From the transient diffraction signals one can see little evidence of periodic structure after the first pulse, but on subsequent pulses there is strong *transient* diffraction associated with the formation of the liquid metallic strips on a background dielectric. After about two or three shots there is *permanent*, albeit weaker, diffraction from the modulated dielectric surface. Strong non-linear growth is noted from these transient characteristics. It might have been concluded from the intermediate value of the reflectivity that the surface during the pulse consisted of an optically thin but uniform layer of liquid on top of the solid and the difference in the transient and permanent diffraction signals was due to differences in a metallic versus a dielectric grating [6]. Having performed calculations of these diffraction efficiencies, under no circumstance are we able to account for the large ratios seen here. To positively confirm our ideas of inhomogeneous melting, we experimentally determined that the ratio of p- to s-polarized scattering efficiencies was approximately 3.5 for an incidence angle of 60° and this number is confirmed by a detailed theoretical calculation [6] for a surface which consisted of alternating metallic and dielectric strips.

The transient characteristics associated with the four different regimes are summarized in fig. 16. In regime B it can be seen that the reflectivity attains the bulk liquid value of 0.75 and remains there for tens of nanoseconds before returning to the value characteristic of the solid (0.50). The diffraction signal, on the other hand, shows two transient peaks. These are associated with the formation of the metallic liquid strips near the beginning of the pulse because of inhomogeneous energy deposition in the solid, and because of the return of the nonuniform melt front to the surface near the end of the pulse. In regime C the duration of the liquid layer is considerably longer and there is no definite transient behaviour. Rather, permanent diffraction signals are observed as inhomogeneous energy occurs in the liquid layer from light scattered near the periphery of the pulse and permanent structure is observed as the non-uniform melt front resolidifies. In the highest fluence regime D, the liquid layer is observed to endure for hundreds of nanoseconds after the pulse is turned off. The transient diffraction signals show strong oscillations characteristic of the excitation of surface capillary waves. These are observed to decay away over the duration of the pulse in the particular case shown here and therefore leave little permanent diffraction signal.

The capillary waves can be definitively identified by their dispersion relation and their decay characteristics. The dispersion relation of capillary waves, in the small wavelength limit where gravitational effects can be neglected, is given by

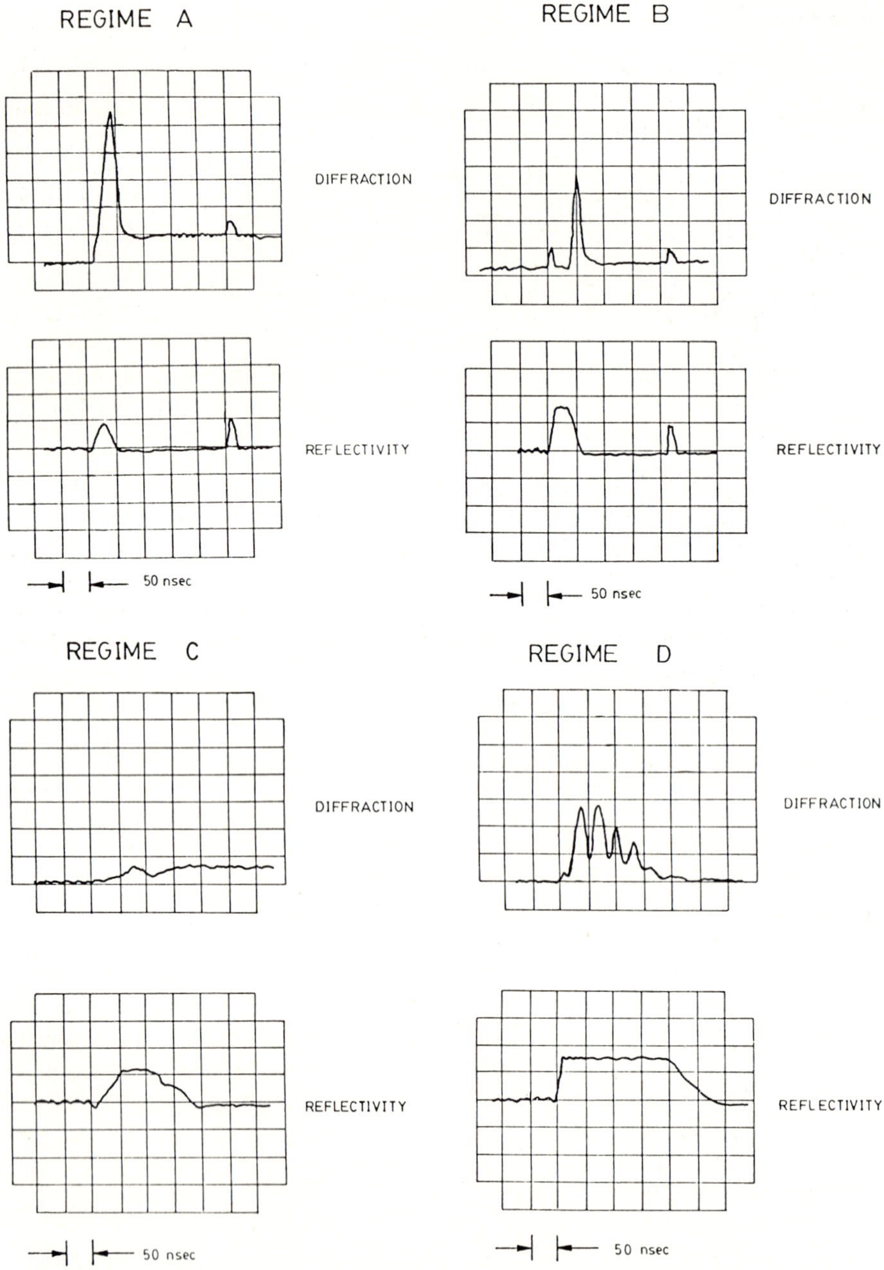

Fig. 16. A sequence of oscilloscope traces representing the transient diffraction and reflectivity characteristics of a cw, Ar^+ laser beam from a Ge surface while illuminated by 1.06 μm pulses in the four (A–D) fluence regimes. The small delayed spikes seen in regimes A and B are timing pulses superimposed on the beam.

$$\omega^2=(\alpha/\rho)k^3, \tag{3}$$

where ω is the frequency, α is the surface tension of the liquid, ρ is the liquid density and k is the wavevector. We have excited capillary waves of various wavevectors by varying the angle of incidence, and hence the value of k_i of the incident beam. Figure 17 shows the results obtained for the oscillation frequencies. The straight line is a theoretical curve based on the known [24] density (5.6 g/cm^3) and surface tension (0.557×10^3 g/s^2). The agreement is excellent.

The corresponding decay times as a function of capillary wavevector are shown in fig. 18. Theoretically, the amplitude decay rate is given by

$$\gamma=1/\tau=2\nu k^2,$$

where τ is the decay time and ν is the kinematic viscosity of the liquid. The theoretical curve in fig. 18 is based on a value of ν of 1.02×10^{-3} cm^2/s [24], which is well within the range of measured values near the melting point. It is also interesting to note that the decay characteristics we observed for capillary waves on Ge are in much better agreement with theory than Keilmann [22] has obtained for capillary waves on mercury, which is already a liquid at room temperature. This is possibly due to the fact that it is extremely difficult to maintain a clean liquid surface, especially for mercury which has such a high density and surface tension.

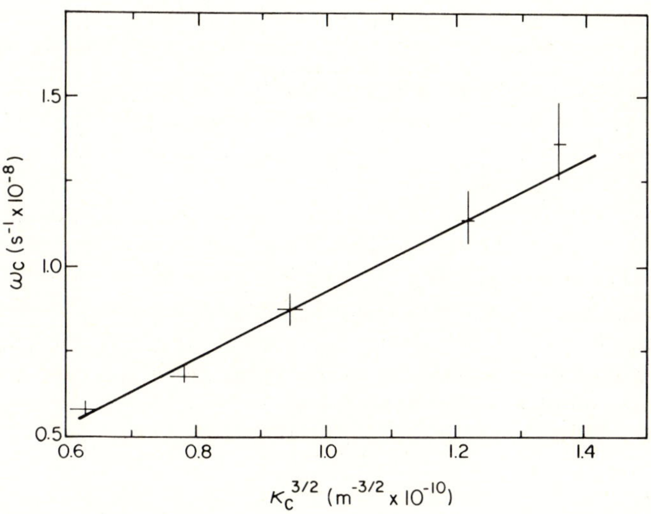

Fig. 17. The measured natural oscillation frequency, $\omega_c=\pi/T$, of the capillary waves generated on molten germanium plotted as a function of their wavevector to the three halves power, $\kappa_c^{3/2}$. The straight line represents the best linear fit to the data which also passes through the origin.

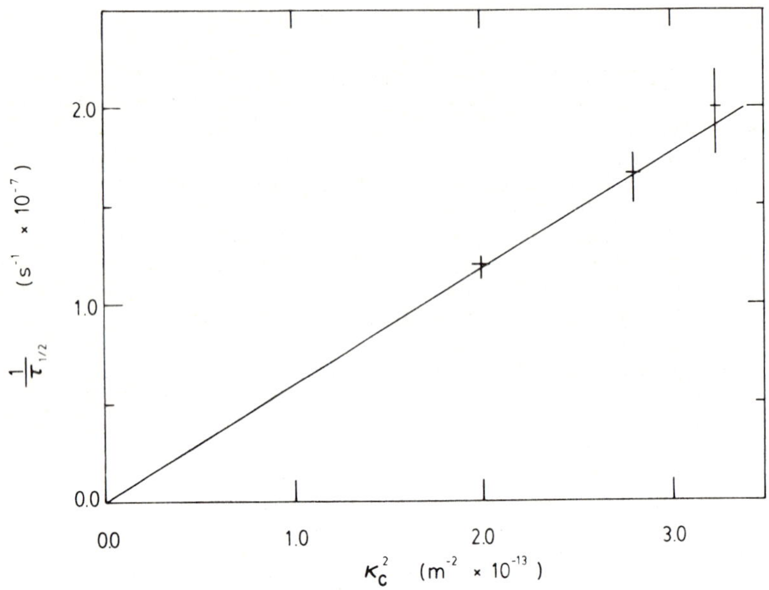

Fig. 18. The measured inverse half life, $1/\tau_{1/2}$, of the squared amplitude of the decaying capillary waves generated on molten germanium as a function of the squared magnitude of the wavevector, κ_c^2. The straight line represents the best linear fit to the data which also passes through the origin.

5. Conclusions

The phenomenon whereby single intense polarized laser beams can induce periodic structure on surfaces is an intriguing one which points out the richness of the physics associated with laser material interactions. The phenomenon is directly related to having a material interface and a beam which is polarized and spatially coherent. The effect of these parameters is not often appreciated in laser physics phenomena. From the point of material processing one may wish to reduce or enhance the effects depending on applications. Apart from this however, there is a wealth of physics to be learned from laser induced periodic surface structure. Besides the detailed electrodynamics of interfaces one can obtain information about the dynamics of an interface which is forced to undergo a spatially inhomogeneous phase transition.

6. Acknowledgements

We gratefully acknowledge financial support of this work from the Natural Sciences and Engineering Research Council of Canada.

References

[1] J.F. Young, J.E. Sipe, M.I. Gallant, J.S. Preston and H.M. van Driel, in: Laser and Electron Beam Interactions with Solids, eds. B.R. Appleton and G.K. Celler (North-Holland, Amsterdam, 1982) p. 233.

[2] J.E. Sipe, J.F. Young, J.S. Preston and H.M. van Driel, Phys. Rev. B 27 (1983) 1141.

[3] J.F. Young, J.S. Preston, H.M. van Driel and J.E. Sipe, Phys. Rev. B 27 (1983) 1155.

[4] S.R.J. Brueck and D.J. Ehrlich, Phys. Rev. Lett. 48 (1982) 1678.

[5] H.M. van Driel, J.E. Sipe and J.F. Young, Phys. Rev. Lett. 49 (1982) 1955.

[6] J.F. Young, J.E. Sipe and H.M. van Driel, Phys. Rev. B 30 (1984) 2001.

[7] M. Birnbaum, J. Appl. Phys. 36 (1965) 3688.

[8] L.J. Willis and D.C. Emmony, Opt. Laser Technol. 7 (1975) 222.

[9] M. Oron and G. Sorensen, Appl. Phys. Lett. 35 (1979) 782.

[10] R.M. Walser, M.F. Becker, D.Y. Sheng and J.G. Ambrose, in: Laser and Electron Beam Interactions and Material Processing, eds. T.J. Gibbons, W. Hess and T. Sigmon (Elsevier, New York, 1981).

[11] N.R. Isenor, Appl. Phys. Lett. 31 (1977) 148.

[12] P.A. Temple and M.J. Soileau, IEEE J. Quant. Elec. QE-17 (1981) 2067.

[13] Z. Guosheng, P.M. Fauchet and A.E. Siegman, Phys. Rev. B 26 (1982) 5366.

[14] F. Keilmann and Y.H. Bai, Appl. Phys. A 29 (1982) 9.

[15] R.H. Ritchie, Surf. Sci. 34 (1973) x.

[16] A. Sommerfeld, Ann. Phys. (Leipzig) 28 (1909) 665; Ann. Phys. 81 (1926) 1135.

[17] L.M. Brekhovskikh, Waves in Layered Media, second ed. (Academic Press, New York, 1980).

[18] J.N. Hodgson, Philos. Mag. 6 (1961) 509.

[19] J.F. Young, J.E. Sipe and H.M. van Driel, in: Laser Controlled Chemical Processing of Surfaces, eds. A.W. Johnson and D.J. Ehrlich (Elsevier, New York, 1984).

[20] T. Surek and B. Chalmers, J. Crystal Growth 29 (1975) 1.

[21] Jeff F. Young, Ph.D. Thesis (University of Toronto) 1983 (unpublished).

[22] W.G. Hawkins and D.K. Biegelsen, Appl. Phys. Lett. 42 (1983) 358.

[23] F. Keilmann, Phys. Rev. Lett. 51 (1983) 2097.

[24] Liquid Semiconductors, V.M. Glazov, S.N. Chinzhevskaya and N.N. Glagoleva (Plenum Press, New York, 1969).

Journal of Luminescence 30 (1985) 472–487
North-Holland, Amsterdam

ELECTRONIC STRUCTURE OF TWO-DIMENSIONAL SEMICONDUCTOR SYSTEMS

M. ALTARELLI

Max-Planck-Institut für Festkörperforschung, Hochfeld-Magnetlabor, 166X, F-38042 Grenoble–Cédex, France

The motion of two-dimensional carriers in semiconductor heterojunctions, quantum wells and superlattices is discussed, with emphasis on subband dispersion parallel to the interfaces and on quantization in a perpendicular magnetic field. The envelope function approximation is shown to provide an efficient and accurate method for the calculation of electronic states. It is shown that for states derived from degenerate or coupled bulk bands (valence bands, $\boldsymbol{k}\cdot\boldsymbol{p}$-coupled conduction and valence bands in narrow-gap semiconductors) striking non-parabolicities of the subband dispersion and non-linearities of the Landau levels versus magnetic field occur. Results for GaAs–AlGaAs and InAs–GaSb systems are presented and compared with experiment.

1. Introduction

The investigation of two-dimensional electronic systems at semiconductor interfaces, which in the past involved mostly inversion or accumulation layers in Si–MOS structures, has in recent years received great impulse from the fabrication of extremely high-quality III–V compound heterostructures by means of MBE and MOCVD techniques [1,2].

In this paper we shall consider three different types of heterostructures (heterojunctions, quantum-well structures and superlattices) which lead to formation of nearly ideal two-dimensional electron and hole states. The motion of the carriers perpendicular to the interfaces is quantized in discrete levels, with energy spacings typically in the 10–100 meV order of magnitude, resulting in a two-dimensional "subband" structure for the unrestricted motion parallel to the interfaces. There are at least two reasons that make the problem of calculating the subband dispersion in the interface plane very interesting. On the one hand, the dispersion is very far from trivial for states derived from coupled bulk bands, such as the p-like valence bands of all III–V semiconductors or the $\boldsymbol{k}\cdot\boldsymbol{p}$ coupled conduction and valence bands of narrow-gap materials. On the other, the dispersion of subbands along the interfaces is essential in

00-2313/85/$03.30

determining the density of states and therefore the optical properties, the structure of excitons and of impurity states, and, of most interest to us, the Landau level quantization in an external magnetic field. The non-parabolicities in the subband dispersion that we mention do indeed take place on a scale of momenta and energies relevant for these phenomena.

It is well-known that a distinctive feature of a two-dimensional system is the complete quantization, i.e., the appearance of an entirely discrete spectrum in an external magnetic field. This is also the reason why magneto-optical and magneto-transport experiments are particularly revealing for two-dimensional systems, and best suited for a detailed test of theoretical results. This is the main reason for the interest of a calculation of the Landau levels.

The plan of the paper is as follows. After a brief statement of the requirements that a satisfactory procedure to calculate the subband dispersion must fulfill, in section 2 we shall describe the multi-band envelope-function approximation, introduced by Sham and coworkers [3] and Bastard [4]. As recently shown, this method can be improved to provide an efficient and realistic description of electronic states in semiconductor heterostructures [5–8], also in presence of an external magnetic field. We shall then proceed, in section 3, to illustrate results for the valence subbands of GaAs–AlGaAs superlattices, quantum wells and heterojunctions, as well as for the strongly coupled conduction- and valence-band derived states of InAs–GaSb superlattices. The GaAs–AlGaAs heterojunction [9] and the InAs–GaSb superlattices are especially interesting in that significant charge transfer across the interfaces takes place and is accounted for self-consistently in the Hartree approximation. In section 4 the calculation of Landau levels is briefly described and comparison with magneto-optical experiments is discussed. Section 5 is devoted to a summary of the results and concluding remarks.

2. Envelope-function description of electronic states

There are some rather stringent requirements a satisfactory procedure to compute electronic states in heterostructures must satisfy. First, it should be suitable to handle the heterojunction, the quantum well and the superlattice geometry. Second, it must be able to deal with a realistic description of the bulk band structure of the constituent materials, including band couplings and degeneracies. Third, it should be suitable for self-consistent calculations in the cases in which a substantial charge rearrangement across the interfaces takes place. Finally, it is also highly desirable to have a method which can incorporate external fields, e.g. magnetic fields.

A truly microscopic calculation with the methods of modern band structure theory can immediately be ruled out. A superlattice with ~ 100 Å thick layers has a unit cell containing ~ 100 atoms, far beyond the present possibilities, and similar problems occur for other geometries. Semi-empirical tight-binding theory is a much better candidate, and indeed interesting results have been obtained by this method [10,11]. However, it is not easy to see how such a method could be used to account for charge rearrangements or external fields; furthermore, if one wants to keep the number of orbitals and range of interactions limited, it is not always possible to describe the band edges accurately [11]. The envelope-function approximation in its multi-band version fulfills all requirements and has been shown to lead to satisfactory results [3,4,5–7,9].

In the envelope-function approximation the wavefunction in each layer of the heterostructure is expressed through the products of slowly varying envelope functions and of the $k=0$ Bloch functions of the corresponding semiconductor. The envelope functions satisfy effective-mass-like equations and are joined at each interface with appropriate boundary conditions, which must be derived from the smoothness conditions for the total wavefunction. In the method presented here, the boundary conditions are embodied in a variational principle, which automatically produces correctly matched wavefunctions. The numerical work is thus greatly simplified and reduces to the usual diagonalization of matrices. It is then possible to handle the problem with a realistic description of the constituent bands, including, when necessary, coupled heavy-hole, light-hole, conduction and split-off bands, and to calculate the subband dispersion in the direction parallel to the layers in a realistic way.

Consider the standard second order $\boldsymbol{k}\cdot\boldsymbol{p}$ description of a set of N coupled band edges at the Γ point of a direct-gap semiconductor [12]. The band structure at a given $\boldsymbol{k}$-value is given by the eigenvalues of an $N\times N$ matrix

$$H_{jj'}=\sum_{\alpha,\beta} D^{\alpha\beta}_{jj'}k_\alpha k_\beta+P^{\alpha}_{jj'}k_\alpha+E_j\delta_{jj'},\qquad (j,j'=1,\ldots,N) \tag{1}$$

where α and β run over x, y and z. A well-known example of such a $\boldsymbol{k}\cdot\boldsymbol{p}$ Hamiltonian is Luttinger's 4×4 valence band Hamiltonian [13] describing the light- and heavy-hole bands. The $D_{jj'}$ and $P_{jj'}$ matrices give the representation of, respectively, second-order and first-order $\boldsymbol{k}\cdot\boldsymbol{p}$ interactions in the basis of the N Γ-point Bloch functions $u_j(\boldsymbol{r})$ of the bands in question, with energies E_j at $\boldsymbol{k}=0$. The two materials, A and B, forming the heterostructure will be each characterized by specifying the D and P matrices and the band edge positions E_j. We can now, following the standard many-band effective-mass theory, as used in magneto-optics [14] or in the theory of acceptor impurities [15], write the wavefunction in the A layer in terms of slowly varying envelope functions F^{A}_j as

$$\psi^{\mathrm{A}}(\boldsymbol{r})=\sum_{j=1}^{N}\left[F^{\mathrm{A}}_j(\boldsymbol{r})u^{\mathrm{A}}_j(\boldsymbol{r})+\sum_{i\neq1,\ldots,N}\frac{\mathrm{i}\boldsymbol{k}\cdot\boldsymbol{P}_{ij}}{E_j-E_i}F^{\mathrm{A}}_j(\boldsymbol{r})u^{\mathrm{A}}_i(\boldsymbol{r})\right], \tag{2}$$

where the second term represents a correction (to first order in $\boldsymbol{k}\cdot\boldsymbol{p}$) due to the higher bands ($i\neq1,\ldots,N$) not included in the original set, and can, for most purposes, be neglected. In eq. (2) and in the following, k represents the operator $-i\hbar\nabla$ acting on envelope functions only. Periodicity parallel to the interfaces always allows us to write

$$F^{A}(\boldsymbol{r}) \rightarrow \exp(i\boldsymbol{k}_{\parallel} \cdot \boldsymbol{r}_{\parallel}) F^{A}(x), \tag{3}$$

with $k_{\parallel}$ a good quantum number, if x denotes the $\langle 100 \rangle$ direction normal to the interfaces. The envelope functions in A or B satisfy an effective-mass equation

$$[H_{jj'}^{A,B} + V^{A,B}(x)\delta_{jj'}]F_{j'}^{A,B}(\boldsymbol{r}) = EF_{j'}^{A,B}(\boldsymbol{r}). \tag{4}$$

In eq. (4) $V(x)$ is the potential arising from charge redistributions in the system, averaged over the unit cell parallel to the interface. Equation (4) is valid in the A or B layers and the boundary condition at the A–B interface must be written in order to determine the eigenvalues. We know that the total wavefunctions, eq. (4), must be continuous with continuous derivatives. In order to express the boundary conditions in terms of the envelope functions only we need some information on the behavior of the Bloch functions u_j of the A and B materials. Given the structural and chemical similarity of all III–V compounds, we shall assume for simplicity that [3]

$$u_j^{A} = u_j^{B}, \qquad j = 1, \ldots, N. \tag{5}$$

Any attempt to improve upon eq. (5) must necessarily introduce some parameters of unknown value into the theory, and, as long as no obvious conflict arises with experimental facts, we shall retain this simple assumption.

Imposing the continuity of the charge density and the divergence-free character of the probability current, both quantities being evaluated to first-order in $\boldsymbol{k} \cdot \boldsymbol{p}$ and averaged over a microscopic unit cell, we obtain [5] with the help of eq. (5):

$$F_j^{A}(x=0^-) = F_j^{B}(x=0^+), \qquad j = 1, \ldots, N,$$

$$\sum_{j'=1}^{N} \left[(D_{jj'}^{xy} + D_{jj'}^{yx})k_y - 2iD_{jj'}^{xx} \frac{\partial}{\partial x} \right] F_{j'} \text{ continuous} \tag{6}$$

In the second condition it was assumed that the y direction is defined to lie along the parallel wavevector $k_{\parallel}$ (see eq. (3)). Equation (6) provides the $2N$ boundary conditions necessary to specify completely the solution of eq. (4). It is worthwhile to remark that the second boundary condition is the generalization, to a multi-band system, of that used by Bastard [4] in the case of simple bands

$$\frac{1}{m_A^*} \frac{\partial}{\partial x} F^{A}(0^-) = \frac{1}{m_B^*} \frac{\partial}{\partial x} F^{B}(0^+) \tag{7}$$

to which it reduces if the $D_{jj'}^{\alpha\beta}$ matrices are diagonal in α and β.

To determine the eigenvalues and the eigenfunctions of the system of differential eqs. (4) with the boundary conditions (6), we resort to a standard method of band structure theory [16]. One can define an "interface functional" I embodying the boundary conditions (6), such that if $\phi=(F^{\mathrm{A}}, F^{\mathrm{B}})$ is a solution with energy E of eq. (4) written as

$$H\phi=\begin{pmatrix} H^{\mathrm{A}} & 0 \\ 0 & H^{\mathrm{B}} \end{pmatrix}\begin{pmatrix} F^{\mathrm{A}} \\ F_{\mathrm{B}} \end{pmatrix}=E\begin{pmatrix} F^{\mathrm{A}} \\ F^{\mathrm{B}} \end{pmatrix}, \tag{7}$$

and if F^{A} and F^{B} match with the right boundary conditions, then

$$\langle\phi|H+I|\phi\rangle=E \tag{8}$$

is an extremum value with respect to arbitrary variations $\delta\phi$ of the wavefunctions [5]. I is of course not diagonal in the representation (7) and ensures the correct "coupling" between the A and B functions. Application of this method allows a simple and efficient generation of eigenvalues and wavefunctions, via a standard matrix diagonalization, thus making iteration to self-consistency feasible.

It will be shown in the following that the many-band envelope function method yields satisfactory results and in some cases very good agreement with experiment. One should nonetheless keep in mind the possible sources of uncertainty and of error. First of all, the interfaces are assumed to be perfect, i.e., no role is assumed to be played by interface defects, at least on the coarse-grained length-scale of an effective-mass calculation. Second, the assumption expressed by eq. (5) is chosen as the only simple and parameter-free answer to a difficult question. Third, the bandgap discontinuities, which are an input parameter to the calculations, where they enter through the relative positioning of the E_j values characterizing semiconductors A and B in eq. (1), are not known very well. They cannot be determined very accurately either by theory or by experiments [17]. On the other hand, other corrections such as the presence of strains due to lattice mismatch between the constituents could be easily accommodated into the formalism.

3. Two-dimensional subbands in semiconductor heterostructures

In this section application of the envelope-function method to the calculation of electronic states in semiconductor heterojunctions, quantum-well and superlattice structures will be discussed, with reference to GaAs–AlGaAs and InAs–GaSb systems. It is evident that once the boundary condition at one interface is chosen (see eq. (6)) it is possible to set up the problem also when two interfaces are present, as in a quantum-well structure, or for a periodic repetition of interfaces, as in a superlattice.

Let x denote the direction perpendicular to the interfaces and let y, z denote two orthogonal directions in the $x=0$ plane. Following Marques and Sham [3] we choose z as the quantization axis of angular momenta and consider $\boldsymbol{k}$-vectors in the xy plane. The six states of the conduction and upper spin–orbit split valence bands (we shall have to take some account of the split-off band when a magnetic field is present, see section 4) decouple into two sets. Our basis is $u_1=|s\uparrow\rangle$, $u_2=|3/2,3/2\rangle$, $u_3=|3/2,-1/2\rangle$ and the analogous u_4, u_5, u_6, for the other set. For $\boldsymbol{k}$ in the xy plane we take the description of the band dispersion in each material given by the eigenvalues of the following matrix (in atomic units):

$$H_{jj'}=\begin{vmatrix} E_c+\frac{1}{2}k^2 & iPk_+/\sqrt{2} & -iPk_-/\sqrt{6} \\ -iPk_-/\sqrt{2} & E_v-\frac{1}{2}(\gamma_1+\bar{\gamma})k^2 & \sqrt{3}\bar{\gamma}k_-^2/2 \\ iPk_+/\sqrt{6} & \sqrt{3}\bar{\gamma}k_+^2/2 & E_v-\frac{1}{2}(\gamma_1-\bar{\gamma})k^2 \end{vmatrix}. \tag{9}$$

Here $k_\pm=k_x\pm ik_y$, $E_c(E_v)$ is the conduction- (valence-)band edge energy, P is Kane's [12] momentum matrix element $P=-i\langle s|P_x|x\rangle$ and γ_1 and $\bar{\gamma}$ are valence-band parameters of the Luttinger type [13] in the spherical ($\gamma_2=\gamma_3=\bar{\gamma}$) approximation. In the larger-gap materials like GaAs and AlGaAs, the valence–conduction coupling can be neglected; we then let $P=0$, so that γ_1 and $\bar{\gamma}$ are the usual Luttinger parameters [13]; for the smaller-gap cases like InAs and GaSb P is important and the direct coupling to the conduction band must then be subtracted from the definition of γ_1 and $\bar{\gamma}$ [18].

The most severe approximation in eq. (9) is the spherical one, although the parameters γ_2 and γ_3 are rather close for the semiconductors of interest; it would be possible to eliminate this approximation should it appear evident that it is necessary for the interpretation of some experiment. The very small $\boldsymbol{k}$-linear terms introduced by the violation of inversion symmetry in III–V compounds [19] are also neglected. At the wavevectors of interest here, order-of-magnitude estimates show that they would lead to effects so small to be hardly observable.

The matrix corresponding to states u_4, u_5 and u_6 of our basis is not identical but equivalent to eq. (9) and yields the same eigenvalues (Kramers degeneracy). However, when we insert eq. (9) into eq. (4) together with a potential $V(x)$, this degeneracy is lifted whenever $V(x)$ does not have inversion symmetry, i.e., when

$$V(x)\neq V(-x). \tag{10}$$

This will be the case in heterojunctions, but not in quantum wells or superlattices (choosing the origin at the center of a layer). In the latter two cases the Kramers degeneracy remains.

We begin our discussion of results by the quantum well case, which, because of its similarity to the one-dimensional potential well of elementary quantum mechanics, is most suitable to display the striking effects brought about by coupled bands. In fig. 1(a), the computed valence subband structure of a GaAs quantum well with thickness $L=14$ nm between $Al_xGa_{1-x}As$ barriers with

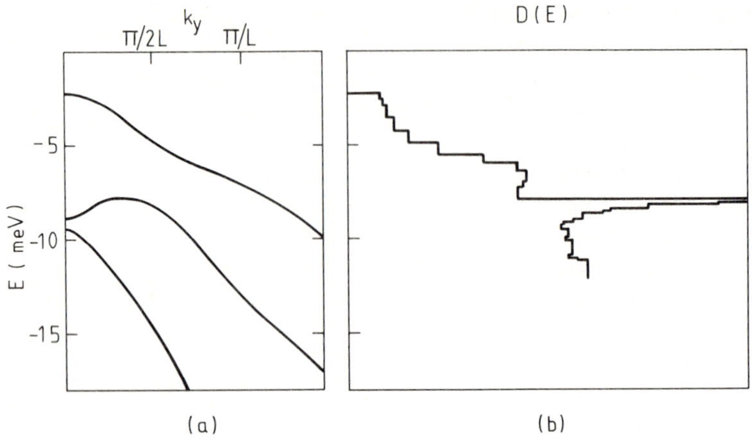

Fig. 1. (a) Valence subband structure for a GaAs quantum well of thickness $L=14$ nm between $Al_{0.21}Ga_{0.79}As$ barriers. The zero of energy is the top of the bulk valence band. (b) Density of states $D(E)$ (arbitrary units) corresponding to this two-dimensional subband structure.

$x=0.21$ is shown. With the band-gap parameters we use [20] and the empirical "85% rule" for the band-gap discontinuity [21], this corresponds to a valence band well 36 meV deep. It is worth remembering that the valence-band part of Hamiltonian (9) (for $P=0$) has eigenvalues corresponding to light and heavy holes

$$E(\boldsymbol{k})=E_v-\tfrac{1}{2}(\gamma_1\pm 2\bar{\gamma})k^2. \tag{11}$$

In the quantum well, light and heavy holes give separate quantized levels when $k_y=0$, i.e., when there is no momentum parallel to the interfaces. For $k_y\neq 0$ (motion along the quantum well) this is no longer possible. Light- and heavy-hole waves have to be mixed in order to satisfy the boundary conditions. This mixing [22] is the reason for the conspicuous non-parabolicities to be seen in fig. 1(a). Notice in particular the second subband that starts off with a positive (i.e., electron-like) mass, and has a maximum at a finite k_y value of the order $\pi/2L$. In fig. 1(b) the density of states corresponding to this subband structure is shown. Notice in particular the sharp singularity corresponding to the maximum in the second subband. If the spherical approximation is released the very sharp peak is somewhat broadened, but it remains very prominent. It is quite evident from fig. 1(b) that the density of states of the valence bands is rather different from the normally assumed "staircase" corresponding to parabolic two-dimensional subbands.

It is important to notice the momentum and energy scales over which the deviations from parabolicity occur. The well thickness, $L=14$ nm is of the same order of the exciton radius in GaAs, and the energy scale is also comparable to the exciton binding energy. This means that the subbands shown in fig. 1(a) will all be mixed in the formation of exciton states, as well as in forming acceptor-bound states. To the author's best knowledge these band structure effects have never been included in a theoretical treatment. It appears, however, that they should have a significant influence on exciton and acceptor spectra in quantum wells (see, note added in proof). For their influence on interband transition matrix elements, a tight-binding treatment was given in ref. [11].

To conclude the discussion of quantum wells, it is worth mentioning that conduction subbands in GaAs–AlGaAs behave in a simpler way, since they are derived from a non-degenerate, very nearly parabolic band.

The bands of fig. 1(a) are Kramers degenerate (apart from the very small splitting which would arise from the inclusion of k-linear terms). If we now consider a single heterojunction between lightly doped n-type GaAs and heavily p-type $Al_{0.5}Ga_{0.5}As$, a two-dimensional hole gas [23] is formed in the inversion layer near the interface. Recently, Ekenberg and Altarelli [9] performed self-consistent calculations (in the Hartree approximation) of the subband structure (fig. 2) which show a spin-splitting at $k_y \neq 0$ very similar to the experimentally observed one. Analogous calculations were also reported by Landwehr and Bangert [24], which replaced the $\sim$90 meV barrier with an infinite one, but did not make the spherical approximation. As one can see from fig. 2, the spin-splitting is of the order of a few meV at $k_y \sim 2 \times 10^6 \text{cm}^{-1}$, at least one order of magnitude bigger than that resulting from the lack of inversion symmetry in bulk GaAs. Needless to say, also in this case the non-parabolicities are quite easily recognized in fig. 2.

As an example of calculation for the superlattice geometry, we shall discuss InAs–GaSb superlattices [6]. These systems are extremely interesting because the top of the GaSb valence band lies higher in energy than the bottom of the InAs conduction band [25], by about 0.15 eV. Charge transfer across the inter-

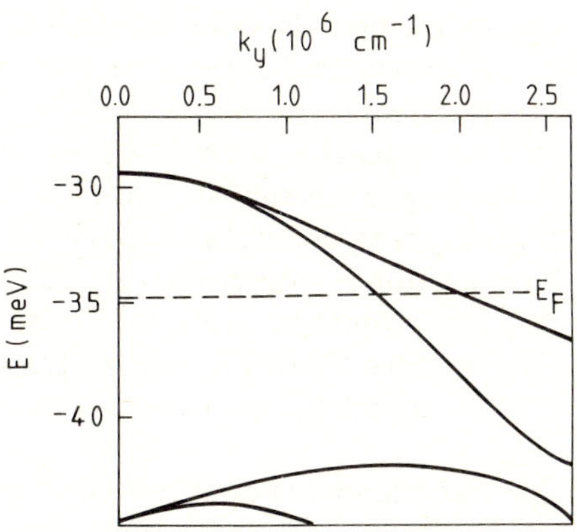

Fig. 2. Two-dimensional energy band structure for a hole system with density 5×10^{11} cm^{-2} at the GaAs–$Al_{0.5}Ga_{0.5}As$ interface (after Ekenberg and Altarelli [9]). E_F denotes the Fermi energy.

faces is therefore to be expected and experiments show a semimetallic behavior for superlattice periods exceeding ~18 nm. Self-consistent calculations of the band structure of those systems in the envelope-function approximation were reported in ref. [6]. In fig. 3 results are shown for the case of InAs thickness $d_A = 12$ nm, GaSb thickness $d_B = 8$ nm. The superlattice period $d = 20$ nm exceeds the critical value so that a semimetallic band structure is expected. In reality, a small hybridization gap of a few meV opens up at the Fermi energy, producing a very small-gap semiconductor rather than a semimetal, at least for an intrinsic sample. The bands obey indeed a "non-crossing" rule for $k_y \neq 0$, the same which is easily recognized in the GaAs–AlGaAs quantum well results of fig. 1. If the lack of inversion symmetry in the bulk materials were taken into account all subbands would be spin-split, and the two components would be allowed to cross. Apart from this small splitting, other crossings are not allowed as simple group-theoretical considerations show. It is hard to conceive

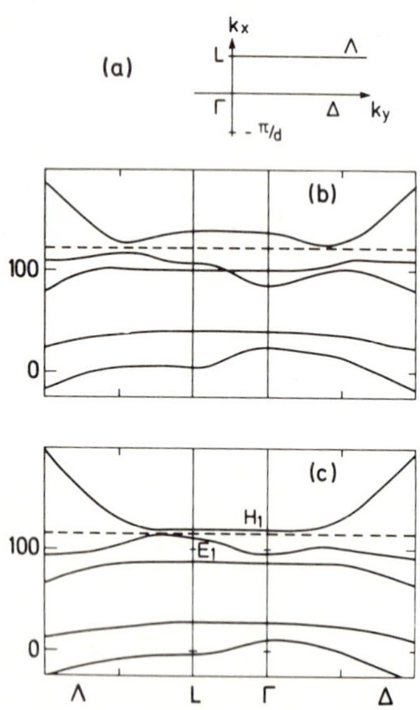

Fig. 3. Band structure of a $d = 12 + 8$ nm InAs–GaSb superlattice (b) non-self-consistent (c) in the self-consistent Hartree approximation. The Brillouin zone nomenclature is defined in (a).

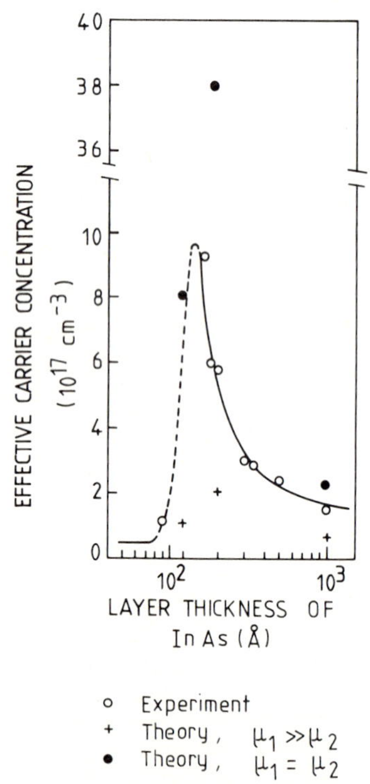

Fig. 4. Hall coefficient, expressed as an effective carrier concentration, as a function of InAs layer thickness. Empty circles: experiment (from ref. [26]), crosses and full circles: theory with $\mu_1 \gg \mu_2$ or $\mu_1 \simeq \mu_2$ (see text).

an experiment which would explicitly show the small gap in InAs–GaSb. The band structure shown in fig. 3 is, however, consistent with a large body of data on this system. We shall discuss later the far-infrared magneto-optics [7,8], and discuss here the results of Hall measurements [26]. In fig. 4 the open circles are experimental results showing the inverse of the Hall coefficient (expressed as an "effective" carrier concentration) as a function of layer thickness of InAs. We assume light n-type doping, $n \simeq 5\times10^{16}$ cm^{-3}, as suggested by the results on the thinner samples, and predict in our model occupation of a conduction band "valley" at some $k \neq 0$. Neglecting the small dispersion of the conduction band in the x-direction (see fig. 3(c)), i.e., assuming a purely two-dimensional model, this means a cylindric ring-shaped Fermi surface as shown in fig. 5. We have an outer, electron-like surface with wavefunctions mostly derived from the InAs conduction band, and an inner one with mostly GaSb valence-band states. We then use the standard result for the Hall coefficient of two-carrier systems,

$$R=(n_1\mu_1^2-n_2\mu_2^2)/(n_1\mu_1+n_2\mu_2)^2, \tag{12}$$

where

$$n_1=\frac{1}{2\pi^2}\,\pi k_{F_1}^2, \qquad n_2=\frac{1}{2\pi^2}\,\pi k_{F_2}^2,$$

and μ_1, μ_2 are the mobilities of carriers on the two portions of the Fermi surface. We expect μ_1 to be somewhat larger than μ_2, as electrons in InAs have higher

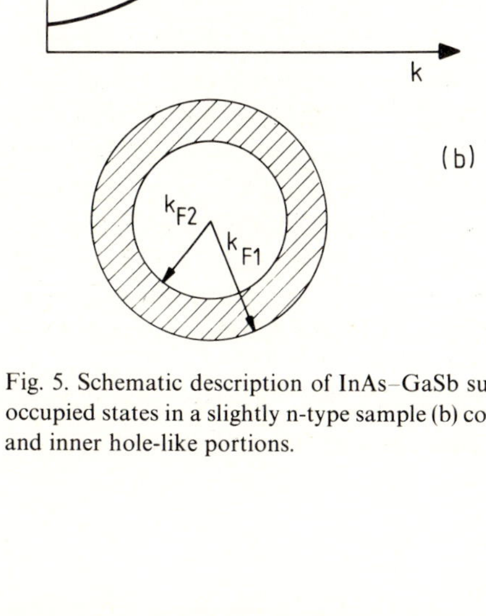

Fig. 5. Schematic description of InAs–GaSb superlattice band structure in the interface plane (a) occupied states in a slightly n-type sample (b) corresponding Fermi surface, with outer electron-like and inner hole-like portions.

mobility than holes in GaSb. In fig. 4 we show results obtained in the two limiting cases $\mu_1 \gg \mu_2$ or $\mu_1 \simeq \mu_2$, using the k_{F_1} and k_{F_2} values obtained, for each thickness, by self-consistent calculations [27]. It is evident that experimental results lie in between, confirming that the mobility ratio is intermediate, $\mu_1 \simeq 2\mu_2$, in good agreement with estimates based on conduction and valence wave-function admixture in our calculations. It is important to notice how calculations in this system are crucially dependent on a formalism which allows inclusion and mixing of conduction and valence states and self-consistent treatment of charge rearrangements.

We conclude this section by repeating, in view of the results described, our main conclusion: degenerate and/or coupled bands in bulk materials result in very non-trivial and non-parabolic subband dispersion in heterostructures.

4. Landau levels in quantum wells and superlattices

There are good experimental reasons to investigate the electronic states of superlattices and quantum wells in an external magnetic field, and there is also the interesting theoretical question of the Landau quantization for the strongly non-parabolic two-dimensional subbands just described. In fact whenever energy vs. k_y dispersion curves are non-parabolic we can anticipate energy vs. magnetic field curves to deviate from linearity. For high oscillator quantum numbers this can be seen from the semiclassical limit, according to which Landau levels of a subband $E(k_y)$ are given by the correspondence

$$E_n(B) = E(k_y = (2eB/c)^{1/2} n) \quad \text{for} \quad n \gg 1. \tag{13}$$

Interesting experiments, however, concern the quantum limit of low n, and for their interpretation we need realistic calculations. The inclusion of an external magnetic field in the envelope-function formalism was described in ref. [7,8]. The starting point is the description of the bulk band structure given by eq. (9). Although the split-off valence band is not included in our basis set, we can no longer neglect its effect on the conduction-band effective g-factor. We therefore explicitly include a corresponding correction to the conduction-band mass m^* and g^* following Roth et al. [28]. In the absence of magnetic fields it was possible to decouple the bulk Hamiltonian into two 3×3 systems. When a field is present this cannot be done and there is therefore no advantage in the somewhat unconventional coordinate system described in section 3. We go back to the usual notation and take the z-axis to be (a) the normal to the interfaces, along which the external field $\boldsymbol{B} = (0,0,B)$ lies and at the same time (b) the axis of quantization of angular momenta. We therefore have to work from now on with 6×6 matrices.

Following the work of Fasolino and Altarelli [7,8], the inclusion of the magnetic field in the envelope function formalism is accomplished by:

(i) Replacing $\boldsymbol{k}$ by $\boldsymbol{k}-(e/c)\boldsymbol{A}$ in the Hamiltonian of each constituent,

(ii) Adding new terms arising from electron or hole spin coupling to the field. For the conduction band this means that the diagonal terms become

$$H_{cc}=E_c+\frac{1}{2m^*}(\boldsymbol{k}-(e/c)\boldsymbol{A})^2+\frac{e}{2m_0c}g^*s_zB, \tag{14}$$

where $\boldsymbol{s}$ is the electron spin and

$$\frac{1}{m^*}=\frac{1}{m_0}\left(1+\frac{2P^2}{3(E_c-E_v+\Delta)}\right), \qquad g^*=2m_0/m^*, \tag{15}$$

Δ being the spin–orbit splitting of the valence band, and P Kane's momentum matrix element. Equation (15) embodies the split-off band contribution mentioned above [28]. For the valence band, this means adding the term [13]

$$(e/c)\kappa J_zB+(e/c)qJ_z^3B, \tag{16}$$

where J_z is the spin 3/2 matrix and κ and q are material parameters. Actually, q is very small for the semiconductors of interest here and the second term can be neglected.

(iii) Modifying the boundary conditions at the interfaces which determine the eigenvalues. Indeed, the new terms introduced above modify the probability current operator and therefore the conditions enforcing its conservation at the interfaces. Before writing the new matching conditions, we must specify the structure of the solutions in each material. Introducing the harmonic oscillator representation for motion in the xy plane based upon the commutation rules for the x and y components of $\boldsymbol{k}-(e/c)\boldsymbol{A}$ [18], the envelope wavefunction can be written:

$$\psi_n=(c_1(z)\phi_n,\ c_2(z)\phi_{n-1},\ c_3(z)\phi_{n+1},\ c_4(z)\phi_{n+1},\ c_5(z)\phi_n,\ c_6(z)\phi_{n+2}), \tag{17}$$

where ϕ_n are harmonic oscillator wavefunctions, $n=-2,-1,0,1$, and the c coefficients are automatically vanishing for those components which have a negative oscillator index, if $n<1$. For the superlattice case all six $c_j(z)$ functions have the common prefactor $\exp(\mathrm{i}k_sz)$ where k_s is a quasi-momentum in the superlattice Brillouin zone, parallel to the external field. In terms of the ψ_n states (17) the effective Hamiltonian can be written as a 6×6 matrix with the structure

$$H_{jj'}=(D_{jj}^{zz}k_z^2+E_j)\delta_{jj'}+(P_{jj'}^z+A_{jj'}(n,B))k_z+C_{jj'}(n,B), \tag{18}$$

where j and j' label the 6 basis states, $P_{jj'}^z$ is the coefficient of k_z in the terms present at zero field via the $\boldsymbol{k}\cdot\boldsymbol{p}$ interaction, and the A and C matrices represent the "new" magnetic field induced terms. With the replacement $k_z\to-\mathrm{i}\,\partial/\partial z$, eq. (18) gives a differential system for the $c_j(z)$ functions. Deriving from eq. (18) an expression for the current operator and imposing its divergence-free character at the interfaces we obtain, besides the continuity of the c functions, the following conditions, in analogy to eqs. (6):

$$-\mathrm{i}D_{jj}^{zz}\frac{\partial}{\partial z}c_j(z)+\tfrac{1}{2}\sum_{j'}A_{jj'}c_{j'}(z)\quad\text{continuous}\quad j=1,\ldots,6. \tag{19}$$

We then apply to eqs. (18) and (19) the method of solution explained in section 2 to obtain the Landau levels as a function of the field B and, in a superlattice geometry, of the quasi-momentum k_s.

This procedure can be applied to the GaAs–AlGaAs or to the InAs–GaSb system. The computed Landau levels are directly comparable to magneto-optical experiments once the selection rules for optical transitions have been established. In the Faraday configuration (σ^+ or σ^- circular polarization) we easily obtain from eq. (17), in terms of eigenstates ψ_n, $\Delta n=+1$ or -1 and, in the superlattice case, also $\Delta k_s=0$. These rules are not very stringent as they do not take into account possible reduction or actual vanishing of the matrix

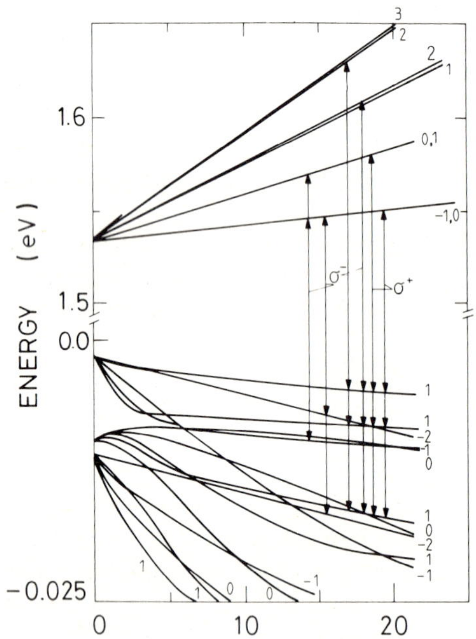

Fig. 6. Landau level diagram at $k_s=0$ for a superlattice with GaAs layers of thickness $d_A=12.5$ nm between $Al_{0.21}Ga_{0.79}As$ layers of thickness $d_B=8$ nm. The period d is then 20.5 nm. The quantum number n (see text) is indicated for each level, as well as allowed valence–conduction transitions for σ^+ and σ^- polarization (after ref. [29]).

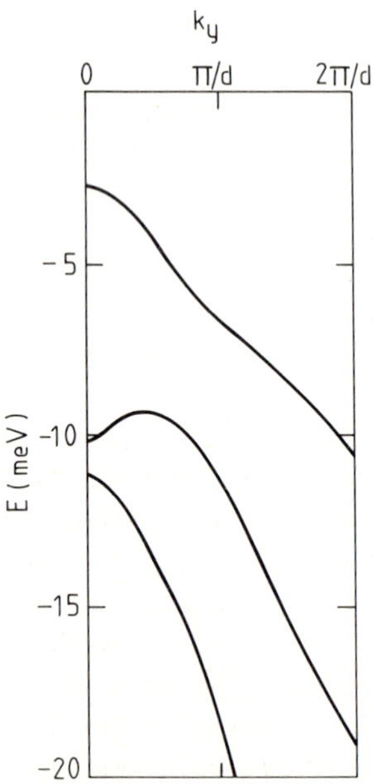

Fig. 7. Two-dimensional band structure at zero field for the superlattice of the previous figure.

elements due to the explicit wavefunction shape in the z direction ("electrical" selection rules). They tend to predict more transitions than actually observable.

In ref. [7] Landau levels for InAs–GaSb systems were discussed and compared favorably with magneto-optical experiments. The satisfactory agreement provided support to the band structure calculations at zero field, shown in fig. 3. Here we concentrate on the GaAs–AlGaAs system, for which one can also anticipate a very complex Landau level structure for the valence-band related states. In fig. 6, the computed Landau levels at $k_s=0$ for a superlattice with $d_A=12.5$ nm GaAs layers between $d_B=8$ nm $Al_xGa_{1-x}As$ layers with $x=0.21$ are shown [29]. One can see that the first three valence-band related subbands (fig. 7) result in the very intricate pattern of fig. 6. It is very interesting to notice how some Landau levels, more specifically the topmost group with $n=0, \pm 1$, have an extremely weak field dependence over a very large range of field values. In the customary language this would correspond to a very large cyclotron mass, but its relation to the very non-parabolic zero-field band structure is of course quite involved. From an experimental investigation of the exciton spectrum of quantum wells in high magnetic fields [29] it was indeed observed that the hole contribution to the reduced mass of the electron–hole pair corresponds to an unexpectedly large mass. These superlattice and quantum well results lead to the suspicion that the large cyclotron masses observed by Störmer et al. [23] in the heterojunction to which calculations of fig. 2 pertain may be related to a similar behavior of the Landau levels. We hope to test this hypothesis soon by a calculation.

In fig. 6 also the allowed optical transitions in the σ^+ and σ^- polarizations are reported. The experimental spectra [29] show a somewhat smaller number of transitions, as well as very prominent exciton effects which are not at all considered in our calculations.

5. Summary and conclusions

It was our aim to make a few points concerning the electronic structure of two-dimensional systems in semiconductors. First of all, it was shown that the envelope-function approximation is an efficient procedure to compute the two-dimensional subband structure. We supported this claim by giving examples of calculations for quantum wells and heterojunctions for GaAs–AlGaAs, and superlattices for InAs–GaSb. Some of these calculations were carried out self-consistently. The agreement with experiment is consistently satisfactory, although we acknowledge that some of the input parameters are known only approximately and some of the underlying assumptions may not be verified by real samples.

It was also shown that the method lends itself to the calculation of energy levels in presence of an external magnetic field, a situation of great importance for the experimental implications. Also in this case, comparison with experiment is rather encouraging. It seems fair to conclude that a reliable and flexible theoretical tool is available for the determination of one-particle properties of semiconductor heterostructures.

As far as the physics emerging from the results is concerned, it should be quite evident that the motion of carriers along interfaces in a semiconductor heterostructure displays many interesting and intriguing aspects. This is certainly the case for all subbands deriving their characters at least in part from the valence bands of the constituents. The complex nature of the bulk band structure are reflected in extremely non-parabolic bands, and in unexpected features like the positive effective masses so commonly encountered among hole subbands. These effects are bound to be of importance in the physics of excitons and of acceptor states in heterostructures. It was also explicitly shown how the motion in an external magnetic field is complicated considerably.

In view of the interest and of the importance of these aspects of band structure in two-dimensional systems, it would be highly desirable to develop some analytic solution for simplified cases as a first step towards a deeper understanding and prediction power without resort to numerical calculations.

Note added in proof

After completion of the manuscript, I became aware of a paper by Masselink et al. (Phys. Rev. B 28 (1983) 7373) in which the acceptor problem is treated with full inclusion of valence band coupling. Results confirm the importance of this coupling emphasized above.

Acknowledgement

I am pleased to thank my coworkers U. Ekenberg, A. Fasolino and J.C. Maan for their invaluable help. I also acknowledge a very useful correspondence with L.L. Chang. The numerical results were obtained through the facilities of the Centre de Calcul Vectoriel pour la Recherche, Palaiseau.

References

[1] For a comprehensive review of the physical properties of two-dimensional electron systems, see, T. Ando, A.B. Fowler and F. Stern, Rev. Mod. Phys. 54 (1982) 437;
K. Ploog and G.H. Döhler, Adv. Phys. 32 (1983) 285.

[2] See, e.g., K. Ploog, Ann. Rev. Mater. Sci. 12 (1982) 123;
J.F. Bonfils, S.J.C. Irvine and J.B. Mullin, eds., Metalorganic Vapor Phase Epitaxy (North-Holland, Amsterdam, 1981).

[3] S.R. White and L.J. Sham, Phys. Rev. Lett. 47 (1981) 879;
G.E. Marques and L.J. Sham, Surf. Sci. 113 (1982) 131;
S.E. White, G.E. Marques and L.J. Sham, J. Vac. Sci. Tech. 21 (1982) 544.
[4] G. Bastard, Phys. Rev. B 24 (1981) 5693; Phys. Rev. B 25 (1982) 7584.
[5] M. Altarelli, in: Applications of High Magnetic Fields in Semiconductor Physics, ed., G. Landwehr (Springer, Berlin, 1983) p. 174.
[6] M. Altarelli, Phys. Rev. B 28 (1983) 842.
[7] A. Fasolino and M. Altarelli, Surf. Sci. 142 (1984) 322.
[8] A. Fasolino and M. Altarelli, in: Two-Dimensional Systems, Heterostructures, and Superlattices, eds., G. Bauer, F. Kuchar and H. Heinrich (Springer, Berlin, 1984) p. 176.
[9] U. Ekenberg and M. Altarelli, Phys. Rev. B 30 (1984) 3569.
[10] J.N. Schulman and Y.-C. Chiang, Phys. Rev. B 24 (1981) 4445.
[11] Y.-C. Chiang and J.N. Schulman, Appl. Phys. Lett. 43 (1983) 536.
[12] E.O. Kane, in: Semiconductors and Semimetals, eds., K. Willardson and A.C. Beer (Academic, New York, 1966) vol. 1, p. 75.
[13] J.M. Luttinger, Phys. Rev. 102 (1956) 1030.
[14] L.M. Roth, B. Lax and S. Zwerdling, Phys. Rev. 114 (1959) 90.
[15] See, e.g., M. Altarelli and F. Bassani, in: Handbook on Semiconductors, ed., W. Paul (North-Holland, Amsterdam, 1982) vol. 1, p. 269 and references therein.
[16] H. Schlosser and P.M. Marcus, Phys. Rev. 131 (1963) 2529, especially Appendix 1.
[17] For a critical overview see H. Kroemer, Surf. Sci. 132 (1983) 543.
[18] C.R. Pidgeon and R.N. Brown, Phys. Rev. 146 (1966) 575.
[19] E.O. Kane, in: Handbook on Semiconductors, ed., W. Paul (North-Holland, Amsterdam, 1982) vol. 1, p. 193;
C.R. Pidgeon and S.H. Groves, Phys. Rev. 186 (1969) 824.
[20] "Landolt-Börnstein Numerical Data and Functional Relationships in Science and Technology" Group III, vol. 17, ed., O. Madelung (Springer, Berlin, 1982).
[21] R. Dingle, A.C. Gossard and W. Wiegmann, Phys. Rev. Lett. 34 (1975) 1327.
[22] Experimental evidence for light-hole heavy-hole mixing was reported by R. Sooryakumar, D.S. Chemla, A. Pinczuk, A.C. Gossard and W. Wiegmann, Bull. Am. Phys. Soc. 29 (1984) 256.
[23] H.L. Störmer and W.T. Tsang, Appl. Phys. Lett. 36 (1980) 685;
H.L. Störmer, Z. Schlesinger, A. Shang, D.C. Tsui, A.C. Gossard and W. Wiegmann, Phys. Rev. Lett 51 (1983) 126.
[24] G. Landwehr and E. Bangert, in: Two-Dimensional Systems, Heterostructures, and Superlattices, eds., G. Bauer, F. Kuchar and H. Heinrich (Springer, Berlin, 1984) p. 40.
[25] See for example L.L. Chang, J. Phys. Soc. Jpn. 49 Suppl. A (1980) 997 and references therein.
[26] L.L. Chang, N. Kawai, G.A. Sai-Halasz, R. Ludeke and L. Esaki, Appl. Phys. Lett. 35 (1979) 939.
[27] For the thickest sample in fig. 4 we used the Thomas–Fermi method as described by J.C. Maan, in: Applications of High Magnetic Fields in Semiconductor Physics, ed., G. Landwehr (Springer, Berlin, 1982) p. 163.
[28] L.M. Roth, B. Lax and S. Zwerdling, Phys. Rev. 114 (1959) 90.
[29] J.C. Maan, A. Fasolino, G. Belle, M. Altarelli, K. Ploog, Physica B + C, to appear.

Journal of Luminescence 30 (1985) 488–501
North-Holland, Amsterdam

COULOMBIC BOUND STATES IN SEMICONDUCTOR QUANTUM WELLS

G. BASTARD

Groupe de Physique des Solides de l'Ecole Normale Supérieure, 24 rue Lhomond, 75231 Paris Cedex 05, France

Recent theoretical works on Coulombic bound states in semiconductor quantum wells (Q.W.) are reviewed. Due to carrier confinement along the growth axis the bound impurity or exciton states display enhanced binding energies over the bulk values. The presence of free carriers in modulation-doped quantum wells decreases the impurity binding energies. However the quasi-bidimensionality of the carrier motion prevents a complete vanishing of impurity bound states. The photoluminescence line of high-quality quantum well is often Stokes-shifted with respect to the absorption or excitation spectra. This Stokes shift can be correlated with interface defects in a qualitative fashion.

1. Introduction

The modern growth techniques such as Molecular-Beam-Epitaxy or Metal-Organic-Chemical-Vapor-Deposition have made possible the realization of high quality semiconductor heterojunctions, e.g., quantum wells (Q.W.) or superlattices (S.L.). In these systems quantum size effects become a key feature [1,2]. For instance, GaAs–$Ga_{1-x}Al_xAs$ Q.W.'s have an effective bandgap which can be blue-shifted by as much as 445 meV from bulk GaAs bandgap [3] due to electron and hole confinements in GaAs slabs. In a defectless Q.W. the carriers move freely in the layer planes but are bound along the growth axis. This leads to the formation of quasi-two-dimensional (2D) subbands and to a step-like density of states (D.O.S.). Despite their high quality the Q.W. structures contain residual charged impurities. Also a Q.W. can be intentionally doped and the dopant can be placed either in the barrier (modulation doping) or in the well (anti-modulation doping). As a result of doping the Q.W. structure may contain free carriers which will screen the impurity potential Since the Q.W. D.O.S. strongly differs from the bulk one, one may expect modifications of the Coulombic bound states (unscreened or screened) in such structures.

The absorption and photoluminescence spectra of high quality single and multiple GaAs–$Ga_{1-x}Al_xAs$ Q.W. structures are dominated by excitonic effects [4]. In fact the room temperature absorption spectra of GaAs Q.W.'s display very strong excitonic lines [5] whereas bulk GaAs hardly shows an exciton-related shoulder under the same conditions. This is a clear indication that the exciton binding energy, another Coulombic problem, is markedly increased by the electron and hole confinement.

In this paper we present a review of recent theoretical results obtained on the problem of Coulombic binding in quasi-2D semiconductor heterostructures. Section 2 is devoted to Coulombic impurities (unscreened or screened) in single Q.W.'s whereas section 3 will deal with the exciton problem. The exciton trapping on a model interface defects is discussed in section 4.

2. Coulombic impurities in single Q.W. structures [6–16]

Consider an idealized Q.W. and assume that (i) a single spherical parabolic band is involved in the carrier kinematics in each layer; (ii) the carrier effective mass m^* and the relative dielectric constant κ are constant throughout the whole structure. These assumptions are rather well justified in the case of the bound states supported by Coulombic donors in GaAs–$Ga_{1-x}Al_xAs$ Q.W. Thus the effective mass Hamiltonian we have to investigate is

$$\mathscr{H} = p^2/2m^* + V_b(z) - e^2/\kappa[\rho^2 + (z-z_i)^2]^{1/2}, \quad (1)$$

$$V_b(z) = V_b Y[z^2 - L^2/4]. \quad (2)$$

In eqs. (1) and (2) the z origin is taken at the center of the Q.W.; $Y(x)$ is the step function which is equal to 1 if $x>0$ and vanishes if $x<0$; L is the Q.W. thickness; V_b the barrier height; $\rho^2 = x^2 + y^2$ and z_i is the position of the impurity along the growth (z) axis. The characteristic length associated with the sole Coulombic potential ($V_b = 0$) is the three-dimensional (3D) Bohr radius a_0^* ($a_0^* = \kappa\hbar^2/m^*e^2 \sim$ 100 Å for donors in bulk GaAs) and the characteristic energy is the 3D Rydberg R_0^* ($R_0^* = m^*e^4/2\kappa^2\hbar^2 \sim 5$ meV for donors in bulk GaAs). Two dimensionless parameters govern the behaviour of the Coulombic bound states in Q.W.: L/a_0^* and $|z_i|/a_0^*$. The first parameter gives a measure of the dimensionality of the motion: if V_b is finite and L/a_0^* is $\gg 1$ or $\ll 1$ the problem is almost three dimensional. If $L/a_0^* \gg 1$ the size quantization becomes negligible compared to R_0^*, $\hbar^2\pi^2/2m^*L^2 = \pi^2 R_0^*(a_0^*/L)^2 \ll 1$: many quasi-2D subbands are admixed by the Coulombic potential and participate in the building of the impurity bound state. For on-center impurity the binding energy extrapolates to R_0^* when $L \to \infty$ and the shape of the ground bound state wavefunction approaches that of a 1s hydrogenic wavefunction. If on the other hand $L/a_0^* \ll 1$ the ground bound level of the Q.W. E_1 raises up to V_b and its associated wavefunction $\chi_1(z)$ increasingly leaks in the barrier material. $\chi_1(z)$ is more and more coupled to the Q.W. continuum states by the Coulombic potential. In turn the Q.W. continuum increasingly contributes to the impurity wavefunction. For $L/a_0^* \ll 1$, the Q.W. becomes a perturbation and at $L=0$ the impurity binding energy is R_0^* (since the well and the barrier materials are assumed to have identical effective mass and dielectric constant) [7–9]. For finite V_b and intermediate thickness ($L/a_0^* \sim 1$) the binding energy is larger than that found at both $L=0$ and $L=\infty$. The z motion becomes forced by the Q.W. effect and the impurity binding energy is larger than in 3D situations because the electron is held close to the attractive center by the Q.W. walls. The positive energy associated with this extra-localization is taken up by E_1 (the onset energy of the Coulombic continuum). Summarizing the previous hand-waving arguments we note that the Coulombic binding energy first increases from R_0^* by decreasing L, then reaches a maximum value and then decreases to R_0^* at $L=0$.

If however V_b is infinite E_1 is liable to diverge and is increasingly separated from the excited states when L decreases. The situation is qualitatively opposite at infinite V_b to that found at finite V_b: decreasing L to zero reinforces the two-dimensional character if $V_b=\infty$. Consequently the impurity binding energy monotically increases reaching $4R_0^*$ at $L=0$ which is the binding energy of a two-dimensional hydrogenic atom.

The second parameter relevant to the discussion of impurity binding energy in heterostructures is the position of the impurity along the growth axis. In a bulk material the ground bound state is degenerate with respect to the impurity position: two impurity sites separated by a lattice vector are equivalent for impurity binding energy. That is no longer true in Q.W. structures which lack for translational invariance along z. Consider for instance a thick well. The on-center impurity binding energy is close to R_0^*. If on the other hand $z_i=\frac{1}{2}L$ the Q.W. wall prevents a quasi-1S state to be an acceptable solution of the on-edge impurity problem. Instead a quasi-$2P_z$ state which nearly vanishes at the impurity site will be an acceptable solution. Thus for thick wells the on-edge impurity binding energy should approach $R_0^*/4$. A dispersion of the binding energy upon z_i is a characteristic feature of the impurity problem in Q.W. [6]. If the impurities are uniformly distributed in the structure the on-center impurities are statistically dominant. An impurity placed in the barrier material ($|z_i|>L/2$) can still bind a state (i.e., a state whose energy is lower than the onset of the ground Q.W. subband E_1). At large $|z_i|$ its binding energy approaches zero [6,9]. There exists a second impurity state attached to the edge of the barrier material (energy $\sim V_b-R_0^*$). That state is not a true bound state since it interferes with the continuum associated with the 2D subbands $E_1, E_2, \ldots$. However its lifetime τ roughly given by

$$\frac{1}{\tau}=\sum_n \frac{1}{\tau_n}; \qquad \frac{1}{\tau_n}=\frac{4\pi m^* a_0^*}{\hbar^2\kappa^2}\chi_n^2(z_i)\left[\frac{R_0^*}{V_b-R_0^*}\right]^2 \tag{3}$$

can be very long for remote impurities.

Various variational solutions of (1) have been proposed. For instance [6,9,10],

$$\psi_{\mathrm{I}}=N\chi_1(z)\exp\left\{-\frac{1}{\lambda}[\rho^2+(z-z_{\mathrm{i}})^2]^{1/2}\right\}; \qquad \psi_{\mathrm{II}}=N\chi_1(z)\exp(-\rho/\lambda), \tag{4}$$

ψ_{I} is exact in both $L=0$ and $L=\infty$ limits. For Q.W. of current interest ($50\,\text{Å} \lesssim L \lesssim 250\,\text{Å}$) ψ_{II} is as good as ψ_{I} for on-center impurities and allows for simpler calculations when the Q.W. is perturbed (e.g., by an electric field applied parallel to z; when one wants to discuss the effect of effective mass mismatch at the interfaces or when the impurity is screened by free carriers). In fig. 1 we present the L dependence of on-edge donor binding energy in GaAs–$Ga_{1-x}Al_xAs$ Q.W. calculated with ψ_{I} and in fig. 2 the z_{i} dependence of the binding energy of the bound level attached to E_1 and created by a donor ion placed in the barrier material ($V_{\mathrm{b}}=\infty$). Note that Tanaka et al. [9] have shown that finite V_{b} sensitively increases the binding energy of the in-barrier impurities compared to the results shown on fig. 2.

Suppose that the Q.W. barriers are selectively doped with donors. At thermal equilibrium some of these donors have released their electrons (fig. 3) which are transferred in the Q.W. leaving fixed positive charges in the barrier. This spatial separation between electrons and their parent donors creates band bending on the one hand and strongly decreases the ionic scattering suffered by the quasi-2D gas on the other hand. We are interested in calculating the bound states supported by a single residual impurity, which we assume to be placed

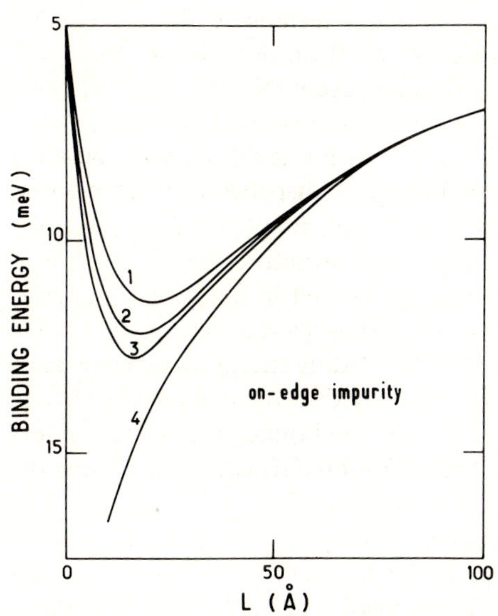

Fig. 1. Thickness dependence of the on-edge donor binding energy in GaAs–$Ga_{1-x}Al_xAs$ Q.W. The curves labelled 1, 2, 3, 4 correspond to $V_{\mathrm{b}}=212$ meV, 318 meV, 424 meV and $V_{\mathrm{b}}=\infty$, respectively.

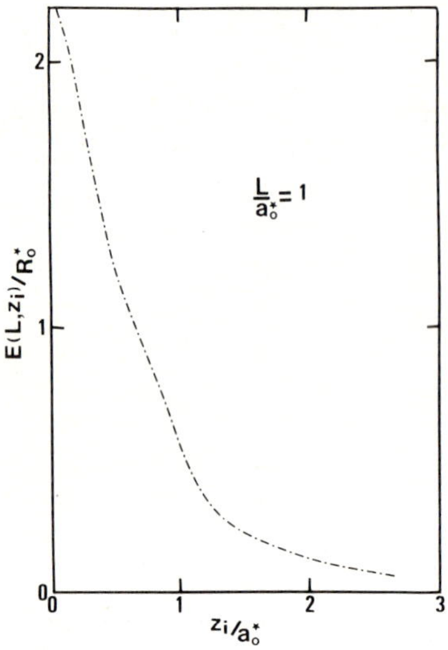

Fig. 2. Position dependence of the donor binding energy (in dimensionless units) in a Q.W. of infinite height. The donor is assumed to be placed in the barrier. $L = a_0^*$.

at the center of the well ($z_i = 0$) [11]. Due to the presence of free carriers the electrostatic interaction between the electron and the on-center impurity is screened. In the random-phase approximation, if a single Q.W. subband is occupied and if $T = 0$, this quasi-2D screening is expressed by [15]

$$\langle \chi_1 | V_s(q, z, z_i = 0) | \chi_1 \rangle = \frac{1}{\varepsilon(q)} \left\langle \chi_1 \left| -\frac{2\pi e^2}{\kappa q} \exp(-q|z|) \right| \chi_1 \right\rangle, \tag{5}$$

where $\boldsymbol{q}$ is a 2D (in-plane) wavevector; $V_s(q, z, z_i = 0)$ is the coefficient of the 2D Fourier expansion of the screened Coulombic potential and $\varepsilon(q)$ the dielectric constant:

$$\varepsilon(q) = 1 + (q_{TF}/q) f_s(q) g(q), \tag{6}$$

$$f_s(q) = \int \chi_1^2(z) \chi_1^2(z') \exp(-q|z - z'|) \, dz \, dz', \tag{7}$$

$$g(q) = 1 - Y(q - 2k_F)[1 - (2k_F/q)^2]^{1/2}. \tag{8}$$

In eqs. (6)–(8) $q_{TF} = 2/a_0^*$ is the 2D Thomas–Fermi wavevector and k_F is the Fermi wavevector related to the 2D concentration n_e by $k_F = (2\pi n_e)^{1/2}$. The quasi-2D screening differs from the bulk one by two important factors.

(i) Only the averages over χ_1 of the 2D Fourier coefficients of the screened potential are easily evaluated. In the bulk instead, one knows the 3D Fourier coefficients $V_s(\boldsymbol{q})$ exactly. Thus $V_s(\boldsymbol{r})$ is easily calculated. In the Thomas–Fermi approximation, for instance, one gets $V_s(\boldsymbol{r}) = -(e^2/\kappa r) \exp(-Q_{TF} r)$.

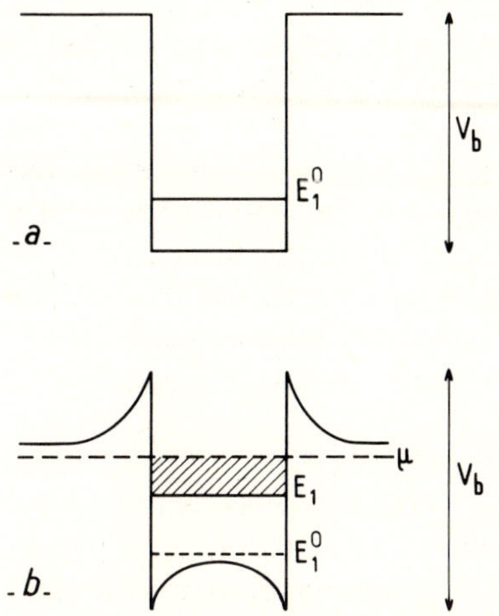

Fig. 3. Conduction band profile of a modulation-doped Q.W. The donors are placed in the barrier and some of them have released their electrons which accumulate in the well. E_1^0 labels the Q.W. ground state in the absence of charge transfer and E_1 is the Q.W. ground state taking into account the band bending effects.

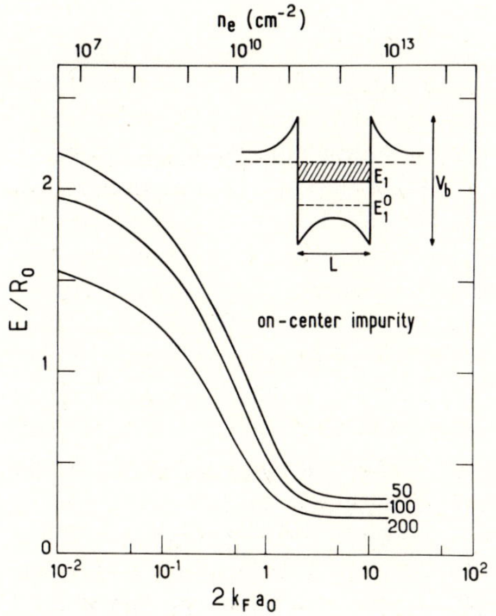

Fig. 4. Binding energy of a on-center screened donor in GaAs–$Ga_{1-x}Al_xAs$ Q.W. (V_b = 0.2 eV) plotted versus the 2D electron concentration n_e for three Q.W. thicknesses L.

(ii) The 2D Thomas–Fermi wavevector q_{TF} is constant whereas the bulk quantity Q_{TF} is concentration-dependent. The latter feature implies that a screened Coulombic potential may have no bound state if the 3D-concentration N_e is larger than a critical value N_e^c such that $Q_{TF}a_0^* \sim 1$. In quasi-2D systems the constant q_{TF} implies that there always exists a bound state: the binding energy saturates to a finite value at large n_e. The binding energy of a screened on-center donor states has been recently calculated by Brum et al. [11]. The results of these calculations are presented on fig. 4 for GaAs–$Ga_{1-x}Al_xAs$ Q.W. ($V_b = 0.2$ eV). A sizeable binding energy remains at large n_e which should allow the observation of impurity-related features at low temperature in modulation-doped Q.W.'s.

Various impurity features have been reported [12,13,16] in photoluminescence spectra of GaAs–$Ga_{1-x}Al_xAs$ Q.W. structures. The impurity lines are much weaker than in bulk GaAs and have been associated either to donors or to acceptors. The observed donor binding energies reported by Shanabrook et al. [12] are well described by Mailhiot et al.'s or Greene and Bajaj's calculations [7,8]. The acceptor binding energies reported by Miller et al. [13] have been quantitatively interpreted by Masselink et al. [14]. Note that acceptor calculations are more complicated than donor calculations due to the coupling between light and heavy hole subbands.

3. Exciton binding energy in single quantum well [17–24]

When a Q.W. structure is shined by near- (effective) bandgap light the lowest lying electron–hole state corresponds to a bound state for the reduced electron–hole motion: the exciton. To its energy one should add the kinetic energy of the center-of-mass (C.M.). Since the electron and hole z motions are confined by Q.W. walls one deals with quasi-2D excitons and the C.M. motion occurs only in the layer plane. Thus, to a great deal, the calculations of exciton binding energy resemble those of Coulombic impurities. There are however two important new features brought by the mobile nature of the positive charge.

(i) The electron and the hole can be mostly localized within the same slab (type I Q.W., e.g., GaAs–$Ga_{1-x}Al_xAs$) or within adjacent layers (type II Q.W., e.g., InAs–GaSb) (fig. 5). Clearly the exciton binding should be larger in type I Q.W. than in type II Q.W. [18], all other parameters being identical. This difference is reminiscent of the much larger binding energies found for in-well impurities compared to those of in-barrier impurities.

(ii) The in-plane hole kinematics is complicated because of the coupling between light and heavy hole subbands at finite wavevector $\boldsymbol{k}_\perp$ in the layer plane [19–21]. In bulk materials the Γ_8 valence Hamiltonian does not reduce to a scalar like the Γ_6 conduction Hamiltonian. To the scalar part one should add a $(\boldsymbol{k}\cdot\boldsymbol{J})^2$ term where $J=\frac{3}{2}$. Quantizing $\boldsymbol{J}$ along the growth axis and adding the

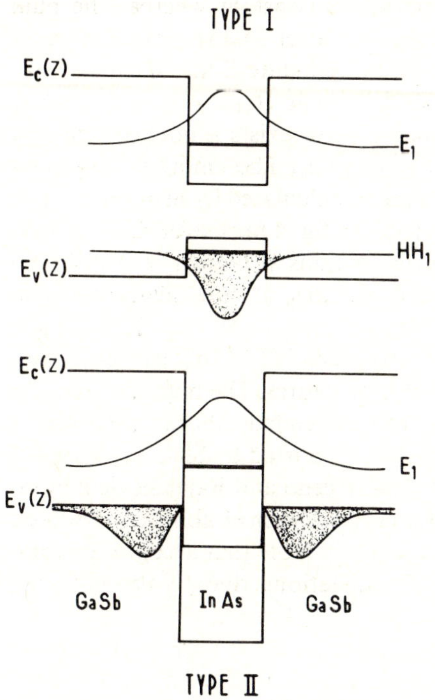

Fig. 5. Schematic representation of electron (E_1) and hole wavefunction in type I and type II (GaSb–InAs–Sb, $L_{InAs} \lesssim 100$ Å) quantum wells, $E_c(z)$ and $E_v(z)$ denote the position-dependent conduction and valence band edges respectively.

valence barrier term $V_v(z)$ leads at $\boldsymbol{k}_\perp = \mathbf{0}$ to decoupled light (LH_n) and heavy (HH_n) subbands whose confinement energies involve the bulk heavy and light hole effective masses. At $\boldsymbol{k}_\perp \neq \mathbf{0}$ these subbands are admixed and for instance anti-crossings take place [19–21]. The in-plane effective mass of heavy hole and light hole subbands have no obvious relationships with the bulk hole masses.

Experimentally the existence of two kinds of excitons formed between electron and heavy hole on the one hand and electron and light hole on the other hand is very well documented in GaAs–$Ga_{1-x}Al_xAs$ Q.W.'s [2,4,5,17]. That is due to a sizeable difference between HH_n and LH_n confinement energies when $L \lesssim 150$ Å. No direct information on the exciton binding energy itself are available (except in [17] where a small hump adjacent to the main E_1–HH_1 exciton line has been interpreted as the "2S" state of the E_1–HH_1 exciton).

In principle the exciton binding energy is obtained by finding the eigenenergies of a 4×4 matrix which represents the reduced motion of a Γ_6 electron and Γ_8 hole. In bulk materials Baldereschi and Lipari [23] have shown that the isotropic parts of the diagonal terms of this matrix are much larger than the other terms. Thus the exciton problem reduces to a donor-like problem if only the isotropic terms are retained and if the C.M. wavevector is assumed to be zero.

In Q.W. structures model exciton calculations have been proposed [18,22]. If one neglects the valence band intricacies and thus deals with idealized isotropic valence bands the exciton binding energies vary with the Q.W. thickness in a very similar way as the donor binding energy does [18]. Greene et al. [24] and Miller et al. [17] have used part of the bulk procedure to solve the Q.W. exciton problem, i.e., they have neglected the off-diagonal terms of the exciton matrix. The diagonal terms involve reduced masses which are different for the z and the in-plane motions and also different for electron–light hole, E_n–LH_n, and electron–heavy hole, E_n–HH_n, excitons. Greene et al. [24] showed that finite barrier effects are essential in exciton calculations. They found that the binding energies of E_1–HH_1 and E_1–LH_1 excitons are influenced by two effects.

(i) The electron and hole leakages in the $Ga_{1-x}Al_xAs$ barrier. The LH_1 state is less localized in GaAs than the HH_1 state. Thus the heavy hole exciton is more two dimensional, i.e., more tightly bound than the light hole exciton for a given L.

(ii) However the in-plane light hole exciton effective mass is *larger* than the one of the heavy hole exciton. Thus the LH_1–E_1 exciton has a tendency to a larger binding than the HH_1–E_1 exciton. The mass effect prevails at large L whereas at small L the E_1–HH_1 exciton is more bound than the E_1–LH_1 exciton. It would be desirable to calculate the effects of the off-diagonal terms

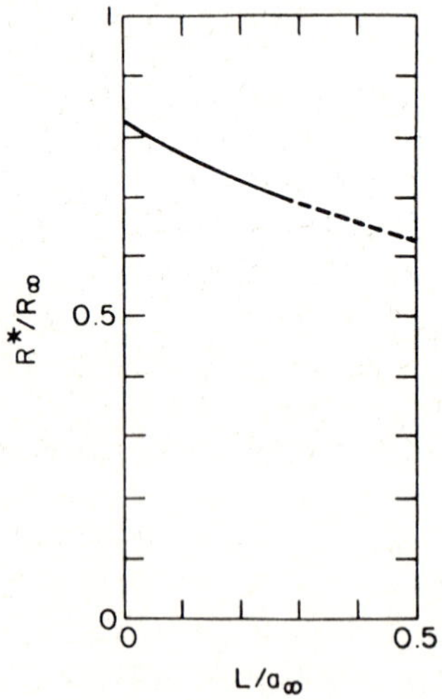

Fig. 6. Thickness dependence of the exciton binding energy in GaSb–InAs–GaSb quantum wells ($L_{InAs} \lesssim 100$ Å).

of the exciton matrix and also of the effective mass mismatch between GaAs–$Ga_{1-x}Al_xAs$ before drawing definite conclusions on the Q.W. exciton problem.

As far as excitons in type II Q.W.'s are concerned, only a model calculation has been proposed [18]. As expected, the spatial separation between electron and hole drastically reduces the exciton binding energy (fig. 6).

4. Low temperature exciton trapping on interface defects [25,26]

The interfaces between two semiconductors, even lattice-matched, are not perfect. In SiO_2–Si metal–oxide-semiconductor structures the Si–SiO_2 interface roughness is known to contribute to a limitation of channel mobility [15]. In GaAs–$Ga_{1-x}Al_xAs$ modulation-doped heterojunctions whose interface quality is far superior to that found in Si–SiO_2 the interface roughness has been calculated to be a scattering mechanism of secondary importance [27]. In GaAs–$Ga_{1-x}Al_xAs$ Q.W.'s the optical probes (luminescence, excitation spectroscopy) may provide a finer access to the interface roughness than transport measurements. This is due to the influence of interface defects on the optical properties of Q.W.'s at low temperature. Suppose an exciton has acquired by some means a finite in-plane C.M. wavevector $\boldsymbol{K}$ and assume that Coulombic impurities are too scarce to play a significant part. The interfaces being imperfect deviate locally from a plane (here as in previous sections we are only concerned with envelope functions which makes possible the definition of planar interfaces). When this deviation corresponds to an attractive potential for both electron and hole, i.e., when extra GaAs protrudes in the $Ga_{1-x}Al_xAs$ barriers, the moving exciton is scattered and even get trapped in this quasi-2D potential well. Little is known on $\mathscr{H}_{def}$ the potential energy associated with interface defects and to a great deal we have to define it. In an empirical approach an interface defect can be characterized by a depth b (along z) and a lateral (in-plane) extension a. Many $\mathscr{H}_{def}$ fulfill these requirements. We took [26] a semi-Gaussian model

$$\mathscr{H}_{def} = -V_c \exp(-\rho_e^2/2a^2)\exp(-z_e^2/2b^2) \times Y[-z_e] - V_v \exp(-\rho_h^2/2a^2)\exp(-z_h^2/2b^2)Y[-z_h] \qquad (9)$$

for a defect protruding in the left-hand side barrier. In eq. (9) V_c (V_v) is the height of the electron (hole) confining barriers; the subscripts e (h) refers to the electron (hole) and L is the average Q.W. thickness.

High quality Q.W.'s are characterized by intense, narrow (few meV) photoluminescence lines which are little Stokes-shifted (0 to ~5 meV for $L \lesssim 100$ Å) from the maximum of the E_1–HH_1 free exciton line seen in absorption or exci-

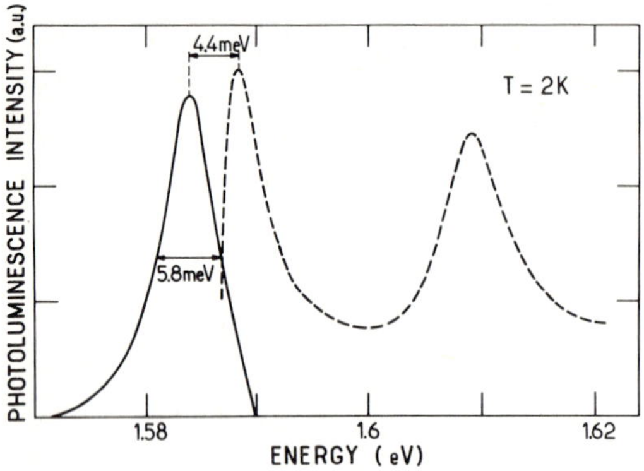

Fig. 7. Photoluminescence (solid line) and excitation (dashed line) spectra of a 70 Å thick GaAs–$Ga_{1-x}Al_xAs$ single Q.W. ($x=0.52$).

tation spectra (X.S.) (fig. 7). Qualitatively one may define two limiting regimes characterizing the optical properties of the excitons interacting with interface defects.

(i) Either a is very large compared to a_0^* (say $a \sim 10^3$ Å). Then each of the defect acts like a micro-sample having its own thickness L'. Inside such an island the exciton is free, the C.M. wavevector is a good quantum number. The absorption or excitation spectrum is essentially governed by the states of high D.O.S. Thus it will reflect the existence of many micro-samples, each of different L' and thus of different effective bandgap $HH_1(L')+E_1(L')+\varepsilon_g-R^*_{exc}(L')$. Suppose one deals with a set of samples of varying average thickness L and assume that, irrespective of L, the lateral island sizes are very large ($a \to \infty$) and the defect depth b is constant. The width of the $E_1 \to HH_1$ exciton peak seen in X.S. will increase like b/L^3. This behaviour was observed in high quality MBE-grown samples by Weisbuch et al. [25]. The width of the P.L. line and its (eventual) Stokes-shift with respect to X.S. peak can hardly be correlated with the X.S. linewidth. This is because P.L. involves excited carriers which, during their lifetime before recombination, may relax or not through all the quantum states of low energy which exist in the excited band. The P.L. emphasizes the quantum state of low energy (which may have a low D.O.S.) provided the capture processes in these low energy states are faster than the radiative lifetime.

(ii) The other limiting regime occurs for $a \sim a_0^*$. A single defect has few bound states (often a single one) because of the in-plane confinement. The C.M. wavevector is no longer a good quantum number. The exciton states consist of a 2D continuum associated with moving excitons scattered by attractive or repulsive interface defects. At lower energy bound states of exciton trapped on attractive interface defects occur. Defects of various (a, b) give rise to different binding energies (fig. 8). Thus a whole band of trapped excitons is generated in

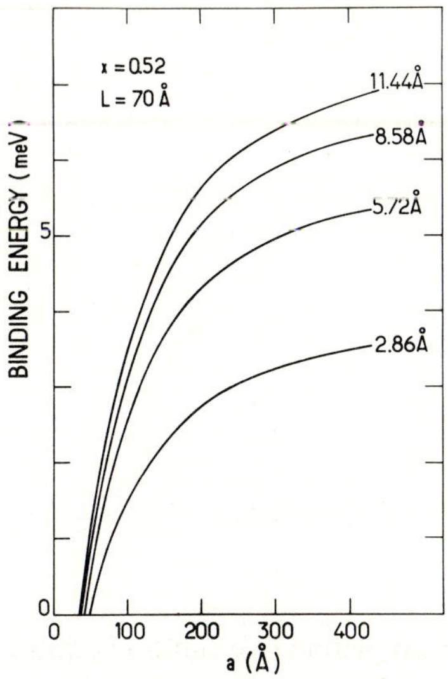

Fig. 8. Trapped exciton binding energy versus the lateral size a of attractive interface defects for different defect depth b in GaAs–$Ga_{1-x}Al_xAs$ quantum well ($x=0.52$). The average quantum well thickness is $L=70$ Å.

the limit of small defects. The X.S. probes the exciton continuum near $\boldsymbol{K}=\mathbf{0}$. The P.L. line is more sensitive to trapped excitons and must be structureless unless a definite pair (a, b) is statistically dominant, which seems unlikely. As far as thermalization of trapped excitons is concerned it appears difficult in a given site since most of the defects have a single bound level. It may occur through phonon-assisted hopping from site to site. To be efficient, that mechanism requires the interface defects to be numerous since they have small radii.

In addition to optical experiments there have been interface characterizations [28] which provide independent evidence that in high quality samples the mean defect depth b is $\sim$ one monolayer (2.83 Å in GaAs) and its lateral extension is $\sim$200–300 Å. If we take these numbers into consideration model (ii) seems preferable to model (i).

The optical experiments performed by the E.N.S. group on MOCVD-grown GaAs–$Ga_{1-x}Al_xAs$ single Q.W.'s ($L=70$ Å, $x=0.52$) reveal that the P.L. line of a representative high quality sample was 4.4 meV Stokes-shifted from the E_1–HH_1 exciton peak seen in the X.S. The P.L. line was structureless and did not change shape when the laser intensity varied from 1 mW to 1 W: neither carrier heating nor saturation effects were detectable. We have qualitatively evaluated the characteristics of the trapped exciton band, (i) by calculating the exciton binding energy on the semi-Gaussian interface defects (eq. (9)) versus defect sizes a and b, (ii) by assuming a Gaussian distribution of defect sizes.

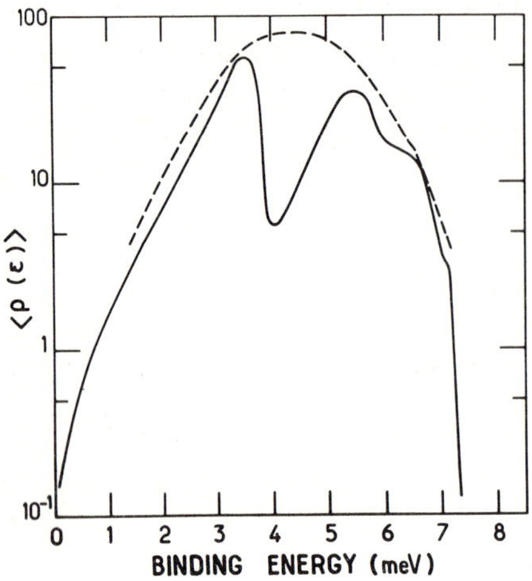

Fig. 9. Trapped exciton density-of-states versus binding energy. The structures are associated to discrete b whereas the smooth curve would correspond to a continuous variation of b in a GaAs $Ga_{1-x}Al_xAs$ ($x=0.52$) Q.W. of average thickness 70 Å.

The trapped exciton D.O.S. is shown on fig. 9. It peaks ~4 meV below the edge of the unperturbed C.O.M. continuum and is 2.5 meV broad at half maximum. We did not attempt to optimize the parameters of the defect sizes distribution. To identify the D.O.S. to the P.L. line one needs to assume that the optical matrix element governing recombination is constant and that there is no thermalization inside the trapped exciton band. At low temperature ($T=1.7$ K) once an exciton is trapped in a given defect it can only emit acoustical phonons to tunnel to another defect. We showed that for reasonable defect concentrations that process lasts longer than the measured luminescence decay time [29], supporting the contention that trapped excitons are not thermalized.

5. Conclusion

Coulombic problems in semiconductor Q.W. have rapidly diversified. The gross features of basic problems (such as the determination of the binding energy of a Coulombic impurity placed somewhere in a Q.W.) are now understood. The calculations have now to address to more difficult situations (perturbed Q.W.'s, interplay between interfaces and impurities, etc.) to match the challenge of present experiments.

Acknowledgements

I am much indebted to Drs J.A. Brum, C. Delalande, C. Guillemot, M.H.

Meynadier, P. Voisin, M. Voos at the E.N.S. and Drs L.L. Chang, L. Esaki, E.E. Mendez at IBM for their active participation in the works described in this paper. The Groupe de Physique de l'E.N.S. is "Laboratoire associé au C.N.R.S.". The work at the E.N.S. has been partly supported by C.N.E.T.

References

[1] L. Esaki and R. Tsu, IBM J. Res. Develop. 14 (1970) 61.
[2] R. Dingle, in: Festkörperprobleme: Advances in Solid State Physics, ed., H.J. Queisser (Pergamon–Vieweg, Braunschweig, 1975) Vol. XV, p. 21.
[3] B.A. Vojak, W.D. Laidig, N. Holonyak Jr., M.D. Camras, J.J. Coleman and P.D. Dapkus, J. Appl. Phys. 52 (1981) 621.
[4] C. Weisbuch, R.C. Miller, R. Dingle, A.C. Gossard and W. Wiegmann, Solid State Commun. 37 (1981) 219.
[5] D.A.B. Miller, D.S. Chemla, D.J. Eilenberger, P.W. Smith, A.C. Gossard and W.T. Tsang, Appl. Phys. Lett. 41 (1982) 679.
[6] G. Bastard, Phys. Rev. 24 (1981) 4714.
[7] C. Mailhiot, Yia-Chung Chang and T.C. McGill, Phys. Rev. B 26 (1982) 4449.
[8] R.L. Greene and K.K. Bajaj, Solid State Commun. 45 (1983) 825.
[9] K. Tanaka, M. Nagaoka and T. Yamabe, Phys. Rev. B 28 (1983) 7068.
[10] C. Priester, G. Allan and M. Lannoo, Phys. Rev. B 28 (1983) 7194.
[11] J.A. Brum, G. Bastard and C. Guillemot, Phys. Rev. B 30 (1984) 905.
[12] B.V. Shanabrook and J. Comas, in: Electronic Properties of Two-Dimensional Systems, Oxford 1983; Surf. Sci. 142 (1984) 504.
[13] R.C. Miller, A.C. Gossard, W.T. Tsang and O. Munteanu, Phys. Rev. B 25 (1982) 3871.
[14] W.T. Masselink, Yia-Chung Chang and H. Morkoç, Phys. Rev. B 28 (1983) 7373.
[15] See, e.g., T. Ando, A.B. Fowler, F. Stern, Rev. Mod. Phys. 54 (1982) 437.
[16] B. Lambert, B. Deveaud, A. Regreny and G. Talalaeff, Solid State Commun. 43 (1982) 443.
[17] R.C. Miller, D.A. Kleinman, W.T. Tsang and A.C. Gossard, Phys. Rev. B 24 (1981) 1134.
[18] G. Bastard, E.E. Mendez, L.L. Chang and L. Esaki, Phys. Rev. B 26 (1982) 1974.
[19] S.S. Nedorezov, Fis. Tverd. Tela 12 (1970) 2269, Sov. Phys. Solid State 12 (1971) 1814.
[20] M. Altarelli, Phys. Rev. B 28 (1983) 842.
[21] A. Fasolino and M. Altarelli, in: Heterostructures and Two-Dimensional Electronic Systems in Semiconductors; Mauterndorf, Springer Series in Solid State Sciences 43 (1984) 176.
[22] S. Satpathy and M. Altarelli, Phys. Rev. B 23 (1981) 2977.
[23] A. Baldereschi and N.O. Lipari, Phys. Rev. B 3 (1971) 439.
[24] R.L. Greene, K.K. Bajaj and D.E. Phelps, Phys. Rev. B 29 (1984) 1807.
[25] C. Weisbuch, R. Dingle, A.C. Gossard and W. Wiegmann, Solid State Commun. 38 (1981) 709.
[26] G. Bastard, C. Delalande, M.H. Meynadier, P. Frijlink and M. Voos, Phys. Rev. B 29 (1984) 7042.
[27] T. Ando, J. Phys. Soc. Japan 51 (1982) 3900.
[28] P.M. Petroff, J. Vac. Sci. Technol. 14 (1977) 973.
[29] E.O. Göbel, H. Jung, J. Kuhl and K. Ploog, Phys. Rev. Lett. 51 (1983) 1588.
[30] C. Priester, G. Bastard, G. Allan and M. Lannoo, Phys. Rev. B (1984) in press.

Journal of Luminescence 30 (1985) 502–519
North-Holland, Amsterdam

TWO-DIMENSIONAL SEMICONDUCTORS: RECENT DEVELOPMENT

D.S. CHEMLA

AT&T Bell Laboratories, Holmdel, N.J. USA

The present article reviews the physics and the applications of the nonlinear optical and electro-optical effects associated with room temperature quasi-two-dimensional excitons in GaAs/AlGaAs quantum well structures.

Introduction

Recently analog and digital optical processing have attracted much attention. However the operation of optical devices such as switches or modulators requires very large optical or electro-optical nonlinearities. These can be found in semiconductors at excitonic transition resonances [1]. In bulk semiconductors excitonic resonances are only observed at low temperatures. It has recently been shown that excitonic effects are enhanced in ultra-thin semiconductor layers due the quasi-two-dimensional carrier confinement [2]. The combination of an increased exciton binding energy [3] and a slightly reduced overall exciton–phonon interaction results in the observation of clear and well resolved exciton peaks in the absorption spectra of GaAs/AlGaAs multiple quantum well structures (MQWS) at room temperature [4].

The nonlinear optical effects involving room temperature excitons are extremely interesting to investigate for practical as well as fundamental reasons. From the point of view of applications MQWS exhibit at room temperature very large nonlinear optical and electro-optical effects at wavelengths, power levels and speeds compatible with laser diodes and electronics. In addition MQWS are prepared by advanced semiconductor technology [5], and therefore they have the potential to be incorporated in integrated systems. From a more fundamental viewpoint, MQWS form a model system to study the effects of reduced dimensionality on excitonic behavior. Furthermore, the room temperature excitons found in MQWS, are in the very unusual situation where

$\hbar\Omega_{LO} > kT > E_b$.

0022-2313/85/$03.30

In this equation, $\hbar\Omega_{LO} \sim 36$meV is the GaAs LO-phonon energy, $kT \sim 25$ meV at room temperature and $E_b \sim 10$ meV is the exciton binding energy. Under these conditions excitons generated by resonant optical excitation are ionized in a very short time by phonon collisions. This provides a unique opportunity to study the dynamics of the transformation of excitons into an electron–hole (e–h) plasma and to investigate the nonlinear optical effects associated with both species.

The article is organized as follows: in section 1 we review the nonlinear absorption and refraction observed with picosecond and cw lasers, and we give a brief summary of the recent studies using femtosecond spectroscopic techniques; in section 2 we review the electro-absorption effects observed by application of static electric fields perpendicular and parallel to the layers, and we present some devices which have already been demonstrated.

1. Optical nonlinearities of GaAs/AlGaAs multiple quantum well structures

The low temperature linear absorption in MQWS has been reviewed in ref. [2] and the room temperature linear absorption has been discussed in ref. [6]. The confinement of an e–h pair in ultra-thin semiconductor layers increases the exciton binding energy[3,7,8] resulting in the observation of well resolved resonances in the room temperature absorption spectra of MQWS [4,9,10,11]. This is shown in fig. 1(a) where the 300 K absorption spectra of a 3.2 μm thick high quality GaAs sample is compared to that of a MQWS sample. The MQWS consists of 77 periods of 102Å GaAs quantum wells alternating with 207Å $Al_{0.28}Ga_{0.72}As$ barrier layers. A small bump is seen at the edge of the GaAs spectrum, it corresponds to the remains of the excitonic resonance. The MQWS spectrum show two clear resonances for the heavy and the light hole excitons, followed by the flat two-dimensional continuum and the onset of the next subband transitions. This type of absorption spectrum is very well described by two Gaussian absorption lines corresponding to the two excitons and a two-dimensional continuum with the appropriate Sommerfeld enhancement factor for the interband transitions [6]. In fig. 1(b) is shown the variation as a function of the temperature of the width of the heavy hole exciton peak, it is well described by a semi-empirical fit, shown by the solid line, involving two terms: a constant inhomogeneous line width, $\Gamma_0 \sim 2$ meV [12,13,14] and a term proportional to the density of LO phonons [9,6]. The coefficient giving the temperature-dependent broadening, $\Gamma_{ph} \sim 5$ meV, is smaller than that of bulk GaAs, $\Gamma_{ph} \sim 7$ meV [15]. From this dependence one can deduce a mean time for ionization by thermal LO-phonon collision of about 400 fs. Since at room temperature $kT \sim 2.5 \times E_b$ almost no excitons can reform from the electrons and holes created by the thermal phonons ionization [6]. *Therefore, although excitonic resonances are observed at room temperature the gas of excitons*

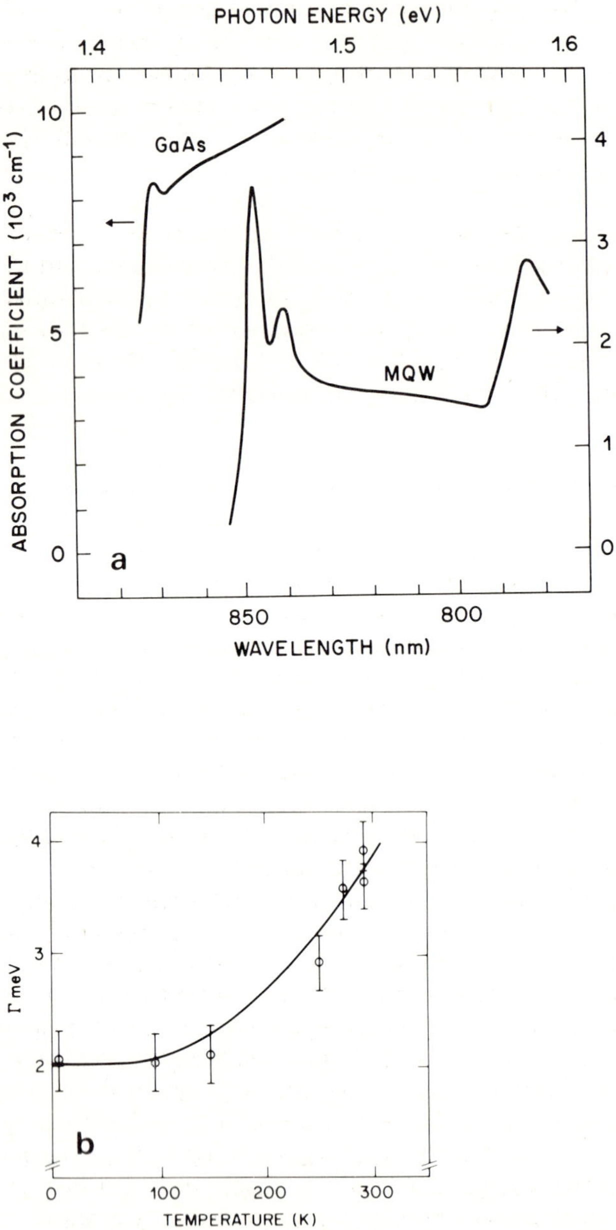

Fig. 1. (a) Room temperature absorption spectra of a 3.2 μm thick high purity GaAs sample and of a MQW sample consisting of 77 periods of 102 Å GaAs layers and 207 Å $Al_{0.28}Ga_{0.72}As$ layers. (b) Temperature dependence of the line width of the $n=1$ heavy hole exciton. The solid line corresponds to fit discussed in the text.

resonantly generated through photon absorption is very short lived; it transforms into a free e–h *plasma in less than* 0.5 ps.

Under optical excitation, as free e–h are generated they screen the Coulomb interaction causing the renormalization of the single-particle states, with however very little shift of the exciton energy because of its neutrality [16,17]. Nevertheless, the binding energy of the exciton with respect to the normalized band gap decreases, the e–h correlation diminishes progressively and the height of the absorption peak decreases. Eventually the correlation is completely lost and the exciton peak disappears leaving a step-like absorption continuum edge near the exciton original position. This edge is much more difficult to saturate and can be shifted further to low energy if more carriers are generated. The large optical nonlinearity observed in MQWS with cw and ps sources results from conjunction of two factors; the concentration of a large oscillator strength in a narrow energy band and the screening of the Coulomb interaction by a charged plasma that modifies and eventually destabilizes the excitonic states.

Saturation of excitonic absorption in MQWS has been observed with cw laser excitation at resonance [4,9] and well above the exciton energy well in the continuum [6]. The intensity dependence of the exciton peaks of the two samples (whose absorption spectra are given in fig. 1) are shown in fig. 2(a) and 2(b), respectively, for the GaAs sample and the MQWS. The solid curves are semi-empirical fits based on the model discussed in the previous paragraph. For GaAs the saturation is well described by a constant ($\sim$not saturable) background plus a saturable species with a saturation intensity $I_s \sim 4.4$ kW/cm^2. The fit for the MQWS involve two saturable species, one with a very small saturation intensity $I_s \sim 580$ W/cm^2 and the other one more difficult to saturate. $I_s \sim 44$ kW/cm^2 [9]. According to our model these results can be interpreted in terms of the bleaching of the excitonic absorptions for the two samples, followed, in the case of the MQWS, by the saturation of the renormalized two-dimensional continuum as it is shifted down in energy. In the case of GaAs, the interband transitions in the three-dimensional continuum show no saturation, at least up to the intensities reached in our experiments. This interpretation is well supported by the shapes of the spectra observed with a low intensity tunable cw dye laser with or without the excitation of a cw diode laser delivering about 800 W/cm^2, and emitting at 42 meV above the exciton peak [6]. Examples of these spectra are shown in fig. 3. The fast ripples are Fabry–Perot fringes in the thin glass plate covering the sample. Under excitation the exciton peaks have disappeared, and the absorption spectrum observed is a step-like edge which has exactly the same shape as the two-dimensional continuum used to describe the interband transition in the unexcited spectrum.

The absorptive and dispersive part of the optical nonlinearity at low excitation has been measured by forward degenerate four-wave mixing (DFWM) using a picosecond dye laser [18]. In these experiments the nonlinear transmission

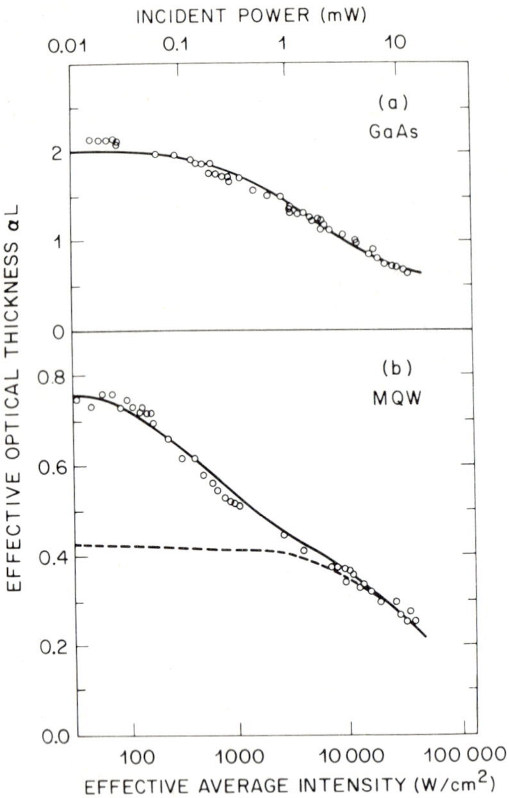

Fig. 2. Intensity dependence of the absorption at the peak of the exciton resonance for the GaAs (curve (a)) and MQW (curve (b)) samples whose absorption spectra are shown in fig. 1. The solid line corresponds to the empirical fits discussed in the text.

induced by the pump beam on the probe beam and the diffraction efficiency of the coherent gratings generated by pump-probe interference can be simultaneously measured. In addition to the linear transmission the same set-up gives access to the recombination life time, $\tau \sim 30$ ns, and to the diffusion coefficient, $D \sim 13\ \mathrm{cm}^2/\mathrm{s}$. An example of the experimental data is shown in fig. 4(a),(b),(c) for the linear transmission, the differential nonlinear transmission, and the DFWM signal, respectively. From the spectra of figs. 4(a) and 4(b) and an adapted Kramers–Krönig inversion it is possible to obtain the spectra of the changes of absorption coefficient per e–h pair σ_{eh} and of refractive index per e–h pair n_{eh} [6] these are shown in fig. 5. They determine complètely the optical nonlinearity in the fundamental absorption region. A severe test of the accuracy of this determination is that the two complex spectra of fig. 5 should reproduce the measured DFWM diffraction efficiency of fig. 4(c). The comparison of the

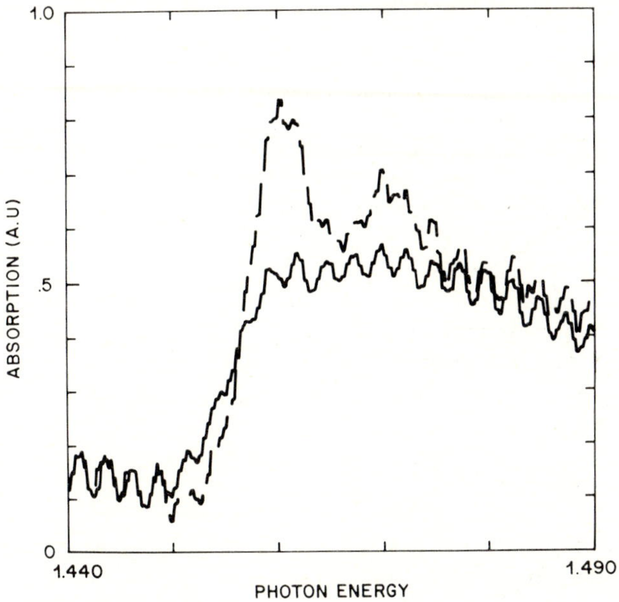

Fig. 3. Absorption spectra of a MQW sample consisting of 65 periods of 96Å GaAs layers and 98Å $Al_{0.3}Ga_{0.7}As$ layers with (solid line) and without (dashed line) a pumping beam from a cw laser diode operating 42 meV above the first exciton resonance. The intensity of the pumping beam is $I = 800$ W/cm^2. The fast oscillations in these spectra are Fabry–Perot fringes in a thin glass plate covering the sample.

experimental data of fig. 4 with the theoretical predictions based on the spectra of fig. 5 is shown in fig. 6, the agreement is excellent [6]. The measured non-linearities are very large, more than one order of magnitude larger than that of GaAs and six order of magnitude than that of Si at room temperature.

To demonstrate the potential use of MQWS for nonlinear optics with cw laser diodes; DFWM experiments were performed in a 1.25 μm MQWS using a commercial diode laser as the sole light source [19]. Typical results are shown on fig. 7. Internal diffraction efficiency of about 5×10^{-4} was measured with pump power as low as 2.5 mW ($I_p \sim 20$ W/cm^2) and probe power of 0.1 mW ($I_t \sim 0.8$ W/cm^2). It should be mentioned that such experiments are usually performed on thick samples with high power lasers.

Optical bistability using MQWS as the nonlinear element in a Fabry–Perot etalon, has been demonstrated [20].

More recently saturation of excitonic absorption in MQWS has been used to mode-lock laser diodes. Pulses as short as 1.6 ps and reliable operation have been obtained [21].

The dynamics of the exciton ionization by thermal phonons has been studied with the recently developed femtosecond spectroscopic techniques [22]. The

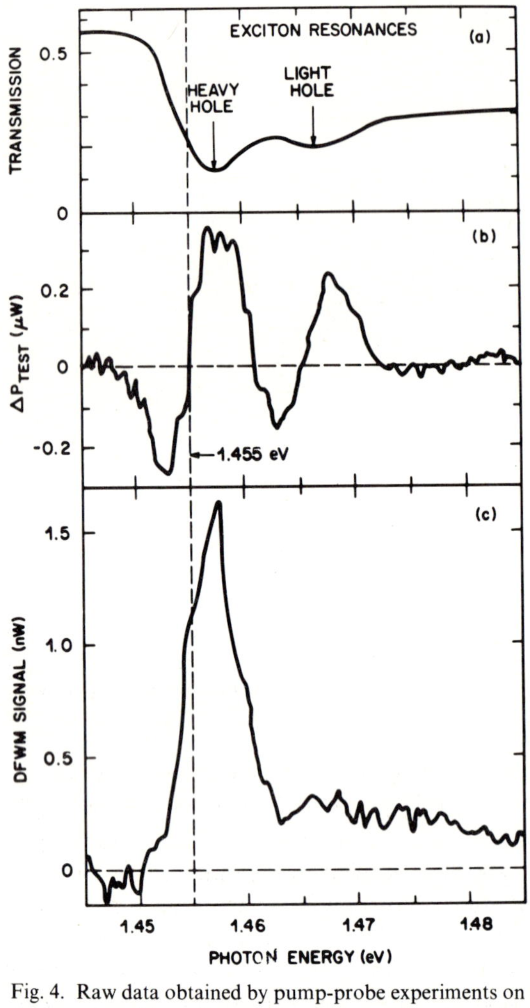

Fig. 4. Raw data obtained by pump-probe experiments on the sample described in fig. 3. (a) linear transmission; (b) nonlinear transmission; diffraction efficiency on the light induced grating. For (b) and (c) the pump and probe beam powers were $\sim 300\,\mu W$ and $\sim 60\,\mu W$, respectively.

changes in the absorption spectrum of MQWS samples under excitation resonant with the heavy hole exciton or under nonresonant excitation in the continuum, have been measured with a resolution of 150 fs [23,24]. In order to avoid any interference with band gap renormalization effects a moderate excitation was used in both cases. It was found that after a transient evolution which last approximately 0.5 ps, the changes are very similar to those observed with cw or ps lasers. However the transient regime ($t < 0.5$ ps) is highly dependent on the nature of the excitation. For the nonresonant case a smooth evolution of the spectrum takes place, from the unpumped spectrum to that observed

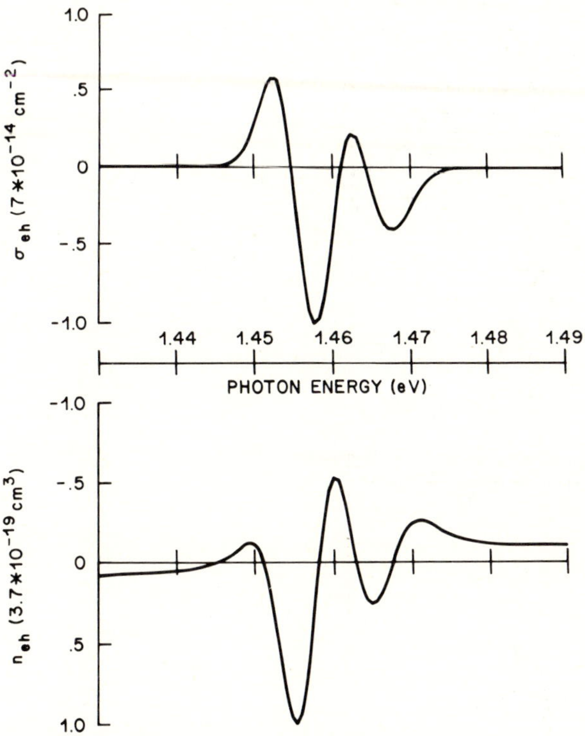

Fig. 5. Spectra of the imaginary and real parts of nonlinearity of the MQW sample. σ_{eh} and n_{eh} describe, respectively, the change of absorption and the change of refractive index induced by one electron–hole pair.

at long times ($t > 1$ ps). Conversely in the case of resonant pumping one observes a very efficient bleaching of the exciton absorption that recovers partially in 300 ± 100 fs. Typical spectra under resonant pumping and at various probe delays, are compared to the unexcited spectrum on fig. 8. The time dependence of the height of the heavy hole exciton peak for the two excitation conditions is shown in fig. 9. The smooth variation found for nonresonant pumping follows quite well the integral of the pump pulse, showing that the effects are proportional to the number of e–h pairs generated in the continuum. The behavior in the resonant pumping case corresponds to the instantaneous generation of a species that, in 300 ± 100 fs, transforms into free e–h pairs, which have the same effects as in the nonresonant pumping case [23,24]. The experiment is interpreted as the first observation of the exciton saturation due to the generation of excitons that occupy the lowest lying energy states of the system, followed by the thermal phonon ionization of these excitons into free e–h pairs. The saturation induced by the resonantly generated excitons is indeed large, and can be used for applications in ultra-fast optical devices [24].

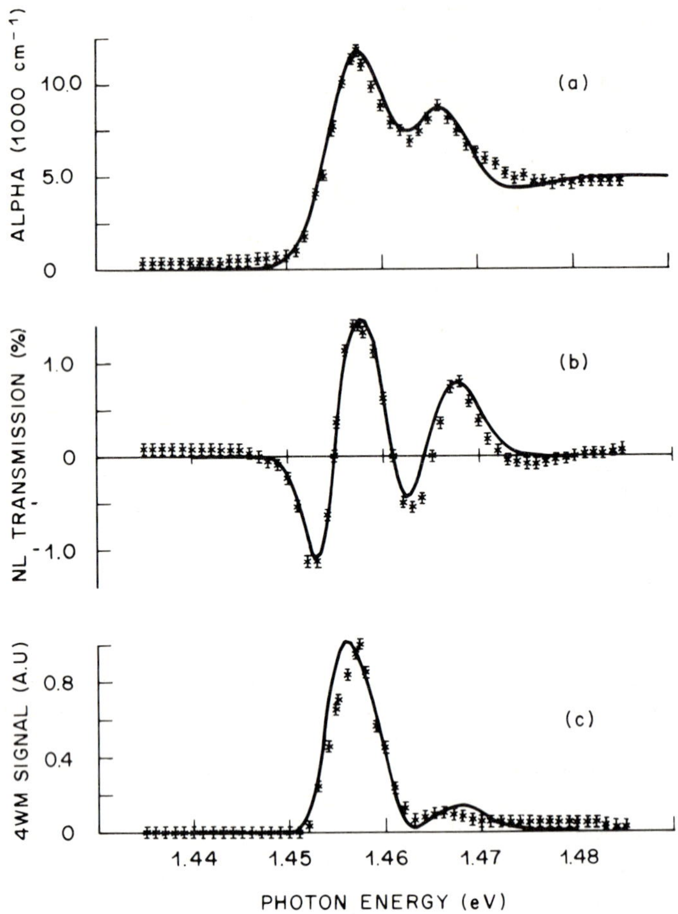

Fig. 6. Comparison of the experimental data of fig. (4), corrected for the small variations of the laser power, with the semi-empirical fit for the linear absorption (a) and nonlinear absorption (b). The solid curve in (c) is obtained by the Kramers–Kronig inversion discussed in the text.

2. Electro-absorption in GaAs/AlGaAs multiple quantum well structures

MQWS absorption is also very sensitive to perturbation by electrostatic fields. The reason for such sensitivity is that the energies associated with the carrier confinement in the layers or with the Coulomb attraction between electrons and holes are small, typically ~ 10 meV, and that the corresponding envelope wavefunctions are large, of the order of ~ 100 Å. Therefore the application of rather small fields, of the order of 10 mV/100 Å, i.e., 10^{+4} V/cm, well within the range that can be applied to semiconductors will cause large

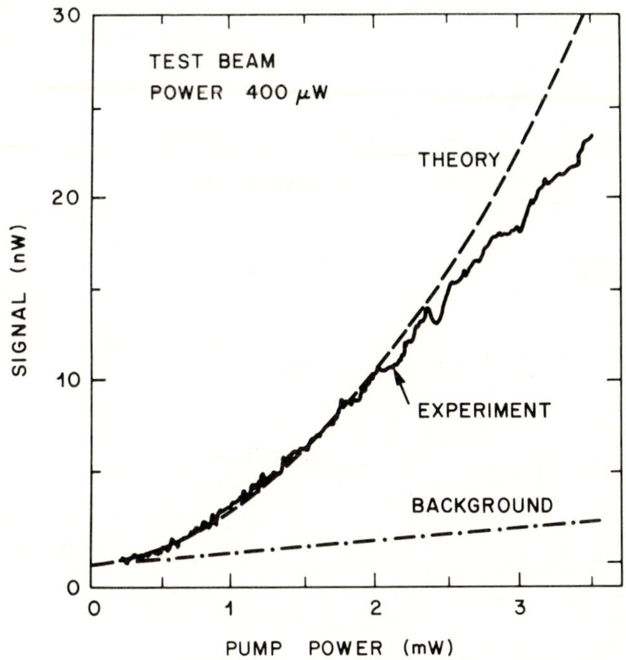

Fig. 7. Degenerate four-wave mixing signal found using a diode laser source on a MQW sample. The dashed line is a theoretical fit showing quadratic fit behavior up to ~2 mW pump power with a small linear background (shown separately as a dotted–dashed line). Saturation of the nonlinearity at higher pump power shows up as a deviation from the theoretical curve.

changes of the absorption spectra [25–28]. Large changes of the luminescence efficiency with applied field have also been observed at low temperatures [29,30].

By symmetry it is expected that fields parallel and perpendicular to the layers will produce different effects. For a field parallel to the layers there is no effect on the confinement but the relative motion of the electron and the hole is perturbed as in the case of the Stark effect, with however some changes due to the lowered dimensionality of excitons in MQWS [31]. For a field perpendicular to the layers the confinement energy of the electron and the hole are affected, and since the two particles are pushed away from one another and squeezed against the walls of the potential well the exciton binding energy is also changed.

The design of samples to distinguish between the two geometries is quite important. In the case of field parallel to the layers photolithography and strong Zn diffusion is used to define proper electrode geometry [32]. The main observed effect is a broadening of the exciton peak which is due to the reduction of the exciton life time by the field ionization. This shown in fig. 10. At an

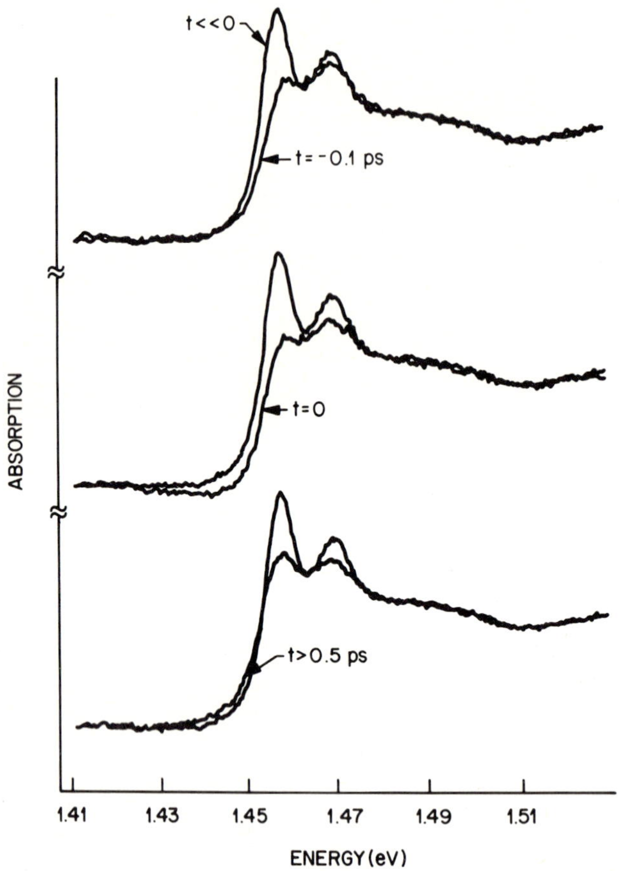

Fig. 8. Comparison of the absorption spectrum of an unexcited MQWS to the spectra after excitation by an ultra short pump pulse at resonance with the heavy hole exciton.

applied field of $1.6 \times 10^{+4}$ V/cm the two peaks have substantially broadened, and they are no more resolvable at $5 \times 10^{+4}$ V/cm. The low field ionization rate has been calculated for two-dimensional excitons by the same method as in the case of the hydrogen atom [32]. It is smaller than in three dimensions because of the increased binding energy and because of other effects related to the lowered dimensionality, the results are in qualitative agreements with the measurements.

To apply fields perpendicular to the layers special samples have been prepared; they consist of a MQWS in the intrinsic region of a p–i–n diode [26]. The p and n regions are made of AlGaAs transparent at the energy of the fundamental gap of the MQWS. The diode is operated back-biased to vary the field across the MQWS and yet avoid large currents and thermal effects.

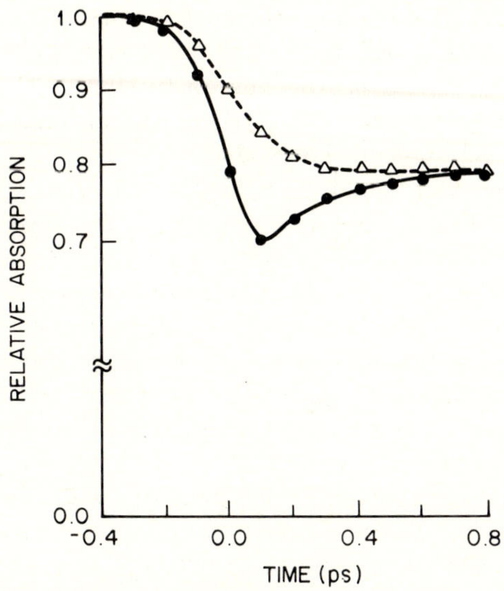

Fig. 9. Dynamic of the absorption of the heavy hole exciton peak for resonant and non-resonant femtosecond pumping.

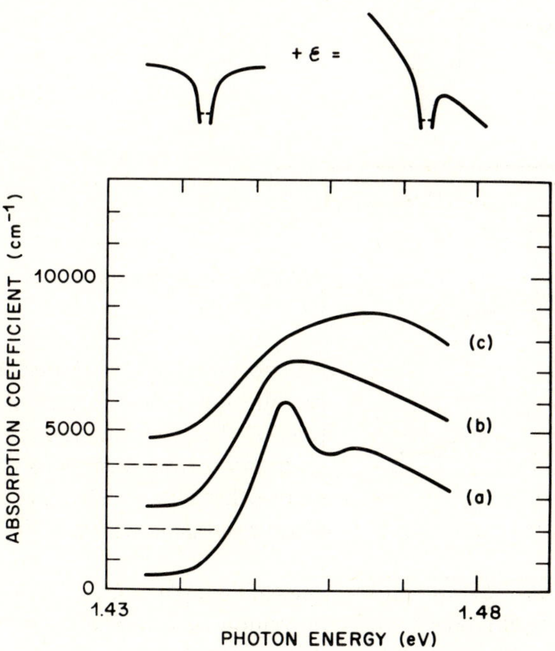

Fig. 10. Absorption spectra for electric fields E parallel to the quantum well layers. (a) $E \simeq 0$; (b) $E \simeq 1.6 \times 10^4$ V/cm; (c) $E \simeq 4.8 \times 10^4$ V/cm. The insert shows figuratively the distortion of the electron–hole Coulomb potential with the applied field.

The dominant effect is a shift of the whole spectrum with a small broadening of the exciton peaks. However these remain well resolved at fields as large as 10^{+5} V/cm, i.e., more than fifty times the classical ionization field. At such field shifts 2.5 times the binding energy are observed. Typical spectra under various static fields are shown in fig. 11. At fields larger than 10^{+5} V/cm the excitons are no longer resolvable. The interpretation of these observations is the following. As the static field is applied the potential seen by the electrons and the holes in the GaAs layers becomes that of a skewed well and the energy levels of the individual particles are shifted down. Simultaneously the wave-functions are compressed against the walls of the well; the charge distribution is thus modified resulting in a variation of the binding energy of the excitons. If the field is further increased, the less-confined particles, i.e., the holes finally tunnel out owing to the finite depth of the well, so that the resonances can no longer be resolved. The results of a calculation based on this model are shown in fig. 12. No adjustable parameters are used and the agreement with the measured shifts is very good [27]. In this calculation the usual discontinuities at the band edges (i.e., $0.15 \times E_g$ for the valence band and $0.85 \times E_g$) for the conduction band have been used [2]; an almost perfect agreement is obtained

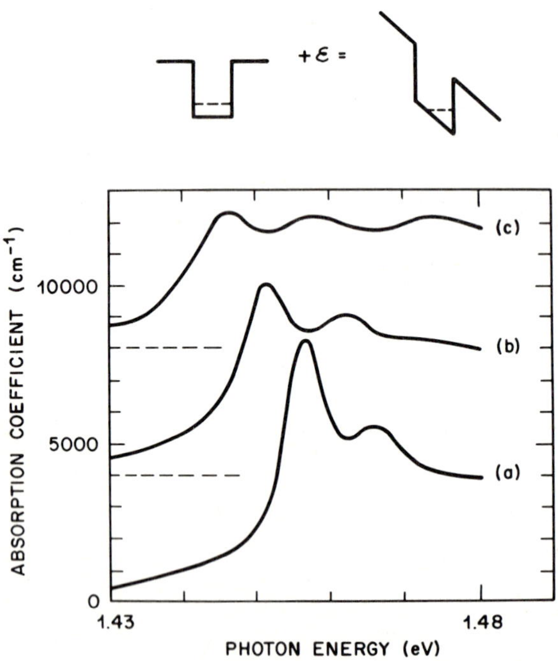

Fig. 11. Absorption spectra for electric fields E perpendicular to the quantum well layers. (a) $E \simeq 1 \times 10^4$ V/cm; (b) $E \simeq 4.7 \times 10^4$ V/cm; (c) $E \simeq 7.3 \times 10^4$ V/cm. The insert shows figuratively the distortions of the quantum well potentials with the applied field.

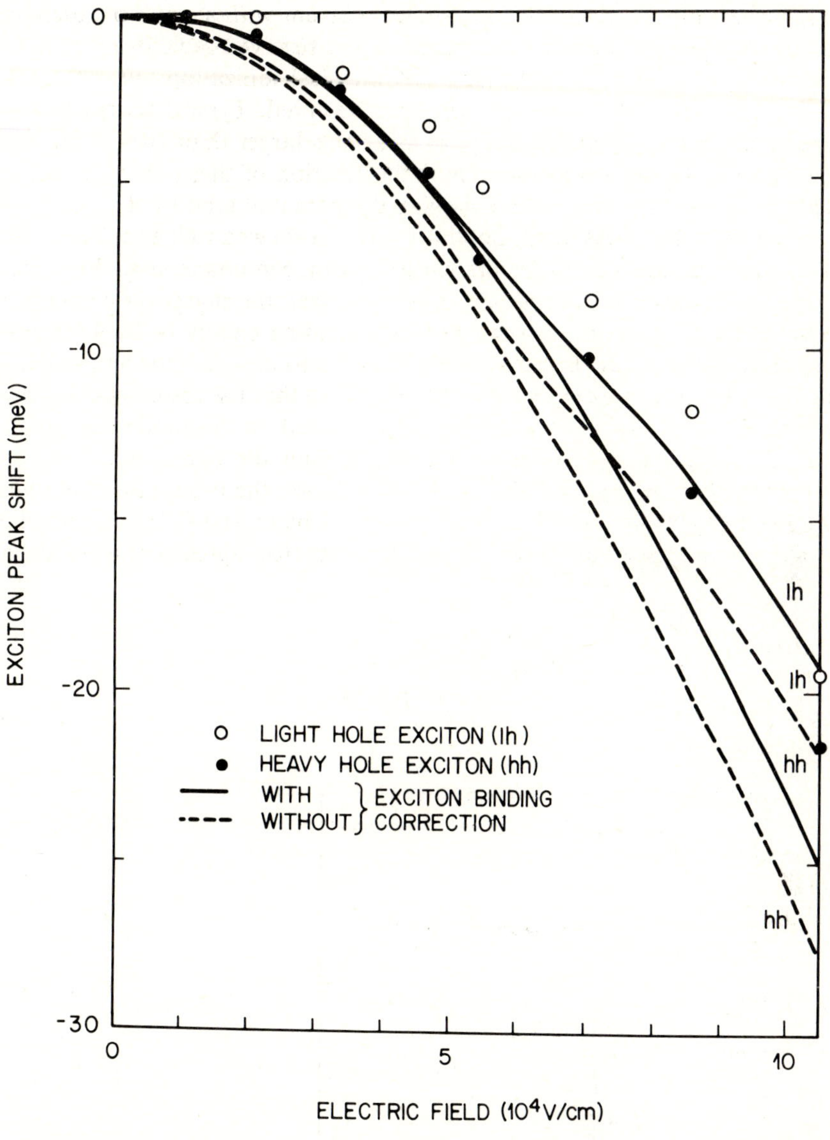

Fig. 12. Shift of the position of light hole (lh) and heavy hole (hh) exciton peaks with applied field perpendicular to 95Å GaAs quantum wells. Points are experimental, lines are theoretical. The dashed lines are calculations including only the shift of the electron and hole subbands. The solid lines include the correction to the exciton binding energy due to the movement of electrons and holes to opposite sides of the wells. There are no fitted parameters in the theory.

if a larger discontinuity at the valence edge is taken ($\sim 0.40 \times E_g$) as recently suggested [33]. The reason why excitons are still observed at fields much larger than the ionization field is that the large potential barriers prevent the particles from escaping from one another. This remarkable property is specific to MQWS; it allows us to move an abrupt absorption edge into a region of the spectra where the sample is transparent in the absence of field.

The intrinsic speed of the effect is very high since it corresponds to the response of the electronic envelope functions. In practice the limits are set by the *RC* constant of the circuit in which the MQWS is incorporated [26]. Recently high speed modulation of laser diode has been achieved using a p–i–n MQWS 0.965 μm thick in form of a 95 μm mesa with a 25 μm optical window. When the device is driven by electrical pulses of -8 V and 125 ps duration, optical pulses ~ 131 ps are produced [28]. A schematic of the device and the optical

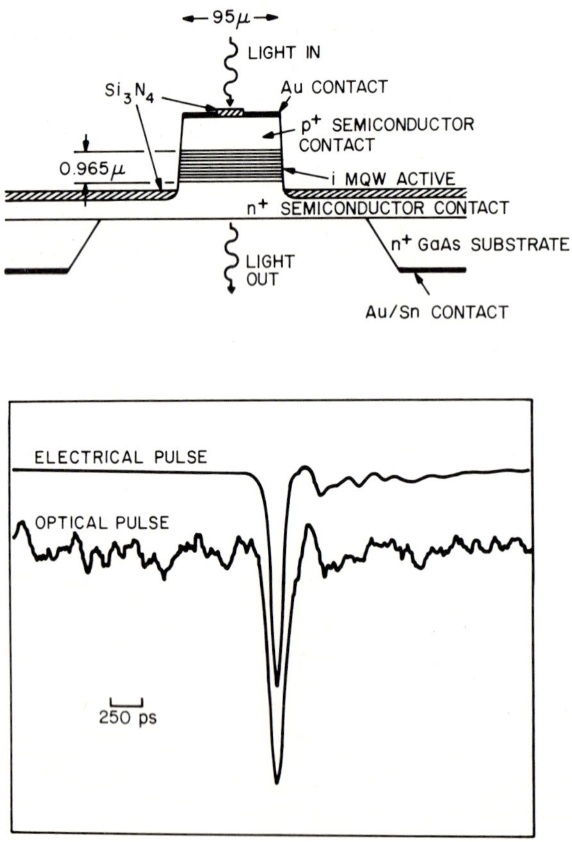

Fig. 13. Schematic of the p–i–n MQWS modulator. Comparison of the electric pulse applied to the modulator and the optical pulse generated by the modulator.

response compared to the electrical excitation pulse are shown in fig. 13. A modulation of 2.3 dB was measured. The overall response time of the device is well accounted for by the characteristics of the electric circuit. Deeper modulation and faster responses should be obtained with thicker MQWS and smaller device area.

A bistable optical switch based on a novel principle, has been demonstrated using the p–i–n MQWS devices; it is called Self-Electro-Optic-Effect-Device or SEED [34]. It presents several attractive features: it requires no cavity, it has a very small switching energy, it operates over a broad band of wavelengths and eventually with incoherent light sources. When a reverse bias is applied to the p–i–n MQWS, the structure also acts as an efficient photodetector. The bistable optical switch consists of a series resistor and a constant voltage supply that are used to bias a p–i–n MQWS. A light beam at a wavelength close to the peak of the exciton at zero bias is sent through the sample and the bias is applied. The absorption spectrum is shifted so that the transmission is high. As the intensity of the incident beam is increased, the photocurrent passing through the resistor induces a reduction of the voltage applied to the device, which in turn increases the absorption by shifting the spectrum thus further increasing the photocurrent and so on. This process provides the feedback mechanism for optical bistability [34]. An example of the optical hysteresis loop obtained with a SEED is shown in fig. 14. The SEED used in these experiments was rather large (about 600 μm in diameter), nevertheless the measured switching energy per unit area is remarkably low, $W_{so} \sim 4$ fJ/μm^2

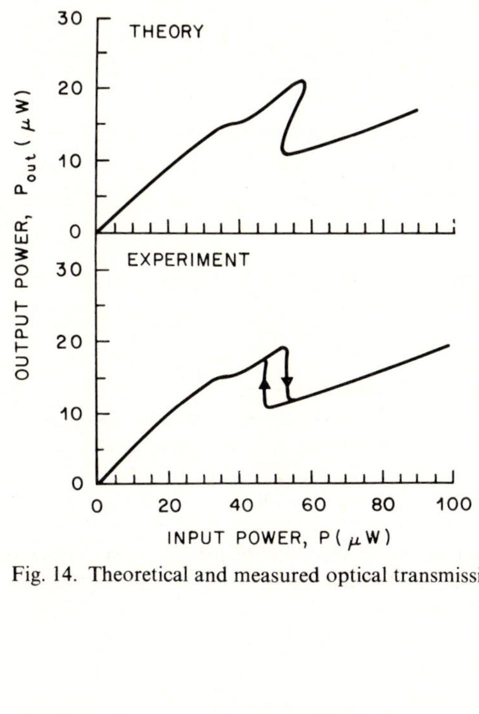

Fig. 14. Theoretical and measured optical transmission of the optical bistable SEED.

for optical incident energy and $W_{st} \lesssim 18$ fJ/μm^2 for the total switching energy (electrical plus optical) substantially lower than any previously reported bistable optical device. Even lower switching energy should be possible for smaller devices, the physical limit for a $(\lambda/n)^2$ device is ~ 1 fJ/μm^2, comparable to that of electronic switches.

Conclusion

We have presented some recent progress made in the investigation and the utilization of the excitonic resonances observed at room temperature in GaAs/AlGaAs multiple quantum well structures. A number of novel nonlinear optical and electro-optical effects are observed which result from the lowered dimensionality of the electrons in ultra-thin semiconductor layers. Room temperature excitonic peak are not only interesting for the applications, they also exhibit the most unusual properties owing to their extremely short life time and their transformation into free e–h pairs.

Finally, we would like to stress that although the GaAs/AlGaAs alloys form a model system from which extremely high quality MQWS have already been made, MQWS made with other semiconductors should soon be available with comparable quality. In addition, quantum wells represent only one type of microstructure in which quantum size effects are expected, soon it will be possible to prepare novel microstructures and investigate their even more exciting physical properties.

Acknowledgements

The work that has been reviewed in this article has been performed at the AT&T Bell Laboratories in close collaboration with D.A.B. Miller, A.C. Gossard, P.W. Smith, T.C.Damen, T.H. Wood, C.A. Burrus, W.H. Knox, R.L. Fork and C.V. Shank.

References

[1] D.S. Chemla and A. Maruani, Progr. Quant. Electron. 8 (1982) 1.

[2] R. Dingle, Confined carrier quantum states in ultra-thin semiconductor heterostructures, in: Festkörperprobleme XV, ed., H.J. Queiser (Pergamon, Vieweg, Braunschweig, 1975).

[3] R.C. Miller, D.A. Kleinman, W.T. Tsang and A.C. Gossard, Phys. Rev., B24 (1981) 1134.

[4] D.A.B. Miller, D.S. Chemla, P.W. Smith, A.C. Gossard and W.T. Tsang, Appl. Phys. B28 (1982) 96.

[5] A.C. Gossard, Molecular beam epitaxy of superlattices in thin films, in: Thin Films Preparation and Properties, eds., K.N. Tu and R. Rosenberg (Academic Press, New York, 1983).

[6] D.S. Chemla, D.A.B. Miller, P.W. Smith, A.C. Gossard and W. Wiegmann, IEEE JQE 20 (1984) 265.
[7] G. Bastard, E.E. Mendez, L.L. Chang and L. Ezaki, Phys. Rev. B26 (1982) 1974.
[8] R.L. Greene and K.K. Bajaj, Solid State Commun. 45 (1983) 831.
[9] D.A.B. Miller, D.S. Chemla, D.J. Eilenberger, P.W. Smith, A.C. Gossard and W.T. Tsang, Appl. Phys. Lett. 41 (1982) 679.
[10] Y. Ishibashi, S. Taruitha, H. Okamoto, Inst. Phys. Conf. Ser. No. 63, Proc. Int. Symp. GaAs and Related Compounds, Japan 1981, pp. 587–588 (1982).
[11] S.W. Kirchoefer, N. Holonyak, K. Hess, D.A. Gulino, H.G. Drickamer, J.J. Coleman and P.D. Dapkus, Appl. Phys. Lett. 40 (1982) 821.
[12] C. Wiesbuch, R.C. Miller, R. Dingle and A.C. Gossard, Solid State Commun. 37 (1981) 219.
[13] R.C. Miller, D.A. Kleinman, W.A. Norland and A.C. Gossard, Phys. Rev. B22 (1980) 863.
[14] C. Weisbuch, R. Dingle, A.C. Gossard and W. Wiegmann, J. Vac. Sci. Technol. 17 (1980) 1128.
[15] V.I. Alperowich, V.M. Zalekin, A.F. Kranchinko and A.S. Terekhev, Phys. Status Solidi B77 (1976) 466.
[16] H. Haug and S. Schmitt-Rink, Progr. Quant. Electron. (in press).
[17] See for example the article by H. Haug in these proceedings and the references therein.
[18] D.A.B. Miller, D.S. Chemla, D.J. Eilenberger, P.W. Smith, A.C. Gossard and W. Wiegmann, Appl. Phys. Lett. 42 (1983) 925.
[19] D.A.B. Miller, D.S. Chemla, P.W. Smith, A.C. Gossard and W. Wiegmann, Opt. Lett. 8 (1983) 477.
[20] See for example H.M. Gibbs, S.S. Tarng, J.L. Jewell, D.A. Weinberger, K. Tai, A.C. Gossard, S.L. McCall, A. Passner and W. Wiegmann, Appl. Phys. Lett. 41 (1982) 221.
[21] Y. Sipberberg, P.W. Smith, D.J. Eilenberger, D.A.B. Miller, A.C. Gossard and W. Wiegmann (submitted to Optic Letters).
[22] R.L. Fork, C.V. Shank, R. Yen, C.A. Hirlimann, IEEE JQE 19 (1983) 500.
[23] W.H. Knox, R.L. Fork, M.C. Downer, D.A.B. Miller, D.S. Chemla, C.V. Shank, A.C. Gossard and W. Wiegmann, Proc. Ultrafast Phenomena, Monterey, CA (1984).
[24] W.H. Knox, R.L. Fork, M.C. Downer, D.A.B. Miller, D.S. Chemla, C.V. Shank, A.C. Gossard and W. Wiegmann (to be published).
[25] D.S. Chemla, T.C. Damen, D.A.B. Miller, A.C. Gossard and W. Wiegmann, Appl. Phys. Lett. 42 (1983) 864.
[26] T.H. Wood, C.A. Burrus, D.A.B. Miller, D.S. Chemla, T.C. Damen, A.C. Gossard and W. Wiegmann, Appl. Phys. Lett. 44 (1984) 16.
[27] D.A.B. Miller, D.S. Chemla, T.C. Damen, A.G. Gossard, W. Wiegmann, T.H. Wood and C.A. Burrus (to be published).
[28] T.H. Wood, C.A. Burrus, D.A.B. Miller, D.S. Chemla, T.C. Damen, A.C. Gossard and W. Wiegmann (to be published).
[29] E.E. Mendez, G. Bastard, L.L. Chang, L. Esaki, H. Morkoc and R. Fisher, Phys. Rev. B26 (1982) 7101.
[30] R.C. Miller and A.C. Gossard, Appl. Phys. Lett. 43 (1983) 954.
[31] J.D. Dow and D. Redfield, Phys. Rev. B1 (1970) 3358.
[32] D.A.B. Miller, D.S. Chemla, T.C. Damen, A.C. Gossard, W. Wiegmann, T.H. Wood and C.A. Burrus (to be published).
[33] R.C. Miller, A.C. Gossard, D.A. Kleinman and O. Munteanu, Phys. Rev. B29 (1984) 3740.
[34] D.A.B. Miller, D.S. Chemla, T.C. Damen, A.C. Damen, A.C. Gossard, W.Wiegmann, T.H. Wood and C.A. Burrus, Appl. Phys. Lett. 45 (1984) 13.

Journal of Luminescence 30 (1985) 520–540
North-Holland, Amsterdam

EXCITONS IN GaAs QUANTUM WELLS

R.C. MILLER and D.A. KLEINMAN

AT&T Bell Laboratories, Murray Hill, New Jersey, 07974, USA

This paper attempts to summarize some of the salient properties of excitons in GaAs quantum wells and in doing so it will emphasize work at AT&T Bell Labs with which the authors have been associated. Although the text relies heavily on published material, an effort has been made to stress new material, and where feasible, unpublished aspects, e.g., figures, related to earlier work. Topics discussed on the quasi-2D excitons in GaAs quantum wells include: their inherent tendency for intrinsic free-exciton emission, exciton binding energies, bound and localized excitons including biexcitons and excitons bound to neutral impurities, effects of n- and p-type modulation and antimodulation doping, and the developments leading to a proposed set of quantum well parameters that results in acceptable fits to the observed exciton transitions for GaAs quantum wells with both square and parabolic potential profiles.

1. Introduction

It was evident from the earliest work by Dingle et al. [1] on the optical properties of GaAs quantum wells grown by molecular beam epitaxy (MBE) with the GaAs–AlGaAs system that excitons play a much more significant role in these quasi-2D systems than in bulk GaAs. The photoluminescence at low temperatures from a few microns of high-quality MBE-grown undoped GaAs is usually characterized by emission at $\approx$1.49 eV due to donor–acceptor pair recombination (D^0–A^0) and free electron–neutral acceptor recombination (e–A^0) plus emission at the exciton edge, $\cong$1.515 eV, due mainly to excitons bound to D^0 and A^0, and hole–neutral donor (h–D^0) recombination. Small contributions due to free-exciton (polariton) emission are also frequently observed. However, the main point is that for the bulk GaAs grown by MBE the emission at low excitation levels is usually predominantly extrinsic in nature [2].

This is in marked contrast to GaAs quantum wells with similar purity to the bulk GaAs material mentioned above. Absorption, emission, and excitation spectra of high-quality GaAs quantum wells with good interfaces demonstrate that intrinsic free-exciton recombination dominates the emission even at very

0022-2313/85/$03.30

low temperatures where bound exciton and impurity processes dominate for the bulk GaAs [2,3].

Extensive absorption spectra obtained by Dingle et al. [1,4] on multiquantum well samples from which the GaAs substrate had been etched off (window samples) revealed sharp peaks at the onset of the band to band transitions showing that exciton effects are also pronounced in absorption. A classic example of this is shown in fig. 12 of ref. [4]. In the absence of such exciton effects, the onset of the various band-to-band transitions should be step-like reflecting the step-like changes in the 2D density-of-states. Excitation spectra taken later also lead to the same conclusions [3,5].

The splitting of the heavy and light hole bands in GaAs quantum wells due to their different effective masses leads to well resolved heavy and light hole exciton transitions in absorption [4] and emission [3,5]. The assignments of these transitions were made initially on the basis of the good agreement between the observed and calculated transition energies. However, linear and circular optical polarization measurements provided direct confirmation of these light and heavy hole assignments [3,5].

Since optical electron spin orientation using circular polarization excitation and detection techniques [6] will be mentioned frequently throughout this paper, some of the underlying principles will be discussed briefly with the aid of fig. 1(b). The relevant circular polarization phenomena are determined by the relative transition strengths (3 and 1), the m_j values ($\pm 3/2$ and $\pm 1/2$), and the effective hole masses $m_h^* = 0.45m_0$ [7] and $m_\ell^* = 0.088m_0$ [8] for the heavy and light holes, respectively. The absorption and emission processes considered are for detection and excitation at normal incidence to the plane of the layers. Since the electron states are $m_j = \pm 1/2$, the resonant absorption of $\sigma^+ (\Delta m_j = +1)$ polarized light by heavy and light hole transitions will generate electrons with $m_j = -1/2$ and $+1/2$, respectively. With $m_h^* > m_\ell^*$, the emission at low temperatures will be via the heavy hole ground state. Thus assuming a spin relaxed hole population [9] and incomplete electron spin relaxation, resonant excitation of the lowest energy heavy or light hole exciton transitions should lead to emission from the ground-state heavy hole exciton polarized σ^+ and σ^-, respectively. Likewise, at elevated temperatures where the lowest light hole level is thermally populated, the above resonant excitation conditions should give rise to emission from the ground-state light hole exciton with polarization σ^- and σ^+, respectively. As there is always some relaxation of the electron spins, the luminescence is never 100% polarized. In any event, a decreased, or a negative polarization, i.e., a polarization opposite to the incident polarization, is the hallmark of light hole transitions in excitation with detection set at the heavy hole emission which is the usual way of obtaining an excitation spectrum. For nonresonant excitation, σ^+ emission at low temperatures dominates since the heavy hole transitions are 3 times the strength of the light hole transitions. Thus circular polarization techniques can be very useful in identifying transitions

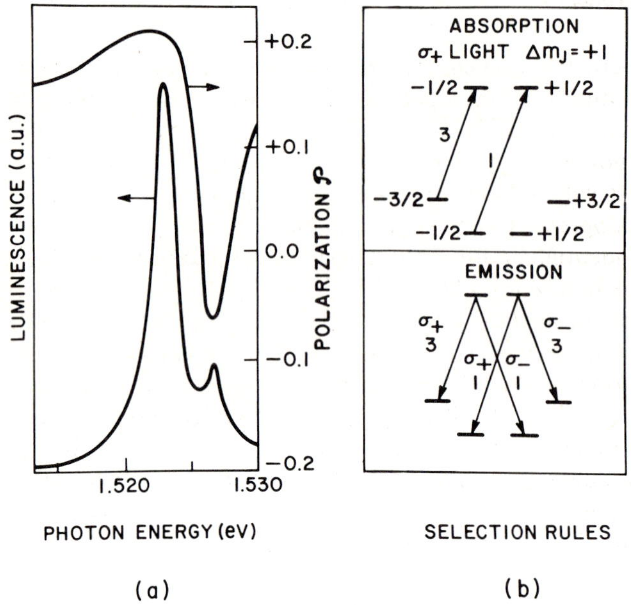

Fig. 1. (a) Photoluminescence and its circular polarization at 50 K for excitation at 1.65 eV with circular polarized light. The sample has 25 GaAs wells with $L = 188$ Å and 19 Å $x = 0.3$ alloy barriers. For this nonresonant excitation, $m_j = -1/2$ electron states are preferentially created which leads to heavy and light hole exciton emission with positive and negative polarizations, respectively. (b) Absorption and emission transitions for GaAs quantum wells with circular polarized light. The transitions involving heavy ($m_j = \pm 3/2$) and light ($m_j = \pm 1/2$) holes have relative strengths of 3 and 1, respectively.

(fig. 1(a)) and indeed have been utilized to verify the original assignments by Dingle et al. [3,5]. Although not discussed herein, the decrease in polarization of the emission with a transverse magnetic field (Hanle effect) can be used to estimate the electron lifetime and the electron spin relaxation time [5,6].

The intrinsic free-exciton nature of the recombination mentioned above for GaAs quantum wells is deduced largely from the observed small Stokes shift (a fraction of the line width) between the heavy hole photoluminescence emission and the heavy hole ground-state exciton absorption peak [3,5]. Likewise, when the photoluminescence emission investigated involves the light hole, there is at most a small Stokes shift between emission and absorption. Weak extrinsic photoluminescence is frequently observed from undoped GaAs quantum wells of the same quality as that mentioned above [10]. The most easily identified extrinsic emission has been that due to e–A^0 where carbon is thought to be the neutral acceptor.

2. Exciton binding energies

Dingle et al. [1,4] were able to estimate the binding energy of the quasi-2D excitons for GaAs quantum wells of width L by extrapolating to zero the confinement energies of the various exciton peaks measured in absorption. In this manner, the exciton binding energy was estimated to be ≈ 9 meV for $L \lesssim 100$ Å. This is to be compared with 4.2 meV for bulk GaAs. However, a number of GaAs quantum well samples showed explicit structure in the excitation spectrum which could be identified with excited and/or continuum states of the 2D excitons. One of the best examples of this is shown in fig. 2 which gives the excitation spectrum for an unusually good single well sample with $L = 50$ Å and alloy barriers of width $L_b = 0.65$ µm with $x = 0.37$. Throughout this work, allowed transitions involving electrons and holes of quantum number n, where $n = 1, 2, 3$, etc. with $\Delta n = 0$, will be identified by E_{ni} where $i = h$ for heavy holes and ℓ for light holes. Forbidden transitions where $\Delta n \neq 0$ will be identified by E_{ijk} where i is the electron quantum number, j the hole quantum number, and k the type of hole, either h or ℓ. Thus the parity allowed $\Delta n = 2$ transition between the $n = 1$ electron and $n = 3$ heavy hole is designated E_{13h}.

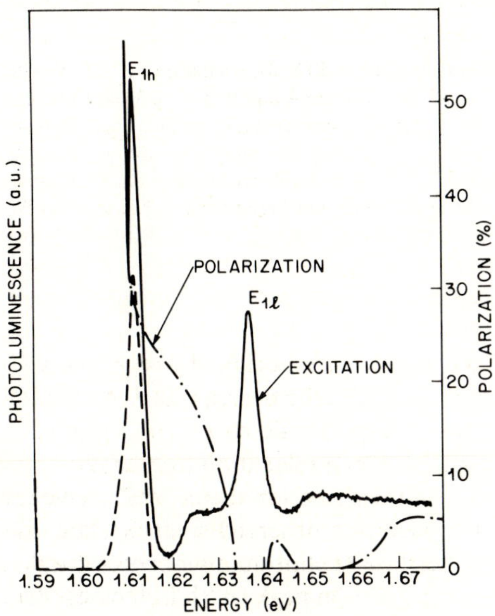

Fig. 2. Photoluminescence (dashed), excitation (solid) and smoothed polarization (dashed–dot) spectra at 5 K for a single GaAs well sample with $L = 50$ Å. For the excitation spectrum detection was at 1.611 eV. The polarization actually goes negative in the 1.637 eV and 1.652 eV spectral regions. The data show that the shoulders in the photoluminescence at 1.624 eV and 1.652 eV are heavy and light hole character, respectively. It is proposed that these shoulders represent the onset of excited exciton states.

Of most interest in fig. 2 are the distinct shoulders on the high-energy side of the E_{1h} and $E_{1\ell}$ exciton peaks which the polarization shows are heavy and light hole character, respectively. It is concluded that these shoulders represent the onset of heavy and light hole excited states and that the discrete excited states in this case are not resolved from the continuum. Also it should be noted that the luminescence level at the light hole continuum edge is about 4/3 that at the heavy hole continuum edge which is consistent with heavy and light hole band relative strengths of 3 and 1, respectively. The apparent continuum edge is taken to be the 2S excited state of the exciton. In this interpretation the true continuum edge corresponding to the ionization threshold is not seen directly in these spectra. The midpoint of the rising portion of the shoulder is taken as the 2S level while the 1S level is the exciton peak itself. This therefore provides a direct measurement of the term value B_{1S}–B_{2S}. Since the in-plane heavy hole mass is less than that of the light hole, the binding energy of the light hole exciton is expected to be greater than that of the heavy hole exciton as is observed [11]. Similar results have been obtained in the same manner for samples with L ranging 42 Å to 340 Å. However for $L = 340$ Å, a distinct peak in the excitation spectrum is attributed to the 2S heavy hole exciton excited state. The term values B_{1S}–B_{2S} increase with decreasing L and for example, for $L = 50$ Å (fig. 2), $B_{1S}(\ell)$–$B_{2S}(\ell) = 10.7$ meV.

These data have been compared with a variational calculation of the exciton binding B_{1S} and the term value $B_{1S} - B_{2S}$ as a function of L for GaAs quantum wells with infinite potential barriers [11]. The formulation contains no adjustable parameters other than the variational parameter and it is exact for $L \to 0$. Since the calculated term values are found to be in good agreement with the data, the calculated B_{1S} values can be considered reliable. It is interesting to note that the 9 meV value for $B_{1S}(\mathrm{h})$ estimated by Dingle et al. [4] is in good agreement with these results. The integrated strengths of the light and heavy hole excitons were also calculated and found to be in reasonable agreement with the data where the continuum thresholds were sufficiently well resolved [11]. The calculated strength increases by about a factor of 2 as L decreases from 140 Å to 42 Å.

This calculation [11], and a subsequent one by Bastard [12], assume no penetration into the alloy barriers, i.e., infinitely high potential wells. Greene et al. [13] have taken finite barriers into account and show that, as expected, the B_{1S} values are reduced for small L due to penetration. As a result, the B_{1S} values from Greene et al. for small L are somewhat less than those determined with no penetration and hence lead to $B_{1S} - B_{2S}$ term values somewhat less than the data as interpreted in ref. [11]. Also Greene et al. find that the effects due to penetration for the light hole exciton are sufficiently more pronounced than those for the heavy hole exciton to give $B_{1S}(\ell) < B_{1S}(\mathrm{h})$ for $L < 50$ Å and $x = 0.3$ alloy barriers. The data in fig. 2 for $L = 50$ Å and $x = 0.37$ alloy barriers, suggest that $B_{1S}(\ell) > B_{1S}(\mathrm{h})$.

3. Bound and localized excitons

3.1. Biexcitons

Biexcitons are perhaps the simplest form of bound excitons and there are data that suggest that this species does indeed exist at low temperatures ($\approx$5 K) in high-quality GaAs quantum wells [14]. Samples grown under conditions which lead to intense, sharp, photoluminescence peaks, and excitation spectra which indicate abrupt interfaces and wells of uniform thicknesses usually exhibit a luminescence peak in the region of the free exciton which consists of two sharp components. An example of this is shown in fig. 3. At the lowest excitation intensities, $I_p \approx 1$ mW cm^{-2}, one peak is observed which excitation spectra show is due to the E_{1h} free exciton. As I_p is increased, a shoulder appears on the low-energy side, and finally at sufficiently high I_p, two distinct peaks separated by $\approx$1 meV are observed. The splitting and the energy of the E_{1h} peak are independent of I_p. When T is increased to $\approx$11 K, $kT \approx 1$ meV, the free-exciton peak always dominates and its intensity is enhanced. This temperature dependence was observed for all samples which exhibited the two sharp

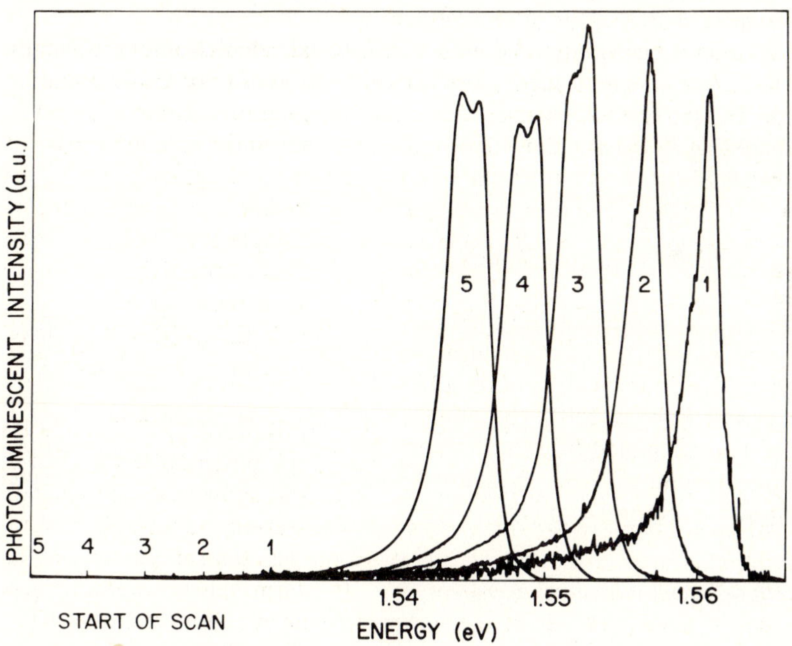

Fig. 3. Photoluminescent intensity at 5 K from a sample with 81 Å GaAs quantum wells and 200 Å $x = 0.26$ alloy barriers versus emission energy for excitation at 1.695 eV with 1.4 mW/cm^2 (1), 3.1 mW/cm^2 (2), 21.2 mW/cm^2 (3), 45.3 mW/cm^2 (4), and 187 mW/cm^2 (5). The peak at lower energy attributed to biexcitons comes in clearly at about 20 mW/cm^2.

peaks. Also, the change from one peak to two peaks with increasing I_p was observed with samples where L ranged from 81 Å to 327 Å. The "threshold" I_p for the second peak tends to decrease with increasing L, and two samples with $L > 327$ Å exhibited two peaks even at the lowest intensities, ≈ 1 mW/cm^2.

With excitation by polarized light, a pronounced minimum in the polarization at the lower-energy peak has been observed [14]. This observation is consistent with zero polarization for the process responsible for this peak. If this peak is indeed due to biexcitons as proposed, it would be expected to radiate with zero polarization. However the fact that this distinct minimum in the polarization is not always observed was tentatively attributed to the need for a detailed line-shape analysis.

Thus, the dependence of the lower-energy peak on excitation intensity, temperature, and polarization is consistent with a 2D biexciton with a binding energy of ≈ 1 meV. Kleinman [15] has made a variational calculation of the binding energy B_{xx} of the biexciton in GaAs quantum wells. This calculation contains no adjustable parameters other than the variational parameters and makes use of the six-parameter wavefunction of Brinkman, Rice, and Bell which yields $B_{xx} = 0.13$ meV for bulk GaAs [16]. For GaAs quantum wells it is found that the calculated biexciton binding energy obeys Haynes's rule [17] with a Haynes constant $f_H \cong 0.1$, i.e., $B_{xx} \cong 0.1 B_{1S}(h)$, and that in the 2D limit B_{xx}/B_{1S} is 3–4 times larger than the corresponding quantity in the 3D case. The calculated values of B_{xx} are in good agreement with the ≈ 1 meV splittings mentioned above for $L < 250$ Å. However for $L > 250$ Å, the observed splittings are too large, suggesting as noted below, that some other mechanism might be relevant for this range of L.

3.2. Excitons bound to neutral donors

Kleinman has also estimated the binding energy B_{XD} of excitons bound to neutral donors for the GaAs quantum well 2D system [15]. These values are larger than B_{xx} and are also found to obey Haynes's rule with $f_H \cong 0.1$ as above for biexcitons. For $L > 250$ Å the observed splittings discussed above tend to agree with the calculated B_{XD} but this agreement should be viewed with caution since the development of this "biexciton" peak with increasing I_p is not expected for excitons on neutral centers.

3.3. Excitons bound to neutral acceptors

A photoluminescence peak attributed to ground-state E_{1h} excitons bound to neutral acceptors (E_{1h}–A^0) has been observed from GaAs quantum wells doped with [Be] $\approx 10^{17}$ cm^{-3} or zinc diffused [18]. Both single and multiple quantum wells have exhibited this extrinsic luminescence. In the four cases where this luminescence peak has been reported the wells were p-type doped, free E_{1h} excitons were in evidence, the luminescence peak exhibited evidence of saturation, and it decreased rapidly in intensity with increasing T. Also the Stokes shift of this peak from E_{1h} was too small to be attributed to e–A^0 recombination which in fact could be identified as additional structure in the photoluminescence. The dissociation energy E_d of the E_{1h}–A^0 complex into A^0 and a free E_{1h} exciton was obtained directly from the observed separation of this extrinsic peak from the intrinsic E_{1h} peak. For GaAs quantum wells with $L=46$ Å, $E_d=6.5$ meV. It was noted that E_d decreases with increasing L obeying Haynes's rule with $f_H \approx 0.13$. No convincing evidence of this E_{1h}–A^0 emission had been seen in the 100 or so undoped GaAs quantum well samples examined. The photoluminescence from two quantum well structures similar to the Be-doped ones mentioned above but with [Be]= $\approx 3\times10^{16}$ cm^{-3} exhibited two e–Be^0 peaks, a strong E_{1h} peak, but no discernible E_{1h}–Be^0 recombination peak. Thus in contrast to bulk GaAs, [Be]$\cong 10^{17}$ cm^{-3} would seem to be required with the GaAs quantum wells studied to produce an easily observed E_{1h}–Be^0 luminescence. This is regarded as evidence that E_{1h} excitons in GaAs quantum wells do not migrate easily so that only when the average separation of the Be^0 centers is comparable to the exciton diameter $\cong 300$ Å [11] will this emission be observed.

3.4. Excitons localized by weak disorder

The exciton transition at low temperature in relatively pure high quality GaAs quantum wells often exhibits a strong resonant Rayleigh scattering [19]. When the incident laser frequency is well outside the exciton resonance a weak Rayleigh scattering is observed due to defect scattering. When it is close to the resonance the Rayleigh scattering increases by as much as 200 times the defect scattering. The scattering has the same spectral and temporal profile as the laser light, and therefore can be distinguished from the ordinary luminescence. Hegarty et al. [19], have attributed this scattering to spatial fluctuations of the exciton resonance frequency due to fluctuations of the effective well width L. We note that fluctuations in L which are multiples of half the lattice constant can arise from a step-like growth during MBE, while smaller fluctuations with a continuous distribution (weak disorder) may arise from random concentration fluctuations of Al in the barriers. On the basis of an assumed Lorentzian model for the exciton line shape, Hegarty et al. [19] have analyzed the absorption and resonant Rayleigh scattering spectra of several samples to obtain the frequency variation of the homogeneous width across the exciton line. The results show that the broadening is inhomogeneous on the low-frequency side of the absorption peak, but the homogeneous contribution rises (and the inhomogeneous contribution falls) rapidly with increasing frequency near the absorption peak. Also the homogeneous part increases with increasing temperature. These results are consistent with the following picture of the free-exciton transition: (a) the energy spectrum is continuous, but due to the weak disorder there exists a "mobility edge" near the absorption peak separating localized states below the edge from delocalized states above the edge; (b) below the edge the line shape (inhomogeneously broadened) represents the density of localized states; (c) above the edge the line shape (homogeneously broadened) represents the scattering of freely migrating excitons by the weak disorder.

4. Excitons in modulation doped GaAs quantum wells

It is to be expected that quasi-2D electron or hole gases in sufficient concentration in GaAs quantum wells will affect the exciton characteristics of these structures. Some interesting effects on the quantum-well excitons believed due to charge carriers was noted during studies of single GaAs quantum wells grown by metalorganic chemical vapor deposition (MOCVD) [20]. While the better samples grown by MOCVD were comparable exciton-wise to the best MBE-grown material, many exhibited an anomalous excitation spectrum. In particular, the E_{1i} transitions were weakened compared to E_{2h}, especially E_{1h}, all exciton peaks were broadened, the emission was Stokes shifted from E_{1h}, and the photoluminescence increased gradually as the excitation photon energy E_p approached $E_{1\ell}$ from the high-energy side. In some cases, as shown in fig. 4, the E_{1h} transition was missing altogether while the other exciton transitions including $E_{1\ell}$ were very much in evidence. Furthermore as T was increased from 5 K to 77 K a distinct peak due to E_{1h} appeared in the excitation spectrum near the expected energy. These observations led to the speculation the anomalous behavior may be due to holes in the GaAs quantum wells [20].

4.1. p-type modulation doped quantum wells

Of much interest from a device as well as a physics point of view are the "modulation doped" structures in which the n- or p-type dopants are restricted during growth to the alloy barriers [21]. In order to maintain a constant Fermi level in such a quantum well structure, the dopants in the alloy barriers ionize with the free carriers migrating to the lower-energy undoped quantum wells. Thus one has a situation where the ionized impurities are spatially separated from the carriers and the resultant carrier mobilities high [21].

Figure 5 shows the photoluminescence spectra for a p-type modulation doped sample grown by MBE with 20 periods of $L = 115$ Å GaAs wells and 150 Å $x = 0.44$ alloy barriers. The middle 50 Å of each barrier was doped with [Be] $\approx 10^{18}$ cm^{-3} which led to a 2D-concentration $N_h = 5.4 \times 10^{10}$ cm^{-2} at 5 K. The excitation spectrum for this sample is anomalous in the same sense as

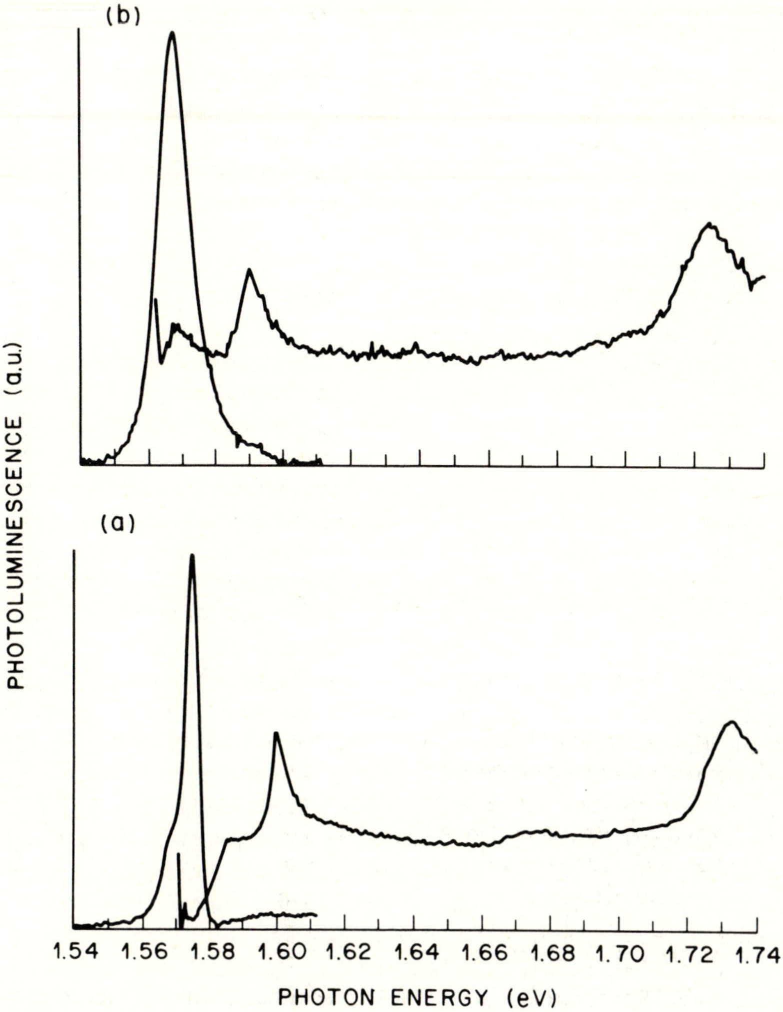

Fig. 4. Photoluminescence and excitation spectra from a single GaAs quantum well with $L=70$ Å grown by metalorganic chemical vapor deposition. In (a) at 6 K, the E_{1h} peak is missing while $E_{1\ell}$ (1.60 eV) and E_{2h} (1.73 eV) are clearly seen in the excitation spectrum. At 77 K (b) a peak attributed to E_{1h} appears (1.567 eV) and is shifted from the free exciton emission peak by $\cong 1$ meV. The detector gain in (b) is four times that of (a).

described above for the MOCVD single-well samples. Such characteristics have been observed with other Be-modulation doped multiquantum well samples with $N_h \approx 10^{11}$ cm^{-2}.

Another sample also grown by MBE had 20 periods of $L=90$ Å GaAs wells and 550 Å $x=0.45$ alloy barriers with the center 43 Å of each doped with Be=

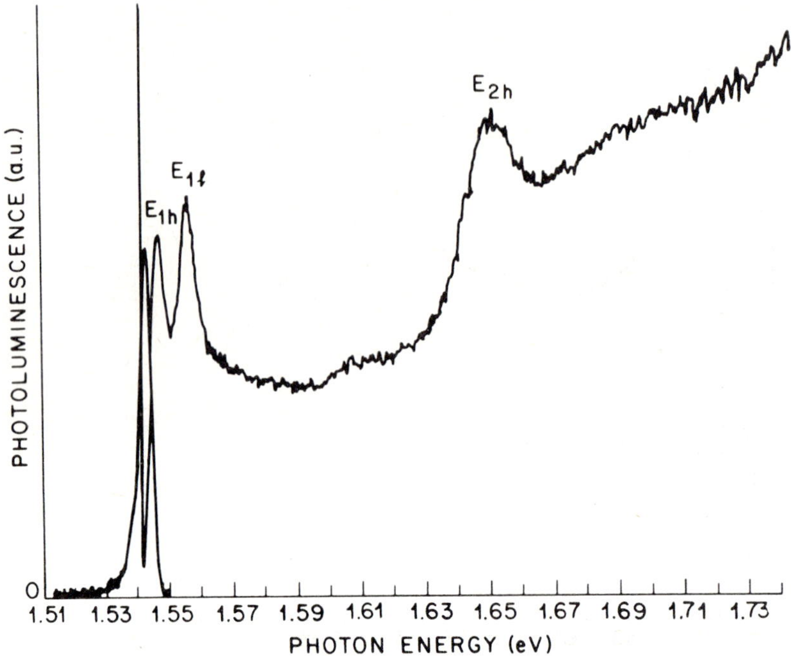

Fig. 5. Photoluminescence and excitation spectra at 5 K for a p-type modulation doped multi-quantum well sample with 115 Å wells and 150 Å $x=0.44$ alloy barriers. The low-temperature 2D carrier density is $N_h=5.4\times10^{10}$ cm^{-2}. The main effects on the excitons due to the holes are in the region of E_{1h} and $E_{1\ell}$ as discussed in the text.

2×10^{18} cm^{-3} to yield $N_h=5.3\times10^{11}$ cm^{-2} at low temperatures. In this case, as shown in fig. 6, at 5 K the E_{1h} transition is missing and the circular polarization shows that the lowest energy peak is light hole in character. However, as T is increased a shoulder appears on the low-energy side of the peak assigned to $E_{1\ell}$ and finally at 43 K a distinct peak due to E_{1h} is seen as shown in fig. 7. There is a sizeable Stokes shift between the emission and the absorption peak due to E_{1h} contributions.

It is most interesting that the E_{1h} exciton peak can be completely absent from the excitation spectrum at low temperatures while other excitons, both light ($E_{1\ell}$) and heavy excitons (E_{2h}), still retain significant strength. The absence of E_{1h} at low temperatures and its reappearance as T is increased has been observed for a number of different samples and for N_h as low as $\approx 3\times10^{11}$ cm^{-2}. Screening of the excitons by the holes would seem to be ruled out since screening should affect all excitons in roughly the same manner which is clearly not the case. It has been suggested [22] that at low temperatures heavy hole band filling prevents the formation of E_{1h} excitons due to the Pauli exclusion principle, i.e., the hole states required to form excitons are already filled. Then as the

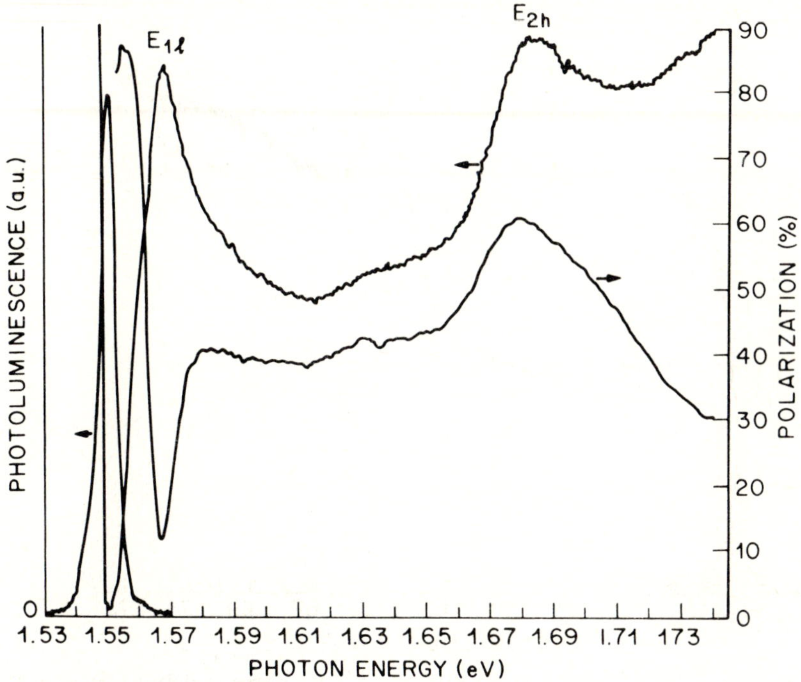

Fig. 6. Photoluminescence, excitation, and polarization spectra at 5 K for p-type modulation doped multiquantum well sample with 90 Å wells and 550 Å $x=0.45$ alloy barriers with the center 43 Å doped to give $N_h=5.3\times10^{11}$ cm^{-2}. The exciton peak E_{1h} is missing and the dip in the polarization at the peak labelled $E_{1\ell}$ shows it is of light hole character.

temperature is increased, holes can boil out of the bottom of the valence band and hence allow the formation of the correlated electron–hole pairs required for E_{1h} excitons. However, it has not been demonstrated theoretically that the Pauli exclusion principle should come into play at smaller hole densities than that required for significant screening of the excitons.

4.2. n-type modulation and antimodulation doped quantum wells

Prior to the investigation of n-type modulation-doped structures, n-type antimodulation doped GaAs quantum wells were investigated to compliment earlier studies on p-type antimodulation doped wells [10,18]. In the antimodulation doped configuration, the impurity is restricted to the quantum well during growth. Most striking with these n-type structures was the observation that with increasing n-type doping (Si), the circular polarization of the luminescence at 5 K with resonant excitation of the $n=2$ heavy hole exciton decreases markedly and, in fact, becomes opposite to that of the excitation for [Si] $\approx10^{18}$ cm^{-3} [23].

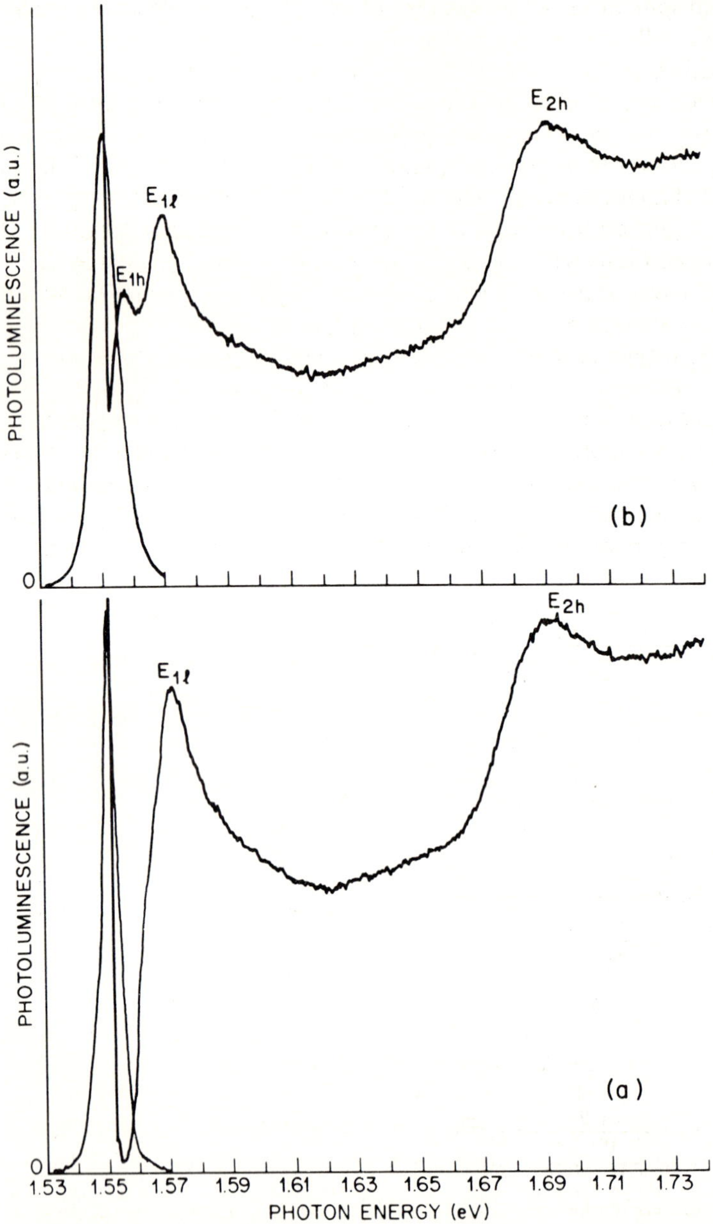

Fig. 7. Photoluminescence and excitation spectra at 5 K (a) and 43 K (b) for the same sample described in connection with fig. 6. At 43 K, a distinct peak due to E_{1h} can be seen. The emission is Stokes shifted from this peak by $\cong 6.5$ meV.

Furthermore, in spite of the large background of unpolarized electrons, the Si-doped quantum wells can exhibit highly polarized luminescence, especially for resonant excitation of E_{1h}. These anomalous characteristics of the polarization were greatly reduced, or even absent, when the sample temperature was increased to ≈ 20 K. At the same time, the polarization of the extrinsic luminescence at energies lower than that of the main peak is not anomalous and shows relatively small changes in magnitude over the 5–20 K temperature range. The mechanism that can result in a negative polarization for resonant excitation of E_{2h} in the Si-doped GaAs wells would seem to require a spin flip of the optically excited electrons and/or holes, or excitons, that give rise to the main luminescence peak. Such a mechanism has not been proposed.

Subsequently, n-type modulation doped GaAs quantum wells were investigated. One such structure incorporated 20 periods of $L = 104$ Å GaAs well and 227 Å thick superlattice barriers doped in the center 24 Å with Si to give $N_e = 1.4 \times 10^{11}$ cm^{-2}. The photoluminescence for this sample shown in fig. 8 exhibits all the anomalous characteristics mentioned above for the Si antimodulation doped structures and hence suggests that it is the electron gas which gives rise to the negative polarization effect. Although this would seem to be a simpler

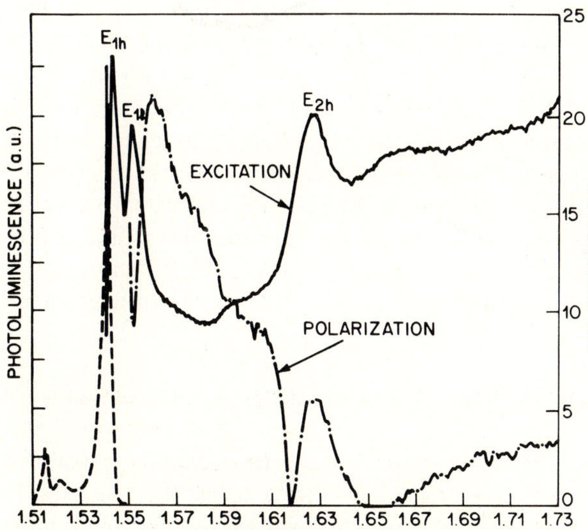

Fig. 8. Photoluminescence (dashed), excitation (solid), and polarization (dashed–dotted) spectra at 5 K for an n-type modulation doped GaAs quantum well sample $L = 104$ Å and superlattice barriers $x \approx 0.3$, doped in the center 24 Å with Si to give $N_e = 1.4 \times 10^{11}$ cm^{-2}. The phase of the detected polarization signal has been changed by 180° between $E_p \cong 1.617$ eV and 1.652 eV so that the polarization detected is actually negative for excitation between these two energies, i.e., for resonant excitation of E_{2h}.

system to investigate from a theoretical point of view than the Si-doped wells, a viable mechanism has yet to be proposed.

The excitation spectra of the GaAs wells doped with electrons or Si donors exhibit some of the same characteristics as p-type modulation doped GaAs wells. In particular, one observes a broadening and weakening of E_{1i} relative to E_{2h}, the gradual rise in the luminescence level as E_p approaches $E_{1\ell}$ from above, and a Stokes shift between emission and absorption (excitation). Although some data suggest that E_{1h} is affected more than $E_{1\ell}$ in the electron containing GaAs wells, the effects are not nearly as pronounced as those for holes in the wells as discussed above. In contrast to the p-type quantum wells, with $N_e \cong 5\times10^{11}$ cm^{-2}, there is no evidence of E_{1i} and perhaps other low lying exciton transitions [24]. However there is a weak modulation, ≈15%, on the excitation spectrum for higher quantum number transitions indicating weakened electron–hole final state interactions, presumably due to screening.

4.3. Stokes shifts

It has been noted that Stokes shifts between emission and absorption are observed for the modulation doped GaAs quantum wells. Attempts to explain the size of the Stokes shifts with a simple model which assumes equal exciton contributions in absorption and emission have not met with success. This model also includes; effects that arise from the electrostatic potential (band bending) due to the charge transfer from the alloy to the quantum well [21], calculations of the resultant electron and hole energy levels, and computations of the Fermi level. As an example of the failure of this simple model, for electron and hole densities in the range 1–2 × 10^{11} cm^{-2}, the observed Stokes shifts are significantly smaller than that predicted by the simple model, typically by about 5 meV. These results may be indicating that exciton effects are significantly less in emission than in absorption. There are also some difficulties with the sample parameters although it appears that this cannot be the only problem.

5. Fitting GaAs quantum well exciton transitions – energy gap discontinuties

Dingle et al. [4] were quite successful in fitting the series of exciton transitions for GaAs quantum wells as observed in absorption. The parameters used in this early work have by and large been accepted until very recently. Of most importance is the division of the energy-gap discontinuity $\Delta E_g = E_g(\mathrm{AlGaAs}) - E_g(\mathrm{GaAs})$ between the conduction and valence band wells. It was concluded that the $\Delta n = 0$ allowed transitions could be fit with a conduction band offset, $\Delta E_c = Q_e \Delta E_g$, with $Q_e = 0.85 \pm 0.03$ which then leads to a valence band offset $\Delta E_v = Q_h \Delta E_g = (1 - Q_e)\Delta E_g$ with $Q_h = 0.15 \pm 0.03$. The effective masses employed were; $m_e^* = 0.0665 m_o$, $m_h^* = 0.45 m_o$, and $m_l^* = 0.080 m_o$. Dingle et al. did

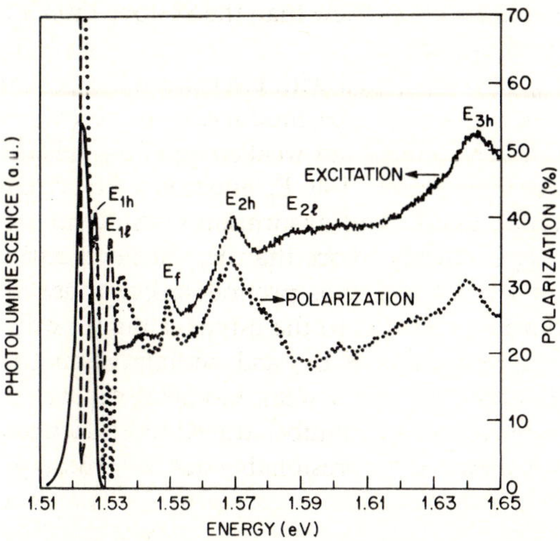

Fig. 9. Photoluminescence (dashed), excitation (solid), and polarization (dotted) spectra at 5 K for an undoped GaAs multiquantum well sample with $L=180$ Å wells and 211 Å $x=0.29$ alloy barriers. The phase of the polarization has been changed by 180 Å for excitation at $E_p=1.53$ eV to show the negative polarization due to $E_{1\ell}$. For E_p equal to the energy of the forbidden transition E_f, a distinct increase in polarization shows this transition is of heavy hole character leading to the assignment E_{13h}.

not claim to fit the forbidden exciton transitions which appeared in the data. Polarization measurements [5], including the data in fig. 9, show that this forbidden transition is heavy hole in character so that it has been assigned to the $\Delta n=2$ parity allowed exciton involving the $n=1$ electron and the $n=3$ heavy hole, i.e., E_{13h}. For GaAs wells with $L=100$ Å and $x=0.3$ alloy barriers, the calculated (above or "usual" parameters) and observed E_{13h} exciton transitions energies measured from the E_{1h} peak are 38 meV and 64 meV, respectively. This serious discrepancy has been known for some time but has not emphasized [4].

5.1. Parabolic GaAs quantum wells

GaAs quantum wells with a parabolic potential profile have been grown by MBE and their photoluminescence properties investigated at low temperatures [25]. Parabolic potential wells are of special interest since the Q's appear directly in the eigenvalue expressions. For square wells with infinite barriers

$$E_{ni}=(1/2m_i^*)[n\pi\hbar/L_z]^2, \tag{1}$$

where $n=1,2,3$, etc. With parabolic wells

$$E_{ni}=(n-1/2)\hbar\omega_{0i}, \tag{2}$$

where again $n=1,2,3$, etc. and $\omega_{0i}=\sqrt{K_i/m_i^*}$ with K_i equal to the curvature of the parabolic well. Defining K_i by the potential height of the finite parabolic well with $z=\pm L/2$, namely, $Q_i\Delta E_g$, one has

$$E_{ni}=2(n-1/2)\frac{\hbar}{L}\left[\frac{2Q_i\Delta E_g}{m_i^*}\right]^{1/2}. \tag{3}$$

Parabolic profiles of aluminum content were generated by chopping an aluminum molecular beam with a computer controlled shutter. The duty ratio during which the aluminum beam was turned on was made proportional to the desired concentration and potential profile, reaching unity for the barriers. Each well consisted of 41 layers. Three samples were grown by this technique.

Photoluminescence measurements at 5 K shown in fig. 10 reflect the expected harmonic oscillator like electron and hole energy levels. Analyses of the energies of the various exciton transitions and their polarization characteristics lead to their identification. Differences in the energies of these transitions are then used to determine the electron, heavy hole, and light hole energy level ladders, ΔE_e, ΔE_h, and ΔE_ℓ, respectively. For these estimates, the binding energies of all the excitons were assumed equal.

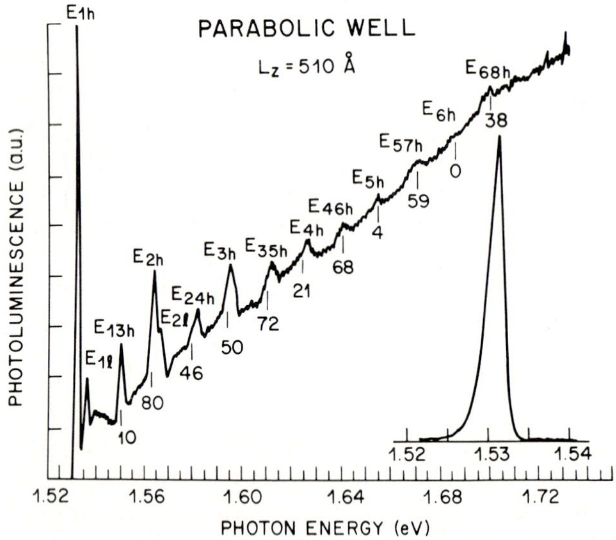

Fig. 10. Photoluminescence (insert) and excitation spectra at 5 K for GaAs quantum wells with a parabolic profile. The wells were of width $L=510\pm35$ Å measured at the 225 Å wide $x=0.30\pm0.06$ alloy barriers. Short vertical bars show the calculated energies of the various heavy hole exciton transitions with $Q_e=0.51$ and the usual effective masses. The integers below the peaks represent the calculated strengths normalized to 100 for E_{1h} (without the resonant enhancement) and demonstrate, in agreement with the data, that the $\Delta n=2$ transitions gain strength relative to the $\Delta n=0$ transitions as n increases.

The agreement between these experimental ΔE_i and those calculated from eq. (3) was very poor. Removing the L dependence by taking ratios gives, for example, $\Delta E_e/\Delta E_h = 2.6$ for the experimental value to be compared with the calculated value

$$[Q_e m_h^*/(1-Q_e)m_e^*]^{1/2} = 6.0, \tag{4}$$

using the usual parameters. This discrepancy is not significantly improved using an exact calculation for E_{ni} which takes into account finite barrier heights and the variation of the effective mass with z. With the usual effective masses, eq. (4) requires that $Q_e \approx 0.5$ to fit the experimental ratio 2.6, in marked contrast to the usual value of 0.85 [1,4]. The many allowed and forbidden heavy hole transitions observed with the parabolic well samples can be fit accurately with $Q_e \approx Q_h$ using the usual effective masses as shown in fig. 10. However, $Q_e \approx 0.5$ with the m_i^* cited above does not lead to satisfactory fits of observed series of exciton transitions for square wells, and conversely, $Q_e = 0.85$ does not lead to acceptable fits of the parabolic well data.

5.1. Unified picture

It is clear that what is needed to fit the square well forbidden transitions is a smaller heavy hole mass and/or a larger Q_h than usually assumed. As a result, a heavy hole mass of $0.34m_0$ has been proposed for these quantum well structures [26]. This value is consistent with that utilized in some theoretical work [27,28] and also obtained by Kleinman in some earlier attempts to fit square quantum well data for wide wells [15]. With this m_h^* and the usual m_e^*, $Q_e \approx 0.57$ is required to fit the parabolic well $\Delta E_e/\Delta E_h$ data via eq. (4). Similar arguments with $Q_e = 0.57$ for the light hole transitions for parabolic wells requires that $m_\ell^* = 0.94m_0$ which is about 7% larger than the value ($0.088m_0$) preferred earlier for square wells. Thus the new parameters proposed [26] for both parabolic and square wells are; $Q_e = 0.57$, $m_e^* = 0.0665m_0$, $m_h^* = 0.34m_0$, and $m_\ell^* = 0.094m_0$.

The energies of the various light and heavy hole exciton transitions have been calculated with these new parameters as a function of L for square wells. As shown in fig. 11 these calculated energies are found to be in good agreement with the corresponding experimental energies including those for the forbidden transitions determined from excitation spectra for ten different high-quality multiquantum square well samples grown by MBE with L ranging from 51 Å to 521 Å and alloy barriers with $x \cong 0.3$.

The new parameters have been obtained in such a manner that the ratios involving the ΔE_i for the parabolic wells agree with the parabolic well data. However the values of L required via eq. (3), or the exact calculation, for the calculated ΔE_i to agree with the data using the new parameters are in agreement with the values of L estimated from the growth parameters. This is the case for all three parabolic well samples [26].

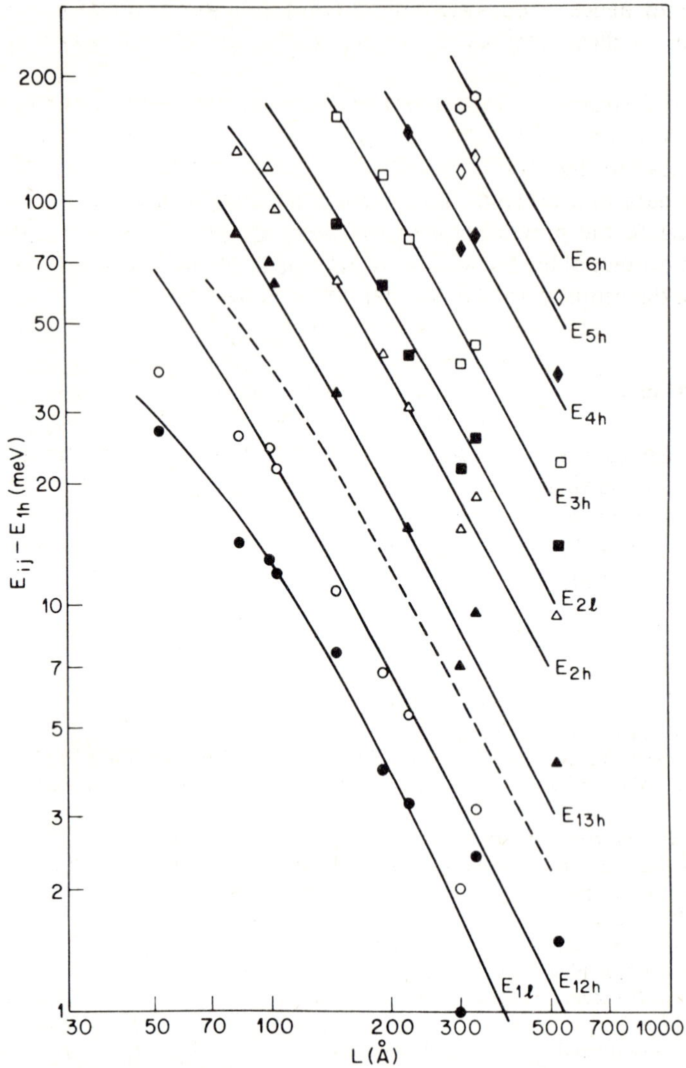

Fig. 11. Logarithm of the observed energies of the exciton transitions E_{ij} measured from the $n=1$ heavy hole exciton E_{1h} versus the logarithm of L for a series of high-quality square GaAs multi-quantum well samples with $x=0.30\pm0.05$ alloy barriers. The solid lines are the calculated $E_{ij}-E_{1h}$ using the new parameters: $Q_e=0.57$, $m_e^*/m_0=0.0665$, $m_h^*/m_0=0.34$, and $m_\ell^*/m_0=0.094$. The dashed line represents $E_{13h}-E_{1h}$ calculated with the usual parameters, $Q_e=0.85$ and $m_h^*/m_0=0.45$.

The calculations discussed above do not include effects due to the non-parabolicity (NP) of the conduction and light hole bands. For the heavy hole, NP is negligible. The effects of NP on the eigenvalues for both square and parabolic wells have been estimated and are found to be of order 7% for the highest energies and hence nontrivial, but generally too small compared to other uncertainties to effect significantly the comparison between theory and experiment [26].

Thus a single set of parameters leads to acceptable fits to the observed exciton transitions for both square and parabolic potential wells. A fine tuning of these parameters could lead to slightly different Q values, but the extensive data cited herein, and other data of a different nature not cited, lend strong support to $Q_e \approx 0.6$ in contrast to the previously accepted value $Q_e \approx 0.85$. These results emphasize the importance of the forbidden exciton transitions and the parabolic quantum wells for determining the GaAs quantum well parameters.

6. Acknowledgements

The authors wish to acknowledge the most important contributions of R.D. Dupuis, A.C. Gossard, Won T. Tsang and W. Wiegmann in providing the many unique structures whose properties have been discussed herein. In addition, the data acquisition and manipulation skills of O. Munteanu greatly facilitated the accumulation of the recent results presented.

References

[1] R. Dingle, W. Wiegmann and C.H. Henry, Phys. Rev. Lett. 33 (1974) 827.
[2] R. Dingle, C. Weisbuch, H.L. Stormer, H. Morkoc and A.Y. Cho, Appl. Phys. Lett. 40 (1982 507.
[3] C. Weisbuch, R.C. Miller, R. Dingle, A.C. Gossard and W. Wiegmann, Solid State Commun. 37 (1981) 219.
[4] R. Dingle, Festkorperprobleme 15 (1975) 21.
[5] R.C. Miller, D.A. Kleinman, W.A. Nordland, Jr. and A.C. Gossard, Phys. Rev. B22 (1980) 863.
[6] See, for example, G. Lampel, Proc. XIIth Int. Conf. Physics of Semiconductors, Stuttgart, 1974, ed., M.H. Pilkuhn (Teubner, Stuttgart, 1974) p. 743.
[7] Q.H.F. Vrehen, J. Phys. Chem. Solids 29 (1968) 129.
[8] A. L. Mears and R.A. Stradling, J. Phys. C 4 (1971) L22.
[9] B.P. Zakharchenya, V.I. Zemskii and D.N. Mirlin, Zh. Eksp, Teor. Fiz. 70 (1976) 1092 [Sov. Phys. JETP 43 (1976) 569].
[10] R.C. Miller, A.C. Gossard and W.T. Tsang, Physica 117B and 118B (1983) 714.
[11] R.C. Miller, D.A. Kleinman, W.T. Tsang and A.C. Gossard, Phys. Rev. B 24 (1981) 1134.
[12] G. Bastard, E.E. Mendez, L.L. Chang and L. Esaki, Phys. Rev. B 26 (1982) 1974.
[13] R.L. Greene, K.K. Bajaj and D.E. Phelps, Phys. Rev. B 29 (1984) 1807.
[14] R.C. Miller, D.A. Kleinman, A.C. Gossard and O. Munteanu, Phys. Rev. B 25 (1982) 6545.
[15] D.A. Kleinman, Phys. Rev. B 28 (1983) 871.
[16] W.F. Brinkman, T.M. Rice, and B. Bell, Phys. Rev. B 8 (1973) 1570.
[17] J.R. Haynes, Phys. Rev. Lett 4 (1960) 361.
[18] R.C. Miller, A.C. Gossard, W.T. Tsang and O. Munteanu, Solid State Commun. 43 (1982) 7, 519.
[19] J. Hegarty, M.D. Sturge, C. Weisbuch, A.C. Gossard and W. Wiegmann, Phys. Rev. Lett. 49 (1982) 930.

[20] R. C. Miller, R.D. Dupuis and P.M. Petroff, Appl. Phys. Lett. 44 (1984) 508.
[21] R. Dingle, H.L. Stormer, A.C. Gossard and W. Wiegmann, Appl. Phys. Lett. 33 (1978) 665.
[22] D.A. Kleinman (unpublished).
[23] R.C. Miller and A.C. Gossard, Phys. Rev. B 28 (1983) 3645.
[24] A. Pinczuk, J. Shah, R.C. Miller, A.C. Gossard, and W. Wiegmann, Solid State Commun. 50 (1984) 735.
[25] R.C. Miller, A.C. Gossard, D.A. Kleinman and O. Munteanu, Phys. Rev. B 29 (1984) 3740.
[26] R.C. Miller, D.A. Kleinman and A.C. Gossard, Phys. Rev. B 29 (1984) 7085.
[27] P. Lawaetz, Phys. Rev. B 4 (1971) 3460.
[28] A. Baldereschi and N.O. Lipari, Phys. Rev. B 3 (1971) 439.

Journal of Luminescence 30 (1985) 541–550
North-Holland, Amsterdam

SHORT PULSE PHYSICS OF QUANTUM WELL STRUCTURES

E.O. GÖBEL, J. KUHL and R. HÖGER

Max-Planck-Institut für Festkörperforschung, Heisenbergstr. 1, D-7000 Stuttgart 80, FRG

This paper reports some recent results of time-resolved studies of the carrier dynamics in GaAs/GaAlAs quantum well structures with picosecond and subpicosecond time resolution. These experiments have provided insight into carrier trapping, energy relaxation, and carrier recombination processes. Carrier trapping into the quantum well layers is very efficient and determines the decay of the GaAlAs luminescence even for 1 μm thick cladding layers. Carrier recombination is enhanced particularly at low temperatures. This effect has been attributed to the increased overlap of electron and hole (exciton) wavefunctions in the quasi-two-dimensional carrier system.

Introduction

The electrical and optical properties of the quasi-two-dimensional (2D) electronic system in quantum well structures (QWS) have gained considerable interest because of their importance for fundamental semiconductor physics as well as technical applications [1–5].

The optical, in particular the luminescence properties of QWS can be summarized as follows: The low temperature luminescence of GaAs/GaAlAs QWS with thickness $L_z \leqslant 25$ nm at moderate excitation intensities is dominated by intrinsic free exciton recombination [6,7] and exciton processes may be present even up to room temperatures [8]. The energetic position of the lowest intrinsic electron–hole excitation depends on barrier height and L_z as expected on the base of fundamental quantum mechanics and verified experimentally by absorption [1–5] or excitation spectroscopy [1–5,9], and reflectivity measurements [10]. The binding energy of the free exciton increases [9,11] and the Bohr radius decreases with decreasing L_z reflecting the transition from the 3D to 2D exciton [9,12–14]. The decrease of the Bohr radius is accompanied by a respective increase of the oscillator strength of the quasi-2D exciton [9,10]. The exciton line is inhomogeneously broadened due to inherent fluctuations in L_z [6,7]. The homogeneous linewidth of the exciton decreases strongly at the low-energy side of the inhomogeneously broadened line [15], which may indicate the localization of excitons in the random potential fluctuations [15,16].

0022-2313/85/$03.30

At high excitation intensities filling of the quantum well states is possible resulting in luminescence also from higher subbands. The optical properties under these conditions consequently should be described within a free-carrier model taking into account the 2D exciton enhancement within the continuum states [17,18].

Experimental techniques

Several experimental techniques have been employed so far to investigate the carrier dynamics in QWS. Indirect information on radiative and non-radiative carrier lifetime has been obtained by stationary luminescence studies of optically pumped QWS [19]. The spontaneous carrier lifetime at laser threshold of GaAs/GaAlAs quantum well lasers has been measured by using the lasing delay technique [20]. The first direct study of carrier dynamics in QWS with 0.2 ps time resolution has been performed by Shank et al. [21,22] by means of excite- and probe-absorption experiments. A picosecond nonlinear hot-luminescence correlation technique [23] has been employed by Xu et al. to investigate hot carrier relaxation in QWS [24].

The most widely used technique so far, however, is time-resolved luminescence spectroscopy [25–29]. A typical experimental setup using a synchronously mode-locked dye laser and a synchroscan streak camera together with a 25 cm grating monochromator (600 grooves/mm) is shown in fig. 1a. The pump laser is an actively mode-locked Ar^+ laser with a stabilized mode-lock driver [30]. We have used Rhodamine 6G and Pyridine 1 as laser dyes for the wavelength region of about 560–750nm. The width of the synchronously mode-locked dye laser pulses was inbetween 2–5 ps for different emission

a

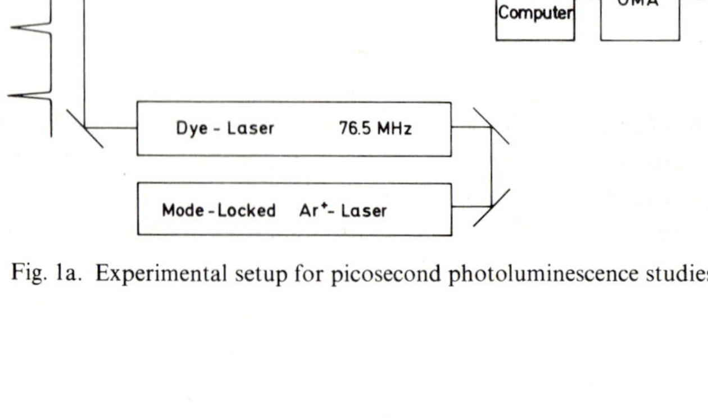

Fig. 1a. Experimental setup for picosecond photoluminescence studies.

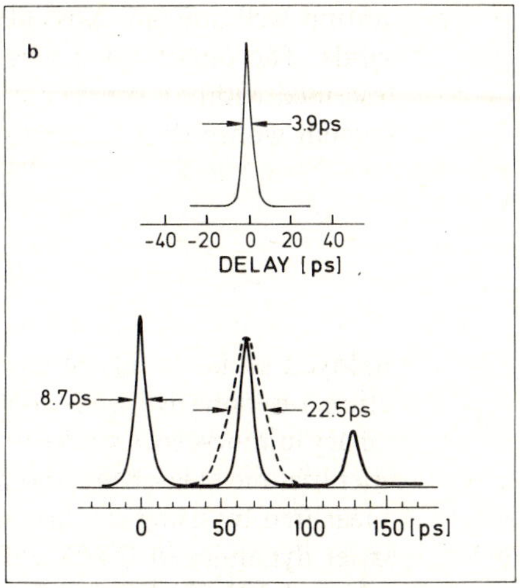

Fig. 1b. A typical autocorrelation trace of the synchronously mode-locked dye laser pulses is shown in the upper part of the figure. The lower part represents the time resolution of the streak camera (full trace) and of the entire experimental setup (dashed trace).

wavelengths. The average output power is typically 50 to 100 mW at about 80 MHz repetition rate and an average pump power of 600 mW. The synchroscan streak camera (Imacon 500) has a S-20 spectral response and is read out with a Si-diode array. The temporal resolution of the system is demonstrated in fig. 1b. The upper part shows an autocorrelation trace of the synchronously mode-locked dye laser output (Rh 6G, two plate birefringent filter) measured by noncollinear frequency doubling in KDP. The autocorrelation full width at half maximum (FWHM) of 3.9 ps corresponds to an actual pulse width of 2.7 ps for a Gaussian pulse shape. The lower part in fig. 1b depicts streak camera traces obtained from the mode-locked dye laser after multiple reflection in an etalon with 1 cm mirror separation before (full line) and after (dashed line) the beam has passed the 25 cm grating spectrometer. The actual time resolution in the present luminescence experiments thus is determined to be about 25 ps.

Results

Carrier *relaxation and thermalization* in QWS has been first studied by Shank et al. [21,22] using excite and probe techniques. The authors conclude from their data that at high excitation intensities ($n \geqslant 5 \times 10^{11}$ cm^{-2} corresponding to an electron–hole pair density of 2.5×10^{17} cm^{-3}) exciton contributions to the absorption are immediately screened (within less than 0.2 ps) and that at the highest excitation ($n \sim 10^{13}$ cm^{-2}) gain occurs below and above the $n=1$ subband due to band gap renormalization and band filling, respectively. From the shape of the induced absorption (gain) spectra the respective values for the carrier density and temperature are obtained employing a free-carrier recombination model taking into account the 2D exciton enhancement. The results show that thermalization is very similar in QWS and bulk GaAs samples. The carrier cooling rate slows down at high excitation intensities in QWS similar to bulk GaAs [31,32]. This reduced cooling, which has been originally attributed to free-carrier screening of the Fröhlich interaction [31,32], seems to be mainly caused by nonequilibrium phonon effects [33].

Pronounced differences in the width of the hot-luminescence correlation peaks of bulk GaAs and GaAs/GaAlAs QWS have been observed by Xu et al. [24] at high excitation intensities ($n \sim 3.8 \times 10^{18}$ cm^{-3}), which have been attributed to differences in the carrier cooling rates. A reduced cooling of free carriers in QWS at high excitation intensities may be inferred also from the time-integrated luminescence data of Xu et al. [24] as well as from time-resolved luminescence studies (c.f. fig. 4 and also ref. 28).

Energy relaxation of excitons in QWS has been investigated by Masumoto et al. [27]. According to their results the exciton system loses its energy at a constant rate of 1×10^6 eV/s, which is much slower than expected for an ideal 2D exciton. The reduced energy loss is attributed to the fact that the exciton is inhomogeneously broadened and exciton localization in the random potential fluctuations may occur. After that, excitons lose their energy slowly, migrating towards lower-energy positions in the spatially inhomogeneous well with the emission of acoustic phonons [27] similar to the findings for heavily doped bulk GaAs [34].

The dynamics of resonantly excited excitons in QWS at room temperature has been investigated very recently by Knox et al. [35] with a time resolution of 150 fs. Due to this high time resolution, which is faster than the optical phonon scattering times, nonequilibrium distributions of excitons at 300 K could be studied for the first time.

The dynamics of carrier *trapping and recombination* can be inferred from, e.g., time-resolved luminescence studies as shown in figs. 2 and 3 for a bath temperature of 10 K and 77 K, respectively [25]. These data have been obtained for a double QWS with $L_z = 14$ nm. The stationary photoluminescence, excitation spectra and reflectivity data of this particular sample clearly demonstrate the intrinsic 2D free-exciton nature of the optical transition [7,10]. The sample has been excited by the optical pulse train with a photon energy of 2.17 eV which is well above the band gap of the GaAlAs ($x=0.21$). The excitation pulses thus are completely absorbed within the upper 1 μm thick GaAlAs layer (indirect excitation). The time-integrated luminescence spectrum at a bath temperature of $T=10$ K (inset in fig. 2) reveals the recombination of carriers within the GaAlAs cladding layers at 1.78 eV and a weak band at 1.49 eV

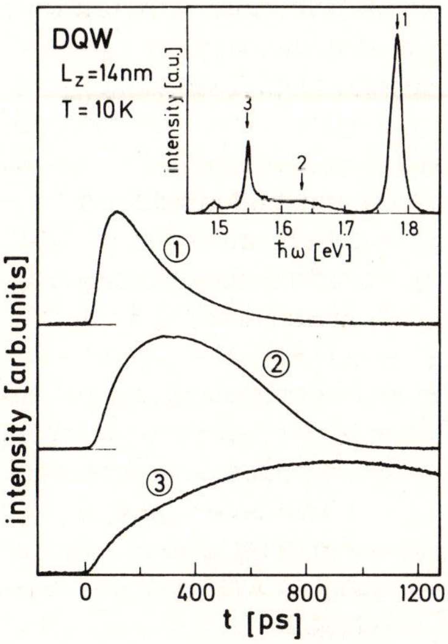

Fig. 2. Time-integrated emission spectrum (inset) and time behavior of the luminescence of a double QWS with $L_z = 14$ nm at $T = 10$ K for indirect excitation.

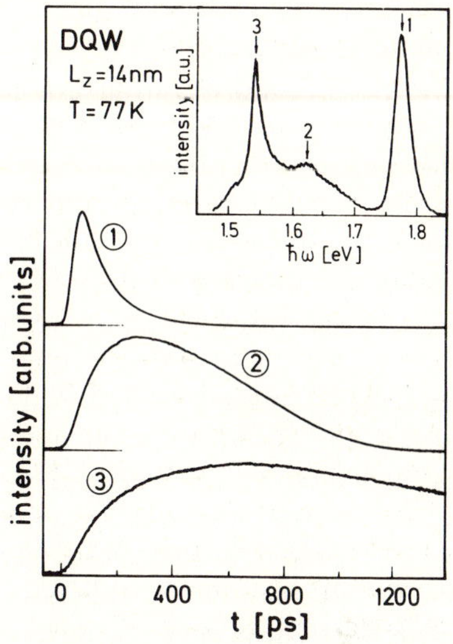

Fig. 3. Time-integrated emission spectrum (inset) and time behavior of the luminescence of a double QWS with $L_z = 14$ nm (same sample as in fig. 2) at $T = 77$ K for indirect excitation.

resulting from recombination within the GaAs substrate. The narrow and broad features with maxima at about 1.55 eV and 1.63 eV are attributed to the $n=1$ and $n=2$ subband transitions within the QW, respectively. The time behavior of the photoluminescence corresponding to transitions within the GaAlAs (curve (1)) and within the QW (curves (2) and (3)) is displayed by the original streak camera traces in the lower part of fig. 2. The luminescence from the GaAlAs shows an immediate response, whereas the QW recombination is delayed and the intensity increase follows approximately the decrease of the GaAlAs luminescence. The decay of the GaAlAs luminescence is faster than in bulk samples, indicating that the fast luminescence decay is at least partly due to trapping of carriers into the QW. This interpretation is further supported by the data obtained for the same sample at 77 K as shown in fig. 3. The time-integrated spectrum shows the same features as observed at 10 K, however, the intensity of the QW luminescence relative to the GaAlAs luminescence is higher at 77 K. The decay of the GaAlAs luminescence (curve (1) in fig. 3), on the other hand, is faster at 77 K than at 10 K. These two experimental facts, the faster decay of the GaAlAs luminescence together with the increase of the QW luminescence intensity definitely proves that the trapping of carriers contributes essentially to the luminescence decay of the GaAlAs. The faster trapping at higher temperatures may be attributed to an increase of the diffusion coefficient with temperature. The mean carrier velocities which would be required if carrier trapping alone is responsible for the fast decay of the GaAlAs luminescence are in the order of 10^6 cm/s. Such high carrier velocities have also been found in quasi-stationary photoluminescence experiments of bulk GaAs samples [36]. Even higher values have been claimed for multiple quantum wells with very thin GaAlAs layers [26].

The transient photoluminescence spectra of the same double QWS are shown in fig. 4 again for indirect excitation. Broad spectra of more than 100 meV width are observed in the early time regime ($t<600$ ps). The spectral width initially increases reflecting again the trapping of carriers from the GaAlAs into the QW. The spectra can be described within a free-carrier recombination model [37] taking into account some broadening of the density-of-states as well as the 2D excitonic enhancement. The full line in the curve for $t=250$ ps is calculated with a density of $n=5\times10^{12}$ cm^{-2} and an effective temperature of $T_{eff}=180$ K. This effective temperature is higher than for bulk GaAs [32,38], which indeed may be due to a slower energy relaxation in 2D QWS as compared to 3D bulk material. Further information on inter- and intra-subband relaxation is needed, however, to confirm this definitely. A relatively narrow spike close to the $n=1$ subband dominates the spectra for $t>600$ ps indicating the formation of 2D excitons. In the early time regime some exciton recombination may come from the outer regions of the excited area.

Exciton recombination is observed in the entire time regime for lower carrier densities, i.e., in particular for direct excitation of the QW. Experimental

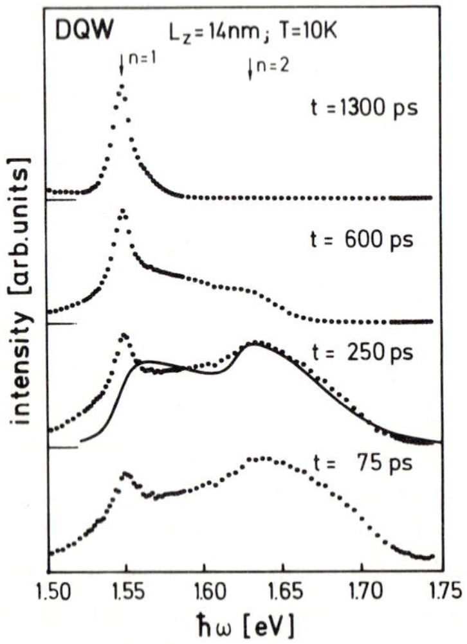

Fig. 4. Transient emission spectra of the double QWS with $L_z=14$ nm at $T=10$ K for indirect excitation. The arrows indicate the position of the $n=1$ and $n=2$ heavy hole exciton, respectively, as determined from excitation spectroscopy. The full line for $t=250$ ps represents a calculated curve as described in the text.

results for two different excitation intensities ($P=3$ mW and $P=55$ mW) at $T=10$ K are depicted in fig. 5. The photon energy of the optical pulses was 1.675 eV and thus no absorption was possible in the GaAlAs. The strong emission for $\hbar\omega < 1.52$ eV seen in the time-integrated spectra in fact shows that most of the excitation light is absorbed in the GaAs substrate. The lowest subband free-exciton emission is observed for both the lower as well as the higher excitation. At the low excitation intensity the exciton emission spectrum shows a pronounced tail to lower energies which is characteristic also for the stationary photoluminescence spectra at low excitation intensities [10]. This low-energy tail again may reflect localization of free excitons [15,16]. At higher excitation the low-energy part of the exciton line saturates [10]. The time behavior of the exciton luminescence shows a fast onset, because trapping is not present for the direct excitation. The decay is slightly different for the low and high excitation intensity and the time at which the intensity has dropped to 1/e of its maximum value, $\tau_{1/e}$, amounts to 410 ps and 570 ps, respectively [39]. This increase in decay time with excitation intensity may indicate saturation of a fast process, which could be responsible also for the low-energy

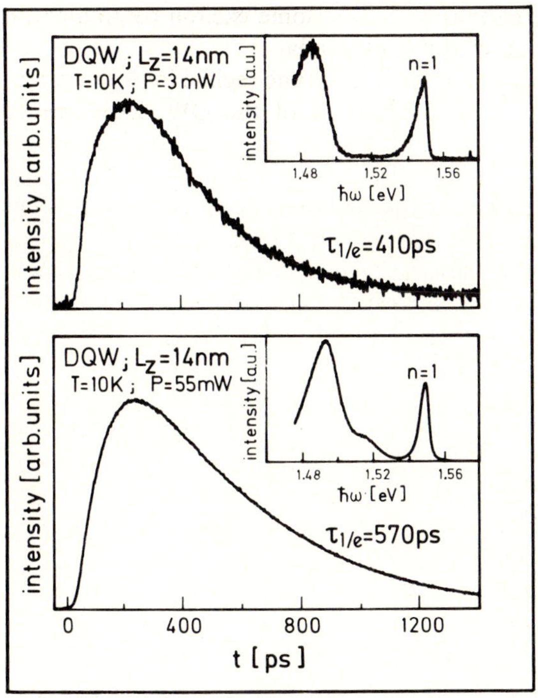

Fig. 5. Time-integrated emission spectra (insets) and time behavior of the luminescence for the double QWS with $L_z=14$ nm at 10 K for an average excitation intensity of 3 mW (upper part) and 55 mW (lower part) in the case of direct excitation.

tail of the low-intensity emission line. The photoluminescence decay is definitely faster than in 3D bulk samples [25,26].

The low-temperature ($T=10$ K) results for $\tau_{1/e}$ of several single, double, and multiple QWS and a double heterostructure with a 0.5 μm thick (bulk) GaAs layer are summarized in fig. 6 for two different excitation intensities [37]. The large difference of $\tau_{1/e}$ of the bulk sample for the two different excitation intensities arises from the fact that at the low excitation level recombination via bound excitons dominates, whereas at the higher excitation recombination occurs mainly via free excitons due to a possible heating of the crystal. The 1/e decay times for all the different QWS are shorter than for the bulk sample and they decrease slightly with decreasing well thickness L_z. A decrease of the QW luminescence decay time with decreasing L_z is indeed expected independent on whether the recombination is dominated by radiative or nonradiative processes. The transition probability for radiative free-exciton recombination is proportional to the electron and hole wavefunction overlap, i.e., is inversely proportional to the "exciton volume". In a QWS the exciton volume is approximately given by the product of the square of the exciton Bohr radius within the plane of the quantum well $a_0(L_z)$ and the well thickness L_z (this simple consideration neglects penetration of the exciton wavefunction into the GaAlAs). We thus obtain for the radiative free exciton lifetime τ_r

$$\tau_r \sim 1/|\Phi(0)|^2 \sim a_0^2(L_z)\cdot L_z \qquad (0<L_z\leqslant a_0^{3D}). \tag{1}$$

The Bohr radius $a_0(L_z)$ itself decreases with decreasing L_z [12–14] and varies between a_0^{3D} for large L_z and $a_0^{3D}/2$ for $L_z\to 0$ (a_0^{3D} is the 3D exciton Bohr radius; $a_0^{3D}=14.5$ nm for GaAs). A superlinear dependence of τ_r on L_z is thus expected on the base of eq. (1).

Nonradiative transitions may be due to Auger as well as defect recombination. Auger recombination in QWS has been recently discussed by several authors

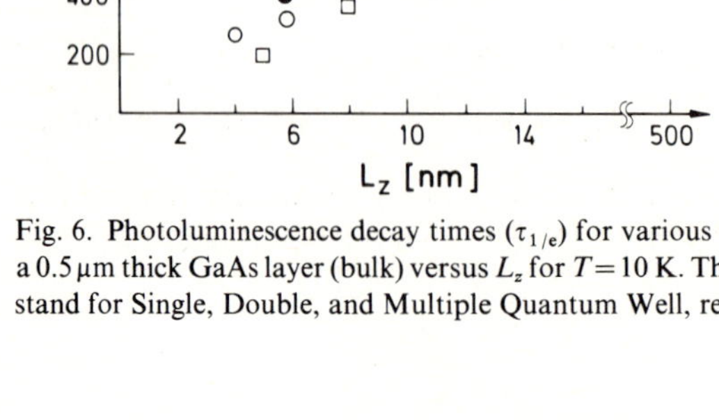

Fig. 6. Photoluminescence decay times ($\tau_{1/e}$) for various QWS and a double heterostructure with a 0.5 μm thick GaAs layer (bulk) versus L_z for $T=10$ K. The abbreviations SQW, DQW, and MQW stand for Single, Double, and Multiple Quantum Well, respectively.

[40–43], in relation to quantum well lasers. Auger recombination, however, may be neglected at low carrier densities where the optical properties are governed by excitonic effects. Defect recombination in QWS will be most likely determined by interface recombination [44]. The nonradiative lifetime τ_{nr} then can be approximated by [45]

$$\tau_{nr} \sim L_z/2S_i, \tag{2}$$

where S_i is the surface recombination velocity. For high quality GaAs/GaAlAs interfaces S_i has been determined to be about 300 cm/s at room temperature which on the base of eq. (2) would yield $\tau_{nr} \sim 2.3$ ns for $L_z = 14$ nm. At low temperatures τ_{nr} is expected to be even larger, because S_i decreases with temperature at least for low excitation intensities, where a potential barrier at the interface exists [45].

The subnanosecond luminescence decay times measured at low temperatures for high quality QWS thus are very likely due to a decrease of the radiative lifetime because of exciton localization. This interpretation is further supported by the result that the low-temperature data follow the same L_z-dependence for all the different structures investigated. Presently, however, it is still an open question whether additional localization of excitons, e.g., within potential fluctuations due to L_z-fluctuations, affects the low-temperature photoluminescence decay.

Acknowledgements

We are very much indepted to K. Ploog, MPI für Festkörperforschung, und G. Weimann, Forschungsinstitut der Deutschen Bundespost Darmstadt, for the high quality QWS and for many helpful discussions. Expert experimental assistance by K. Rother and A. Jonietz is also gratefully acknowledged.

References

[1] R. Dingle, Confined carrier quantum states in ultrathin semiconductor heterostructures, in: Festkörperprobleme (Advances in Solid State Physics) vol. XV, H.J. Queisser, ed. (Pergamon-Vieweg, Braunschweig, 1975) p. 21.

[2] N. Holonyak, R.M. Kolbas, R.D. Dupuis, P.D. Dapkus, IEEE J. Quant. Electr. QE-16 (1980) 170.

[3] See e.g. in: Electronic Properties of Two-Dimensional Systems, F. Stern, ed. (North-Holland, Amsterdam, 1982).

[4] N.T. Linh, The two-dimensional electron gas and its technical applications, in: Festkörperprobleme (Advances in Solid State Physics) vol. XXIII, P. Grosse, ed. (Pergamon-Vieweg, Braunschweig, 1983) p. 227.

[5] D.S. Chemla, Helvetica Physica Acta 56 (1983) 607.

[6] C. Weisbuch, R.D. Miller, R. Dingle, A.C. Gossard and W. Wiegmann, Solid State Commun. 37 (1981) 219.
[7] H. Jung, A. Fischer and K. Ploog, Appl. Phys. A 33 (1984) 97.
[8] P. Dawson, G. Duggan, H.I. Ralph and K. Woodbridge, Phys. Rev. B 28 (1983) 7381.
[9] R.C. Miller, D.A. Kleinman, W.T. Tsang and A.C. Gossard, Phys. Rev. B 24 (1981) 1134.
[10] L. Schultheis and K. Ploog, Phys. Rev. B 30 (1984) 1090.
[11] J.C. Maan, G. Belle, A. Fasolino, M. Altarelli and K. Ploog Phys. Rev. B 30 (1984) 2253.
[12] G. Bastard, E.E. Mendez, L.L. Chang and L. Esaki, Phys. Rev. B 26 (1982) 1974.
[13] R.L. Greene, K.K. Bajaj and D.E. Phelps, Phys. Rev. B 29 (1984) 1807.
[14] Y. Shinozuka and M. Matsuura, Phys. Rev. B 28 (1983) 4878.
[15] J. Hegarty, M.D. Sturge, C. Weisbuch, A.C. Gossard and W. Wiegmann, Phys. Rev. Lett. 49 (1982) 930.
[16] G. Bastard, C. Dalalande, H.M. Meynadier, P.M. Frijlink and M. Voos, Phys. Rev. B 29 (1984) 7042.
[17] M. Shinada and S. Sugano, J. Phys. Soc. Jpn. 21 (1966) 1936.
[18] F.L. Ledermann and J.D. Dow, Phys. Rev. B 13 (1976) 1633.
[19] R.C. Miller, D.A. Kleinmann, W.A. Nordland, Jr. and A.C. Gossard, Phys. Rev. B 22 (1980) 863.
[20] N.K. Dutta, R.L. Hartmann, W.T. Tsang, IEEE J. Quant. Electr. QE-19 (1983) 1243.
[21] C.V. Shank, R.L. Fork, B.I. Greene, C. Weisbuch and A.C. Gossard, Surf. Sci. 113 (1982) 108.
[22] C.V. Shank, R.L. Fork, R. Yen, J. Shah, B.I. Greene, A.C. Gossard and C. Weisbuch, Solid State Commun. 47 (1983) 981.
[23] D. von der Linde, J. Kuhl and E. Rosengart, J. Lumin. 24/25 (1981) 675.
[24] Z.Y. Xu and C.L. Tang, Appl. Phys. Lett. 44 (1984) 692.
[25] E.O. Göbel, H. Jung, J. Kuhl and K. Ploog, Phys. Rev. Lett. 51 (1983) 1588.
[26] J. Christen, D. Bimberg, A. Steckenborn and G. Weimann, Appl. Phys. Lett. 44 (1984) 84.
[27] Y. Masumoto, S. Shionoya and H. Kawaguchi, Phys. Rev. B 29 (1984) 2324.
[28] J.F. Ryan, R.A. Taylor, A.C. Maciel, J.M. Worlock, A.C. Gossard and W. Wiegmann, Bull. Am. Phys. Soc. 29 (1984) 214; Proc. 17th Int. Conf. Phys. Semicond., San Francisco, 1984; Phys. Rev. Lett. 53 (1984) 1841.
[29] P. Dawson, G. Duggan, H.I. Ralph and K. Woodbridge, Proc. 17th Int. Conf. Phys. Semicond., San Francisco, 1984.
[30] H. Klann, J. Kuhl and D. von der Linde, Opt. Commun. 38 (1981) 390.
[31] R.F. Leheny, J. Shah, R.L. Fork, C.V. Shank and A. Migus, Solid State Commun. 31 (1979) 809.
[32] W. Graudszus and E.O. Göbel, Physica 117 B & 118 B (1983) 55.
[33] W. Pötz and P. Kocevar, Phys. Rev. B 28 (1983) 7040.
[34] E.O. Göbel and W. Graudszus, Phys. Rev. Lett. 48 (1982) 1277.
[35] W.H. Knox, R.L. Fork, M.C. Dower, D.A.B. Miller, D.S. Chemla, C.V. Shank, A.C. Gossard and W. Wiegmann, in: Ultrafast Phenomena, D.H. Auston and K.B. Esenthal, eds. (Springer, Berlin, 1984) p. 162.
[36] A. Forchel, H. Schweizer, H. Nather, K.M. Romanek, J. Fischer and G. Mahler, Physica (Utrecht) 117 B & 118 B (1983) 336.
[37] R. Höger, E.O. Göbel, J. Kuhl, K. Ploog, Verhdlg. DPG (VI) 19 (1984) 213; R. Höger, E.O. Göbel, J. Kuhl, K. Ploog, G. Weimann, Proc. 17th Int. Conf. Phys. Semicond., San Francisco, 1984.
[38] D. von der Linde and R. Lambrich, Phys. Rev. Lett. 42 (1979) 1090.
[39] The decay itself is nonexponential; see also refs. [26] and [27].
[40] L.C. Chiu and A. Yariv, IEEE J. Quant. Electr. QE-18 (1982) 1406.
[41] C. Smith, R.A. Abram and M.G. Burt, J. Phys. C 16 (1981) L171.
[42] A. Sugimura, Appl. Phys. Lett. 42 (1983) 17.
[43] N.K. Dutta, J. Appl. Phys. 54 (1983) 1236.
[44] D.E. Aspnes, Surf. Sci. 132 (1983) 406.
[45] R.J. Nelson and R.G. Sobus, Appl. Phys. 49 (1978) 6103; R.J. Nelson, J. Vac. Sci. Techn. 15 (1978) 1475.

Journal of Luminescence 30 (1985) 551–561
North-Holland, Amsterdam

QUANTUM WELL LASERS

L.C. CHIU

ITT Advanced Technology Center, Shelton, CT 06484, USA

A. YARIV

California Institute of Technology, Pasadena, CA 91125, USA

Basic issues of quantum well lasers, such as the gain, temperature dependence of the threshold current and modulation characteristics are reviewed and discussed.

1. Introduction

The field of semiconductor lasers has witnessed tremendous advancements in the last decade, primarily due to the importance of these devices in optical communication. The most common semiconductor lasers are of the double heterostructure (DH) type which consists of a thin active layer (0.1–0.3 μm) sandwiched between two layers of higher gap materials. The rapid advancement in crystal growth techniques has made possible a new class of semiconductor lasers in which the thickness of the active layer is on the order of or smaller than the de Broglie wavelength of the carriers. When the thickness approaches the de Broglie wavelength, quantized size effect emerges, and the material properties gradually deviate from the more familiar bulk properties as the size is reduced.

In these structures, the carrier motion in the plane perpendicular to the heterojunction (z direction) is restricted by the potential barrier and very different from the motion in the plane parallel to the junction (x-y plane). In fact, the carrier motion in the z direction can be viewed as a particle in a finite square well potential which is treated in all quantum mechanics text books, and these devices are appropriately called quantum well lasers. Devices which have a single low bandgap active layer are known as single quantum well lasers, and those with active layer consisting of many alternating layers of low and high bandgap materials forming a superlattice type structure are generally called multiple quantum well lasers. Quantum well (QW) lasers have received much attention in recent years, as they possess several novel characteristics different

0022-2313/85/$03.30

from conventional DH lasers, among which are extremely low threshold current densities [1,2], less temperature dependence of the threshold currents [3,4], wavelength tailoring through variation of the well sizes, and improved dynamical properties [5,6]. Apart from the most common GaAs–GaAlAs devices, single and multiple quantum well lasers have been fabricated in a host of other material systems such as InGaAsP–InP [7], InGaAs–InP [8,9], GaSb–GaAlSb [10], InGaAs–InAlAs [11], and InGaAs–GaAs [12]. This paper reviews some of the basic issues of quantum well lasers, and parameters pertaining to the GaAs–GaAlAs system will be used. Section 2 addresses the gain in QW structures. Section 3 illustrates the effects of gain, mode gain, confinement factor on laser threshold. Section 4 deals with the T_0 of QW lasers. Section 5 discusses the modulation characteristics of QW lasers. Section 6 concludes the discussion and findings.

2. Gain in quantum well structures

The gain in QW lasers has been treated in several works using simple models [13,14,15]. Normally, the wave function of the carriers is assumed to be of the form

$$\psi_n=(2/L_z A)^{1/2} f_n(z) \exp(\mathrm{i}\boldsymbol{k}\cdot\boldsymbol{\rho}) u_k(\boldsymbol{\rho}),$$

where L_z is the thickness of the quantum well, A is the area, $\boldsymbol{k}$ and $\boldsymbol{\rho}$ are the momentum and position vectors in the x-y plane respectively, u is the modulation part of the Bloch function, and n denotes the nth subband in which the carriers reside.

The transition matrix element $H_{k'k}$ is then given by

$$H_{k'k}=-\frac{eA_0}{2m_0} M\times I_{n'n},$$

where e is the electronic charge, A_0 is the amplitude of the vector potential of the radiation field, m_0 is the mass of the electron, and M is the momentum matrix element given by

$$M=-\mathrm{i}\hbar\int u_{k'}^{*}\nabla_{\rho} u_k\, \mathrm{d}\boldsymbol{\rho}, \qquad I_{n'n}=\frac{2}{L_z}\int_0^{L_z} f_{n'} f_n\, \mathrm{d}z.$$

For simplicity, most authors assume either an infinite square well [15] or that the carriers reside only in the lowest subband, and the spatial dependence of f_n is assumed to be sin $n\pi z/L_z$. Moreover, the band in the x-y plane is taken to be parabolic, and the energy spectrum is simply given by

$$E_{\mathrm{c,v}}=\hbar^2 k^2/2m_{\mathrm{c,v}}+E_z,$$

where $m_{\mathrm{c,v}}$ is the effective mass of the electrons and holes in the conduction and valence bands respectively. Applying the above approximations and the assumption of a constant matrix element M enables an analytical expression to be derived for the gain which results in

$$g=\frac{\hbar}{\bar{n}c\varepsilon_0(\hbar\omega)}\left(\frac{e}{m_0}\right)^2\left(\frac{m_r}{L_z\hbar^2}\right)M^2(f_c-f_v)\,\mathrm{Int}\left(\frac{\hbar\omega-E_g}{E_{1c}+E_{1v}}\right)^{1/2}\delta_{n'n},$$

where $\bar{n}$ is the refractive index, ε_0 is the permittivity of free space, $\hbar\omega$ is the energy of the photon, m_r is the reduced mass of electrons and holes, E_g is the bandgap, $E_{1c,v}$ are the lowest subband energies of the conduction and valence band respectively, and the f's are the Fermi functions. The Kronecker delta denotes the allowed transitions with $\Delta n=0$. The momentum matrix element M is

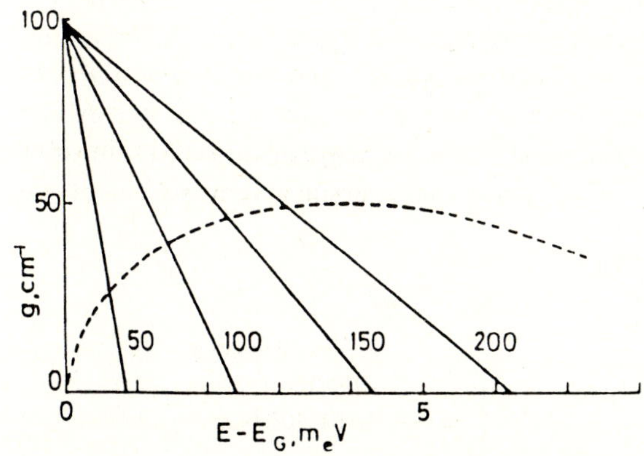

Fig. 1. Gain spectra of quantum-well lasers from ref. [13].

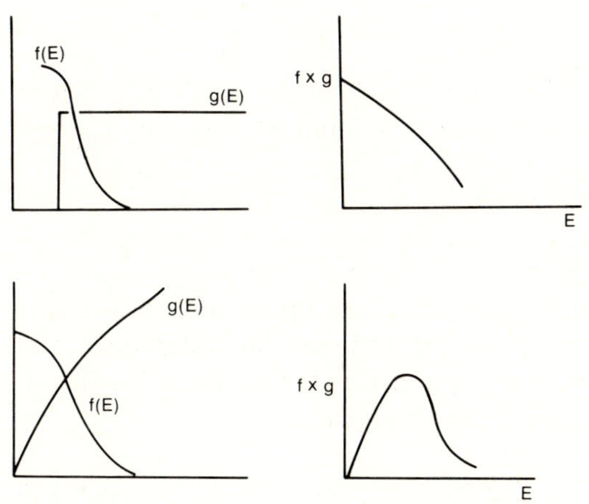

Fig. 2. Densities of states and Fermi functions for conventional and quantum well lasers.

assumed to be identical to or scale directly with the bulk value. It should be noted that even if M scales with the bulk value, the magnitude of the scaling factor is unknown, and is usually assumed to be unity in most publications [13,14,15].

Figure 1 shows the calculated gain from ref. [13]. Note that the gain always assumes its maximum value at the bottom of the well. This is due to the quasi-two-dimensional nature of the density of states, and can be understood from the simple drawing of fig. 2. Figure 3(a) is the measured gain of a single QW laser with $L_z = 210$ Å [16], and fig. 3(b) is that of a conventional DH laser [17]. Comparing figs. 1 and 3, it can be seen that the measured gain profile for the QW laser resembles more in shape to the bulk case. However, the gain profile is narrower than the conventional case, and has much smaller shift of the peak gain, properties indicative of the two-dimensional nature. This implies that the actual density of states is not exactly step-like, but is probably broadened (could be due to variation in layer thickness) as depicted in fig. 4. Thus, simple models, such as parabolic band and infinite square well approximations, provide a quick way to estimate the general trends and properties of QW and MQW lasers, but is nevertheless inadequate for obtaining accurate quantitative estimates important to device optimization and design. This area probably warrants further fundamental research.

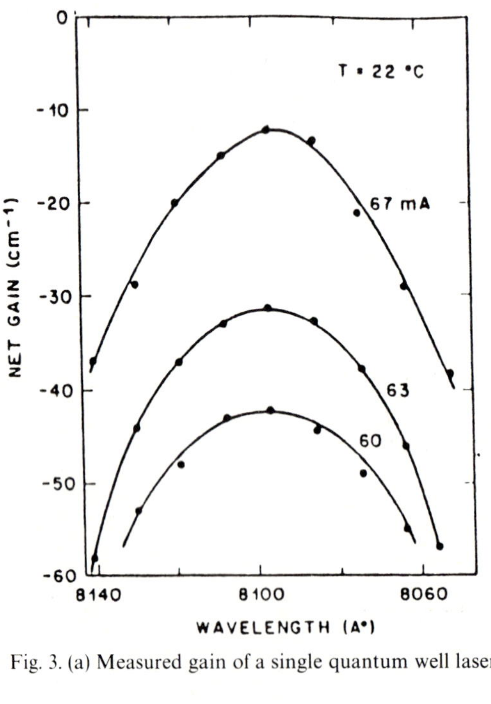

Fig. 3. (a) Measured gain of a single quantum well laser from ref. [16].

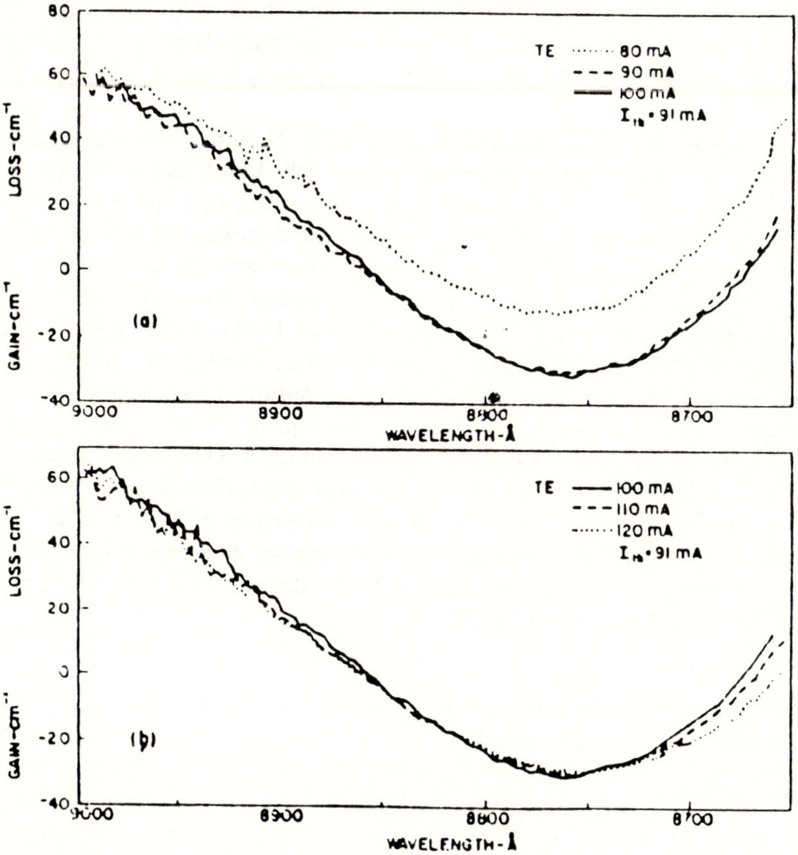

Fig. 3. (b) Measured gain of a conventional DH laser from ref. [17].

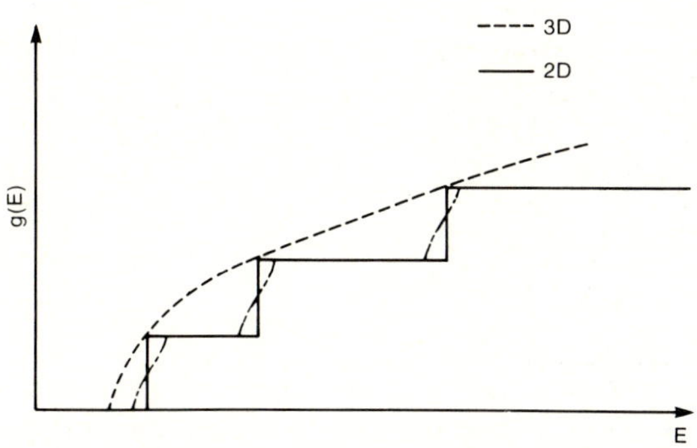

Fig. 4. Broadening of two-dimensional densities of states as indicated by the dash–dot line.

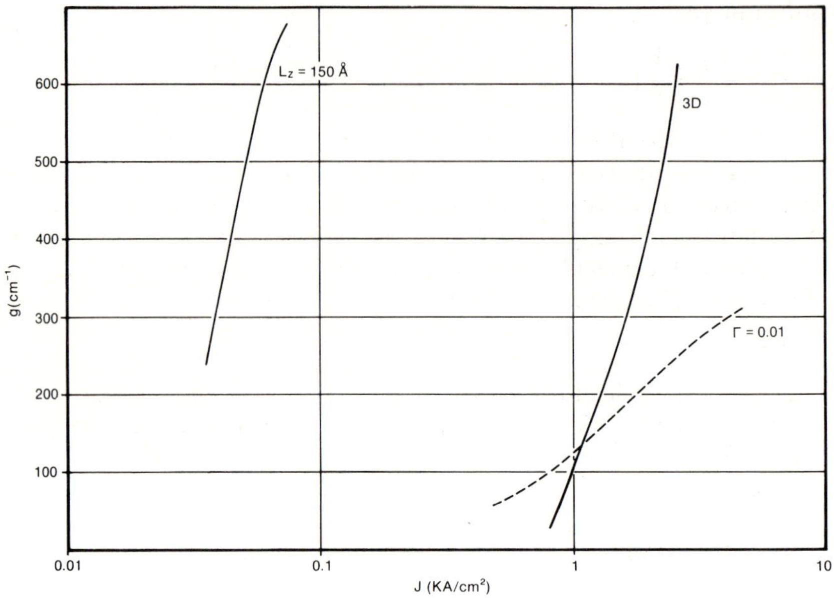

Fig. 5. Gain as a function of current density J for QW and conventional lasers.

Figure 5 shows the calculated (using simple models) gain peak as a function of current density J for a well with $L_z = 150\,\text{Å}$. From fig. 5, we have under normal lasing conditions, the gain peak can be expressed as $g_{\max} = 12.5\,(J - 19.2)$ for $L_z = 150\,\text{Å}$ QW laser. For conventional GaAs laser, the same relation can be expressed as $g_{\max} = 0.045\,(J/d - 4222)$ where d is the active layer thickness. Thus, at the same current density, the peak gain of the QW laser is much higher than that of a conventional laser. However, it should be noted that in actual lasers, it is the "mode gain" which is the relevant parameter. The dotted line of fig. 5 shows the effect when the confinement factor is taken into account. Thus, due to the small confinement factor and high gain needed in single QW lasers, the threshold current density is a very sensitive function of net losses. For losses of about 30–60 cm^{-1}, the threshold current density of a QW laser is comparable to that of a conventional laser. The use of MQW structure is advantageous. This topic will be discussed in the next section.

3. Mode gain

The mode gain G is defined as $G \equiv \Gamma g$, where Γ is the optical confinement factor. The mode gain required at lasing is determined by the total absorption and mirror losses

$$G=\Gamma g=\alpha+\frac{1}{L}\ln\left(\frac{1}{R}\right),$$

where L is the cavity length and R the reflectivity of the mirror. Though the gain g increases as the active layer thickness is reduced, the confinement factor Γ decreases rapidly as L_z is reduced. Thus, there is a trade-off between the two effects. To illustrate the impact on device design, we will consider the threshold current densities for a single and 3 well laser with $L_z=150\,\text{Å}$. The threshold current density J_{th} can be expressed as

$$J_{\text{th}}=n_{\text{th}}L_z(e/\tau),$$

where τ is the carrier lifetime. Assuming the barrier and the well thicknesses are the same for the three well laser, the ratio of the threshold current densities is approximately given by

$$J_1/J_3 \approx n_1/3n_3.$$

The ratio of the threshold carrier densities will be determined by G, g and Γ.

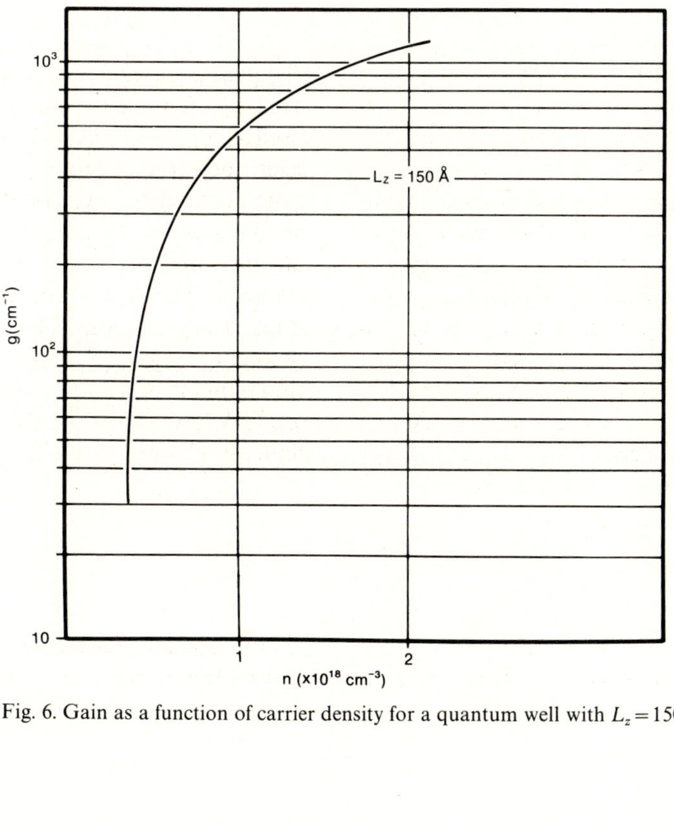

Fig. 6. Gain as a function of carrier density for a quantum well with $L_z=150\,\text{Å}$.

The confinement factors for the 1 and 3 well lasers will be approximately 0.01 and 0.23 respectively. Assuming a mode gain of 30 cm^{-1} is required for lasing, then the gain g needed will be 3000 and 130 cm^{-1} for the 1 and 3 well lasers respectively. The threshold carrier densities can then be determined (e.g. through fig. 6), and are approximately $\sim 8\times10^{18}$ and 0.5×10^{18} cm^{-3}. Thus the threshold current density for the 1 well laser will be about five times higher than that for the 3 well laser. The simple example here is only to illustrate the factors involved in the actual design of devices, and shows the advantage of using multiple quantum well structure. However, as stated in the last section, the use of simple models is inadequate for arriving at accurate values of well thickness, number of wells, etc., for actual devices, and values obtained should only be regarded as general guidelines.

4. Temperature dependence of threshold current

The temperature dependence of the threshold current of semiconductor lasers is given by the empirical expression

$$I = I_0 \exp(T/T_0),$$

where T_0 is a characteristic temperature which measures the temperature sensitivity of the threshold current. An interesting property of the QW lasers is that they possess higher T_0's than conventional DH lasers. Hess et al. [4] derived

$$J_{th} \sim kT_c \exp(-T^*/T_c)$$

for the threshold current density of QW lasers in the case where Fermi–Dirac distribution applies. In the above, T^* is a parameter related to the densities of states and transition matrix element, and T_c is the carrier temperature, which may be different from the lattice temperature T_1. For large T_c, J_{th} varies linearly with T_c, rather than exponentially. It has been found that $T_c > T_1$ at high carrier densities [18], and depends only weakly on T_1. Assuming $T_c = T_1$, QW lasers would still be less temperature sensitive, as can be understood qualitatively from the expression for $f_c - f_v$. To achieve optical gain, one requires $f_c - f_v > 0$, which for QW lasers is given (using simple models) by

$$\frac{\exp(n/N_c)-1}{\exp\left\{\left[\frac{m^r}{m_c}(\hbar\omega-E_g)-E_{1c}\right]\Big/kT\right\}+\exp\left(\frac{n}{N_c}\right)-1}$$

$$-\frac{1}{\left[\exp\left(\frac{n}{N_v}\right)-1\right]\exp\left\{-\left[\frac{m_r}{m_v}(\hbar\omega-E_g)+E_{1v}\right]\Big/kT\right\}+1} > 0.$$

Since $N_c \sim T$ and $N_c \sim T^{3/2}$ for two- and three-dimensional systems respectively, the condition to achieve optical gain is thus less sensitive to temperature for QW lasers.

Experimentally, some QW lasers do exhibit higher T_0's [3,19], whereas others have T_0's of the same magnitude as conventional lasers [2,9]. The low value of T_0 can probably be attributed to carrier leakage over the heterobarrier and other nonradiative recombinations. The work of the Thomson-CSF group [2] has indicated inefficient collection of electrons for wells below 150 Å, and the graded index structure (GRIN) has been demonstrated to improve carrier collection and increase T_0. J. Tang et al. [20] has shown that carrier scattering decreases rapidly for $L_z \leqslant 100$ Å. To increase carrier collection efficiency, GRIN or multiple quantum well structures are desirable.

5. Dynamical characteristics

An important and rather unique characteristic of semiconductor laser is that it can be directly modulated by simply varying the applied current. The capability and the ease at which semiconductor lasers can be directly modulated have made these devices indispensable components in modern optical communication links. It has been found that QW lasers have superior dynamical behavior of suppressed relaxation oscillation and reduced quantum noise peak than conventional DH lasers. In addition, due to the narrower gain spectra (fig. 3), the number of longitudinal modes can be reduced [6]. Another advantage that can be expected for QW well lasers is the extended bandwidth for modulation. The highest frequency at which one can modulate semiconductor lasers is determined by the relaxation oscillation frequency. From simple rate equation analysis, the relaxation oscillation frequency is given by

$$f_r = (1/2\pi)(AP/\tau_p)^{1/2},$$

where A is the differential gain coefficient, P is the photon density in the laser cavity, and τ_p the photon lifetime. Figure 7 shows the ratio of the differential gain coefficient for QW and conventional lasers [13], and the square root of the ratio gives directly the possible increase in modulation bandwidth at the *same* P and τ_p. As is evident from fig. 7, significant improvement in speed performance can be obtained in quantum well lasers. Another important parameter related to A is the linewidth enhancement factor α [21]. Because of the higher A, one would intuitively expect the linewidth enhancement factor α to be smaller in QW lasers. Though α for QW lasers is smaller at the *same* excitation level $\hbar\omega - E_g$, no significant advantage is gained in QW lasers as lasing occurs much closer to the bandgap than conventional lasers, where lasing at a few kT above E_g leads to smaller α [21]. Thus, no significant reduction in α is expected for QW lasers [21, 22].

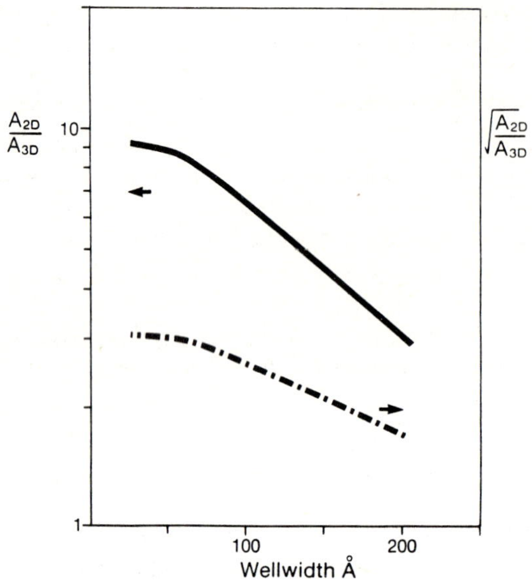

Fig. 7. Comparison of the differential gain constant for quantum well and conventional lasers.

6. Summary

Quantum well lasers have very different characteristics from conventional lasers due to the two-dimensional nature of the devices. The higher gain attainable in QW lasers has made possible lasers with extremely low threshold current densities. These would lead to much lower power dissipation, and made these lasers viable candidates for integrated optoelectronics. The gain of QW lasers has been discussed in some depth, and it is shown that though simple models are useful to predict general properties and trends, more sophisticated models should be employed in order to obtain meaningful quantitative values important for the design and optimization of QW lasers. The topic of QW lasers will continue to be an area of active research, and many more fruitful results can be expected in the near future.

References

The reference articles are chosen as representatives for the topics discussed, and the list is by no means exhaustive. The authors regret for the omission of any deserving articles which may have been left out.

[1] W.T. Tsang, Extremely low threshold (ALGa)As modified multiquantum well heterostructure lasers grown by molecular-beam epitaxy, in: Appl. Phys. Lett. 39 (1981) pp. 786–788.

[2] S.D. Hersee, B. de Cremoux and J.P. Duchemin, Some characteristics of the GaAs/GaAlAs graded-index separate-confinement heterostructure quantum well laser structure, in: Appl. Phys. Lett. 44 (1984) pp. 000–000.

[3] W.T. Tsang and R. L. Hartman, cw narrow beam (AlGa)As multiquantum-well heterostructure lasers grown by molecular beam epitaxy, in: Appl. Phys. Lett. 38 (1981) pp. 502–504.

[4] K. Hess, B. A. Vojak, N. Holonyak, Jr. and R. Chin, Temperature dependence of threshold current for a quantum-well heterostructure laser, in: Solid State Elect. 23 (1980) pp. 585–589.

[5] E. R. Anderson, B. A. Vojak, N. Holonyak, Jr., G.E. Stillman, J.J. Coleman and P.D. Dapkus, Transient and noise characteristics of quantum-well heterostructure lasers, in: Appl. Phys. Lett. 38 (1981) pp. 585–587.

[6] H. Iwamura, T. Saku, T. Ishibashi, K. Otsuka and Y. Horikoshi, Dynamic behaviour of a GaAs–AlGaAs MQW laser diode, in: Elect. Lett. 19 (1983) pp. 180–181.

[7] E.A. Rezek, R. Chin, N. Holonyak, Jr., S.W. Kirchoefer and R.M. Kolbas, Quantum-well InP–$In_{1-x}GasbxP_{1-z}As_z$ heterostructure lasers grown by liquid phase epitaxy (LPE), in: J. Elect. Mater. 9 (1980) pp. 1–27.

[8] K. Kodama, J. Komeno and M. Ozeki, in: Optically Pumped $In_{0.53}Ga_{0.47}As$/InP MQW lasers grown by chloride vapour-phase epitaxy, Elect. Lett. 20 (1984) pp. 48–50.

[9] W.T. Tsang, Gasb0.47Insb0.53As/InP multiquantum well heterostructure lasers grown by molecular beam epitaxy operating at 1.53 μm, in: Appl. Phys. Lett. 44 (1984) pp. 288–290.

[10] H. Temkin and W.T. Tsang, Optically pumped GaSb/$Al_{0.6}Ga_{0.4}Sb$ multiquantum well lasers operating in the $\lambda = 1.5$–1.6 μm region, in: J. Appl Phys. 55 (1984) pp. 1413–1415.

[11] H. Temkin, K. Alavi, W.R. Wagner, T.P. Pearsall and A.Y. Cho, 1.5–1.6 μm Gasb.47$In_{.53}$as/$Al_{.48}In_{.52}As$ multiquantum well laser grown by molecular beam epitaxy, in: Appl. Phys. Lett. Lett. 42 (1983) pp. 845–847.

[12] W.D. Laidig, P.J. Caldwell, Y.F. Lin and C.K. Peng, Strained-layer quantum-well injection laser, in: Appl. Phys. Lett. 44 (1984) pp. 653–655.

[13] M.G. Burt, Gain spectra of quantum-well lasers, in: Elect. Lett. 19 (1983) pp. 210–211.

[14] N.K. Dutta, Calculated threshold current of GaAs quantum well lasers, in: J. Appl. Phys. 53 (1982) p. 7211–7214.

[15] A. Sugimura, Threshold currents for AlGaAs quantum well lasers, in: IEEE J. Quant. Elect. QE-20 (1984) pp. 336–343.

[16] N.K. Dutta, R.L. Hartman and W.T. Tsang, Gain and carrier lifetime measurements in AlGaAs single quantum well lasers, in: IEEE J. Quant. Elect. QE-19 (1983) pp. 1243–1246.

[17] B.W. Hakki and T.L. Paoli, Gain spectra in GaAs double heterostructure injection lasers, in: J. Appl. Phys. 46 (1975) pp. 1299–1306.

[18] Z.Y. Xu and C.L. Tang, Picosecond relaxation of hot carriers in highly photoexcited bulk GaAs and GaAs–ALGaAs multiple quantum wells, in: Appl. Phys. Lett. 44 (1984) pp. 692–694.

[19] K. Kodama, J. Komeno and K. Ozeki, Photoexcited $In_{0.53}Ga_{0.47}As$/InP quantum-well lasers with high characteristic temperatures, in: Elect. Lett. 20 (1984) pp. 42–44.

[20] J.Y. Tang, K. Hess, N. Holonyak, Jr., J.J. Coleman and P.D. Dapkus, The dynamics of electron–hole collection in quantum well heterostructures, in: J. Appl. Phys. 53 (1982) pp. 6043–6046.

[21] K. Vahala, L.C. Chiu, S. Margalit and A. Yariv, On the linewidth enhancement factor α in semiconductor injection lasers, in: Appl. Phys. Lett. 42 (1983) pp. 631–633.

[22] M.G. Burt, Linewidth enhancement factor for quantum-well lasers, in: Elect. Lett. 20 (1984) pp. 27–29.

Journal of Luminescence 30 (1985) 562–579
North-Holland, Amsterdam

INJECTION, INTERSUBBAND RELAXATION AND RECOMBINATION IN GaAs MULTIPLE QUANTUM WELLS

D. BIMBERG and J. CHRISTEN

Institut für Festkörperphysik I, Technische Universität Berlin, Strasse des 17. Juni 135, 1000 Berlin 12, FRG

A. STECKENBORN

Heinrich-Hertz-Institut für Nachrichtentechnik Berlin GmbH, Einsteinufer 37, 1000 Berlin 10, FRG

G. WEIMANN and W. SCHLAPP

Forschungsinstitut der Deutschen Bundespost beim Fernmeldetechnischen Zentralamt, 6100 Darmstadt, FRG

Results of continuous excitation, time-delayed and time-resolved cathodoluminescence experiments on undoped and Be-doped GaAs multiple quantum wells of thickness 5–11 nm are reported for temperatures 5–300 K and excitation intensities 1–10^3 W/cm^2. Comparison is made with similar investigations of bulk epitaxial GaAs layers. Increasing structural localisation of the carriers is identified to lead to a qualitative change of the type of recombination. Increasingly faster radiative excitonic recombination leads to a bypass of impurity and trap capture processes in undoped as well as in doped material, at low temperatures as well as at room temperature. Despite the exponential character of excitonic decay, the luminescence transients are found to be very complex due to an interplay of intersubband scattering and recombination processes and time-dependent carrier temperature. Transients are analysed in detail, excitonic lifetimes and intersubband scattering times are derived. It is argued that both the lifetime reduction in quantum wells and the novel process of recombination heating lead to a strongly increased quasiequilibrium temperature of excited carriers in the wells as compared to bulk material. Injection of carriers from 18 nm GaAlAs barriers to GaAs wells is found to occur in less than 10^{-12} s without loss.

1. Introduction

The invention and refinement of novel techniques – like molecular beam epitaxy (MBE) and metal organic chemical vapour phase deposition (MOCVD) [1,2] for the deposition of thin single-crystalline layers of a thickness of a few lattice constants of high crystallographic and chemical perfection lead to a new

0022-2313/85/$03.30 © Elsevier Science Publishers B.V.
(North-Holland Physics Publishing Division)

type of "structural architecture" in semiconductor and device physics called "wavefunction engineering". Structural localisation, e.g., in two-dimensional quantum wells or in enhancement layers at heterointerfaces enforces a quantisation of all electronic levels [3–5]. Consequently electronic and optical properties of such structures are size dependent and differ fundamentally from those of the three-dimensional "parent" material.

Investigations of the radiative recombination of localized electrons and holes yield information about the position of the lowest 2-D energy levels and partly about the type of the recombination process involved. Although recent extensive photoluminescence experiments of many groups [6] show that the radiative recombination in two dimensions does not only differ in its energetic position but also by its type from that in three dimensions no clear insight exists about the physical origin of the latter fundamental difference. It is the purpose of this paper to provide the missing insight: The increase of structural localisation leads to an increase of the radiative recombination rate of intrinsic excitonic recombination processes as compared to extrinsic ones and to trapping. Thus charge carriers do not relax and recombine stepwise via impurities as in a three-dimensional material, but in one step, emitting radiation of a photon energy of the order of the band gap: Impurity and trap states are bypassed by the charge carriers. Time-delayed and time-resolved cathodoluminescence experiments presented here show this unambiguously. The kinetics of the decay of the carrier population in the wells and in addition to that the dynamics of charge carrier transfer from barriers via interfaces to quantum wells are studied in detail.

The paper is organised as follows: In the next two sections we describe briefly our new experimental technique of time-resolved cathodoluminescence and some of the properties of the superlattices used for our experiments. Results of continuous excitation experiments and surprising conclusions on speed and efficiency of charge carrier transfer from the $Al_xGa_{1-x}As$ barriers to the GaAs quantum wells are presented in section 4. A comparative study of carrier recombination and trapping after the end of an exciting pulse in MQW's and bulk material is given in section 5. A detailed analysis of cathodoluminescence decay times given in section 6 yields information about the kinetics of interband and intersubband recombination and relaxation processes. Some remarks on energy relaxation of hot carriers given in section 7 and a brief summary (section 8) conclude the paper.

A preliminary report on some results was presented recently [7]. Picosecond time resolved photoluminescence studies using streak camera techniques complement and extend our results to shorter times and to the range of inhomogeneous excitation [8]. The unique properties of structurally localised charge carriers in quantum wells described here can be used in many respects to create novel and to improve present opto-electronic devices [9].

2. A setup for time-resolved cathodoluminescence experiments

The cathodoluminescence (CL) system – a block diagram is shown in fig. 1 – is implemented on a commercial scanning electron microscope (SEM). The vacuum system is modified and cryogenic shielding is introduced to suppress contamination of the samples at low temperatures [10]. A beam-blanking unit is implemented directly below the electron gun. Electron pulses of rise and decay time $\leqslant 200$ ps, varying width of $\leqslant 1$ ns up to 10 μs and a repetition rate of 1 kHz up to 1 MHz are used for excitation. A miniaturized continuous flow He cryostat manufactured by Cryovac similar to the one described in refs. [10,11] is fitted to a standard specimen stage. The temperature is continuously variable in the range $5\ \mathrm{K} \leqslant T \leqslant 300$ K. Lattice and charge carrier or exciton temperatures of typical samples are determined independently as described earlier [10,12] to control or avoid heating of the lattice as compared to the cryostat temperature. 30 keV electrons at varying currents yielding intensities $0.8\ \mathrm{W/cm^2} \leqslant I_{exc} \leqslant 1800\ \mathrm{W/cm^2}$ are used for excitation. The emitted light is focussed by an elliptical mirror on to the entrance slit of a McPherson 30 cm vacuum monochromator. The electron beam is always scanned over a small square of 7 μm × 7 μm and the luminescence light presents a mean value of the properties of this area. The light is detected by a cooled photomultiplier with GaAs cathode. The pulse generator which triggers the beam blanking unit also triggers a photon counting system via a delay line. The photon counting system operates according to the method of delayed coincidence in an inverted

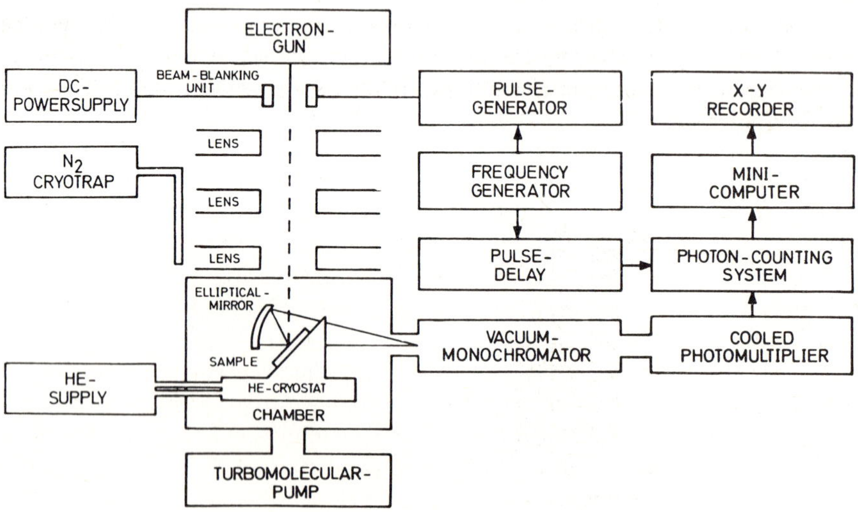

Fig. 1. Scanning electron microscope with low temperature cathodoluminescence setup, schematically. Temperature of sample, $5\ \mathrm{K} \leqslant T \leqslant 300$ K; time resolution $\leqslant 250$ ps.

mode, however, thus yielding an improved S/N ratio. The dynamic range of our detection system is $\cong 10^5$. The pulse height distribution generated by the photon counting system is evaluated by a computer controlled (DEC LSI 11/23) multichannel analyser, a system usually employed for energy-dispersive X-ray analysis. The data are stored on a magnetic disc. Our data acquisition technique has the unique advantage that simultaneously up to 12 time windows with varying widths – to optimize independently the S/N ratio – can be electronically set. Thus up to 12 spectra taken at different times with respect to the start of the exciting pulse can be recorded simultaneously. The time resolution of the *total* CL setup including photomultiplier and electronics is experimentally found to be $\leqslant$250 ps.

Figure 2 presents an illustration of this novel procedure. A 85 ns electron pulse is employed here to excite the luminescence. The onset of the luminescence and its kinetics, the quasisteady state and the decay and its kinetics are shown. There are two important advantages of this method:

All spectra taken at different delay times are generated from the same pulses under the same experimental conditions. No repetitive wavelength scans are necessary.

The data acquisition time is largely reduced as compared to a sequential measurement. Beyond that a time-resolved cathodoluminescence experiment as described here is superior to a typical time-resolved photoluminescence experiment having similar time resolution from three different points of view:

(a) The depth of excitation can be easily varied by varying the electron acceleration voltage between 200 V and 40 keV (Jeol JSM 840). The projected range of such electrons, e.g., in GaAs varies from a few nm only to 5.3 μm, respectively. Thus predominantly surface or surface-free excitation, respectively, can be chosen depending on the requirements of the actual experiments. At any penetration depth the excitation is much more homogeneous in volume than after laser illumination.

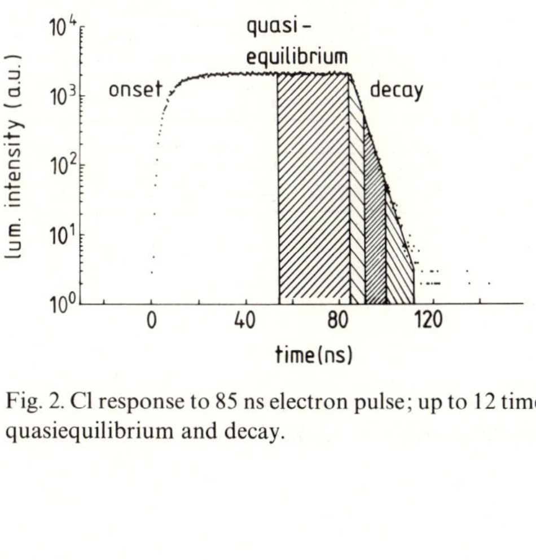

Fig. 2. Cl response to 85 ns electron pulse; up to 12 time windows can be arbitrarily set during onset, quasiequilibrium and decay.

(b) The width of the exciting pulses can be varied by more than four orders of magnitude without altering the pulse rise and decay time. Thus the temporal resolution of the experiment is independent of the pulse width.

(c) The electron beam can be easily scanned over the surface of the crystal using the standard scanning coils of the SEM. Two-dimensional pattern of the luminescence intensity and lifetime can be taken in a few minutes [11].

3. Properties of the samples

Figure 3 gives a cross section of a typical sample. Each multiple quantum well system consists of up to 60 GaAs layers, sandwiched between $Ga_{0.6}Al_{0.4}As$ layers, having both a constant but different thickness. The width of the GaAlAs layers is 17.6 nm, to ensure decoupling of the individual GaAs wells. The composition of GaAlAs is such that the material is still direct, but the depth of the QW is close to maximum. Charge carriers which are transferred from the GaAlAs to the GaAs do not need to be scattered from an indirect to a direct minimum. The width of the GaAs layers varies from sample to sample between 5.2 nm and 11.3 nm. Bulk material is investigated additionally. The MQW layers are either high purity undoped p-type ($p \approx 10^{14}$ cm^{-3}) or uniformly Be-doped ($p \approx 10^{16}$ cm^{-3}). All samples used for the work reported here are grown under similar conditions from the same sources on top of semi-insulating Cr-doped GaAs substrates. 0.5 μm thick GaAs and 0.13 μm thick $Ga_{0.6}Al_{0.4}As$ buffer layers are grown between the substrate and the MQW's. The surfaces of all crystals are passivated by a 0.13 μm $Ga_{0.6}Al_{0.4}As$ cap layer. Using 30 keV electrons for excitation the penetration depth is large enough that the substrate

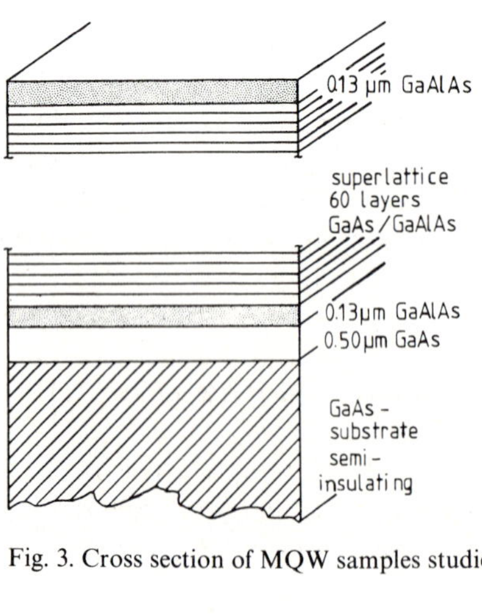

Fig. 3. Cross section of MQW samples studied in this paper.

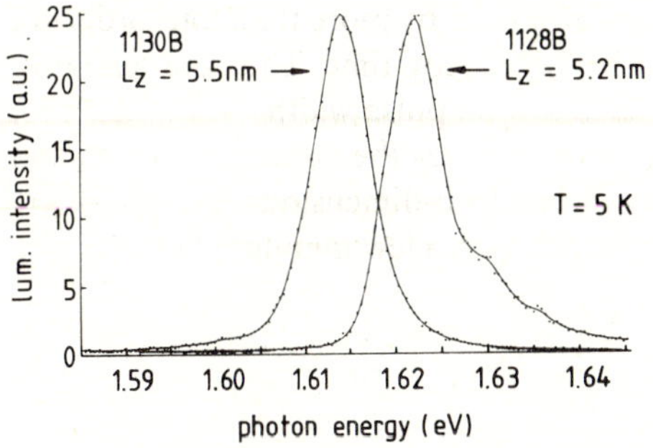

Fig. 4. Low temperature Cl spectra of doped and undoped sample; note: the well thickness differs by ≈1 atomic layer only.

is still excited to a depth of ≈1 μm. It is obvious that neighbouring GaAs wells and GaAlAs barriers are excited with similar intensity.

The halfwidth of the luminescence line can be used as a measure of the abruptness of the heterointerface and of the reproducibility of the thickness of subsequent wells. Figure 4 shows on a linear scale the luminescence spectra of an undoped sample with $L_z = 5.2$ nm and a Be-doped sample with $L_z = 5.5$ nm. The L_z values given here are based on a numerical solution of the one-dimensional Schrödinger equation of a particle in a box with finite barrier height, using the known GaAs electron and hole masses [13]. Although the mean difference of L_z of the two QW systems is only ≈1 atomic layer both lines are clearly separate from each other. The $L_z = 5.2$ nm line has a halfwidth of 6.6 meV.

Let us assume that the intrinsic line width of the recombination process and its thermal broadening be zero. A variation of the well thickness L_z either laterally or of subsequent wells leads then to a finite line width $\Delta\hbar\omega$ of the luminescence. The variation ΔL_z in the excitation volume is

$$|\Delta L_z| = \tfrac{1}{2}(\Delta\hbar\omega/\Delta E_g)L_z, \tag{1}$$

if an energy dependence of the lowest conduction and valence bands $\sim L_z^{-2}$ is assumed. ΔE_g is the difference between the two-dimensional and the three-dimensional fundamental energy gap. We find $|\Delta L_z| = 0.16$ nm for sample 1128 B. This value presents an upper limit. Since L_z can have only discrete values this $|\Delta L_z|$ value can be interpreted by the appearance of one monolayer step of the length of 36.0 nm in every 100.0 nm interval.

4. Dynamics of charge transfer from the barriers to the quantum wells

Figure 5 shows on a logarithmic scale the cathodoluminescence intensity of three different samples over a large range of photon energies. Light emitted from bulk GaAs, from GaAs quantum wells and from $Ga_{0.6}Al_{0.4}As$ layers is

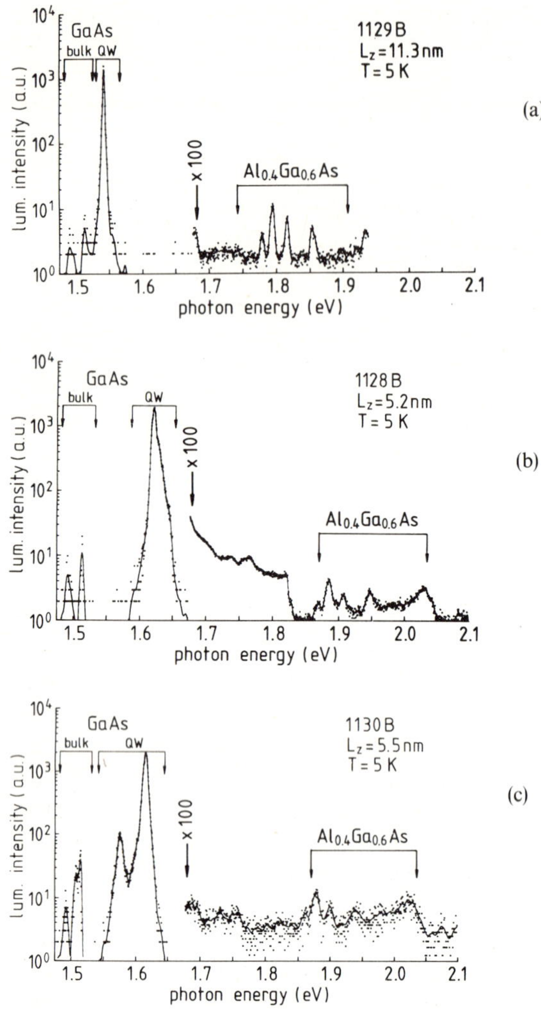

Fig. 5. Comparison of spectra of three different MQW samples: (a) $L_z = 11.3$ nm, undoped $p = 1 \times 10^{14}$ cm^{-3}; (b) $L_z = 5.2$ nm, undoped $p = 1 \times 10^{14}$ cm^{-3}; (c) $L_z = 5.5$ nm, Be-doped $p = 1 \times 10^{16}$ cm^{-3}.

clearly observed at its different energetic positions. It is surprising, however, that the intensity of the light from the GaAlAs layers is so much weaker than the intensity of the light from the GaAs quantum wells. For sample, 1130 B $I^{QW}/I^{GaAlAs} = 2000$ although the ratio of total layer thickness $d^{QW}/d^{GaAlAs} = 1/4$ which is equal to the ratio of the total excited volumes. This typical result will now be used to derive the *effective* lifetime of charge carriers in the GaAlAs barriers.

The luminescence intensity I depends on the generation or injection rate g of charge carriers, the excited volume V and the quantum efficiency η.

$$I = g \cdot V \cdot \eta, \tag{2}$$

$\eta = \tau_{eff}/\tau_r$, where τ_{eff} is the effective lifetime of the charge carriers and τ_r is the radiative lifetime. $1/\tau_{eff} = \sum_i 1/\tau_i$, where i represents the different recombination, trapping, etc., processes. The generation rate of charge carriers by high energy electrons is inversely proportional to the band gap. Thus $g^{QW}/g^{GaAlAs} = 1.27$. If there appears an additional flux g_f of charge carriers from the GaAlAs barriers to the QW's being equal to $g^{GaAlAs} \cdot V^{GaAlAs}$ – which will be shown to be correct below –, then

$$I^{GaAs}/I^{GaAlAs} = [(g^{QW} d^{QW} + g^{GaAlAs} d^{GaAlAs})/g^{GaAlAs} d^{GaAlAs}](\eta^{QW}/\eta^{GaAlAs}). \tag{3}$$

It is certainly reasonable to assume $\eta^{GaAs} \approx 1$. Then

$$I^{GaAs}/I^{GaAlAs} = [1 + (d^{QW} g^{QW}/d^{GaAlAs} g^{GaAlAs})](\tau_r^{GaAlAs}/\tau_{eff}^{GaAlAs}). \tag{4}$$

Inserting the values given above we find

$$\tau_{eff}^{GaAlAs} = 6.6 \times 10^{-4} \tau_r^{GaAlAs}. \tag{5}$$

Using a typical value $\tau_r^{GaAlAs} \approx 10^{-9}$ s, we find $\tau_{eff}^{GaAlAs} \approx 6.6 \times 10^{-13}$ s. The effective lifetime of the charge carriers in GaAlAs is reduced by more than three orders of magnitude as compared to the radiative lifetime. This large reduction can be only explained as a consequence of carrier transfer into the quantum wells. The transfer can occur via scattering by longitudinal optical phonons. The transfer time which we find here is indeed ≈ 4 times longer than typical LO emission times of electrons (1.3×10^{-13} s) and holes (1.8×10^{-13} s) in GaAs [14]. This transfer time is also by orders of magnitude shorter than any capture or nonradiative recombination time which can be reasonably assumed for the relatively perfect and pure GaAlAs layers used here. We conclude therefore that the transfer or injection efficiency of carriers from GaAlAs barriers to GaAs quantum wells is close to or equal 100%, which justifies a posterio the additional generation term assumed in eq. (3). Thus carrier injection should occur with large efficiency for multiple quantum well lasers which have typically structures similar to those discussed here [15–17].

5. Carrier recombination and trapping in two- and three-dimensional GaAs

The recombination channels of nonlocalized and structurally localized charge carriers in GaAs will be shown in this section to differ fundamentally. In fig. 6 three luminescence spectra of a high purity ($[p] \approx 10^{15}$ cm^{-3}) bulk sample taken at 5 K 0,20 and 50 ns after the end of a long excitation pulse are depicted [12]

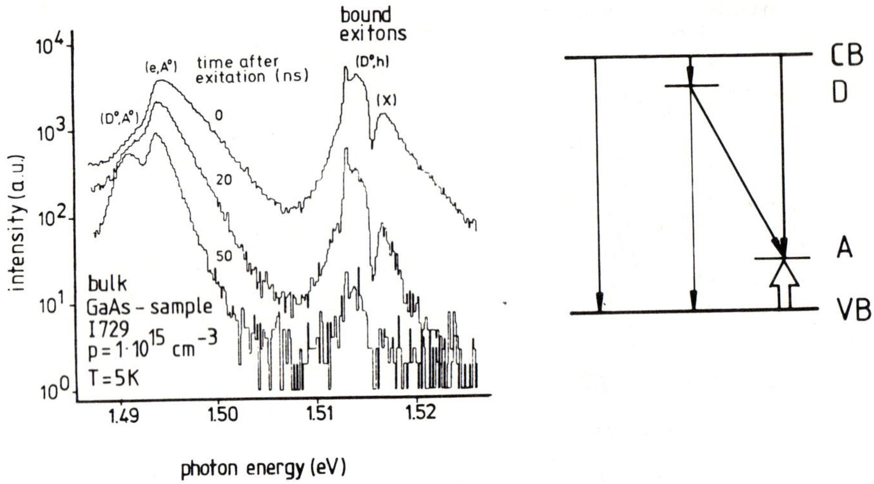

Fig. 6. Left: time-delayed spectrum of high purity GaAs bulk sample; right: the dominant recombination and trapping processes are shown schematically.

on a semilogarithmic scale. The spectrum at 0 ns consists of the well known bound exciton (D^0, h), (e, A^0), (D^0, A^0) recombinations and shows at the high energy side a much weaker free exciton band. The dominant radiative recombination processes are impurity induced. With increasing time after the end of the excitation the center of gravity of the recombination shifts towards longer wavelengths. Lines caused by free holes or excitons disappear since free holes are trapped first.

The return to the thermal equilibrium number of charge carriers is a steplike process as indicated schematically on the right-hand side of the figure. The capture of holes by traps is apparently faster than the recombination of the free exciton.

Figure 7 shows time delayed spectra of a MQW sample with $L_z = 5.5$ nm. This sample has an acceptor concentration (Be) which is a factor 10 larger than in the sample of fig. 6. One might expect a still more rapid disappearance of all free hole or excitonic features with increasing delay time. The reverse is true. Despite the larger contents of acceptors the impurity-induced recombination is faint under continuous excitation and remains faint at any time after the end of excitation. Apparently capture of free charge carriers is bypassed here by fast free-excitonic recombination. The free exciton recombination dominates the luminescence at any time and at temperatures up to room temperature. Figure 8 shows the temperature dependence of the recombination of the same sample as shown in fig. 7 on a logarithmic scale. The faint acceptor induced feature due to (e, A_c) recombination, A_c is the center acceptor [18,19], disappears completely with increasing temperature. Figure 9 compares the temperature dependence

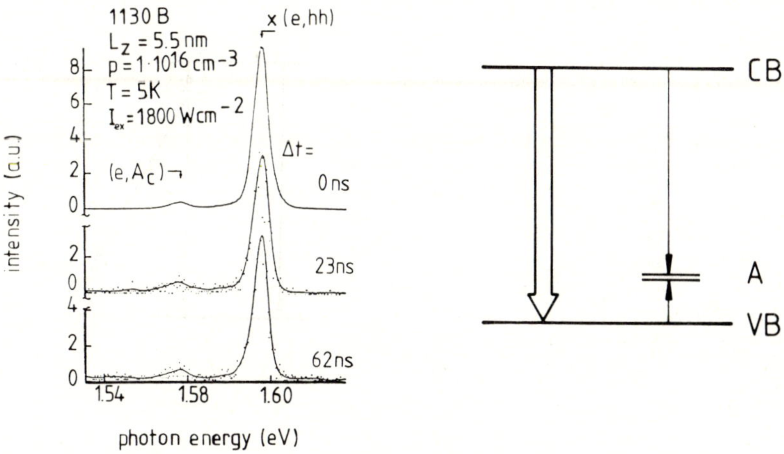

Fig. 7. Left: time-delayed spectrum of doped MQW sample, $L_z = 5.5$ nm; right: recombination via intrinsic (excitonic) processes, bypass of trapping by impurities (schematic).

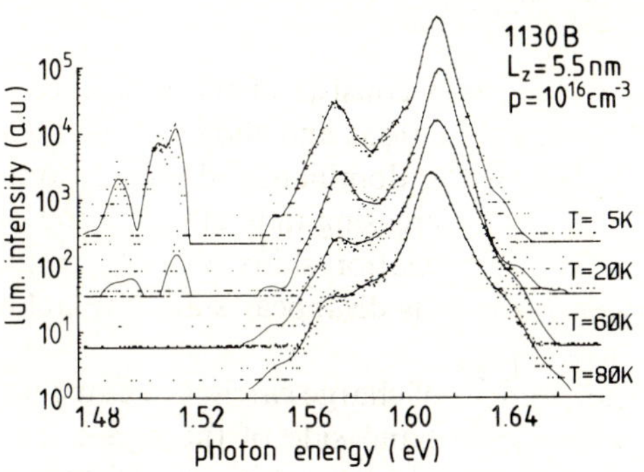

Fig. 8. Temperature dependence of emission (5 K < T < 80 K) of doped sample.

of the peak positions of the heavy X(e, hh) and light X(e, lh) excitons with the temperature variation of the three-dimensional gap. All three temperature variations are exactly identical. The three lines are displaced by each other by a fixed, temperature independent energy difference. Apparently the room temperature recombination from a quantum well has the same free-excitonic origin as the low temperature recombination, otherwise there would be a high-energy-shift of the order of the 2-D exciton binding energy. The identity of the slope $dE/dT = 0.4$ meV/K for 2-D and 3-D at temperatures larger than 100 K indicates that the phonon coupling in the quantum wells investigated here is not different from the coupling in 3 dimensions.

The temperature dependence of the integrated luminescence of quantum wells, however, is very different from that of bulk material. Figure 10 compares

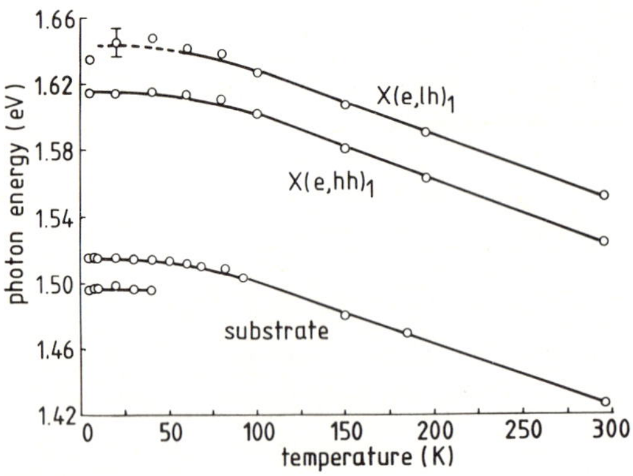

Fig. 9. Temperature dependence of the energetic position of QW and bulk luminescence; energy shift for $T > 100$ K: -0.4 meVK^{-1} for all three cases.

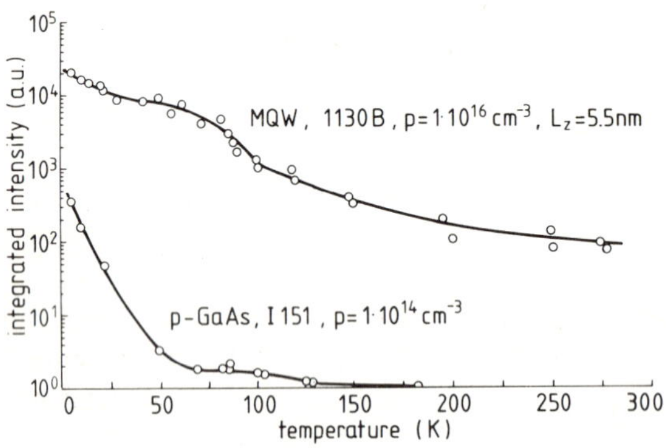

Fig. 10. Temperature dependence of integrated Cl intensity of a QW and a bulk GaAs sample.

the same samples as above. Already at 5 K the bulk sample has a much lower yield. Its intensity drops strongly up to 70 K by 2.5 orders of magnitude. The intensity of the MQW sample drops in the same range only by a factor 4. The delayed decrease found at 60–100 K is due to the ionisation of the neutral acceptors which leads to a larger population of hole subbands higher in energy and in turn to an increased excitonic recombination probability as will be discussed in the next section.

6. Kinetics of interband and intersubband recombination in quantum wells

The dominating processes which govern the return from an excited state to a thermal equilibrium population in a quantum well are much simpler by their nature than in a bulk semiconductor. They are of purely excitonic origin.

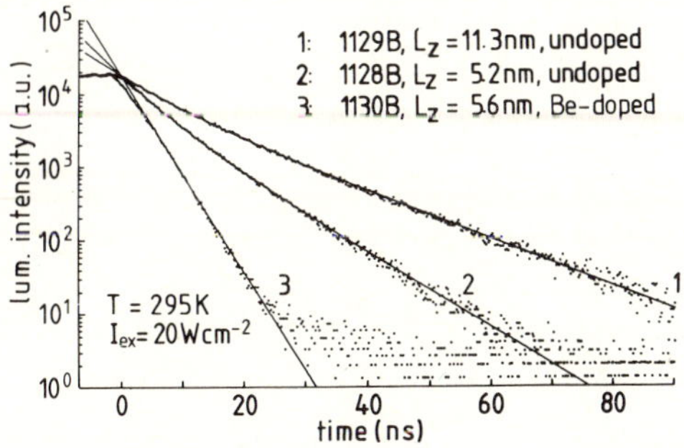

Fig. 11. Room temperature Cl decay after long pulse excitation.

Figure 11 illustrates the three most important qualitative features already reported earlier by us [7,9]. In this figure the transients of three different MQW samples taken at room temperature are superimposed. The width L_z of the wells of the undoped sample 1 is roughly a factor 2 larger than the width of the wells of samples 2 and 3. The acceptor concentration of sample 3 is larger by two orders of magnitude than the acceptor concentrations in samples 1 and 2. We note that:

(a) the transients are apparently exponential, as it is expected for the decay of free excitons. Bimolecular e–h recombination is thus on the other hand excluded as a recombination channel.

(b) the lifetime decreases with decreasing L_z. Increased localisation of the exciton increases its radiative recombination probability. The experimentally observed enhancement is found to agree quantitatively with theoretical predictions [7,20,21].

(c) p-doping further reduces the lifetime without introducing additional impurity related recombination channels. The time-delayed spectra presented in fig. 7 show indeed no influence of the additional acceptors. Instead hole subbands lying higher in energy are additionally populated by free holes and new radiative excitonic recombination channels open. A comparison of the luminescence spectra of the undoped and doped $L_z = 5$ nm samples (fig. 12) shows indeed a stronger $n = 1$ light hole exciton in the doped sample. The same mechanism, increased hole population, leads to the delayed drop of intensity between 60 K and 100 K in the MQW I (T) dependence shown in fig. 10.

A closer examination of the transients shows, however, that they cannot be fitted by one simple exponential over the whole dynamic range of close to 5 orders of magnitude of intensity. At least a sum of 2, often a sum of three functions is needed.

For a given sample and temperature the transients still depend on the wavelength at which they are taken and on excitation intensity. Figure 13 shows transients taken after excitation with pulses of the same length but an excitation

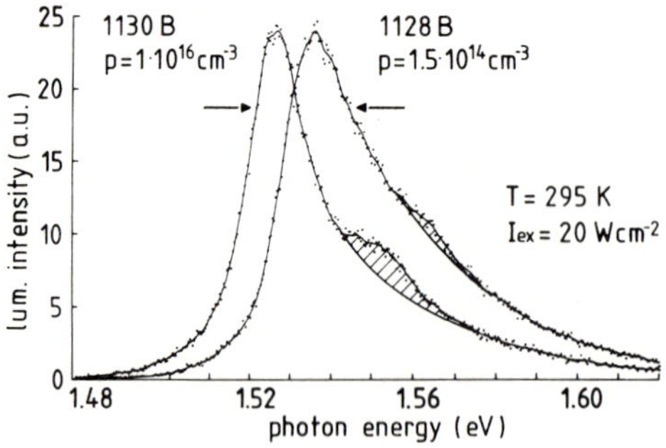

Fig. 12. Cl spectra of same samples as in fig. 4 but at RT. Note: doping opens no additional impurity channel, but increases population of higher hole subbands.

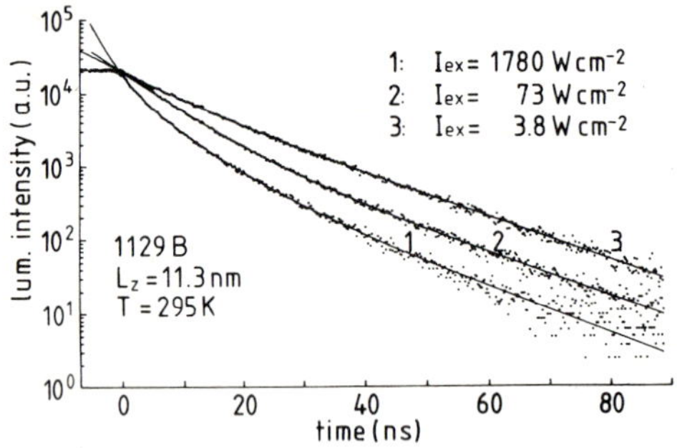

Fig. 13. Cl decay of $X(e,lh)_1$ at 3 different excitation densities.

intensity varying from 3.8 W/cm^2 to 1780 W/cm^2 at the wavelength of the $n=1$ X(e, lh) recombination. At larger excitation the initial decay is much faster, since excitonic recombination channels at higher energies are activated. With increasing time these higher energetic excitons are depleted and contribute less to the recombination.

Figure 14(a) compares cw spectra of a $L_z = 11.3$ nm MQW system taken at various excitation intensities. The relative increase of the $X(e, hh)_2$ exciton luminescence is clearly visible. Figure 14(b) shows time-delayed spectra after 1780 W/cm^2 excitation. The lineshape changes strongly during the first 30 ns due to depopulation of higher energetic exciton states. After that time no such large change occurs any more.

Apparently the distribution function has changed its character. The exciton system then obeys Maxwell statistics with time-dependent temperature. Indeed for low excitation little variation of time-delayed spectra is found already at short delay times.

Although the transients shown in fig. 13 look pretty different, a numerical analysis shows that they can be all described by the same three time constants.

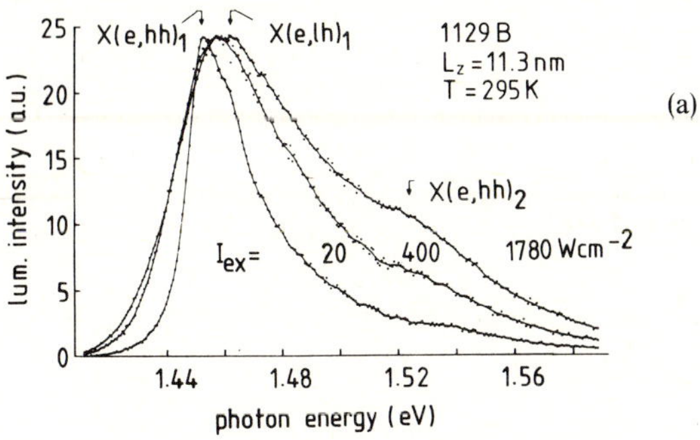

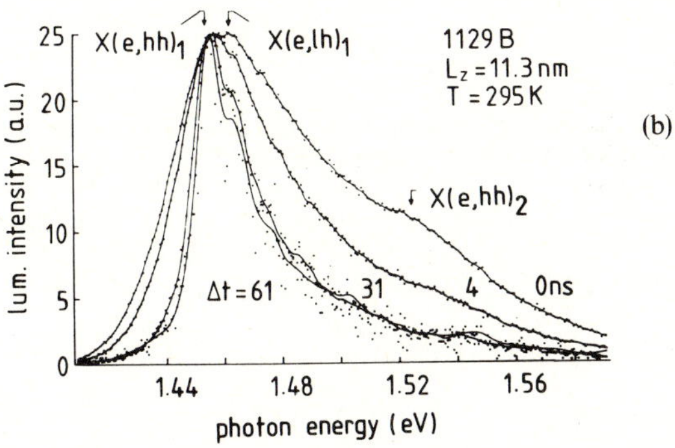

Fig. 14. RT Cl spectra of a MQW with $L_z = 11.3$ nm: (a) variation of excitation density; (b) time delayed spectra at high excitation density.

$\tau_1 = 15$ ns, $\tau_2 = 7$ ns and $\tau_3 = 2$ ns. Only the weight factors by which these time constants contribute differ. At larger excitation the relative contributions of the shorter decay times are increased, thus causing different transients. This brief discussion also shows that great care must be taken when transient experiments are interpreted. Too straight-forward conclusions based on observations of differences of slopes might lead to invalid conclusions.

The results of figs. 13 and 14 will now be used as an example to gain further insight in the recombination kinetics. Figure 14 shows that 5 different subbands apparently participate in the recombination: the $n = 1$ and 2 electron bands, the $n = 1$ and 2 heavy hole bands and the $n = 1$ light hole band. Figure 15 shows their relative energetic ordering and the 5 different time constants which are characteristic for the various intersubband relaxation (τ_e, τ_h) and excitonic recombination processes ($\tau(\mathrm{e,hh})_1$, $\tau(\mathrm{e,hh})_2$, $\tau(\mathrm{e,lh})_1$). The rate equations of the different populations $e_1, e_2, (\mathrm{hh})_1, (\mathrm{hh})_2$ and $(\mathrm{lh})_2$ are (assuming no re-excitation):

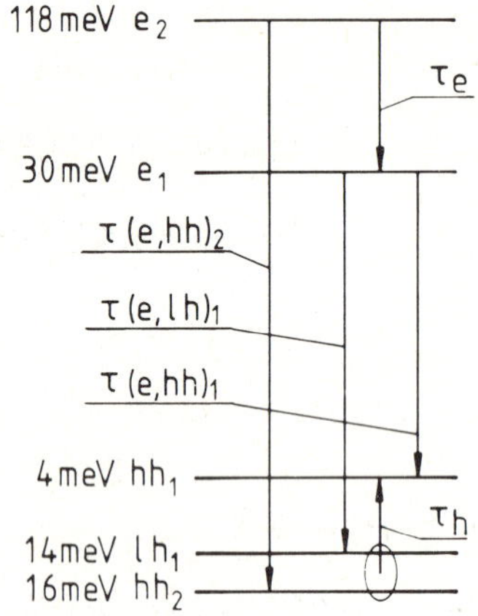

Fig. 15. The different intersubband relaxation and recombination processes which govern the Cl of the MQW sample with $L_z = 11.3$ nm (see fig. 14) are shown schematically. The energies give the high energy shifts of the various levels with respect to the 3-D conduction and valence bands.

$$\mathrm{d}e_2/\mathrm{d}t = -[1/\tau_e + 1/\tau(\mathrm{e, hh})_2]e_2, \tag{6}$$

$$\mathrm{d}e_1/\mathrm{d}t = e_2/\tau_e - (\mathrm{hh})_1/\tau(\mathrm{e,hh})_1 - (\mathrm{lh})_1/\tau(\mathrm{e,lh})_1, \tag{7}$$

$$\mathrm{d(hh)}_2/\mathrm{d}t + \mathrm{d(lh)}_1/\mathrm{d}t = -[(\mathrm{hh})_2 + (\mathrm{lh})_1]/\tau_\mathrm{h} - e_2/\tau(\mathrm{e,hh})_2 - (\mathrm{lh})_1/\tau(\mathrm{e,lh})_1, \tag{8}$$

$$\mathrm{d(hh)}_1/\mathrm{d}t = [(\mathrm{hh})_2 + (\mathrm{lh})_1]/\tau_\mathrm{h} - (\mathrm{hh})_1/\tau(\mathrm{e,hh})_1. \tag{9}$$

The $(\mathrm{lh})_1$ and the $(\mathrm{hh})_2$ level almost coincide in energy and are treated together. Deriving eqs. (6)–(9) we have made the assumption that the concentration of a certain type of exciton is equal to the smaller one of the two subband populations which make up this particular state (e.g. $X(\mathrm{e,hh})_2 = e_2$). Group theoretical selection rules predict that

$$\tau = \tau(\mathrm{e, hh})_1 = \tau(\mathrm{e, hh})_2 \tag{10}$$

and

$$3\tau(\mathrm{e, hh})_1 = \tau(\mathrm{e, lh})_1. \tag{11}$$

Inserting (10) and (11) into (6)–(8) we get for the time dependence of the luminescence intensity

$$I(t) = c_1 \exp[-t/\tau] + c_2 \exp[-t(1/3\tau + 1/\tau_\mathrm{h})] + c_3 \exp[-t(1/\tau + 1/\tau_e)]. \tag{12}$$

Thus the experimentally observed time constants are not identical with the true decay times of, e.g., different excitons or with intersubband relaxation times but can be used to determine them, with one exception: 'long time' after excitation the higher excited states are depopulated. Only one recombination channel having the minimum photon energy is still active. Thus the decay time at long time is identical to the lifetime τ of the $(\mathrm{e,hh})_1$ exciton. On the other hand an identification of the *initial* decay time with an exciton lifetime as is frequently done is a priori incorrect. We find

$\tau_e = 2.3$ ns, $\tau_h = 8$ ns and $\tau = 15$ ns.

Results for other samples and the whole range of temperatures covered by our experiments will be given elsewhere. It should be noted here, however, that the analysis proves the three important observations noted at the beginning of this section to remain valid.

7. Relaxation of hot carriers in quantum wells

A very brief remark on energy relaxation processes (intraband relaxation) will conclude the paper. At low temperatures we find that the excitons are not in thermal equilibrium with the lattice – independent of excitation intensity or delay time after excitation. The exciton temperature is higher than the lattice temperature and higher than the carrier temperature in bulk GaAs under the same experimental conditions. This observation could lead to the premàture conclusion that the electron–phonon coupling in quantum wells is reduced as compared to bulk GaAs – in contrast to theoretical expectations [22]. A well-founded analysis, however, should take into account two effects which both increase strongly the experimentally observed carrier temperature, even if the phonon coupling is unaltered or increases:

(a) The lifetime of the charge carriers and excitons decreases due to localisation, as shown here. Thus less and less excess energy can be dissipated during the lifetime of a carrier or an exciton.

(b) The recombination of the excitons occurs preferentially at small $\boldsymbol{k}$ values. This recombination leads to a distortion of the thermal distribution of the excitons and – after rapid reestablishment of a distribution function – to an increased temperature of the remaining excitons. This recently discovered "recombination heating" [23] can balance the dissipation via acoustic phonons at low temperatures since the heating rate

$$\mathrm{d}(kT)/\mathrm{d}t = kT/\tau \qquad (14)$$

is of the same order as the dissipation rate. τ is the lifetime of the transition.

A detailed analysis of the energy relaxation of excitons in MQWs taking into account both effects will be presented elsewhere.

8. Summary

Structural localisation of charge carriers in quantum wells leads to a strongly increased excitonic recombination probability and ultimately to a complete change of the type of recombination: it changes from extrinsic to intrinsic. This conclusion is based on results of continuous excitation, time-delayed and time-resolved cathodoluminescence experiments of GaAs quantum wells of varying thickness and p-type doping which are reported for the temperature range 5–300 K and various excitation levels.

The novel experimental setup used for these experiments which were described in detail has an overall time resolution of better than 250 ps independent of the width of the exciting pulse. Charge carriers excited in the 18 nm thick $Ga_xAl_{1-x}As$ barriers surrounding the wells are found to be scattered or injected into the wells in 0.6 ps with $\approx$100% efficiency. The localisation induced enhancement of excitonic recombination probabilities which manifests itself in decreased lifetimes leads to predominant intrinsic excitonic recombination still at room temperature. Carrier capture by impurities or traps is increasingly bypassed with decreasing quantum well width. A comparison of time-delayed recombination spectra in bulk material and quantum well illuminates these qualitative differences. The recombination in bulk material occurs stepwise via impurities. The recombination in QWs is intrinsic of nature and impurities show negligible contribution to recombination still in 10^{16} cm^{-3} p-doped material. p-doping further reduces the luminescence lifetime since additional excitonic radiative recombination channels start to participate. A detailed analysis of the transients show that despite the exponential law which governs the decay of a single level exciton population, the actual situation is complex. Using a detailed rate equation model for the first time interband and intersubband recombination and relaxation times are derived. Comparatively larger carrier or exciton temperatures in quantum wells than in bulk material at low lattice temperatures are understood in terms of the reduced lifetime leading to reduced energy relaxation. Additionally there occurs recombination heating due to the $\boldsymbol{k}$-dependent recombination which partly balances dissipation via electron–acoustic phonon interaction.

Acknowledgements

We are indebted to R. Gerhardts and J. Mycielski for some helpful comments and discussions. Part of this work is supported by the DFG in the framework of SFB 6.

References

[1] A.Y. Cho and J.R. Arthur, in: Progress in Solid State Chemistry. 10, J.O. McColdin and G. G. Somorjai, eds. (Pergamon, Oxford, 1976) p. 157;
G. Weimann and W. Schlapp, in: Springer Series in Solid-State Sciences 53, G. Bauer, F. Kuchar and H. Heinrich, eds. (Springer, Berlin, 1984) p. 88.

[2] M. Razeghi and J.P. Duchemin, in: Springer Series in Solid-State Sciences 53, G. Bauer, F. Kuchar and H. Heinrich, eds. (Springer, Berlin, 1984) p. 100.

[3] L. Esaki and R. Tsu, IBM J. Res. Dev. 14 (1970) 61.

[4] R. Dingle, in: Advances in Solid State Physics XVII, J. Treusch, ed. (Vieweg, Braunschweig, 1977) 21.
[5] T. Ando, A.B. Fowler and F. Stern, Rev. Mod. Phys. 54 (1982) 437.
[6] See e.g. R.C. Miller, J. Lumin. 30 (1985) 520.
[7] J. Christen, A. Steckenborn, D. Bimberg and G. Weimann, Verhandl. DPG (VI) 18 (1983) 668;
J. Christen, D. Bimberg, A. Steckenborn and G. Weimann, Appl. Phys. Lett. 44 (1984) 84.
[8] E.O. Goebel, H. Jung, J. Kuhl and K. Ploog, Phys. Rev. Lett. 51 (1983) 1588.
[9] D. Bimberg, J. Christen and A. Steckenborn ref. [2] p. 136.
[10] A. Steckenborn, H. Münzel and D. Bimberg, BEDO 13 (1980) 157.
[11] A. Steckenborn, H. Münzel and D. Bimberg, J. Lumin. 24/25 (1981) 351.
[12] H. Münzel, D. Bimberg and A. Steckenborn, Physica 117B + 118B (1983) 214.
[13] D. Bimberg, in ref. [4] p. 195.
[14] D. von der Linde and R. Lambrich, Phys. Rev. Lett. 42 (1979) 1090.
[15] W.T. Tsang, Appl. Phys. Lett. 39 (1981) 786.
[16] R. Fischer, J. Klein, T.J. Drummond, W. Kopp, H. Morkoc, E. Anderson and M. Pion, Appl. Phys. Lett. 44 (1984) 1.
[17] H. Sakaki, J. Yoshino, Y. Sekiguchi and K. Sakai, Electron. Lett. (1984) in print.
[18] G. Bastard, Phys. Rev. B 24 (1981) 4714.
[19] R.C. Miller, A.C. Gossard, W.T. Tsang and O. Munteanu, Phys. Rev. B 25 (1982) 3871.
[20] R.C. Miller, D.A. Kleinman, W.T. Tsang and A.C. Gossard, Phys. Rev. B 24 (1981) 1134.
[21] G. Bastard, E.E. Mendez, L.L. Chang and L. Esaki, Phys. Rev. B 26 (1982) 1974.
[22] K. Hess, Appl. Phys. Lett. 35 (1979) 484.
[23] D. Bimberg and J. Mycielski, Proc. of the 17th Int. Conf. Physics of Semic., J. Chadi, ed. (Springer, New York, 1984) in print and J. Phys. C in print.

Journal of Luminescence 30 (1985) 580–584
North-Holland, Amsterdam

THE EFFECTS OF PHOTOEXCITATION ON EXCITONS IN SEMICONDUCTOR DOPING SUPERLATTICES

T.L. REINECKE

Naval Laboratory, Washington, D.C. 20375, USA

F. CROWNE

Martin Marietta Research Laboratories, Baltimore, MD 21227, USA

Calculations of the properties of excitons in doping superlattices have been made as a function of doping density using a variational approach. Interesting new features are obtained when the exciton energy becomes comparable to the superlattice potential energy. These results are compared to recent experimental data on GaAs doping superlattices.

Semiconductor doping superlattices are especially attractive for photoluminescence studies because their properties can be modified over wide ranges by photoexcitation [1]. These systems are formed by alternating layers of n and p doping in a bulk material. Electrons are transferred from the n layers to the p layers creating a modulated electric field and a modulated electrostatic potential for the carriers. This potential confines the carrier motion in the direction perpendicular to the layers and produces quasi-two-dimensional subbands for the electrons and holes which have an indirect gap in real space and consequently long electron–hole recombination times.

Under photoexcitation both the net charge in the layers and the band bending due to the superlattice potential are reduced. It is possible to reduce the superlattice potential almost to zero by photoexcitation, and in particular the band bending can become comparable to the exciton energy. In this case interesting new effects can occur as a result of the competition between the superlattice potential which favors spatial separation of the electron and hole and the Coulomb attraction between the electron and hole.

We have made a series of calculations of the properties of excitons in doping superlattices. These include excitons formed from electrons and holes in the superlattice subbands, excitons with an electron or a hole in a hydrogenic state at a donor or an acceptor, and excitons in hetero-doping superlattices [2]. In the present note we summarize the results for the case of an exciton formed from an electron and a hole in the superlattice subbands, and we point out the relevance of these results for photoluminescence studies of doping superlattices. Fuller details will be presented elsewhere.

Results

The system is composed of alternating layers of uniform positive charge $|e|N_D$ and negative charge $-|e|N_A$ whose centers are separated by distance d along the $\hat{z}$-axis. The electron and hole are taken to be in adjacent layers and are described by the wavefunction

$$\Psi(\boldsymbol{r}_e, \boldsymbol{r}_h) = N\zeta_e(z - z_D/2)\zeta_h(z - z_A/2)\exp\{-[\alpha^{-1}((x_e - x_h)^2 + (y_e + y_h)^2) + \beta^{-1}(z_e - z_h)^2]^{1/2}\}. \tag{1}$$

Here ζ_i, i=e, h are the subband wavefunctions, α and β are the variational parameters representing the extent of the wavefunction parallel and perpendicular to the layers, and N is a normalizing constant.

The superlattice potential energy in the absence of the Coulomb interaction is parabolic, and the subband functions ζ_i are given straightforwardly in terms of harmonic oscillator functions with subband energies ε_i, i=e, h where $\varepsilon_e = \frac{1}{2}\hbar(4\pi e^2 N_D/\kappa m_e)^{1/2}$ and similarly for ε_h. κ is the background dielectric constant, and the conduction and valence band edges are taken to be described by single isotropic mass parameters. When the Coulomb interaction is present the problem is solved variationally. Lengths are given in terms of the exciton Bohr radius $a_x = \hbar^2\kappa/\mu e^2$ and energies in terms of the effective exciton Rydberg $R_x = e^2/2\kappa a_x$ where $\mu^{-1} = m_e^{-1} + m_h^{-1}$.

Results for the exciton binding energy are shown in fig. 1. The binding energy is measured relative to the subband energies as $E_b = \langle\hat{H}\rangle - \varepsilon_e - \varepsilon_h$. The results are expressed in terms of a dimensionless parameter $\Delta = d/a_x$ giving the separation between the layers and a parameter $\Sigma = \frac{1}{4}(\Gamma_e^{-1} + \Gamma_h^{-1})a_x^{-2}$ where $\Gamma_e = (m_e/\hbar)(4\pi e^2 N_D/\kappa m_e)^{1/2}$ and similarly for Γ_h. Σ is a measure of the (inverse of the) doping density, and it varies substantially for systems of physical interest. For example, for GaAs with $N_A = N_D$, $\Sigma = 0.03$ for $N_D = 10^{19}$ cm^{-3}, and $\Sigma = 10$ for $N_D = 10^{15}$ cm^{-3}.

For $\Delta = 0$ the correct three- and two-dimensional binding energies are obtained respectively in the limit of high Σ ($E_b/R_x = -1$ and $\alpha/a_x = \beta/a_x = 1$) and of low Σ ($E_b/R_x = -4$ and $\alpha/a_x = \frac{1}{2}$). For nonzero Δ in the high doping limit E_b approaches a constant which depends on Δ [3]. In addition, E_b develops a minimum as a function of Σ. This minimum results from an interplay between the effects of reduced dimensionality which increase $|E_b|$ and the spatial separation of the electron and the hole which reduces $|E_b|$.

The novel physics of the present system results from a competition between

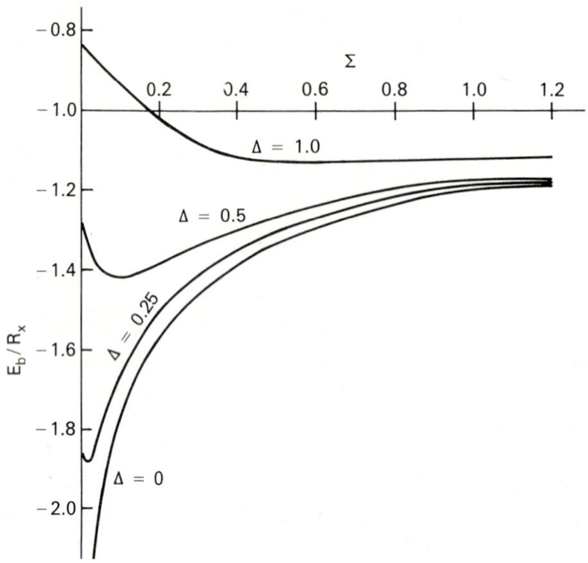

Fig. 1. Binding energy vs doping parameter Σ for several separations Δ.

the superlattice potential energy and the Coulomb attraction between the electron and the hole. By direct calculation we have shown that the difference between the expectation values of the electron and hole positions goes to zero approximately exponentially with decreasing doping. As the electron and hole move together their wavefunction overlap increases, and their recombination probability increases. For a correlated system like the present one the factor in the recombination probability which represents the wavefunction

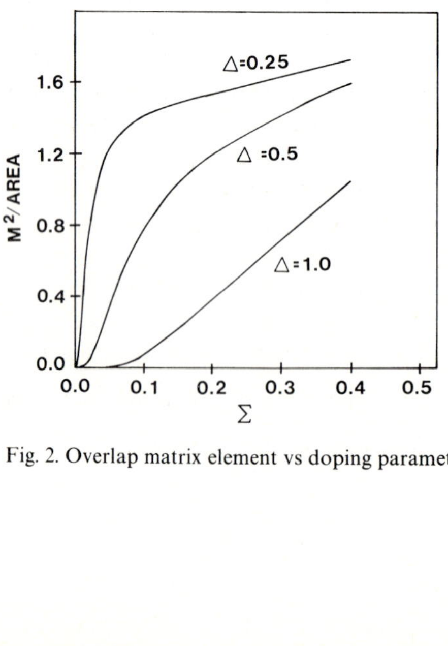

Fig. 2. Overlap matrix element vs doping parameter Σ for several separations Δ.

overlap is M^2 where

$$M = \int d^3r \Psi(r = r_e; r = r_h). \tag{2}$$

This is shown in fig. 2 where it is seen that M^2 is exponentially small for large doping and increases with decreasing doping.

Discussion

The present results provide a useful picture of the effects of photoexcitation on the behavior of excitons and of photoluminescence in doping superlattices. After photoexcitation the space charge of the layers is partly neutralized, and the superlattice potential decreases. An electron and a hole excited into the subbands then will move closer together under the influence of the Coulomb interaction, and their recombination probability will increase. The corresponding features in the luminescence spectra will occur at higher energies with increasing photoexcitation because of the decrease of the superlattice potential variation.

The present picture is consistent with a new feature seen in recent photoluminescence studies [4] of a doping superlattice in GaAs. For low and moderate photoexcitation intensity a broad recombination band like that seen previously [5] was observed. It shifted to higher energy with increasing intensity and is attributed to recombination of electrons from the electron subband with holes at acceptors. For the highest excitation, however, a second line occurred in the spectrum at a higher energy than the first. Its energy is consistent with recombination involving an interacting electron and hole in their respective subbands or with an exciton bound at an impurity. In the picture developed here the recombination probability of electrons and holes from the subbands is increased by increasing photoexcitation because of the increased wavefunction overlap, and the probability of formation of a bound exciton complex at an impurity also is increased. As a result the recombination from both of these processes increases with photoexcitation which is consistent with what is seen experimentally.

This work was supported in part by an ONR contract (T.L.R.).

References

[1] G.H. Döhler, H. Künzel, D. Olega, K. Ploog, P. Ruden, H.J. Stolz and G. Abstreiter, Phys. Rev. Lett. 47 (1981) 864.
[2] G.H. Döhler and P.P. Ruden, Surf. Sci. 142 (1984) 474.

[3] Yu.E. Lozvik and N.V. Nishanov, Sov. Phys. Solid State 18 (1976) 905.
[4] F. Crowne, T.L. Reinecke and B.V. Shanabrook, Proc. XVIIth Int. Conf. Physics of Semiconductors, San Francisco, 1984, eds., D.J. Chadi and W.A. Harrison (to be published).
[5] H. Jung, G.H. Döhler, H. Künzel, K. Ploog, P. Ruden and H.J. Stolz, Solid State Commun. 43 (1982) 291.

Journal of Luminescence 30 (1985) 585–596
North-Holland, Amsterdam

EXCITONS AND ELECTRON–HOLE PLASMA IN QUASI-TWO-DIMENSIONAL SYSTEMS*

S. SCHMITT-RINK and C. ELL

Institut für Theoretiscne Physik der Universität Frankfurt, Robert-Mayer-Str. 8, D-6000 Frankfurt-Main, FRG

A theoretical study of many-body effects in quasi-two-dimensional electron–hole systems is presented. The renormalized single-particle energies and the exciton binding energy are calculated as functions of the carrier density and temperature. A simple model for the non-linear excitonic absorption and refraction is proposed.

1. Introduction

In the last few years, much attention has been focused on quasi-two-dimensional electron–hole systems in semiconductors, particularly with regard to possible applications in integrated optics [1]. The most prominent feature of these systems is the fact that excitons are observable even at room temperature, due to the strong enhancement of their binding energy under quantum confinement.

Experimentally, such systems are best realized in thin semiconductor films or in thin semiconductor quantum wells with very large (infinite) barrier thicknesses and heights. In both cases, the quantization in the direction perpendicular to the confining layer gives rise to discrete energy levels. If the thickness d of the confining layer is much smaller than the bulk exciton radius a_∞ the system becomes quasi-two-dimensional. Keldysh has shown that in this limit the Coulomb interaction between two charges e and e' has the form [2]

$$V(r)=\frac{ee'}{\varepsilon}\int_0^\infty \mathrm{d}q\, J_0(qr)\left[\varepsilon_0 \tanh\left(\frac{qd}{2}\right)+\varepsilon\right]\Big/\left[\varepsilon \tanh\left(\frac{qd}{2}\right)+\varepsilon_0\right]. \tag{1}$$

Here, $r \gg d$ is the distance between the charges and q is the two-dimensional momentum in the plane of the confining layer. ε and ε_0 are the dielectric constants of the confining layer and the surrounding medium, respectively.

* This work has been supported by the Deutsche Forschungsgemeinschaft through the Sonderforschungsbereich Frankfurt/Darmstadt and by the Commission of the European Communities (under the experimental phase of the European Stimulation Action).

0022-2313/85/$03.30

For $\varepsilon = \varepsilon_0$ (no image forces) eq. (1) reduces to the familiar expression

$$V(r) = ee'/\varepsilon_0 r. \tag{2}$$

For the more realistic case $\varepsilon \gg \varepsilon_0$ one obtains

$$V(r) \simeq \frac{2ee'}{\varepsilon d} \int_0^\infty \mathrm{d}x \, \frac{J_0(x)}{x + 2\varepsilon_0 r/\varepsilon d}$$

$$= \frac{\pi ee'}{\varepsilon d} \left[H_0\left(\frac{2\varepsilon_0 r}{\varepsilon d}\right) - Y_0\left(\frac{2\varepsilon_0 r}{\varepsilon d}\right) \right], \tag{3}$$

where H_0 and Y_0 are the Struve function and the Bessel function of the second kind, respectively. For $\lambda = 2\varepsilon_0 r/\varepsilon d \gg 1$ this potential reduces to the potential (2), whereas for $\lambda \ll 1$ it behaves like a logarithmic potential [2,3]. Throughout this paper we will consider the limit $\lambda \gg 1$, so that eq. (2) holds.

The low-temperature properties of electrons and holes interacting via the potential (2) have been studied in a number of papers [4–9]. The present paper deals mainly with their high-temperature properties, which are very interesting with regard to possible applications. Nevertheless, we make contact with the earlier calculations in order to check our theoretical model.

The paper is organized as follows. In section 2 we study the ground state of a quasi-2D electron–hole plasma (EHP) within the plasmon-pole approximation. Our results agree well with the genuine RPA calculation of Kuramoto and Kamimura [4], which confirms our simple model. In section 3 we consider the problem of an exciton embedded in a gas of free carriers. We solve the homogeneous Bethe–Salpeter equation variationally, taking into account exchange, bandfilling and screening corrections. Screening is treated both statically and dynamically and the respective critical densities for exciton ionization are determined. Finally, in section 4, we utilize the results of section 3 and calculate the nonlinear optical susceptibility of quasi-2D excitons.

2. Ground-state properties

In the quasi-2D limit one can directly apply the theoretical methods which have been developed for the description of electrons and holes in 3D systems (for a recent review see, e.g., ref [10]). In the following we will use the familiar plasmon-pole approximation in order to calculate the renormalized single-particle energies and the ground-state energy of a quasi-2D EHP. This approximation, which is originally due to Lundquist [11] and Overhauser [12], has been successfully applied to the 3D electron gas and the 3D EHP. It has been extended to the quasi-2D electron gas by Vinter [13]. It will be shown below that for a quasi-2D EHP the plasmon-pole approximation yields results which are close to the results of the genuine RPA calculation of Kuramoto and

Kamimura [4]. This confirms that free-carrier screening in quasi-2D electron–hole systems can be approximately treated within this simple model.

At zero temperature we write for the renormalized single-particle energies of a quasi-2D EHP ($\hbar=1$)

$$e_i(k)=E_g/2+\varepsilon_i(k)+\mu_{ixc}, \qquad i=e, h, \tag{4}$$

where μ_{ixc} is the exchange-correlation contribution to the chemical potential μ_i and $\varepsilon_i(k)=k^2/2m_i$. E_g is the effective bandgap, which differs from the corresponding bulk value by the confinement energies of the electrons (e) and holes (h). Within the RPA μ_{ixc} reads

$$\mu_{ixc}=\text{Re}\,\Sigma_i(k_F), \tag{5}$$

where $k_F=\sqrt{2\pi n}$ is the Fermi wavenumber and

$$\Sigma_i(k)=-\sum_{\boldsymbol{q}} V_s(q, e_i(k)-e_i(\boldsymbol{k}-\boldsymbol{q}))\theta(k_F-|\boldsymbol{k}-\boldsymbol{q}|) -\sum_{\boldsymbol{q}}\int_0^{\infty}\frac{d\omega}{\pi}\frac{\text{Im}\,V_s(q,\omega)}{e_i(k)-e_i(\boldsymbol{k}-\boldsymbol{q})-\omega+i\delta} \tag{6}$$

is the single-particle self-energy, which is separated into a screened-exchange term (sx) and a Coulomb-hole term (ch). Here, $V_s(q, \omega)$ is the dynamically screened Coulomb interaction. Within the plasmon-pole approximation the total RPA excitation spectrum is replaced by a single mode $\omega(q)$, so that $V_s(q, \omega)$ reduces to the simple expression

$$V_s(q, \omega)=V(q)\{1+\omega_p^2(q)/[(\omega+i\delta)^2-\omega^2(q)]\}. \tag{7}$$

Here, $V(q)=2\pi e^2/\varepsilon_0 q$ is the Fourier transform of eq. (2) and $\omega_p(q)=(2\pi ne^2q/\varepsilon_0 m)^{1/2}$ is the 2D plasma frequency. m is the reduced electron–hole mass, $m^{-1}=m_e^{-1}+m_h^{-1}$. For the effective plasmon frequency $\omega(q)$ we use

$$\omega^2(q)=\omega_p^2(q)(1+q/\kappa)+(\varepsilon_e(q)+\varepsilon_h(q))^2/4, \tag{8}$$

where $\kappa=2e^2(m_e+m_h)/\varepsilon_0$ is the 2D screening wavenumber. This dispersion relation reproduces the correct screening behaviour at long wavelengths and simulates the individual particle–hole excitations at short wavelengths.

In fig. 1 we have plotted the resulting exchange-correlation potential $\mu_{xc}=\mu_{exc}+\mu_{hxc}$ and the various contributions to it for $m_e=m_h$ as a function of the plasma density n. The 2D exciton Rydberg $E_0=2me^4/\varepsilon_0^2$ and the 2D exciton radius $a_0=\varepsilon_0/2me^2$ are chosen as the units of energy and length, respectively. Figure 1 shows, that for low densities ($k_F\ll\kappa$) exchange effects are screened efficiently and μ_{xc} approaches the Coulomb-hole potential μ_{ch}, whereas for high densities ($k_F\gg\kappa$) exchange becomes the dominant feature. The density dependence of μ_{ch} is close to $-3.1\,(na_0^2)^{1/3}E_0$. A similar behaviour ($\mu_{ch}\propto -n^{1/4}$) has been found for the 3D EHP [14]. Using the methods of ref. [14], one obtains

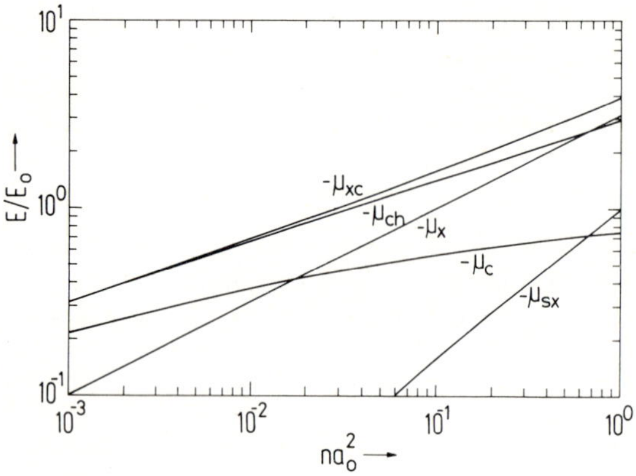

Fig. 1. Exchange-correlation contribution to the chemical potential as a function of the density ($\mu_{xc}=\mu_x+\mu_c=\mu_{sx}+\mu_{ch}$).

analytically a value of $\mu_{ch}=-3.68\,(na_0^2)^{1/3}E_0$, which is somewhat too low when compared to the empirical result.

In fig. 2 we have plotted the mean effective electron mass $\bar{m}_e$ defined by $k_F^2/2\bar{m}_e=k_F^2/2m_e+\Sigma_e(k_F)-\Sigma_e(0)$ for some typical plasma densities. This quantity is a measure for the deviation of $e_e(k)$ from the parabolic behaviour (4). Figure 2 shows, that the renormalization of the bandwidth is small. As in three dimensions, the k-dependences of the exchange and correlation self-energies cancel to a large extent, so that $\Sigma_i(k)$ is almost independent of k.

By integrating the chemical potential $\mu=\mu_e+\mu_h$ over the density n we obtain the free energy per electron–hole pair:

$$F(n)=\frac{1}{n}\int_0^n \mu(n')\,\mathrm{d}n'. \tag{9}$$

Figure 3 shows the resulting ground-state energy for $m_e=m_h$. Its minimum value is $-0.87\,E_0$ at $r_s\simeq1.4$, where $r_s=1/(\pi na_0^2)^{1/2}$ is the 2D density parameter. This result is very close to the result of Kuramoto and Kamimura [4] ($-0.86\,E_0$

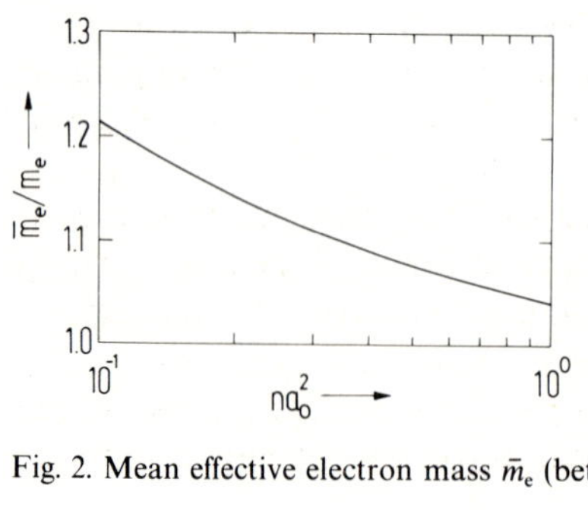

Fig. 2. Mean effective electron mass $\bar{m}_e$ (between $k=0$ and k_F) as a function of the density.

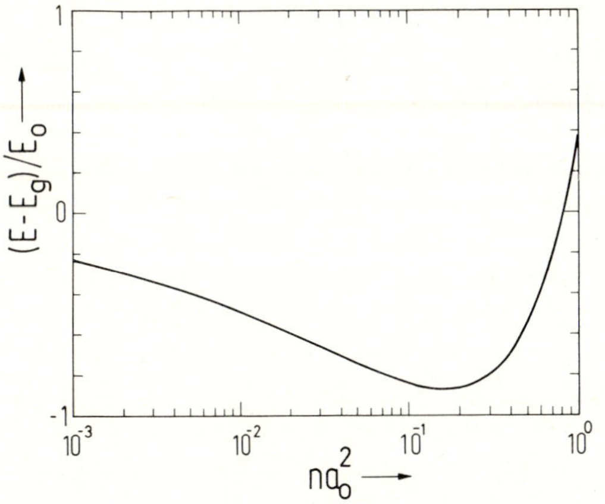

Fig. 3. Ground-state energy as a function of the density.

at $r_s \simeq 1.4$). These authors calculated the ground-state energy within the RPA (including the 2D Hubbard modification) using the method of Brinkman and Rice [15]. Similar results have been obtained in ref. [9], whereas in refs. [5] and 6 somewhat lower values of the ground-state energy have been reported. The influence of electron–hole pairing on the ground-state energy, i.e., the formation of a dielectric electron–hole system, has been studied in ref. [7].

The good agreement of our results with the RPA results of ref. [4] shows, that dynamical free-carrier screening in a quasi-2D EHP can be approximately treated within the plasmon-pole approximation. Therefore, we will use this simple model in our further calculations.

3. Screening of the exciton

In this section we will consider how electron–hole pair states, which determine the optical properties of a semiconductor in a very decisive way, are influenced by many-body effects. It is well known, that in a two-component Coulomb system bound pair states, i.e., excitons, exist only up to a critical pair density n_c, which is determined by the vanishing of their binding energy due to phase space occupation and screening effects. This density is usually called Mott density [16].

At low temperatures, $T < E_0$ ($k_B = 1$), excitons are thermally stable, so that the screened pairs and the screening pairs are almost identical. Hence, if $T < E_0$, a fully self-consistent treatment is needed in order to describe the ionization of the exciton. First attempts in this direction are reviewed in ref. [10].

At high temperatures, $T > E_0$, excitons are thermally ionized and the screening and phase space occupation processes are governed by free carriers. In that case one can consider a given electron–hole pair embedded in a gas of free electrons and holes. For 3D excitons this problem has been studied within the framework of the so-called Shindo approximation [17,18], which yields a consistent description of dynamical screening effects. This approximation, which is valid for shallow excitons, can be simplified further by considering the limit of a vanishing exciton binding energy. It reduces then to the so-called Haken approximation [19].

In the following we will use the latter approximation together with the plasmon–pole approximation in order to calculate the renormalized exciton energy and the critical density for exciton ionization in the nondegenerate limit (similar calculations have been performed for 3D excitons; see, e.g., refs [20] and [21]). Additionally, we perform the same calculation using the static RPA screening function. We compare the respective results and show that they are similar to the results, which have been obtained for 3D excitons.

The appropriate basis for the discussion of electron–hole pair states in an EHP is the equation of motion for the two-particle Green's function, the so-called Bethe–Salpeter equation. In the screened ladder approximation and in the nondegenerate limit its homogeneous part yields the effective exciton Schrödinger equation [10,18]

$$\sum_{\boldsymbol{k}'} H_{\boldsymbol{k},\boldsymbol{k}'}\phi(\boldsymbol{k}')=E\phi(\boldsymbol{k}'), \tag{10}$$

where

$$H_{\boldsymbol{k},\boldsymbol{k}'}=\left(E_{\mathrm{g}}+\frac{k^2}{2m}+\Delta(k)\right)\delta_{\boldsymbol{k},\boldsymbol{k}'}-(1-f_{\mathrm{e}}(k)-f_{\mathrm{h}}(k))V(\boldsymbol{k}-\boldsymbol{k}')$$
$$-(I(\boldsymbol{k},\boldsymbol{k}')-V(\boldsymbol{k}-\boldsymbol{k}')). \tag{11}$$

Here,

$$f_{\mathrm{i}}(k)=\frac{\pi n}{m_{\mathrm{i}}T}\exp\left(-\frac{k^2}{2m_{\mathrm{i}}T}\right) \tag{12}$$

is the single-particle distribution function and $I(\boldsymbol{k},\boldsymbol{k}')$ is an effective nonlocal electron–hole interaction,

$$\Delta(k)=\sum_{\boldsymbol{k}'}\{I(\boldsymbol{k},\boldsymbol{k}')-(1+f_{\mathrm{e}}(k')+f_{\mathrm{h}}(k'))V(\boldsymbol{k}-\boldsymbol{k}')\} \tag{13}$$

being the corresponding self-energy shift.

The first term in eq. (11) is the energy of an uncorrelated electron–hole pair including Coulomb-hole and (unscreened) exchange self-energy corrections. The second and the last term in eq. (11) describe the attractive electron–hole interaction and its renormalization due to band filling and screening effects, respectively.

Using the plasmon–pole approximation (7) together with the Haken approximation one obtains in the nondegenerate limit

$I(\boldsymbol{k}, \boldsymbol{k}') \equiv I(\boldsymbol{k}-\boldsymbol{k}')$,

where

$$I(q)=V(q)\left\{1-\frac{\omega_{\mathrm{p}}^2(q)}{2\omega(q)}\left(\frac{1+N(q)}{\varepsilon_{\mathrm{e}}(q)+\omega(q)}+\frac{N(q)}{\varepsilon_{\mathrm{e}}(q)-\omega(q)}+(\mathrm{e}\leftrightarrow\mathrm{h})\right)\right\}. \tag{14}$$

Here, $N(q)=\{\exp(\omega(q)/T)-1\}^{-1}$ is the Bose distribution function. The screening wavenumber κ, which enters the plasmon dispersion (8), is now $\kappa=4\pi ne^2/\varepsilon_0 T$.

For the discussion of static screening we use the correct RPA expression (see, e.g., ref. [22])

$$I(q)=\frac{2\pi e^2}{\varepsilon_0}\Bigg/\left[q+\frac{\kappa}{2}\sum_{\mathrm{i}=\mathrm{e,h}}\frac{\sqrt{8m_{\mathrm{i}}T}}{q}F\left(\frac{q}{\sqrt{8m_{\mathrm{i}}T}}\right)\right] \tag{15}$$

instead of

$$I(q)=V(q)[1-\omega_{\mathrm{p}}^2(q)/\omega^2(q)], \tag{16}$$

which follows from eq. (14) in the limit $m_{\mathrm{e,h}}\to\infty$. Here, $F(x)$ is Dawson's integral.

In order to calculate the renormalized exciton energy E_x we solve the effective Schrödinger equation (10) variationally. For that purpose we minimize the expectation value of the effective Hamiltonian (11)

$$E_x=\sum_{\boldsymbol{k},\boldsymbol{k}'}\phi^*(\boldsymbol{k})H_{\boldsymbol{k},\boldsymbol{k}'}\phi(\boldsymbol{k}'), \tag{17}$$

using the 1s-trial function

$$\phi(k)=\sqrt{8\pi a^2}/(1+a^2k^2)^{3/2}. \tag{18}$$

The renormalized exciton binding energy can be obtained by subtracting E_{x} from the renormalized bandgap

$$E_{\mathrm{g}}'=E_{\mathrm{g}}+\Delta(0). \tag{19}$$

Figure 4 shows the renormalized exciton energy E_{x} and the renormalized bandgap E_{g}' for the statically screened Coulomb interaction (15) and $m_{\mathrm{e}}=m_{\mathrm{h}}$. As in three dimensions, the exciton energy is almost independent of the plasma density n, whereas the continuum edge moves down rapidly, when n increases. The remarkable constancy of E_{x} can be explained by a strong compensation between the changes of the exciton binding energy and of the bandgap up to high densities. Physically, it reflects the charge neutrality of an exciton.

Figure 5 shows E_{x} and E_{g}' for the dynamically screened Coulomb interaction (14). Since screening is now less effective, the values of the renormalized bandgap and of the exciton binding energy are considerably larger than those of fig. 4. The dashed lines in fig. 5 represent results for E_{g}', which include screening of the exchange interaction and Fermi statistics. For the densities of interest and for

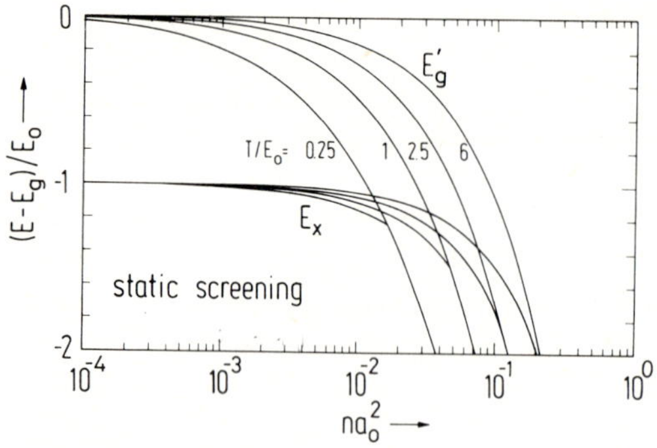

Fig. 4. Renormalized bandgap E_g' and renormalized exciton energy E_x for the statically screened Coulomb interaction (15).

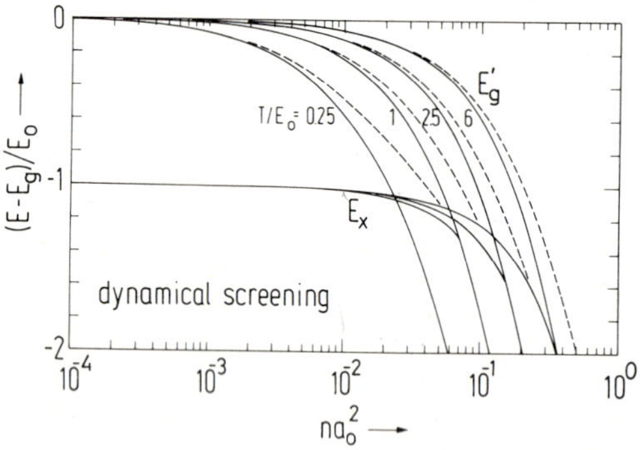

Fig. 5. Renormalized bandgap E_g' and renormalized exciton energy E_x for the dynamically screened Coulomb interaction (14).

$T > E_0$ they are close to the asymptotic curves according to eqs. (13) and (19), which confirms our low-density expansion.

The point, where the exciton vanishes in the continuum, yields the value of the Mott density n_c. In fig. 6 we have plotted n_c for $m_e = m_h$ as a function of the temperature. Since screening is overestimated in the static approximation, the static Mott density is always lower than the dynamical one. At high temperatures, $T > E_0$, the numerical n_c is close to the value, which results from the simple criterion $\kappa a_0 \simeq 0(1)$, where $\kappa = 4\pi n e^2/\varepsilon_0 T$ is the 2D screening wavenumber. At low temperatures, this criterion becomes meaningless, because κ reduces to a constant, when $T \to 0$.

Note again, that our calculation yields only asymptotic results for n_c. In particular, these results are not valid for $T < E_0$. Calculations of the low-temperature Mott density $n_c(T \gtrsim 0)$ have been performed in refs. [8] and [23] (note, that at $T \equiv 0$ the system is always an insulator). Starting from the metallic side Andryushin and Silin [8] determined $n_c(T \gtrsim 0)$ within the static RPA

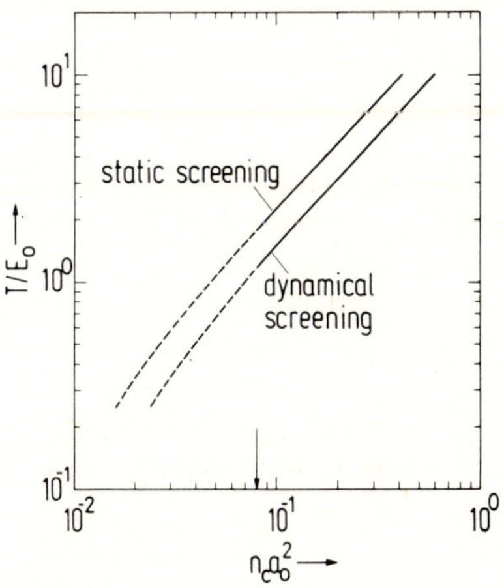

Fig. 6. Mott density as a function of the temperature. The arrow marks the low-temperature result of ref. [23].

(including the 2D Hubbard modification). For $m_e = m_h$, they obtained a value of $n_c a_0^2 \simeq 0.06$ ($r_{sc} \simeq 2.3$). Stern and Howard [23] estimated $n_c(T \gtrsim 0)$ from the overlap of the bound-state wavefunctions. They obtained a value of $n_c a_0^2 \simeq 0.08$ ($r_{sc} = 2$). In fig. 6 this value is marked by an arrow. For $T \simeq E_0$, it becomes equal to our asymptotic result.

A general shortcoming of our model is the fact that we did not take short-range correlations properly into account. The RPA and thus the screened ladder approximation yield an incomplete description of the low-density domain. For 3D systems, this incompleteness is not very significant, because in the non-degenerate limit the RPA reduces to the Debye–Hückel theory, which is known to be valid in leading order in the plasma parameter. For 2D systems, however, the situation is quite different. In two dimensions short-range correlations are more important and the RPA is less satisfactory than in three dimensions [24]. A refined calculation is thus expected to yield results, which are somewhat different from ours. In particular, we expect larger values of E_g', i.e., smaller values of $-\Delta(0)$, and somewhat larger values of the Mott density. Going beyond the screened ladder approximation is however not worth the effort. The inclusion of finite thickness effects seems to be more promising. These effects will greatly improve the validity of our approximations, because for a finite layer thickness short-range correlations are less important [22].

4. Nonlinear optical susceptibility

The possibility of observing excitonic effects at room temperature makes quasi-2D electron–hole systems very interesting, particularly with regard to possible applications in integrated optics. In the last few years, extensive experimental studies of the nonlinear optical properties of quasi-2D excitons have been performed [1]. Furthermore, the observation of room-temperature excitonic optical bistability has been reported [25].

Physically, the observed room-temperature nonlinearities originate from the bleaching of the exciton resonance by free-carrier screening and bandfilling effects, the free carriers being generated through thermal ionization of the photoexcited excitons. For bulk materials, which exhibit a similar behaviour, the corresponding nonlinear optical susceptibility has been calculated only numerically [26]. In the present paper, a simple analytical expression for the nonlinear optical susceptibility of quasi-2D excitons is derived. Our theoretical model is an appropriate extension of the 3D zero-temperature blocking model of Goll and Haken [27]. It fits the existing experimental data quite well.

The linear optical properties of quasi-2D excitons have been investigated in detail by Shinada and Sugano [28]. As in three dimensions, the linear exciton oscillator strength f_0 is related to the probability of finding the electron near the hole, i.e., $f_0 \propto |\phi_0(r=0)|^2 = 2/\pi a_0^2$, where $\phi_0(r)$ is the unperturbed exciton orbital wavefunction. If the exciton is embedded in an EHP, its oscillator strength is strongly reduced. Keeping only the leading terms, the renormalized exciton oscillator strength f can be approximately expressed as $f/f_0 = 1 - n/n_{sat}$, where n_{sat} is the critical plasma density, at which the excitonic susceptibility saturates. In general, the inverse saturation density n_{sat}^{-1} consists of two terms, $n_{sat}^{-1} = n_1^{-1} + n_2^{-1}$. The first term n_1^{-1} is due to the renormalization of the exciton orbital wavefunction. Utilizing the results of the previous section, it can be approximately expressed as $n_1^{-1} = 2/n \lim_{n\to 0}(1 - a_0/a)$, where $a \geqslant a_0$ is the variational exciton Bohr radius. The second term n_2^{-1} is due to the familiar bandfilling (blocking) modification of the transition probability. In the case of free-carrier bandfilling one obtains $n_2^{-1} = \sum_k (f_e(k) + f_h(k))\phi_0(k)/n/\phi_0(r=0)$, where $f_i(k)$ is the single-particle distribution function (12). In ref. [27] only n_2^{-1} has been considered (at zero temperature and in the dilute limit $f_i(k) = n/2|\phi_0(k)|^2$).

In fig. 7 we have plotted the resulting saturation density for $m_e = m_h$ as a function of the temperature. It varies in much the same way as the Mott density n_c. In particular, the static saturation density is smaller than the dynamical one, due to the overestimate of screening effects. The experimental point in fig. 7 is taken from ref. [1]. It has been obtained by measuring the room-temperature absorption saturation of the electron–heavy hole exciton in a $GaAs/Al_xGa_{1-x}As$ multiple quantum well structure. It is in good agreement with our theoretical results.

Since the exciton energy remains practically constant, the nonlinear optical susceptibility χ varies in the same way as the oscillator strength f:

$$\chi/\chi^0 = 1 - n/n_{sat}, \qquad (20)$$

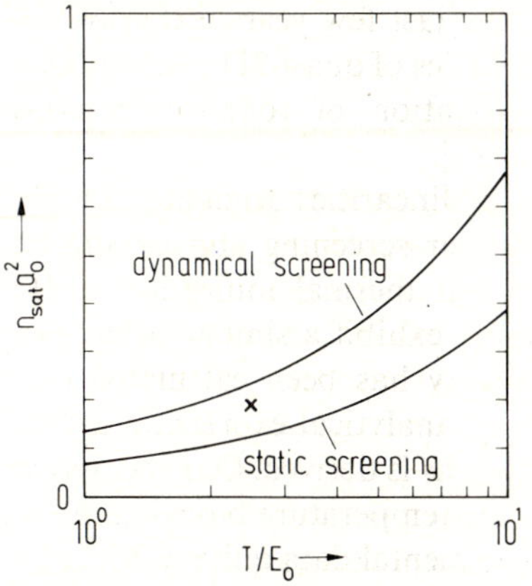

Fig. 7. Saturation density as a function of the temperature. The experimental point is taken from ref. [1].

where χ^0 is the unperturbed optical susceptibility. Substituting the stationary solution of the rate equation

$$\dot{n}=\chi_2\,\mathrm{d}|E|^2/2-n/\tau \tag{21}$$

into eq. (20), we obtain the simple saturation form

$$\chi/\chi^0=\frac{1}{1+|E|^2/|E|^2_{\mathrm{sat}}}, \tag{22}$$

where

$$|E|^2_{\mathrm{sat}}=2n_{\mathrm{sat}}/\mathrm{d}\tau\chi_2^0 \tag{23}$$

is the saturation intensity. Such a saturation behaviour has been observed indeed [1,29]. Thus, our simple theory explains not only the low saturation density of quasi-2D excitons at room temperature but also the experimentally observed Bloch-like shape of the nonlinear optical susceptibility.

Conclusions

We have discussed some many-body effects in quasi-2D electron–hole systems. Treating dynamical free-carrier screening within the plasmon-pole approximation, we have studied the ground-state of a metallic EHP and the room-temperature ionization and saturation of the exciton. Our results are qualitatively similar to the corresponding results in three dimensions and agree with the existing experimental data.

Acknowledgements

We are grateful to Professor H. Haug for bringing this subject to our attention.

References

[1] D.S. Chemla, Helv. Phys. Acta 56 (1983) 607.
[2] L.V. Keldysh, JETP Lett. 29 (1979) 658.
[3] J. Katriel, Phys. Lett. A 101 (1984) 158.
[4] Y. Kuramoto and H. Kamimura, J. Phys. Soc. Japan 37 (1974) 716.
[5] E.A. Andryushyn and A.P. Silin, Solid State Commun. 20 (1976) 453.
[6] E.A. Andryushyn and A.P. Silin, Sov. Phys. Solid State 18 (1976) 1243.
[7] A.V. Klyuchnik and Yu.E. Lozovik, Sov. Phys. Solid State 20 (1978) 364.
[8] E.A. Andryushyn and A.P. Silin, Sov. Phys. Solid State 21 (1979) 129.
[9] P.K. Isihara, Y. Nakane and A. Isihara, J. Phys. C 15 (1982) 2929.
[10] H. Haug and S. Schmitt-Rink, Progr. Quant. Electron. (in press).
[11] B.I. Lundquist, Phys. Kondens. Mat. 6 (1967) 193 and 206.
[12] A.W. Overhauser, Phys. Rev. B 3 (1970) 1888.
[13] B. Vinter, Phys. Rev. B 13 (1976) 4447.
[14] F. Thuselt, Phys. Lett. A 94 (1983) 93.
[15] W.F. Brinkman and T.M. Rice, Phys. Rev. B 7 (1973) 1508.
[16] N.F. Mott, Metal-Insulator Transitions (Taylor and Francis, London, 1974).
[17] K. Shindo, J. Phys. Soc. Japan 29 (1970) 287.
[18] R. Zimmermann, K. Kilimann, W.D. Kraeft, D. Kremp and G. Röpke, Phys. Status Solidi (b) 90 (1978) 175.
[19] H. Haken, in: Polarons and Excitons, eds., C.G. Kuper and G.D. Whitfield (Oliver and Boyd, Edinburgh, 1963).
[20] H. Schweizer, A. Forchel, A. Hangleiter, S. Schmitt-Rink, J.P. Löwenau and H. Haug, Phys. Rev. Lett. 51 (1983) 698.
[21] G. Mahler and J.L. Birman, Phys. Rev. B 4 (1977) 1552.
[22] T. Ando, A.B. Fowler and F. Stern, Rev. Mod. Phys. 54 (1982) 437.
[23] F. Stern and W. E. Howard, Phys. Rev. 163 (1967) 816.
[24] H. Totsuji, J. Phys. Soc. Japan 40 (1976) 857.
[25] H.M. Gibbs, S.S. Tarng, J.L. Jewell, D.A. Weinberger, K. Tai, A.C. Gossard, S.L. McCall, A. Passner and W. Wiegmann, Appl. Phys. Lett. 41 (1982) 221.
[26] J.P. Löwenau, S. Schmitt-Rink and H. Haug, Phys. Rev. Lett. 49 (1982) 1511.
[27] J. Goll and H. Haken, Phys. Status Solidi (b) 101 (1980) 489.
[28] M. Shinada and S. Sugano, J. Phys. Soc. Japan 21 (1966) 1936.
[29] The experimental point in fig. 7 corresponds to a saturation intensity of 580 W/cm^2. Here, the following parameters have been used: excitation frequency $\omega \simeq 1.463$ eV, recombination lifetime $\tau \simeq 21$ ns, exciton Bohr radius $a_0 = 60$ Å and unperturbed optical layer thickness $D \simeq 10^{-2}$.

AUTHOR INDEX

SUBJECT INDEX

Bistability

Cadmium sulphide

Cadmium telluride

Copper chloride

Electron–hole plasmas

Excitons

Gallium arsenide, gallium aluminium arsenide

Germanium

Hot carriers

Screening, Mott transition

Short pulses

Silicon

Superlattices

Transport

Two-dimensional semiconductors, quantum wells